Blood Cell Biochemistry

Volume 3
Lymphocytes and
Granulocytes

Blood Cell Biochemistry

Series Editor

J. R. Harris, *North East Thames Regional Transfusion Centre, Brentwood, Essex, England*

A Continuation Order Plan is available for this series. A continuation order will bring delivery of each new volume immediately upon publication. Volumes are billed only upon actual shipment. For further information please contact the publisher.

Blood Cell Biochemistry

Volume 3
Lymphocytes and Granulocytes

Edited by
J. R. Harris
North East Thames Regional Transfusion Centre
Brentwood, Essex, England

Springer Science+Business Media, LLC

Library of Congress Cataloging-in-Publication Data

Lymphocytes and granulocytes / edited by J.R. Harris.
     p.   cm. -- (Blood cell biochemistry ; v. 3)
    Includes bibliographical references and index.
    ISBN 978-0-306-43546-1   ISBN 978-1-4615-3796-0 (eBook)
    DOI 10.1007/978-1-4615-3796-0
    1. Lymphocytes. 2. Granulocytes. I. Harris, James R.
  II. Series.
    [DNLM: 1. Granulocytes. 2. Lymphocytes. WH 200 L9862]
  QR185.8.L9L89 1991
  612.1'12--dc20
  DNLM/DLC
  for Library of Congress   90-14317
                                   CIP

Contributors

Futwan A. Al-Mohanna   University Department of Surgery, University of Wales College of Medicine, Cardiff CF4 4XN, United Kingdom

Bernard M. Babior   Department of Molecular and Experimental Medicine, Research Institute of Scripps Clinic, La Jolla, California 92037, USA

Antonio Bonati   Institute of Medical Pathology, University of Parma, 43100 Parma, Italy

John C. Cambier   Department of Pediatrics, Divisions of Basic Sciences and Basic Immunology, National Jewish Center for Immunology and Respiratory Medicine, Denver, Colorado 80206, USA

Arnold S. Freedman   Division of Tumor Immunology, Dana-Farber Cancer Institute and the Department of Medicine, Harvard Medical School, Boston, Massachusetts 02115, USA

Giulio Gabbiani   Department of Pathology, University of Geneva, Geneva, Switzerland

Renato Gennaro   Institute of Biology, University of Udine, 33100 Udine, Italy

Maurice B. Hallett   University Department of Surgery, University of Wales College of Medicine, Cardiff CF4 4XN, United Kingdom

Michael A. Horton   Imperial Cancer Research Fund Haemopoiesis Research Group, Department of Haematology, St. Bartholomew's Hospital, London EC1A 7BE, United Kingdom

James D. Katz   Department of Medicine, The Western Pennsylvania Hospital, Pittsburgh, Pennsylvania 15224, USA; *present address*: Division of Rheumatic Diseases, Department of Medicine, The University of Connecticut Health Center, Farmington, Connecticut 06030, USA

Kwang-Myong Kim   Department of Pediatrics, Faculty of Medicine, Kyoto University, Sakyo-ku, Kyoto 606, Japan

H. Phillip Koeffler   Division of Hematology/Oncology, UCLA School of Medicine, Los Angeles, California 90024, USA

Byoung S. Kwon   Department of Microbiology and Immunology and Walther Oncology Center, Indiana University School of Medicine, Indianapolis, Indiana 46223, USA

Kathrin L. Lehmann   Department of Pediatrics, Divisions of Basic Sciences and Basic Immunology, National Jewish Center for Immunology and Respiratory Medicine, Denver, Colorado 80206, USA

Chau-Ching Liu   Laboratory of Cellular Physiology and Immunology, The Rockefeller University, New York, New York 10021, USA

Mitsufumi Mayumi   Department of Pediatrics, Faculty of Medicine, Kyoto University, Sakyo-ku, Kyoto 606, Japan

Haruki Mikawa   Department of Pediatrics, Faculty of Medicine, Kyoto University, Sakyo-ku, Kyoto 606, Japan

Eric F. Rimmer   Imperial Cancer Research Fund Haemopoiesis Research Group, Department of Haematology, St. Bartholomew's Hospital, London EC1A 7BE, United Kingdom

Domenico Romeo   Department of Biochemistry, Biophysics and Macromolecular Chemistry, University of Trieste, 34127 Trieste, Italy

Annette Schmitt-Gräff   Department of Pathology, University of Geneva, Geneva, Switzerland

Barbara Skerlavaj   Department of Biochemistry, Biophysics and Macromolecular Chemistry, University of Trieste, 34127 Trieste, Italy

Robert M. Smith   Division of Pulmonary and Critical Care Medicine, Department of Internal Medicine, University of California Medical Center, San Diego, California 92103, USA

Arthur K. Sullivan   McGill Cancer Centre, and Division of Hematology, Royal Victoria Hospital, Montreal, Quebec H3G 1Y6 Canada

Andreas Tobler   Central Hematology Laboratory, University of Berne, Inselspital, Switzerland

Joseph A. Trapani   Laboratory of Human Immunogenetics, Memorial Sloan-Kettering Cancer Center, New York, New York 10021, USA

William F. Wade   Department of Pediatrics, Divisions of Basic Sciences and Basic Immunology, National Jewish Center for Immunology and Respiratory Medicine, Denver, Colorado 80206, USA

Richard C. Woodman   Department of Medicine, University of Calgary, Calgary, Alberta T2N 2T9, Canada

John Ding-E Young Laboratory of Cellular Physiology and Immunology, The Rockefeller University, New York, New York 10021, USA

Lucy H. Y. Young The Massachusetts Eye and Ear Infirmary, Harvard Medical School, Boston, Massachusetts 02114, USA

Margherita Zanetti Department of Biochemistry, Biophysics and Macromolecular Chemistry, University of Trieste, 34127 Trieste, Italy

Preface

This, the third volume of the *Blood Cell Biochemistry* series, follows the pattern established in the two previous volumes by containing up-to-date specialist reviews of topics of current interest within the field of study defined by the subtitle. Thus, the topics included can be loosely classified under the broad subtitle "Lymphocytes and Granulocytes," but this does not indicate the full scope of content, scientific interest, and emphasis of the present volume.

The opening chapter, by Antonio Bonati, surveys the currently available biochemical, immunological, and molecular markers of hemopoietic precursor cells. This is followed, appropriately, by a contribution from Arnold S. Freedman on the cell surface markers in leukemia and lymphoma. In a detailed chapter, Annette Schmitt-Gräff and Giulio Gabbiani discuss the cytoskeletal organization of normal and leukemic lymphocytes and lymphoblasts. John C. Cambier and his colleagues then present a discussion of the signaling events in T-lymphocyte-dependent B-lymphocyte activation. Lymphocyte IgE receptors and IgE-binding factors are dealt with by Kwang-Myong Kim and his colleagues, and the role of granule mediators in lymphocyte-mediated cytolysis is covered by John Ding-E Young and his associates. A short contribution from James D. Katz deals with the intricacies and difficulties of studies on the complement C3b (CR1) receptor and its cytoskeletal interactions in neutrophils. Arthur K. Sullivan then presents an in-depth survey of the membrane biochemistry surrounding the flow of granule organelles in leukocyte differentiation. A thorough discussion of the respiratory burst oxidase of human phagocytes under the intriguing title "The Elusive Oxidase" comes from Bernard M. Babior and his colleagues. The consideration of cytolytic systems is taken further within the chapter on the localization, structure, and function of myeloperoxidase, by Andreas Tobler and H. Phillip Koeffler, which together with a consideration of the importance of intracellular calcium and cytoskeletal interactions in neutrophil oxidase activation by Futwan A. Al-Mohanna and Maurice B. Hallet, provides a central thematic core to the volume. Indeed, the following chapter, by Renato Gennaro and his colleagues on neutrophil and eosinophil "defense" proteins, continues this consideration of bacteriocidal activity.

The final chapter deals with membrane glycoproteins of mast cells and basophils. This comes from Eric F. Rimmer and Michael A. Horton and places emphasis on the various membrane-bound receptors for complement anaphylatoxins and immuno-globulins.

The thoroughly international authorship of this volume highlights the prominent worldwide interest in and importance of white blood cell biochemistry. It is hoped that this volume will be of value to hematologists, immunologists, cell biologists and others interested in the biochemistry of lymphocytes and granulocytes.

Robin Harris

Brentwood, Essex, England

Contents

Chapter 3

Cytoskeletal Organization of Normal and Leukemic Lymphocytes and Lymphoblasts

Annette Schmitt-Gräff and Giulio Gabbiani

Chapter 4

Signaling Events in T-Lymphocyte-Dependent B-Lymphocyte Activation

John C. Cambier, Kathrin L. Lehmann, and William F. Wade

Chapter 5

IgE Receptors on Lymphocytes and IgE-Binding Factors

Kwang-Myong Kim, Mitsufumi Mayumi, and Haruki Mikawa

Chapter 6

Lymphocyte-Mediated Cytolysis: Role of Granule Mediators

John Ding-E Young, Byoung S. Kwon, Joseph A. Trapani, Chau-Ching Liu,
and Lucy H. Y. Young

Chapter 7

CR1–Cytoskeleton Interactions in Neutrophils

James D. Katz

Chapter 8

The Flow of Granular Organelles in Leukocyte Differentiation

Arthur K. Sullivan

Chapter 9

The Elusive Oxidase: The Respiratory Burst Oxidase of Human Phagocytes

Robert M. Smith, Richard C. Woodman, and Bernard M. Babior

Chapter 10

Myeloperoxidase: Localization, Structure, and Function

Andreas Tobler and H. Phillip Koeffler

Chapter 11

Mechanisms of Oxidase Activation in Neutrophils: Importance of Intracellular Calcium and Cytoskeletal Interactions

Futwan A. Al-Mohanna and Maurice B. Hallett

Chapter 12

Neutrophil and Eosinophil Granules as Stores of "Defense" Proteins

Renato Gennaro, Domenico Romeo, Barbara Skerlavaj, and Margherita Zanetti

Chapter 13
Membrane Glycoproteins of Mast Cells and Basophils
Eric F. Rimmer and Michael A. Horton

Chapter 1

Biochemical, Immunological, and Molecular Markers of Hemopoietic Precursor Cells

Antonio Bonati

1. INTRODUCTION

Hematology is a discipline that is undergoing improvements regarding technical applications and treatment of blood diseases and also is reaching a better understanding of the nature of different normal and hemopoietic precursors.

For several years, hematologists have faced the problem of identifying blood cells only by morphological criteria. From the first observations of leukemia carried out by Virchow in 1845 and Ehrlich in 1898 until the new era of monoclonal antibodies (MAbs), which developed in the late 1970s and early 1980s, no substantial progress was reached in discriminating clearly between the variability of hemopoietic cells, particularly at the most immature stages of differentiation. Ehrlich introduced the theory of the dualistic origin of leukemic cells by considering the existence of two main types of leukocytes, "nongranular or lymphoid" and "granular or myeloid." In 1900, Naegeli reformed this concept by identifying an immature myeloid cell that is nongranular, the myeloblast; and by describing a corresponding leukemia that is acute myeloid leukemia (reviewed in Ferrata and Storti, 1958).

Other eminent hematologists then developed the view of a unique hemopoietic precursor from which all other blood cells derive. This cell, named the hemocytoblast, was described as a nucleolated early immature precursor with a high nuclear/cytoplasmic ratio, thin chromatin, and lack of granules. The authors thought that the most immature precursors have the ability to differentiate along different hemopoietic lines in response to different microenvironmental factors. Therefore, no strict lineage fidelity could be present throughout the hemopoietic maturative process (Ferrata and Storti, 1958). This view

Antonio Bonati   Institute of Medical Pathology, University of Parma, 43100 Parma, Italy.

preceded some concepts that today are sustained on the basis of a new technical approach provided by molecular biological analysis. For example, in acute lymphocytic leukemia (ALL), both in fresh leukemic cells and in cell lines with B-cell or T-cell precursor immunophenotype, there is a cross-lineage immunoglobulin heavy-chain (IgH) or T-cell receptor (TCR) beta gene rearrangement in approximately 25% of cases. Both types of rearrangement or TCR gamma gene rearrangement are also detectable in acute myeloid leukemia (AML) (Furley et al., 1987). As has been convincingly demonstrated, acute leukemic cells are the counterparts of normal hematopoietic precursors frozen at an early stage of differentiation; therefore, it could be suggested that hemopoietic precursors of mixed molecular phenotype are present even during the normal hemopoietic maturative process, although in small percentage and with the significance of a transient population (Greaves, 1986; Greaves et al., 1986). It is expected that future experiments combining molecular biological techniques with single-cell analysis (such as in situ hybridization) will confirm this point of view.

In this chapter, we describe the biochemical, immunological, and molecular markers that have led to a better understanding of hemopoietic cells. In addition, we discuss the main application of these markers to human hemopoietic ontogenesis analysis, to diagnosis of leukemia, and to bone marrow purging.

2. BIOCHEMICAL MARKERS

Considerable progress was made in the characterization of normal and leukemic cells by the study of leukocyte enzymes. The enzymes of the myeloid series were widely investigated, but the existence of characteristic enzymatic markers of lymphocytes and of lymphoid precursors was recognized only in the late 1970s and 1980s (Hoffbrand and Janossy, 1981). Leukocyte enzymes can be studied by biochemical assays of whole-cell extracts, by cytochemical methods in blood, bone marrow, lymph node imprints, or cytospins, or in frozen sections of different tissues by histochemistry. Some enzymes can also be studied by immunological methods using polyclonal or monoclonal antibodies by techniques such as immunofluorescence, immunoperoxidase, or immunoalkaline phosphatase labeling (APAAP). Certain enzymes can be analyzed by various methods, for example, both by biochemical assay of whole-tissue extracts and by immunological detection at the single-cell level [e.g., terminal deoxynucleotidyl transferase (TdT)] (Coleman and Hutton, 1981). In this section we describe enzymatic markers of normal and leukemic hemopoietic cells detectable by biochemical assay.

2.1. Terminal Deoxynucleotide Transferase

One of the most important enzymes of the hemopoietic system is TdT. TdT is a DNA polymerase that has the unique property of synthesizing DNA molecules in the absence of a template. TdT is significant in hematology because of its occurrence in white blood cells and marrow, particularly in patients with leukemia (Bollum, 1979). The enzyme was first described in the early 1960s by Bollum (1960) and Krakow et al. (1962) in calf thymus gland extracts.

TdT purified from calf thymus gland has two peptide subunits with molecular weights (MW) of 8,000–10,000 (the alpha subunit) and 24,000–30,000 (the beta subunit) (Chang, 1971). The human enzyme purified from lymphoblasts was recognized to be constituted of a single polypeptide chain of 62,000 MW, but the human enzyme from thymus was found to have a subunit structure like that of calf thymus (Deibel and Coleman, 1979). The mapping patterns of the different subunits suggested that a single polypeptide weighing 60 kDa could be the original form of mammalian TdT and that the alpha and beta subunits could be proteolytic products derived from the 60-kDa form via an intermediate 42-kDa form. The putative process of TdT degradation described by Nakamura *et al.* (1981) is represented in Figure 1. Similar results were reported by Bollum and Chang (1981), who identified a single peptide of 58,000–60,000 Da from thymus of different species and in both mouse and human lymphoblastoid cells. Some authors (Deibel *et al.*, 1981) suggested that the proteolytic form of the enzyme might originate during the differentiation of hemopoietic precursor cells.

TdT functions in the absence of a template, property that distinguishes this enzyme from the usual DNA polymerases. TdT requires an initiator molecule containing a free 3′-hydroxyl group to which the 5′-deoxynucleotides are added. If the initiator used is a short polydeoxynucleotide such as $poly(dA)_{12-18}$ or $poly(dA)_{50}$ and the substrate is dGTP, this reaction can be used to detect TdT in tissue extacts (Bollum, 1979).

Use of the $poly(dA)_{50}$ initiator with acid precipitation of the product is the most convenient and rapid method for quantitative detection of TdT, because the enzyme is extracted quickly by sonication and ultracentrifugation. Methods that use column chromatography are important for investigating different migratory forms of TdT. Biochemical assay procedures for detecting TdT by both ion-exchange chromatography and acid precipitation were extensively reported by Coleman and Hutton (1981).

Hematologists were provided a good opportunity to investigate TdT in normal and leukemic cells after an antibody against calf TdT was obtained in the rabbit by Bollum (1975). The antiserum to the calf enzyme cross-reacts with human, rodent, and chicken TdT. The use of anti-calf TdT antiserum allows rapid detection of TdT by indirect immunofluorescence. Furthermore, by immunofluorescence analysis, it is possible to detect the phenotype of TdT^+ cells by combining rabbit anti-calf TdT with different murine MAbs (Bollum, 1979). The important information concerning the nature of the hemopoietic precursors obtained by this technique is discussed below.

The first information that TdT provided hematolologists was based on biochemical assay. Interpretation of TdT activity requires knowledge of the range of values obtained when assaying specific types of tissues in a particular laboratory. TdT activities in nucleated cells from the bone marrow and peripheral blood (PB) of normal patients are used to define a normal range. In a large series of normal cases (160 adults and 190 children), the means and standard deviations of TdT activities (1 unit = 1 nmol of dGTP polymerized per hr) were 2.7 ± 3.7 and 5.9 ± 1.9 units/10^8 cells. TdT activity is present in large amounts in malignant cells from certain patients with leukemia and lymphoma. Leukemic

FIGURE 1. Putative process of TdT degradation. A single polypeptide of 60 kDa could be the original form of mammalian TdT, with the alpha- and beta-subunit proteolytic products derived from the 60-kDa form via an intermediate 42-kDa form.

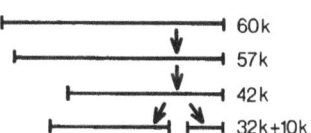

cells from approximately 300 patients with ALL were assayed and reported as having high levels of TdT in 90% of cases (Coleman and Hutton, 1981).

There are two variants of ALL that express very low levels of TdT. In one variant, the cells have immunoglobulins on their surface, and cytologic features of the blasts are undistinguishable from Burkitt's lymphoma. In the other variant, the leukemic cells express cytoplasmic IgM (cyIgM) but not surface immunoglobulin (pre-B-cell leukemias). In patients with AML, TdT is generally within the normal range, although 5–10% of patients may have high TdT levels. Chronic myeloid leukemia (CML) patients have low levels of TdT activity both in bone marrow and PB, whereas 35% of blastic crisis CML patients have high TdT levels. These cases were reported as lymphoid blast crisis and were good responders to vincristine and prednisone (Marks et al., 1978). Lymphoblastic leukemia of T-cell or null cell type (the latter leukemia is derived from B-cell precursors) shows high TdT activity.

TdT is absent in chronic lymphoproliferative disorders. In 11 Hodgkin's disease and 5 hairy cell leukemia cases, TdT activity was detected at convincing levels only in one patient in each of these categories (Srivastava et al., 1978). High TdT levels that disappeared after splenectomy were detected in a case of hairy cell leukemia (Bonati et al., 1983b). This finding could indicate that even in the course of chronic lymphoid disorders a significant wave of immature precursors may be present, as recently found in common ALL antigen-positive (CALLA[+]) myeloma (Durie and Grogan, 1985).

The chromatographic difference between leukemia-associated TdT (TdT elutes as two forms, one at 0.3 M salt and one at 0.4 M salt) and normal marrow-associated TdT (TdT elutes as a single peak at 0.3 M) suggested that phosphocellulose TdT patterns could be useful for monitoring residual marrow leukemic disease in TdT[+] leukemias (Bell et al., 1982). This analysis was performed in 376 untreated TdT[+] ALL patients, but in 8 patients studied in early relapse, the blast cell TdT pattern was of the single-peak 0.3 M type. Therefore, leukemia cell TdT cannot be differentiated from normal marrow TdT. A similar pattern was observed in a case of Ph[+] AML (Casoli et al., 1983). Moreover, the same physicochemical characteristics of TdT were present in bone marrow and blood in the acute phase. Blood and bone marrow TdT activity at first and second relapse showed chromatographic forms of TdT different from that of the acute phase and from that of the cerebrospinal fluid, which conserved a form like that of the acute phase.

Considerable attention has been given to the biological significance of this unique DNA polymerase. There is convincing evidence that TdT acts as a somatic mutagen by the insertion of extra, random nucleotides (N regions) at the D-J joining region during both immunoglobulin and TCR beta gene rearrangement, thus contributing to the generation of antibody and TCR diversity (N diversity) (Desiderio et al., 1984; Yancopoulos et al., 1986).

2.2. Enzymes in Nucleotide Metabolism

The enzymes involved in the purine salvage pathway have an important function in lymphoid cell metabolism. These enzymes are adenosine deaminase (ADA), purine nucleotide phosphorylase (PNP), and 5'-nucleotidase (5'NT). 5'NT is localized in the cell membrane and precedes ADA and PNP in the metabolic pathway; ADA and PNP are mainly localized in the cytoplasm (Van Laarhoven and De Bruyn, 1983).

Much effort was devoted to understanding the role of purine metabolism in immune function after a causal relationship between genetic deficiency of ADA and a combined deficiency of T- and B-cell function was observed (Giblett *et al.*, 1972). These studies involved both analysis of human lymphoblastoid cell lines *in vitro* and analysis of human and animal cell extracts after collection of tissues *in vivo*. From these analyses, it was concluded that purine metabolism plays an essential role in the development of the immune system; it is of interest also that different activity levels of purine enzymes are present in different stages of lymphoid cell differentiation (Van Laarhoven and De Bruyn, 1983).

2.2.1. Adenosine Deaminase

ADA catalyzes the deamination of deoxyadenosine and adenosine to deoxyinosine and inosine, respectively (Figure 2). ADA activity in T lymphocytes is 10–12 times greater than in B lymphocytes, and increased activity was observed in thymocytes but not in tonsil lymphocytes. In contrast to TdT, which decreases from early immature cells, ADA tends to increase at least in the thymic cells and the leukemias from which they are derived (Hoffbrand and Janossy, 1981; Van Laarhoven and De Bruyn, 1983).

Coleman *et al.* (1978) examined 26 cases of children 2–15 years old, combining TdT and ADA biochemical activity detection with cell surface marker analysis. TdT activity was high in blasts from the 20 children with either null or T-cell ALL. The activity of ADA was higher than that of TdT and was lower in T-cell than in null cell leukemias. TdT and ADA were detectable at very low levels in three cases of B-cell ALL that were positive for surface-associated immunoglobulins. The authors suggested that the levels of

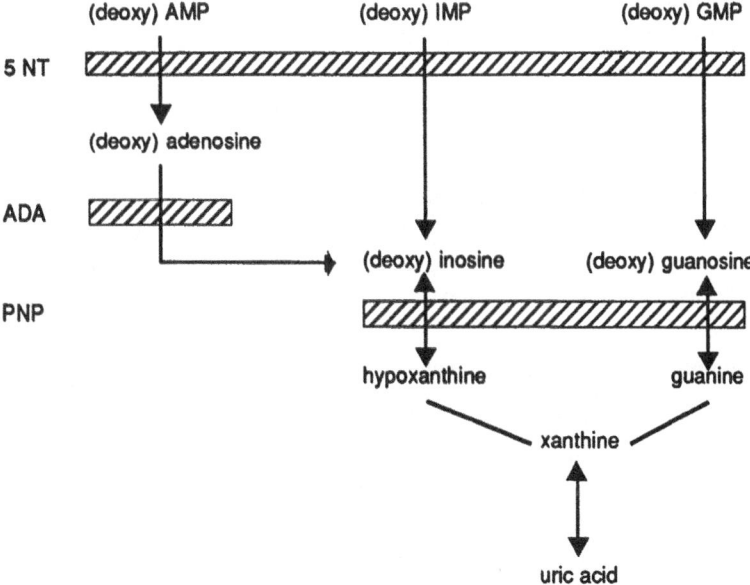

FIGURE 2. Pathways of 5'NT, ADA, and PNP purine metabolism. The level at which the main inborn errors occur is represented.

both enzymes might be keys to a better understanding of the different forms of acute leukemias: in non-B, non-T ALL (including common and null ALL), TdT is high and ADA is intermediately raised; in T-ALL, both TdT and ADA are high; in B-ALL, both TdT and ADA are low; in AML, TdT is low but ADA is high. Others confirmed these data by examining large numbers of ALL patients (Brusamolino et al., 1984).

The prognostic significance of ADA in acute and chronic myeloid leukemias was studied (Bertazzoni et al., 1982). Survival was significantly longer for AML patients with low levels of ADA; in the chronic phase of CML, ADA values were higher than in normal controls only in patients with early signs of transformation.

In B-cell chronic lymphocytic leukemia (B-CLL), ADA activity is extremely low, but in contrast to patients with severe combined immunodeficiency, the enzyme is normally expressed in the erythrocytes (Tung et al., 1976). Moreover, it was found that B-CLL lymphocytes have about half the activity of normal B lymphocytes. These data notwithstanding the diagnostic value of the enzyme assay in blood malignancies is limited. In fact, the overlap in ADA levels among the various forms of acute and chronic leukemias (ADA is higher in T-cell than in B-cell chronic lymphoproliferative disorders) is too great for the test to be reliable (Hoffbrand and Janossy, 1981).

Some attempts to produce "nucleoside toxicity" for killing of leukemic cells were carried out by using 2'-deoxycoformicin (dCF), which selectively inhibits ADA activity. Since T-cells are less able than B cells to degrade dATP, ADA inhibition may lead to the accumulation of dATP; dATP inhibits ribonucleotide reductase, producing an interruption of the supply of dTTP, dATP, and dCTP and so inhibiting DNA synthesis (Van Laarhoven and De Bruyn, 1983). On the basis of these studies on biochemical mechanisms, therapeutic trials were carried out; a considerable decrease in bone marrow lymphoblasts of two patients with T-ALL resistant to conventional chemotherapy was induced (Prentice et al., 1980). However, the toxicity observed in these and some other cases was too high to permit introduction of dCF in T-ALL treatment.

It is of speculative interest that dCF therapy produced a phenotypic conversion of acute leukemia from T lymphoblastic to myeloblastic (Murphy et al., 1983). This finding would suggest that dCF determined a selection pressure blocking T-cell differentiation and proliferation, permitting the emergence of a myeloid subclone of a pluripotential leukemic progenitor with the ability to differentiate along both T-lymphocytic and myeloid lineages.

Recently, two patients with hairy cell leukemia with massive splenomegaly and severe pancytopenia, after failure of alpha interferon therapy, were treated with dCF with satisfactory results (Foon et al., 1986).

2.2.2. Purine Nucleoside Phosphorylase

PNP acts sequentially (Figure 2) with ADA in purine degradation, catalyzing the conversion of inosine or deoxyinosine to hypoxanthine and guanosine or deoxyguanosine to guanine. A deficiency of PNP was detected in several patients with defective T-cell immunity but with normal B-cell function (Giblett et al., 1975). PNP was shown to be significantly reduced in eight cases of T-ALL compared with non-T, non-B ALL or with normal peripheral blood (Blatt et al., 1980). The detection of PNP activity was recently found to be a reliable method for distinguishing between B-CLL (low PNP levels) and prolymphocytic leukemia (high PNP levels) (Ratech et al., 1985).

2.2.3. 5'-Nucleotidase

5'NT catalyzes the first enzymatic step in the degradation of purine nucleotides leading to the formation of the corresponding nucleosides (Figure 2). 5'NT acts in part at the surface of the cell membrane (ecto-5'NT) and in part at the intracellular level. Decreased 5'NT activity was detected both in B and T cells in patients with X-linked agammaglobulinemia (Cohen *et al.*, 1980).

Since 5'NT and PNP reach the lowest values seen during lymphocyte development at the maturative stage of cortical thymocytes, it was suggested that the loss of 5'NT and PNP activity may contribute to determining the "biochemical suicide" of these cells (Ma *et al.*, 1982). It is well known that most TdT$^+$ cortical thymocytes are cells committed to die in order to permit the destruction of cell clones active against self-antigen proteins (Rothemberg and Triglia, 1983). The particularly effective deoxyribonucleoside kinase characteristics of these cells may rapidly lead to toxic concentrations of purine deoxynucleotides, especially dATP and dGTP, in the absence of the enzymes that catalyze purine metabolism. The fact that ADA activity is low in cortical thymocytes may also contribute to this effect. Only a few cortical thymocytes are rescued, by a still unknown mechanism.

2.2.4. Isoenzymes

A series of analyses was carried out on carboxylic esterase, acid phosphatase, β-hexosaminidase, and lactate dehydrogenase expression in various B- and T-cell leukemia–lymphoma cell lines (Drexler *et al.*, 1985a–d). The authors showed by an isoelectric focusing method that distinct isoenzyme profiles were characteristic of different stages of differentiation. Moreover, non-leukemia–lymphoma-specific isoenzyme or an additional isoenzyme undetectable in normal hemopoietic cells could be seen. The results confirm that leukemia–lymphoma cells are apparently not different from their normal counterparts.

The enzyme profiles of eight Hodgkin's disease cell lines were examined by immunoelectrofocusing or chromatographic techniques (Scott *et al.*, 1988). The results showed that all of the cell lines analyzed had enzymatic features typical of lymphoid cells so that a monocyte/histiocyte origin of these cell lines could be excluded. These results, together with the data provided by immunology and molecular biology (Jones, 1987; Drexler *et al.*, 1988), suggest a lymphoid origin of *in vivo* Hodgkin's Reed-Sternberg cells.

2.2.5. Cyclic AMP

Cyclic AMP (cAMP) is a metabolite important in the control of various intracellular processes. It interacts with cAMP-dependent protein kinases, which are represented by cAMP-binding regulatory subunits and catalytic subunits capable of phosphorylating other proteins. The regulatory units are recognizable as two distinct classes, I and II, identifiable in hemopoietic cells of different lineages (Elias and Papayannopoulou, 1985). For example, B-cell transformed lines express higher levels of type II kinase than do those of T-cell origin, as shown by column chromatographic patterns (Elias and Papayannopoulou, 1985).

A marked increase was found in type I cAMP-dependent protein kinase in HL-60 cells exposed to agents that induce myeloid differentiation but not when HL-60 differentiates along the monocyte/macrophage pathway (Fontana et al., 1984). More recently, among cells of human hemopoietic origin, four binding protein subtypes (a, b, c, and d) were identified by photoaffinity labeling and gel electrophoresis. Human erythroblasts taken from methylcellulose cultures of fetal liver and adult blood and from murine and human erythroleukemic cells cultured with and without chemical inducers of erythrodifferentiation were examined. An isozymic cAMP-binding protein pattern was identified as characteristic of the erythroid lineage, and the relative amounts of the three components identified (a, b, and c) expressed modification during induced differentiation (Elias and Papayannopoulou, 1985). Thus, cAMP-binding protein analysis may be useful for investigation of hemopoietic differentiation.

2.2.6. Thymidine Kinase

dTTP, one of the precursors for DNA synthesis, can be synthesized by a de novo pathway and a salvage pathway; the rate-limiting step in the salvage pathway is catalyzed by thymidine kinase (Munch-Petersen and Tyrsted, 1977).

There is much evidence for a positive correlation between thymidine kinase activity and DNA synthesis (Tyrsted and Munch-Petersen, 1977). In human tissues, two isoenzymatic forms of thymidine kinase (TK1 and TK2), distinguishable by chromatographic, electrophoretic, and kinetic properties, were reported (Taylor et al., 1972). Isoenzyme detection in leukemic cells led to the recognition of three new forms of thymidine kinase: TK1, found in B-CLL; TK3, found in ALL and acute monocytic leukemia (AMOL); and TK4, found in AMOL (Munch-Petersen and Tyrsted, 1986). However, the practical value of this knowledge for hematologists was low until an assay for detection of thymidine kinase in serum was developed. By this assay, a striking correlation between high serum thymidine kinase levels and poor prognosis in AML was found (Archimbaud et al., 1988).

A new aspect of biochemical analysis of the cells is the study of cell cycle-specific proteins. The group of Julio Celis recently identified, by two-dimensional gel electrophoresis, two human phosphoproteins (dividin and IEF 59dl) that are first detected late in G_1 near the G_1/S transition border of the cell cycle (Nielsen et al., 1987). Work is now in progress by the same group to prepare MAbs against dividin. This effort will be of value for purifying this protein and will provide a rapid probe to screen normal and tumor tissues.

Moreover, Celis et al. (1988) established data bases of protein information derived by analyzing, by two-dimensional gel electrophoresis, transformed human amnion cells and PB mononuclear cells. A total of 1781 [^{35}S]methionine-labeled amnion proteins and a total of 1311 proteins from PB were resolved and recorded by using computerized two-dimensional gel electrophoresis. The authors compared the overall pattern of protein synthesis of normal human lymphocyte subpopulations after separation by fluorescence-activated cell sorting (FACS) into CD4$^+$ helper T cells, CD8$^+$ suppressor T cells, CD20$^+$ B cells, and NHK-1$^+$ cells. Of about 1000 proteins detected in each case, most were found to be common to all subpopulations (Madsen et al., 1988).

3. IMMUNOLOGICAL MARKERS

Immunology has provided many means of investigating the hemopoietic system, particularly since the revolution introduced by the development of MAbs and their subsequent use in the diagnosis of leukemia and related diseases. In 1978, the leukemic marker conferences and World Health Organization-sponsored committees defined the following reagents as important in this area: (1) anti-ALL antiserum reacting with CALLA, constituted of a single glycosylated polypeptide of a MW 95,000–100,000; (2) antisera to common core determinants of HLA-DR molecules, which are glycoproteins of MW 33,000 and 28,000 (p28 and p33) controlled by the HLA-DR locus of the major histocompatibility (MHC) gene complex (HLA system in humans); (3) anti-human T-lymphocyte antigen; (4) anti-surface membrane immunoglobulins (smIg); and (4) the handful of antimyeloid antisera reacting to AML, AMML, (acute myelo-monocitic leukemia), AMOL, and myeloid blast crisis of CML that were available at that time (Janossy et al., 1986).

The technique discovered by Kohler and Milstein (1975) to produce antibodies reactive to specific clonal determinants (MAbs) provided hematologists and immunologists with a rich source of reagents. The majority of these reagents were classified into clusters of differentiation (CD). International workshops held in Paris (1982), Boston (1984), and Oxford (1986) were organized for this purpose, and a later workshop was held in Vienna (February 1989). The large series of reagents available might create some confusion and the wrong impression that many reagents are required in the hematological field. In fact, laboratory experts need only a handful of selected MAbs for a correct diagnosis of leukemia. In addition to MAbs, some heterologous reagents are important for defining the phenotypes of hemopoietic precursors. This is the case for a particular antiserum produced in rabbits against calf nuclear TdT. Also, the use of heterologous antisera reactive against smIg of different isotypes and against cytoplasmic mu chains may be helpful in the diagnosis of leukemia (Janossy et al., 1986).

Different methods can be used to detect the antigens. Immunofluorescence is the fastest method, particularly when directly labeled antibodies are used. The fluorochromes most frequently used are fluorescein isothiocyanate (FITC; green), tetraethyl rhodamine isothiocyanate (TRITC; red), and the recently introduced phycoerythrin (PE). Directly labeled antibodies are used when the antigens are expressed on the cell surface at relatively high density, such smIg, HLA-DR, antigens associated with natural killer cells, and antigens characteristic of certain T-cell subsets (CD8$^+$ cells). Indirect immunofluorescence assay using a two-step reaction, such as a mouse MAb followed by an FITC-or TRITC-conjugated goat anti-mouse antiserum, may also enable detection of antigens weakly expressed and thus not easily recognizable by direct staining. For example, some cases of B-CLL express only small amounts of smIg, and in the common form of ALL, the intensity of the specific CALLA may be extremely variable within the malignant clone. Therefore, we could erroneously judge as CALLA$^-$ a lymphoblastic leukemia if the blasts are examined by a direct immunofluorescence test only. Double-cell immunofluorescence may be a source of additional information for analysis of particular cell subpopulations. In fact, the combination of different reagents may lead to recognition of different cells in an apparently homogeneous population (for example, in the case of biphenotypic leukemia) (Bonati *et al.*, 1986) or to definition of the phenotype of a particular subset of cells (for example, the few TdT$^+$ cells of normal bone marrow that are CALLA$^+$) (Janossy *et al.*, 1979).

Double-cell immunofluorescence can be carried out in four different ways: (1) with directly labeled antibodies used as a single layers; (2) with MAbs of different immunoglobulin subclasses in the first step and fluorochrome-labeled MAbs specific for these in the second step; (3) with an MAb (and the corresponding fluorochrome-labeled second layer, such as goat anti-mouse immunoglobulin) together with a heterologous antiserum directly labeled (e.g., goat anti-human IgM for detecting cyIgM) or indirectly labeled (e.g., rabbit anti-TdT and goat anti-rabbit FITC or TRITC-labeled antiserum); and (4) with hapten-labeled MAbs together with fluorochrome-labeled heterologous antihapten antisera such as in the case of biotin- or arsanilic acid-conjugated reagents. The latter is a system feasible for amplifying the expression of a particular antigen. Even systems with triple-layer amplifications are feasible.

Recently triple combination of first-step reagents was employed (Van Dongen *et al.*, 1985a,b) by using FITC and TRITC excitation combined with immunogold staining. Visualization of the cells stained by fluorochromes is possible by using a microscope equipped with an epifluorescence condenser and selective filters for FITC and TRITC excitation. PE, which fades quickly with filters for TRITC, can be better observed with filters for FITC. To see the orange color of PE with FITC filters, the selective green barrier filter must be removed from the epifluorescence reflector housing. The morphological characteristics of the stained cells may be visualized by a phase contrast objective.

Detection of antigens by flow cytometry is widely used today. This method leads to quantitatively positive detection of antigens and to easier detection of weak antigens. Small subsets of cells may be separated for analysis in culture or for other research purposes by FACS, which, like flow cytometry, uses fluorescence laser properties to separate cells positive for different antibodies. Immunoperoxidase labeling, and in particular APAAP, have recently been recognized as tests that may parallel immunofluorescence detection in importance. The advantages provided by these techniques are (1) easier Giemsa counterstaining of cells, with better morphological resolution and examination; (2) ease of examination of cells on cytospins to identify particular subsets that could be lost in mononuclear cell suspension; and (3) use for analysis of tissue sections. Particularly APAAP leads to reliable results when bone marrow or lymph node biopsies are examined (Janossy *et al.*, 1986).

The reagents available for detecting hemopoietic cells may be classified as stem cell associated, myeloid, common ALL and B-lineage associated, and T-lineage associated.

Table I lists MAbs reacting with B, T, and myeloid (nonlymphoid) cells, classified by CD, with some of the most frequently used reagents for detecting the phenotypes of normal and leukemic hemopoietic precursors.

The stem cell-related MAbs such as My10 (CD34) recently gained importance not only for diagnosing leukemia but also for isolating blood populations (e.g., by FACS, magnetic beads, or panning) enriched with early precursors in order to carry out differentiation studies (Strauss *et al.*, 1986). The anti-My10 MAb binds to a 115-kDa membrane glycoprotein that is expressed on the cell surface membrane of 1–4% of normal

Table I

Abs Reacting with B, T, and Myeloid Cells Classified by CD with Some of the Most Frequently Used Reagents for Detecting the Phenotypes of Normal and Leukemic Precursors

CD group	Mol. wt. (kDa)	Reactivity[a]	Examples of antibodies	Antigen designation[a]
CD1a	49	Thy, LC	NA1/34, OKT6	T6
CD1b	47	Thy, LC	NU-T2, 4A76	T6
CD1C	43	Thy, LC, some B	M241	T6
CD2	50	T	OKT11, NU-T1	T11, ER, LFA-2
CD3	19–29	T	Anti-Leu-4UCHT1, OKT3	T3
CD4	55	$T_{H/I}$, M	Anti-Leu-3a, OKT4	T4
CD5	67	T, B-CLL	Anti-Leu-1, Tu71	T1
CD6	120	T (mature)	Tu33, OKT17	T12
CD7	41	T	Tu14, 3A1	T2
CD8	32–33	$T_{S/C}$	Anti-Leu-2a, OKT8	T8
CD10	100	Early B pr.	VIL-A1, J5, NU-N1	CALLA
C11b	160, 95	G, M, DRC, LGL	OKM1	Mac-1 α chain (C3biR)
CD13	150	G, M	My7, MCS2	
CD14	55	G, M, IDR, DRC	UCHM1, Mo2, My4	
CD16	50–60	G, NK, M	Anti-Leu-11	FcR
CD19	95	B	Anti-B4, HD37	
CD20	37, 32	B, M	Anti-B1, NU-B2, G28-2	
CD21	140	B, DRC	HB5, B2	C3dR
CD22	140, 130	B	To15, HD39, HD6	
CD24	42	B, G, M	BA-1, HB8	
CD25	55	T, B, M, NK	Anti-Tac, Tu69	IL2R
CD33	67	Myeloid progenitors	L4F3, L1B2, My9	
CD34	115	Early G, B pr.	My10, BI-3C5	Immature G
CD35	220	G, M, B, RBC, DRC, glomeruli	E11, To5, J3B11	C3bR
CDw41	130, 115	Th	J15	gpllb-llla
CDw42	150	Th	HPL14, AN51	gplb

[a]Abbreviations: B, B lymphocyte; B-CLL, B-type chronic lymphatic leukemia; B pr., B precursors; CALLA, common acute lymphatic leukemia antigen; DRC, dendritic reticulum cell; ER, sheep erythrocyte receptor; FcR, Fc receptor (for IgG); G, granulocyte and precursors; IDR, interdigitating reticulum cell; IL2R, interleukin 2 receptor; LC, Langerhans cell; LFA, lymphocyte function-associated antigen; LGL, large granular lymphocyte; M, monocyte/macrophage; NK, natural killer cell; RBC, red blood cell; T, T lymphocyte; T_{act}, T-cell activation antigen; Th, thrombocyte; $T_{H/I}$, T helper/inducer cell; Thy, thymocyte; $T_{S/C}$, T suppressor/cytotoxic cell.

adult bone marrow cells (Strauss *et al.*, 1986). This antigen may be identified in both AML and ALL, and a large number of positive cells in a blood sample immediately indicates a preponderance of immature leukemic cells (Janossy *et al.*, 1986). The same epitope is recognized by MAb 3C5. Anti-HLA-DR MAb may recognize normal myeloblasts, lymphoid precursors, and the leukemias corresponding to these normal subsets, but also a mature population such as PB B lymphocytes and B-CLL lymphocytes (Janossy *et al.*, 1986).

My9 (CD33) produced by Griffin *et al.* (1984), detects over 80% of cases of AML, and My7 (CD13) reacts with 70% of AML; both reagents react with less 1% of ALLs. Mo2 (CD14), My4, and UCHM1 stain leukemia cells with monocytic differentiation (Griffin *et al.*, 1983a,b). Antibodies specific for normal monocytes and not granulocytes have a preferential reactivity against M4-M5 leukemias. Mo2 detects 46% of cases of M4-

M5 leukemias but only 15% of M1-M2 leukemias. My4 reacts with over 60% of M4-M5 leukemias, but only with 14% of cases of M1-M2. A greater degree of discrimination was reported with the antibodies UCHM1 and UCHALF, which are more selective for monocytes in normal blood and bone marrow. UCHM1 reacts with 95% of cases of M4-M5 leukemias and 5% of cases of M1-M2 leukemias; similarly, UCHALF reacts with 95% of cases of M4-M5 leukemias and 5% of cases of M1-M2 leukemias (Linch et al., 1984). Another series of interesting reagents to detect myeloid antigens was produced in the W. Knapp laboratory (Vienna) (Majdić et al., 1981; Paietta et al., 1983.)

Antiglycophorin antibody R10 was found to be active against blasts from 80% cases of erythroleukemia (Greaves et al., 1983b). Some of the antiplatelet antibodies (J15) may be of value in detecting M7 megakaryoblastic leukemia (Vainchenker et al., 1982; Griffin et al., 1983b).

It is important to add that the immunological phenotype reflects in general the level of differentiation of the AML cells, but the correlation with the French-American-British (FAB) classification is imperfect. Prediction of the FAB classification by immunological phenotyping is possible only after computer analysis in 65% of cases (Linch and Griffin, 1986). Moreover, with respect to leukemia, classification is not particularly helpful for determining whether the immunological phenotype corresponds to conventional morphological subtyping. The crucial question is whether the immunological phenotype provides useful information, such as prognostic values (Linch and Griffin, 1986).

Civin et al. (1983) analyzed the expression of My1 (CD15) and My10 (CD34) MAbs in 33 cases of AML. All of the My1$^+$ patients reached a complete remission with a percentage significantly higher than that of the My1$^-$ patients. Only a few My1$^-$ patients were My10$^+$; 87% of the My10$^+$ patients achieved a complete remission.

An important group of antibodies is represented by the CD10 cluster. These reagents include antibodies reactive against an antigen of 100 kDa characteristic of most cases of common ALL. A few T-cell acute lymphoid leukemias (Greaves et al., 1983a) and a few TdT$^+$ lymphoid precursors in normal adult bone marrow may express this antigen (Janossy et al., 1979). A B-cell-related antigen of 95 kDa, which is expressed along the B-cell lineage from early precursors to mature B lymphocytes, is identified by the CD19 cluster of antibodies. This antigen is also expressed in cases of very immature ALL negative for CALLA antigen (Nadler et al., 1983).

CD20 (35 kDa) and CD22 (135 kDa) antigens are present on the membrane of B cells after the lack of CALLA antigen (mature B cells). CD22 identifies an antigen that is expressed early on the cytoplasm of B cells and so is easily detectable on cytospins or tissue sections of B-cell tissues. Polyclonal or monoclonal antibodies active against smIg of different isotypes and polyclonal or monoclonal antibodies that identify the μ chain of immunoglobulin in the cytoplasm of B cells broaden the range of useful reagents for analyzing hemopoietic cell lineages (Janossy et al., 1986).

smIg are weakly expressed in B-CLL but strongly expressed in other chronic lymphoproliferative disorders such as prolymphocytic leukemia. Moreover, the monoclonality test for kappa and lambda light-chain expression is very important for diagnosing B-cell chronic lymphoproliferative malignancy (Janossy et al., 1986).

Early thymic precursors and corresponding leukemias are stained by surface membrane CD7 (40 kDa), CD5 (65 kDa) (weakly), and intracytoplasmic CD3 antigens (Furley et al., 1986a; Campana et al., 1987). CD3 is expressed on the surface membrane of medullary thymocytes and PB T lymphocytes, although recent studies by flow cytometry analysis have found its weak expression also in cortical thymocytes (Lanier et al., 1986). CD3 is constituted by three polypeptide invariable chains of 25 kDa (gamma chain) and 20 kDa (delta and epsilon chains) (Pessano et al., 1985). It is of interest that CD5 is characteristically weakly expressed on B-CLL cells; since in normal PB it is strongly expressed on T lymphocytosis, by using this reagent one can easily distinguish between a reactive lymphocytosis and a B-CLL proliferation (Janossy et al., 1986). CD2 (50 kDa) identifies the E-rosette receptor; the intrathymic population that lacks this antigen represents the most immature T-cell population (Furley et al., 1986a), but the most immature prothymocytic leukemias usually express it (Furley et al., 1986a; Van Dongen et al., 1987). CD1a (49 kDa) is the characteristic antigen expressed by cortical thymocytes which is not present in earlier thymocytes and is lost in medullary and PB T lymphocytes. It plays an important but uncompletely known role in the mechanism that regulates acquisition of self- and non-self-immunocompetence (Lanier et al., 1986). CD1a, CD1b (45 kDa), and CD1c (43 kDa) are members of the immunoglobulin supergene family and share a low degree of homology to both human and mouse MHC class I alpha and class II beta chains (Martin et al., 1986). CD1b has a distribution in human tissues like that of CD1a but different from that of CD1c. CD1c is expressed on the surface of cortical thymocytes but also on PB and spleen B cells, on the follicular mantle B cells of lymph nodes, and on about 40% of cases of B-CLL and about 30% of cases of B-cell non-Hodgkin's lymphomas of both follicular and diffuse histotypes (Delia et al., 1988).

CD25 (55 kDa) reacts with interleukin 2 receptors (Tac antigen) on activated T cells, whereas resting B and T cells either do not express detectable Tac antigen or display only small number of receptors per cell as compared with stimulated T cells. CD25 is present on the surface membrane of hairy cell leukemia cells and may help to distinguish this disorder from other chronic lymphoproliferative diseases (Janossy et al., 1986; Uchiyama et al., 1981).

Another series of reagents is of particular interest for studying T-cell ontogenesis and immunoregulation of the T-cell system. These reagents recognize epitopes localized on different chains of the TCR. BetaF1 (Brenner et al., 1987), a framework MAb that reacts with an epitope localized on the TCR beta chain at the "hidden" part of the T-lymphocyte surface membrane, is of particular interest (see below). MAb WT31 may recognize the beta chain only when linked with the alpha chain to form a complete CD3 alpha/beta heterodimer on the cell surface membrane (Spits et al., 1985). Recently, reagents reactive with the gamma/delta heterodimer present on a small subset of PB cytotoxic T lymphocytes were also identified (Haynes et al., 1989). At the Oxford workshop, a series of reagents reactive against the products of variable regions of the TCR was presented (Boylston et al., 1986).

A rabbit anti-calf TdT antiserum is an important reagent for detecting immature normal precursors both on B- and T-cell series. It is particularly useful in double indirect immunofluorescence assays to identify small subsets of immature cells (Bollum, 1979). TdT was found also in a small percentage (about 10%) of AML, but a normal counterpart of this TdT$^+$ AML cell has not been until now identified (Bonati and Starich, 1986).

Mouse MAbs against human TdT are now available. However, they do not have substantial advantages over polyclonal reagents and require the use of a goat anti-mouse serum of a different class when double-marker analysis is necessary. It is better to use a combination of three reagents, HTdT-1, HTdT-3, and HTdT-4 (Supertechs, Bethesda, Md.), since these reagents on their own give moderately strong staining not comparable to that provided by the polyclonal rabbit anti-TdT (Janossy *et al.,* 1986).

In George Janossy's laboratory at Royal Free Hospital, London, to develop the UK ALL (United Kingdom acute lymphocytic leukemia) project, a rapid micropletes method for leukemia diagnosis was developed. Briefly, the most relevant antibodies are used for rapid screening by immunofluorescence assay using microplates with C wells. After staining, the cells are pelleted on multitest slides coated with poly-L-lysine and examined. Depending on the results, it is possible to perform some other analyses (1) by staining the residual cells in the microplate with directly labeled second antibody, (2) by preparing cytospins from selected microwells and by staining these with anti-TdT or anti μ-chain antiserum, and finally (3) by analyzing cells from selected microwells on the cell sorter (Janossy *et al.,* 1986).

4. MOLECULAR MARKERS

The recent progress in molecular biology has provided hematologists with new tools to develop additional markers for recognizing hemopoietic precursors. The ability to easily extract DNA from eukaryotic cells, digest it by restriction endonucleases, and analyze it by Southern blot (Southern, 1975) allowed one to detect the amplification or rearrangement of various genes during the maturative process of hemopoietic cells (Greaves, 1986).

Clonal neoplastic tissues are a good model for analyzing hemopoietic precursor cell genes, since the same gene rearrangement is repeated in a multiplicity of cells and thus it may be easier to detect than the rearrangements that occur at the level of polyclonal (nonneoplastic) tissues (Greaves, 1986; Furley *et al.,* 1986a). RNA is studied after extraction by the so-called Northern blot technique (Thomas, 1983). A limitation of RNA investigation was the necessity for a large number of hemopoietic cells, but a recent single-step method of RNA isolation by acid guanidinium thiocyanate–phenol–chloroform extraction has overcome this difficulty (Chomczynski and Sacchi, 1987). In addition, techniques such as *in situ* hybridization permits one to identify both DNA and RNA at the single-cell level.

B-cell receptor for antigen is represented by smIg, and TCR is represented by a 90 kDa heterodimer composed of disulfide-bounded alpha and beta chains, glycoproteins that express clonal variability in their primary structure (Meuer *et al.,* 1983). Recently, a second CD3-associated heterodimer composed of gamma and delta gene products was identified on the surface of a small subset of PB T-cell cytotoxic lymphocytes (Brenner *et al.,* 1986). B-cell and T-cell receptors are encoded by genes constructed from variable (V), joining (J), diversity (D), and constant (C) regions that undergo rearrangement during T-cell differentiation (Greaves, 1986). The rearrangement and expression of immunoglobulins and TCR genes is a prerequisite for the selection of a variable repertoire of B- and T-cell antigen receptors that recognize antigen diversity in the context of self-MHC restriction (Greaves, 1986). The use of probes to genes encoding immunoglobulins and TCR glycoproteins has become a sensitive mean to assess clonality and lineage in lymphoid malignancies (Greaves, 1986a,b; Furley *et al.,* 1986a,b).

The relationship among immunoglobulin gene rearrangement, cytoplasmic immunoglobulin production, and cell surface antigen expression in cases of ALL has been explored (Korsmeyer *et al.*, 1983). All cases of the non-B, non-T classification, which lacked both definitive T-cell markers and surface immunoglobulins, had rearranged immunoglobulin genes, indicating that they represent precursor cells already committed to the B-cell lineage at the gene level. Therefore, we can consider the so-called immunologically null ALLs as leukemias that belong to the B-cell lineage. The presence of HLA-DR but not CALLA in many of the cases that had rearranged heavy-chain genes but retained germ line light-chain genes offers evidence that expression of HLA-DR may precede expression of CALLA during B lymphoid differentiation. In addition, some ALLs examined that lacked BA-1 antigen (CD24, which identifies an antigen, p30, on cells at multiple stages of B-cell development) were HLA-DR$^+$ CALLA$^-$. Intracytoplasmic μ chains were detectable in leukemias at a more advanced stage of maturation. These findings suggest a coordinate sequence of immunoglobulin gene rearrangement and B-cell surface antigen expression in acute leukemias of B-cell lineage.

A parallel examination carried out on T-cell leukemias found a developmentally regulated rearrangement and expression of genes encoding the TCR–T3 (CD3) complex (Furley *et al.*, 1986a). The CD3 complex is represented by invariant proteins (delta, epsilon, and gamma), non covalently associated with the alpha/beta complex, which are implicated in membrane signal transduction or regulation (Meuer *et al.*, 1983).

Different types of T leukemias and normal thymic cells were examined. T-cell leukemias were distinguished in precursor cortical thymocyte phenotype (CD3$^-$ CD1$^-$ CD7$^+$ TdT$^+$), immature cortical thymocyte phenotype (CD3$^-$ CD1$^+$ CD7$^+$ TdT$^+$), relative mature cortical thymolyte phenotype (CD3$^+$ CD1$^+$ CD7$^+$ TdT$^+$), and mature T-cell subset phenotype (CD3$^+$ CD1$^-$ CD7$^+$ or $^-$ TdT$^-$ CD4$^+$ or CD8$^+$). The results showed that human leukemic cells corresponding to the earliest identifiable stages of intrathymic T-cell differentiation lack cell surface expression of the TCR alpha/beta–CD3 complex but transcribe TCR beta on RNA from either germ line (1 of 13 cases) or partially (DJ) or fully (VDJ) rearranged (12 of 13 cases) genes. These cells do not express TCR alfa on RNA that appears in leukemias with a relatively mature cortical thymocyte phenotype but express CD3 delta and epsilon RNA and accumulate CD3 proteins in the cytoplasm. Experiments performed on equivalent normal T cells isolated from CD2$^-$ and CD1$^-$ fractions of human thymuses showed that the TCR beta gene was in a predominantly germ line configuration in these immature T-cell fractions. Expression of the CD3 gene is a very early event in T-cell differentiation also in these normal equivalents of leukemic cells and is one of the earliest signs of T-cell commitment. TCR alpha-chain production appears to be the limiting maturation-linked event in the transport, assembly, and cell surface membrane insertion of the TCR alpha/beta–CD3 complex.

Others, by investigating TCR gamma gene rearrangement, found a germ line aspect of this gene in prothymocytic leukemias but rearrangement of the gene in the other T-ALL (Van Dongen *et al.*, 1987). Moreover, distinct levels of TCR gamma transcript were present only in some thymocytic T-ALL, i.e., some precursor cortical thymocyte leukemias and immature cortical phenotype leukemias, but not in the most mature form of T-cell leukemias, suggesting that TCR gamma gene expression may occur only in the early phases of thymic differentiation, whereas TCR beta gene transcription continues during further differentiation. These data suggest a predetermined order of rearrangement and expression of the TCR genes during development. Ontogenic studies carried out in mice confirmed that rearrangement and expression of the TCR genes is a regulated process, since the gamma genes are rearranged and expressed first, followed by the beta and then the alpha genes (Haars et al., 1986).

A frequence occurrence of IgH gene rearrangement in human T-cell leukemias and of TCR gene rearrangement in B-cell leukemias was found (Furley et al., 1987). It was calculated that an inappropriate or cross-lineage IgH or TCR beta gene rearrangement occur in approximately 25% of the cases. Inappropriate gamma gene rearrangement was less frequently observed, whereas rearrangements of TCR alpha genes are difficult to investigate by standard Southern blotting techniques. AML occasionally shows clonal rearrangement of IgH genes and, more rarely, of TCR genes. It is of interest that IgH and TCR beta and gamma gene rearrangements in AML are frequently associated with the expression of TdT (Foa et al., 1987; Seremetis et al., 1987), a nuclear enzyme that may induce somatic mutations.

Analysis of inappropriate rearrangements showed the involvement of the DJ region but not of the V region, although rearrangements apparently involving the V region have been reported (Pelicci et al., 1985).

Mature T- or B-lymphoid leukemias and lymphomas show a "cross-lineage" rearrangement in 5–10% of cases (Greaves et al., 1987). Rearrangements of kappa and lambda chains are very rare in T cells, perhaps because these would require a prior fully functional VDJ rearrangement. The inappropriate rearrangements are usually not accompanied by detectable mRNA transcript, but paradoxically, transcripts from unrearranged genes in the "wrong" lineage can be occasionally detected, for example, IgH in the myeloid cell line KG1 (Furley et al., 1986b) and TCR beta in normal B cells (Calman and Peterlin, 1986). Inappropriate or cross-lineage rearrangements could also be expressed in a few normal early precursors that are the putative counterparts of these leukemic cells (Greaves, 1986). The existence of these rare populations within normal hemopoietic tissues could be tested by new methodological approaches such as in situ hybridization (Pardoll et al., 1987) or polymerase chain reaction (PCR) amplification (Lee et al., 1987a,b).

The discovery of cellular oncogenes opened the possibility of obtaining new markers of leukemic cells. The transforming genes of retroviruses are derived from normal cellular genes (c-onc) that are conserved among vertebrates (Bishop, 1983). The function(s) of these cellular gene products is not clearly defined. However, there is evidence that virus-induced neoplastic transformation is correlated with enhanced levels of expression of these genes (Duesberg, 1983). Over 25 oncogenes were identified, and many were cloned and sequenced (Bishop, 1983). Changes in either the coding or control regions of these genes were implicated in the development of cancer. Several molecular mechanisms resulting in the amplification of normal oncogene products or the development of aberrant proteins that modify the normal growth control process may be implicated. It is of interest that several cellular oncogenes are located at the breakpoint of chromosomal translocation, as is the case of c-abl in t(9;2) of Ph1$^+$ CML (Dalla Favera et al., 1983) and c-myc in t(8;14) t(8;22) t(2;8) of Burkitt's lymphoma or B-ALL (Hamlyn and Rabbitts, 1983; De Klein et al., 1982).

Alteration of cellular oncogenes was reported in neoplasias of the hemo-lymphopoietic system (Westin et al., 1982; Blick et al., 1984), although many analyses were carried out on cell lines that may not exactly reproduce the genotype and phenotype of the primary neoplastic clone.

A systematic study of the expression of 11 cellular oncogenes was carried out in fresh tumoral cells from 51 ALL and AML patients (Mavilio *et al.*, 1986). The results of this extensive study showed that (1) c-*myc* and c-*myb* are consistently expressed in acute leukemias, whereas c-*fos* is selectively expressed in M4-M5 AML; (2) c-*fes* transcripts are more abundant in AML than in ALL, c-*abl* is variably express at low levels, and c-*src* and c-*erbB* transcripts are not detectable; (3) expression of c-Ha-*ras* and N-*ras* is low or undetectable in all acute leukemias, whereas c-Ki-*ras* is observed only in T- ALL; and (4) the c-*sis* gene is never expressed in AML or ALL.

Some problems such as the lack of uniformity in the level of expression of different cases of the same leukemic subtype and the expression at detectable levels of most cellular oncogenes also in normal hemopoietic cells made the practical use of the oncogenes as hemopoietic cell markers difficult (Birnie *et al.*, 1986; Ferrari *et al.*, 1985; Garin Chesa *et al.*, 1987).

Recently, some interesting practical aspects have been found with respect to CML. In 90% of the patients affected by this disease is present the t(9;22)(q34;q11) chromosome translocation (Rowley, 1973). In this alteration, the breakpoint on chromosome 22 occurs within a 5.8-kb segment of DNA referred to as the "breakpoint cluster region (bcr) (Groffen *et al.*, 1984). The hybrid gene gives rise to a hybrid 8.5-kb bcr–*abl* RNA and encodes a novel p210 bcr–*abl* protein with greater tyrosine kinase activity than the normal p145 c-*abl* protein (Stam *et al.*, 1985; Neria *et al.*, 1986; Konopka *et al.*, 1984). Moreover, the Ph chromosome translocation is demonstrable in some ALL with B-cell precursor phenotype some of which have bcr rearrangement (bcr+) and some of which do not (bcr−) (Erikson *et al.*, 1986). There is now evidence that the Ph+ bcr− leukemias are associated with a novel p190 *abl* protein. Therefore, the bcr+ p210+ ALL are probably lymphoid blast crisis of a CML clinically silent until the blastic phase, whereas the bcr− p190+ cases are *de novo* ALL (Kurzrock *et al.*, 1987; Chan *et al.*, 1987).

Studies were carried out to determine whether differences in the structure of the Ph1 chromosome could be correlated with the clinical stage of the disease. Restriction mapping of the bcr has shown that most chronic-phase CML patients have a breakpoint located in the 5′ region of the bcr, whereas patients in the blast crisis exhibit a breakpoint in the 3′ region (Schaefer-Rego *et al.*, 1987). Although other reports had indicated that there is no correlation between the phase of the disease and the size of breakpoint within the bcr, the breakpoint in a 5′ region can result in a fourfold-longer chronic phase than one in the 3′ region, suggesting that mapping of the breakpoint in the bcr is of considerable prognostic value (Mills *et al.*, 1988). An analysis at the DNA level of few CML long-term survivors confirmed the presence of a breakpoint in the 5′ segment of the bcr; this configuration was preserved also in the blastic phase without a shift to 3′ breakpoint (Selleri *et al.*, 1989).

The degree of variability of the molecular defect present on CML cells is higher than initially thought. Of 80 patients examined, 7 showed rearrangement falling outside the bcr (Saglio *et al.*, 1988). Three cases expressed a rearrangement upstream (two) and downstream (one) of the bcr, whereas in four others the breakpoint of chromosome 22 was not detected. Others reached similar results and found the lack of 8.5-kb RNA transcript in cases with heterogeneous rearrangement (Selleri *et al.*, 1987).

The underlying genetic mechanism in Ph+ bcr− ALL was recently explained (Hermans *et al.*, 1987). Chromosomal breaks still occur within the bcr gene, but they take place in the putative first intron of the gene, which is 5′ of the bcr involved in CML. An alternative bcr–c-*abl* fusion gene is created which is transcribed into a chimeric 7-kb bcr–

abl mRNA encoding the novel p190 fusion protein. Final proof of the existence of p190 bcr–*abl* awaits the development of antibodies active against peptides encoded by the first bcr gene exon. In accordance with the proposed structure for p190 bcr–*abl* is the fact that the available antisera against the bcr gene do not precipitate p190 bcr-*abl* since they are not directed against peptides derived from the first exon of the bcr gene (Van Denderen *et al.*, 1989). The availability of these antibodies not only for immunoprecipitation of cell lysates but also for staining of single cells would lead to a better understanding of the subpopulations involved in CML.

The availability of a molecular probe for the myeloperoxidase (MPO) gene prompted some authors to investigate the levels of expression of this gene in several blast cell populations, clearly identified as lymphoid by immunological and molecular criteria (Ferrari *et al.*, 1988). An unexpectedly high level of MPO mRNA was detected in three of eight patients in whom no MPO protein was detected. This could mean that the normal pattern of lineage-associated gene expression is intrinsically flexible in uncommitted progenitors; this normally transient period of "lineage promiscuity" could then be retained when cells are transformed, arrested at the bi- or multipotential stage of hematopoiesis, and clonally expanded by a leukemogenic factor(s) (Greaves, 1986). This hypothesis needs to be tested in normal tissues by techniques such as separation of early precursors, e.g., by FACS or magnetic beads, and detection of gene expression by *in situ* hybridization (Pardol *et al.*, 1987) or PCR (Lee *et al.*, 1987a,b), as we have suggested above. Alternatively, leukemic cells with a mixed phenotype could be derived by genetic misprogramming, producing the so-called aberrant phenotypes. These apparently aberrant phenotypes, characterized both by intramyeloid infidelity (for example, single cells with erythropoietic and granulopoietic markers) or interlineage infidelity (single cells with myeloid and lymphoid markers) have been identified in several patients (Smith *et al.*, 1983; Bonati *et al.*, 1986).

5. ANALYSIS OF HUMAN HEMOPOIETIC ONTOGENESIS BY IMMUNOLOGICAL AND MOLECULAR MARKERS

New insights into human hemopoietic ontogenesis were gained by using immunological and molecular techniques. In particular, B- and T-cell lineages were widely investigated. The immunofluorescence technique was used to examine several aspects of B-cell ontogeny in humans (Gathings *et al.*, 1977). Large lymphoid cells positive for intracytoplasmic IgM (pre-B cells) were detected in fetal liver as early as 7 weeks of gestation, 2 weeks before the appearance of smIgM$^+$ B lymphocytes. In fetuses after 13 weeks of gestation, pre-B cells and smIgM$^+$ B lymphocytes were represented in equal percentages in fetal liver and bone marrow, and lymphocytes bearing smIgG were detected earlier than those that express smIgD or smIgA. By double immunofluorescence assay, the percentage of B lymphocytes bearing only smIgM as opposed to those expressing both smIgM and smIgD was found to be much higher in fetal liver and bone marrow than in spleen, blood, and lymph nodes. Moreover, smIgG, smIgA, and smIgD are expressed independently on smIgM$^+$ lymphocytes of adult blood, whereas neonatal lymphocytes may simultaneously bear three or more smIg isotypes.

An orderly expression similar to that of surface immunoglobulins was found for rearrangement and expression of TCR and gamma genes during thymic development. A study carried out in the mouse found that gamma genes are rearranged and expressed first, followed by the beta and then the alpha genes (Haars *et al.*, 1986), as discussed above. TCR gene rearrangement occurs exclusively in the thymus, although some gamma gene rearrangement may be observed also in fetal liver, probably originated in committed T-cell progenitors.

We discussed above developmentally regulated rearrangement and expression of genes encoding the TCR–T3 complex found in T-leukemic cells and in thymic cells that are their normal counterparts (Furley *et al.*, 1986a). The existence of a coordinate expression of TdT and TCR beta-chain genes at the protein level was recently investigated in fetal and pediatric thymocytes (Bonati *et al.*, 1989). It was shown that TCR beta chains are not expressed in the $CD2^-$ fraction and in the large TdT^+ thymic blasts that are $CD2^-$ or $CD2^+$ and represent the most immature thymic fractions. The beta chains appear in $CD2^+$ $CD1^+$ TdT^+ thymic subsets prevalently represented by cortical thymocytes, and their expression increases in the medullary thymocytes along with loss of TdT. A similar orderly expression of TdT and cytoplasmic μ chains was observed in B-cell precursors of pediatric and regenerating bone marrow, where TdT expression declines coincident with cytoplasmic μ-chain synthesis (Janossy *et al.*, 1979). This sequence of events along B-cell lineage is detectable even at 15–19 weeks of gestation in fetal liver and bone marrow (Bodger *et al.*, 1983; Hoklawd, *et al.*, 1983; Bonati *et al.*, 1984a,b).

As summarized in Figure 3, μ-chain and TCR beta-chain expression follows TdT expression both during B- and T-cell lineage maturation and in leukemic cells that are the putative counterparts of these normal maturative stages.

Figure 4 shows "lymphoid-like" TdT^+ precursors identified in fetal liver at around 15 weeks of gestation (Bonati *et al.*, 1983a,b).

This orderly sequence of TdT and TCR beta chain is disrupted in early fetal thymocytes (before 20 weeks of gestation), since about 58% of thymocytes at 15 weeks of gestation are beta-chain positive when TdT^+ cells are not present (Bonati *et al.*, 1989). The disruption of TdT and TCR beta chain in fetal thymic ontogeny implies that these proteins are unlikely to have N-region diversifications at this stage of development. A similar situation may apply to early B-cell precursor development in fetal liver with respect to μ chain and TdT, since μ chains appear in fetal liver as early as seven weeks of gestation (Gathings *et al.*, 1977), suggesting that neither μ- nor beta-chain protein expresses N-region diversification at an early stage of development. In fact, it was found that the presence of N-region sequences of the TCR delta gene in mouse adult, but not fetal thymocytes, is associated with the appearance of TdT (Elliott *et al.*, 1988).

The phenotype of both B- and T-cell precursors was widely studied during ontogenesis by the use of MAbs. Most TdT^+ cells that are present at early stages of fetal liver and bone marrow development (Bonati *et al.*, 1983a,b) may express CALLA antigen on their surface and B-cell precursor-associated antigens so that these immature cells are indistinguishable from those of $CALLA^+$ leukemias (Bodger *et al.*, 1983; Bonati *et al.*, 1984a,b; Hokland *et al.*, 1983). Conversely, most fetal spleen B cells at the same gestational stages lack TdT and reflect more advanced stages of B-cell development (Delia *et al.*, 1985).

Fetal spleen at 22 weeks and peripheral lymph nodes at 16–18 weeks contain a lymphoid subset that expresses IgM and IgD weakly and the T-cell-associated antigens

PUTATIVE STEM CELL

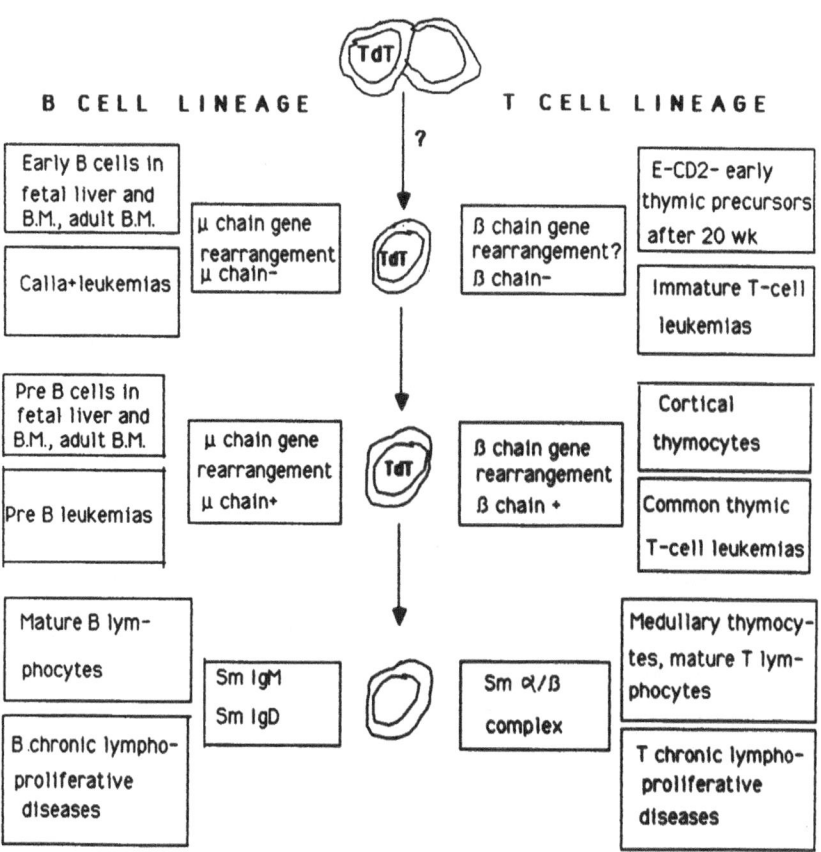

FIGURE 3. Diagram showing that μ- and β-chain protein expression follow TdT both during B- and T-cell normal maturation and in leukemic cells that are the putative counterparts of these normal stages. The figure depicts only a schematic model that is not representative of all aspects of the process; e.g., most immature T-cell leukemias express CD2 and rearrange the TCR beta gene (Furley *et al.*, 1986a; Van Dongen *et al.*, 1987); 25% of CD2[−] normal thymocytes can rearrange the TCR beta gene (Furley *et al.*, 1986a).

CD5 and Tu33 (Bofill *et al.*, 1985; Nadler *et al.*, 1984). These cells are the putative normal counterparts of B-CLL cells and were recognized also in a very small subset of the adult spleen (Caligaris-Cappio *et al.*, 1983).

A recent paper has led to new consideration of the role of fetal lymph nodes during the ontogenesis of the human system (Cattoretti *et al.*, 1989). In 16–22 weeks fetuses, TdT[+] B-cell precursors are nonrandomly allocated in the connective tissue around developing abdominal organs and populate newly formed fetal lymph nodes. TdT[+] B-cell precursors of the lymph nodes differentiate, giving rise to the full range of maturation-associated phenotypes up to the mature follicular B cell. Therefore, the role of the fetal lymph nodes as Bursa-equivalent sites, such as fetal liver and bone marrow, is strongly supported. Moreover, this analysis confirmed the role of the fetal spleen as a site of recovery of more mature B-cell compartments.

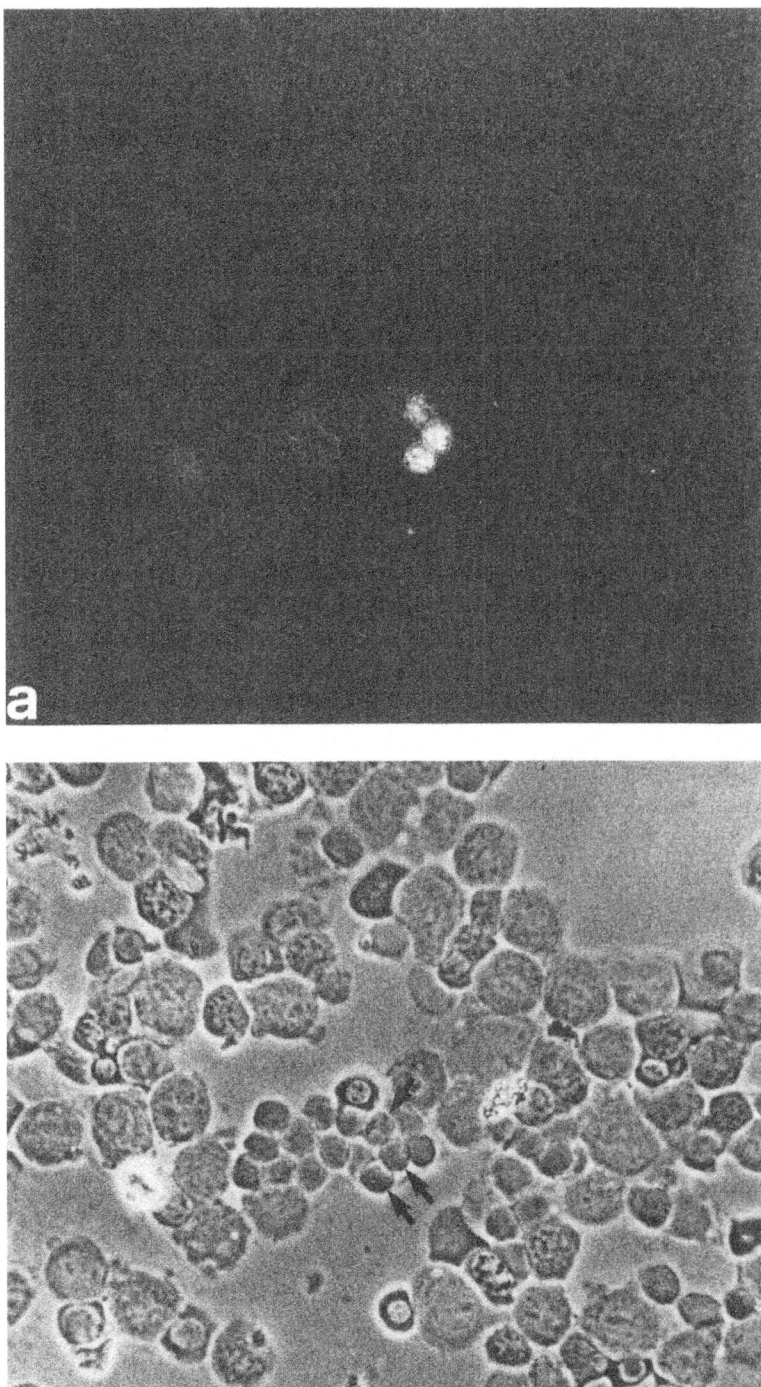

FIGURE 4. TdT⁺ cells in fetal liver at 17 weeks of gestation. (a) Three cells with a clear nuclear positivity; (b) the same field, with arrows pointing to three "lymphoid-like" cells that are TdT⁺ (see panel a) (Bonati *et al.*, 1983a; 1984b).

It was suggested that the thymus is colonized during ontogenesis by waves of pro-thymocytic cells that are able to migrate from bursa-equivalent sites to thymic rudiments and give rise to T lymphocytes (Bodger *et al.*, 1983; Van Dongen *et al.*, 1985a,b). Whereas colonization of the thymus occurs late in the final third of gestation in the mouse, in birds and humans the thymus is colonized by hematopoietic stem cell precursors during the first third of gestation (Lobach *et al.*, 1985). However, there is no certain evidence of the existence of a prothymocytic cell outside the thymus. A very rare HLA-DR$^+$ TdT$^+$ CD7$^+$ cell (putative prothymocyte) was identified in postnatal human bone marrow by a three-step immunoassay using immunogold particles (Van Dongen *et al.*, 1985a,b), but this interesting result was not reproduced by others (Campana *et al.*, 1987). Moreover, T-cell-associated antigens were detected as very rare cells early during ontogenesis in fetal liver, bone marrow, yolk sack (only CD7), and mesenchyme (only CD7), but these cells did not express TdT as a marker of immaturity or TdT testing was not done (Bonati *et al.*, 1984a,b; Lobach *et al.*, 1985; Haynes *et al.*, 1988; Rosenthal *et al.*, 1983).

After lymphoid colonization of the thymic rudiment at 10 weeks of fetal gestation, fetal thymic cells reacted with CD7, CD5, CD4, and CD8 antibodies. The reactivity to membrane surface CD3, CD2, and CD1 appeared at 12 weeks of gestation (Lobach *et al.*, 1985). CD1a, CD1b, and CD1c molecules are expressed early, but in thymuses obtained from younger fetuses (around 12 weeks), a large subpopulation of cells was stained by one and not by the other CD1 MAbs (Amiot *et al.*, 1987). Around 15–19 weeks of gestation, the percentage of cells expressing individual T-cell-associated antigens suggested that many of these antigens are coexpressed on individual cells (Lobach *et al.*, 1985; Bonati *et al.*, 1984a).

The analysis of myeloid lineage during ontogenesis was carried out by tissue culture of precursor cells rather than by immunological detection. By this method, both circulating blood as well as fetal liver and bone marrow were found to be a rich source of CFU-GM around 15 weeks of gestation (Moore and Williams, 1973; Linch *et al.*, 1982). Surface marker analysis of myeloid-associated antigens in fetal tissues at around 15–19 weeks of gestation indicated low-level expression in fetal liver and a percentage like that of the adult in bone marrow (Bonati *et al.*, 1984b).

6. NEW OPPORTUNITIES FOR IMMUNOLOGICAL AND MOLECULAR MARKERS: MINIMAL RESIDUAL DISEASE DETECTION AND THERAPEUTICAL APPROACHES

Immunological and molecular markers have attained great importance in attempts to detect leukemic minimal residual disease. Monoclonal antibodies are currently used also for purging bone marrow in both autologous and allogeneic bone marrow transplantation.

It was suggested that the most sensitive method for detecting lymphoid cells that show the phenotype of ALL of B or T lineage is double immunofluorescence staining for TdT and B- or T-lineage antigens (Janossy *et al.*, 1988). In dilution experiments, double-labeled leukemic cells (TdT$^+$ CD10$^+$ or TdT$^+$ CD7$^+$) were regularly detected on all cytospins of peripheral blood containing 0.01% leukemia cells with the expected phenotype. Also, a method for testing the cytolytic efficacy against leukemic cells of the relevant MAbs in the presence of complement, by mixing TdT$^+$ lymphoblasts and an equal proportion of "inert" cells and then examining on cytospin the residual percentages of TdT $^+$ cells, was developed (Campana and Janossy, 1986). By this method, more than a 4-log cytoreduction of leukemic TdT $^+$ cells was detected. With the help of sensitive tests in the presence of rabbit complement (C1), MAbs CD10 and CD19 and their cocktails were found to be capable of killing >3 log blast cells in 84, 75.5, and 90% of cases of ALL, respectively. CD7 had the same effect on T-ALL cells, killing >3 log leukemic cells in 73% of cases (Janossy *et al.*, 1988).

The encouraging results of these *in vitro* tests prompted the selection of 36 patients in first remission (10 considered at high risk), 23 in second remission, 2 in third remission, and 1 without remission for injection of autologous bone marrow after immunological purging (Janossy *et al.*, 1988). After an observation period of 5–35 months, 24 of the 36 patients are alive and in complete remission. After a complete follow-up of the patients, it will be possible to have a reliable evaluation of the efficacy of these methods of purging in ALL.

Bone marrow purging by using MAbs and complement has been suggested also in AML patients (Lange *et al.*, 1984), but the results until now are not as encouraging as those obtained in ALL (Ball *et al.*, 1986, 1990).

Immunological purging is also used for purging bone marrow from T cells in allogenic bone marrow transplantations to decrease the incidence and gravity of graft-versus-host disease. In particular, Campath1, which was produced in Cambridge, is used on a large scale (Hale *et al.*, 1983); however, the efficacy of T-cell depletion in allogenic bone marrow transplantation is controversial, since this treatment produces a decreased mortality for graft-versus-host disease but the cases of recurrent leukemia are more frequent in the group of purged than of nonpurged patients. The explanation of this behavior is the lack of graft-versus-leukemia reaction in the treated bone marrow (Gale, 1987).

New instruments have become available to hematologists for detecting minimal residual disease since the introduction of molecular biology techniques. DNA sequence polymorphism studies provide a new tool for the observation of engraftment after bone marrow transplantation, for evaluation of posttransplant lymphoma or leukemic relapse, and for analysis of mixed hematopoietic and lymphoid chimeric states (Ginsburg *et al.*, 1985). A technique recently described for diagnosing sickle cell anemia prenatally was used for detection of minimal residual disease (Lee *et a.*, 1987a,b). This technique exploits the sequence amplification produced by the PCR, which can amplify copies of target DNA exponentially and requires small amounts of DNA (less than 1 μg). By means of the PCR, DNA sequences were amplified that flank the crossover sites of a characteristic chromosomal translocation for follicular lymphoma (t(14;18)(q32;q21) (Lee *et al.*, 1987a). The remission marrow and blood samples of a patient with follicular lymphoma and t(14;18) were apparently normal by both morphological and conventional Southern blot analysis, but the t(14;18) hybrid DNA sequences were observed by the PCR method, which permits the detection of cells carrying pathological DNA sequences at a dilution 1 : 100,000. An assay for detecting minimal residual BCR–*abl* transcripts by a modified PCR that uses RNA amplification was described (Lee *et al.*, 1987a,b).

It is of interest that an improved DNA probe assay using a Ph1–BCR-3 probe can identify as few as 1% leukemic cells in a specimen after mixing experiments with normal cells (Blennerhassett *et al.*, 1988). A clinical evaluation of CML patients in remission after allogenic transplantation revealed by this methodology a small fraction of residual

Ph1-positive leukemic cells. It is expected that this assay will permit detection of minimal residual disease in CML patients after autologous and allogenic bone marrow transplantation.

New opportunities for hematological research could be opened by molecular analysis of the mechanism(s) that may induce drug resistance. It is well known that leukemia can be easily treated at first acute phase, but the same drugs that induce complete remission are usually partially or not active at relapse. An overexpressed plasma membrane glycoprotein of molecular mass 170,000 (p-glycoprotein) is detectable in different multidrug-resistant human cell lines and in transplantable tumors; consequently, it was suggested that p-glycoprotein mediates multidrug resistance (Riordan *et al.*, 1985). Chromosomal *in situ* hybridization localized the amplified p-glycoprotein gene that accompanies the development of multidrug resistance in the J774.2 murine cell line (Slovak *et al.*, 1988). It is expected that similar research will be carried out on fresh human leukemic cells for analyzing in both whole-tissue extracts and single cells the mechanisms that induce drug resistance.

7. REFERENCES

Amiot, M., Dastot, H., Schmid, M., Bernard, A., and Boumsell, L., 1987, Analysis of CD1 molecules on thymus cells and leukemic T lymphoblasts identifies discrete phenotypes and reveals that CD1 intermolecular complexes are observed only on normal cells, *Blood* **70**:676–685

Archimbaud, E., Vigreux, B., Tigaud, J. D., Maupas, J., Guyotat, D., Viala, J. J., and Fiere, D., 1988, Serum thymidine kinase in acute non lymphoblastic leukemia, *Leukemia* **2**:245–246

Ball, E. D., Mills, L. E., Coughlin, C. T., Beck, J. R., Cornwell, G. G. III, 1986, Autologous bone marrow transplantation in acute myelogenous leukemia. In vitro treatment with myeloid cell specific monoclonal antibodies, *Blood* **68**:1311–1315

Ball, E. D., Mills, L. E., Cornwell, G. G. III, Davis, B. H., Coughlin, C. T., Howell, A. L., Stukel, T. A., Dain, B. J., McMillan, R., Spruce, W., Miller, W. E., and Thompson, L., 1990, Autologous Bone Marrow Transplantation for Acute Myeloid Leukemia Using Monoclonal Antibody-Purged Bone Marrow, *Blood* **75**:1199–1206

Bell, R., Lillquist, A., Abelson, H., and McCaffrey, R., 1982, Chromatographic forms of terminal deoxynucleotidyl transferase lymphoid cells and in leukemia cells at presentation and relapse, *Leuk. Res.* **6**:775–780

Bertazzoni, U., Brusamolino, E., Isernia, P., Scovassi, I., Torsello, S., Lazzarino, M., and Bernasconi, C., 1982, Prognostic significance of terminal transferase and adenosine deaminase in acute and chronic myeloid leukemia, *Blood* **60**:685–692

Birnie, G. D., Warnock, A. M., Burns, J. H., and Clark, P., 1986, Expression of myc gene locus in populations of leukocytes from leukaemia patients and normal individuals, *Leuk. Res.* **10**:515–526

Bishop, J. M., 1983, Cellular oncogenes and retroviruses, *Annu. Rev. Biochem.* **52**:301–306

Blatt, J., Reaman, G. H., Levin, N., and Poplack, D. G., 1980, Purine nucleoside phosphorylase activity in acute lymphoblastic leukemia, *Blood* **56**:380–382

Blennerhassett, G. T., Furth, M. E., Anderson, A., Burns, J. P., Chaganti, R. S. K., Blick, M., Talpaz, M., Dev, V. G., Chan, L. C., Wiedemann, L. M., Greaves, M. F., Hagemeijer, A., Van Der Plas, D., Skuse, G., Wang, N., and Stam, K., 1988, Clinical evaluation of a DNA probe assay for the Philadelphia (Ph[1]) translocation in chronic myelogenous leukemia, *Leukemia* **2**:648–657

Blick, M., Westin, E., Gutterman, J., Wong-Staal, F., Gallo, R. C., McCredie, K., Keating, M., and Murphy, E., 1984, Oncogene expression in human leukemia, *Blood* **64**:1234–1239

Bodger, M. P., Janossy, G., Bollum, F. J., Burford, G. D., and Hoffbrand, A. V., 1983, The ontogeny of terminal deoxynucleotidyl transferase positive cells in the human fetus, *Blood* **61**:1125–1134

Bofill, M., Janossy, G., Janossa, M., Burford, G. D., Seymour, G. J., Wernet, P., and Kelemen, E., 1985, Human B cell development. II. Subpopulations in the human fetus, *J. Immunol.* **134**:1531–1538

Bollum, F. J., 1960, Oligodeoxyribonucleotide primers for calf thymus polymerase, *J. Biol. Chem.* **235:**18–20

Bollum, F. J., 1975, Antibody to terminal deoxynucleotidyl transferase, *Proc. Natl. Acad. Sci. USA* **72:**4119–4122

Bollum, F. J., 1979, Terminal deoxynucleotidyl transferase as a hemopoietic cell marker, *Blood* **54:**1203–1215

Bollum, F. J., and Chang, L. M. S., 1981, Immunological detection of conserved structure for terminal deoxynucleotidyl transferase, *J. Biol. Chem.* **256:**8767–8770

Bonati, A., and Starcich, B., 1986, Terminal deoxynucleotidyl transferase: A nuclear marker of hemopoietic precursors. Biochemical, immunological and clinical aspects, *Haematologica* **71:**419–429

Bonati, A., Casoli, C., Starcich, B., and Buscaglia, M., 1983a, Terminal deoxynucleotidyl transferase (TdT) in human foetus, *Scand. J. Haematol.* **31:**447–453

Bonati, A., Ferrari, C., Starchich, B., and Tedeschi, F., 1983b, Terminal deoxynucleotidyl transferase (TdT) levels in hairy cell leukemia, *Haematologica* **68:**114–118

Bonati, A.,Delia, D., Rignanese, G., and Buscaglia, M., 1984a, Phenotype of human fetal thymus mono-nuclear cells. Analysis with a panel of monoclonal antibodies in *New Trends in Experimental Hematology* (C. Peschle and C. Rizzoli, eds.), pp. 387–389, Ares Serono Symposia, Rome

Bonati, A., Delia, D., Starcich, B., and Buscaglia, M., 1984b, Phenotype of the terminal transferase-positive cells in human fetal liver and bone-marrow: Analysis with monoclonal antibodies, *Scand. J. Haematol.* **33:**418–424

Bonati, A., Delia, D., and Starchich, R., 1986, Progression of a myelodysplastic syndrome to pre-B acute lymphoblastic leukaemia with unusual phenotype, *Br. J. Haematol.* **64:**487–491

Bonati, A., Cattoretti, G., and Starcich, R., 1989, Discordant expression of terminal transferase and T cell receptor β chain in fetal and pediatric thymocytes, *Leukemia* **3:**130–132

Boylston, A. W., Borst, J., Yssel, H., Blanchard, D., Spits, H., and De Vries, J. E., 1986, Properties of a panel of monoclonal antibodies which react with the human T cell antigen receptor on the leukemic line HPB-ALL and a subset of normal peripheral blood T lymphocytes, *J. Immunol.* **137:**741–744

Brenner, M. B., McLean, J., Dialynas, D. P., Strominger, J. L., Smith, J. A., Owen, F. L., Seidman, J. G., Ip, S., Rosen, F., and Krangel, M. S., 1986, Identification of a putative second T-cell receptor, *Nature (London)* **322:**145–150

Brenner, M. B., McLean, J., Scheft, H., Warnke, R. A., Jones, N., and Strominger, J. L., 1987, Characterization and expression of the human αβ T-cell receptor by using a framework monoclonal antibody, *J. Immunol.* **138:**1502–1509

Brusamolino, E., Isernia, P., Lazzarino, M., Scovassi, I., Bertazzoni, U., and Bernasconi, C., 1984, Clinical utility of terminal deoxynucleotidyl transferase and adenosine deaminase determinations in adult leukemia with a lymphoid phenotype, *J. Clin. Oncol.* **2:**871–880

Caligaris-Cappio, F., Gobbi, M., and Janossy, G., 1983, Infrequent normal B cells express features of B chronic lymphocytic leukemia, *J. Exp. Med.* **155:**623–628

Calman, A. F., and Peterlin, B. M., 1986, Expression of T cell receptor genes in human B cells, *J. Exp. Med.* **164:**1940–1957

Campana, D., and Janossy, G., 1986, Leukemia diagnosis and testing of complement-fixing antibodies for bone marrow purging in acute lymphoid leukemia, *Blood* **68:**1264–1271

Campana, D., Thompson, J. S., Amlot, P., Brown, S., and Janossy, G., 1987, The cytoplasmic expression of CD3 antigens in normal and malignant cells of the T lymphoid lineage, *J. Immunol.* **138:**648–655

Casoli, C., Bonati, A., and Starchich, B., 1983, Ph1 positive acute myelocytic leukemia with high TdT levels, *Cancer* **52:**1210–1214

Cattoretti, G., Parravicini, C., Bonati, A., Buscaglia, M., Zuliani, G., Plebani, A., Delia, D., and Rilke, F., 1989, Terminal deoxynucleotidyl transferase positive B cell precursors in fetal lymph nodes and extra-hemopoietic tissues, *Eur. J. Immunol.* **19:**493–500

Celis, J. E., Petersen Ratz, G., Celis, A., Madsen, P., Gesser, B., Kwee, S., Madsen, P. S., Nielsen, H. V., Yde, H., Lauridsen, J. B., and Basse, B., 1988, Towards establishing comprehensive databases of cellular proteins from transformed human epithelial amnion cells (AMA) and normal peripheral blood mononuclear cells, *Leukemia* **2:**561–602

Chan, L. C., Karhi, K. K., Rayter, S. I., Heisterkamp, N., Eridani, S., Powles, R., Lawler, S. D., Groffen, J., Foulkes, J. G., Greaves, M. F., and Wiedemann, L. M., 1987, A novel *abl* protein expressed in Philadelphia chromosome positive acute lymphoblastic leukaemia, *Nature (London)* **325:**635–637

Chang, L. M. S., 1971, Deoxynucleotyde polymerizing enzymes of calf thymus gland. V. Homogenous terminal deoxynucleotidyl transferase, *J. Biol. Chem.* **246:**909–916

Chomczynski, P., and Sacchi, N., 1987, Single-step method of RNA isolation by acid guanidinium thiocyanate-phenol-chloroform extraction, *Anal. Biochem.* **162:**156–159

Civin, C. I., Vaughan, W. P., Strauss, L. C., Schwartz, J. F., Karp, J. E., and Burke, P. J., 1983, Diagnostic and prognostic utility of cell surface markers in acute non-lymphocytic leukaemia (ANLL), *Exp. Haematol.* **11**(Suppl. 14):152a

Cohen, A., Mansour, A., Dosch, H. M., and Gelfand, E. W., 1980, Association of a lymphocyte purine enzyme deficiency (5′ nucleotidase) with combined immunodeficiency, *Clin. Immunol. Immunopathol.* **15:**245–248

Coleman, M. S., and Hutton, J. J., 1981, Terminal transferase, in *The Leukemic Cell* (D. Catovsky, ed.), pp. 203–219, Churchill Livingston, London

Coleman, M. S., Greenwood, M. F., Hutton, J. J., Holland, P., Lampkin, B., Crill, C., and Kastelic, J. E., 1978, Adenosine deaminase, terminal deoxynucleotidyl transferase (TdT), and cell surface markers in childhood acute leukemia, *Blood* **52:**1125–31

Dalla Favera, R., Martinotti, S., Gallo, R. C., Erickson, J., and Croce, C. M., 1983, Translocation and rearrangements of the c-myc oncogene locus in human undifferentiated B-cell lymphomas, *Science* **219:**963–967

Deibel, M. R., Jr., and Coleman, M. S., 1979, Purification of a high molecular weight human terminal deoxynucleotidyl transferase, *J. Biol. Chem.* **254:**8634–8640

Deibel, M. R., Jr., Coleman, M. S., Acree, K., and Hutton, J. J., 1981, Biochemical and immunological properties of human terminal deoxynucleotidyl transferase purified from blasts of acute lymphoblastic and chronic myelogenous leukemia, *J. Clin. Invest.* **67:**725–734

De Klein, A., Van Kessel, A. G., Grosveld, G., Bartram, C. R., Flagemeijer, A., Bootsma, D., Spurr, N. K., Heisterkamp, N., Groffen, J., and Stephenson, J. R., 1982, A cellular oncogene is translocated to the Philadelphia chromosome in chronic myelocytic leukemia, *Nature (London)* **300:**765–767

Delia, D., Cattoretti, G., Bonati, A., Villa, S., De Braud, F., and Buscaglia, M., 1985, Detection of the common acute lymphoblastic leukaemia antigen (CALLA) on B cells from human fetal tissues. A multiple phenotypic characterization, *Clin. Exp. Immunol.* **59:**305–314

Delia, D., Cattoretti, G., Polli, N., Fontanella, E., Aiello, A., Giardini, R., Rilke, F., and Della Porta, G., 1988, CD1c but neither CD1a nor CD1b molecules are expressed on normal, activated, and malignant human B cells: Identification of a new B-cell subset, *Blood* **72:**241–247

Desiderio, S. V., Yancopoulos, G. D., Paskind, M., Thomas, E., Boss, M. A., Landau, N., Alt, F. W., and Baltimore, D., 1984, Insertion of N regions into heavy-chain genes is correlated with expression of terminal deoxytransferase in B cells, *Nature (London)* **311:**752–755

Drexler, H. G., Gaedicke, G., and Minowada, J., 1985a, Isoenzyme studies in human leukemia-lymphoma cell lines. I. Carboxylic esterase, *Leuk. Res.* **9:**209–229

Drexler, H. G., Gaedicke, G., and Minowada, J., 1985b, Isoenzyme studies in human leukemia-lymphoma cell lines. II. Acid phosphatase, *Leuk. Res.* **9:**537–548

Drexler, H. G., Gaedicke, G., and Minowada, J., 1985c, Isoenzyme studies in human leukemia-lymphoma cell lines. III. β-Hexosaminidase (E.C. 3.2.1.30), *Leuk. Res.* **9:**549–559

Drexler, H. G., Gaedicke, G., and Minowada, J., 1985d, Isoenzyme studies in human leukemia-lymphoma cell lines. IV. Lactate dehydrogenase, *Leuk. Res.* **9:**561–571

Drexler, H. G., Leber, B. F., Norton, J., Yaxley, J., Tatsumi, E., Hoffbrand, A. V., And Minowada, J., 1988, Genotypes and immunophenotypes of Hodgkin's disease-derived cell lines, *Leukemia* **2:**371–376

Duesberg, P. H., 1983, Retroviral transforming genes in normal cells, *Nature (London)* **304:**219–226

Durie, B. G. M., and Grogan, T. M., 1985, CALLA-positive myeloma: An aggressive subtype with poor survival, *Blood* **66:**229–232

Elias, L., and Papayannopoulou, 1985, A cyclic AMP binding protein pattern useful as a biochemical differentiation marker of erythroleukemic cell lines and normal cloned erythroblasts, *Leuk. Res.* **9:**1457–1461

Elliott, J. F., Rock, E. P., Patten, P. A., Davis, M. M., and Chien, Y. H., 1988, The adult T-cell receptor δ-chain is diverse and distinct from that of fetal thymocytes, *Nature (London)* **331:**627–631

Erikson, J., Griffin, C. A., Rushdi, A., Valtieri, M., Hoxie, J., Finan, J., Emanuel, B. S., Rovera, G., Nowell, P. C., and Croce, C. M., 1986, Heterogeneity of chromosome 22 breakpoint in Philadelphia-positive acute lymphocytic leukemia, *Proc. Natl. Acad. Sci. USA* **83:**1807–1811

Ferrari, S., Torelli, U., Selleri, L., Donelli, A. Venturelli, D., Narni, F., Moretti, L., and Torelli, G., 1985, Study of the levels of expression of two oncogenes, c-myc and c-myb, in acute and chronic leukemias of both lymphoid and myeloid lineage, *Leuk. Res.* **9:**833–842

Ferrari, S., Mariano, M. T., Tagliafico, E., Sarti, M., Ceccherelli, G., Selleri, L., Merli, F., Narni, F., Donelli, A., Torelli, U., and Torelli, G., 1988, Myeloperoxidase gene expression in blast cells with a lymphoid phenotype in cases of acute lymphoblastic leukemia, *Blood* **72:**873–876

Ferrata, A., and Storti, E., 1958, *Le Malattie del Sangue,* F. Vallardi, Milan

Foa, R., Casorati, G., Giubellino, M. C., Basso, G., Schirò, R., Pizzolo, G., Lauria, F., Lefranc, M. P., Rabbitts, T. H., and Migone, N., 1987, Rearrangements of immunoglobulin and T-cell receptor β and γ genes are associated with terminal deoxynucleotidyl transferase expression in acute myeloid leukemia, *J. Exp. Med.* **165:**879–890

Fontana, J. A., Emler, C., Ku, K., McClung, J. K., Butcher, F. R., and Durham, J. P., 1984, Cyclic AMP-dependent and -independent protein kinases and protein phosphorylation in human promyelocytic leukemia (HL60) cells induced to differentiate by retinoic acid, *J. Cell. Physiol.* **120:**49–60

Foon, K. A., Nakano, G. M., Koller, C. A., Longo, D. L., and Steis, R. G., 1986, Response to 2'-deoxycoformicin after failure of interferon-α in nonsplenectomized patients with hairy cell leukemia, *Blood* **68:**297–300

Furley, A. J., Mizutani, S., Weilbaecher, K., Dhaliwal, H. S., Ford, A. M., Chan, L. C., Molgaard, H. V., Toyonaga, B., Mak, T., Van Den Elsen, P., Gould, D., Terhorst, C., and Greaves, M. F., 1986a, Developmentally regulated rearrangement and expression of genes encoding the T-cell receptor-T3 complex, *Cell* **46:**75–87

Furley, A. J., Reevers, W. B. R., Mizutani, S., Altass, L. J., Watt, S. M., Jacob, M. C., Van Den Elsen, P., Terhorst, C., and Greaves, M. F., 1986b, Divergent molecular phenotypes of KG1 and KG1a myeloid cell lines, *Blood* **68:**1101–1107

Furley, A. J. W., Chan, L. C., Mizutani, S., Ford, A. M., Weilbaecher, K., Pegram, S. M., and Greaves, M. F., 1987, Lineage specificity of rearrangements and expression of genes encoding the T cell receptor-T3 complex and immunoglobulin heavy chain in leukemia, *Leukemia* **1:**644–652

Gale, R. P., 1987, T-cells, bone marrow transplantation and immunotherapy: Use of monoclonal antibodies. Immune interventions in disease, *Ann. Intern. Med.* **106:**263–267

Garin Chesa, P., Retting, W. J., Melamed, M. R., Old, L. J., and Niman, H. L., 1987, Expression of p21[ras] in normal and malignant human tissues: Lack of association with proliferation and malignancy, *Proc. Natl. Acad. Sci. USA* **84:**3234–3238

Gathings, W. E., Lawton, A. R., and Cooper, M. D., 1977, Immunofluorescent studies of the development of pre-B cells B lymphocytes and immunoglobulin isotype diversity in humans, *Eur. J. Immunol.* **7:**804–810

Giblett, E. R., Anderson, J. E., Cohen, F., Pollara, B., and Meuwissen, H. J., 1972, Adenosine-deaminase deficiency in two patients with severely impaired cellular immunity, *Lancet* **ii:**1067–1069

Giblett, E. R., Amman, A. J., Wara, D. W., Sandmann, R., and Diamond, L. K., 1975, Nucleoside phosphorylase deficiency in a child with severely defective T-cell immunity and normal B-cell immunity, *Lancet* **i:**1010–1013

Ginsburg, D., Joseph, H. A., Smith, B. R., Orkin, S. H., and Rappeport, J. M., 1985, Origin of cell population after bone marrow transplantation. Analysis using DNA sequence polymorphisms, *J. Clin. Invest.* **75:**596–603

Greaves, M. F., 1986, Differentiation-linked leukaemogenesis in lymphocytes, *Science* **234:**697–704

Greaves, M. F., Hariri, G., Newman, R. A., Sutherland, D. R., Ritter, M. A., and Ritz, J., 1983a, Selective expression of the common acute lymphoblastic leukemia (gp 100) antigen on immature lymphoid cells and their malignant counterparts, *Blood* **61:**628–639

Greaves, M. F., Sief, C., and Edwards, P. A. W., 1983b, Monoclonal antiglycophorin as a probe for erythroleukaemias, *Blood* **61:**645–646

Greaves, M. F., Chan, L. C., Furley, A. J. W., Watt, S. M., and Molgaard, H. V., 1986, Lineage promiscuity in hemopoietic differentiation and leukemia, *Blood* **67:**1–11

Greaves, M. F., Furley, A. J. W., Chan, L. C., Ford, A. M., and Molgaard, H. V., 1987, Inappropriate rearrangement of immunoglobulin and T-cell receptor genes, *Immunol. Today* **8:**115–116

Griffin, J. D., Mayer, R. J., Weinstein, H. J., Rosenthal, D. S., Coral, F. S., Beveridge, R. P., and Schlossman, S. F., 1983a, Surface marker analysis of acute myeloblastic leukaemia: Identification of differentiation associated phenotypes, *Blood* **62:**557–563

Griffin, J. D., Ritz, J., Beveridge, R. P., Lipton, J. M., Daley, J. F., and Schlossman, S. F., 1983b, Expression of MY 7 antigen on myeloid precursor cells, *Int. J. Cell. Cloning* **1:**33–48

Griffin, J. D., Todd, R. J., Ritz, J., Nadler, L. M., Canellos, G. P., Rosenthal, D., Gallivan, M., Beveridge, R.

P., Weinstein, H., Karp, D., and Schlossman, S. F., 1983c, Differentiation patterns in the blastic phase of chronic myeloid leukaemia, *Blood* **61**:95–91

Griffin, J. D., Linch, D., Sabbath, K., Larcom, P., and Schlossman, S. F., 1984, A monoclonal antibody reactive with normal and leukemic human myeloid progenitor cells, *Leuk. Res.* **8**:521–534.

Groffen, J., Stephenson, J. R., Heisterkamp, N., De Klein, A., Bartram, C. R., and Grosveld, G., 1984, Philadelphia chromosomal breakpoints are clustered within a limited region, bcr, on chromosome 22, *Cell* **36**:93–99

Haars, R., Kronenberg, M., Gallatin, W. M., Weissman, I. L., Owen, F. L., and Hood, L., 1986, Rearrangement and expression of T-cell antigen receptor and γ genes during thymic development, *J. Exp. Med.* **164**:1–24

Hale, G., Bright, S., Chumbley, G., Hoang, T., Metcalf, D., Munro, A., and Waldmann, H., 1983, Removal of T-cells from bone marrow for transplantation: A monoclonal anti-lymphocyte antibody which fixes human complement, *Blood* **62**:873–882

Hamlyn, P. H., and Rabbitts, T. H., 1983, Translocation joins c-myc and immunoglobulin 1 genes in a Burkitt lymphoma revealing a third exon in the c-myc oncogene, *Nature (London)* **304**:135–139

Haynes, B. F., Martin, M. E., Kay, H. H., and Kurtzberg, J., 1988, Early events in human T cell ontogeny. Phenotypic characterization and immunohistiologic localization of T cell precursors in early human fetal tissues, *J. Exp. Med.* **168**:1061–1080

Haynes, B. F., Denning, S. M., Singer, K. H., and Kurtzberg, J., 1989, Ontogeny of T-cell precursors: A model for the initial stages of human T-cell development, *Immunol. Today* **10**:87–91

Hermans, A., Heisterkamp, N., Von Linden, M., Van Baal, S., Meijer, D., Van Der Plas, D., Wiedeman, L. M., Groffen, J., Bootsma, D., and Grosveld, G., 1987, Unique fusion of bcr and c-abl genes in Philadelphia chromosome positive acute lymphoblastic leukemia, *Cell* **51**:33–40

Hoffbrand, A. V., and Janossy, J., 1981, Enzyme and membrane markers in leukaemia: Recent developments, *J. Clin. Pathol.* **34**:254–262

Hokland, P., Rosenthal, P., Griffin, J. D., Nadler, L. M., Daley, J., Hokland, M., Schlossman, S. F., and Ritz, J., 1983, Purification and characterization of fetal hematopoietic cells that express the common acute lymphoblastic leukemia antigen (CALLA), *J. Exp. Med.* **157**:114–127

Janossy, G., Bollum, F. J., Bradstock, K. F., McMichael, A., Rapson, N., and Greaves, M. F., 1979, Terminal transferase-positive human bone marrow cells exhibit the antigenic phenotype of common acute lymphoblastic leukemia, *J. Immunol.* **123**:1525–1529

Janossy, G., Bollum, J., and Campana, D., 1986, Immunofluorescence studies in leukaemia diagnosis, in *Monoclonal Antibodies* (P. C. L. Beverley, ed.), pp. 97–131, Churchill Livingstone, London

Janossy, G., Campana, D., Burnett, A., Coustan-Smith, E., Timms, A., Bekassy, A. N., Hann, I., Alcorn, M. J., Totterman, T., Simonsson, B., Bengtsson, M., Laurent, J. C., and Poncelet, P., 1988, Autologous bone marrow transplantation in acute lymphoblastic leukemia-preclinical immunologic studies, *Leukemia* **2**:485–495

Jones, D. B., 1987, The histogenesis of Reed-Sternberg cell and its mononuclear counterparts, *J. Pathol.* **151**:191–196

Kohler, G., and Milstein, C., 1975, Continuous cultures of fused cells secreting antibody of predetermined specificity, *Nature (London)* **256**:495–497

Konopka, J. B., Watanabe, S. M., and Witte, O. N., 1984, An alteration of the human c-abl protein in K562 leukemia cells unmasks associated tyrosine kinase activity, *Cell* **37**:1035–1042

Korsmeyer, S. J., Arnold, A., Bakhshi, A., Ravetch, J. V., Siebenlist, U., Hieter, P. A., Sharrow, S. O., LeBien, T. W., Kersey, J. H., Poplack, D. G., Leder, P., and Waldmann, T. A., 1983, Immunoglobulin gene rearrangement and cell surface antigen expression in acute lymphocytic leukemias of T cell and B cell precursor origins, *J. Clin. Invest.* **71**:301–313

Krakow, J. S., Coutsogeorgopoulos, C., and Canellakis, E. S., 1962, Studies on the incorporation of deoxynucleotides and ribonucleotides into deoxyribonucleic acid, *Biochem. Biophys. Acta* **55**:639–650

Kurzrock, R., Shtalrid, M., Romero, P., Kloetzer, W. S., Talpaz, M., Trujillo, J. M., Blick, M., Beran, M., and Guttermann, J. U., 1987, A novel c-abl protein product in Philadelphia-positive acute lymphoblastic leukemia, *Nature (London)* **325**:631–635

Lange, B., Ferrero, D., Pessano, S., Hubbell, H., Palumbo, A., Lai, S. K., and Giovanni, R., 1984, Discrimination between normal hemopoietic stem cells and myeloid leukaemia cells using monoclonal antibodies, in *Minimal Residual Disease in Acute Leukaemia* (B. Lowenberg and A. Hagenbeek, eds.), pp. 55–65, Martinus Nijhoff Publishers, The Hague, Netherlands

Lanier, L. L., Allison, J. P., and Phillips, J. H., 1986, Correlation of cell surface antigen expression on human thymocytes by multi-color flow cytometric analysis: Implications for differentiation, *J. Immunol.* **137:**251–2507

Lee, M. S., Chang, K. S., Cabanillas, F., Freireich, E. J., Trujillo, J. M., and Stass, S. A., 1987a, Detection of minimal residual cells carrying the (t14;18) by DNA sequence amplification, *Proc. Nat. Acad. Science USA* **237:**175–178

Lee, M. S., Chang, K. S., Freireich, E. J., Kantarjian, H. M., Talpaz, M., Trujillo, J. M., and Stass, S. A., 1987b, Detection of minimal residual *bcr/abl* transcripts by a modified polymerase chain reaction, *Blood* **72:**893–897

Linch, D. C., and Griffin, J. D., 1986, Monoclonal antibodies reactive with myeloid associated antigens, in *Monoclonal Antibodies* (P. C. L. Beverley, ed.), pp. 222–246, Churchill Livingstone, London.

Linch, D. C., Knott, L. J., Rodek, C. H., and Huehns, E. R., 1982, Studies of circulating haemopoietic progenitor cells in human fetal blood, *Blood* **59:**976–979

Linch, D. C., Allen, C., Beverley, P. C. L., Bynoe, A. G., Scott, C. S., and Hogg, N., 1984, Monoclonal antibodies differentiating between monocytic and nonmonocytic variants of AML, *Blood* **63:**556–573

Lobach, D. F., Hensley, L. L., Ho, W., and Haynes, B. F., 1985, Human T cell antigen expression during the early stages of fetal thymic maturation, *J. Immunol.* **135:**1752–1759

Ma, D. D. F., Sylwestrowicz, T. A., Granger, S., Massaia, M., Franks, G., Janossy, G., and Hoffbrand, A. V., 1982, Distribution of terminal deoxynucleotidyl transferase, purine degradative and synthetic enzymes in subpopulations of human thymocytes, *J. Immunol.* **129:**1430–1435

Madsen, P. S., Hokland, M., Ellegaard, J., and Hokland, P., Petersen Ratz, G., Celis, A., and Celis, J. E., 1988, Major proteins in normal human lymphocyte subpopulations separated by fluorescence-activated cell sorting and analyzed by two-dimensional gel electrophoresis, *Leukemia* **2:**602–615

Majdić, O., Liszka, K., Lutz, D., Knapp, W., 1981, Myeloid differentiation antigen defined by a monoclonal antibody, *Blood* **58:**1127–1133

Marks, S. M., Baltimore, D., and McCaffrey, R., 1978, Terminal transferase as a predictor of initial responsiveness to vincristine and prednisone in blastic chronic myelogenous leukemia, *N. Engl. J. Med.* **298:**812–814

Martin, L. H., Calabi, F., and Milstein, C., 1986, Isolation of CD1 genes: A family of major histocompatibility complex-related differentiation antigens, *Proc. Natl. Acad. Sci. USA* **83:**9154–9158

Mavilio, F., Sposi, N. M., Petrini, M., Bottero, L., Marinucci, M., De Rossi, G., Amadori, S., Mandelli, F., and Peschle, C., 1986, Expression of cellular oncogenes in primary cells from human acute leukemias, *Proc. Natl. Acad. Sci. USA* **83:**4394–4398

Meuer, S. C., Acuto, O., Hussey, R. E., Hodgdon, J. C., Fitzgerald, K. A., Schlossman, S. F., and Reinherz, E. L., 1983, Evidence for the T3 associated 90Kd heterodimer as the T-cell antigen receptor, *Nature (London)* **303:**808–810

Mills, K. I., MacKenzie, E. D., and Birnie, G. D., 1988, The site of the breakpoint within the bcr is a prognostic factor in Philadelphia-positive CML patients, *Blood* **72:**1237–1241

Moore, M. A. S., and Williams, N., 1973, Analysis of proliferation and differentiation of foetal granulocyte macrophage progenitor cells in haemapoietic tissues, *Cell Tissue Kinet.* **6:**461–476

Munch-Petersen, B., and Tyrsted, G., 1986, Thymidine kinase isoenzymes in human acute and chronic lymphatic leukemia, *Leuk. Res.* **10:**637–642

Murphy, S. B., Stass, S., Kalwinsky, D., and Rivera, G., 1983, Phenotypic conversion of acute leukaemia from T-lymphoblastic to myeloblastic induced by therapy with 2′-deoxycoformicin, *Br. J. Haematol.* **55:**285–293

Nadler, L. M., Anderson, K. C., Marti, G., Bates, M., Park, E., Daley, J. F., and Schlossman, S. F., 1983, B4, a human B lymphocyte associated antigen expressed on normal, mitogen activated and malignant B lymphocytes, *J. Immunol.* **131:**244–250

Nadler, L. M., Korsmeyer, S. J., Anderson, K. C., Boyd, A. W., Slaughenhoupt, B., Park, E., Jensen, J., Coral, F., Meyer, R. J., Sallan, S. E., Ritz, J., and Schlossman, S. F., 1984, B cell origin of non-T cell acute lymphoblastic leukaemia. A model for discrete stages of neoplastic and normal pre-B-cell differentiation, *J. Clin. Invest.* **74:**332–340

Nakamura, H., Tanabe, K., Yoshida, S., and Morita, T., 1981, Terminal deoxynucleotidyl transferase of 60,000 daltons from mouse, rat and calf thymus, *J. Biol. Chem.* **256:**8745–8751

Neria, Y. B., Daley, G. Q., Mes-Masson, A. M., Witte, O. N., and Baltimore, D., 1986, The chronic myelogenous leukemia-specific P210 protein is the product of the *bcr*/c-*abl* hybrid gene, *Science* **233**:212–214

Nielsen, S., Celis, A., Petersen Ratz, G., and Celis, J. E., 1987, Identification of two human phosphoprotein (dividin and IEF 59dl) that are first detected late in G_1 near the G_1/S transition border of the cell cycle, *Leukemia* **1**:68–77

Paietta, E., Bettelheim, P., Schwarzemeier, J. D., Lutz, D., Majdic, O., Knapp, W., 1983, Distinct lympho-blastic and myeoblastic populations in TdT positive acute myeoblastic leukemia: evidence by double-fluorescence staining, *Leuk. Res.* **7**:301–307

Pardoll, D. M., Fowlkes, B. J., Lechler, R. I., Germain, R. N., and Schwartz, R. H., 1987, Early genetic events in T cell development analyzed by in situ hybridization, *Exp. Med.* **165**:1624–1638

Pelicci, P. G., Tabilio, A., Knowles, D. M., II, and Dalla-Favera, R., 1985, Lymphoid tumors displaying rearrangements of both immunoglobulin and T cell receptor genes, *J. Exp. Med.* **162**:1015–1024

Pessano, S., Oettgen, H., Bhan, A. K., and Terhorst, C., 1985, The T3/T cell receptor complex: Antigenic distinction between the two 20-kd T3 (T3-δ and T3-ε) subunits, *Embo. J.* **4**:337–344

Prentice, H. G., Smyth, J. F., Ganeshaguru, K., Wonke, B., Bradstock, K. F., Janossy, G., Goldstone, A. H., and Hoffbrand, A. V., 1980, Remission induction with adenosine-deaminase inhibitor 2′-deoxicoformycin in T-lymphoblastic leukemia, *Lancet* **ii**:170–172

Ratech, H., Borer, W. Z., Winberg, C. D., and Rappaport, H., 1985, Enzymatic differences between chronic lymphocytic leukemia and prolymphocytic leukemia, *Leuk. Res.* **9**:1271–1275

Riordan, J. R., Deuchars, K., Kartner, N., Alon, N., Trent, J., and Ling, V., 1985, Amplification of P-glycoprotein genes in multidrug resistant mammalian cell lines, *Nature (London)* **316**:317–319

Rosenthal, P., Rimm, I. J., Umiel, T., Griffin, J. D., Osathanondh, R., Schlossmann, S. F., and Nadler, L. M., 1983, Ontogeny of human hematopoietic cells: Analysis utilizing monoclonal antibodies, *J. Immunol.* **31**:232–237

Rothemberg, E., and Triglia, D., 1983, Clonal proliferation unlinked to terminal deoxynucleotidyl transferase synthesis in thymocytes of young mice, *J. Immunol.* **130**:1627–1633

Rowley, J. D., 1973, A new consistent chromosomal abnormality in chronic myelogenous leukaemia identified by quinacrine fluorescence and Giemsa staining, *Nature (London)* **243**:290–293

Saglio, G., Guerrasio, A., Tassinari, A., Ponzetto, C., Zaccaria, A., Testoni, P., Celso, B., Rege Cambrin, G., Serra, A., Pegoraro, L., Avanzi, G. C., Attadia, V., Falda, M., and Gavosto, F., 1988, Variability of the molecular defects corresponding to the presence of Philadelphia chromosome in human hematologic malignancies, *Blood* **72**:1203–1208

Schaefer-Rego, K., Dudek, H., Popenoe, D., Arlin, Z., Mears, J. G., Bank, A., and Leibowitz, D., 1987, CML patients in blast crisis have breakpoints localised to a specific region of the bcr, *Blood* **70**:448–455

Scott, C. S., Stark, A. N., Jones, D. B., Minowada, J., Roberts, B. E., and Drexer, H. G., 1988, Quantitative and qualitative enzyme studies of Hodgkin's disease-derived cell lines, *Leukemia* **2**:447–452

Selleri, L., Narni, F., Emilia, G., Colò, A., Zucchini, P., Venturelli, D., Donelli, A., Torelli, U., and Torelli, G., 1987, Philadelphia-positive chronic myeloid leukemia with a chromosome 22 breakpoint outside the breakpoint cluster region, *Blood* **70**:1659–1664

Selleri, L., Emilia, G., Temperani, P., Grassilli, E., Zucchini, P., Tagliafico, E., Bonati, A., Venezia, L., Ferrari, S., Torelli, U., and Torelli, G., 1989, Philadelphia-positive chronic myelogenous leukemia with typical bcr/abl molecular features and atypical, prolonged survival, *Leukemia* **3**:538–542

Seremetis, S. V., Pelicci, P. G., Tabilio, A., Ubriaco, A., Grignani, F., Guttner, J., Winchester, T. J., Knowles, D. M., II, and Dalla-Favera, R., 1987, High frequency of clonal immunoglobulin or T-cell receptor gene rearrangements in acute myelogenous leukemia expressing terminal deoxyribonucleotidyl transferase, *J. Exp. Med.* **165**:1703–1712

Slovak, M. L., Lothstein, L., Horwitz, S. B., and Trent, J. M., 1988, Molecular/cytogenetic alterations accompanying the development of multidrug resistant in the J774.2 murine cell line, *Leukemia* **2**:453–458

Smith, L. J., Curtis, J. E., Messner, H. A., Senn, J. S., Furthmayr, H., and McCulloch, E. A., 1983, Lineage infidelity in acute leukemia, *Blood* **61**:1138–1145

Southern, E. M., 1975, Detection of specific sequences among DNA fragments separated by gel electrophoresis, *J. Mol. Biol.* **98**:503–517

Spits, H., Borst, J., Tax, W., Capel, P. J. A., Terhorst, and De Vries, J. E., 1985, Characteristics of a monoclonal antibody (WT-31) that recognizes a common epitope on the human T cell receptor for antigen, *J. Immunol.* **135**:1922–1928

Srivastrava, B. I., Khan, S. A., and Song, S. Y., 1978, Terminal deoxynucleotidyl transferase in hairy-cell leukaemia and Hodgkin's disease, *Br. J. Cancer* **38**:643–644

Stam, K., Heisterkamp, N., Grosveld, G., De Klein, A., Verma, R. S., Coleman, M., Dosik, H., and Groffen, J., 1985, Evidence of a new chimeric *bcr/c-abl* mRNA in patients with chronic myelocytic leukemia and the Philadelphia chromosome, *N. Engl. J. Med.* **313**:1429–1433

Strauss, L. C., Rowley, S. D., La Russa, V. F., Sharkis, S. J., Stuart, R. K., and Civin, C. I., 1986, Antigenic analysis of hematopoiesis. V. Characterization of My-10 antigen expression by normal lymphohematopoietic progenitor cells, *Exp. Hematol.* **14**:878–886

Taylor, A. T., Stafford, M. A., and Jones, O. W., 1972, Properties of thymidine kinase partially purified from human fetal and adult tissue, *J. Biol. Chem.* **247**:1930–1935

Thomas, P., 1983, Hybridization of denaturate RNA transferred or dotted onto nitrocellulose paper, *Methods Enzymol.* **100**:255–266

Tung, R., Silber, R., Quagliata, F., Conckyn, M., Gottesman, J., and Hirschhorn, R., 1976, Adenosine deaminase activity in chronic lymphocytic leukemia. Relationship to B- and T-cell subpopulations, *J. Clin. Invest.* **57**:756–761

Tyrsted, G., and Munch-Petersen, B., 1977, Early effects on phytohemagglutinin on induction of DNA polymerase, thymidine kinase, deoxyribonucleoside triphosphate pools and DNA synthesis in human lymphocytes, *Nucleic Acids Res* **8**:2713–2716

Uchiyama, T., Broder, S., and Waldmann, T. A., 1981, A monoclonal antibody (anti-Tac) reactive with activated and functionally mature human T cells, *J. Immunol.* **126**:1393–1397

Vainchenker, W., Deschamps, J. F., Bastin, J. M., Guichard, J., Titeux, M., Breton Gorius, J., and McMichael, A. J., 1982, Two monoclonal anti-platelet antibodies as markers of human megakarocyte maturation: Immunofluorescent staining and platelet peroxidase detection in megakaryocyte colonies and in in vivo cells from normal and leukaemic patients, *Blood* **59**:514–521

Van Denderen, J., Hermans, A., Meeuwsen, T., Troelstra, C., Zegers, N., Boersma, W., Grosveld, G., and Van Ewijk, W., 1989, Antibody recognition of the tumor-specific *bcr-abl* joining region in chronic myeloid leukemia, *J. Exp. Med.* **169**:87–89

Van Dongen, J. J. M., Hooijkaas, H., Comans-Bitter, W. M., Benne, K., Van Os, T. M., and De Josselin de Jong, J., 1985a, Triple immunological staining with colloidal gold fluorescein and rhodamine as labels, *J. Immunol. Methods* **80**:1–6

Van Dongen, J. J. M., Hooijkaas, H., Comans-Bitter, W. M., Hählen, K., De Klein, A., Van Zanen, G. E., Van't Veer, M. B., Abels, J., and Benner, R., 1985b, Human bone marrow cells positive for terminal deoxynucleotidyl transferase (TdT), HLA-DR, and T cell marker may represent prothymocytes, *J. Immunol.* **135**:3144–3150

Van Dongen, J. J. M., Quetermous, T., Bartram, C. R., Gold, D. P., Wolvers-Tettero, I. L. M., Comans-Bitter, W. M., Hooijkaas, H., Adriaansen, H. J., De Klein, A., Raghavachar, A., Ganser, A., Duby, A. D., Seidman, J. G., Van Den Elsen, P., and Terhorst, C., 1987, T-cell receptor-CD3 complex during early T-cell differentiation: Analysis of immature T-cell acute lymphoblastic leukemias (T-ALL) at DNA, RNA, and cell membrane level, *J. Immunol.* **138**:1260–1269

Van Laarhoven, J. P. R. M., and De Bruyn, C. H. M. M., 1983, Purine metabolism in relation to leukemia and lymphoid cell differentiation, *Leukemia Res.* **7**:451–480

Westin, E. H., Wong-Staal, F., Gelmann, E. P., Dalla Favera, R., Papas, T. S., Lautenberger, J. A., Eva, A., Premkumar Reddy, E., Tronik, S. R., Aaronson, S. A., and Gallo, R. C., 1982, Expression of cellular homologues of retroviral *onc* genes in human hematopoietic cells, *Proc. Natl. Acad. Sci. USA* **79**:2490–2494

Yancopoulos, G. D., Blackwell, T. K., Suh, H., Hood, L., and Alt, F. W., 1986, Introduced T-cell receptor variable region gene segments recombine in pre-B cells: Evidence that pre-B and T-cells use a common recombinase, *Cell* **44**:251–259

Chapter 2

Cell Surface Markers in Leukemia and Lymphoma

Arnold S. Freedman

1. INTRODUCTION

The hematologic malignancies have long been recognized to be both morphologically and clinically heterogeneous. In the past decade, there have been significant advances in the identification and characterization of the cell surface molecules that are expressed on normal lymphoid and myeloid cells. The application of these markers to the study of leukemias and lymphomas has initiated an understanding of the heterogeneity and both the biologic and clinical behavior of these tumors. It is now possible to assign a lineage derivation to a neoplastic hematopoietic cell in virtually all cases by the expression of lineage-restricted antigens (Ags). More importantly, by using markers that define stages of normal T, B, and myeloid differentiation, it is possible to begin to identify subsets of patients within histologically defined subgroups of leukemias and lymphomas that demonstrate unique clinical presentations, disease courses, and responses to therapy.

Until the advent of monoclonal antibodies (MAbs) directed against lineage-restricted surface Ags on lymphoid and myeloid cells, the assignment of cellular derivation often proved difficult. B cells have been traditionally defined as cells that express surface immunoglobulin (sIg) or can be induced to secrete Ig. T cells, although morphologically indistinguishable from B cells, have been classically defined by the expression of sheep red blood cell (RBC) receptors (E-rosette receptor). In contrast, myeloid cells express non-lineage-restricted markers like major histocompatibility complex (MHC) class II (HLA-DR/Ia) Ags, receptors for the cleavage products of the C3 component of complement, and the Fc portion of IgG. Through the efforts of numerous laboratories worldwide, the vast number of MAbs that define lymphoid and myeloid differentiation Ags have been

Arnold S. Freedman Division of Tumor Immunology, Dana-Farber Cancer Institute and the Department of Medicine, Harvard Medical School, Boston, Massachusetts 02115, USA.

characterized. The International Workshops on Human Leukocyte Differentiation Antigens have facilitated the classification of hundreds of MAbs. This has led to the clustering of cell surface Ags defined by MAbs [CD—(cluster designation)] on the basis of cellular expression and biochemical studies.

It is within the context of a biologic understanding of lymphoid and myeloid ontogeny that I will review the expression of lymphoid and myeloid Ags on leukemias and lymphomas. Normal T, B, and myeloid cell ontogeny and the associated expression of cell surface Ags will be discussed first. Then, in the context of normal cellular differentiation, I will relate the phenotype of the malignant cell to that of its normal cellular counterpart. We will also examine the heterogeneity of Ag expression within histologically defined subgroups.

2. B CELLS

2.1. Normal B-Cell Ontogeny

The *sine qua non* of the B cell is the expression of cytoplasmic and/or integral cell sIg (Preud'homme and Seligman, 1972; Gathings *et al.*, 1977). B-cell ontogeny has been operationally divided into stages including pre-B cell, resting B cell, activated/proliferating B cell, differentiating B cell, and plasma cell. These stages can be defined by the expression of unique cytoplasmic and cell surface markers. In this regard, human B-cell Ags can be subgrouped into four broad categories: (1) Ags that span ontogeny (the so-called pan-B-cell Ags); (2) Ags that appear on the resting B cell and are lost with activation; (3) Ags that are not expressed on the resting B cell but appear after activation; and (4) Ags that appear at the terminal stages of differentiation (Figure 1 and Table I).

The earliest pre-B cells have been defined by their expression of cell surface Ags, including MHC class II (Ia) and CD19, and rearrangements of immunoglobulin μ heavy-chain genes (Korsmeyer *et al.*, 1981, 1983a; Nadler *et al.*, 1983, 1984; Brouet *et al.*,

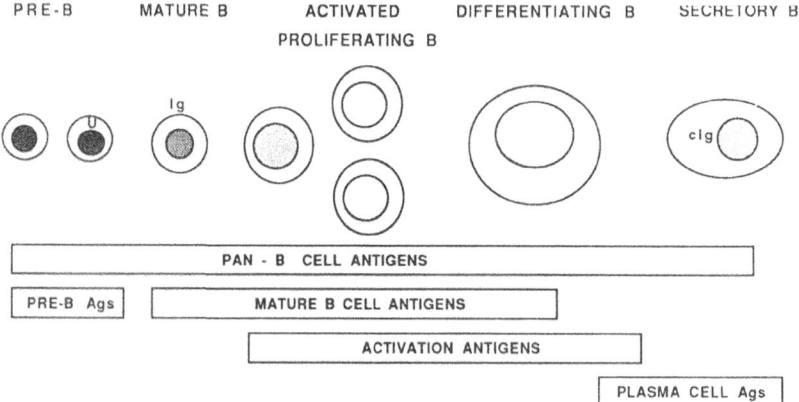

FIGURE 1. Stages of normal B-cell differentiation. The expression subgroups of cell surface antigens characterize these stages.

Table I
Cell Surface Antigen Expression in B-Cell Ontogeny

Antigen or CD no.	Mol. wt. (kDa)	Lineage restriction	Expression in B-cell ontogeny
Ia	29, 34	B, activated T, myeloid	Pan-B
19	95	B	Pan-B
20	35	B	Pan-B
24	42	B	Pan-B
10	100	B, granulocyte	Pre-B
9	26	B, activated T, platelet	Pre-B
45R	220/200	B, T subset, myeloid	Pan-B
40	50	B	Pan-B
72	43/39	B	Pan-B
73	69	B, T	Pan-B
74	41/35/33	B, monocyte	Pan-B
w78		B, monocyte	Pan-B
22	130	B	Mature B lost with activation (cytoplasmic pre-B)
sIg			
IgM	900	B	Mature B lost with activation
IgG	150		
IgD	1150		
21	140	B, DRC[a]	Mature B lost with activation
35	220	B, myeloid, RBC, DRC[a]	Mature B, decrease at plasma cell stage
44	90	B, T, myeloid	Mature B, decrease at plasma cell stage
37	40–45	B, T, myeloid	Mature B, decrease at plasma cell stage
39	80	B, T, monocyte	Mature B, decrease at plasma cell stage
w75	53(?)	B, T	Mature B, decrease at plasma cell stage
76	85, 67	B, T	Mature B, decrease at plasma cell stage
1c	43	B, thymocyte	Mature B, decrease at plasma cell stage
11b/18	180/95	B, T	Mature B, decrease at plasma cell stage
70		B activated, T	Activated B
71	90	All proliferating cells	Activated B
4F2	140	B activated, T, monocyte	Activated B
54	90	B activated, T, myeloid	Activated B
Blast-1	45	B, activated T	Activated B
25	55	B, activated T, monocyte	Activated B
5	67	B, T	Activated B
23	45	B, T	Activated B
77		B	Activated B
B5	75	B	Activated B
BB-1/B7	46	B	Activated B
Bac-1		B	Activated B
38	45	Prothymocyte, plasma cell	Plasma cell
PCA-1	26	Plasma cell, myeloid	Plasma cell

[a]DRC = dendritic reticulum cell.

1979). By virtue of its B-lineage restriction, CD19 is the most reliable cell surface marker of B lineage at the pre-B-cell level. It is not known whether there are Ia$^+$ μ Ig heavy-chain rearranged CD19$^-$ pre-B cells. However, the identification of CD19$^+$ Epstein-Barr virus (EBV)-transformed cytoplasmic (c) μ^- pre-B-cell lines and cytoplasmic μ^- acute lymphoblastic leukemias (ALLs) suggests that CD19 is expressed prior to Ig heavy-chain rearrangement. CD19 is a 95-kDa glycoprotein for which the cDNA has been recently isolated, and sequence homologies suggest that it is a member of the Ig supergene family, with the extracellular portion of CD19 having three Ig-like domains (Stamenkovic and Seed, 1988a; Tedder et al., 1989b). Studies by several investigators have begun to examine the function of the CD19 Ag. In vitro studies with normal mature B cells and pre-B cells demonstrate that anti-CD19 MAbs inhibit B-cell proliferation in response to anti-Ig (Pezzutto et al., 1986). In contrast, malignant pre-B cells are stimulated to grow by anti-CD19 MAbs (Uckun and Ledbetter, 1989a). The effects of anti-CD19 MAbs on Ig secretion are controversial, with both inhibition and enhancement of Ig secretion observed (Dorken et al., 1989; Pezzutto et al., 1989). Additional studies suggest that CD19 may be a receptor involved in the regulation of intracellular Ca^{2+} levels.

In addition to these cell surface Ags, a B-cell-restricted Ag, CD22, is present in the cytoplasm (cCD22) at the earliest stages of pre-B-cell differentiation but is not exported to the cell surface until the mature resting B-cell stage, when sIgD is first expressed (Dorken et al., 1986, 1987). CD22 is a 135-kDa glycoprotein with a 100-kDa protein core (Boue and LeBien, 1988). Recent isolation of the cDNA demonstrates that CD22 has five Ig domains and has significant homology to neutral cellular adhesion molecule and myelin-associated glycoprotein (Stamenkovic and Seed, 1990). Functional studies with MAbs against CD22 have demonstrated augmentation of B-cell proliferation following stimulation with anti-Ig, and it has been hypothesized that CD22 is involved in signal transduction following crosslinking of sIg (Pezzutto et al., 1987).

The stages of normal pre-B-cell differentiation have been examined by isolating pre-B cells from fetal hematopoietic tissues and adult bone marrow (Hokland et al., 1983, 1984, 1985). A sequence of pre-B-cell differentiation was reported by the isolation of distinct populations of pre-B cells and then inducing them to differentiate in vitro (Nadler et al., 1982; Cossman et al., 1982). Following the appearance of CD19 and cytoplasmic CD22, cells express CD10 [common ALL antigen (CALLA)]. CD10 is a 100-kDa glycoprotein that is also expressed on granulocytes (Ritz et al., 1980; Greaves et al., 1983). Recent isolation of the cDNA for CD10 suggests that it is a zinc metalloprotease identical to enkephalinase, although its precise function on lymphoid cells is unknown (Shipp et al., 1988, 1989). Pre-B cells next express the pan-B-cell-restricted Ag CD20 (Stashenko et al., 1980; Nadler et al., 1981b), which is a 35-kDa nonglycosylated phosphoprotein. In vitro B-cell functional studies support the hypothesis that CD20 is involved in the regulation of B-cell activation. Different MAbs directed against CD20 can both inhibit B-cell activation and proliferation or induce B cells to go from the G$_0$ phase of the cell cycle to the G$_1$ phase, depending on the source and state of activation of B cells (Clark et al., 1985; Tedder et al., 1985). The isolation of the cDNA for CD20 has permitted a theoretical model of the structure of CD20 (Tedder et al., 1988; Stamenkovic and Seed, 1988b; Einfeld et al., 1988). CD20 has a very small extracellular domain, several hydrophobic regions that traverse the cell membrane multiple times, and a large cytoplasmic domain. This structure is reminiscent of rhodopsin and the β-adrenergic receptor, both of which are GTP-binding proteins. A recent report by Tedder *et al.* (1989a) suggests that CD20 is involved in transmembrane ion flux and may regulate intracellular Ca^{2+} levels.

The levels of expression of these cell surface Ags define stages of pre-B-cell differentiation that include Ia^+ $CD19^+$ $cCD22^+$ $CD10^-$ $CD20^-$, Ia^+ $CD19^+$ $cCD22^+$ $CD10^+$ $CD20^-$, and Ia^+ $CD19^+$ $cCD22^+$ $CD10^+$ $CD20^+$ (Figure 2). Finally, the last stage of pre-B-cell ontogeny appears with the expression of $c\mu$ Ig heavy chain without the expression of light chains. The expression of $c\mu$ further divides the Ia^+ $CD19^+$ $cCD22^+$ $CD10^+$ $CD20^+$ stage into $c\mu^-$ and $c\mu^+$ stages. A recent report has suggested a somewhat different scheme of cell surface Ag expression in pre-B-cell ontogeny (Hurwitz *et al.*, 1988). These differences include the expression of CD34, a 115-kDa glycoprotein present on myeloid progenitor cells, along with Ia, CD19, cCD22, and CD10 at the earliest pre-B-cell stage. With further maturation, cells no longer express CD34 and have decreased CD10 expression.

Several other B-cell Ags are expressed during pre-B-cell development, including CD9, CD24, CD40, and CD45R (leukocyte common Ag/T200). CD9 is a 26-kDa glycoprotein present on thymocytes, activated B and T cells, and platelets (Kersey *et al.*, 1981; Hercend *et al.*, 1981). The only associated function of CD9 is that anti-CD9 MAbs induce platelet aggregation (Favier *et al.*, 1989). CD24 is a 45 kDa sialoglycoprotein that is also expressed on granulocytes (Abramson *et al.*, 1981). The expression of CD24 on pre-B cells has not been related to the previously discussed stages of pre-B-cell ontogeny. To date, *in vitro* studies have not identified a function for this molecule. CD40 is a 50 kDa phosphoprotein expressed on a subset of pre-B cells (Uckun and Ledbetter, 1989b; Braesch-Anderson *et al.*, 1989). The cDNA for CD40 has been isolated and found to be homologous to the cDNAs of both nerve and epidermal growth factor receptors (Inui *et al.*, 1989). The proliferation of B cells that have been previously activated is enhanced by anti-CD40 MAbs, suggesting that CD40 can mediate a growth signal and is involved in signal transduction. The leukocyte common Ag family of molecules (CD45) is widely

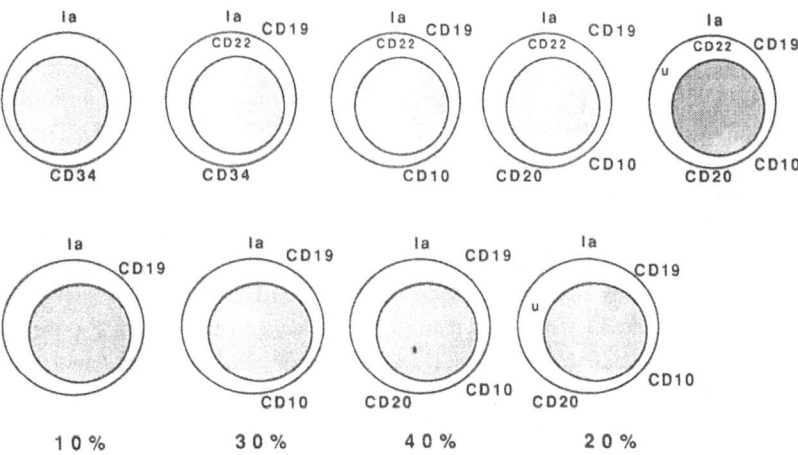

FIGURE 2. Stages of normal pre-B-cell ontogeny (top row) and corresponding subgroups of non-T-cell ALL (bottom row). The percentage of all cases which each subgroup represents is indicated.

expressed on hematopoietic cells (Ritter *et al.*, 1985). MAbs directed against CD45 precipitate four chains of 220, 205, 190, and 180 kDa. The generation of these different polypeptide chains is through differential mRNA splicing of a single transcript (Streuli *et al.*, 1987). The cytoplasmic domain of CD45 has been recently shown to have tyrosine phosphotase activity, suggesting a regulatory function of this molecule (Charbonneau *et al.*, 1988). The 220- and 200-kDa chains are preferentially expressed by B lymphocytes (termed CD45R), and a subset of pre-B cells express CD45R; however, the precise populations have not been defined. Several other pan-B-cell Ags were recently characterized at the 4th International Workshop of Human Leukocyte Differentiation Antigens. These include CD72 (43/39) (Schwarting and Stein, 1989), CD73 (ecto-5' nucleotidase) (Thompson *et al.*, 1989), CD74 (invariant chain of Ia), and CDw78 (Funderud *et al.*, 1989). All of these Ags are present at the pre-B-cell stage and decrease prior to the plasma cell stage.

In adults, pre-B-cell development appears to take place in the bone marrow. As pre-B cells mature, they are exported to the peripheral blood and lymphoid tissues, where they reside until activated by Ag. At present, the definition of the true resting B cell is controversial. Traditionally, this cell was considered to be a small dense cell that expresses sIgM and sIgD. These cells continue to express Ia, CD19, CD20, CD24, and CD22 but no longer express CD10. Coincident with the maturation to the sIg^+ stage, cells express the CD11a/CD18 LFA-1 complex (Campana *et al.*, 1986). CD11a is noncovalently associated with CD18 and is a member of a glycoprotein family involved in cell–cell adhesion. Resting B cells also express a 90-kDa glycoprotein, CD44, which has recently been shown to be involved in lymphocyte homing and in the adhesion of lymphocytes to high-endothelial venules. The recent isolation of the cDNA for CD44 has demonstrated homology to cartilage link and proteoglycan core proteins (Goldstein *et al.*, 1989; Stamenkovic *et al.*, 1989a). Resting B cells also now express CD21, which is the receptor for the C3d cleavage fragment of complement and for EBV (Nadler *et al.*, 1981a, 1986; Iida *et al.*, 1983; Fingeroth *et al.*, 1984), and CD35, which is the C3b complement receptor. CD21 is a 140-kDa nonphosphorylated glycoprotein that is relatively B-cell restricted within the hematopoietic system, with follicular dendritic cells expressing the Ag. In contrast, CD35 is expressed on granulocytes, monocytes, RBCs and follicular dendritic cells (Hogg and Horton, 1987). CD21 appears to be capable of generating a transmembrane stimulatory signal to B cells, as studies with MAbs against CD21 demonstrate stimulation of B-cell proliferation (Frade *et al.*, 1985; Nemerow *et al.*, 1985; Wilson *et al.*, 1985). The 40–45-kDa glycoprotein CD37 and the 80-kDa Ag CD39 are similar in expression to several of these other mature B-cell Ags (CD21, CD22, and CD44), both being lost during B-cell maturation to the plasma cell stage (Ling *et al.*, 1987). The recently described CDw75 (Erikstein *et al.*, 1989) and CD76 (85/67 kDa) (Moller *et al.*, 1989) both appear at the $sIgM^+$ stage and decrease prior to the plasma cell stage. To date, these Ags have not been functionally characterized. A subset of mature B cells have recently been described which express the T-cell-associated Ag CD1. CD1 is a member of the Ig supergene family and shares homology to both MHC class I and class II. CD1 exists as three molecules, CD1a (49 kDa), CD1b (45 kDa), and CD1c (43 kDa), which are noncovalently associated with β_2-microglobulin. Approximately 50% of normal adult splenic B cells and a proportion of peripheral blood B cells are $CD1c^+$, and after activation, CD1c is up regulated (Small *et al.*, 1989).

Recent studies suggest that resting B cells are heterogeneous by virtue of metabolic, genetic, and cell surface markers of activation. As early as seconds after binding of Ag or mitogen, resting B cells demonstrate one or more events of activation. These include increases in both intracellular Ca^{2+} concentration and inositol phospholipid metabolism, expression of the proto-oncogene c-*myc*, synthesis of RNA, increases in cell size, and then entrance into the G_1 phase of the cell cycle (Paul *et al.*, 1986). Therefore, in this context, resting B cells are small dense cells that are sIgM/D$^+$, are in the G_0 phase of the cell cycle, and have not yet begun to undergo any of the above events of activation.

After triggering with Ag or various polyclonal mitogens, resting B cells are activated and subsequently proliferate. In addition to mitogens such as anti-Ig or *Staphylococcus aureus* Cowan 1, which activate B cells by crosslinking sIg, or EBV, which activates through interaction with CD21, several cytokines have been shown to have a stimulatory effect on resting B cells. Both gamma interferon-γ (INF-γ) (Boyd *et al.*, 1987) and interleukin 1 (IL1) (Freedman *et al.*, 1988) prime resting B cells to have an enhanced proliferative response following activation with anti-Ig. The T-cell-derived growth factor IL4 activates B cells from G_0 to G_1, induces the B-cell activation Ag CD23, and hyperinduces MHC class II Ag (Rabin *et al.*, 1985; Defrance *et al.*, 1987). *In vitro*, DNA synthesis begins at between 30 and 48 hr and peaks at 72 hr. Data from *in vitro*, *in vivo*, and *in situ* studies indicate that the activation of resting B cells is accompanied by a sequence of cell surface antigenic changes. Within 24 hr of activation, resting B cells begin to lose sIgD, CD21, and CD22, and this process is complete by 72–95 hr (Boyd et al., 1985a). As these Ags are lost, a number of B-cell-restricted and -associated Ags sequentially appear (Kehrl *et al.*, 1984). These activation Ags are excellent candidates for growth factor receptors, molecules that regulate proliferation and differentiation, structures involved in cell–cell interaction, and molecules that play a role in the localization and binding of activated B cells within a microenvironment.

The activation Ags can be divided into those which are B-cell associated, including CD70 (Stein *et al.*, 1989), CD71 (Haynes *et al.*, 1981b), 4F2 (Haynes *et al.*, 1981a), CD54 (Rothlein *et al.*, 1986), Blast-1 (Thorley-Lawson *et al.*, 1982), CD25 (Tsudo *et al.*, 1984; Boyd *et al.*, 1985b), CD5 (Freedman *et al.*, 1987a; and CD23 (Thorley-Lawson *et al.*, 1985), and those which are B-cell restricted, including CD77 (Murray *et al.*, 1985), B5 (Freedman *et al.*, 1985a), BB-1/B7 (Yokochi *et al.*, 1982; Freedman *et al.*, 1987b), and Bac-1 (Suzuki *et al.*, 1986). The majority of these activation Ags demonstrate peak expression by 72 hr and are no longer expressed on the cell surface at 120 hr. The expression of these Ags is heterogeneous, following different *in vitro* stimuli. Although the functions of most of these Ags are unknown, recent studies have provided information on four of them. CD71 is the transferrin receptor (see Chapter 13 in *Subcellular Biochemistry*, Vol. 17) and is expressed on all proliferating cells. MAbs reactive with CD71 inhibit *in vitro* B-cell proliferation and differentiation (Neckers *et al.*, 1984). CD25, which is the 55-kDa Tac Ag, is the low-affinity IL2 receptor and is present on activated B cells, T cells, and monocytes. CD23 is a 45-kDa glycoprotein that is preferentially expressed on B cells following activation with EBV, phorbol esters, and IL4; it is also present on monocytes, eosinophils, platelets, and activated T cells. CD23 is the low-affinity receptor for IgE and may be associated with a low-molecular-weight B-cell growth factor (BCGF) receptor (Kikutani *et al.*, 1986; Gordon *et al.*, 1986). From the isolation of the cDNA, CD23 is homologous to the asialoglycoprotein receptor (Ikuta *et al.*, 1987).

The fourth activation Ag (CD54) that has been functionally characterized was first identified on activated B cells by the MAb LB-2 (Clark and Yokochi, 1984). This MAb defines the 90-kDa glycoprotein ICAM 1, which is the ligand for LFA-1 (Marlin and Pringer, 1987) and is involved in homotypic adhesion of activated B cells (Patarroyo *et al.*, 1987) as well as the binding of activated B cells to T cells, monocytes, and endothelial cells (Patarroyo *et al.*, 1988; Prieto *et al.*, 1988).

Following activation, B cells are competent to proliferate in response to a variety of cytokines (Table II) (Kishimoto, 1985). There is evidence that IL2, IL4, low- and high-molecular-weight BCGF, INF-γ, and IL1 can induce B-cell proliferation. Proliferating B cells then are induced to secrete Ig and undergo Ig heavy-chain class switching in response to various cytokines. IL2 in high concentrations, IL6, and IL1 have been shown in various systems to induce IgG secretion, whereas IL5 appears to specifically induce IgA secretion. IL4 promotes isotype switching to IgE and IgG1 but inhibits IgM, IgG3, IgG2a, and IgG2b (Snapper and Paul, 1987). In contrast, INF-γ stimulates the IgG2a isotype but inhibits IgG3, Ig2b, IgG1, and IgE production. Accompanying this differentiative stage is the gradual loss of the B-cell activation Ags as well as pan-B-cell Ags, including Ia, CD19, CD20, and CD24. This terminal differentiation stage is also characterized by the appearance of several other Ags, including CD38 and PCA-1, which are expressed on plasma cells (Terhorst *et al.*, 1981; Stamenkovic and Seed, 1989; Anderson *et al.*, 1983). Both of these Ags are B-cell associated; CD38 is expressed on activated T cells and myeloid progenitor cells, whereas PCA-1 is present on activated T cells and granulocytes.

2.2. B-Cell Leukemias and Lymphomas

2.2.1. Non-T-Cell ALL

Leukemic cells from approximately 80% of patients with ALL lack sIg and T-cell Ags and have been considered to be of non-B-, non-T-cell derivation. Although the expression of a number of cell surface Ags, including Ia, CD10 (CALLA), CD9, CD24, and CD71 identifies most non-B-, non-T-cell ALLS, these do not define lineage (Sobol *et al.*, 1985). Several lines of evidence have subsequently supported the conclusion that the majority of these leukemias are derived from stages of pre-B-cell differentiation. Initially, cμ heavy chains were noted in the tumor cells from approximately 20% of patients (Brouet *et al.*, 1979; Greaves *et al.*, 1980; Vogler *et al.*, 1978). Further studies demonstrated that 50% of these leukemias expressed the B-cell-restricted Ag CD20, further suggesting that these ALLs were of B-cell origin (Nadler *et al.*, 1981b). Finally, the observation that greater than 95% of non-B-, non-T-cell ALLs expressed the B-cell-restricted Ag CD19 suggested that these ALLs were derived from pre-B cells (Nadler *et al.*, 1984; Flug *et al.*, 1985; Chen *et al.*, 1986).

More recent studies have suggested that approximately one-third of ALLs express myeloid markers (Sobol *et al.*, 1987; Davey *et al.*, 1988). These myeloid Ags include CD13 and CD33, which are present on the majority of acute myelogenous leukemias (AMLs). Approximately 60% of the myeloid Ag-positive ALLs coexpress B-cell Ags. From a clinical viewpoint, these CD13$^+$ CD33$^+$ ALLs have a lower complete remission (CR) rate than do other phenotypically defined groups. It has also been recently reported that there are ALL cells (5 of 336 cases) and normal fetal liver and bone marrow cells that

Table II
Cell Surface Antigen Expression in T-Cell Ontogeny

Antigen or CD no.	Mol. wt. (kDa)	Lineage restriction	Expression in T-cell ontogeny				
			Prothymocyte, stage I	Thymocyte, stage II	Thymocyte, stage III	Mature T	Activated T
2	50	T (?B)	+	+	+	+	+
7	40	T (some ANNL)	+	+	+	+	+
71	90	All proliferating cells	+				+
38	45	T, myeloid progenitors B (plasma cells)	+				+
1	49	T, subpopulation B		+			
4	55–62	T, monocyte		+	+	+	+
8	34	T		+	+	+	+
3	20–25	T			+	+	+
5	67	T, subpopulation B			+	+	+
6	120	T, activated B			+	+	+
TCR (α/β)	90	T			+	+	+
11a/18	180/95	T, B, myeloid			+	+	+
45R	220, 200	T subset B			+	+	+
44	90	T, B			±	+	+
29	130	T, B, myeloid		+	+	+	+
25	55	Activated T					+
		Activated B monocyte					
26	105	Activated T					+
70		Activated T					+
		Activated B					
9	26	Activated T, pre-B, platelet					+
Ia	29, 34	Activated T, B, myeloid, monocyte					+

coexpress the T-cell Ag CD2 (E-rosette receptor) and CD19 (Uckun *et al.*, 1989). Moreover a subset of murine B cells are also reported to express CD2 (Yagita *et al.*, 1989).

The expression of Ia, CD19, CD10, and CD20 has led to the definition of four subgroups of ALL that correspond to the stages of normal pre-B-cell ontogeny (Figure 2). Nadler *et al.* (1984) examined 138 cases of non-T-cell ALL and noted that tumor cells could be assigned to one of these four subgroups: Ia alone (4%); Ia CD19 (14%); Ia CD19 CD10 (33%); and Ia CD19 CD10 CD20 (49%). It is controversial as to whether the identification of subgroups of pre-B-cell ALL has not yet had an impact on either prognosis or therapy (Hoelzer *et al.*, 1988); however, it is suggested that cell surface phenotype may correlate with age at presentation. Most very young children (less than 2 years old) appear to develop ALL from the earliest pre-B cells (Ia$^+$ CD19$^+$) (Nadler *et al.*, 1984; Pui *et al.*, 1986; Dinndorf and Reaman, 1986; Katz *et al.*, 1988). In contrast, the commonest pre-B ALL in adults express the most mature phenotype (Ia$^+$ CD19$^+$ CD10$^+$ CD20$^+$). This hypothesis is consistent with the hypothesis that the neoplastic event affects the major pre-B-cell population found at each level of development. For example, a large proportion of cells in the fetal bone marrow express Ia and B4, whereas the most common pre-B-cell population in adults is Ia$^+$ CD19$^+$ CD10$^+$ CD20$^+$. A more recent study has reported that ALLs do not as closely correspond to stages of normal pre-B-cell ontogeny (Hurwitz *et al.*, 1988).

Precursor B-cell ALLs express additional B-cell Ags, including CD72 (50% of cases), CDw78 (50% of cases), and CD40 (23% of cases). CD9 is present on 83% of CD10$^+$ ALLs and 68% of CD10$^-$ cases. It has recently been reported that over 50% of these ALLs express CD25 and high-affinity IL2 receptors; however, similar to hairy cell leukemia, they do not proliferate *in vitro* in the presence of IL2 (Woermann *et al.*, 1987). Generally, these leukemias do not express the adhesion molecules CD11a, CD11b, CD18 (Miedema *et al.*, 1985) and CD54, as well as other mature B-cells Ags.

2.2.2. Chronic B-Cell Leukemias

Approximately 95% of chronic lymphocytic leukemias (CLLs) are of B-cell origin (Koziner *et al.*, 1982; Anderson *et al.*, 1984). Whereas 80% of CLLs expressed monoclonal sIg, the remaining sIg$^-$ non-T-cell CLLs expressed B-cell-associated determinants, including receptors for mouse RBCs (MRBCs) and complement components, suggesting that these were also of B-cell derivation. More recent demonstration that CLLs express B-cell-restricted Ags and have rearranged Ig heavy- and light-chain genes provides definitive evidence for a B-cell origin of 95% of CLLs (Arnold *et al.*, 1983). Although the lineage of CLLs has been defined, the state of differentiation of B-CLL is unclear. Morphologically and by conventional cell surface markers, B-CLL resembles the small peripheral blood B cell. This conclusion was based on the observation that B-CLLs express sIgM/D, complement receptors, as well as Ia, CD19, CD20, CD24, CD40, CD44, CD75, and CDw78. However, in contrast to peripheral blood B cells, which express CD11a, CD22, CD35 (C3b receptors), and CD21 (C3d/EBV receptors), B-CLLs are generally only CD21$^+$, only 25% express CD22, and they have less intense expression of CD45R than do normal B cells (Brown *et al.*, 1987). CLLs are heterogeneous for the expression of adhesion molecules. Although CLLs do not express LFA-1 (CD11a), unlike peripheral blood B cells they express the other members of this glycoprotein family, CD11b and CD11c noncovalently associated with CD18 (De la Hera *et al.*, 1988; Morabito *et al.*, 1987a; Pinto *et al.*, 1989). Whereas 50% of peripheral blood B cells express CD1c, only 20–40% of CLLs are CD1c$^+$ (Delia *et al.*, 1988; Orazi *et al.*, 1989). Moreover, virtually all B-CLLs express the CD5 Ag, which is expressed on mature T cells but is not detectable on normal unstimulated B cells (Koziner *et al.*, 1982; Martin *et al.*, 1980, 1981; Boumsell *et al.*, 1980; Kamoun *et al.*, 1981). These studies suggest that B-CLL cells are not derived from small resting B cells. More recently, CLLs have been shown to express several Ags that are not detected on small resting B cells but appear with activation. These include the B-cell activation Ags B5, Blast-1, CD23, and CD25 (Freedman *et al.*, 1987a; Sheibana *et al.*, 1987).

Several studies have demonstrated that cells which phenotypically resemble B-CLL cells can be detected in normal lymphoid populations (Caligaris-Cappio *et al.*, 1982; Ault *et al.*, 1985; Bofill *et al.*, 1985; Freedman *et al.*, 1988). Small numbers of cells that coexpress CD5 and weak sIg and form MRBC rosettes have been observed in normal adult lymph nodes. In addition, CD5$^+$ B cells have been observed at the periphery of the germinal center of adult lymph nodes. CD5$^+$ B cells have been observed to be present in increased numbers in the peripheral blood from patients following allogeneic and autologous bone marrow transplantation, patients with rheumatoid arthritis, and patients with systemic lupus erythematosis. Very small numbers of CD20$^+$ CD5$^+$ cells have also been isolated from normal adult peripheral blood and tonsil but not bone marrow. Further studies have shown that CD5$^+$ B cells are a major subset of fetal B cells and that they closely resemble most B-CLLs by virtue of their expression of Ia, CD19, CD20, CD21, weak sIg, and CD5 but lack of CD35. The demonstration of expression of activation Ags on B-CLL cells has led to the examination of normal *in vitro* activated B cells for a normal cellular counterpart of B-CLL. Whereas anti-Ig-activated B cells express a variety of B-cell-restricted and -associated activation Ags, including B5, Blast-1, and CD25, they do not express CD5. However, a subset of normal B cells stimulated with phorbol myristic acetate (Miller and Gralow, 1984; Freedman *et al.*, 1988, 1989a), which directly activates cells via protein kinase C, express CD5, B5, and CD25. In addition, various other B-cell stimuli, cytokines, and mitogens were examined for their effects on CD5 induction on normal B cells and only IL4 affected CD5 expression by inhibiting the expression of CD5 mRNA (Freedman *et al.*, 1989a and b). These studies suggest that B-CLLs phenotypically resemble several subsets of normal B cells, including *in vitro* activated B cells.

Prolymphocytic leukemia (PLL), although considered to be a chronic leukemia of predominantly B-cell origin, differs from B-CLL both morphologically and clinically. Similar to most B-CLLs, PLL cells are Ia$^+$ CD19$^+$ CD20$^+$ CD21$^+$. However, in contrast to most CLLs, PLL cells express LFA-1, sIg, and CD45R more intensely and generally lack MRBC receptors and CD5 (less than 50% of cases) (Catovsky *et al.*, 1976; Stark *et al.*, 1986). The FMC7 MAb, which defines a subset of normal B cells, is reactive with most PLLs but largely unreactive with CLLs (Brooks *et al.*, 1981; Catovsky *et al.*, 1981). In a preliminary study from our laboratory in which PLLs were examined with a panel of MAbs that define B-cell activation Ags, similar to B-CLLs, most PLLs (four of five cases) expressed B5 and Blast-1 (five of five cases), with less frequent expression of CD25 and CD23 (two of five cases). A report of 24 patients with PLL has confirmed the expression of Ia, CD19, and CD20 but noted CD21 and CD35 on half of the cases (Berribi *et al.*, 1990). The expression of several activation Ags and plasma cell-associated Ags (CD38 and PCA-1) was noted in two-thirds of the cases. The expression of activation Ags is therefore quite similar to that seen with B-CLL; however, by virtue of other differences in cell surface Ags, the normal cellular counterpart of PLL is different from that of CLL.

2.2.3. Hairy Cell Leukemia

Hairy cell leukemias (HCLs) are of B-lineage derivation by the expression of sIg (Catovsky *et al.*, 1974) and the demonstration of rearranged Ig genes (Korsmeyer *et al.*, 1983b). However, the normal B cell(s) from which HCL is derived is controversial. Generally, HCLs express the pan-B-cell Ags Ia, CD19, CD20, CD24, CD45R, and CD72 as well as Ags more limited in their expression on B cells, including CD22, CD40, CD76, and sIg (Anderson *et al.*, 1985; Divine *et al.*, 1984). Although lineage has been clarified by the expression of these Ags, insight into the state of differentiation has come about by studies with addition reagents. The expression in most cases of HCL of several B-cell activation Ags, including B5, CD23, CD25, Bac-1, and HC2 (Posnett *et al.*, 1985, 1987), would lead to the hypothesis that HCL correspond to a subset of activated B cells. However, the demonstration that most HCLs express the monocyte-associated Ags CD11b and CD11c (Schwarting *et al.*, 1985; Kristensen *et al.*, 1987) and PCA-1 suggest that HCL may correspond to a unique minor subpopulation of B cells. Of recent interest is the observation that although HCLs express CD25, they do not respond *in vitro* to IL2, implying that these are low-affinity receptors or are down regulated by some other mechanism (Robb *et al.*, 1984). However, the same cells will proliferate in response to BCGF, and they may in fact produce autostimulatory BCGF (Ford *et al.*, 1985).

2.2.4. Burkitt's Lymphoma

Burkitt's lymphoma is a tumor of B-lineage derivation by the expression of a variety of B-cell-restricted Ags, including CD19, CD20, and sIgM/D as well as the B-cell-associated Ags Ia and CD24. Further examination of these lymphomas has demonstrated that these tumors express CD10 (Ritz *et al.*, 1981) and variably express CD22 as well receptors for the Fc portion of IgG. Burkitt's lymphomas are usually CD1c⁻, and most lack the adhesion molecules LFA-1 (CD11a/CD18) (Clayberger *et al.*, 1987; Inghirami *et al.*, 1988), p150,95 (CD11c), and CD44 (Pals *et al.*, 1989; Picker *et al.*, 1988). When these tumors were examined for the expression of B-cell activation Ags, they were B5⁺ BB-1/B7⁺ CD71⁺ (Porwit-Ksiazek *et al.*, 1983; Kvaloy *et al.*, 1984; Freedman *et al.*, 1987; Salter *et al.*, 1987), and some Burkitt's lymphoma cell lines express CD25 and Bac-1. The normal B lymphocyte from which these lymphomas are derived is controversial. Studies of normal lymphoid tissues suggest that Burkitt's lymphomas may be derived from a subset of germinal center B cells (Mann *et al.*, 1976). The observation that CD10, several B-cell activation Ags, and the Burkitt's lymphoma-associated glycolipid Ag CD77 are detected in the germinal centers of lymph nodes suggests that Burkitt's lymphoma may be the neoplastic counterpart of a subset of normal activated B cells (Harris *et al.*, 1984; Murray *et al.*, 1985).

2.2.5. Nodular Lymphomas

Morphologically, the nodular lymphomas resemble and have been believed to be derived from normal lymphoid follicles. Early studies suggested that nodular lymphoma and normal follicular cells were related by the expression of cell surface Ags, specifically C3 receptors (Jaffe *et al.*, 1974; Cossman and Jaffe, 1981; Stein *et al.*, 1978; Cossman *et al.*, 1984a). Additional immunologic studies of normal lymph node and nodular lymphomas have further supported this hypothesis. Virtually all nodular lymphomas [nodular poorly differentiated lymphocytic (NPDL), nodular mixed lymphocytic/histiocytic, and nodular histiocytic] express monoclonal sIg (Anderson *et al.*, 1984). In most cases, the Ig isotype is $\mu \pm \delta$ or $\mu \pm \gamma$ although occasional cases lack sIg (usually nodular histiocytic) (Aisenberg *et al.*, 1983). Virtually all cases express Ia, CD19, CD20, CD21, CD24, and CD10 but lack CD5. Follicular lymphomas are variably expression for the cell adhesion molecules, being LFA-1$^+$ CD44$^+$ (in 30–50% of cases), but p150,95$^-$ (Pallesen *et al.*, 1989). Several groups have examined normal lymph nodes by immunoperoxidase and have found that germinal centers are stained with MAbs directed against CD20, Ia, CD21, CD71, and CD10 (Bhan *et al.*, 1981; Stein *et al.*, 1982; Greil *et al.*, 1986). Occasional sIg$^-$ cells are also seen in the germinal center. Moreover, a recent examination of the expression of B-cell activation Ags on NPDLs revealed that most cases expressed B5 and BB-1/B7 and half expressed CD71, Blast-1, and CD25 (Freedman *et al.*, 1987). These studies further support the relationship between the normal activated germinal center B cells and nodular lymphomas.

2.2.6. Diffuse Lymphomas

The diffuse lymphomas, including well-differentiated lymphocytic lymphoma (DWDLL), intermediate lymphocytic lymphoma, poorly differentiated lymphocytic lymphoma (DPDLL), and large-cell (DLCL) lymphoma, are very heterogeneous phenotypically as they are clinically. The DWDLL are phenotypically nearly identical to B-CLL by virtue of the expression of Ia, CD19, CD20, CD21, CD24, CD44, p150,95 (CD11c/CD18), weak sIg, and CD5 and the formation of MRBC rosettes. The presence of the adhesion molecule CD11a/CD18 (LFA-1) on DWDLL, which is in contrast to CLL, may account for the differences in anatomic sites of involvement of these two diseases. When examined for the expression of B-cell activation Ags, DWDLL is identical to B-CLL (B5$^+$ CD25$^\pm$ Blast-1$^\pm$) (Freedman *et al.*, 1987; Grant *et al.*, 1986). The normal cellular counterparts of DWDLL cells, as previously discussed for CLL, may be a subset of normal activated cells. The intermediate lymphocytic lymphomas, which correspond phenotypically to normal follicular mantle zone cells, closely resemble DWDLLs. The only difference is the lack of CD21 on intermediate lymphomas (Weisenberger *et al.*, 1987a,b). In contrast to DWDLL, the DPDLLs do not express CD5 and are intensely sIg$^+$. This histologic subtype was very heterogeneous with respect to the expression of B-cell activation Ags. The heterogeneity of DPDLLs makes the correlation of cell surface phenotype to normal B cells difficult. The immunophenotype of the DLCLs has been intensively studied (Freedman *et al.*, 1985b; Van der Valk *et al.*, 1983; Pinkus and Said, 1978; Horning *et al.*, 1984). The vast majority (75–85%) of DLCLs are of B-cell derivation by virtue of the expression of monoclonal sIg and/or the B-cell-restricted Ag CD20.

Approximately 15–20% of cases are of T-cell origin, while true histiocytic (myeloid/ monocytic) DLCLs are rare. The cell surface phenotypes from 57 cases of DLCL have been reported, and virtually all cases expressed Ia, CD20, CD19, and sIg. Approximately 40% weakly expressed CD21, and few cases expressed sIgD or the plasma cell Ag PCA-1. Previous studies of normal B cells have shown that after activation, the expression of CD21 and sIgD decreases and by 4 days they are not detected (Stashenko *et al.*, 1981). A minor population of large cells from normal spleen was isolated that had weak to absent CD21 expression and histologically resembled DLCL cells. These studies suggested that DLCL cells correspond to normal transformed B cells. They are further supported by the demonstration of a common activation Ag phenotype (B5$^+$ CD71$^+$ BB-1/B7$^+$ Blast-1$^+$ 4F2$^+$) for most DLCLs, suggesting that DLCLs may correspond to a subpopulation of activated B cells. The DLCLs are heterogeneous for the expressions of the family of adhesion structures, with 50–75% of tumors expressing LFA-1 and CD44; however, DLCLs are rarely p150,95$^+$. Although several studies have attempted to correlate the immunologic heterogeneity of DLCLs with clinical characteristics and survival, these results have been controversial (Horning *et al.*, 1984; Habeshaw *et al.*, 1983; Rudders *et al.*, 1983).

2.2.7. Disorders of Secretory B Cells

Functionally, multiple myeloma and Waldenström's macroglobulinemia correspond to more terminally differentiated normal B cells. The correlation of cell surface phenotype of these disorders with normal B-cell ontogeny stems from the antigenic changes observed 4–7 days after B cells are stimulated with anti-Ig. Antigens including CD38 and PCA-1 are acquired, and by 7 days the pan-B-cell Ags Ia, CD19, and CD20 are no longer expressed. Cells isolated from patients with Waldenström's macroglobulinemia express Ia, CD19, CD20, and PCA-1 (Ellegaard *et al.*, 1987), whereas myelomas express CD38 and PCA-1, lack Ia, CD19, CD20, CD37, and CD44, and variably express CD11a/CD18, CD24, and CD39 (Anderson *et al.*, 1984). These studies further support the hypothesis that myelomas correspond to more differentiated B cells. Although the terminally differentiated myeloma cell has a unique phenotype, pre-B cells (CD10$^+$) have been detected in bone marrow and peripheral blood from myeloma patients (Pilarski *et al.*, 1985; Goldstein *et al.*, 1985; Caligaris-Cappio *et al.*, 1985). The demonstration of cells that appear to be progenitors of the terminally differentiated neoplastic cells suggests that the neoplastic event may occur at an early stage of B-cell ontogeny.

3. T CELLS

3.1. Normal T-Cell Ontogeny

A large number of MAbs have been developed that define cell surface structures expressed on human T cells. These MAbs have been used to characterize the stages of T-cell ontogeny and differentiation, identify subsets of functionally distinct T cells, and elucidate the functions of some of these cell surface Ags.

During embryonic and early postnatal life, bone marrow precursor cells migrate to the thymus. The thymic microenvironment provides a setting for the processing and eventual development of functionally competent T cells. These cells are subsequently exported into peripheral lymphoid tissues and the circulation (Moore and Owen, 1967; Owen and Ritter, 1969; Owen and Raff, 1970; Stutman and Good, 1971). A sequence of changes in cell surface Ags identified by MAbs is observed to accompany intrathymic differentiation (Reinherz and Schlossman, 1980) (Table II and Figure 3). The cells in the earliest stage (I) of intrathymic differentiation, which constitute 10% of the thymic lymphocytes, express CD2 (E-rosette receptor) (Howard *et al.*, 1981), CD71 (the transferrin receptor) (Terhorst *et al.*, 1981; Haynes *et al.*, 1981b), CD38, and CD7 (Haynes *et al.*, 1979; Vodinelich *et al.*, 1983; Mossalayi *et al.*, 1989). CD2 is a 50-kDa glycoprotein whose natural ligand is LFA-3 (CD58) (Selvaraj *et al.*, 1987). Peripheral blood T cells can be induced to proliferate by the binding of MAbs directed against two different epitopes of the CD2 molecule (Meuer *et al.*, 1984). This triggers T-cell activation, IL2 production, and IL2 receptor expression. Similarly, thymocytes and natural killer cells can be activated via the CD2 alternative pathway. CD7 is a 40-kDa glycoprotein of unknown function; however, CD7 has been proposed to be an Fc receptor for IgM, and anti-CD7 MAb inhibits one-way mixed-lymphocyte reactions.

Stage II thymocytes are characterized by the loss of CD71, the acquisition of CD1 (Terhorst *et al.*, 1981; Amiot *et al.*, 1987), and the coexpression of CD4 and CD8 (McMichael and Gotch, 1987). Both CD4 and CD8 are members of the Ig supergene family. CD4 is a 55–62-kDa glycoprotein that has been implicated in the regulation of T-

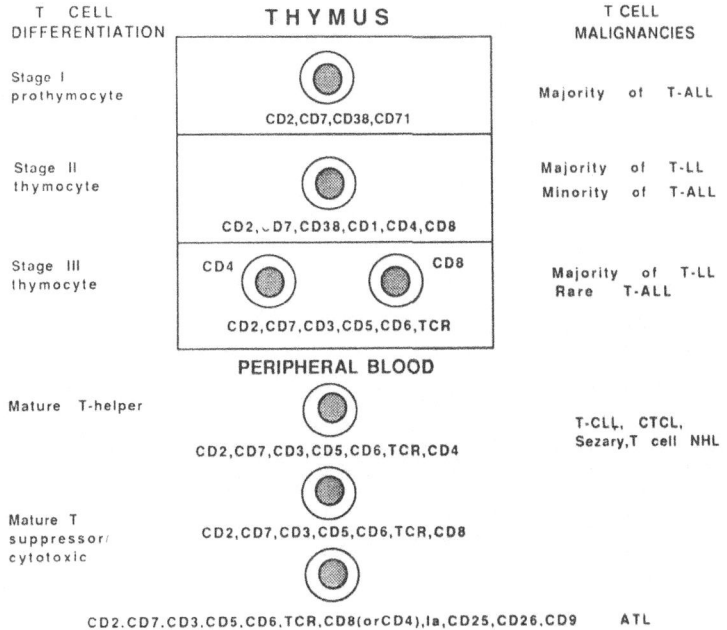

FIGURE 3. Stages of normal T-cell differentiation. Thymic differentiation can be divided into stages on the basis of expression of T-cell antigens. The T-cell leukemias and lymphomas phenotypically resemble stages of normal T-cell ontogeny.

cell activation and is involved in the recognition of MHC class II Ags. CD4 regulatory function may be through its association with a T-cell-specific protein tyrosine kinase (p58^{lck}) (Rudd et al., 1988). CD4 also serves as the receptor of human immunodeficiency virus type 1. CD8 a 34-kDa glycoprotein which, like CD4, may interact with products of the MHC. More recently, it has been suggested that CD8 is involved in regulating T-cell growth; like CD4, CD8 is associated with the p58^{lck} tyrosine kinase (Barber et al., 1989). The population, coexpressing CD1, CD2, CD4, CD7, CD8 and CD38, constitute 70% of thymocytes.

With further maturation, cells lose CD1 and acquire mature T-cell Ags CD3 (Campana et al., 1987), CD5, and CD6 (stage III). CD3 is a complex of three chains, γ, δ, and ε (20–25 kDa) that are noncovalently associated with the T-cell antigen receptor (McMichael and Gotch, 1987). Anti-CD3 MAbs are mitogenic for T cells and induce a Ca^{2+} flux. These and other studies suggest that CD3 is involved in transmembrane signal transduction. CD5, a 67-kDa glycoprotein, from its predicted structure and homologies may mediate a transmembrane signal (McMichael and Gotch, 1987). This view is further supported by observations that anti-CD5 MAbs can augment T-cell proliferation and IL2 production. In parallel with the expression of CD3, cells express the T-cell antigen receptor (TCR), which is noncovalently associated with CD3 (Acuto and Reinherz, 1985). The TCR can exist as a 90-kDa α/β heterodimer that is a member of the Ig supergene family; unlike the antigen receptor for B cells, sIg, the CD3/TCR complex recognizes Ag in the context of MHC. A second TCR, termed γ/δ (70–80 kDa), is also associated with CD3 (Brenner et al., 1986). Cells that express the γ/δ TCR appear earlier in ontogeny than α/β TCR$^+$ cells, are CD4 and CD8 negative, and are associated with natural killer cell activity (non-MHC-restricted cytotoxicity).

When cells leave the thymus, they no longer express CD38 and are segregated into cells expressing CD4 or CD8, constituting 60–70% or 30–40% of peripheral blood T cells, respectively. More recently, a series of MAbs have been developed that further subdivide T4 helper cells into inducers of help (CD45R) and inducers of suppression (CD29) (Morimoto et al., 1985a,b). Peripheral T cells, when activated by Ag, mitogen, anti-CD2, anti-CD3, or anti-TCR MAbs, undergo additional changes in cell surface Ags. During the first 2 days after activation, T cells express the IL2 receptor (CD25) (Uchiyama et al., 1981) and CD26 (Fox et al., 1984). The transferrin receptor (CD71), CD70, CD9 (Hercend et al., 1981), as well as CD38 reappear by 4 days of in vitro culture, and Ia Ags are present by 6–8 days.

3.2. T-Cell Leukemias and Lymphomas

Studies of cell surface molecules have demonstrated that a minority of acute and chronic leukemias as well as non-Hodgkin's lymphomas (NHLs) are of T-cell lineage. With the definition of a series of differentiation Ags, T-cell malignancies generally correspond to distinct stages of normal T-cell ontogeny (Figure 3).

3.2.1. T-Cell ALL

The T-cell acute lymphoblastic leukemias (T-ALLs) constitutes 15–20% of all cases of ALL. The initial observation of heterogeneity of ALL and that some cases were of T lineage was by the fact that those cells formed rosettes with sheep RBCs (Kersey *et al.*, 1973; Dow *et al.*, 1977). MAbs directed against the pan-T-cell Ag CD7 have demonstrated highly specific and sensitive detection of T-ALLs (Borowitz *et al.*, 1985). Monoclonal antibodies to CD7 react with nearly 100% of T-ALLs and 75% of Ia$^-$ E rosette$^-$ ALLs, which also lack a variety of T-cell Ags (CD2, CD5, CD3, CD4, and CD8). However, a small number of AMLs (6%) and chronic myelogenous leukemia (CML) blast crisis of myeloid type are also CD7$^+$. This Ag, although highly conserved in T-cell ontogeny from early thymocytes to and including mature T cells, does not identify heterogeneity of T-ALLs. CD6 has been reported to identify approximately half of the cases of T-ALL (Matutes *et al.*, 1985). In a study that compared various T-cell markers with CD6, the CD6$^-$ cases expressed CD7. Similar to the recent report of ALLs that coexpress B-cell and myeloid Ags, 20% of the ALLs that are CD14$^+$ or CD33$^+$ (myeloid specific) also react with T-cell-specific MAbs.

Studies with a panel of MAbs defining T-cell differentiation Ags have demonstrated that generally T-ALL cells correspond phenotypically to cells in early stages of intrathymic differentiation. Reinherz *et al.* (1979a,b) examined cells from 21 patients with T-ALL and found that the majority correspond to stage I thymocytes (CD38$^+$ and/or CD71$^+$ CD38$^+$) (Figure 3). It has been reported that up to 50% of T-ALLs are Ia$^+$, and these largely correspond to the stage I thymocytes. Only 20% express the phenotype of stage II thymocytes (CD4$^+$ CD1a$^+$ CD8$^+$). One case resembled stage III thymocytes by the expression of CD3. T-ALLs also express additional Ags present on thymocytes, including CD28, LFA-1 (CD11a/CD18), and CD29. Similar to most thymocytes, T-ALLs are CD25$^-$ CD26$^-$ CD45R$^-$. T-ALLs have been reported to be more heterogeneous, with additional cases expressing the phenotypes of mature thymocytes. Link *et al.* (1985) demonstrated in 10 cases of T-ALL that all were essentially CD5$^+$ CD7$^+$ CD38$^+$, with four cases expressing CD3. Interestingly, all of the CD3$^-$ cases had intracytoplasmic CD3, suggesting that cytoplasmic CD3 expression may occur early in T-cell ontogeny (Van Dongen *et al.*, 1988). The cytoplasmic CD3$^+$ T-ALLs have a rearranged TCR β chain, whereas the surface CD3$^+$ have both α and β chains rearranged. As previously discussed, the CD3 structure has been noted to not be expressed on most thymocytes (stage I and early stage II). Similarly, on both early thymocytes and the T-ALLs that are CD3$^-$, the 90-kDa α and β chains of the TCR are not detected (Mirro *et al.*, 1987; Morabito *et al.*, 1987b). In contrast, T-ALL cell lines that express CD3 also coexpress the noncovalently associated TCR structure. Of interest, it has been observed that CD3 on T-ALLs is a functionally active molecule. Stimulation of T-ALL with anti-CD3 induces both colony formation and intracellular Ca^{2+} flux (Ledbetter and Uckun, 1989).

3.2.2. Lymphoblastic Lymphoma

Lymphoblastic lymphoma (LL), a subtype of NHL, is most commonly seen in children and young adults (Jaffe and Berard, 1978). The vast majority of lymphoblastic lymphomas are of T-cell lineage, as determined by formation of E rosettes and reactivity with anti-CD2. Bernard *et al.* (1981) noted that most (49%) T-LLs correspond to common (stage II) thymocytes (CD4$^+$ CD1a$^+$ CD8$^+$), with the remaining cases corresponding to early (30%) or late (20%) thymocytes (CD38$^+$ or CD3$^+$ CD4$^\pm$ CD8$^\pm$, respectively). Recently studies have observed that most T-LLs also lack surface CD3 but express CD5,

CD1, CD71, CD38, and CD7 (Link *et al.*, 1985; Weiss *et al.*, 1986). In addition, one-third of cases are Ia $^+$, and 50% express LFA-1 (CD11a/CD18). Again, as observed in T-ALL, the majority of the CD3 $^-$ cases expressed cytoplasmic CD3. In contrast to these findings, Cossman *et al.* (1983) have observed T-LL to be highly heterogeneous, with CD2 and CD38 ($\pm$CD1) expression the most consistent cell surface phenotype. From a clinical point of view as well as by cell surface markers, T-ALL and T-LL have considerable overlap. However, one difference that has been noted is that approximately 40% of T-LLs express CD10; in contrast, less than 10% of T-ALLs are CD10 $^+$ (Nadler *et al.*, 1980; Ritz *et al.*, 1981). Approximately 50% of T-LLs express the homing receptor/cell adhesion molecule CD44, and this expression correlates with stages of intrathymic differentiation. The CD4 $^-$ CD8 $^-$ T-LLs are CD44 $^+$, whereas the CD4 $^+$ CD8 $^+$ cells are CD44 $^-$. The presence of CD44 did not, however, correlate with leukemic presentation (Picker *et al.*, 1988).

3.3.3. T-Cell Non-Hodgkin's Lymphomas

The vast majority of NHLs are of B-cell origin. T-cell diffuse NHLs, referred to as peripheral T-cell lymphomas to distinguish them from T-LL, are classified according to a modified Rappaport classification as predominantly DLCL or DPDL. Only about 15% of DLCLs are of T-cell origin (Freedman *et al.*, 1985b). Generally, these tumors are either CD3 $^+$ and/or CD2 $^+$ and lack the thymocyte-associated Ag CD1. Most cases, when examined for the expression of mature T-cell Ags, are CD3 $^+$ CD5 $^+$ CD7 $^+$ CD4 $^+$ CD70 $^+$ Ia $^+$, with a minority being CD8 $^+$ (Cossman *et al.*, 1984b; Brouet *et al.*, 1984; Winberg *et al.*, 1985; Grogan *et al.*, 1985; Horning *et al.*, 1986; Stein *et al.*, 1989). Occasional cases are reported that coexpress CD4 and CD8 or lack both Ags. Lennert's lymphoma, the lineage of which has been previously uncertain, appears to be of T-cell derivation. These tumors have rearranged TCR β chains and are CD2 $^+$ CD3 $^+$ CD4 $^+$ (Feller *et al.*, 1986; Spier *et al.*, 1988; Stonesifer *et al.*, 1986). Unlike T-ALLs and T-LLs, the patterns of expression of cell surface antigens on peripheral T-cell NHLs have less consistency and resemble the phenotypes of mature T cells, with a much greater variability and heterogeneity of Ag expression.

3.3.4. Cutaneous T-Cell Lymphomas

Cutaneous T-cell lymphomas (CTCLs) include mycosis fungoides and Sézary syndrome (Broder and Bunn, 1980). Most cases of these malignancies are of the T-helper cell phenotype (CD4 $^+$), but occasional cases are reported to be CD8 $^+$ (Boumsell *et al.*, 1981; Nasu *et al.*, 1985; Jimbow *et al.*, 1985; Haynes *et al.*, 1981c). Functionally, these CD4 $^+$ CTCL cells can provide help in a pokeweed mitogen-driven system with normal B cells. This is further supported by the cell surface phenotype of the helper inducer CD4 $^+$ T cell, CD29 $^+$ CD45R $^-$. These CTCL cells generally also express the mature T-cell Ags CD2, CD3, CD5, CD6, CD11a/CD18, and CD44. Differences in cell surface phenotype have been noted between circulating CTCL cells (Sézary cells) and neoplastic cells infiltrating the skin (Haynes *et al.*, 1982). Sézary cells generally lack CD7 and Ia, whereas the cells infiltrating the skin are CD7 $^+$ Ia $^+$ and variably express CD25 and CD71, suggesting that they may be neoplastic counterparts of activated T-helper cells. It has also been reported that CD7 is not present on CTCL cells that are aggressive or of advanced stage (Sterry and Mielke, 1989).

3.3.5. Adult T-Cell Leukemia/Lymphoma

Adult T-cell leukemia/lymphoma (ATL) has been shown to be associated with the human T-cell leukemia virus type I, with clustering of cases in southern Japan, the southeastern United States, and the Caribbean (Uchiyama et al., 1977). An initial examination of cell surface markers found three cases to express the phenotype of mature activated T-helper cells (CD3$^+$ CD4$^+$ CD5$^+$ CD11a$^+$ CD30$^+$ CD38$^+$) (Hattori et al., 1981; Dallenbach et al., 1989). The expression of CD25 has been variable in some reports, often less intense or absent on freshly isolated cells, and increasing in Ag density with in vitro culture (Waldmann et al., 1984; Uchiyama et al., 1985). Additional characterization of surface Ag phenotype has demonstrated these cells to be generally CD71$^+$ Ia$^+$ (both expressed on activated T cells) CD6$^+$ (marker of mature peripheral T cells) but CD1$^-$. Atypical cases have been described that coexpress CD4 and CD8 as well (Yamada et al., 1985; Tsudo et al., 1986). Functional studies of ATL cells have demonstrated that despite their expression of CD25, most of these cells do not proliferate in response to exogenous IL2, whereas a case of T-CLL that expressed CD25 responded to IL2 and the receptor was modulated by IL2 (Yamada et al., 1984). In contrast, although ATL cells express CD4, they lack helper activity in a pokeweed mitogen-driven system of Ig production by normal B lymphocytes. A consistent observation has been that ATL cells actually mediate suppression of normal Ig production, and these leukemic cells appear to correspond to the normal subset of CD4$^+$ peripheral blood T cells that are inducers of suppression (Morimoto et al., 1985c).

3.3.6. Chronic T-Cell Leukemias

CLLs are generally of B-cell derivation; less than 5% of CLLs and about 20% of PLLs are of T-cell origin. Most cases of T-CLL (Reinherz et al., 1979b; Pandolfi et al., 1982, 1985; Simpkins et al., 1985; Baldini et al., 1985) and T-PLL (Volk et al., 1983; Tsai et al., 1984; Catovsky et al., 1982; Woods et al., 1985) have mature T-helper cell phenotypes (CD2$^+$ CD3$^+$ CD4$^+$ CD5$^+$ CD7$^+$ CD11a$^+$), with occasional reported cases of cells coexpressing CD4 and CD8. Both T-CLL and T-PLL have been recently shown to express CD25; the T-CLL case required phytohemagglutinin stimulation to induce CD25. Functionally, the proliferative response to mitogens of T-CLL and T-PLL cells have been generally diminished with occasional cases reported to have intact helper function, with either pokeweed mitogen- or IL2-driven Ig synthesis.

Several cases have been described of T-CLLs that lack CD4 and CD5 and express CD8 or CD11b (Tagawa et al., 1983, 1986a). These cases generally express CD3 and receptors for the Fc portion of IgG (CD16) and are probably part of the T-gamma lymphoproliferative disorder (Rümke et al., 1982; Reynolds et al., 1984; Miedema et al., 1986). Morphologically, these cells often resemble normal large granular lymphocytes (LGL) (Ferrarini et al., 1983; Loiseau et al., 1987; Oshimi et al., 1988; Tagawa et al.,

1986b). These populations include most if not all natural killer cells, which express CD16, CD2, often CD38, CD8, CD3, and CD11b. The neoplastic cells are also generally CD2$^+$, with certain populations being CD3$^+$ CD8$^+$ CD11b$^+$ Ia$^\pm$ and expressing a variety of cell surface Ags characteristic of natural killer cells. *In vitro* assays of functional activity have demonstrated that cells from certain cases of Fcγ-positive T-CLL and T-gamma lymphoproliferative disease can function as natural killer and/or suppressor cells (as measured by suppression of Ig production by normal B cells in a pokeweed mitogen-driven system) after culture with anti-CD3 or IL2 (Palutke *et al.*, 1983; Loughran *et al.*, 1987). Studies of TCR rearrangement in these LGL leukemias have been variable, with rearrangements of γ as well as β chains reported. The characterization with MAbs of subpopulations of peripheral blood lymphocytes with natural killer activity (Hercend *et al.*, 1985) may provide further understanding of this disorder and certain of its manifestations, including anemia, neutropenia, and hypogammaglobulinemia.

4. MYELOID CELLS

4.1. Normal Myeloid Ontogeny

In normal bone marrow, pluripotent stem cells with both self-regenerating and differentiating properties give rise to all mature myeloid cells (Quesenberry and Levitt, 1979). As the progeny of these pluripotent stem cells differentiate, they become committed to a single hematopoietic lineage (Till and McCulloch, 1980; Bradley and Metcalf, 1966; Pike and Robinson, 1970). Colony assays have been developed for committed progenitor cells of the erythroid lineage (BFU-E, CFU-E), granulocyte/monocyte lineage (CFU-GM), and platelet lineage (CFU-mega). In addition, multipotent colony-forming cells can be assayed (CFU-GEMM), and most recently, *in vitro* assays have been described for pluripotent stem cells. In all stages of normal myelopoiesis, proliferation is tightly coupled to differentiation where each maturing cell stage has less proliferative potential. This is in contrast to AML, where proliferation proceeds in the absence of terminal differentiation. Therefore, analysis of cell surface Ags and functional capacity of normal myeloid progenitors may provide insight into aberrant differentiation of myeloid leukemias.

A variety of techniques have been used to isolate myeloid progenitor cells in order to examine the cell surface phenotypes of these cells and the patterns of expression of a variety of Ags during normal myeloid differentiation (Figure 4 and Table III). The earliest myeloid Ag expressed on multipotent progenitor cells is CD34 (Strauss and Civin, 1983). CD34 is a 115-kDa glycoprotein that identifies long-term bone marrow-initiating cells. Cells at an immature level of differentiation, CFU-GEMM, express CD34 and CD33, which is a 67-kDa glycoprotein (Griffin *et al.*, 1984). The CFU-GM is identified by the expression of Ia, CD13 (Griffin *et al.*, 1981, 1983b; Sakai *et al.*, 1987), and CD15. CD13 is a 150-kDa glycoprotein identified as aminopeptidase N (Look *et al.*, 1989), which is homologous to other zinc-binding metalloproteases. This enzyme is believed to metabolize regulatory peptides. Griffin *et al.* (1983) have observed that this Ag identifies a subset of CFU-GM cells that are most actively proliferating. Anti-CD15 MAb reacts with lac-

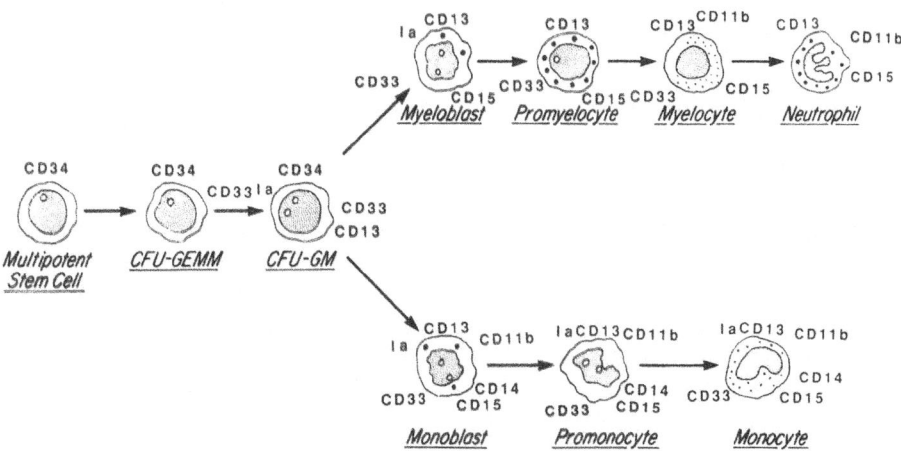

FIGURE 4. Cell surface antigen expression during normal myeloid cell diffentiation.

to-*N*-fucose pentosyl III (Tetteroo *et al.*, 1984). These MAbs exert a variety of functions on granulocytes, including suppression of zymosan-stimulated oxygen consumption, increase in intracellular Ca^{2+}, stimulation of locomotion, and suppression of phagocytosis. With differentiation along the granulocyte pathway, CD11b is acquired. CD11b is noncovalently associated with CD18 and is the receptor for the C3bi cleavage fragment of the third component of complement (Arnaout *et al.*, 1983). As granulocyte maturation continues, cells lose Ia at the promyelocyte level and CD11b increases in Ag density. Initially, monocyte maturation from the CFU-GM is characterized by acquisition of PM-81, similar to granulocyte differentiation. Further maturation of monocytes is characterized by the acquisition of CD11b followed by CD14 (Todd *et al.*, 1981). CD14 is a 55-kDa glycoprotein, the gene for which maps to chromosome 5 in the area encoding other hematopoietic growth factors and receptors. In studies using anti-CD14 MAbs, it has been suggested that CD14 may be involved in monocyte phagocytic function. In contrast to granulocyte precursors and granulocytes, where Ia and CD33 are lost, the monocyte pathway is

Table III
Cell Surface Antigen Expression in Myeloid Cell Ontogeny

Antigen or CD no.	Mol. wt. (kDa)	Stages of myeloid differentiation
34	115	Pluripotent stem cell, CFU-GEMM, CFU-GM
Ia	29, 34	CFU-GEMM, CFU-GM, myeloblast, monocyte
33	67	CFU-GEMM, CFU-GM, promyelocyte, monocyte
13	150	CFU-GEMM, CFU-GM, granulocyte, monocyte
15		Myeloblast, monoblast, granulocyte, monocyte
116/18	180/95	Monoblast, monocyte, myelocyte, granulocyte
14	55	Monoblast, monocyte

characterized by the persistent expression of Ia and CD33 through to and including differentiated monocytes.

4.2. Myeloid Leukemias

4.2.1. ANNL

Although most acute leukemias can be classified by morphology and cytochemical staining as being of lymphoid or myeloid lineage, a small number are unclassifiable. As in the case of B- and T-cell-derived leukemias, where expression of a series of cell surface Ags has defined subgroups of non-T-cell and T-cell ALLs, anti-myeloid MAbs have been useful in both defining the lineage of acute leukemias and identifying subgroups of acute nonlymphocytic leukemia (ANNL) (Griffin et al., 1983a–d; van der Reijden et al., 1983; Sabbath et al., 1985; Katz et al., 1985; Tindle et al., 1985; Drexler et al., 1986; Dinndorf et al., 1986; Hanson et al., 1987). Further studies have permitted correlation of myeloid leukemias with normal myeloid progenitor cells. Finally, recent studies will be discussed which correlate cell surface markers with prognosis.

A large number of MAbs have been extensively screened on large numbers of patients, which permits analysis of the phenotypes of ANNLs. MAbs that define Ia and CD45 Ags, although identifying essentially all cases of ANNL, are not restricted to myeloid leukemias. Several antibodies have been extensively screened and have excellent ability to distinguish ANNL from ALL. These antibodies include anti-CD13, anti-CD33, and PM-81, which are myeloid restricted. All three have been noted to identify over 80% of ANNLs [except French-American-British (FAB) classification M6], and 73% have been reported to coexpress CD13 and CD33. Another antibody, termed VIM-2 (Majdi et al., 1984), identifies over 90% of ANNLs of M1–5 FAB classification. Monoclonal antibodies directed against CD34 react with normal myeloid and mixed colony-forming cells in 85% of M1 ANNLs; however, the Ag is also expressed by some ALLs and cells from patients with CML in lymphoid blast crisis (Katz et al., 1985; Tindle et al., 1985).

The immunologic characterization of ANNL cells can distinguish essentially all cases of ANNL from ALL. However, a small number of leukemias are reported to be of dual lineage. Many of these cases consist of two different populations of leukemic cells, for example, $CD10^+$ lymphoblasts intermingled with blasts expressing myeloid Ags. There have also been cases of lymphoid and myeloid markers on the same cell. These include a patient with acute leukemia whose cells coexpressed Ia and CD10 and also reacted with anti-CD13 MAb, which detects myeloid cells (Pui et al., 1984). Two cases of infant null cell ALL have been reported in which cells coexpressing CD19 and CD33 (as well as CD13 and CD15) were identified. Another case of CML in blast crisis revealed cells reported to coexpress T-cell Ags (CD3) and myeloid Ags CD11b and CD13 (Griffin et al., 1983c). In a recent study of 74 cases of adult ALL, 31% of cases were found to coexpress lymphoid and myeloid cell surface Ags (Sobol et al., 1985). These bi-phenotypic cases tended to be older (45 years versus 34 years) and had a statistically significant poorer response to therapy (CR, 30% versus 73%; $P = 0.01$) and shorter survival. Generally, it has been reported that adult ANNLs (usually less than 4% of cases) rarely express T- or B-cell Ags (Griffin et al., 1986).

The definition of lineage of acute leukemias has become a major use of MAbs as diagnostic reagents. Further studies have identified subgroups of ANNL and attempted to correlate cell surface phenotype with the FAB classification. The M1 ANNLs, which correspond to immature myeloid progenitors, express CD13, CD33, and CD34. The M1 and M2 ANNLs are reactive with MAb MA1, which defines a myeloid-associated Ag, whereas other FAB subtypes are generally negative (Nogiwa *et al.*, 1988). Anti-CD11b MAb has been reported to react with approximately 60% of M1 and M2 ANNLs. This is in contrast to the normal cellular counterparts of M1 ANNLs, which are CD11b⁻. Ball and Fanger (1983) and Neame *et al.* (1986) have examined cells from patients with ANNL by using three monoclonal antibodies: PMN-6 and PMN-29, which are granulocyte specific, and CD14, which is monocyte specific. They observed that less differentiated ANNLs of M1 and M2 FAB classification lacked all three Ags, whereas M4 (myelomonocytic) ANNLs expressed all three Ags. The monocytic leukemias (M5) were reactive with CD14. These studies demonstrated that subsets defined by these three MAbs correlated highly with FAB classification. Linch *et al.* (1984) examined 70 cases of ANNL and found that the reactivity of a series of MAbs that define Ags on monocytes (CD14 and CD35) as well as monocytes and granulocyte precursors (TG1) correlated well with FAB subgroups M4 and M5. The other members of the LFA family of cell adhesion molecules, CD11a and CD11c, were largely expressed on only M4 and M5 ANNLs. These MAbs were essentially unreactive with M1, M2, and M3 leukemias. Two reports of 191 and 105 cases of ANNL revealed that the MAb CD15 was generally very weakly or unreactive with promyelocytic leukemias (M3), whereas most myeloid leukemias of FAB classifications M2, M4, and M5 were CD15⁺ (Bettelheim *et al.*, 1985a; Neame *et al.*, 1986). Although not as extensively characterized as other FAB subtypes, MAbs against glycophorin identify M6 ANNLs (San Miguel *et al.*, 1986). The M7, acute megakaryoblastic leukemias have been shown in several studies to express Ia, GPIIb-IIIa, CD33, CD34, and CD11c but not CD13 or CD14 (Erber *et al.*, 1987; Koike *et al.*, 1987; Varon *et al.*, 1989). Griffin *et al.* (1983a) have examined 70 cases of ANNL with a panel of MAbs, including anti-Ia, CD14, CD13, CD11b, and CD33, which had previously been used to characterize stages of normal myeloid differentiation. Four phenotypes characteristic of normal immature myeloid cells have been characterized. The cell surface phenotypes of 62 of the 70 patients were identical to one of these four groups. The most immature, group I (13 patients), corresponding to the CFU-GM, expressed Ia, CD13, and CD33. The majority (79%) of these patients were classified as M1 or M2 subtypes. Group II included 16 cases expressing Ia, CD13, CD33, and detectable levels of CD11b, phenotypically resembling normal myeloblasts. Similar to group I, 73% were considered to have M1 or M2 leukemias and 27% were of the M4 subtype. The third group (III, promyelocytes), characterized by loss of Ia, included all cases of acute promyelocytic leukemia and some patients with AML (acute myeloblastic leukemia) or AMML (acute myelomonocytic leukemia). The largest group (IV, promonocyte), expressing Ags acquired by cells destined to be monocytes (CD11b, CD14, CD13, CD33, and Ia) included 28 cases. The vast majority (81%) of these had monocyte morphology (M4/M5), and 19% were considered to have ANNL. Although cell surface phenotype was related to the FAB classification in this study, phenotypic groups did contain patients with differing morphologies. These studies relate surface Ag expression to stages of normal myeloid differentiation; as will be discussed, the use of this panel of MAbs may provide important clinical correlations.

The identification of subgroups with poor-prognosis ANNL has been difficult. Although in certain studies factors such as a history of prior myelodysplastic syndrome, prior chemotherapy or radiotherapy, and monocytic features within the FAB classification have identified patients with a poor prognosis, these findings are controversial (Keating, 1982; Cadman et al., 1977). With the observation of heterogeneity of myeloid Ag expression on ANNL, attempts have been made to identify clinically relevant subgroups by surface Ag expression. Civin et al. (1983) examined 33 patients for expression of CD15 and CD34. They observed that the expression of CD15 and lack of CD34 identified patients with greater likelihood of obtaining CR. A recent investigation (Griffin et al., 1986) examined the surface Ag expression of 196 patients with ANNL with a panel of 16 MAbs. All patients were uniformly treated in this study. It was observed that the expression of two Ags, CD14 and CD13, predicted for a low CR rate. CD14, as previously discussed, is expressed by cells with monocytic features, whereas CD13 is expressed on multipotent stem cells as well as more mature myeloid cells (granulocytes and monocytes). $CD14^+$ cases had a CR rate of 53%, whereas $CD14^-$ cases had a CR rate of 69% ($P = 0.03$). Cases that were $CD13^+$ had a CR rate of 55%, whereas $CD13^-$ cases had a CR rate of 73% ($P = 0.01$). The absence of both markers was associated with a CR rate of 82%, while the CR rate for all other cases were 54%. Two other groups, identified by the expression of Ia and CD11b, were associated with a decreased continuous CR rate in 1 year and lower 1-year survival rate, respectively. A more limited study confirmed that $CD13^+$ cases had poor response to treatment but did not confirm similar findings for CD11b and CD14 expression (Pilkington et al., 1989; Schwarzinger et al., 1989). These studies suggest that immunologically defined subgroups may have clinical significance beyond that predicted by standard criteria such as the FAB classification.

4.2.2. CML

CML in chronic phase is characterized by the proliferation of mature granulocytes and their relatively mature precursors. After a variable period of time, most patients enter a terminal phase resembling acute blastic leukemia. Morphologically, about one-third of these blast crisis leukemias resemble lymphoblasts and express CD10. The majority of cases, however, are heterogeneous and felt to be related to ANNL. The clarification of lineage of CML in blast crisis is clinically relevant, as the subgroup of patients with lymphoid blast crisis frequently respond to therapy with vincristine and prednisone. Outside of the characterization of the lymphoid blast crisis with MAbs that define CD10, the remaining cases have until recently been characterized only by their lack of lymphoid markers. Using a panel of MAbs reactive with myeloid, erythroid, lymphoid, or megakaryocyte lineage cells, Griffin et al. (1983d) determined the cell surface phenotypes of 30 patients with CML blast crisis. Eleven cases expressed a phenotype similar to that of ALL cells, expressing Ia and CD10, with five of the cases expressing the pan-B-cell Ag CD20. One-third of the cases were $CD13^+$ Ia^+, with variable expression of CD11b and CD14, and this group resembled early myeloid cells, commonly seen in ANNL. One patient had an erythroleukemia phenotype (expressing glycophorin), and one expressed Plt-1, representing a megakaryoblastic phenotype. Other studies have reported a higher frequency (30–40%) of megakaryoblastic cells by the expression of GPIIb-IIIa. One case had a mixed population of $CD13^+$ $CD10^+$ cells, and six cases were termed undifferentiated because they did not express markers characteristic of any lineage. Bettelheim *et al.* (1985b) examined 45 cases of CML blast crisis and observed that the majority (28 patients) were of myeloid derivation (Ia$^+$ CD13$^+$ VIM-2$^+$). Similar to the study by Griffin *et al.*, about one-third were lymphoid (Ia$^+$ CD10$^+$ CD24$^+$). Two other cases were of mixed myeloid and lymphoid cells, and one was unclassifiable. Although CML blast crisis cells of T lineage were not observed in these studies, several cases have been reported of lymphoid blast crisis with T-cell markers (Allouche *et al.*, 1985, Griffin *et al.*, 1983c). These studies have demonstrated in most cases that a dominant population of cells could be identified of only one lineage. Moreover, CML blast crisis cells represent limited differentiation of the pluripotent stem cell, believed to be the target of leukemic transformation.

5. SUMMARY

Cell surface markers have identified considerably more heterogeneity within human T, B, and myeloid neoplasms than was evident by standard morphologic and histochemical techiques. By using markers specific for the lineage and state of differentiation, it is now possible to correlate malignant lymphoid and myeloid cells with their normal cellular counterparts. Considering the complexity of the normal hematapoietic system with regard to ontogeny, differentiation, and function, it is not surprising that these malignancies reflect this diversity. Hopefully, with increasing characterization of the normal function of cell surface molecules, as well as the subpopulations of normal cells to which these malignancies correspond, we will have a better understanding of the biologic and clinical behavior of these malignancies.

ACKNOWLEDGMENT. This work was supported by U.S. Public Health Service grant K08 CA 01105 awarded by the National Cancer Institute, DHHS.

6. REFERENCES

Abramson, C., Kersey, J. H., and LeBien, T. W., 1981, A monoclonal antibody (BA-1) primarily reactive with cells of human B lymphocyte lineage, *J. Immunol.* **126**:83–88

Acuto, O., and Reinherz, E. L., 1985, The human T-cell receptor: Structure and function, *N. Engl. J. Med.* **312**:1100–1111

Aisenberg, A. C., Wilkes, B. M., and Harris, N. L., 1983, Monoclonal antibody studies in non-Hodgkin's lymphoma, *Blood* **61**:461–475

Allouche, M., Bourinbaiar, A., Georgoulias, V., Consolini, R., Salvatore, A., Auclair, H., and Jasmin, C., 1985, T cell lineage involvement in lymphoid blast crisis of chronic myeloid leukemia, *Blood* **66**:1155–1161

Amiot, M., Dastot, H., Schmid, M., Bernard, A., and Boumsell, L., 1987, Analysis of CD1 molecules on thymus cells and leukemic T lymphoblasts identifies discrete phenotypes and reveals that CD1 intermolecular complexes are observed only on normal cells, *Blood* **70**:676–685

Anderson, K. C., Park, K., Bates, M., Leonard, R. C. F., Hurdy, R., Schloosman, S. F., and Nadler, L. M., 1983, Antigens on human plasma cells identified by monoclonal antibodies, *J. Immunol* **130**: 1132–1138

Anderson, K. C., Bates, M. P., Slaughenhoupt, B. L., Pinkus, G. S., O'Hara, C., Schlossman, S. F., and Nadler, L. M., 1984, Expression of human B cell associated antigens on leukemias and lymphomas: A model of B cell differentiation, *Blood* **63**:1424–31

Anderson, K. C., Boyd, A. W., Fisher, D. C., Leslie, D., Schlossman, S. F., and Nadler, L. M., 1985, Hairy cell leukemia: A tumor of pre-plasma cells, *Blood* **65**:620–629

Arnaout, M. A., Todd, R. F., Dana, N., Melamed, J., Schlossman, S. F., and Colter, H. R., 1983, Inhibition of phagocytes of complement C3 or IgG coated particles and of C3Gi binding by monoclonal antibodies to a monocyte-granulocyte membrane glycoprotein (Mol), *J. Clin. Invest.* **72**:171–179

Arnold, A., Cossman, J., Bakhshi, A., Jaffe, E. S., Waldmann, T. A., and Korsmeyer, S. J., 1983, Immunoglobulin-gene arrangements as unique clonal markers in human lymphoid neoplasms, *N. Engl. J. Med.* **309**:1593–1599

Ault, K. A., Antin, J. H., Ginsburg, D., Orkin, S. H., Rappeport, J. M., Keohan, M. L., Martin, P., and Smith, B. R., 1985, Phenotype of recovering lymphoid cell populations after marrow transplantation, *J. Exp. Med.* **161**:1483–1502

Baldini, L., DiPadova, F., Cortelezzi, A., Nori, A., Nobili, L., Lavezzi, A. M., Maiolo, A. T., and Polli, E. E., 1985, Functional and multimarker analysis of T cell chronic lymphocytic leukemia, *Scand. J. Haematol.* **34**:88–96

Ball, E. D., and Fanger, M. W., 1983, The expression of myeloid-specific antigens on myeloid leukemia cells: Correlations with leukemia subclasses and implications for normal myeloid differentiation, *Blood* **61**:456–463

Barber, E., K., Dasgupta, J. D., Schlossman, S. F., Trevillyan, J. M., and Rudd, C. E., 1989, The CD8 antigen is coupled to a protein-tyrosine kinase (p58Lck) that phosphorylates the CD3 complex in vitro, *Tissue Antigens* **33**:93

Bernard, A., Boumsell, L., Reinherz, E. R., Nadler, L. M., Ritz, J., Coppin, H., Richard, Y., Valensi, F., Dausset, J., Flandrin, G., Lemerle, J., and Schlossman, S. F., 1981, Cell surface characterization of malignant T cells from lymphoblastic lymphoma using monoclonal antibodies: Evidence for phenotypic differences between malignant T cells from patients with acute lymphoblastic leukemia and lymphoblastic lymphoma, *Blood* **57**:1105–1110

Berribi, A., Dagan, S., Bassous-Guedj, L., Vorst, E., and Shtalrid, M., 1989, Further characterization of prolymphocytic leukemia cells as a tumor of activated B cells, *Tissue Antigens* **33**:324

Bettelheim, P., Panzer, S., Majdic, O., Stockinger, H., Roithner, A., Köller, U., Meryn, S., Lechner, K., and Knapp, W., 1985a, Unexpected absence of a myeloid surface antigen (3-fucosyl-n-acetyllactosamine) in promyelocytic leukemia, *Leuk. Res.* **9**:1323–1327

Bettelheim, P., Lutz, D., Majdic, O., Paietta, E., Haas, O., Linkesch, W., Neumann, E., Lechner, K., and Knapp, W., 1985b, Cell lineage heterogeneity in blast crisis of chronic myeloid leukaemia, *Br. J. Haematol.* **59**:395–409

Bhan, A. K., Nadler, L. M., Stashenko, P., and Schlossman, S. F., 1981, Stages of B cell differentiation in human lymphoid tissues, *J. Exp. Med.* **154**:737–749

Bofill, M., Janossy, G., Janossa, M., Burford, G. D., Seymour, G. J., Wernet, P., and Kelemen, E., 1985, Human B cell development. II. Subpopulations in the human fetus, *J. Immunol.* **134**:1531–38

Borowitz, M. J., Dowell, B. L., Boyett, J. M., Falletta, J. M., Pullen, D. J., Crist, W. M., Humphrey, G. B., and Metzgar, R. S., 1985, Monoclonal antibody definition of T cell acute leukemia: A pediatric oncology group study, *Blood* **65**:785–788

Boue, D. R., and LeBien, T. W., 1988, Expression and structure of CD22 in acute leukemia, *Blood* **71**: 1480–1486

Boumsell, L., Coppin, H., Pham, D., Raynal, B., Lemerle, J., Dausset, J., and Bernard, A., 1980, An antigen shared by a human T cell subset and B cell chronic lymphocytic leukemic cells, *J. Exp. Med.* **152**:229–234

Boumsell, L., Bernard, A., Reinherz, E. R., Nadler, L. M., Ritz, J., Coppin, H., Richard, Y., Dubertret, L., Valensi, F., Degos, L., Lemerle, J., Flandrin, G., Dausset, J., and Schlossman, S. F., 1981, Surface antigens on malignant Sezary and T-CLL cells correspond to those of mature T cells, *Blood* **57**:526–530

Boyd, A. W., Anderson, K. C., Freedman, A. S., Fisher, D. C., Slaughenhoupt, B. L., Schlossman, S. F., and Nadler, L. M., 1985a, Studies of in vitro activation and differentiation of human B lymphocytes. I. Phenotypic and functional characterization of the B cell population responding to anti-Ig antibody, *J. Immunol.* **134**:1516–1523

Boyd, A. W., Fisher, D. C., Fox, D., Schlossman, S. F., and Nadler, L. M., 1985b, Structural and functional characterization of IL-2 receptors on activated B cells, *J. Immunol.* **134**:2387–92

Boyd, A. W., Tedder, T. F., Griffin, J. D., Freedman, A. S., Fisher, D. C., Daley, J., and Nadler, L. M., 1987, Pre-exposure of resting B cells to gamma interferon enhances their proliferative response to subsequent activation signals, *Cell Immunol.* **106**:355–65

Bradley, T. R., and Metcalf, D., 1966, The growth of mouse bone marrow cells in vitro, *Aust. J. Exp. Biol. Med.* **44**:287–300

Braesch-Anderson, S., Paulie, S., Aspenstrom, P., Koho, H., and Perlmann, P., 1989, Biochemical characteriztics of the human B-cell and carcinoma antigen CDw40, *Tissue Antigens* **33**:129

Brenner, M. B., McLean, J., Dialynas, D. P., Strominger, J. L., Smith, J. A., Owen, F. L., Seidman, J. G., Ip, S., Rosen, F., and Krangel, M. S., 1986, Identification of a putative second T-cell receptor, *Nature (London)* **322**:145–149

Broder, S., and Bunn, P. A., 1980, Cutaneous T cell lymphomas, *Semin. Oncol.* **7**:310–331

Brooks, D. A., Beckman, I. G. R., Bradley, J., McNamara, P. J., Thomas, M. E., and Zola, H., 1981, Human lymphocyte markers defined by antibodies derived from somatic cell hybrids. IV. A monoclonal antibody reacting specifically with a subpopulation of human B lymphocytes, *J. Immunol.* **126**:1373–1377

Brouet, J. C., Preud'homme, J. L., Penit, C., Valensi, F., Rouget, P., and Seligmann, M., 1979, Acute lymphoblatic leukemia with pre-B cell characteristics, *Blood* **54**:269–273

Brouet, J. C., Rabian, C., Gisselbrecht, C., and Flandrin, G., 1984, Clinical and immunological study of non-Hodgkin T cell lymphomas (cutaneous and lymphoblastic lymphomas excluded), *Br. J. Haematol.* **57**:315–327

Brown, V., Smith, S. Dewar, E., and Maddy, A., 1987, The correlation between surface immunoglobulin expression and the leucocyte-common antigen in B-cell chronic lymphocytic leukaemia, *Leuk. Res.* **11**:903–910

Cadman, E. C., Capizzi, R. L., and Bertino, J. R., 1977, Acute nonlymphocytic leukemia. A delayed complication of Hodgkin's disease therapy. Analysis of 109 cases, *Cancer* **40**:1280–1296

Caligaris-Cappio, F., Gobbi, M., Bofill, M., and Janossy, G., 1982, Infrequent normal B lymphocytes express features of B chronic lymphocytic leukemia, *J. Exp. Med.* **155**:623–627

Caligaris-Cappio, F., Bergui, L., Tesio, L., Pizzolo, G., Malavasi, F., Chilosi, M., Campana, D., van Camp, B., and Janossy, G., 1985, Identification of malignant plasma cell precursors in the bone marrow of multiple myeloma, *J. Clin. Invest.* **76**:1243–1251

Campana, D., Sheridan, B., Tidman, N., Hoffbrand, A. V., and Janossy, G., 1986, Human leukocyte function-associated antigens on lympho-hematopoietic precurser cells, *Eur. J. Immunol.* **16**:537–542

Campana, D., Thompson, J. S., Amlot, P., Brown, S., and Janossy, G., 1987, The cytoplasmic expression of CD3 antigens in normal and malignant cells of T lymphoid lineage, *J. Immunol.* **138**:648–655

Catovsky, D., Pettit, J. E., Galetto, J., Okos, A., and Galton, D. A. G., 1974, The B-lymphocyte nature of the hairy cell of leukemic reticuloendotheliosis, *Br. J. Haematol.* **26**:29–37

Catovsky, D., Cherchi, M., Okos, A., Hedge, U., and Galton, D. A. G., 1976, Mouse red blood cell rosettes in B lymphoproliferative disorders, *Br. J. Haematol.* **33**:173–177

Catovsky, D., Cherchi, M., Brooks, D., Bradley, J., and Zola, H., 1981, Heterogeneity of B-cell leukemias demonstrated by the monoclonal antibody FMC7, *Blood* **58**:406–408

Catovsky, D., Wechsler, A., Matutes, E., Gomez, R., Bourikas, G., Cherchi, M., Pepys, E. O., Pepys, M. B., Kitani, T., Hoffbrand, A. V., and Greaves, M. F., 1982, The membrane phenotype of T-prolymphocytic leukemia, *Scand. J. Haematol.* **29**:398–404

Charbonneau, H., Tonks, N. K., Walsh, K. A., and Fischer, E. H., 1988, The leukocyte common antigen (CD45): A putative receptor-linked protein tyrosine phosphatase, *Proc. Natl. Acad. Sci. USA* **85**:7182–7186

Chen, Z., Sigaux, F., Miglierina, R., Valensi, F., Daniel, M. T., Ochoa-Noguera, M. H., and Flandrin, G., 1986, Immunological typing of acute lymphoblastic leukemia: Concurrent analysis by flow cytofluorometry and immunocytology, *Leuk. Res.* **10**:1411–1417

Civin, C. I., Vanghan, W. P., Straa, L. C., Schwartz, J. F., Karp, J. E., and Burke, P. J., 1983, Diagnostic and prognostic utility of cell surface markers in acute non-lymphocytic leukemia (ANLL), *Exp. Hematol.* **11**(Suppl. 14):152a

Clark, E. A., and Yokochi, T., 1984, Human B cell and B cell blast-associated surface molecules defined with monoclonal antibodies, in *Leukocyte Typing I* (A. Bernard *et al.*, eds.), pp. 339–346, Springer-Verlag, Berlin

Clark, E. A., Shu, G., and Ledbetter, J. A. 1985, Role of the Bp35 cell surface polypeptide in human B-cell activation, *Proc. Natl. Acad. Sci. USA* **82**:1766–1770

Clayberger, C., Medeiros, L. J., Link, M. P., Warnke, R. A., Wroght, A., Koller, T. D., Smith, S. D., and Krensky, A. M., 1987, Absence of cell surface LFA-1 as a mechanism of escape from immunosurveillance, *Lancet* **ii**:533–536

Cossman, J., and Jaffe, E. S., 1981, Distribution of complement receptor subtypes in non-Hodgkins lymphomas of B cell origin, *Blood* **58**:20–26

Cossman, J., Neckers, L. M., Arnold, A., And Korsmeyer, S. J., 1982, Induction of differentiation in a case of common acute lymphoblastic leukemia, *N. Engl. J. Med.* **307**:1251–1254

Cossman, J., Chused, T. M., Fisher, R. I., Magrath, I., Bollum, F., and Jaffe, E. S., 1983, Diversity of immunological phenotypes of lymphoblastic lymphoma, *Cancer Res.* **43**:4486–4490

Cossman, J., Neckers, L. M., Jones, T., Hsu, S. M., Jaffe, E. S., 1984a, Low grade lymphomas: Expression of developmentally regulated B cell antigens, *Am. J. Pathol.* **115**:117–124

Cossman, J., Jaffe, E. S., and Fisher, R. I., 1984b, Immunologic phenotypes of diffuse, aggressive, non-Hodgkin's lymphomas: Correlations with clinical features, *Cancer* **54**:1310–1317

Dallenbach, F., Josimovic-Alasevic, O., Durkop, H., Schwarting, R., Pizzolo, G., Diamantstein, T., Takasuki, K., and Stein, H., 1989, Soluble Ki-1 (CD30) antigen in the sera of patients with adult T cell lymphoma/leukemia: A marker of disease activity, *Tissue Antigens* **33**:309

Davey, F. R., Mick, R., Nelson, D. A., MacCallum, J., Sobol, R. E., Royston, I., Cuttner, J., Ellison, R. R., and Bloomfield, C. D., 1988, Morphologic and cytochemical characterization of adult lymphoid leukemias which express myeloid antigen, *Leukemia* **2**:420–428

Defrance, T., Aubry, J. P., Rousset, F., Vanberuliet, B., Bonnefoy, J. Y., Arai, N., Takebe, Y., Yokota, T., Lee, F., Aral, K., DeVries, J., and Banchereau, J., 1987, Human recombinant interleukin 4 induces FcE receptors (CD23) on normal human B lymphocytes, *J. Exp. Med.* **165**:1459–1467

De la Hera, A., Alvarez-Mon, M., and Sanchez-Madrid, F., Martinez A. C., and Durantez, A., 1988, Co-expression of Mac-1 and p150,95 on CD5$^+$ B cells. Structural and functional characterization in a human chronic lymphocytic leukemia, *Eur. J. Immunol.* **18**:1131–1134

Delia, D., Cattoretti, G., Polli, N., Fontanella, E., Aiello, A., Giardini, R., Rilke, F., and Della Porta, G., 1988, CD1c but neither CD1a nor CD1b molecules are expressed on normal activated, and malignant human B cells: Identification of an new B-cell subset, *Blood* **72**:241–247

Dinndorf, P. A., and Reaman, G. H., 1986, Acute lymphoblastic leukemia in infants: Evidence for B cell origin of disease by use of monoclonal antibody phenotyping, *Blood* **68**:975–978

Dinndorf, P. A., Andrews, R. G., Benjamin, D., Ridgway, D., Wolff, L., and Bernstein, I. D., 1986, Expression of normal myeloid-associated antigens by acute leukemia cells, *Blood* **67**:1048–1053

Divine, M., Farcet, J. P., Gourdin, M. F., Tabilio, A., Vasconcelos, A., Andre, C., Jouault, A., Bouguet, J., and Reyes, F., 1984, Phenotype study of fresh and cultured hairy cells with the use of immunologic markers and electron microscopy, *Blood* **64**:547–552

Dorken, B., Moldenhauer, G., Pezzutto, A., Schwartz, R., Feller, A., Kiesel, S., and Nadler, L. M., 1986, HD39 (B3), a B lineage-restricted antigen whose cell surface expression is limited to resting and activated human B lymphocytes, *J. Immunol.* **136**:4470–4479

Dorken, B., Moldenhauer, G., Pezzutto, A., and Emmrich, F., 1989, Analysis of functional effects of B cell antibodies using an antigen specific assay: CD19 antibodies enhance Ig secretion, *Tissue Antigens* **33**:142

Dorken, P., Pezzutto, A., and Hunstein, W., 1987, Expression of cytoplasmic CD22 in B cell ontogeny, in *Leukocyte Typing III* (A. J. McMichael, ed.), pp. 474–475 Oxford University Press, Oxford

Dow, L. W., Borella, Sen, L., Aur, R. J. A., George, S. L., Mauer, A. M., and Simone, J. V., 1977, Initial prognostic factors and lymphoblast-erythrocyte rosette formation in 109 children with acute lymphoblastic leukemia, *Blood* **50**:671–682

Drexler, H. G., Sagawa, K., Menon, M., and Minowada, J., 1986, Reactivity pattern of "myeloid monoclonal antibodies" with emphasis on MCS-2, *Leuk. Res.* **10**:17–23

Einfeld, D. A., Brown, J. P., Valentine, M. A., Clark, E. A., and Ledbetter, J. A., 1988, Molecular cloning of the human B cell CD20 receptor predicts a hydrophobic protein with multiple transmembrane domains, *EMBO J.* **7**:711–717

Ellegaard, J., Kucharska-Pulczynska, M., and Hokland, P., 1987, Analysis of leukocyte differentiation antigens in blood and bone marrow from patients with Waldenstrom's macroglobulinemia, in *Leukocyte Typing III* (A. J. McMichael, ed.), pp. 512–515, Oxford University Press, Oxford

Erber, W. N., Breton-Gorius, J., Villeval, J. L., Oscier, D. G., Bai, Y., and Mason, D. Y., 1987, Detection of cells of megakaryocyte lineage in hematological malignancies by imuno-alkaline phosphatase labelled smears with a panel of monoclonal antibodies, *Br. J. Haematol.* **65**:87–94

Erikstein, B. K., Asheim, H. C., Smeland, E. B., Beiske, K., and Funderud, S., 1989, Characterization of a new monoclonal antibody HH2 which recognizes a cell cycle regulated antigen specific for human B-lymphocytes, *Tissue Antigens* **33**:150

Favier, R., Morel, M. C., Potevin, F., Kaplan, C., and Lecompte, T., 1989, CD9 and GPIIa-IIIa: Their potential association and role on platelet activation, *Tissue Antigens* **33**:350

Feller, A. C., Griesser, G. H., Mak, T., and Lennert, K., 1986, Lymphoepithelioid lymphoma (Lennert's lymphoma) is a monoclonal proliferation of helper/inducer T cells, *Blood* **68**:663–667

Ferrarini, M., Romagnani, S., Montesoro, E., Zicca, A., Del Prete, G. F., Nocera, A., Maggi, E., Leprini, A., and Grossi, E., 1983, A lymphoproliferative disorder of the large granular lymphocytes with natural killer activity, *J. Clin. Immunol.* **3**:30–41

Fingeroth, J. D., Weis, J., Tedder, T. F., Strominger, J. L., Biro, P. A., and Fearon, D. T., 1984, Epstein-Barr virus receptor of human B lymphocytes is the C3d receptor CR2, *Proc. Natl. Acad. Sci.* USA **81**:4510–4514

Flug, F., Dodson, L., Wolff, J., Guarini, L., Rausen, A., Wang, C. Y., and Knowles, D. M., 1985, B-lymphocyte associated antigen expression by non-B, non-T acute lymphoblastic leukemia, *Leuk. Res.* **9**:1051–1058

Ford, R. J., Yoshimura, L., Morgan, J., Quesada, J., Montagna, R., and Maizel, A., 1985, Growth factor-mediated tumor cell proliferation in hairy cell leukemia, *J. Exp. Med.* **162**:1093–1098

Fox, D. A., Hussey, R. E., Fitzgerald, K. A., Acuto, O., Poole, C., Palley, L., Daley, J. F., Schlossman, S. F., and Reinherz, E. L., 1984, Ta1, a novel 105 kd human T cell activation antigen defined by a monoclonal antibody, *J. Immunol.* **133**:1250–1256

Frade, R., Crevon, M. C., Barel, M., Vazquez, A., Kirkorian, L., Charriaut, C., and Galahand, P., 1985, Enhancement of human B cell proliferation by an antibody to the C3d receptor, the gp 140 molecule. *Eur. J. Immunol.* **15**:73–76

Freedman, A. S., Boyd, A. W., Berrebi, A., Horowitz, J. C., Rosen, K. J., Slaughenhoupt, B., Levy, D., Daley, J., Levine, H., and Nadler, L. M., 1987c, Expression of B cell activation antigens on normal and malignant B cells, *Leukemia* **1**:9–15

Freedman, A. S., Boyd, A. W., Anderson, K. C., Fisher, D. C., Schlossman, S. F., and Nadler, L. M., 1985a, B5, a new B cell restricted activation antigen, *J. Immunol.* **134**:2228–2235

Freedman, A. S., Boyd, A. W., Anderson, K. C., Fisher, D. C., Pinkus, G. S., Schlossman, S. F., and Nadler, L. M., 1985b, Immunologic heterogeneity of diffuse large cell lymphoma, *Blood* **65**:630–637

Freedman, A. S., Boyd, A. W., Bieber, F., Daley, F., Rosen, K., Horowitz, J., Levy, D., and Nadler, L. M., 1987a, Normal cellular counterparts of B cell chronic lymphocytic leukemia, *Blood* **70**:418–27

Freedman, A. S., Freeman, G., Horowitz, J. C., Daley, J., and Nadler, L. M., 1987b, B7, a B cell restricted antigen which identifies pre-activated B cells, *J. Immunol.* **137**:3260–3267

Freedman, A. S., Freeman, G., Whitman, J., Segil, J., Daley, J., and Nadler, L. M., 1988, Pre-exposure of human B cells to recombinant interleukin I enhances subsequent proliferation, *J. Immunol.* **141**:3398–404

Freedman, A. S., Freeman, G., Whitman, J., Segil, J., Daley, J., and Nadler, L. M., 1989a, Studies of in vitro activated CD5+ B cells, *Blood* **73**:202–208

Freedman, A. S., Freeman, G., Whitman, J., Segil, J., Daley, J., Levine, H., and Nadler, L. M., 1989b, Expression and regulation of CD5 on in vitro activated human B cells, *Eur. J. Immunol.* **19**:849–855

Funderud, S., Blomhoff, H. K., Asheim, H. C., Beiske, K., Totterman, T., and Smeland, E. B., 1989, Characterization of a new non-clustered B cell specific antigen (FN1) preferentially expressed on resting cells, *Tissue Antigens* **33**:151

Gathings, W. E., Lawton, A. R., and Cooper, M. D., 1977, Immunofluorescent studies on the development of pre-B cells, B lymphocytes and immunoglobulin isotype diversity in humans, *Eur. J. Immunol.* **7**:804–810

Goldstein, L. A., Zhou, D. F. H., Picker, L. J., Minty, C. N., Bargatze, R. F., Ding, J. F., and Butcher, E. C., 1989, A human lymphocyte homing receptor, the Hermes antigen, is related to cartilage proteoglycan core and link proteins, *Cell* **56**:1063–1072

Goldstein, M., Hoxie, J., Zembryki, D., Matthews, D., and Levinson, A. I., 1985, Phenotypic and functional analysis of B cell lines from patients with multiple myeloma, *Blood* **66**:444–446

Gordon, J., Rowe, M., Walker, L., and Guy, G., 1986, Ligation of the CD23 p45 antigen triggers cell-cycle progression of activated B lymphocytes, *Eur. J. Immunol.* **16**:1075–1080

Grant, B. W., Platt, J. L., Jacob, H. S., and Kay, N. E., 1986, Lymphocyte populations and tac-antigen in diffuse B-Cell lymphomas, *Leuk. Res.* **10**:1271–1278

Greaves, M. F., Verbi, W., Vogler, L. B., Cooper, M., Ellis, R., Ganeshguru, K., Hoffbrand, V., Janossy, G., and Bollum, F. J., 1980, Antigenic and enzymatic phenotypes of the pre-B subclass of acute leukemia, *Leuk. Res.* **3**:353–362

Greaves, M. F., Hariri, G., Newman, R. A., Sutherland, D. R., Ritter, M. A., and Ritz, J., 1983, Selective expression of the common acute lymphoblastic leukemia (gp100) antigen on immature lymphoid cells and their malignant counterparts, *Blood* **61:**628–639

Greil, R., Gattringer, C., Schultz, T., Knapp, W., Radaskiewicz, T., Dierich, M. P., and Huber, H., 1986, Receptors for the third component of complement: Their association with maturation stage in non-Hodgkin's lymphoma and their possible implication with the development of follicular structure, *Clin. Exp. Immunol.* **64:**423–431

Griffin, J. D., Ritz, J., Nadler, L. M., and Schlossman, S. F., 1981, Expression of myeloid differentiation antigens on normal and malignant myeloid cells, *J. Clin. Invest.* **68:**932–941

Griffin, J. D., Mayer, R. J., Weinstein, H. J., Rosenthal, D. S., Coral, F. S., Beveridge, R. P., and Schlossman, S. F., 1983a, Surface marker analysis of acute myeloblastic leukemia: Identification of differentiation-associated phenotypes, *Blood* **62:**557–563

Griffin, J. D., Ritz, J., Beveridge, R. P., Lipton, J. M., Daley, J. F., and Schlossman, S. F., 1983b, Expression of MY7 antigen on myeloid progenitor cells, *Int. J. Cell Cloning* **1:**33–48

Griffin, J. D., Tantravahi, R., Canellos, G. P., Wisch, J. S., Reinherz, E. L., Sherwood, G., Beveridge, R. P., Daley, J. F., Lane, H., and Schlossman, S. F., 1983c, T cell surface antigens in a patient with blast crisis of chronic myeloid leukemia, *Blood* **61:**640–644

Griffin, J. D., Todd, R. F., Ritz, J., Nadler, L. M., Canellos, G. P., Rosenthal, D., Gallivan, M., Beveridge, R. P., Weinstein, H., Karp, D., and Schlossman, S. F., 1983d, Differentiation patterns in the blastic phase of chronic myeloid leukemia, *Blood* **61:**85–91

Griffin, J. D., Linch, D., Sabbath, K. D., Larcom, P., and Schlossman, S. F., 1984, A monoclonal antibody reactive with normal and leukemic human myeloid progenitor cells, *Leuk. Res.* **8:**521–534

Griffin, J. D., Davis, R., Nelson, D. A., Davey, F. R., Mayer, R. J., Schiffer, C., McIntyre, O. R., and Bloomfield, C. D., 1986, Use of surface marker analysis to predict outcome of adult acute myeloblastic leukemia, *Blood* **68:**1232–1241

Grogan, T. M., Fielder, K., Rangel, C., Jolley, C. J., Wirt, D. P., Hicks, M. J., Miller, T. P., Brooks, R., Greenberg, B., and Jones, S., 1985, Peripheral T cell lymphoma: Aggressive disease with heterogeneous immunotypes, *Am. J. Clin. Pathol.* **83:**279–288

Habeshaw, J. A., Lister, T. A., and Starsfeld, A. G., 1983, Correlation of transferrin receptor expression with histologic class and outcome in non-Hodgkin's lymphoma, *Lancet* **i:**498–501

Hanson, C. A., Gajl-Peczalska, J., Parkin, J. L., and Brunning, R. D., 1987, Immunophenotyping of acute myeloid leukemia using monoclonal antibodies and the alkaline phosphatase-antialkaline phosphatase technique, *Blood* **70:**83–89

Harris, N. L., Nadler, L. M., and Bhan, A. K., 1984, Immunologic characterization of two malignant lymphomas of germinal center type (centroblastic/centrocytic and centrocytic) with monoclonal antibodies, *Am. J. Pathol.* **117:**262–272

Hattori, T., Uchiyama, T., Toibana, T., Takatsuki, K., and Uchino, H., 1981, Surface phenotype of Japanese adult T-cell leukemia cells characterized by monoclonal antibodies, *Blood* **58:**645–647

Haynes, B. F., Eisenbarth, G. S., and Fauci, A. S., 1979, Human lymphocyte antigens: Production of a monoclonal antibody that defines functional thymus derived lymphocyte subsets, *Proc. Natl. Acad. Sci. USA* **76:**5829–5833

Haynes, B. F., Hemler, M. E., Mann, D. L., Eisenberth, G. S., Shelhamer, J., Mostowski, H. S., Thomas, C. A., Strominger, J. L., and Fauci, A. S., 1981a, Characterization of a monoclonal antibody (4F2) that binds to human monocytes and to a subset of activated lymphocytes, *J. Immunol.* **126:**1409–1420

Haynes, B., Hemler, B. F., Cotner, T., Mann, D. L., Eisenberth, G. S., Strominger, J. L., and Fauci, A. S., 1981b, Characterization of a monoclonal antibody (5E9) that defines a human cell surface antigen of cell activation. *J. Immunol.* **127:**347–351

Haynes, B. F., Metgar, R. S., Minna, J. D., and Bunn, P. A., 1981c, Phenotypic characterization of cutaneous T cell lymphoma: Use of monoclonal antibodies to compare with other malignant T cells, *N. Engl. J. Med.* **304:**1319–1323

Haynes, B. F., Hensley, L. L., and Jegasothy, B. U., 1982, Differentiation of human T lymphocytes. II. Phenotypic difference in skin and blood malignant T cell in cutaneous T cell lymphoma, *J. Invest. Dermatol.* **78:**323–326

Hercend, T., Nadler, L. M., Pesando, J. M., Reinherz, E. L., Schlossman, S. F., and Riz, J., 1981, Expression of a 26,000 dalton glycoprotein on activated human T cells, *Cell. Immunol.* **64:**192–199

Hercend, T., Griffin, J. D., Bensussan, A., Schmidt, R. E., Edson, M. A., Brennan, A., Murray, C., Daley, J.

F., Schlossman, S. F., and Ritz, J., 1985, Generation of monoclonal antibodies to a human natural killer clone: Characterization of two natural killer cell-associated antigens, NKH1$_A$ and NKH2, expressed on subsets of large granular lymphocytes, *J. Clin. Invest.* **75**:932–943

Hoelzer, D., Thiel, E., Loeffer, H., Buechner, T., Ganser, A., Heil, G., Koch, P., Freud, M., Diedrich, H., Ruehl, H., Machmeyer, G., Lipp, T., Nowrousian, M. R., Burkert, M., Gerecke, D., Pralle, H., Mueller, U., Lunscken, C., Fuelle, H., Ho, A. D., Kuechler, R., Busch, F. W., Schneider, W., Goerg, C., Emmerich, B., Braumann, D., Vaupel, H. A., von Paleske, A., Bartels, H., Neiss, A., and Messerer, D., 1988, Prognostic factors in a multicenter study for treatment of acute lymphoblatic leukemia in adults, *Blood* **71**:123–131

Hogg, N., and Horton, M. A., 1987, Myeloid antigens: New and previously defined clusters, in *Leukocyte Typing III* (A. J. McMichael, ed.), pp. 576–602, Oxford University Press, Oxford

Hokland, P., Rosenthal, P., Griffin, J. D., Nadler, L. M., Daley, J. F., Hokland, M., Schlossman, S. F., and Ritz, J., 1983, Purification and characterization of fetal hematopoietic cells that express the common acute lymphoblastic leukemia antigen (CALLA), *J. Exp. Med.* **157**:114–129

Hokland, P., Nadler, L. M., Griffin, J. D., Schlossman, S. F., and Ritz, J., 1984, Purification of the common acute lymphoblastic leukemia antigen (CALLA) positive cells from normal bone marrow, *Blood* **64**:662–666

Hokland, P., Ritz, J., Schlossman, S. F., and Nadler, L. M., 1985, Orderly expression of B cell antigens during the in vitro differentiation of non-malignant human pre-B cell, *J. Immunol.* **135**:1746–1751

Horning, S. J., Doggert, R. S., Warnke, R. A., Dorfman, R. F., Cox, R. S., and Levy, R., 1984, Clinical relevance of immunologic phenotype in diffuse large cell lymphoma, *Blood* **63**:1209–1215

Horning, S. J., Weiss, L. M., Crabtree, G. S., and Warnke, R. A., 1986, Clinical and phenotypic diversity of T cell lymphomas, *Blood* **67**:1578–1582

Howard, F. D., Ledbetter, J. A., Wong, J., Bieber, C. P., Stinson, E. B., and Herzenberg, L. A., 1981, A human T lymphocyte differentiation marker defined by monoclonal antibodies that block E-rosette formation, *J. Immunol.* **126**:2117–2122

Hurwitz, C. A., Loken, M. R., Graham, M. L., Kzarp, J. E., Borowitz, M. J., Pullen, D. J., and Civin, C. I., 1988, Asynchronous antigen expression in B lineage acute lymphoblastic leukemia, *Blood* **72**:299–307

Iida, K., Nadler, L. M., and Nussenzweig, V., 1983, The identification of the membrane receptor for the complement fragment C3d by means of a monoclonal antibody, *J. Exp. Med.* **158**:1021–1033

Ikuta, K., Takami, M., Kim, C. W., Honjo, T., Miyoshi, T., Tagaya, Y., Kawabe, T., and Yodoi, J., 1987, Human lymphocyte Fc receptor for IgE: Sequence homology of its cloned cDNA with animal lectins, *Proc. Natl. Acad. Sci. USA* **84**:819–823

Inghirami, G., Wieczorek, R., Zhu, B., Silber, R., Dalla-Favera, R., and Knowles, D. M., 1988, Differential expression of LFA-1 molecules in non-Hodgkin's lymphoma and lymphoid leukemia, *Blood* **72**:1431–1434

Inui, S., Kaisho, T., Clark, E. A., Seed, B., Kikutani, H., and Kishimoto, T., 1989, Expression of intact and mutant CDw40 on murine lymphocytes (cytoplasmic portion is essential for signal transduction through CDw40), *Tissue Antigens* **33**:133

Jaffe, E. S., and Berard, C. W., 1978, Lymphoblastic lymphoma, a term rekindled with new precision, *Ann. Int. Med.* **89**:415–417

Jaffe, E. S., Shevach, E. M., Frank, M. M., Berard, C. W., and Green, I., 1974, Nodular lymphoma—evidence for origin from follicular B lymphocytes, *N. Engl. J. Med.* **290**:814–819

Jimbow, K., Maeda, K., Ito, Y., Ishida, O., and Takami, T., 1985, Heterogeneity of cutaneous T-cell lymphoma: Phenotypic and ultrastructural characterization of four unusual cases, *Cancer* **56**:2458–2469

Kamoun, M., Kadin, M. F., Martin, P. J., Nettleton, J., and Hansen, J. A., 1981, A novel humen T cell antigen preferentially expressed on mature T cells and also on (B type) chronic lymphatic leukemic cells, *J. Immunol.* **127**:987–996

Katz, F. E., Tindle, R., Sutherland, D. R., and Greaves, M. F., 1985, Identification of a membrane glycoprotein associated with haemopoietic progenitor cells, *Leuk. Res.* **9**:191–198

Katz, F., Malcolm, S., Gibbons, B., Tilly, R., Lam, G., Robertson, M. E., Czepulkowski, B., and Chessells, J., 1988, Cellular and molecular studies on infant null acute lymphoblastic leukemia, *Blood* **71**:1438–1447

Keating, M. J., 1982, Early identification of potentially cured patients with acute myelogenous leukemia—a recent challenge, in *Adult Leukemias* 1 (C. D. Bloomfield, ed.), pp. 237–263, Martinus Nijhoff, Boston

Kehrl, J. H., Muraguchi, A., and Fauci, A. S., 1984, Differential expression of Oell activation markers after stimulation of resting human B lymphocytes, *J. Immunol.* **132**:2857–2861

Kersey, J. H., Sabad, A., Gajl-Peczalska, F., Hallgen, H. M., Yunis, E. J., and Nesbit, M. E., 1973, Acute lymphoblastic leukemic cells with T (thymus-derived) lymphoma markers, *Science* **182**:1355–1356

Kersey, J. H., LeBien, T. W., Abramson, C. S., Newman, R., Sutherland, R., and Greaves, M., 1981, p24: A human hemopoietic progenitor and acute lymphoblastic leukemia-associated cell surface structure identified with a monoclonal antibody, *J. Exp. Med.* **153**:726–731

Kikutani, H., Inui, S., Sato, R., Barsumain, L. E., Owaki, H., Yamaski, K., Kaisho, T., Uchibayashi, N., Hardy, R. R., Hirano, T., Tsunasawa, S., Sakiyama, F., Suemura, M., and Kishimato, T., 1986, Molecular structure of human lymphocyte receptor for immunoglobulin E, *Cell* **47**:657–665

Kishimoto, T., 1985, Factors affecting B cell growth and differentiation, *Annu. Rev. Immunol.* **3**:133–152

Koike, T., Aoki, S., Maruyama, S., Narita, M., Ishizuka, T., Imanaka, H., Adachi, T., Maeda, H., and Shibata, A., 1987, Cell surface phenotyping of megakaryoblasts, *Blood* **69**:957–960

Korsmeyer, S. J., Hieter, P. A., Ravetch, J. V., Poplack, D. G., Waldmann, T. A., and Leder, P., 1981, Developmental hierarchy of immunoglobulin gene rearrangements in human leukemic pre-B cells, *Proc. Natl. Acad. Sci. USA* **78**:301–313

Korsmeyer, S. J., Arnold, A., Bakhshi, A., Ravetch, J. V., Siebenlist, V., Hieter, P. A., Sharrow, S. O., LeBien, T. W., Kersey, J. H., Poplack, D. G., Leder, P., and Waldman, T. A., 1983a, Immunoglobulin gene rearrangement and cell surface antigen expression in acute lymphocytic leukemias of T cell and B cell precursor origins, *J. Clin. Invest.* **71**:301–313

Korsmeyer, S. J., Greene, W. C., Cossman, J., Hsu, S. M., Jensen, J. P., Neckers, L. M., Marshall, S. L., Bakhshi, A., Depper, J. M., Leonard, W. J., Jaffe, E. S., and Waldmann, T. A., 1983b, Rearrangement and expression of immunoglobulin genes and expression of Tac antigen in hairy cell leukemia, *Proc. Natl. Acad. Sci. USA* **80**:4522–4526

Koziner, B., Gebhard, D., Denny, T., and Evans, R. L., 1982, Characterization of B cell type chronic lymphocytic leukemia cells by surface markers and a monoclonal antibody, *Am. J. Med.* **73**:802–807

Kristensen, J. S., Ellegaard, J., and Hokland, P., 1987, A two-color flow cytometry assay for detection of hairy cells using monoclonal antibodies, *Blood* **70**:1063–1068

Kvaloy, S., Langholm, R., Kaalhus, O., Michaelsen, T., Funderud, S., Foss Abrahamsen, A., and Godal, T., 1984, Transferrin receptor and B-lymphoblast antigen—their relationship to DNA synthesis, histology and survival in B-cell lymphomas, *Int. J. Cancer* **33**:173–1777

Ledbetter, J. A., and Uckun, F. M., 1989, Expression of a functional CD3 T-cell receptor complex on leukemic human T-cell precursers, *Tissue Antigens* **33**:88

Linch, D. C., Allen, C., Beverley, P. C. L., Bynoe, A. G., Scott, C. S., and Hogg, N., 1984, Monoclonal antibodies differentiating between monocytic and nonmonocytic variants of AML, *Blood* **63**:566–573

Ling, N. R., MacLennan, I. C. M., and Mason, D. Y., 1987, B-cell antigens: New and previously defined clusters, in *Leukocyte Typing III* (A. J. McMichael, ed.), pp. 302–335, Oxford University Press, Oxford

Link, M. P., Stewart, S. J., Warnke, R. A., and Levy, R., 1985, Discordance between surface and cytoplasmic expression of the Leu-4 (T3) antigen in thymocytes and in blast cells from childhood T lymphoblasts malignancies, *J. Clin. Invest.* **76**:248–253

Loiseau, P., Divine, M., Le Paslier, D., Marolleau, J. P., Farcet, J. P., Flandrin, G., Cohen, D., Degos, L., Sigaux, F., and Reyes, F., 1987, Phenotypic and genotypic heterogeneity in large granular lymphocyte expansion, *Leukemia* **1**:205–209

Look, A. T., Ashmun, R. A., Shapiro, L. H., and Peiper, S. C., 1989, The human myeloid plasma membrane antigen CD13 (gp150) is identical to aminopeptidase N, *Tissue Antigens* **33**:228

Loughran, T. P., Draves, K. E., Starkebaum, G., Kidd, P., and Clark, E. A., 1987, Induction of NK activity in large granular lymphocyte leukemia: Activation with anti-CD3 monoclonal antibodies, *Blood* **69**:72–78

Majdi, O., Bettelheim, P., Stockinger, H., Aberer, W., Liszka, K., Lutz, D., and Knapp, W., 1984, M2, a novel myelomonocytic cell surface antigen and its distribution on leukemic cells, *Int. J. Cancer* **33**:617–623

Mann, R. B., Jaffe, E. S., Braylen, R. C., Nanba, K., Frank, M. M., Ziegler, J. L., Berard, C. W., 1976, Nonendemic Burkitt's lymphoma: A B cell tumor related to germinal centers, *N. Engl. J. Med.* **295**:685–691

Marlin, S. D., and Pringer, T. A., 1987, Purified intracellular adhesion molecule-1 (ICAM-1) is a ligand for lymphocyte-functional antigen 1 (LFA-1), *Cell* **51**:813–819

Martin, P. J., Hansen, J. A., Nowinski, R. C., and Brown, M. A., 1980, A new human T cell differentiation antigen: Unexpected expression on chronic lymphocytic leukemia cells, *Immunogenetics* **11**:429–439

Martin, P. J., Hansen, J. A., Siadak, A. W., and Nowinski, R. C., 1981, Monoclonal antibodies recognizing normal human T lymphocytes and malignant B lymphocytes: A comparative study, *J. Immunol.* **127:**1920–1923

Matutes, E., Parreira, A., Foa, R., and Catovsky, D., 1985, Monoclonal antibody OKT17 recognizes most cases of T-cell malignancy, *Br. J. Haematol.* **61:**649–656

McMichael, A. J., and Gotch, F. M., 1987, T-cell antigens: New and previously defined clusters, in *Leukocyte Typing III* (A. J. McMichael, ed.), pp. 31–62, Oxford University Press, Oxford

Miedema, F., and Melief, J. M., 1986, Immunobiology of the expanded T cells in T-cell leukemia and T-gamma lymphocytosis, *Leuk. Res.* **10:**469–474

Miedema, F., Tromp, J. F., Veer, M. B., Poppema, S., and Melief, C. J. M., 1985, Lymphocyte function-associated antigen 1 (LFA-1) is a marker of mature (immunocompetent) lymphoid cells. A survey of lymphoproliferative disease in man, *Leuk. Res.* **9:**1099–1104

Miller, R. A., and Gralow, J., 1984, The induction of Leu-1 antigen expression in human malignant and normal B cells by phorbol myristic acetate (PMA), *J. Immunol.* **133:**3408–3414

Mirro, J., Kitchingman, G., Behm, F. G, Murphy, S. B., and Goorha, R. M., 1987, T cell differentiation stages identified by molecular and immunologic analysis of the T cell receptor complex in childhood lymphoblastic leukemia, *Blood* **69:**908–912

Moller, P., Moldenhauer, G., and Dorken, B., 1989, Workshop antibodies B29 (HD66) and B74 (CDIS-4) define a new B-cell antigen, *Tissue Antigens* **33:**160

Moore, M. A. S., and Owen, J. T., 1967, Experimental studies on the development of the thymus, *J. Exp. Med.* **126:**715–725

Morabito, F., Prasthofer, E. F., Dunlap, N. E., Grossi, C. E., and Tilden, A. B., 1987a, Expression of myelomonocytic antigen on chronic lymphocytic leukemia B cells correlates with their ability to produce interleukin 1, *Blood* **70:**1750–1757

Morabito, F., Prasthofer, E. F., Pullen, D. J., Mahoney, D., Downing, J. R., Crist, W. M., and Grossi, C. E., 1987b, Analysis of surface antigen profile, TdT expression, and T cell receptor gene rearrangement for maturational staging of leukemic T cells: A pediatric oncology group study, *Leukemia* **1:**514–517

Morimoto, C., Letvin, N. L., Distaso, J. A., Aldrich, W. R., and Schlossman, S. F., 1985a, The isolation and characterization of the human suppressor inducer T cell subset, *J. Immunol.* **134:**1508–1515

Morimoto, C., Letvin, N. L., Boyd, A. W., Hagan, M., Brown, H. M., Kornacki, M. M., and Schlossman, S. F., 1985b, The isolation and characterization of the human helper inducer T cell subset, *J. Immunol.* **134:**3762–3769

Meuer, S., Hussey, R., Fabbi, M., Fox, D., Acuto, O., Fitzgerald, K., Hodgdon, J., Protentis, J., Schlossman, S., and Reinherz, E., 1984, An alternative pathway of T cell activation: A functional role for the 50 KD T11 sheep erythrocyte receptor protein, *Cell* **39:**897–906

Morimoto, C., Matsuyama, T., Oshige, C., Tanaka, H., Hercend, T., Reinherz, E. L., and Schlossman, S. F., 1985c, Functional and phenotypic studies of Japanese adult T cell leukemia cells, *J. Clin. Invest.* **75:**836–843

Mossalayi, M. D., Bertho, J. M., Dalloul, A. H., Lecron, J. C., and Debre, P., 1989, CD7 is the earliest cell surface antigen detected on T cell lineage, *Tissue Antigens* **33:**108

Murray, L. J., Habeshaw, J. A., Wiels, J., and Greaves, M. F., 1985, Expression of Burkitt lymphoma-associated antigen (defined by the monoclonal antibody 38.13) on both normal and malignant germinal-centre B cells, *Int. J. Cancer* **36:**561–565

Nadler, L. M., Reinherz, E. L., Weinstein, H. J., D'Orsi, C. J., and Schlossman, S. F., 1980, Heterogeneity of T cell lymphoblastic malignancies, *Blood* **55:**806–810

Nadler, L. M., Stashenko, P., Hardy, R., van Agthoven, A., Terhorst, C., and Schlossman, S. F., 1981a, Characterization of a B cell specific (B2) distinct from B1, *J. Immunol.* **126:**1941–1947

Nadler, L. M., Stashenko, P., Ritz, J., Hardy, R., Pesando, J. M., and Schlossman, S. F., 1981b, A unique cell surface antigen identifying lymphoid malignancies of B cell origin, *J. Clin. Invest.* **67:**134–40

Nadler, L. M., Ritz, J., Bates, M. P., Park, E. K., Anderson, K. C., Sallan, S. E., and Schlossman, S. F., 1982, Induction of human B cell antigens in non-T cell acute lymphoblastic leukemia, *J. Clin. Invest.* **70:**433–442

Nadler, L. M., Anderson, K. C., Marti, G., Bates, M., Park, E., Daley, J. F., and Schlossman, S. F., 1983, B4, a human B cell associated antigen expressed on normal, mitogen activated, and malignant B lymphocytes, *J. Immunol.* **131:**244–250

Nadler, L. M., Korsmeyer, S. J., Anderson, K. C., Boyd, A. W., Slaughenhoupt, B., Park, E., Jensen, J.,

Coral, F., Mayer, R. J., Sallen, S. E., Ritz, J., and Schlossman, S. F., 1984, B cell origin of non-T cell acute lymphoblastic leukemia, *J. Clin. Invest.* **74:**332–340

Nadler, L. M., Boyd, A. W., Park, E., Anderson, K. C., Fisher, D., Slaughenhoupt, B., Thorley-Lawson, D. A., and Schlossman, S. F., 1986, The B cell-restricted glycoprotein (B2) is the receptor for Epstein-Barr virus, in *Leukocyte Typing II*, Vol. 2 (E. L. Reinherz, B. F. Haynes, L. M. Nadler, and I. D. Bernstein, eds.), pp. 509–518, Springer-Verlag, New York

Nasu, K., Said, J., and Vonderheid, S., 1985, Immunopathology of cutaneous T-cell lymphoma, *Am. J. Pathol.* **119:**436–447

Neame, P. B., Soamboonsrup, P., Browman, G. P., Meyer, R. M., Benger, A., Wilson, W. E. C., Walker, I. R., Saeed, N., and McBride, J. A., 1986, Classifying acute leukemia by immunophenotyping: A combined FAB-immunologic classification of AML, *Blood* **68:**1355–1362

Neckers, L. M., Yenokida, G., and James, S. P., 1984, The role of the transferrin receptor in human B lymphocyte activation, *J. Immunol.* **133:**2437–1441

Nemerow, G. R., McNaughton, M. E., and Cooper, N. R., 1985, Binding of monoclonal antibody to the Epstein Barr virus (EBV) CR2 receptor induces activation and differentiation of human B lymphocytes, *J. Immunol.* **135:**3068–3073

Nogiwa, E., Tamaki, K., Omine, O., and Maekawa, T., 1988, A monoclonal antibody with reaction spectrum covering acute leukemia and T lineage cells, *Leuk. Res.* **12:**249–256

Orazi, A., Cattoretti, G., Polli, N., and Rilke, F., 1989, CD1c and CD23 expression distinguish chronic B-lymphocytic leukemias from other chronic leukemias and leukemic lymphomas, *Tissue Antigens* **33:**68

Oshimi, K., Hoshino, S., Takahashi, M., Akahoshi, M., Saito, H., Kobayashi, Y., Hirai, H., Takaku, F., Yahagi, N., Oshimi, Y., Horie, Y., and Mizoguchi, H., 1988, Ti (WT31)-negative, CD3-positive, large granular lymphocyte leukemia with nonspecific cytotoxicity, *Blood* **71:**923–931

Owen, J. J. T., and Raff, M. C., 1970, Studies on the differentiation of thymus-derived lymphocytes, *J. Exp. Med.* **132:**1216–1232

Owen, J. J. T., and Ritter, M. A., 1969, Tissue interactions in the development of thymus lymphocytes, *J. Exp. Med.* **129:**431–437

Pallesen, G., Hamilton-Dutoit, S., and Plesner, T., 1989, LFA-1 in lymphomas: Absence in hairy cell leukemia. An analysis of the workshop panel of CD11a/18 clustered antibodies, *Tissue Antigens* **33:**258

Pals, S. T., Horst, E., Ossekoppele, G. J., Figdor, C. G., Scheper, R. J., and Meijer, C. J. L. M., 1989, Expression of lymphocyte homing receptor as a mechanism of dissemination in non-Hodgkin's lymphoma, *Blood* **73:**885–888

Palutke, M., Eisenberg, L., Kaplan, J., Hussain, M., Kithier, K., Tabaczka, P., Mirchandani, I., and Tenenbaum, D., 1983, Natural killer suppressor T cell chronic lymphocytic leukemia, *Blood* **62:**627–634

Pandolfi, F., Rossi, G. D., Semenzato, G., Quinti, I., Ranucci, A., De Sanctis, G., Lopez, M., Gasparotto, G., and Aiuti, F., 1982, Immunologic evaluation of T chronic lymphocyte leukemia cells: Correlations among phenotype, functional activities, and morphology, *Blood* **59:**688–695

Pandolfi, F., De Rossi, G., Ranucci, A., Bonomo, G., Pasqualetti, D., Napolitano, M., and Manzari, V., 1985, Tac-positive, HTLV-negative, T helper phenotype chronic lymphocytic leukemia cells, *Blood* **65:**1531–1537

Paul, W. E., Mizuguchi, J., Brown, M., Nakanishi, K., Hornbeck, P., Rabin, E., and Ohara, J., 1986, Regulation of B-lymphocytes activation, proliferation, and immunoglobulin secretion, *Cell. Immunol.* **99:**7–13

Patarroyo, M., Clark, E. A., Prieto, J., Kantor, C., and Gahmberg, C. G., 1987, Identification of a novel adhesion molecule in human leukocytes by monoclonal antibody LB-2, *FEBS Lett.* **210:**127–131

Patarroyo, M., Prieto, J., Betty, P. G., Clark, E. A., and Gahmberg, C. G., 1988, Adhesion-mediating molecules of human monocytes, *Cell. Immunol.* **113:**278–289

Pezzutto, A., Dorken, B., Moldenhauer, G., and Clark, E. A., 1987, Amplification of human B cell activation by a monoclonal antibody to the B cell-specific antigen CD22, Bp 130/140, *J. Immunol.* **138:**98–103

Pezzutto, A., Dorken, G., Rabinovitch, P. S., Snow, P., Reinherz, E., and Schlossman, S. F., 1986, CD19 monoclonal antibody HD37 inhibits anti-immunoglobulin induced B cell activation and proliferation, *J. Immunol.* **138:**2793–2799

Pezzutto, A., Shu, G. L., Barrett, T. B., Ellingsworth, L., Dorken, B., and Clark, E. A., 1989, Downregulation of B cell activation by CD19 monoclonal antibodies vs inhibition by surface Ig F_c receptor cross-linking or TGF-beta, *Tissue Antigens* **33:**144

Picker, L. J., Medeiros, L. J., Weiss, L. M., Warnke, R. A., and Butcher, E. C., 1988, Expression of lymphocyte homing receptor antigen in non-Hodgkin's lymphoma, *Am. J. Pathol.* **130**:496–504

Pike, B. L., and Robinson, W. A., 1970, Human bone marrow colony growth in agar-gel, *J. Cell. Physiol.* **76**:77–84

Pilarski, L. M., Mant, M. J., and Ruether, B. A., 1985, Pre-B cells in peripheral blood of multiple myeloma patients, *Blood* **66**:416–422

Pilkington, G. R., Wolf, M. M., Matthews, J., Griffiths, J. D., Cooper, I. A., Forster, D. C., and Jose, D. G., 1989, Correlation of CD surface antigen phenotype with FAB subtype, remission induction and survival in AML patients, *Tissue Antigens* **33**:243

Pinkus, G. S., and Said, J. W., 1978, Characterization of non-Hodgkin's lymphomas using multiple cell markers: Immunologic, morphologic, and cytochemical studies of 72 cases, *Am. J. Pathol.* **94**:349–380

Pinto, A., Zagonel, V., Carbone, A., Marotta, G., De Rosa, L., Bullian, P. L., Cirillo, D., Attadia, V., Colombatti, A., and Del Vecchio, L., 1989, Expression of myelomonocytic antigens on chronic lymphocytic leukemia B cells identifies a subset of patients with a varient CLL phenotype and different biologic and clinical features, *Tissue Antigens* **33**:181

Porwit-Ksiazek, A., Christersson, B., Lindemalm, C., Mellstedt, H., Tribukait, B., Biberfeld, G., and Biberfeld, P., 1983, Characterization of malignant and non-neoplastic cell phenotypes in highly malignant non-Hodgkin's lymphomas, *Int. J. Cancer* **32**:667–674

Posnett, D. N., Wang, C. Y., Chiorazzi, N., Crow, M. K., and Kunkel, H. G., 1985, An antigen characteristic of hairy cell leukemia cells is expressed on certain activated B cells, *J. Immunol.* **133**:1635–1640

Posnett, D. N., Folkl, R. J., and Wang, C. Y., 1987, A hairy cell leukemia associated antigen (HC2) has a distinct role in normal B cell differentiation, *Leukemia* **1**:384–386

Preud'homme, J. L., and Seligman, M., 1972, Surface bound immunoglobulins as a cell marker in human lymphoproliferative diseases, *Blood* **40**:777–791

Prieto, J., Beatty, P. G., Clark, E. A., and Patarroyo, M., 1988, Molecules mediating adhesion of T and B cells, monocytes and granulocytes to vascular endothelial cell, *Cell* **63**:631–637

Pui, C. H., Dahl, G. V., Melvin, S., Williams, D. L., Peiper, S., Mirro, J., Murphy, S. B., and Stass, S., 1984, Acute leukaemia with mixed lymphoid and myeloid phenotype, *Br. J. Haematol.* **56**:121–130

Pui, C. H., Williams, D. L., Raimondi, S. C., Melvin, S. L., Behm, F. G., Look, A., T., Dahl, G. V., Rivera, G. K., Kalwinsky, D. K., Mirro, J., Dodge, R. K., and Murphy, S. B., 1986, Unfavorable presenting clinical and laboratory features are associated with Calla-negative non-T, non-B lymphoblastic leukemia in children, *Leuk. Res.* **10**:1287–1292

Quesenberry, P., and Levitt, L., 1979, Hematopoietic stem cells, *N. Engl. J. Med.* **301**:755–760

Rabin, E., Ohara, J., and Paul, W. E., 1985, B-cell stimulatory factor 1 activates resting B cells, *Proc. Natl. Acad. Sci. USA* **82**:2935–2929

Reinherz, E. L., and Schlossman, S. F., 1980, The differentiation and functions of human T lymphocytes: A review, *Cell* **19**:821–827

Reinherz, E. L., Nadler, L. M., Sallen, S. E., and Schlossman, S. F., 1979a, Subset derivation of T-cell acute lymphoblastic leukemia in man, *J. Clin. Invest.* **64**:392–397

Reinherz, E. L., Nadler, L. M., Rosenthal, D. A., Moloney, W. C., and Schlossman, S. F., 1979b, T cell subset characterization of human T-CLL, *Blood* **53**:1066–1075

Reynolds, C. W., and Foon, K. A., 1984, Ty-lymphoproliferative disease and related disorders in man and experimental animals: A review of the clinical, cellular and functional characteristics, *Blood* **64**:1146–1158

Ritter, M. A., Sauvage, C. A., Pegram, S. M., Myers, C. D., Dalchau, R., and Fabre, J. W., 1985, The human leukocyte-common (LC) molecule: Dissection of leukemias using monoclonal antibodies directed against framework and restricted antigenic determinants, *Leuk. res.* **9**:1249–1254

Ritz, J., Pesando, M., Notis-McConarty, J., Lazarus, H., and Schlossman, S. F., 1980, A monoclonal antibody to human acute lymphoblastic leukemia antigen, *Nature (London)* **283**:583–585

Ritz, J., Nadler, L. M., Bhan, A. K., Notis-McConarty, J., Pesndo, J. M., and Schlossman, S. F., 1981, Expression of common acute lymphoblastic leukemia antigen (CALLA) by lymphomas of B cell and T cell lineage, *Blood* **58**:648–652

Robb, R., Greene, W. C., and Rusk, C. M., 1984, Low and high affinity cellular receptors for interleukin-2, *J. Exp. Med.* **160**:1126–1146

Rothlein, R., Dustin, M. L., Marlin, S. D., and Springer, T. A., 1986, A human intracellular adhesion molecule (ICAM-1) distinct from LFA-1, *J. Immunol.* **137**:1270–1274

Rudd, C. E., Trevillyan, J. M., Dasgupta, J. D., Wong, L. L., and Schlossman, S. F., 1988, The CD4 receptor is complexed in detergent lysates to a protein-tyrosine kinase (pp58) from human T lymphocytes, *Proc. Natl. Acad. Sci. USA* **85:**5190–5194

Rudders, R. A., DeLellis, R. A., Ahl, E. T., Bernstein, S., and Begg, C. B., 1983, Adult non-Hodgkin's lymphoma, correlation of cell surface markers phenotype with prognosis, the new working formulation, and the Rappaport and Lukes-Collins histomorphologic schemes, *Cancer* **52:**2289–2299

Rümke, H., Miedema, F., Berge, I. J. M., Terpstra, F., van der Reijden, H. J., van de Griend, R. J., de Bruin, H. G., K. von dem Borne, A. E. G., Smit, J. W., Zeijlemaker, W. P., and Melief, C. J. M., 1982, Functional properties of T cells in patients with chronic T gamma lymphocytosis and chronic T cell neoplasia, *J. Immunol.* **129:**419–426

Sabbath, K. D., Ball, E. D., Larcom, P., Davis, R. B., and Griffin, J. D., 1985, Heterogeneity of clonogenic cells in acute myeloblastic leukemia, *J. Clin. Invest.* **75:**746–753

Sakai, K., Hattori, T., Sagawa, K., Yokoyama, M., and Takatsuki, K., 1987, Biochemical and functional characterization of MCS-2 antigen (CD13) on myeloid leukemic cells and polymorphonuclear leukocytes, *Cancer Res.* **47:**5572–5576

Salter, D. M., Krajewski, A. S., and Cunningham, S., 1987, Activation and differentiation antigen expression in B cell non-Hodgkin's lymphoma, *J. Pathol.* **154:**209–222

San Miguel, J. F., Gonzales, M., Canizo, M. C., Anta, J. P., Zola, H., and Lopez Borrasca, A., 1986, Surface marker analysis in AML and correlation with FAB classification, *Br. J. Haematol.* **64:**547–560

Schwarting, R., and Stein, H., 1989, AS-HCL 2: A monoclonal antibody reacting with a new determinant of human B cells, *Tissue Antigens* **33:**161

Schwarting, R., Stein, H., and Wang, C. Y., 1985, The monoclonal antibodies alpha-S-HCL1 (alpha-Leu-14) and alpha-S-HCL3 (alpha-Leu-M5) allow the diagnosis of hairy cell leukemia, *Blood* **65:**974–983

Schwarzinger, I., Valent, P., Koller, U., Knapp, W., Lechner, K., and Bettelheim, P., 1989, Prognostic significance of surface marker expression on blasts of patients with de novo acute myeloid leukemia, *Tissue Antigens* **33:**245

Selvaraj, P., Plunkett, M. L., Dustin, M., Sanders, M. E., Shaw, S., and Springer, T. A., 1987, The T lymphocyte glycoprotein CD2 binds the cell surface ligand LFA-3, *Nature (London)* **326:**400–403

Sheibana, K., Winberg, C. D., Van de Velde, S., Blaney, D. W., and Rappaport, H., 1987, Distribution of lymphocytes with IL-2 receptors (Tac antigen) in reactive lymphoproliferative processes, Hodgkin's disease, and non-Hodgkin's lymphoma, *Am. J. Pathol.* **127:**27–37

Shipp, M. A., Richardson, N. E., Sayre, P. H., Brown, N. R., Masteller, E. L., Clayton, L. K., Ritz, J., and Reinherz, E. L., 1988, Molecular cloning of the common acute lymphoblastic leukemia antigen (CALLA) identifies a type II integral membrane protein, *Proc. Natl. Acad. Sci. USA* **85:**4819–4823

Shipp, M. A., Vijayaraghavan, J., Schmidt, E. V., Masteller, E. L., D'Adamio, L., Hersh, L. B., and Reinherz, E. L., 1989, Common acute lymphoblastic leukemia antigen (CALLA) is active neutral endopeptidase 24.11 ("enkephalinase"): Direct evidence by cDNA transfection analysis, *Proc. Natl. Acad. Sci. USA* **86:**297–301

Simpkins, H., Kiprov, D. D., Davis, J. L., Morand, P., Puri, S., and Grahn, E. P., 1985, T cell chronic lymphocytic leukemia with lymphocytes of unusual immunologic phenotype and function, *Blood* **65:**127–133

Small, T. N., Keever, C. A., Knowles, R. W., O'Reilly, R. J., and Flomenberg, N., 1989, CD1c expression during normal B cell ontogeny, *Tissue Antigens* **33:**71

Snapper, C. M., and Paul, W. E., 1987, Interferon-γ and B cell stimulatory factor-1 reciprocally regulate Ig isotype production, *Science* **236:**944–947

Sobol, R. E., Roystan, I., LeBien, T., Minowada, J., Anderson, K., Davey, F. R., Cuttner, J., Schiffer, C., Ellison, R. R., and Bloomfield, C. D., 1985, Adult acute lymphoblastic leukemia phenotypes defined by monoclonal antibodies, *Blood* **65:**730–735

Sobol, R. E., Mick, R., Royston, I., Davey, F. R., Ellison, R. R., Newman, R., Cuttner, J., Griffin, J. D., Collins, H., Nelson, D. A., and Bloomfield, C. D., 1987, Clinical importance of myeloid antigen expression in adult acute lymphoblastic leukemia, *N. Engl. J. Med.* **316:**1111–1117

Spier, C. M., Lippman, S. M., Miller, T. P., and Grogan, T. M., 1988, Lennert's lymphoma, a clinicopathologic study with emphasis on phenotype and its relationship to survival, *Cancer* **61:**517–524

Stamenkovic, I., and Seed, B., 1988a, CD19, the earliest differentiation antigen of B cell lineage, bears three extracellular immunoglobulin-like domains and an Epstein-Barr virus related cytoplasmic tail, *J. Exp. Med.* **168:**1205–1210

Stamenkovic, I., and Seed, B., 1988b, Analysis of two cDNA clones encoding the B lymphocyte antigen CD20 (B1, Bp35), a type III integral membrane protein, *J. Exp. Med.* **167**:1975–1980

Stamenkovic, I., Amiot, M., Pesando, J. M., and Seed, B., 1989a, A lymphocyte molecule implicated in lymph node homing is a member of the cartilage link protein family, *Cell* **56**:1057–1062

Stamenkovic, I., and Seed, B., 1990, The B-cell antigen CD22 mediates monocyte and erythrocyte adhesion, *Nature* **345**:74–77.

Stark, A. N., Limbert, H. J., Roberts, B. E., Jones, R. A., and Scott, C. S., 1986, Prolymphocytoid transformation of CLL: A clinical and immunological study of 22 cases, *Leuk. Res.* **10**:1225–1232

Stashenko, P., Nadler, L. M., Hardy, R., and Schlossman, S. F., 1980, Characterization of a new B lymphocyte specific antigen in man, *J. Immunol.* **125**:1678–1685

Stashenko, P., Nadler, L. M., Hardy, R., and Schlossman, S. F., 1981, Expression of cell surface markers following human B cell activation, *Proc. Natl. Acad. Sci. USA* **78**:3848–3852

Stein, H., Siemssen, V., and Lennert, K., 1978, Complement receptor subtypes C3b and C3d in lymphocytic tissue and follicular lymphoma, *Br. J. Cancer* **37**:520–532

Stein, H., Gerdes, J., and Mason, D. Y., 1982, The normal and malignant germinal centre, *Clin. Haematol.* **11**:531–559

Stein, H., Perez-Canto, A., Anagnostopoulis, I., Dallenbach, F., Dienemann, D., and Lemke, H., 1989, Expression of the activation antigen Ki-24 by reactive and neoplastic lymphoid cells, *Tissue Antigens* **33**:322

Sterry, W., and Mielke, V., 1989, In reactive skin diseases and cutaneous T cell lymphomas CD4+ memory cells [CD45R/2H4−, CDw29(4B4)+] are the dominating cell type, *Tissue Antigen* **33**:278

Stonesifer, K. J., Benson, N. A., Ryden, S. E., Pawlinger, D. F., and Braylan, R. C., 1986, The malignant cells in a Lennert's lymphoma are T lymphocytes with a mature helper surface phenotype. A multiparameter flow cytometric analysis, *Blood* **68**:426–429

Strauss, L. C., and Civin, C. I., 1983, MY10, a human progenitor cell surface antigen identified by a monoclonal antibody, *Exp. Hematol.* **11**(Suppl. 14):370a

Strauss, L. C., Skubitz, K. M., August, J. T., and Civin, C. I., 1984, Antigenic analysis of hematopoiesis: II. Expression of human neutrophil antigens on normal and leukemic marrow cells, *Blood* **63**:574–578

Streuli, M., Hall, L. R., Saga, Y., Schlossman, S. F., and Saito, H., 1987, Differential usage of three exons generates at least five different mRNAs encoding human leukocyte common antigens, *J. Exp. Med.* **166**:1548–1566

Stutman, O., and Good, R. A., 1971, Immunocompetence of embyronic hematopoietic cells after traffic to thymus, *Transplant. Proc.* **3**:923–925

Suzuki, T., Sanders, S. K., Butler, J. L., Gartland, G. L., Komiyama, K., and Cooper, M. D., 1986, Identification of an early activation antigen (Bac-1) on human B cells, *J. Immunol.* **137**:1208–1213

Swendeman, S. L., and Thorley-Lawson, D. A., 1987, The activation antigen Blast-2, when shed is a BDGF for normal and transformed B-lymphocytes, in *Leukocyte Typing III* (A. J. McMichael, ed.), pp. 453–454, Oxford University Press, Oxford

Tagawa, S., Konishi, I., Kuratune, H., Katagiri, S., Taniguchi, N., Tamaki, T., Inoue, R., Kanayama, Y., Tsubakio, T., Machii, T., Yonezawa, T., and Kitani, T., 1983, A case of T cell chronic lymphocytic leukemia (T-CLL) expressing a peculiar phenotype (E+, OKM1+, Leu 1+, OKT3−, and IgG EA−), *Cancer* **52**:1378–1384

Tagawa, S., Taniguchi, N., Tokumine, Y., Tamaki, T., Konishi, I., Kanayama, Y., Inoue, R., Machii, T., and Kitani, T., 1986a, OKM1-positive T-cell leukemias: Relationships among morphologic features, phenotype, and functional activities, *Cancer* **57**:1507–151499

Tagawa, S., Tokumine, Y., Ueda, E., Waki, K., Kanayama, Y., Taniguchi, N., Nakanishi, T., Inoue, R., and Kitani, T., 1986b, Leu 11+ Tγ cell chronic lymphocytic leukemia with partially activated natural killer function and its further activation by recombinant IL2 in vitro, *Blood* **68**:846–852

Tedder, T. F., Boyd, A. W., Freedman, A. S., and Nadler, L. M., 1985, The B cell surface molecule B1 is functionally linked with B cell activation and differentiation, *J. Immunol.* **135**:973–979

Tedder, T. F., Streuli, M., Schlossman, S. F., and Saito, H., 1988, Isolation and structure of a cDNA encoding the B1 (CD20) cell surface antigen of human B lymphocytes, *Proc. Natl. Acad. Sci. USA* **85**:208–212

Tedder, T. F., Bell, P. D., Frizzell, R. A., and Bubien, J. K., 1989a, CD20 directly regulates transmembrane ion flux in B lymphocytes, *Tissue Antigens* **33**:145

Tedder, T. F., Penta, A., and Isaacs, C., 1989b, Cloning of CD19, a family of immunoglobulin-like proteins, and use of transfected cell lines to examine the workshop panel of monoclonal antibodies, *Tissue Antigens* **33**:140

Terhorst, C., van Agthoven, A., LeClair, K., Ledbetter, J. A., Moldenhauer, G., and Clark, E. A., 1981, Biochemical studies in the human thymocyte antigens T6, T9, and T10, *Cell* **23**:771–780

Tetteroo, P. A. T., van't Veer, M. B., Tromp, J. F., and von dem Borne, A. E. G., 1984, Detection of the granulocyte-specific antigen 3-fucosyl-N-acetyl-lactosamine on leukemic cells after neuraminidase treatment, *Int. J. Cancer* **33**:355–358

Thompson, L. F., Ruedi, J. M., Glass, A., and Lucas, A. H., 1989, Production of monoclonal antibodies to ecto-5'-nucleotidase: A glycosylphosphatidylinositol anchored differentiation antigen expressed on human T and B lymphocytes, *Tissue Antigens* **33**:163

Thorley-Lawson, D. A., Schooley, R. T., Bhan, A. K., and Nadler, L. M., 1982, Epstein-Barr virus superinduces a new human B cell differentiation antigen (B-LAST-1) expressed on transformed lymphoblasts, *Cell* **30**:415–425

Thorley-Lawson, D. A., Nadler, L. M., Bhan, A. K., Schooley, R. T., 1985, Blast-2 (EBVCS) an early cell surface marker of human B cell activation, is superinduced by Epstein-Barr virus, *J. Immunol.* **134**:3007–3012

Till, J. E., and McCulloch, E. A., 1980, Hematopoietic stem cell differentiation, *Biochim Biophys Acta* **605**:431–459

Tindle, R. W., Nichols, R. A. B., Chan, L., Campana, D., Catovsky, D., and Birnie, G. D., 1985, A novel monoclonal antibody BI-3C5 recognizes myeloblasts and non-B non-T lymphoblasts in acute leukaemias and CGL blast crises, and reacts with immature cells in normal bone marrow, *Leuk. Res.* **9**:1–9

Todd, R. F., Nadler, L. M., and Schlossman, S. F., 1981, Antigens on human monocytes identified by monoclonal antibodies, *J. Immunol.* **126**:1435–1442

Tsai, L. C., Tsai, C. C., Hyde, T. P., Thomas, L. A., and Broun, G. O., 1984, T cell prolymphocytic leukemia with helper-cell phenotype and a review of the literature, *Cancer* **54**:463–470

Tsudo, M., Uchiyama, T., and Uchino, H., 1984, Expression of TAC antigen on activated normal human B cells, *J. Exp. Med.* **160**:612–617

Tsudo, M., Uchiyama, T., Uchino, H., and Yodoi, J., 1986, Failure of regulation of Tac antigen/TCGF receptor on adult T-cell leukemia cells by anti-Tac monoclonal antibody, *Blood* **61**:1014–1016

Uchiyama, T., Yodoi, J., Sagawa, K., Takatsuki, K., and Uchino, H., 1977, Adult T-cell leukemia: Clinical and hematologic features of 16 cases, *Blood* **50**:481–492

Uchiyama, T., Broder, S., and Waldmann, T. A., 1981, A monoclonal antibody (anti-TAC) reactive with activated and functionally mature T cells. I. Production of anti-TAC monoclonal antibody and distribution of Tac (+) cells, *J. Immunol.* **126**:1393–1403

Uchiyama, T., Hori, T., Tsudo, M., Wano, Y., Umadome, H., Tamori, S., Yodoi, J., Maeda, M., Sawami, H., and Uchino, H., 1985, Interleukin-2 receptor (Tac antigen) expressed on adult T cell leukemia cells, *J. Clin. Invest.* **76**:446–453

Uckun, F., and Ledbetter, J. A., 1989a, Altered function of the CD19 receptor on leukemic human B-cell precursers in B-lineage acute lymphoblastic leukemia, *Tissue Antigens* **33**:146

Uckun, F., and Ledbetter, J. A., 1989b, Expression and function of CDw40/Bp50 antigen in early human B-lymphocyte ontogeny, *Tissue Antigens* **33**:146

Uckun, F. M., Waddick, K., Kuebelbeck, V., Ledbetter, J., Kishimoto, T., and Koller, B., 1989, Biphenotypic normal and leukemic lymphocyte precursers, *Tissue Antigens* **33**:183

Van der Reijden, H. J., van Rhenen, D. J., Lansdorp, P. M., van't Veer, M. B., Langenhuijsen, M. M. A. C., Engelfriet, C. P., and K. von dem Borne, A. E. G., 1983, A comparison of surface marker analysis and FAB classification in acute myeloid leukemia, *Blood* **61**:443–448

Van der Valk, P., van den Besselaar-Dingjan, G., Daha, M. R., and Meijer, C. J. L. M., 1983, Analysis of large cell lymphomas using monoclonal and heterologous antibodies, *J. Clin. Pathol.* **36**:44–50

Van Dongen, J. J. M., Krissansen, G. W., Wolvers-Tettero, I. L. M., Comans-Bitter, W. M., Adriaansen, H. J., Hooijkaas, H., van Wering, E. R,. and Terhorst, C., 1988, Cytoplasmic expression of the CD3 antigen as a diagnostic marker for immature T-cell malignancies, *Blood* **71**:603–612

Varon, D., Gittel, S., Linder, S., Bassous-Guedj, L., Vorst, E., and Berrebi, A., 1989, An antiplatelet glycoprotein IIb monoclonal antibody (3B2) interacts with activated platelets and recognizes immature megakaryoblasts, *Tissue Antigens* **33**:347

Vodinelich, L., Tax, W., Bai, Y., Pegram, S., Capel, P., and Greaves, M. F., 1983, A monoclonal antibody (WT1) for detecting leukemias of T cell precursors (T-ALL), *Blood* **62**:1108–1113

Vogler, L. B., Crist, W. M., Bockman, D. E., Pearl, E. R., Lawton, A. R., and Cooper, M. D., 1978, Pre-B cell leukemia; a new phenotype of childhood lymphoblastic leukemia, *N. Engl. J. Med.* **298**:872–878

Volk, J. R., Kjeldsberg, C. R., Eyre, H. J., and Marty, J., 1983, T cell prolymphocytic leukemia: Clinical and immunologic characterization, *Cancer* **52**:2049–2054

Waldmann, T. A., Greene, W. C., Sarin, P. S., Saxinger, C., Blayney, D. W., Blattner, W. A., Goldman, C. K., Bongiovanni, K., Sharrow, S., Depper, J. M., Leonard, W., Uchiyama, T., and Gallo, R. C., 1984, Functional and phenotypic comparison of human T cell leukemia/lymphoma virus positive adult T cell leukemia with human T cell leukemia/lymphoma virus negative Sezary leukemia, and their distinction using anti-Tac, *J. Clin. Invest.* **73**:1711–1718

Weisenberger, D. D., Linder, J., Daley, D. T., and Armitage, J. O., 1987a, Intermediate lymphocytic lymphoma. An immunohistologic study with comparison to other lymphocytic lymphomas, *Human Pathol.* **18**:781–790

Weisenberger, D. D., Sanger, W. G., Armitage, J. O., and Purtilo, D. T., 1987b, Intermediate lymphocytic lymphoma: Immunophenotypic and cytogenetic findings, *Blood* **69**:1617–1621

Weiss, L. M., Bindl, J. M., Picozzi, V. J., Link, M. P., and Warnke, R. A., 1986, Lymphoblastic lymphoma: An immunophenotype study of 26 cases with comparison to T cell acute lymphoblastic leukemia, *Blood* **67**:474–478

Wilson, B. S., Platt, J. L., and Kay, N. E., 1985, Monoclonal antibodies to the 140,000 mol wt glycoprotein of B lymphocyte membranes (CR2 receptor) initiates proliferation of B cells in vitro, *Blood* **66**:824–829

Winberg, C. D., Sheibani, K., Krance, R., and Rappaport, H., 1985, Peripheral T cell lymphoma: Immunologic and cell-kinetic observations associated with morphological progression, *Blood* **66**:980–989

Woermann, B., Anderson, J. M., Ling, Z. D., and LeBien, T. W., 1987, Structure/function analysis of IL-2 binding proteins on human B cell precursor acute lymphoblastic leukemia, *Leukemia* **1**:660–666

Woods, G. M., Sawyer, P. J., Kirov, S. M., Lowenthal, R. M., Jupe, D. M., and Catovsky, D., 1985, Functional and phenotypic analysis of a T cell prolymphocytic leukemia, *Leukemia Res.* **9**:587–596

Yagita, H., Nakamura, T., Karasuyama, and Okumura, K., 1989, Monoclonal antibodies specific for murine CD2 reveal its presence on B as well as T cells, *Proc. Natl. Acad. Sci. USA* **86**:645–649

Yamada, Y., Kamihira, S., Amagasaki, T., Kinoshita, K., Kusano, M., Ikeda, S., Toriya, K., Suzuyama, J., and Ichimaru, M., 1984, Changes of adult T cell leukemia cell surface antigens at relapse or at exacerbation phase after chemotherapy defined by use of monoclonal antibodies, *Blood* **64**:440–444

Yamada, Y., Kamihira, S., Amagasaki, T., Kinoshita, K., Kusano, M., Chiyoda, S., Yawo, E., Ikeda, S., Suzuyama, J., and Ichimaru, M., 1985, Adult T cell leukemia with atypical surface phenotypes: Clinical correlation, *J. Clin. Oncol.* **3**:782–788

Yokochi, T., Holly, R. D., and Clark, E. A., 1982, B lymphoblast antigen (BB1) expressed on Epstein-Barr virus-activated B cell blasts. B lymphoblastoid cell lines, and Burkitt's lymphomas, *J. Immunol.* **128**:823–827

Chapter 3

Cytoskeletal Organization of Normal and Leukemic Lymphocytes and Lymphoblasts

Annette Schmitt-Gräff and Giulio Gabbiani

1. INTRODUCTION

The cytoskeleton of eukaryotic cells is a complex array composed of three sets of filaments: (1) microfilaments, 4 to 7 nm in diameter, made up mainly of actin; (2) intermediate filaments, 8 to 11 nm in diameter, formed by at least five distinct classes of proteins; and (3) microtubules, 25 nm in diameter, consisting of tubulin (for review, see Alberts *et al.*, 1983). Many proteins have been described that interconnect these filaments or link them to various organelles. The cytoskeleton is involved in many basic cellular functions, including maintenance of cell shape, motility, organelle movement, and mitosis. Studies of cytoskeletal protein expression and organization have increased our understanding of the cellular adaptation during differentiation and pathological processes (for a review, see Rungger-Brändle and Gabbiani, 1983).

The goal of this chapter is to describe the cytoskeletal elements of lymphoid cells and to demonstrate how they are related to specific cellular functions and to the status of maturation in normal and leukemic lymphocytes and lymphoblasts.

2. MICROFILAMENTS IN NORMAL AND LEUKEMIC LYMPHOCYTES AND LYMPHOBLASTS

2.1. Actin and Actin-Associated Proteins in Nonmuscle Cells

Since the purification of actin from nonmuscle cells (Hatano and Oosawa, 1966), much has been learned about the mechanisms of actin polymerization and its control by

Annette Schmitt-Gräff and Giulio Gabbiani   Department of Pathology, University of Geneva, Geneva, Switzerland.

actin-binding proteins (for reviews, see Korn, 1982, and Stossel *et al.*, 1985). Globular subunits of 42 kDa (G-actin) assemble reversibly to give polymers several micrometers in length (F-actin). Microfilaments are made up of two polymer chains forming a right-handed helix. Actin monomers bind ATP that is hydrolyzed during the polymerization reaction (Wegner, 1985). Actin filaments are bipolar, as shown by decoration with heavy meromyosin (Huxley, 1963), and have a higher affinity for actin monomers at their plus or barbed end than at their minus or pointed end (Woodrum *et al.*, 1975). This leads to a steady-state flux or "treadmilling" of actin subunits (Neuhaus *et al.*, 1983). Nonmuscle actin filaments are considerably more labile than muscle actin filaments. Small changes in monomer pool and in ionic concentration as well as the action of accessory proteins modulate the polymerization and disassembly of actin (Weeds, 1982).

Different groups of actin-binding proteins have been described according to their function. Examples include (1) profilin, which binds predominantly to actin monomers and inhibits their incorporation into filaments; (2) gelsolin, a multifunctional actin-modulating protein that has at least three effects on actin: severing, nucleation, and end blocking dependent on Ca^{2+} concentration; (3) nonerythroid spectrins and vinculin, which are assumed to mediate the attachment of actin filaments to membranes; (4) myosin, which not only generates contractile forces with actin but also is an actin crosslinker; and (5) α-actinin, which binds, crosslinks, and stabilizes actin filaments (for a review, see Stossel *et al.*, 1985). Moreover, cytochalasins are widely used to study cellular functions in which microfilaments are involved. These fungal metabolites shorten actin filaments by binding to the barbed end (Schliwa, 1982), thus promoting actin depolymerization.

Electron microscopic studies have revealed several types of actin filament organization in the cytoplasm of nonmuscle cells (Stossel, 1984). Microfilaments may be arranged in parallel bundles, in which they have a unidirectional polarity and probably play a role in stabilizing cellular structures such as the brush border microvilli of epithelial cells (Bretscher, 1983). Stress fibers, consisting of a bundled network of actin filaments and associated proteins, represent a supramolecular array of cytoskeletal elements present *in vivo* in myofibroblasts (Gabbiani *et al.*, 1971) and endothelial cells (Gabbiani *et al.*, 1983) and on the ventral surface of many cultured cells (Ramaekers *et al.*, 1980).

In motile nonmuscle cells, the peripheral or cortical zone is mainly composed of microfilaments forming orthogonal networks (for a review, see Hartwig *et al.*, 1985). Variations in the number and length of actin filaments alter the consistency of the peripheral cytoplasm. Many motile events such as locomotion, exocytosis, phagocytosis, organelle distribution, and cytoplasmic streaming are accompanied by actin assembly and disassembly (Weeds, 1982; Stossel *et al.*, 1984). Sol-gel transformation with a higher viscosity of the cytoplasm follows actin polymerization (for a review, see Southwick and Stossel, 1983). It is generally agreed that actin must be in the filamentous form (F-actin) to participate with myosin in the force-generating system of motility according to the sliding-filament hypothesis (Weeds, 1982).

2.2. Actin Isoforms

Actin is one of the most conserved proteins in eukaryotic cells, characterized by a tissue-specific rather than a species-specific microheterogeneity (Vandekerckhove and Weber, 1978). By means of two-dimensional PAGE, three isoforms, referred to as α-, β-, and γ-actins, were separated (Garrels and Gibson, 1976). However, sequencing and genetic studies have shown that at least six actin isoforms are expressed in different tissues in mammalians and birds; two are present in every cell (β- and γ-cytoplasmic), two are found in striated muscle (α-skeletal and α-cardiac), and two are specific to smooth muscle (α- and γ-smooth muscle) (Vandekerckhove and Weber, 1979). At present, only a small number of isoactin-specific antibodies recognizing α-smooth muscle actin (Skalli et al., 1986a) or α-sarcomeric actin (Bulinski et al., 1983; Skalli et al., 1988) are available.

Little is known about the function of actin isoforms. Many studies have shown that actin isoform expression may be modulated during development (Woodcock-Mitchel et al., 1988) and in experimental (Skalli et al., 1986b) and pathological (Skalli et al., 1989; Schmitt-Gräff et al., 1989) conditions. Moreover, transformation of cell lines in vitro may result in the expression of new forms of isoactins (Hamada et al., 1981).

2.3. Organization of Actin in Normal Lymphocytes

2.3.1. Actin Content in Normal Lymphocytes

Actin has been identified as a major cellular component of lymphocytes (Barber and Delovitch, 1978). In various lymphocyte cell lines and normal peripheral blood lymphocytes, actin was estimated to constitute 6–16% of total cellular protein (Fechheimer and Cebra, 1979; Leavitt et al., 1980; Stark et al., 1982; Atkins and Anderson, 1982; Schmitt-Gräff et al., 1987a). The discrepant amounts of actin evaluated as a percentage of total cellular protein may in part be explained by the different methods used for actin quantification, including one- and two-dimensional PAGE, densitometric scanning of gels, autoradiography, and DNase I inhibition assay. Recently, an actin content of 3.4 ± 0.12 pg per cell was evaluated in lymphocytes obtained from human blood (Phatak et al., 1988).

Similar amounts of actin were found in T and B-lymphocyte-enriched populations obtained from normal donors (Stark et al., 1982; Schmitt-Gräff et al., 1987a). However, studies done on rat lymphocytes showed that characteristic values were associated with functionally distinct subpopulations (Mely-Goubert and Bellagrau, 1981). Thus, T lymphocytes had significantly higher total actin per cell than did B lymphocytes. When different fractions of sheep thymocytes were separated, the fraction enriched for immunocompetent medullary-type cells showed the highest actin content, but the value was lower than that of peripheral lymphocytes (Miyasaka et al., 1984). In the T-cell population, the values increased with the differentiation state. A correlation between the recirculating properties and the actin content of lymphocytes has been proposed (Mely-Goubert and Bellagrau, 1981).

2.3.2. Actin Isoforms in Lymphocytes

Normal lymphocytes contain the β- and γ-cytoplasmic isoforms of actin (Fechheimer and Cebra, 1979). In normal B and T lymphocytes, β-actin is the predominant form (Stark et al., 1982; Leavitt et al., 1980).

2.3.3. F-Actin in Normal Lymphocytes

In lymphocytes, as in other types of nonmuscle cells, actin exists in a globular (G) and a filamentous (F) form, which is assumed to play an important force-generating and structural role. The extent of actin polymerization in lymphocytes has been assayed by different techniques, including the DNase I inhibition assay (Varani *et al.*, 1983), ultracentrifugation of Triton X-100 cell lysates followed by PAGE (Schmitt-Gräff *et al.*, 1987a; Phatak *et al.*, 1988), and flow cytometry after incubation with 7-nitrobenz-2-oxadiazole-phallacidin (Phatak *et al.*, 1988). In the study of Phatak *et al.* (1988), the Triton-insoluble cytoskeleton of lymphocytes contained 0.95 ± 0.04 pg of actin per cell and the Triton-soluble cytoplasmic fraction contained 2.4 ± 0.08 pg per cell by PAGE, in excellent agreement with the amount of actin directly measured in the whole lymphocytes (see above). According to the results obtained by Varani *et al.* (1983), about 66% of normal lymphocyte actin is found as G-actin. Thus, a majority of normal lymphocyte actin exists as monomers.

Modulation of the polymerization state of actin in lymphocytes was observed under different conditions. Decreasing amounts of G-actin as measured by the DNase I inhibition assay were associated with stimulation of lymphocytes with concanavalin A (ConA) (Rao and Varani, 1982; Varani *et al.*, 1983; Rao, 1984) and with phytohemagglutinin (PHA) and wheat germ agglutinin (Rao, 1984). The decrease of G-actin could be prevented by pretreatment of cells with cytochalasin E, indicating that the decrease was likely due to conversion to F-actin (Rao, 1984). Furthermore, it has been shown that activators of protein kinase C cause a marked increase in F-actin (Phatak *et al.*, 1988). However, most studies of actin conformation have been done by immunofluorescent staining without quantification of F-actin. Thus, compared with neurophils (White *et al.*, 1983; Fechheimer and Zigmond, 1983), less is known about how changes in the quantity of polymerized actin relate to changes in the motility and function of lymphocytes.

2.3.4. Patterns of Actin Distribution in Lymphocytes

Fluorescence microscopic evaluation of actin in lymphocytes has been performed by using polyclonal (Gabbiani *et al.*, 1977; Fagraeus *et al.*, 1980) or monoclonal (Kammer *et al.*, 1988) antiactin antibodies recognizing cytoplasmic actin isoforms. Alternatively, fluorescein or rhodamine conjugates of phallacidin or phalloidin derivatives have been found to be effective stains for polymerized F-actin filaments (Phatak *et al.*, 1988).

Most resting lymphocytes show a smooth configuration with a uniform cytoplasmic actin fluorescence. In addition, a rim of increased submembranous fluorescence corresponding to a fine microfilament meshwork is present at the cell periphery (Figure 1A). In conditions associated with changes in cell surface morphology, the actin network undergoes striking rearrangements. When surface microvilli, pseudopods, and an ameboid morphology with a hyaline veil forward and a tail or uropod at the back are formed, actin staining is predominantly encountered in the cytoplasmic extensions. The increased labeling of surface projections with phallacidin or phalloidin derivatives indicates a localized increase in microfilaments clustered in these active cell edges. Thus, a polarized actin assembly is the hallmark of various conditions characterized by motility, such as lymphocyte locomotion (for a review, see Southwick and Stossel, 1983), binding to target cells (Ryser *et al.*, 1982), and capping of surface receptors (Gabbiani *et al.*, 1977).

Cytochalasins inhibit the polar accumulation of actin beneath the cell membrane, probably by preventing the formation of microfilaments required for motile events (Kammer *et al.*, 1988). Studies on lymphocyte locomotion in relation to fibroblasts in culture

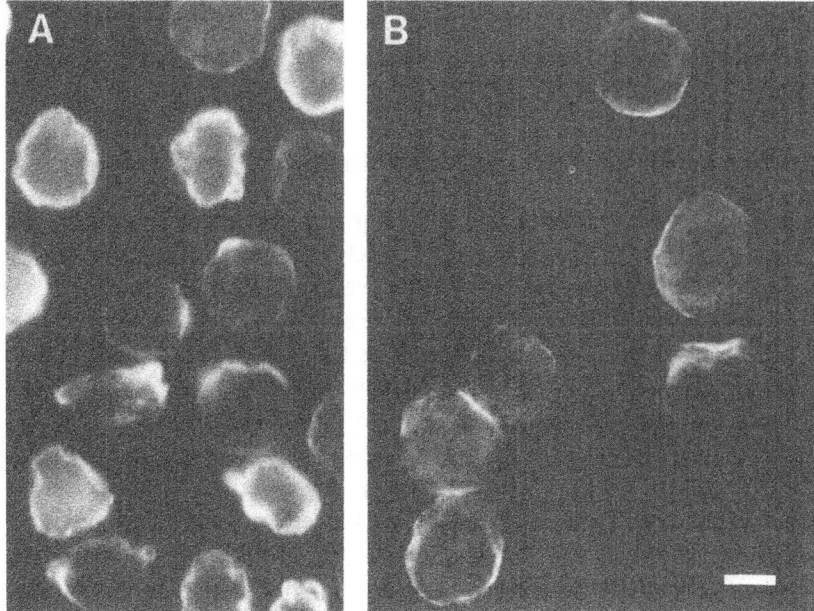

FIGURE 1. (A) Normal peripheral blood lymphocytes exhibiting a pale cytoplasmic fluorescence with a thin rim of increased concentration of F-actin when stained with rhodamine phalloidin. (B) CLL lymphocytes with a decreased concentration of F-actin, as indicated by the decreased staining intensity. Bar = 10 μm.

showed that cytochalasin B abolished lymphocyte shape-changing activity, motility, and crawling (Chang *et al.*, 1979).

Whereas the presence of intracellular actin is well documented, the possibility that some actin is associated with the cell surface is controversial. Owen *et al.* (1978) reported that actin is most probably exposed on the surface of pig, mouse, and human B lymphocytes. Immunofluorescent staining of murine lymph node lymphocytes and thymocytes by Sanders and Craig (1983) revealed two patterns: (1) intensely fluorescent rings representing cytoplasmic actin and (2) a faint punctate pattern probably due to a reaction with surface actin. This staining pattern may correspond to a developmental binding site for actin or to the expression of a protein antigenically related to actin on the lymphocyte surface.

By electron microscopy, a network of microfilaments can be resolved in a narrow submembranous zone of resting lymphocytes (Zucker-Franklin *et al.*, 1979). Microfilament bundles have been observed in surface villi (Fagraeus *et al.*, 1980) and in the uropod region of capped lymphocytes (Zucker-Franklin *et al.*, 1979). In cytotoxic T lymphocytes attached to target cells, the microfilament network is conspicuous under the area of cell contact (Ryser *et al.*, 1982).

When the migration of lymphocytes across the wall of high-endothelial venules was studied, microfilaments were detected in cytoplasmic processes of lymphocytes passing through the vessel wall (Campbell, 1983). It is well accepted that Hermes-defined glycoproteins play a major role in the process of lymphocyte–endothelial cell interaction and lymphocyte extravasation (Jalkanen *et al.*, 1988). In addition, an adenylate cyclase-dependent activation of the cytoplasmic contractile system is probably required for stable lymphocyte adherence to high-endothelim venules of lymph nodes, a process sensitive to cytochalasin B (Spangrude *et al.*, 1984).

2.4. Interactions between Actin and Surface Antigens in Lymphocytes

There is convincing evidence indicating that actin and additional proteins form a matrix associated with the plasma membrane which stabilizes the membrane bilayer and mediates interactions between transmembrane proteins and the cytoskeleton (Mescher *et al.*, 1981). A variety of studies have shown a close association between surface molecules and microfilaments in dynamic processes such as the capping phenomenon in B and T lymphocytes (Gabbiani *et al.*, 1977; Bourguignon and Bourguignon, 1984; Kammer *et al.*, 1983). The crosslinking of surface receptors by ligands induces a series of contractile events in motile cells, including patching, capping, and endocytosis of ligand–receptor complexes (Bretscher, 1984). These events were first described in B lymphocytes in suspension (Taylor *et al.*, 1971). When lymphocytes are exposed to antibodies at 0°C, two-dimensional precipitates of the antigen–antibody complex are formed at the cell surface. At 37°C, these patches move to the tail of the cell and form caps associated with polar aggregates or co-caps of actin and other cytoskeletal proteins beneath the caps (Bourguignon and Bourguignon, 1984). In addition, spontaneous capping of surface molecules and the coincident reorganization of cytoskeletal elements has been observed in lymphocytes dependent on the onset of cellular motility and the formation of a uropod (Schreiner *et al.*, 1976; Braun *et al.*, 1978).

The presence of both actin and myosin beneath the caps (Gabbiani *et al.*, 1977; Schreiner *et al.*, 1977) and the detection of myosin light-chain phosphorylatin in isolated plasma membranes from capped lymphocytes (Kerrick and Bourguignon, 1984) have been used to support the possibility that submembranous microfilament contraction according to the actin/myosin sliding-filament mechanism may be implicated in the capping cascade (Bourguignon *et al.*, 1985). Furthermore, the polymerization of globular or filamentous actin accompanying Thy-1 capping (Laub *et al.*, 1981) and the reduction or inhibition of cap formation by cytochalasins (de Petris, 1974; Kammer *et al.*, 1983, 1988) favor the notion that cytoskeletal microfilaments play a critical role in lymphocyte cap formation. Much has been learned regarding the attachment of surface receptors to the actin-containing cytoskeleton during capping. Braun *et al.* (1982) demonstrated that ligand binding converts surface immunoglobin (sIg) of B lymphocytes from a detergent-soluble (free) to a detergent-insoluble (cytoskeleton-associated) form. Woda and Woodin (1984) showed that binding of both polyvalent and monoclonal antibodies (MAbs) induced 63–77% of sIg to assume a detergent-insoluble form, presumably because of a transmembrane interaction with cytoskeletal structures. In contrast, class II histocompatibility antigens remained predominantly soluble. By flow cytofluorometric assay and biochemical analysis, Albrecht and Noelle (1988) have demonstrated that anti-isotype-specific antibodies induce membrane IgM (mIgM) and mIgD to associate with the cytoskeleton of B lymphocytes in an isotype-specific fashion. An intriguing finding of this study is that less than half of the μ-specific MAbs tested rendered mIgM cytoskeletally associated. A strict correlation between the capacity of MAbs to induce cytoskeletal association and capping was noted.

When the effects of crosslinking conventionally anchored cluster of differentiation (CD) molecules and of glycosyl inositol phospholipid-anchored molecules, which move freely among lipid domains within the plasma membrane, were compared in human T lymphocytes, similar patterns of clustered microfilaments adjacent to the polar caps were observed (Kammer *et al.*, 1988). However, the exact nature of the linkage between membrane surface receptors and the intracellular actin network remains to be elucidated. It has been proposed that serum vitamin D-binding protein may link sIg with actin (Petrini *et al.*, 1983). Hoessli *et al.* (1980) found that α-actinin co-caps with sIg and Thy-1 antigen in mouse spleen lymphocytes and suggested that α-actinin may be involved in the movement of lymphocyte surface receptors. Recently, Gupta and Woda (1988) observed that B-cell sIg becomes detergent insoluble because of its transmembrane association with F-actin via a 70–73-kDa and a 112-kDa protein. The latter was found to be immunologically related to α-actinin. Bourguignon *et al.* (1985) have documented that a T-lymphoma transmembrane glycoprotein (gp180) and fodrin, a nonerythroid type of spectrin, are closely associated in a complex. They have pointed to the importance of this complex for the association of the plasma membrane and the intracellular microfilament network during lymphocyte patching and capping.

The physiological significance of the attachment of surface receptors with the actin-containing cytoskeleton is still being discussed. Apparently, the aggregation of surface receptors plays an important role in the immune response (Monroe and Cambier, 1983). Rao (1984) reported a receptor-mediated actin polymerization following stimulation with the mitogenic lectins ConA and PHA in human peripheral blood lymphocytes. Thus, lectin-induced actin polymerization might be a signal for the initiation of DNA synthesis and mitogenesis.

In human T lymphocytes, activation of protein kinase C, which is a link in a second-messenger system that transduces and amplifies information after engagement of external receptors on the cell surface, causes a marked increase in cellular F-actin and pseudopod formation (Phatak *et al.*, 1988). This observation can be interpreted as an additional indication that actin-conformational changes may be components of signal transduction from the cell surface to the nucleus.

In light of these and other studies (Bourguignon *et al.*, 1984; Goroff *et al.*, 1986; Rothstein, 1986), one could assume that receptor–microfilament association may be implicated in transmembrane signaling, cell activation, and modulation of lymphocyte proliferation.

2.5. Actin in Leukemic Lymphocytes and Lymphoblasts

2.5.1. Total Actin Content of Leukemic Cells

Modifications of actin isoforms, actin content, extent of actin polymerization, and distribution of microfilaments have been reported in transformed lymphoid cell lines *in vitro* and in human leukemias *in vivo*. Several studies have documented a considerable decrease in actin content as a percentage of total cellular protein in chronic lymphocytic leukemia (CLL) cells compared with normal peripheral blood lymphocytes (Stark *et al.*, 1982; Atkins and Anderson, 1982; Schmitt-Gräff *et al.*, 1987a). Moreover, total actin was found to be diminished in transformed cell lines of lymphoid origin (Leavitt *et al.*, 1980; Varani *et al.*, 1983). However, lymphoblasts from children with common acute lymphoblastic leukemia (ALL) did not contain significantly less actin than did controls (Schmitt-Gräff *et al.*, 1987a, b).

2.5.2. Actin Isoforms in Leukemic Cells

According to Leavitt *et al.* (1980), equal amounts of the β- and γ-actins are synthesized in a transformed lymphoid cell type (Molt-4), whereas normal T cells produce β-actin as the predominant form, indicating that the cellular transformation may result in a modulation of actin isoform expression. However, similar ratios of β- to γ-actin were found in lymphocytes from patients with CLL (Stark *et al.*, 1982).

2.5.3. F-Actin Content of Leukemic Cells

Few biochemical studies have examined the content of F-actin in neoplastic lymphoid cells. Varani *et al.* (1983) showed that transformed lymphoid cell lines had a greater level of their actin in a polymerized form than did normal lymphocytes. A T-cell thymoma line was reported to have a lower proportion of G-actin than mature thymocytes (Mely-Goubert and Bellagrau, 1981). However, it must be emphasized that under *in vitro* conditions, actin organization may undergo striking alterations, with a relative increase of actin filaments compared with *in vivo* conditions.

We studied the actin filament content of CLL lymphocytes and ALL blasts freshly isolated from peripheral blood. The amount of F-actin was evaluated by densitometric scanning of SDS-polyacrylamide gels loaded with pellets obtained by ultracentrifugation of total cell lysates. F-actin was diminished by about 50% in both CLL and ALL cells compared with peripheral lymphocytes from healthy donors (Figure 2).

2.5.4. Patterns of Actin Distribution in Leukemic Cells

The quantitative changes of total and F-actin content demonstrated in leukemic cells

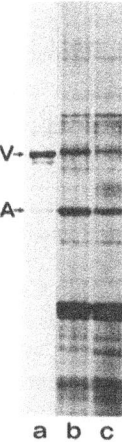

a b c

FIGURE 2. SDS-polyacrylamide gel separation of lymphocyte proteins in ultracentrifugation pellets of Triton X-100 cell lysates. Lane a, Cytoskeletal fraction of cultured baby hamster kidney cells as markers for actin (A) and vimentin (V); lane b, pellet from a control person; lane c, pellet from a patient with CLL. Note that lane b (pellet of control) shows a band of actin larger and more intensely stained than in lane c (pellet of CLL patient). The vimentin band in lane c is clearly weaker than the vimentin band in lane b.

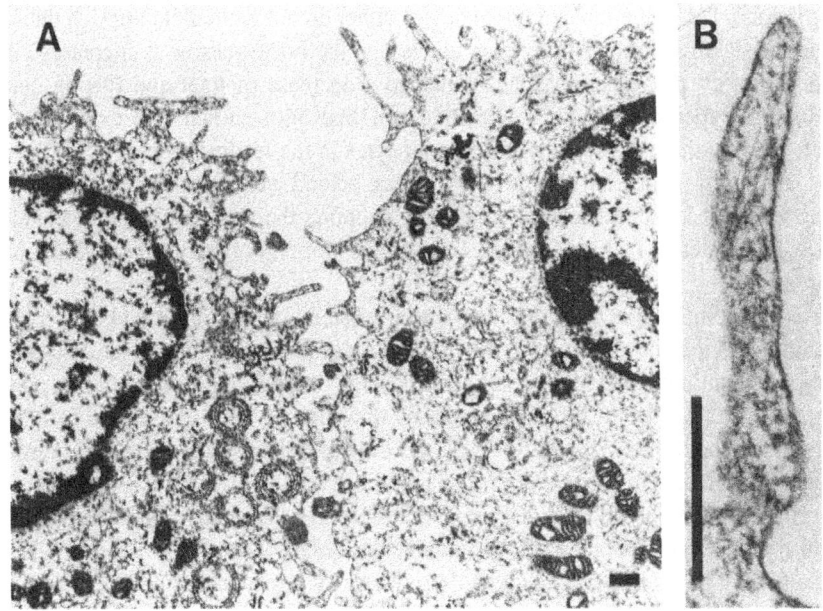

FIGURE 3. (A) Electron micrograph of hairy cells in HCL showing humerous filipodia and ruffled protuberances. (B) Cytoplasmic protrusions filled with a meshwork of microfilaments. Bars = 0.5 μm.

are in agreement with ultrastructural studies of Zucker-Franklin *et al.* (1979), who have found that the marginal cytoplasm of CLL lymphocytes shows a poorly developed microfilamentous network. Labeling of CLL lymphocytes with rhodamine phalloidin shows a weak fluorescence intensity correlating with a low F-actin content (Figure 1B). Caligaris-Cappio *et al.* (1986) have demonstrated by means of fluorescent phalloidin staining that actin is abnormally rearranged in phorbol acetate-stimulated lymphocytes of CLL and hairy cell leukemia (HCL). The majority of hairy cells showed accumulation of F-actin in peripheral protrusions in the form of filipodia and ruffles. On stimulation, hairy cells sprouted long dentritic processes rich in submembranous F-actin. Electron micrographs of hairy cells reveal numerous cytoplasmic protrusions enriched for microfilaments (Figure 3).

2.5.5. Impact of Actin Organization Changes in Leukemic Lymphocytes and Lymphoblasts

Actin organization in leukemic lymphocytes and lymphoblasts is likely to have an important impact on the functional capabilities of neoplastic cells. It has been suggested that the peculiar cytoskeletal features of HCL may account for *in vitro* adhesive properties of hairy cells and the clinical behavior, with a low number of circulating cells but a marked infiltration of bone marrow and spleen (Caligaris-Cappio *et al.*, 1986).

When compared with their normal counterparts, substantial differences in motile and contractile activities and in cell deformability have been noted in lymphocytes from patients with CLL (Stark *et al.*, 1982). CLL cells show an anomalous capping behavior (Cohen and Gilbertsen, 1975; Liebes *et al.*, 1978). Altered migratory properties in CLL lymphocytes as compared with normal cells have been documented (Wagstaff *et al.*, 1981). Indium-111 oxine-labeled lymphocytes either do not recirculate through the spleen or their transit time through this organ as well as the bone marrow is increased. These altered migratory properties may be related to a decrease of total and F-actin content.

Homing experiments with a leukemic cell line have shown that extravasation is characterized by the presence of fibrillar podocytes in the leading front (Azzarelli *et al.*, 1989). Hand mirror forms with a uropod trailing behind were not seen during the transvenular traffic of leukemic cells. This finding supports the notion that the cytoskeleton, which is implicated in cell conformation, may be abnormally organized in lymphoblastic leukemia.

It is tempting to speculate that the pattern of actin expression in leukemic lymphocytes and blasts may be attributed to the immaturity of leukemic cells. Most cases of CLL correspond to an immature counterpart at a midstage of the B-cell differentiation pathway (Anderson *et al.*, 1984). ALL blasts expressing the common ALL antigen are early B-cell precursors (Korsmeyer *et al.*, 1983). Thus, according to the findings reported in myeloid cell lines (Meyer and Howard, 1983; Nagata *et al.*, 1980), an inverse relationship between the level of maturation and total actin content as well as the relative F-actin content may exist in neoplastic lymphoid cells. Moreover, the reduced level of F-actin in CLL and ALL cells may be related to changes in regulatory systems, including Ca^{2+} and actin-associated proteins.

Finally, the changes in actin organization may be related to transformation and leukemogenesis. The actin network rapidly depolymerizes in cells transformed by oncogene-carrying viruses. Furthermore, the β-actin gene was found to be transduced by transforming retroviruses (Nakarro *et al.*, 1984). Thus, oncogenes may be implicated in the abnormal actin expression in neoplastic lymphoid cells. It has been suggested that mutated actins in transformed cells may lead the cells to express the transformed phenotype (Kakunaga *et al.*, 1984).

3. INTERMEDIATE-SIZE FILAMENTS IN NORMAL AND LEUKEMIC LYMPHOCYTES AND LYMPHOBLASTS

3.1. The Intermediate Filament System

Intermediate filaments (IF) constitute an elaborate network coursing throughout the cytoplasma of most but not all eukaryotic cells. These filaments are characterized biochemically by their insolubility in detergents and in low- and high-ionic-strength buffers. They are homo- or heteropolymers assembled from a large family of polypeptides that vary in composition according to the embryological origin of the cell (for a review, see Traub, 1985). Epithelial cells contain keratins (44 to 68 kDa), mesenchymal cells contain vimentin (55 kDa), muscular cells contain desmin, neural cells contain neurofilament proteins (68, 160, and 200 kDa) and glial cells contain glial acid fibrillary protein (51 kDa). IF typing has been widely used for the determination of cellular origin (Osborn and Weber, 1983). However, there are exceptions to the rule of a tissue-specific distribution of IF proteins, such as the simultaneous expression of two IF proteins in the same cell (Czernobilsky *et al.*, 1985) and the absence of detectable IF in certain developing (Jackson *et al.*, 1980) and adult (Dellagi *et al.*, 1983; Möller *et al.*, 1988) tissues.

Despite considerable divergence of amino acid sequence, the secondary structure of IF subunits is conserved (for a review, see Steinert *et al.*, 1985). IF appear to be associated with the other major components of the cytoskeleton. IF organization can be altered by various drugs, such as colcemid, which induces the collapse of vimentin IF into the nuclear region, suggesting a close association between IF and microtubules. The nuclear lamins are also members of the IF protein family (Aebi *et al.*, 1986).

3.2. Distribution of Vimentin in Lymphocytes in Normal Conditions

Immunofluorescence microscopy and gel electrophoresis studies have indicated that the IF of lymphoid cells are made up of vimentin (Franke *et al.*, 1979; Gabbiani *et al.*, 1981, Schmitt-Gräff *et al.*, 1987a). IF from normal human B and T cells isolated from the peripheral blood show a characteristic organization consisting of 10-nm fibrils running diffusely throughout the cytoplasm and extending to the cell periphery. In addition, some peripheral blood lymphocytes contain an aggregate of vimentin at one pole of the cell (Figure 4A, B). IF are prominent in the uropod of cells with a hand mirror configuration (Figure 4C). The formation of ligand–receptor complexes induces vimentin filaments to be reorganized as an aggregate beneath the cap of ligand–receptor complexes (Bourguignon and Bourguignon, 1981). In lymphocytes treated with colcemid, the site of IF collapse lies directly beneath uropods (Dellagi and Brouet, 1982). Loss of a fibrillar vimentin pattern and formation of a spontaneous IF cap take place in lymphocytes after incubation at 37°C on glass cover slips (Stark *et al.*, 1984).

In capping studies of the complement decay-accelerating factor (DAF) in T lymphocytes, it was observed that vimentin IF became localized beneath DAF molecules (Kammer *et al.*, 1988) in a manner identical to that reported for CD-capped cells (Kammer *et al.*, 1983). It is worth noting that cytochalasins B and D blocked migration of vimentin to the cell pole, suggesting an attachment of vimentin to the unpolarized globular actin (Kammer *et al.*, 1988).

Polar expression of IF seems to be a stable characteristic of particular lymphocyte subsets within murine lymphoid tissues (Lee and Repasky, 1987). Submembranous aggregates of vimentin are observed in close association with similarly polarized aggregates of spectrin in cells of B- and T-lymphocyte lineage *in situ*. The differences among lymphocytes regarding the distribution pattern of vimentin and spectrin in anatomically defined areas of lymphoid tissue may reflect the various functional states in which these cells exist (Lee and Repasky, 1987).

Moreover, the expression of vimentin is dependent on lymphocyte development and maturation. Vimentin filaments disappear from B lymphocytes when the cells are maturing toward plasma cells (Dellagi *et al.*, 1983). A complete loss of vimentin occurs in B lymphocytes while entering the follicular center of lymph nodes (Möller *et al.*, 1988). T cells, however, retain IF at all stages of maturation (Dellagi *et al.*, 1983).

3.3. Vimentin in Neoplastic Lymphocytes and Lymphoblasts

An aberrant pattern of IF organization has been observed in peripheral blood lymphocytes from patients with CLL (Liebes *et al.*, 1978; Stark *et al.*, 1984; Schmitt-Gräff *et al.*, 1987a; Zauli *et al.*, 1988). Compared with peripheral blood lymphocytes from healthy donors, CLL lymphocytes showed a significant decrease in IF content when vimentin was

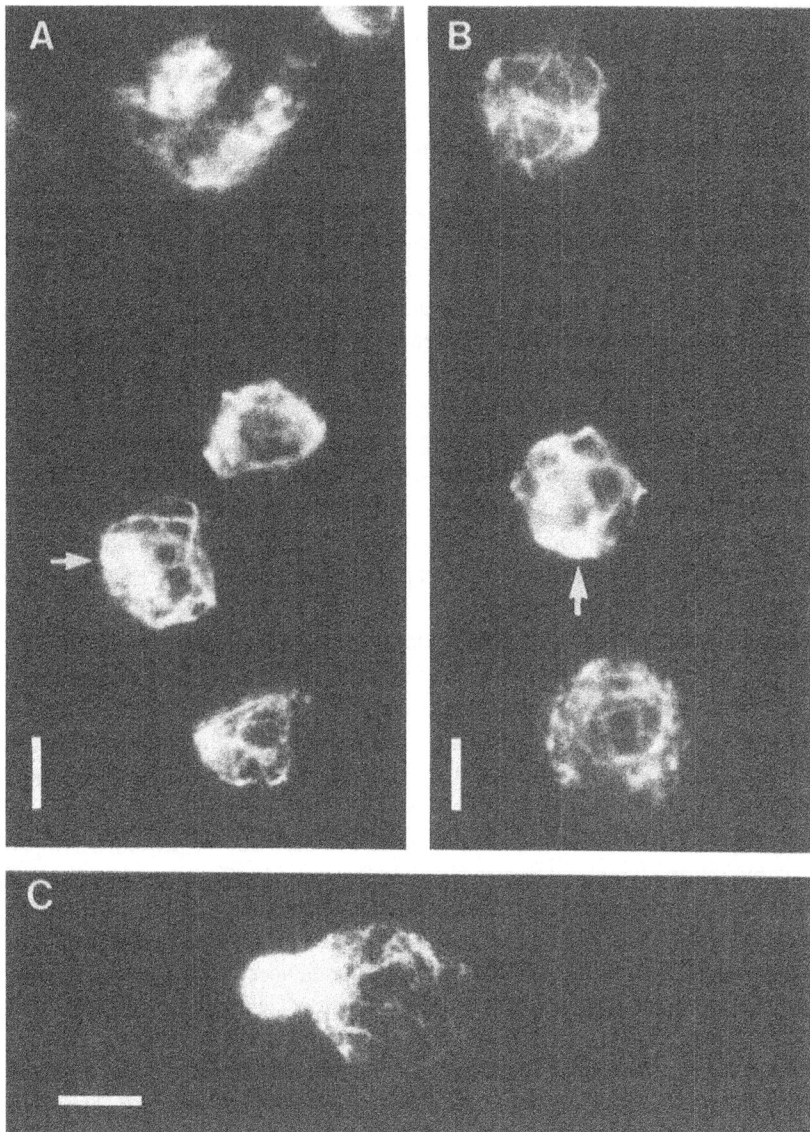

FIGURE 4. Immunofluorescent staining of normal peripheral blood lymphocytes decorated with antivimentin antibodies. (A and B) Intertwining vimentin IF running through lymphocytes with a round shape. Note scanty vimentin patches (arrows). (C) Vimentin filaments aggregated in the uropod of a lymphocyte showing a hand mirror configuration. Bars = 10 μm.

quantified as a percentage of total proteins by densitometric scanning of SDS-poly-acrylamide gels loaded with Triton X-100 cell lysates (Schmitt-Gräff et al., 1987a) (Figure 2). This result is in agreement with indirect immunofluorescence data showing a relatively weak staining of CLL lymphocytes with antivimentin antibodies (Stark et al., 1984; Zauli et al., 1988). According to our observations, lymphocytes from patients with B-CLL show a fine network of IF filaments less organized than that observed in healthy donors. The distribution of vimentin varies from a fine fibrillar network to an irregular cap at the cell periphery (Figure 5A). CLL lymphocytes also differ from normal lymphocytes by an anomalous spontaneous capping of vimentin (Stark et al., 1984) and a diminished capping by multivalent ligands (Liebes et al., 1978). These findings support the concept that the cytoskeletal organization may be impaired in CLL lymphocytes. On peripheral blood smears, CLL are easily traumatized and ruptured, show crushed nuclei, giving the appearance of smudge cells. This higher fragility may result from a defective arrangement of cytoskeletal elements.

Interestingly, some neoplastic lymphoid populations are devoid of IF. An absence of vimentin in Burkitt's lymphomas has been documented in several studies (Dellagi et al., 1984; Möller et al., 1988). It has also been shown that B lymphoblastic, immunoblastic, and centroblastic lymphomas may contain vimentin-negative cells (Ramaekers et al., 1985). The cessation of vimentin expression within a neoplastic population may indicate a follicular center origin or a differentiation stage corresponding to the intrafollicular B-cell transformation (Möller et al., 1988). In this context, it seems reasonable that lympho-blasts bearing the common ALL-associated antigen contain IF, since they appear to repre-sent a monoclonal expansion of early B-cell precursors at an extrafollicular stage of differentiation (Korsmeyer et al., 1983). In lymphoblasts from children with common ALL, we observed a well-developed network of intermediate filaments by immunofluor-escence staining. In ALL blasts with T markers, thick irregular vimentin filament bundles are present, including polar caps, whorls, and rings (Figure 5B,C). Electron micrographs of ALL blasts yield large bundles of IF in close association with the nucleus (Figure 6). It is conceivable that the strong expression of vimentin in common ALL and T-ALL blasts could in part be related to their proliferative activity. It has been described that an increased vimentin content is typical of cells having a high replicating rate, such as cultered mesothelial cells (Connell and Rheinwald, 1983) and vascular smooth muscle cells in vivo (Kocher et al., 1985) and in vitro (Skalli et al., 1986b). Thus, the level of IF expression in lymphoma and leukemia may provide some information on the proliferative activity as well as on the stage of maturation reached by the neoplastic cells.

3.4. Expression of Cytokeratin in Lymphoid Tissues

Examination of human lymph nodes, tonsils, and spleen by using antibodies to cytoskeletal proteins has detected extended arrays of extrafollicular reticulum cells form-ing IF containing cytokeratins 8 and 18 and/or desmin-containing IF, often also in com-bination with vimentin (Franke and Moll, 1987). Moreover, α-smooth muscle actin is synthesized by stroma cells of lymphoid organs (Toccanier-Pelte et al., 1987) and bone marrow (Schmitt-Gräff et al., 1989). In a recent phenotypic analysis of human T-cell lymphomas, it was noted that in contrast to normal lymphocytes, neoplastic lymphoid cells may react positively with cytokeratin antibodies (Feller, 1990). While IF of only the

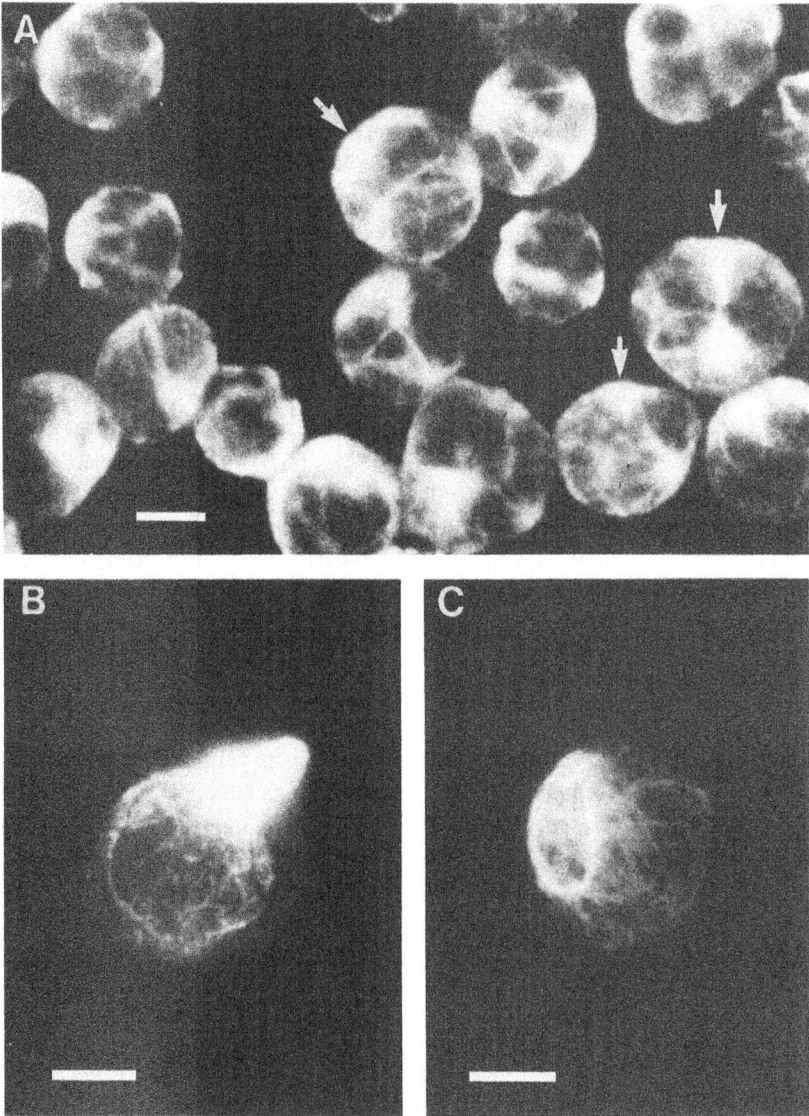

FIGURE 5. Immunofluorescent staining for vimentin of CLL lymphocytes (A) and T-ALL blasts (B and C). (A) Many CLL cells show a focal peripheral concentration of vimentin (arrows) in addition to a fine filamentous network. (B) Vimentin filaments are clustered in the uropod of a T lymphoblast with a spontaneously polarized configuration. (C) Large coils of vimentin filaments are present in the cytoplasm of ALL blasts with a T phenotype. Bars = 10 μm.

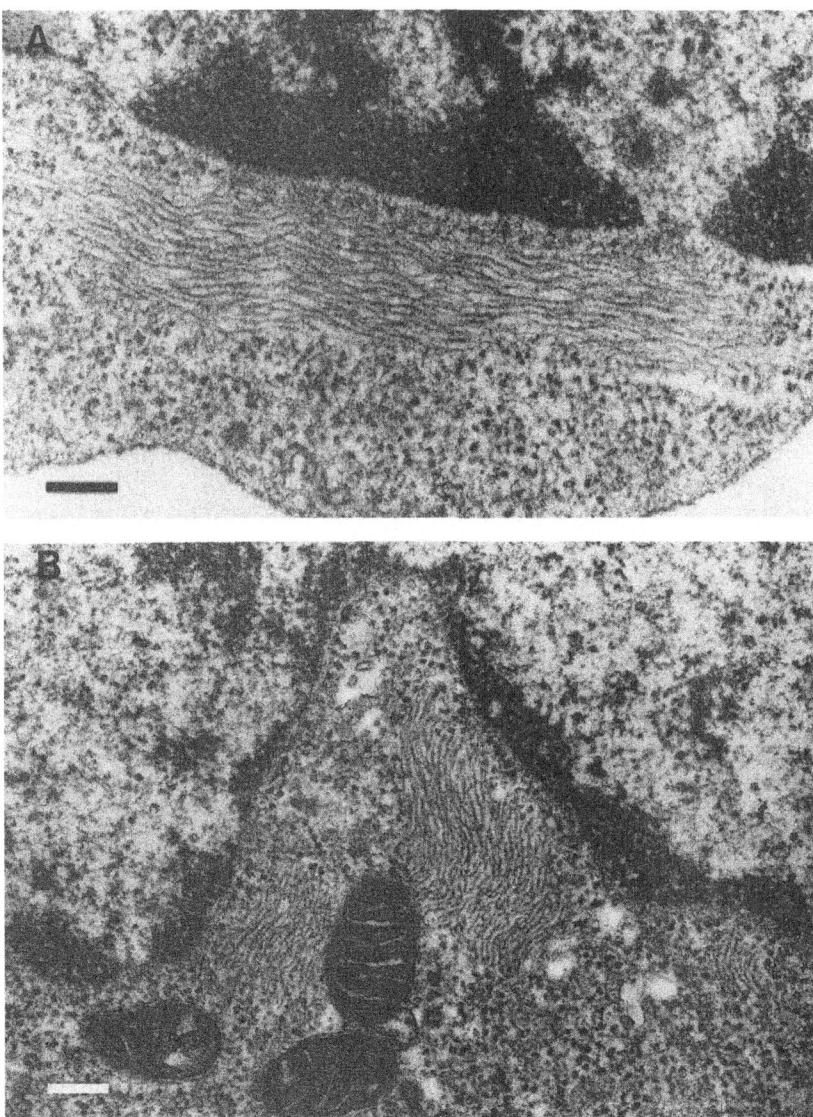

FIGURE 6. Electron micrograph of common ALL blasts showing two typical arrangements of IF bundles in close association with the nucleus and mitochondria. Bars = 0.2 μm.

vimentin type were present in most cases of lymphomas, a small number of large-cell anaplastic lymphomas were shown to express cytokeratins. Examples include lymphomas of high-grade malignancy positive for the K1 antigen (CD30), which may express cytokeratins 8 and 18. (A. C. Feller and O. O. Altmannsberger, personal communication Feller, 1990). Sewell *et al.* (1986) have reported that in a case diagnosed as IgD myeloma/ immunoblastic lymphoma, single tumor cells expressed leukocyte common antigen, cytokeratin, IgD, and lambda chains. The observation may have implications for the interpretation of IF typing in the diagnosis of metastatic carcinomas versus malignant lymphomas.

4. MICROTUBULES IN LYMPHOID CELLS

Microtubules form well-ordered arrays in the cytoplasm of eukaryotic cells. Moreover, they are the main component of specialized structures such as the mitotic spindle, cilia, flagella, and axonema (for a review, see Olmstedt *et al.*, 1984). Their basic subunits are dimers of α- and β-tubulin, which are encoded by a multigene family (for a review, see Raff, 1984). Differences in α- and β-tubulin expression in tissues and species may be related to specific cellular functions (Burland *et al.*, 1983). Tubulin dimers form protofilaments arranged around a central core. Assembly of tubulin molecules is accompanied by the hydrolysis of GTP. The distribution between subunits and polymers within the tubulin pool is regulated by complex mechanisms, including the interaction with associated proteins (MAPs). Since tubulin subunits have sites for the binding of antimitotic drugs such as colchicine, tubulin assembly can be experimentally modified by alkaloids. Centrosomes and kinetochores are cellular components defined as microtubule-organizing centers (Brinkley *et al.*, 1981).

There is evidence indicating that force production for microtubule-dependent movements such as the migration of chromosomes along the mitotic spindle is provided by a contractile system based on microtubule–dynein interactions (Johnson *et al.*, 1984).

Immunofluorescent staining of B and T lymphocytes has demonstrated a generally diffuse labeling present along the cell membrane and in the cytoplasm of cells with rounded configuration (Gabbiani *et al.*, 1977; Kammer *et al.*, 1988). Upon crosslinking of surface molecules after incubation at 37°C, fluorescence for tubulin was restricted to the region of the cap, where actin and vimentin were clustered. Co-capping of actin and tubulin with immunoglobulins in B lymphocytes indicates that not only microfilaments but also microtubules are involved in the regulation of movements of cell surface proteins (Gabbiani *et al.*, 1977). Similar patterns were seen beneath capped CD and DAF molecules in T lymphocytes. Colchicine augmented the capping rate of DAF (Kammer *et al.*, 1988). Relatively little is known about how the cell regulates the interactions of actin filaments and microtubules, but it has been shown that actin filaments and microtubules form three-dimensional networks, with the actin filaments crosslinked to the microtubules by MAPs (Pollard *et al.*, 1984). Studies on mechanisms of human cell-mediated cytotoxicity indicated that natural killing depends on microtubule and microfilament integrity (Katz *et al.*, 1982).

Recently, in a human lymphoma cell line, indirect immunofluorescent staining has shown that the intracellular precursor of interleukin-1β is associated with the microtubule network *in vivo* and *in vitro* (Baldari and Telford, 1989).

By electron microscopy, microtubules originating typically from the region of centrioles and radiating toward the cell periphery may be visualized in both normal and leukemic lymphoid cells (Figure 7).

Tubulin of normal and CLL lymphocytes was quantified by double labeling and isolation of tubulin-derived peptides (Atkins and Anderson, 1982). The tubulin content of

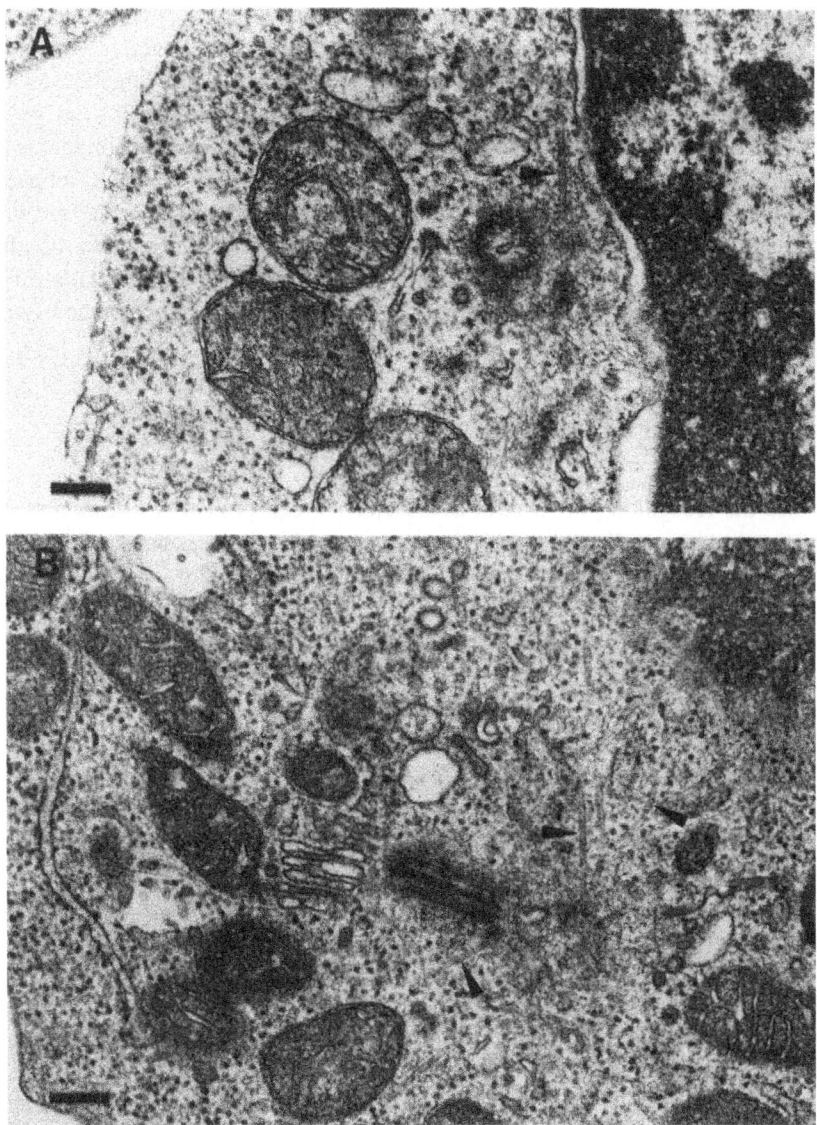

FIGURE 7. Ultrastructure of CLL lymphocytes. Microtubules (arrowheads) are clearly visible, originating from the region of centrioles irrespective of the plane of section (A: transversal; B: longitudinal) and often radiating to the cell periphery. Bars = 0.2 μm.

CLL cells was 4.4 ± 1.5% of total protein, which was significantly less than that of normal lymphocytes (6.1 ± 1.1%). By colchicine binding, no significant difference in the tubulin content of normal and CLL lymphocytes was reported (Liebes *et al.*, 1980).

Increased rates of tubulin synthesis have been reported in polyclonally activated normal human T and B lymphocytes as well as in leukemia cell lines and in Epstein-Barr virus genome-positive lines (Bachvaroff and Rapaport, 1980). Moreover, α_1-, α_2-, and β-tubulins appear on the cell membrane surface of leukemia cells and lymphocytes transformed by mitogens and Epstein-Barr virus (Bachvaroff *et al.*, 1980).

Lymphoblasts from a human leukemia cell line possess surface tubulin that is structurally similar (but with a slightly different isoelectric point) to soluble intracellular tubulin (Quillan *et al.*, 1985). Probably, these leukemia cells possess surface tubulin because they produce a tubulin isomer that has a specific membrane receptor (Quillan *et al.*, 1985). The function of surface or membrane tubulin on lymphoid cells is not yet clear. One could imagine that membrane tubulin interacts with other membrane-associated components.

5. ORGANIZATION OF SPECTRIN IN LYMPHOCYTES

Since the recognition of spectrin as a major red cell protein, 20 years ago, much information has been collected on the membrane-associated cytoskeleton of the erythrocyte (for a review, see Cohen, 1983). Erythrocyte spectrin forms a two-dimensional submembranous framework linking short actin oligomers to the membrane via ankyrin (Branton *et al.*, 1981). In red blood cells, spectrin is a heterodimer made up of an alpha chain of 240 kDa and a beta chain of 220 kDa (for a review, see Goodman and Shiffer, 1983). The spectrin network seems to be important in determining the shape and the mechanical properties of the erythrocyte (Stokke *et al.*, 1986a,b) and in maintaining the membrane phospholipid asymmetry (Williamson *et al.*, 1987).

Several lines of evidence indicate that proteins similar to erythrocyte spectrin and ankyrin are present in a variety of nonerythroid cells, including lymphocytes (Repasky *et al.*, 1982; for a review, see Bennett, 1985). Thus, a nonerythroid type of spectrin referred to as fodrin is present in neural cells and in lymphocytes (Glenney and Glenney, 1983). This form of spectrin is composed of a gamma chain of 235 kDa and an alpha chain of 240 kDa possessing a calmodulin-binding site similar to avian but not to mammalian erythrocyte spectrin (Glenney *et al.*, 1982).

In T and B lymphocytes, crosslinking of cell surface receptors by appropriate ligands induces the coincident aggregation of spectrin (Nelson and Lazarides, 1983). This redistribution of spectrin during the capping of cell surface molecules has led to the hypothesis that lymphocyte spectrin, like erythrocyte spectrin, may be associated with cell surface plasma membrane proteins and restrict their lateral mobility.

Immunofluorescence analysis of mammalian lymphocytes in various lymphoid organs has revealed two different patterns of lymphocyte spectrin organization: spectrin was found (1) symmetrically distributed under the plasma membrane or (2) concentrated in an unipolar submembranous aggregate (Repasky *et al.*, 1984).

Lymphocytes with a naturally capped configuration of spectrin are observed in characteristic locations such as the paracortical and cortical regions of lymph nodes, the periarterial lymphoid sheet of the white pulp in the spleen, and the medulla of the thymus. Cortical thymocytes are mostly noncapped (Repasky *et al.*, 1984). A polar distribution of spectrin associated with aggregates of vimentin has been documented in particular subsets of T and B lymphocytes of murine lymphoid tissues (Lee and Repasky *et al.*, 1987). Moreover, the pattern of spectrin distribution, either aggregated or evenly dispersed, is a feature of various lymphocyte cell lines probably related to the stage of cell maturation (Pauly *et al.*, 1986).

We observed punctate patterns of spectrin staining in cells from CLL and Burkitt's lymphoma, whereas lymphocyte controls showed uniformly bright staining, circumferentially in the region of the plasma membrane (Figure 8).

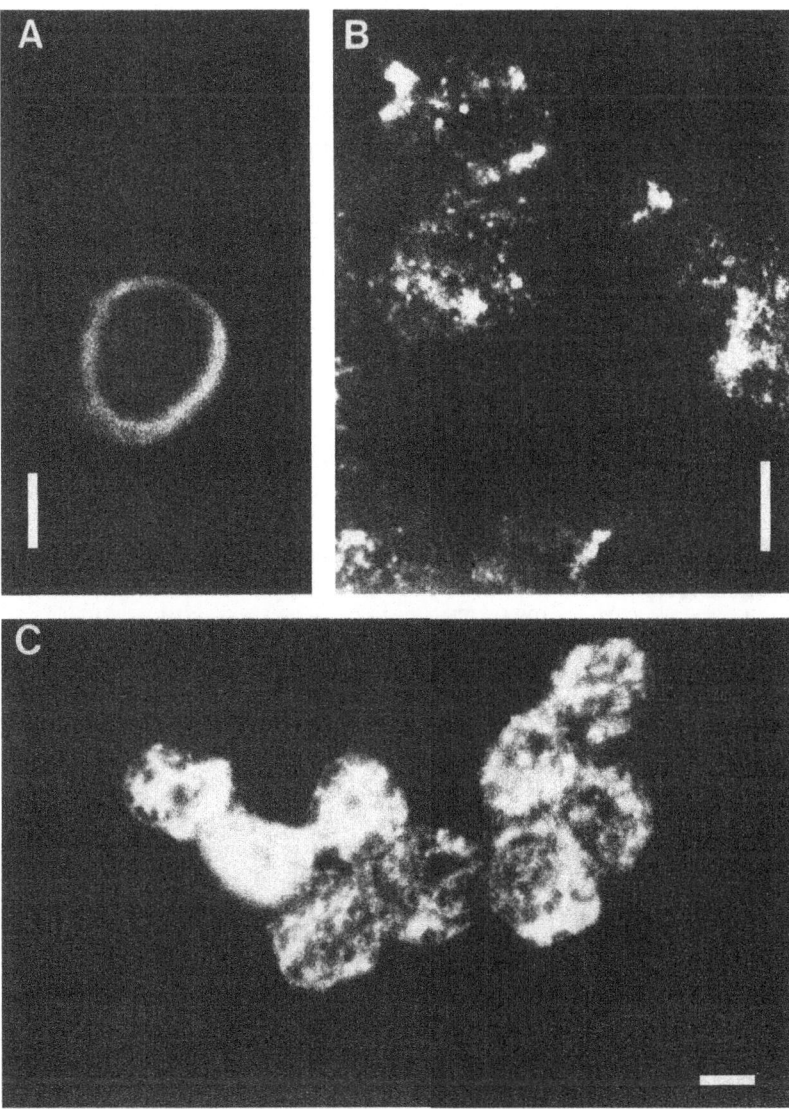

FIGURE 8. Immunofluorescence micrograph of lymphoid cells incubated with anti-chicken spectrin developed in rabbits. (A) A peripheral blood lymphocyte from a control person shows spectrin staining circumferentially in the region of the plasma membrane. (B) In CLL lymphocytes, a punctate pattern of spectrin staining is present. (C) Lymphoblasts from Burkitt's lymphoma contain multiple aggregates of spectrin. Bars = 10 μm.

There is substantial evidence indicating a correlation between the organizational state of spectrin and membrane lipids in lymphocytes (Del Buono *et al.*, 1988). As shown by staining with the fluorescent lipophilic probe meracyanine 540, tightly packed lipids are characteristic of mature lymphocytes with spectrin caps, whereas organized lipids were found in less mature cells with uniformly distributed spectrin (Del Buono *et al.*, 1988). Apparently, modifications of the particular patterns of spectrin distribution in lymphocyte subsets are effected by various signals. Thus, when cells of the functional T-cell hybridoma DO-11.10, showing a polar aggregate of spectrin, are stimulated by mitogen (ConA), the spectrin aggregate redistributes uniformly under the plasma membrane (Gregorio *et al.*, 1989). Black *et al.* (1988) have shown that lymphocyte spectrin is frequently aggregated as a discrete filamentous structure in the Golgi region. T-cell activation results in a rapid fragmentation of the aggregate of spectrin and a movement from the cytoplasmic location to the plasma membrane (Lee *et al.*, 1988). Thus, the association of spectrin with the plasma membrane is dynamic and seems to depend on the activation of lymphocytes. Further studies are needed to assess the role of spectrin in cell-specific functions, including the immune response.

6. CONCLUSIONS

There is no doubt that the cytoskeleton determines the shape and motility of lymphoid cells and is involved in characteristic cellular activities such as recycling (Mely-Goubert and Bellagrau, 1981), binding to target cells (Ryser *et al.*, 1982), and redistribution of surface receptors (Gabbiani *et al.*, 1977).

Reorganization of actin is documented in T and B lymphocytes following crosslinking of surface receptors (Kammer *et al.*, 1988), activation (Phatak *et al.*, 1988), or treatment with mitogens (Rao, 1984). These observations suggest that actin may play an important role in several physiological and pathological processes in lymphocytes, such as signal transduction, activation, and proliferation.

Co-capping of actin, vimentin, tubulin (Kammer *et al.*, 1988), and spectrin (Nelson and Lazarides, 1983) with surface molecules indicates that the control of receptor functions may be achieved by close interactions of different cytoskeletal elements and associated proteins.

An aberrant actin organization has been noted in neoplastic lymphocytes and lymphoblasts (Calligaris-Cappio *et al.*, 1986; Schmitt-Gräff *et al.*, 1987a, b). Changes in actin organization and content may reflect different degrees of differentiation and may be correlated with impaired functions of neoplastic cells.

Whereas microfilaments and microtubules are present in all lymphoid populations studies, vimentin is lost in different subsets, e.g., follicular center cells (Möller *et al.*, 1988). Since the role of intermediate filaments has not been determined, the significance of the repression of vimentin expression is unclear.

Characteristic distribution patterns of vimentin and spectrin have been noted in lymphocyte subsets (Pauly *et al.*, 1986; Lee and Repasky, 1987), which may be related to the stage of maturation or may reflect various functional states of lymphocytes.

Further studies on the distribution and modulation of cytoskeletal components in lymphoid cells may be of use in the elucidation of motile events and membrane protein redistribution. Such studies will contribute to our understanding of pathological phenomena such as defective immune responses and leukemogenesis.

7. REFERENCES

Aebi, U., Cohn, J., Buhle, L., and Gerace, L., 1986, The nuclear lamina is a meshwork of intermediate-type filaments, *Nature* (London) **323**:560–564

Alberts, B., Bray, D., Lewis, J., Raff, M., Roberts, K., and Watson, J. D., 1983, The cytoskeleton, in *Molecular Biology of the Cell*, pp. 549–609, Garland Publishing Inc., New York, New York

Albrecht, D. L., and Noelle, R. J., 1988, Membrane Ig-cytoskeletal interactions. I. Flow cytofluorometric and biochemical analysis of membrane IgM-cytoskeletal interactions, *J. Immunol.* **141**:3915–3922

Anderson, K. C., Bates, M. P., Slaughenhoupt, B. L., Pinkus, G. S., Schlossman, S. F., and Nadler, L. M. 1984, Expression of human B cell-associated antigens on leukemias and lymphomas: A model of human B cell differentiation, *Blood* **63**:1424–1433

Atkins, H., and Anderson, P. J., 1982, Actin and tubulin of normal and leukaemic lymphocytes, *Biochem. J.* **207**:535–539

Azzarelli, B., Easterling, K., and Norton, J. A., 1989, Leukemic cell-endothelial cell interactions in leukemic cell dissemination, *Lab. Invest.* **60**:45–64

Bachvaroff, R. J., and Rapaport, F. T., 1980, Active secretion of cytoskeletal and mechanicochemical proteins in EBV-genome positive human lymphocytes, *Transplant. Proc.* **12**:205–208

Bachvaroff, R. J., Miller, F., and Rapaport, F. T., 1980, Appearance of cytoskeletal components on the surface of leukemia cells and lymphocytes transformed by mitogens and Epstein-Barr virus, *Proc. Natl. Acad. Sci. USA* **77**:4479–4983

Baldari, C. T., and Telford, J. L., 1989, The intracellular precursor of Il-1β is associated with microtubules in activated U 937 cells, *J. Immunol.* **142**:785–791

Barber, B. H., and Delovitch, T. L., 1979, The identification of actin as a major lymphocyte component, *J. Immunol.* **122**:320–325

Bennett, V., 1985, The membrane skeleton of human erythrocytes and its implication for more complex cells, *Annu. Rev. Biochim.* **54**:273–304

Black, J. D., Koury, S. K., Bankert, R. B., and Repasky, E. A., 1988, Heterogeneity in lymphocyte spectrin distribution: Ultrastructural identification of a new spectrin-rich cytoplasmic structure, *J. Cell Biol.* **106**:97–109

Bourguignon, L. Y. W., and Bourguignon, G., 1981, Immunocytochemical localization of intermediate filament proteins during lymphocyte capping, *Cell. Biol. Int. Rep.* **5**:783–789

Bourguignon, L. W., and Bourguignon, G. I., 1984, Capping and the cytoskeleton, *Int. Rev. Cytol.* **87**:195–224

Bourguignon, L. W., Suchard, S. J., Nagpal, M. L., and Glenney, J. R. Jr., 1985, A T-lymphoma glycoprotein (gp 180) is linked to the cytoskeletal protein, fodrin, *J. Cell Biol.* **101**:477–487

Branton, D., Cohen, C. M., and Tylor, J., 1981, Interaction of cytoskeletal proteins of the human erythrocyte membrane, *Cell* **24**:24–32

Braun, J. K., Fujiwara, K., Pollard, T. D., and Unanue, E. R., 1978, Two distinct mechanisms for redistribution of lymphocyte surface marcomolecules. I. Relationship of cytoplasmic myosin, *J. Cell Biol.* **79**:409–418

Braun, J. P., Hochman, P. S., and Unanue, E. R., 1982, Ligand-induced association of surface immunoglobulin with the detergent insoluble cytoskeletal matrix of the B-lymphocyte, *J. Immunol.* **128**:1198–1204

Bretscher, A., 1983, Microfilaments organization in the cytoskeleton of the intestinal brush border, *Cell Muscle Motil.* **4**:239–268

Bretscher, M. S., 1984, Endocytosis: Relation to capping and locomotion, *Science* **244**:681–686

Brinkley, B. R., Cox, S. M., Pepper, D. A., Wible, L., Brenner, S., and Pardue, R. L., 1981, Tubulin assembly sites and the organization of cytoplasmic microtubules in cultured mammalian cells, *J. Cell Biol.* **90**:554–562

Bulinski, J. C., Kumar, S., Titani, K., and Hauschka, S. D., 1983, Peptide antibody specific for the amino terminus of skeletal muscle α-actin, *Proc. Natl. Acad. Sci. USA* **80**:1506–1510

Burland, T., Gull, K., Schedl, R., Boston, R., and Dove, W., 1983, Cell type-dependent expression of tubulins in Physarum, *J. Cell Biol.* **97**:1852–1859

Caligaris-Cappio, F., Bergui, L., Tesio, L., Corbascio, G., Tousco, F., and Marchisio, P. M., 1986, Cytoskeleton organization is aberrantly rearranged in the cells of B chronic lymphocytic leukemia and hairy cell leukemia, *Blood* **67**:233–239

Campbell, F. R., 1983, Intercellular contacts of lymphocytes during migration across high-endothelial venules of lymph nodes: An electron microscopic study, *Anat. Rec.* **207**:643–652

Chang, T. W., Celis, E. C., Eisen, H. N., and Solomon, F., 1979, Crawling movements of lymphocytes on and beneath fibroblasts in culture, *Proc. Natl. Acad. Sci. USA* **76**:2917–2921

Cohen, C. M., 1983, The molecular organization of the red cell membrane cytoskeleton, *Semin. Hematol.* **20**:141–158

Cohen, H. J., and Gilbertsen, B. B., 1975, Human lymphocyte surface immunoglobulin capping: Normal characteristics and anomalous behaviour of chronic lymphocytic leukemic lymphocytes, *J. Clin. Invest.* **55**:84–93

Connell, N. D., and Rheinwald, J. G., 1983, Regulation of the cytoskeleton in mesothelial cells: Reversible loss of keratin and increase in vimentin during rapid growth in culture, *Cell* **34**:245–253

Czernobilsky, B., Moll, R., Levy, R., and Franke, W. W., 1985, Co-expression of cytokeratin and vimentin filaments in mesothelial, granuloas and rete ovarii cells of human ovary, *Eur. J. Cell. Biol.* **37**:175–190

Del Buono, B. J., Williamson, P. L., and Schlegel, R. A., 1988, Relation between the organization of spectrin and of membrane lipids in lymphocytes, *J. Cell Biol.* **106**:697–703

Dellagi, K., and Brouet, J.-C., 1982, Redistribution of intermediate filaments during capping of lymphocyte surface molecules, *Nature* (London) **298**:284–286

Dellagi, K., Vainchenker, W., Vinci, G., Paulin, D., and Brouet, J. C., 1983, Alteration of vimentin intermediate filament expression during differentiation of human hemopoietic cells, *EMBO J.* **2**:1509–1514

Dellagi, K., Brouet, J. C., Portier, M. M., and Lenoir, G. M., 1984, Abnormal expression of vimentin intermediate filaments in human lymphoid cell lines whith deletion or translocation of the distal end of chromosome 8, *J. Natl. Cancer Inst.* **73**:95–99

de Petris, S., 1974, Inhibition and reversal of capping by cytochalasins B, vinblastine and colchicine, *Nature* (London) **250**:54–55

Fagraeus, A., Biberfeld, G., and Norberg, R., 1980, Reaction of anti-actin antibodies with lymphoid cells, *Cell. Mol. Biol.* **26**:129–134

Fechheimer, M., and Cebra, J. J., 1979, Isolation and characterization of actin and myosin from B-lymphocytic guinea pig leukemia cells, *J. Immunol.* **122**:2590–2597

Fechheimer, M., and Zigmond, S. H., 1983, Changes in cytoskeletal proteins of polymorphonuclear leukocytes induced by chemotactic peptides, *Cell. Motil.* **3**:349–361

Feller, A. C., 1990, Phenotypic analysis of human T cell lymphomas, *Progr. Pathol.* **131**: in press

Franke, W. W., and Moll, R., 1987, Cytoskeletal components of lymphoid organs. I. Synthesis of cytokeratins 8 and 18 and desmin in subpopulations of extrafollicular reticulum cells of human lymph nodes, tonsils, and spleen, *Differentiation* **36**:145–163

Franke, W. W., Schmid, E., Winter, S., Osborn, M., and Weber, K., 1979, Widespread occurrence of inter-mediate-sized filaments of the vimentin-type in cultered cells from diverse vertebrates, *Exp. Cell. Res.* **123**:25–46

Gabbiani, G., Ryan, G. B., and Majno, G., 1971, Presence of modified fibroblasts in granulation tissue and their possible role in wound contraction, *Experientia* **27**:549–550

Gabbiani, G., Chaponnier, C., Zumbe, A., and Varsalli, P., 1977, Actin and tubulin co-cap with surface immunoglobulin in mouse B-lymphocytes, *Nature* (London) **269**:697–698

Gabbiani, G., Kapanci, Y., Barazzone, P., and Franke, W. W., 1981, Immunochemical identification of intermediate-sized filaments in human neoplastic cells: A diagnostic aid for surgical pathologists, *Am. J. Pathol.* **104**:206–216

Gabbiani, G., Gabbiani, F., Lombardi, D., and Schwartz, S. M., 1983, Organization of actin cytoskeleton in normal and regenerating arterial endothelial cells, *Proc. Natl. Acad. Sci. USA* **80**:2361–2364

Garrels, J. I., and Gibson, W., 1976, Identification and characterization of multiple forms of actin, *Cell* **9**:793–805

Glenney, J. R., and Glenney, P., 1983, Fodrin is the general spectrin-like protein found in most cells whereas spectrin and the TW protein have a restricted distribution, *Cell* **34**:503–512

Glenney, J. R., Glenney, P., and Weber, K., 1982, Erythroid spectrin, brain fodrin and intestinal brush border proteins (TW-260/240) are related molecules containing a common calmodulin-binding subunit bound to a variant cell type-specific subunit, *Proc. Natl. Acad. Sci. USA* **79**:4002–4005

Goodman, S. R., and Shiffer, K., 1983, The spectrin membrane skeleton of normal and abnormal human erythrocytes: A review, *Am. J. Physiol.* **244:**C121–C141

Goroff, D. K., Stall, A., Mond, J. J., and Finkelman, F. D., 1986, In vitro and in vivo B lymphocyte-activation properties of monoclonal anti-delta antibodies. I. Determinants of B lymphocyte-activating properties, *J. Immunol.* **136:**2382–2392

Gregorio, C. C., Black, J. D., Lee, J. K., and Repasky, E. A., 1989, Organization of spectrin and ankyrin in lymphocytes, *J. Cell Biol.* **107:**26a

Gupta, S. K., and Woda, B. A., 1988, Ligand-induced association of surface immunoglobulin with the detergent insoluble cytoskeleton may involve alpha-actinin, *J. Cell. Immunol.* **140:**176–182

Hamada, H., Leavitt, J., and Kakanuga, T., 1981, Mutated β-actin gene: Coexpression with an unmutated allele in a chemically transformed human fibroblast cell line, *Proc. Natl. Acad. Sci. USA* **78:**3634–3637

Hartwig, J. H., Niederman, R., and Lind, S. E., 1985, Cortical actin structures and their relationship to mammalian cell movements, *Subcell. Biochem.* **11:**1–49

Hatano, S., and Oosawa, F., 1966, Isolation and characterization of plasmodium actin, *Biochim. Biophys. Acta* **127:**488–498

Hoessli, D., Rungger-Brandle, E., Jockusch, B., and Gabbiani, G., 1980, Lymphocyte alpha-actinin: Relationship to cell membrane and co-capping with surface receptors, *J. Cell Biol.* **84:**305–313

Huxley, H. E., 1963, Electron microscope studies on the structure of natural and synthetic protein filaments from striated muscle, *J. Mol. Biol.* **7:**281–308

Jackson, B. W., Grund, C., Schmid, E., Burki, K., Franke, W. W., and Illmensee, K., 1980, Formation of cytoskeletal elements during mouse embryogenesis: Intermediate filaments of the cytokeratin type and desmosomes in preimplantation embryos, *Differentiation* **20:**203–216

Jalkanen, S., Jalkanen, M., Bargatze, R., Tammi, M., and Butcher, E. C., 1988, Biochemical properties of glycoproteins involved in lymphocyte recognition of high endothelial venules in man, *J. Immunol.* **141:**1615–1623

Johnson, K. A., Porter, M. E., and Shimizu, T., 1984, Mechanism of force production for microtubule-dependent movements, *J. Cell Biol.* **99:**132S–136S

Kakunaga, T., Leavitt, J., and Hamada, H., 1984, A mutation in actin associated with neoplastic transformation, *Fed. Proc.* **43:**2274–2275

Kammer, G. M., Smith, J. A., and Mitchel, R., 1983, Capping of human T cell specific determinants: Kinetics of capping and receptor re-expression and regulation by the cytoskeleton, *J. Immunol.* **130:**38–44

Kammer, G. M., Walter, E. J., and Medof, M. E., 1988, Association of cytoskeletal re-organization with capping of the complement decay accelerating factor on T lymphocytes, *J. Immunol.* **141:**2924–2928

Katz, P., Zaytoun, A. M., and Lee, J. H., 1982, Mechanisms of human cell-mediated cytotoxicity. III. Dependence of natural killing on microtubule and microfilament and microfilament integrity, *J. Immunol.* **129:**2816–2825

Kerrick, W. G. L., and Bourguignon, L. Y. W., 1984, Capping of muse T-lymphoma cells is regulated by a calcium activated myosin light chain kinase, *Proc. Natl. Acad. Sci. USA* **81:**165–169

Kocher, O., Skalli, O., Cerruti, D., Gabbiani, F., and Gabbiani, G., 1985, Cytoskeletal features of rat aortic cells during development: An electron microscopic, immunohistochemical and biochemical study, *Circ. Res.* **56:**829–838

Korn, E. D., 1982, Actin polymerization and its regulation by proteins from nonmuscle cells, *Physiol. Rev.* **62:**672–737

Korsmeyer, S. J., Arnold, A., Bakhshi, A., Ravetch, J. V., Siebenlist, U., Hieter, P. A., Sharrow, S. D., Lebien, T. W., Kersey, J. H., Poplack, D. G., Leder, P., and Waldmann, T. A., 1983, Immunoglobulin gene re-arrangement and cell surface antigen expression in acute lymphocytic leukemias of T cell und B cell precursor origins, *J. Clin. Invest.* **71:**301–313

Laub, F., Kaplan, M., and Gitler, C., 1981, Actin polymerization accompanies Thy-1 capping on mouse thymocytes, *FEBS Lett.* **124:**35–38

Leavitt, J., Leavitt, A., and Attallah, A. M., 1980, Dissimilar modes of expression of β- and γ-actin in normal and leukemic human T lymphocytes, *J. Biol. Chem.* **255:**4984–4987

Lee, J. K., and Repasky, E. A., 1987, Cytoskeletal polarity in mammalian lymphocytes in situ, *Cell. Tissue Res.* **247:**195–202

Lee, J. K., Black, J. D., Repasky, E. A., Kubo, R. T., and Bankert, R. B., 1988, Activation induces a rapid reorganization of spectrin in lymphocytes, *Cell* **55:**807–816

Liebes, L. F., Quagliata, F., and Silber, R., 1978, The anomalous capping behaviour of chronic lymphocytic

leukemia lymphocytes: Studied with an anti-lymphocyte antiserum, *Clin. Immunol. Immunopathol.*
10:222–232

Liebes, L. F., Fleit, H., Zucker-Franklin, D., and Silber, R., 1980, Human lymphocyte tubulin: Purification and
characterization in normal and leukemic cells, *Biochim. Biophys. Acta* **633:**245–257

Mely-Goubert, B., and Bellgrau, D., 1981, Actin content in lymphocytes: A proposed correlation with their
recirculating properties, *J. Immunol.* **127:**399–401

Mescher, M. F., Jose, M. J. L., and Balk, S. P., 1981, Actin-containing matrix associated with the plasma
membrane of murine tumour and lymphoid cells, *Nature* (London) **289:**139–144

Meyer, W. H., and Howard, T. H., 1983, Changes in actin content during induced myeloid maturation of human
promyelocytes, *Blood* **62:**308–314

Miyasaka, M., Mely-Goubert, W. M., Dudler, L., and Trnka, Z., 1984, Actin activity is high in the immunocompetent fraction of thymocytes, *Thymus* **6:**57–65

Möller, P., Momburg, F., Hofmann, W. J., and Matthaei-Maurer, D. U., 1988, Lack of vimentin occurring
during the intrafollicular stages of B cell development characterizes follicular center cell lymphoma, *Blood*
71:1033–1038

Monroe, J. G., and Cambier, J. C., 1983, B cell activation. II. Receptor cross-linking by thymus-independent
and thymus-dependent antigens induces a rapid decrease in the plasma membrane potential of antigen-
binding B lymphocytes, *J. Immunol.* **131:**2641–2644

Nagata, K., Sagara, J., and Ichikawa, Y., 1980, Changes in contractile proteins during differentiation of myeloid
leukemia cells. I. Polymerization of actin, *J. Cell Biol.* **85:**273–282

Nakarro, G., Robbins, K. C., and Reddy, E. P., 1984, Gene product of v-fgr onc: Hybrid protein containing a
portion of actin and a tyrosine specific protein kinase, *Science* **223:**63–66

Nelson, W. J., and Lazarides, E., 1983, Expression of the beta-subunit of spectrin in nonerythroid cells, *Proc.
Natl. Acad. Sci. USA* **80:**363–367

Neuhaus, J. M., Wanger, M., Keisler, T., and Wegner, A., 1983, Treadmilling of actin, *J. Muscle Res. Cell.
Motil.* **4:**507–527

Olmsted, J. B., Cox, J. V., Asness, C. F., Parysek, L. M., and Lyon, H. D., 1984, Cellular regulation of
microtubule organization, *J. Cell Biol.* **99:**28S–32S

Owen, M. J., Auger, J., Barber, B. H., Edwards, A. J., Walsh, F. S., and Crumpton, M. J., 1978, Actin may be
present on the lymphocyte surface, *Proc. Natl. Acad. Sci. USA* **75:**4484–4488

Osborn, M., and Weber, K., 1983, Tumor diagnosis by intermediate filament typing: A novel tool for surgical
pathology, *Lab. Invest.* **48:**372–394

Pauly, J. L., Bankert, R. B., and Repasky, E. A., 1986, Immunofluorescent patterns of spectrin in lymphocyte
cell lines, *J. Immunol.* **136:**246–253

Petrini, M., Emerson, D. L., and Galbraith, R. M., 1983, Linkage between surface immunoglobulin and
cytoskeleton of B lymphocytes may involve Gc protein, *Nature* (London) **306:**73–74

Phatak, P. D., Packman, C. H., and Lichtman, M. A., 1988, Protein kinase C modulates actin conformation in
human T lymphocytes, *J. Immunol.* **141:**2929–2934

Pollard, T. D., Selden, S. C., and Maupin, P., 1984, Interaction of actin filaments whith microtubules, *J. Cell
Biol.* **99:**33S–37S

Quillan, M., Castello, C., Krishan, A., and Rubin, R., 1985, Cell surface tubulin in leukemic cells: Molecular
structure, surface binding, turnover, cell cycle expression and origin, *J. Cell Biol.* **101:**2345–2354

Raff, E. C., 1984, Genetics of microtubule systems, *J. Cell Biol.* **99:**1–10

Ramaekers, F. C. S., Osborn, M., Schmid, E., Weber, K., Bloemendal, H., and Franke, W. W., 1980,
Identification of the cytoskeletal proteins in lens-forming cells, a special epitheloid cell type, *Exp. Cell Res.*
127:309–327

Ramaekers, F. C. S., Vroom, T. M., Moesker, O., Kant, A., Scholte, G., and Vooijs, G. P., 1985, The use of
antibodies to intermediate filament proteins in the different diagnosis of lymphoma versus metastatic
carcinoma, *Histochem. J.* **17:**57–79

Rao, K. M. K., 1984, Lectin-induced actin polymerization in human lymphocytes: A possible signal for
mitogenesis, *Cell. Immunol.* **83:**181–188

Rao, K. M., and Varani, J., 1982, Actin polymerization induced by chemotactic peptide and concanavalin A in
rat neutrophils, *J. Immunol.* **129:**1605–1607

Repasky, E. A., Granger, B. L., and Lazarides, E., 1982, Widespread occurrence of avian spectrin in nonerythroid cells, *Cell* **29:**821–833

Repasky, E. A., Symer, D. E., and Bankert, R. B., 1984, Spectrin immunofluorescence distinguishes a population of naturally capped lymphocytes in situ, *J. Cell Biol.* **9**:350–355

Rothstein, T. L., 1986, Stimulation of B cells by sequential addition of antiimmunoglobulin antibody and cytochalasin, *J. Immunol.* **136**:813–816

Rungger-Brändle, E., and Gabbiani, G., 1983, The role of cytoskeletal and cytocontractile elements in pathologic processes, *Am. J. Pathol.* **110**:361–392

Ryser, J. E., Rungger-Brändle, E., Chaponnier, C., Gabbiani, G., and Vassalli, P., 1982, The area of attachment of cytotoxic T lymphocytes to their target cells shows high motility and polarization of actin, but not myosin, *J. Immunol.* **128**:1159–1162

Sanders, S., and Craig, S. W., 1983, A lymphocyte cell surface molecule that is antigenically related to actin, *J. Immunol.* **131**:370–377

Schliwa, M., 1982, Action of cytochalasin D on cytoskeletal networks, *J. Cell Biol.* **92**:79–91

Schmitt-Gräff, A., Chaponnier, C., and Gabbiani, G., 1987a, Cytoskeletal organization of peripheral blood normal and leukemic lymphocytes and lymphoblasts, *J. Submicroscop. Cytol.* **19**:329–335

Schmitt-Gräff, A., Scheulen, M. E., and Gabbiani, G., 1987b, Cytoskeletal organization in acute leukemias, *Haematol. Blood Transfus.* **30**:302–307

Schmitt-Gräff, A., Skalli, O., and Gabbiani, G., 1989, Alpha-smooth muscle actin is expressed in a subset of bone marrow stromal cells in normal and pathological conditions, *Virchows Arch. B* **57**:291–302

Schreiner, G. F., Braun, J., and Unanue, E. R., 1976, Spontaneous redistribution of surface immuno-globulin in the motile B lymphocyte, *J. Exp. Med.* **144**:1683–1688

Schreiner, G. F., Fujiwara, K., Pollard, T., and Unanue, E. R., 1977, Redistribution of myosin accompanying capping of surface Ig, *J. Exp. Med.* **145**:1393–1398

Sewell, H. F., Thompson, W. D., and King, D. J., 1986, IgD myeloma/immunoblastic lymphoma cells expressing cytokeratin, *Br. J. Cancer* **53**:695–696

Skalli, O., Ropraz, P., Trzeciak, A., Benzonana, G., Gillessen, D., and Gabbiani, G., 1986a, A monoclonal antibody against α-smooth muscle actin: A new probe for smooth muscle differentiation, *J. Cell Biol.* **103**:2787–2796

Skalli, O., Bloom, W. S., Ropraz, P., Azzarone, B., and Gabbiani, G., 1986b, Cytoskeletal remodeling of rat aortic smooth muscle cells in vitro: Relationship to culture conditions and analogies to in vivo situations, *J. Submicroscop. Cytol.* **18**:481–493

Skalli, O., Gabbiani, G., Babai, F., Seemayer, T. A., Pizzolato, G., and Schürch, W., 1988, Intermediate filament proteins and actin isoforms as markers for soft tissue tumor differentiation and origin. II. Rhabdomyosarcomas, *Am. J. Pathol.* **130**:515–531

Skalli, O., Schürch, W., Seemayer, T., Lagace, R., Montandon, D., Pittet, B., and Gabbiani, G., 1989, Myofibroblasts from diverse pathologic settings are heterogeneous in their content of actin isoforms and intermediate filament proteins, *Lab. Invest.* **60**:275–285

Southwick, F. S., and Stossel, T. P., 1983, Contractile proteins in leukocyte function, *Semin. Hematol.* **20**:305–321

Spangrude, G. J., Braaten, B. A., and Daynes, R. A., 1984, Molecular mechanisms of lymphocyte extravasation. I. Studies of two selective inhibitors of lymphocyte recirculation, *J. Immunol.* **132**:354–362

Stark, R., Liebes, L. F., Nevrla, D., Conklyn, M., and Silber, R., 1982, Decreased actin content of lymphocytes from patients with chronic lymphocytic leukemia, *Blood* **59**:536–541

Stark, R. S., Liebes, L. F., Shelanski, M. L., and Silber, R., 1984, Anomalous function of vimentin in chronic lymphocytic leukemia lymphocytes, *Blood* **63**:415–420

Steinert, P. M., Steven, A. C., and Roop, D. R., 1985, The molecular biology of intermediate filaments, *Cell* **42**:411–419

Stokke, B. T., Mikkelsen, A., and Elgsaeter, A., 1986a, The human erythrocyte membrane skeleton may be an ionic gel. I. Membrane mechano-chemical properties, *Eur. Biophys. J.* **13**:219–233

Stokke, B. T., Mikkelsen, A., and Elgsaeter, A., 1986b, The human erythrocyte membrane skeleton may be an ionic gel. II. Numerical analyses of cell shapes and shape transformations, *Eur. Biophys. J.* **13**:203–218

Stossel, T. P., 1984, Contribution of actin to the cytoplasmic matrix, *J. Cell Biol.* **99**:15S–21S

Stossel, T. P., Hartwig, J. H., Yin, H. L., Southwick, F. W., and Zaner, K. S., 1984, The motor of leukocytes, *Fed. Proc.* **43**:2760–2763

Stossel, T. P., Chaponnier, C., Ezzell, R. M., Hartwig, J. H., Janmey, P. A., Kwiatkowski, D. J., Lind, S. E.,

Smith, D. B., Southwick, F. S., Yin, H. L., and Zaner, K. S., 1985, Nonmuscle actin-binding proteins, *Annu. Rev. Cell Biol.* **1:**353–402

Taylor, R. B., Duffus, P. H., Raff, M. C., and de Petris, S., 1971, Redistribution and pinocytosis of lymphocyte surface immunoglobulin molecule induced by anti-immunoglobulin antibody, *Nature* (London) *New Biol.* **233:**225–229

Toccanier-Pelte, M. F., Skalli, O., Kapanci, Y., and Gabbiani, G., 1987, Characterization of stromal cells with myoid features in lymph nodes and spleen in normal and pathologic conditions, *Am. J. Pathol.* **129:**109–118

Traub, P., 1985, Intermediate filaments, a review. Springer-Verlag, Berlin

Vandekerckhove, J., and Weber, R., 1978, At least six different actins are expressed in a higher mammal: An analysis based on the amino acid sequence of the amino-terminal tryptic peptide, *J. Mol. Biol.* **126:**783–802

Vandekerckhove, J., and Weber, K., 1979, The complete amino acid sequence from bovine aorta, bovine heart, bovine fast skeletal muscle, and rabbit slow skeletal muscle, a protein chemical analysis of muscle actin differentiation, *Differentiation* **14:**123–133

Varani, J., Wass, J. A., and Rao, K. M. K., 1983, Actin changes in normal human and rat leukocytes and in transformed human leukocytic cells, *J. Natl. Cancer Inst.* **70:**805–809

Wagstaff, J., Gibson, C., Thatcher, N., and Crowther, D., 1981, The migratory properties of indium-111 oxine labelled lymphocytes in patients with chronic lymphocytic leukemia, *Br. J. Haematol.* **49:**283–291

Weeds, A., 1982, Actin-binding proteins—regulators of cell architecture and motility, *Nature* (London) **296:**811–816

Wegner, A., 1985, Subleties of actin assembly, *Nature* (London) **313:**97–98

White, J. R., Naccache, P. H., and Sha'afi, R. I., 1983, Stimulation by chemotactic factor of actin association with the cytoskeleton in rabbit neutrophils, *J. Biol. Chem.* **258:**14041–14047

Williamson, P., Antia, R., and Schlegel, R. A., 1987, Maintenance of membrane phospholipid asymmetry: Lipid-cytoskeletal interactions or lipid pump?, *FEBS Lett.* **219:**316–320

Woda, B. A., and Woodin, M. B., 1984, The interaction of lymphocyte membrane proteins with the lymphocyte cytoskeletal matrix, *J. Immunol.* **133:**2767–2772

Woodcock-Mitchel, J., Mitchel, J. J., Low, R. B., Kieney, M., Sengel, M., Sengel, P., Rubbia, L., Skalli, O., Jackson, B., and Gabbiani, G., 1988, Alpha-smooth muscle actin is transiently expressed in embryonic rat cardiac and skeletal muscles, *Differentiation* **39:**161–166

Woodrum, D. L., Rich, S. A., and Pollard, T. D., 1975, Evidence for biaised unidirectional polymerization of actin filaments using heavy meromyosin by an improved method, *J. Cell Biol.* **67:**231–237

Zauli, D., Gobbi, M., Crespi, C., Tazzari, P. L., Miserocchi, F., and Tassinari, A., 1988, Cytoskeleton organization of normal and neoplastic lymphocytes and lymphoid cell lines of T and B origin, *Br. J. Haematol.* **68:**405–409

Zucker-Franklin, D., Liebes, L. F., and Silber, R., 1979, Differences in the behaviour of the membrane and membrane-associated filamentous structures in normal and chronic lymphocytic leukemia (CLL) lymphocytes, *J. Immunol.* **122:**97–107

Chapter 4

Signaling Events in T-Lymphocyte-Dependent B-Lymphocyte Activation

John C. Cambier, Kathrin L. Lehmann, and William F. Wade

1. INTRODUCTION

The B-cell immune response is initiated *in vivo* by antigen binding to antigen-specific membrane immunoglobulin (mIg) molecules. Binding leads to antigen uptake, processing by proteases, and reexpression, now as immunogenic peptides in association with major histocompatibility complex, (MHC)-encoded class II (Ia) molecules (for a review, see Chesnut and Grey, 1986; Rock *et al.*, 1984). Antigen binding to mIg also leads to signal transduction stimulating alterations in various cellular processes of the B cell (for a review, see Cambier and Ransom, 1987; DeFranco, 1987). Depending on its structure, binding of antigen to B-cell mIg may at one extreme provide a stimulus sufficient to induce expression of a large complex of genes that cause B-cell proliferation and differentiation into antibody-secreting plasma cells or, alternatively, it may stimulate expression of only a limited set of genes (Cambier and Ransom, 1987; DeFranco, 1987; Finkelman *et al.*, 1986).

Antigens that provide all signals required for induction of B-cell proliferation and differentiation are by definition T-lymphocyte independent (TI), meaning that the antibody response to them is independent of T cells. These antigens fall into two categories: TI type 2 antigens, which, presumably because of their rigidity and polyvalency, provide sufficient signal through mIg to drive proliferation and differentiation, and TI type 1 antigens, which drive proliferation and differentiation by virtue of associated carrier moieties that initiate proliferation via interaction with some unknown secondary cellular receptors, i.e., not mIg. In contrast, binding of paucivalent proteinaceious antigens to mIg

John C. Cambier, Kathrin L. Lehmann, and William F. Wade   Department of Pediatrics, Divisions of Basic Sciences and Basic Immunology, National Jewish Center for Immunology and Respiratory Medicine, Denver, Colorado 80206, USA.

does not stimulate B-cell proliferation but does induce increased expression of certain genes, the consequence of which is probably cell death if further signals are not provided (Cambier *et al.*, 1986; Snow *et al.*, 1986; Klemsz *et al.*, 1989a,b). Binding of these antigens appears to prime B cells for subsequent responses which are dependent on T lymphocytes that provide "help" signals (T_h) for both proliferation and differentiation (Cambier and Julius, 1988). These protein antigens are therefore categorized as T lymphocyte dependent (TD).

T cell "help" can take the form of soluble peptide hormones, lymphokines, as well as cell-associated ligands (Cambier and Julius, 1988; O'Garra *et al.*, 1988; Abbas, 1988; DeFranco *et al.*, 1984; Krusemeier and Snow, 1988). Available evidence indicates that B-cell proliferation and differentiation are also regulated by multiple non-T_h-cell-derived cytokines (Plaut, 1987). The complexity of this regulation poses quite a signal-processing problem for the B cell, which must differentially sense, transduce, and integrate the information provided by these factors in order to respond appropriately. In this review, we will discuss B-cell signaling events involved in generation of TD immune responses, with particular reference to the potential receptor–transducer roles of mIg and MHC-encoded Ia molecules. Many of the processes discussed are summarized in Figure 1.

2. B-CELL ANTIGEN RECEPTOR-MEDIATED SIGNALING

Precise determination of the immediate consequences of TD antigen binding to B cells has been hampered by the clonal distribution and thus low frequency within normal B-cell populations of cells specific for a given antigen. As a result, many laboratories have utilized anti-mIg antibodies as antigen surrogates to analyze the effects of antigen binding to mIg. As will be discussed in greater detail below, some anti-Ig antibody preparations, most notably heterologous polyclonal antibodies, act essentially as TI type 2 immunogens (Finkelman *et al.*, 1986; Zitron and Clevinger, 1980; Julius *et al.*, 1984), whereas others, mostly soluble anti-IgM and anti-IgD monoclonal antibodies (MAb), behave similarly to TD immunogens (Cambier *et al.*, 1986; Snow *et al.*, 1986; Klemsz *et al.*, 1989a,b; Leptin, 1985; Myers *et al.*, 1987; Coggeshall and Cambier, 1984; Bijsterbosch *et al.*, 1985; Grupp *et al.*, 1987). Extensive studies using both types of anti-Ig reagents and limited studies of *bona fide* TD antigens have demonstrated convincingly that crosslinking of mIg leads to increased hydrolysis of phosphoinositides, a minor component (~6%) of total B-cell membrane phospholipids, by a phospholipase C (PLC), yielding diacylglycerol (DAG) and inositol mono- and polyphosphates (Myers *et al.*, 1987; Coggeshall and Cambier, 1984; Bijsterbosch *et al.*, 1985; Grupp *et al.*, 1987). Inositol trisphosphate and DAG generated following mIg crosslinking function as second messengers that mediate the activation of protein kinase C (PKC) and mobilization of Ca^{2+} from both intracellular stores and the extracellular space (Bijsterbosch *et al.*, 1986; Ransom *et al.*, 1988; Nel *et al.*, 1986; Chen *et al.*, 1986; MacDougall *et al.*, 1988; Hornbeck and Paul, 1986). Studies using DAG analogs, such as phorbol myristate acetate (PMA), and calcium ionophores suggest that these second messengers are responsible for transcriptional activation of the proto-oncogene c-*fos*, as well as Ia (α and β chains) and invariant-chain genes (Klemsz *et al.*, 1989, 1990; Monroe, 1988), and for induction of

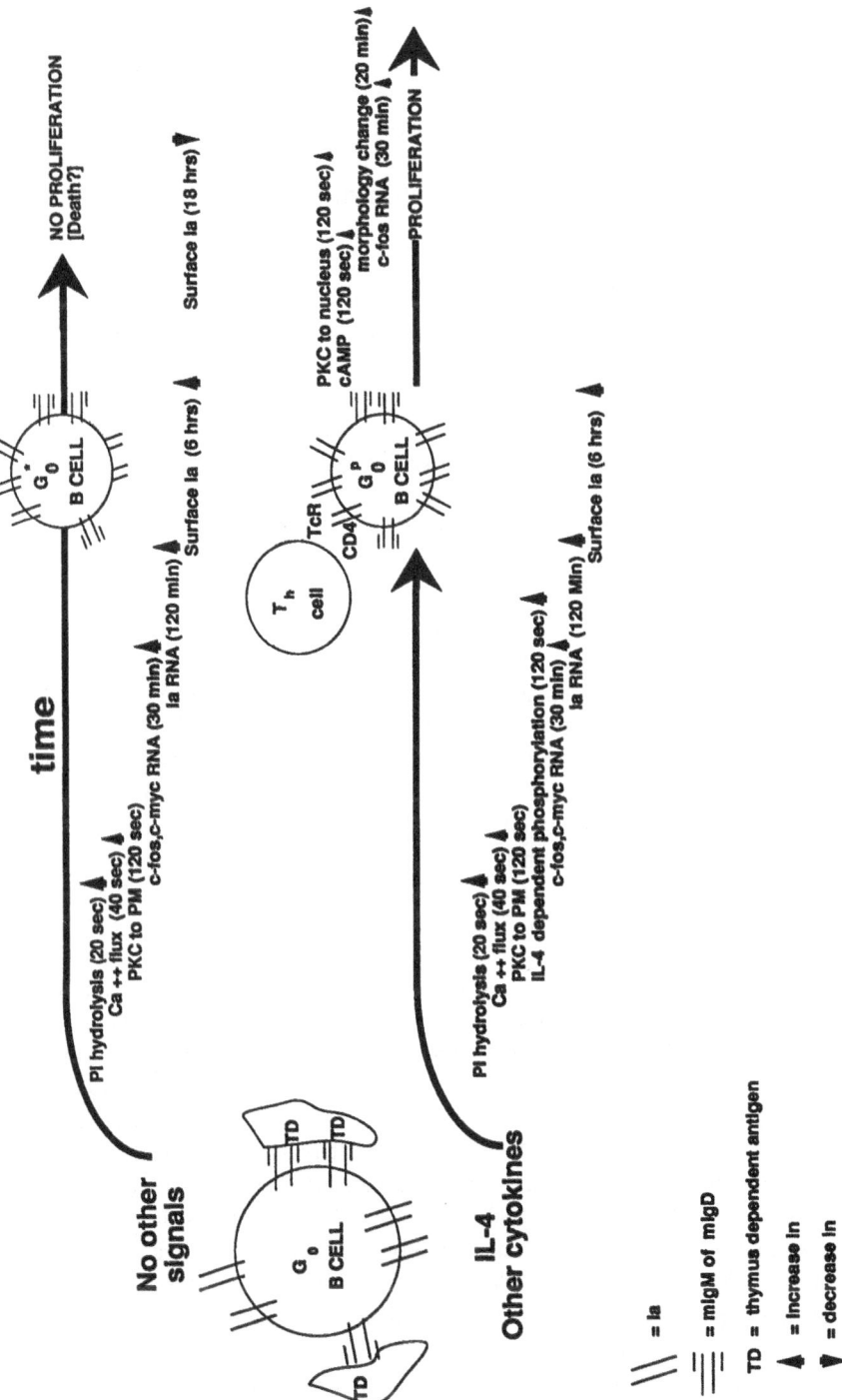

FIGURE 1. B-cell signaling events in T_h-cell-dependent immune response. Changes in the cell physiology of quiescent B cells (G_0) that have bound TD antigen and received no other signals are indicated by the top arrow. These B cells do not proliferate and most probably die in the absence further signals. The bottom arrow represents activation events of quiescent B cells (G_0) that have received IL4 signals in addition to TD antigen-induced signals. If the B cells that received TD antigen and IL4 signals are further signaled by T-cell-associated ligands (antigen receptor or CD4 molecules) that bind Ia, the B cells now proliferate.

c-*myc* mRNA accumulation (Klemsz *et al.*, 1989) following mIg crosslinking. Studies by Klaus *et al.* (1986) and Rothstein *et al.* (1986) indicate that under certain circumstances PMA and calcium ionophores stimulate B-cell proliferation, consistent with observations that extensive crosslinking of mIg by TI type 2 antigens can lead to B-cell proliferation (Cambier *et al.*, 1982). These data suggest that activation of inositol phospholipid hydrolysis by Ig-crosslinking ligands can, under appropriate circumstances, initiate B-cell proliferation. The ability of mIg–antigen interactions to affect the B-cell physiology serves to underscore the signal-transducing role of mIg in the initiation of B-cell immune responses.

It is quite possible that mIg crosslinking leads to stimulation of signaling pathways in addition to that which involves PLC activation. For example, studies by Nel *et al.* (1984) indicate that mIg crosslinking leads to rapid tyrosine phosphorylation of a number of cellular proteins. It should be noted, however, that PMA stimulation also leads to tyrosine kinase activation in B cells (Nel *et al.*, 1985). At this point, it is unclear whether tyrosine kinase activation following mIg crosslinking proceeds via PKC activation or whether mIg is coupled more directly to a tyrosine kinase. Other findings that support a role for an alternate signaling pathway include those by Mond *et al.* (1987), who demonstrated that mIg molecules can transduce proliferative signals in B cells depleted of PKC by high-dose PMA treatment. These findings are consistent with the possibility that mIg-crosslinking ligands may activate multiple signaling pathways.

The precise biologic consequences of mIg crosslinking by TD antigens are a matter of some debate. As noted earlier, these antigens induce hydrolysis of phosphoinositides, leading to Ca^{2+} mobilization, PKC activation, and increased proto-oncogene expression. However, in the absence of additional stimuli, B cells do not proliferate (Cambier *et al.*, 1982; Noelle *et al.*, 1983). In fact, B cells stimulated with TD antigens in the absence of additional stimuli appear to undergo accelerated cell death (J. Cambier, unpublished observation). The earliest point that distinguishes signaling by TD and TI type 2 ligands is marked by the transition of B cells from G_o to G_1 (Snow *et al.*, 1986; Klemsz *et al.*, 1989a; Cambier *et al.*, 1982; Noelle *et al.*, 1983). TD antigens do not induce this response, whereas TI type 2 antigens do. Thus, mIg-binding ligands appear to fall into two categories, those which activate the phosphoinositide cascade but only stimulate cells to progress as far as increased proto-oncogene and Ia expression (TD responses) and those which stimulate the TD responses but also cause cells to enter cycle and proliferate (TI type 2 responses). It is not known why these antigens, which appear to differ only in the relative ability to crosslink mIg, induce such distinct biologic responses. One possible explanation for the lack of a proliferative response to TD antigens is that TD antigens induce generation of smaller Ca^{2+} or DAG responses; alternatively, they may induce these responses for a shorter duration than do TI type 2 antigens. Recent studies by Goroff *et al.* (1986) and Albrecht and Noelle (1988) suggest an additional possibility. These experiments indicate that mitogenic but not nonmitogenic mIg-binding ligands induce association of mIg with the cytoskeleton. Previous studies have shown that cytoskeletal association of mIg leads to subsequent energy-dependent capping and internalization of mIg (Braun and Unanue, 1980; Schreiner and Unanue, 1976). This phenomenon may reflect the activation of an additional signaling mechanism by mitogenic mIg-binding ligands, a mechanism that may require mIg internalization. This hypothesis is consistent with observations of DeFranco *et al.* (1985) which indicate that stimulation of B-cell proliferation by anti-Ig involves two signaling events; the first requires only low antigen receptor (mIg) occupancy, which is sufficient to induce phosphoinositide hydrolysis and Ca^{2+} mobilization, and the second, needed about 24 hr later, requires much higher Ig-binding ligand concentrations that are likely to induce mIg capping and internalization.

Thus, published data are consistent with the possibility that TD antigens initiate the B-cell immune response by activating transduction of signals through mIgM and mIgD, which involves rapid activation of a PLC, leading to increased phosphoinositide hydrolysis and subsequent Ca^{2+} mobilization, PKC activation, and increased expression of proto-oncogenes and Ia. These B cells, however, do not progress into cycle unless additional signals are provided. Progression into cycle is critical to perpetuate the response by enabling generation of daughter cells that will be the effector cells (plasmacytes) and memory cells required for a secondary response. Therefore, it seems likely that B-cell proliferation during TD antigen response is dependent on additional signaling events initiated by T_h-cell-derived ligands.

3. MOLECULAR BASES OF T-CELL-MEDIATED B-CELL SIGNALING

A substantial literature has documented the requirement that T_h cells recognize antigenic peptides in association with Ia molecules for generation of antibody responses to most proteinaceous antigens (for a review, see Singer and Hodes, 1983). This requirement presumably reflects a need to deliver an activating second signal from the T_h cell to the B cell via a soluble or membrane-associated ligand (Julius et al., 1982). Additionally, it may reflect a necessity for delivery of signals from the B cell to the T_h cell. A compelling literature has documented the fact that antigen-presenting B cells and macrophages (collectively APC) internalize, process, and reexpress antigen as peptide associated with Ia and that recognition of this peptide–Ia complex leads to T_h-cell activation and the elaboration of lymphokines (Chesnut and Grey, 1986; Rock et al., 1984). Thus, minimally contact-dependent T_h-cell signaling occurs upon T_h cell–APC interaction.

It is a popular belief that lymphokines elaborated as a result of T_h cell–APC interaction provide all signals necessary to drive B-cell proliferation and differentiation and therefore that B cells need not be signaled by direct contact with T_h cells. However, although low-density (activated) splenic B cells become antibody-secreting cells in response to lymphokines alone, quiescent splenic B cells do not (Whalen et al., 1988). Furthermore, studies by Noelle et al. (1983), Cambier and Julius (1988), and Whalen et al. (1988) have shown that whereas resting B cells proliferate in the presence of intact T_h cells, provided antigen and Ia restriction requirements for T_h-cell activation are met, lymphokine-enriched supernatants from these cultures or other sources do not support proliferation of quiescent B cells. The most notable exception to the requirement for the presence of intact T_h cells to support B-cell proliferation is demonstrated by the studies of B-cell-activating factor (BCAF; a product of some cloned helper T_h cells), which stimulates a proportion of quiescent B cells to proliferate in the absence of T_h cells (Leclerq et al., 1984, 1986). In our hands, BCAF is a much less efficient stimulator of B-cell proliferation than are intact T_h cells (J. C. Cambier and M. H. Julius, unpublished observation). Thus, most available data are consistent with the possibility that cognate T_h cell–B cell interaction results in bidirectional cell contact-mediated signaling and that the

T_h-cell signal to the B cell is important in activating B-cell entry into cell cycle. It should be noted, however, that on the basis of the findings discussed above, it is impossible to distinguish between the possibilities that T_h-cell membrane-associated ligands signal B cells during T_h cell–B cell contact or that signaling is mediated by lymphokines that are active only at very short range. One could argue that high local levels of lymphokines, especially if directionally secreted, could mediate T_h-cell stimulation of B-cell growth very efficiently. Such directional lymphokine secretion is consistent with the reported reorientation of T-cell microtubule-organizing centers and reorganization of membrane-associated cytoskeleton (talin) at cell contact points in T cell–B cell conjugates (Kupfer and Singer, 1989; Kupfer *et al.*, 1987). The concept of T_h-cell antigen receptor-directed focusing of lymphokine release is also supported by results of recent studies by Poo *et al.* (1988), who found that when T_h cells that had been forced into pores of Nuclepore membranes were stimulated with antibodies specific for T_h-cell antigen receptors, interleukin 4 (IL4) was preferentially produced at the pole of the cell where receptors were bound. These findings suggest that upon interaction with antigen-presenting B cells, T_h cells may secrete lymphokines at points of cell apposition, generating high local concentrations of lymphokines such as BCAF which could activate the B cell. Such locally limited lymphokine-mediated B-cell activation could be as effective as cell contact-mediated signaling in preserving the specificity of the immune response.

The above possibility notwithstanding, the requirement that T_h cells recognize antigenic peptides in association with Ia during cognate T_h cell–B cell interactions seems most consistent with the possibility that T_h cells signal B cells by cell-associated ligands and further, that Ia molecules act as transducers of these signals. The possibility that T_h cells signal B cells by direct contact and through Ia is supported by results of Krusemeier and Snow (1988), who demonstrated that T_h cells rendered incapable of producing lymphokines by treatment with cyclosporine A still provide proliferative signal to B cells and that delivery of signal is inhibited by anti-Ia antibodies. These findings were extended and qualified by results of Brian (1988), which indicated that isolated T-cell membranes could stimulate B cells and that this signal could be delivered across Ia haplotype barriers. This finding is consistent with results of studies of DeFranco *et al.* (1984), Cambier and Julius (1988), and Whalen *et al.* (1988) in which the ability of Ia-restricted T_h cells to activate mixed quiescent Ia-compatible and incompatible B cells was assessed. In these studies, it was observed that once activated via interaction with the Ia-compatible antigen-presenting B cells, the T_h cell could activate, although with somewhat less efficiency, proliferation of the Ia-incompatible B cell. Taken together, these data support the possibility that T_h cells provide contact-mediated proliferative signals to B cells via recognition of a monomorphic Ia determinant.

Implicit in this model is the existence on the T_h cell of ligands that bind polymorphic and monomorphic Ia determinants. Clearly, T_h-cell antigen receptors ($\alpha\beta$TCR) bind polymorphic Ia determinants (Chesnut and Grey, 1986; Rock *et al.*, 1984). Two elegant studies have demonstrated the T_h-cell CD4 molecules exhibit specificity of binding monomorphic Ia determinants. Gay *et al.* (1987) transfected human CD4 cDNA into a mouse T-cell hybridoma (3 DT-52.5.8) and found that these cells responded well to P815 cells only if the P815 cells had been cotransfected with the genes for the HLA-DR α and β chains and invariant-chain cDNAs. A more direct approach was taken by Doyle and Strominger (1987), who demonstrated that transfection of CV1 cells with cDNAs encoding CD4 conferred upon CV1 cells the ability to bind class II$^+$ (but not class II$^-$) human B cells. Binding was inhibitable with either anti-CD4 or anti-MHC class II antibodies. Thus, T_h cells can bind B cells via CD4 interaction with monomorphic Ia-determinants.

The results discussed above suggest that cognitive T_h cell–B cell collaboration involves bidirectional signaling and that Ia histocompatibility is required only for T_h-cell activation. Once activated, T_h cells signal B cells by multiple mechanisms involving cell-associated ligand and lymphokines. If this model is correct and T_h-cell contact-mediated B-cell signaling is essential for the B-cell response, one would predict that T_h cell–B cell contact would be required for T_h-cell-dependent B-cell activation even if requirements for T_h-cell activation are obviated by activation of T_h cells using antibodies against the T_h-cell antigen receptor complex. This question has been approached experimentally in a number of recent studies. Owen (1988) demonstrated that anti-$\alpha\beta$TCR monoclonal antibodies (MAb F23)-stimulated T_h clones (E9.D4) activated B cells in T_h cell–B cell mixtures (close proximity) but could not activate B cells if the T_h cells were separated from the B cells by a 0.4-μm membrane. Riedel *et al.* (1988) cultured E9.D4 cells and quiescent B cells in wells that had been coated with MAb F23 and found that B cells responded optimally only when T_h and B cells were in close proximity in one "corner" of the dish. Hirohata *et al.* (1988) conducted experiments using normal human B cells and normal CD4$^+$ CD8$^-$ T cells in conjunction with anti-CD3 MAb. Once again, optimal induction of B-cell proliferation and antibody secretion required culture of T and B cells in close contact. Finally, Julius and Rammensee (1988) demonstrated that MAb F23-stimulated E9.D4 cells helped B cells regardless of Ia allotype. Collectively, these studies support the concept that Ia restriction in T_h cell–B cell collaboration reflects a requirement that T_h cells recognize peptide in association with compatible Ia to become activated. Once activated, T_h cells deliver help to B cells via recognition, possibly by CD4, of a monomorphic Ia determinant (Doyle and Strominger, 1987). Furthermore, the data discussed indicate that the immediate consequence of delivery of this signal is B-cell proliferation.

4. BIOLOGICAL EVIDENCE FOR Ia-MEDIATED SIGNAL TRANSDUCTION

Unambiguous determination of the ability of Ia molecules to act as signal transducers is complicated by the fact that the physiologic Ia-binding ligands are cell associated. When T_h cells are used to signal B cells, it is difficult to ensure that the biochemical or physiologic effects observed reflect changes in the Ia-bearing cell and not the cell that bears its ligand. This is especially true in this case, since transmembrane signal transduction occurs in both cells during T_h cell–B cell interaction. The problem is further complicated by the need, when studying the biochemistry of signal transduction, to synchronously deliver saturating doses of ligand to the responding cells. This is necessary to induce a maximal, and thus optimally detectable, biochemical response. Because of geometric constraints, it is currently impossible to achieve either synchronous receptor ligation or receptor saturation when the ligand is cell associated.

The problems inherent in studying B-cell signaling by T_h-cell-associated ligands have led a number of laboratories to use antibodies to various B-cell surface molecules as surrogates of T_h-cell-associated ligands for such studies. Niederhuber *et al.* (1975) and Forsgren *et al.* (1984) assessed the effect of anti-Ia antibodies on B-cell proliferative responses to lipopolysaccharide (LPS) and observed that anti-Ia antibodies uniformly inhibited these responses. Clement *et al.* (1986) subsequently extended these findings by demonstrating that anti-Ia antibodies inhibited anti-Ig-induced B-cell proliferation. Needless to say, the observed negative effects would not be predicted if Ia is the transducer of T_h-cell signals that initiate proliferation. More predicted effects were seen by Palacios and co-workers (1983), who demonstrated that certain anti-Ia antibodies enhance *in vitro* anti-sheep erythrocyte antibody responses. Finally, Bishop and Haughton (1986) reported that certain anti-Ia antibodies (those specific for I-E) stimulate the differentiation of the CH12 B-cell lymphoma to become antibody-secreting cells. Taken together, these data suggest that Ia molecules transduce signals that promote differentiation of B cells at the apparent expense of proliferation, a result seemingly inconsistent with the ability of T_h cells to stimulate B-cell proliferation.

On the basis of our studies of the biochemistry of Ia signaling (see below), we hypothesize that although anti-Ig and anti-Ia signals are antagonistic if provided simultaneously, they may be complementary if given in sequence, a condition imposed during the immune responses by requirements for antigen processing after mIg ligation but before cognate T_h cell–B cell interaction. Furthermore, the findings of Bishop and Haughton (1986) may be explained if B cells in distinct differentiative states respond differently to Ia ligation; i.e., while antigen-primed quiescent B cells proliferate, postproliferative B cells, presumably analogous to CH12, differentiate.

In a very recent series of studies (Cambier and Lehmann, 1989), we have reexamined the biologic consequences of Ia crosslinking in a model system designed to mimic the *in vivo* immune response in terms of the multiplicity of signals provided and the temporal relationship of their provision. Quiescent B cells were primed with soluble, anti-μ MAb and IL4 for varied time periods before being stimulated by immobilized anti-Ia MAb. The anti-μ MAb Bet2 was used as a substitute for TD antigens because, like these antigens, Bet2 induces phosphoinositide hydrolysis, Ca^{2+} mobilization, and increases in c-*myc* and Ia expression but does not induce entry of quiescent B cells into cell cycle (Cambier, unpublished observation; Kung *et al.*, 1981). The priming effects of IL4 were assessed on the basis of observations of Oliver *et al.* (1985) and Snapper and Paul (1987) that IL4 prestimulation primes cells for a more accelerated progression to S phase following stimulation with LPS or mitogenic anti-Ig antibodies and observations by Sanders *et al.* (1987) that IL4 primes cells for T cell–B cell conjugate formation. Anti-Ia MAb immobilized on polystyrene plates was used to mimic binding of Ia by T_h cells via their $\alpha\beta$TCR or CD4 molecules.

B cells that had been primed for 16 hr with 20 μg/ml of Bet2 and 100 units μg/ml of IL4 and then transferred to anti-Ia MAb (D3.137.5.7)-coated plates underwent a morphologic transformation typified by the extension of long fibrous pseudopods (Cambier and Lehmann, 1989). This response was rapid, being noted in some cells within 20 min of transfer to MAb-coated plates, and was observed in 95% of the B cells. Cells remained in this morphology for 36 to 48 hr. This morphologic change was specific for Ia-binding ligands, since primed cells adhered but remained spherical when cultured on plates coated with anti-H-2K MAb (M1.42.398), anti-Fc receptor MAb (2.4G2), anti-LyB2 MAb (10-1.D.1.), anti-LyB8 MAb (Cy34.1.2), anti-heat-stable antigen MAb (J11D), anti-QA2 MAb (D3.137), or anti-IgD MAb (JA12.5). These findings suggest that Ia molecules transduce a signal that results minimally in the restructuring of cytoskeletal elements.

We subsequently examined the effect of immobilized anti-Ia MAb on the progression of B cells into cycle (Cambier and Lehmann, 1989). Cells were harvested at various times after transfer to anti-Ia MAb-coated plates, stained with acridine orange, and analyzed by flow cytometry. Results indicate that a large proportion (>70%) of cells entered cycle within 48 hr of stimulation. Cells were detected in all stages of cycle, indicating that Ia mediated signals that are sufficient to drive cells through cycle provided anti-μ MAb and IL4 are also present.

Finally, we assessed the ability of primed, anti-Ia MAb-stimulated cells to secrete immunoglobulins (Cambier and Lehmann, 1989). Cells were primed for 16 hr as before and expanded on anti-Ia-coated plates for 48 hr before being washed and cultured on fresh anti-Ia MAb-coated plates with IL4- and IL5-containing supernatants. Ninety-six hours later, culture supernatants were harvested, and their IgM and IgG content was quantified by ELISA. Anti-Ia-stimulated B cells cultured during the final 96 hr with lymphokines produced significant amounts of IgM and IgG compared with the amounts produced by B cells that had been cultured on anti-H-2K-coated plates. B cells stimulated for the entire 160 hr with LPS produced slightly more IgM and IgG than did B cells cultured on anti-Ia MAb-coated plates. Thus, anti-Ia MAb signaling makes B cells competent to differentiate to antibody-secreting cells in response to appropriate lymphokines; furthermore, this response approaches the efficiency of that induced by stimulation of B cells with optimal LPS concentrations.

To ascertain whether the responses observed above were a general feature of Ia-binding ligands, we compared the ability of a number of anti-Ia MAb to activate primed B cells. We observed, using B cells from AKR mice, that immobilized anti-$A^k\alpha$ MAb (2A2), anti-$A^k\beta$ MAb (10.3.6, 10.2.16, 40M, and 39J), and anti-Eα MAb (14.4.4S) are all comparably effective stimulators of proliferation of primed B cells, inducing maximal responses when plates are coated with optimal concentrations of anti-Ia MAb. Thus, the chain or isotype specificity of ligand binding to Ia is not of critical importance in determining the proliferative response. This result is consistent with physiologic signals being delivered by both the T_h-cell antigen receptor, which binds a polymorphic Ia determinant, and CD4, which binds a monomorphic Ia determinant.

We also assessed the degree to which the proliferative responses observed were specific to Ia ligation. Plates were coated with MAb that were specific for a variety of B-cell markers, including LyB8, LyB2, I-A, I-E, H-2K, IgD, and FcR. We assessed the response of primed C57BL/6 × DBA/2 F1 B cells to these matrices on the basis of the MTT assay, which measures mitochondrial metabolic activity (Mossman, 1983). Activity was measured 48 hr after transfer of B cells to MAb-coated plates. Among the ligands studied, primed B cells were activated only by Ia-binding ligands. Findings that Ia ligation leads specifically to changes in morphology, increased proliferation, and immunoglobulin secretion (so long as accessory signals are provided) are consistent with the possibility that Ia is the transducer of a T_h-cell-derived contact-dependent signal required for optimal B-cell responses during TD immune responses.

Having established that the combination of anti-IgM MAb-, IL4-, and anti-Ia MAb-mediated signals was sufficient to stimulate B-cell activation and proliferation, we formally examined, under optimal conditions, the degree to which each is required for the responses. Any combination of two of the three signals was insufficient to stimulate morphologic transformation and activation (MTT assay) or proliferation [^{3}H]thymidine uptake) comparable to that induced by the combination of all three ligands or that induced by LPS (MTT and [³H]thymidine uptake only). These findings, provided in the following sequence, suggest that an antigen signal, an IL4 signal, and an Ia-transduced signal, presumably provided by T_h cells, satisfy all requirements to stimulate B-cell proliferation during T_h-cell-dependent humoral immune responses.

In very recent studies, we have begun to evaluate the usefulness of this experimental protocol for inducing long-term growth of B cells. In a typical experiment, cells were primed with IL4 and anti-μ MAb as before and then transferred to anti-Ia MAb-coated plates for a 48-hr expansion period before being washed and recultured on anti-Ia MAb-coated plates with IL4 and IL5 but without anti-μ MAb. Cells were subsequently washed and recultured every 4 days under the same conditions. Eleven days after initiation of culture, cells were harvested and counted, and expression of Ia, Thy-1, and mIgM was assessed by immunofluorescence. Results demonstrated the initial B-cell population expanded 20-fold during the culture period. When the culture was terminated, the recovered cells were virtually 100% mIgM and Ia positive, indicating that the expanding population was B cells. No T cells were detected among the recovered cells. Thus, the combination of signals used was sufficient to sustain proliferation of B cells for at least 11 days.

5. BIOCHEMICAL EVIDENCE FOR Ia-MEDIATED SIGNAL TRANSDUCTION

The molecular basis of Ia-mediated transmembrane signal transduction in B cells is poorly understood. In 1986, while studying induction of PKC translocation in B cells by mIg-binding ligands, we noted that anti-Ia MAb stimulated an unusual and previously undescribed PKC translocation event (Cambier *et al.*, 1987; Chen *et al.*, 1987). In these studies, anti-IgM and anti-IgD MAb induced translocation of PKC from the cytosolic compartment to a detergent-extractable membrane compartment. Surprisingly, under the conditions used, anti-Ia MAb also induced a rapid loss of detectable PKC from the cytosolic fraction, and this loss was not mirrored by the appearance of PKC in the plasma membrane fraction. Further analysis revealed that anti-Ia MAb induced the association of PKC with particulate material and that this association was refractile to detergent extraction. This translocation involved most (>70%) of the B-cell PKC detectable by assays of enzymatic activity and [³H]phorbol binding or by Western blotting analysis with anti-PKC antibodies (Cambier *et al.*, 1987). Immunohistochemical studies revealed that Ia crosslinking leads to PKC association with the nucleus. At this point, it is unclear whether under these circumstances the PKC actually enters the nucleus or is associated with the nuclear envelope.

Since it occurs in unprimed cells, the PKC translocation response *per se* does not reflect delivery of sufficient signal(s) to stimulate the morphologic changes and B-cell activation described earlier.

The nuclear translocation response can be induced in both resting and IL4-plus-anti-μ-MAb-primed B cells by intact anti-Ia MAb or F(ab')₂ fragments (Z. Z. Chen,

Evidence suggests that anti-Ia MAb induction of PKC translocation to the nucleus is mediated via generation of cyclic AMP (cAMP) (Cambier *et al.*, 1987). Specifically, we observed that anti-Ia MAb induce accumulation of cAMP in B cells and that the response was correlated with PKC translocation. Subsequent experiments showed that dibutyryl-cAMP (dbcAMP) and 8-bromo-cAMP induce PKC translocation, as do agents such as isoproterenol, cholera toxin, and forskolin, which induce cAMP generation. Recent studies of Cohen and Rothstein (1989) have demonstrated that properly timed secondary stimulation of anti-Ig- or PMA- and ionomycin-stimulated cells with dbcAMP or iso-butylmethylxanthine can lead to enhanced proliferation. These data suggest that a causal relationship exists between Ia ligation, cAMP generation, nuclear PKC translocation, and proliferation of B cells.

The basis of coupling of Ia to cAMP-generating systems is unknown. One possibility is that Ia is, or somehow becomes, associated with the β-adrenergic receptor following ligation and the resultant perturbation of the β-adrenergic receptor activates cAMP generation. Consistent with this possibility is the expression of β-adrenergic receptors by B cells (Miles *et al.*, 1984) and the fact that during T_h cell–B cell interactions, Ia molecules become associated with an unidentified 67-kDa B-cell surface molecule (Shivdasani and Thomas, 1988). The molecular weight of this species is similar to that of β-adrenergic receptor (64 kDa). An alternative explanation is that Ia may be directly coupled to G_s or a similar adenyl cyclase-activating GTP-binding protein. Worthy of note is the fact that whereas β-adrenergic agents are efficient activators of adenylate cyclase in isolated B-cell membranes, anti-Ia MAb are not (K. Newell, National Jewish Center, unpublished observation). Therefore, if Ia is somehow coupled to adenylate cyclase activation, that coupling must be disrupted by membrane isolation.

We have recently begun a series of experiments that we hope will begin to better elucidate this coupling mechanism. These studies (Wade *et al.*, 1989) involve the preparation and use of transfected B-cell lines that express mutant Ia molecules. The goal of these studies is to define the Ia structural elements involved in coupling Ia to PKC translocation. Initially, we prepared genomic DNA encoding the α and β chains of I-A^k but containing stop condons at positions appropriate to yield expressed protein products that are truncated at various points in the cytoplasmic domains of either the α or β chain or both. These constructs were transfected into the B-cell lymphoma M12.C3, and clones expressing I-A^k at levels similar to those found in activated B cells were selected for further study.

Using these transfectants, we assessed the ability of wild-type and mutant Ia molecules to transduce signals leading to PKC translocation to the nuclear compartment. Ia molecules are heterodimeric structures composed of an α chain that has two globular extracellular domains connected to a single 29-amino-acid transmembrane-spanning region and a 12-amino-acid carboxy-terminal tail. The β chain also contains two extracellular domains connected to a single 22-amino-acid transmembrane-spanning region and an 18-amino-acid carboxy-terminal cytoplasmic tail. Results of studies of truncation mutants indicate that when coupled with a normal β chain, the entire α-chain cytoplasmic tail may be deleted without significant impairment of Ia signaling ability. However, progressive truncation of the β chain, removing 12 amino acids and pairing it with either a full-length α chain or an α chain missing its 12 carboxy-terminal residues, progressively impairs signaling ability, slowing the PKC translocation response and lessening its magnitude. The ability of Ia to signal is not totally destroyed by β-chain truncation until the most plasma membrane-proximal six amino acids are removed. Thus, among cytoplasmic Ia residues, only the plasma membrane-proximal six amino acids of the β chain seem essential for signal transduction that leads to PKC translocation to the nuclear fraction. Therefore, this region is an obvious candidate for the site of interaction between Ia and a secondary transducer such as the β-adrenergic receptor or a G protein. Consistent with this possibility is the fact that this region is highly charged and highly conserved between mouse I-A and I-E and human DR and DQ (Figueroa and Klein, 1986). An equally tenable possibility is that this region is not a site of interaction with a secondary transducer; rather, its importance lies in maintaining a site in the transmembrane region of the β chain in some conformation necessary for function as a point of contact with accessory molecules. Studies in progress involve production and use of Ia mutants with more conservative changes in the β chain to study this question.

6. CONCLUSIONS

Available evidence is consistent with the possibility that Ia molecules transduce signals in B cells following interaction with T_h-cell antigen receptors and CD4 molecules during cognate T_h cell–B cell conjugate formation. Signals transduced through Ia mediate proliferation of normal B cells provided these cells have been primed by previously reception of signals through mIg and IL4 receptors. This model is consistent with the temporal sequence of signals imposed by requirements that B cells internalize and process native TD antigen and reexpress processed antigen before T_h-cell recognition of that processed antigen is possible. Once "programmed" to proliferate by Ia-binding ligands, B cells become responsive to the proliferation- and differentiation-promoting effects of interleukins.

Signal transduction through Ia appears to involve the activation, via a region of the cytoplasmic domain of the β chain, of a cAMP-generating mechanism, presumably involving adenylate cyclase. The importance of cAMP generation in physiologic Ia-mediated signaling is suggested by recent observations of E. C. Snow and R. J. Noelle (unpublished observation) that T_h cells stimulate cAMP generation in B cells during delivery of T_h-cell help. cAMP appears to mediate the translocation of PKC to a nuclear or nucleus-associated compartment, where it presumably functions in initiation of subsequent processes that lead to proliferation. Although the function of nuclear PKC is unknown, several possibilities are suggested by the current literature. Sequencing and translation of PKC DNA have revealed zinc finger motifs consistent with a possible function as a DNA-binding–transcription-regulating protein (Kikkawa *et al.*, 1987). Additionally, Yamamoto *et al.* (1988) have reported that in PC12 cells, activation of the cAMP-responsive DNA transcription-activating protein CREB requires that the protein be phosphorylated by both PKA and PKC. These findings suggest that cAMP generation stimulates the activity of both PKC and PKA in these cells. Such a mechanism may also be active in B cells. Our future studies will address mechanisms by which early events in Ia-mediated signal transduction are coupled to changes in gene transcription that determine the ultimate biologic response to ligand.

ACKNOWLEDGMENT. This research was supported by National Institutes of Health grants AI 20519 and AI 21768.

7. REFERENCES

Abbas, A. K., 1988, A reassessment of the mechanism of antigen-specific T-cell-dependent B-cell activation, *Immunol. Today* **89**:94

Albrecht, D. L., and Noelle, R. J., 1988, Membrane Ig-cytoskeletal interactions. I. Flow cytofluorometric and biochemical analysis of IgM-cytoskeletal interactions, *J. Immunol.* **141**:3915–3922

Bijsterbosch, M., Meade, J. C., Turner, G. A., and Klaus, G. G. B., 1985, B lymphocyte receptors and polyphosphoinositide degradation, *Cell* **41**:999–1006

Bijsterbosch, M. K., Rigley, K. P., and Klaus, G. G. B., 1986, Crosslinking of surface immunoglobulin on B lymphocytes induces intracellular Ca^{++} release and Ca^{++} influx: Analysis with Indo-1, *Biochem. Biophys. Res. Commun.* **137**:500–506

Bishop, G. A., and Haughton, G., 1986, Induction of differentiation of a transformed clone of $Ly1^+$ B cells by clonal T cells and antibodies, *Proc. Natl. Acad. Sci. USA* **83**:7410–7414

Braun, J., and Unanue, E. R., 1980, B lymphocyte biology studied with anti-Ig antibodies, *Immunol. Rev.* **52**:3–28

Brian, A. A., 1988, Stimulation of B-cell proliferation by membrane-associated molecules from activated T cells, *Proc. Natl. Acad. Sci. USA* **85**:564–568

Cambier, J. C., and Julius, M. H., 1988, Early changes in quiescent B cell physiology subsequent to cognate and bystander interaction with helper T cells, *Scand. J. Immunol.* **27**:59–71

Cambier, J. C., and Lehmann, K. L., 1989, Ia mediated signal transduction leads to proliferation of primed B lymphocyte, *J. Exp. Med.* **170**:877–886

Cambier, J. C., and Ransom, J. T., 1987, Molecular mechanisms of transmembrane signaling in B lymphocytes, *Annu. Rev. Immunol.* **5**:179–199

Cambier, J. C., Monroe, J. G., and Neale, M. J., 1982, Definition of conditions which facilitate antigen specific activation of the majority of isolated trinitrophenol binding B cells, *J. Exp. Med.* **156**:1635–1649

Cambier, J. C., Heusser, C. H., and Julius, M. H., 1986, Abortive activation of B lymphocytes by anti-immunoglobulin antibodies, *J. Immunol.* **136**:3140–3146

Cambier, J. C., Newell, M. K., Justement, L. B., McGuire, J. C., Leach, K. L., and Chen, Z. Z., 1987, Ia binding ligands and cAMP stimulate nuclear translocation of PKC in B lymphocytes, *Nature* (London) **327**:629–632

Chen, Z. Z., Coggeshall, K. M., and Cambier, J. C., 1986, Translocation of protein kinase C during membrane immunoglobulin-mediated transmembrane signaling in B lymphocytes, *J. Immunol.* **136**:2300–2306

Chen, Z. Z., McGuire, J. C., Leach, K. L., and Cambier, J. C., 1987, Transmembrane signaling through B cell MHC class II molecules: Anti-Ia antibodies induce protein kinase C translocation to the nuclear fraction, *J. Immunol.* **138**:2345–3252

Chesnut, R. W., and Grey, H. M., 1986, Antigen presentation by B cells and its significance in T-B interactions, *Adv. Immunol.* **39**:51–94

Clement, L. T., Tedder, T. F., and Gartland, G. L., 1986, Antibodies reactive with class II antigens encoded for by the major histocompatibility complex inhibit human B cell activation, *J. Immunol.* **136**:2375–2381

Coggeshall, K. M., and Cambier, J. C., 1984, B cell activation. VIII. Membrane immunoglobulins transduce signals via activation of phosphatidylinositol hydrolysis, *J. Immunol.* **133**:3382–3386

Cohen, D. P., and Rothstein, T. L., 1989, Adenosine 3′, 5′-cyclic monophosphate modulates the mitogenic responses of murine B lymphocytes, *Cellular Immunol.* **121**:113–124

DeFranco, A. L., 1987, Molecular aspects of B-lymphocyte activation, *Annu. Rev. Cell Biol.* **3**:143–78

DeFranco, A. L., Ashwell, J. D., Schwartz, R. H., and Paul, W. E., 1984, Polyclonal stimulation of resting B lymphocytes by antigen specific T lymphocytes, *J. Exp. Med.* **159**:861–880

DeFranco, A. L., Ravache, E. S., and Paul, W. E., 1985, Separate control of B lymphocyte early activation and proliferation in response to anti-IgM antibodies, *J. Immunol.* **135**:87–94

Doyle, C., and Strominger, J. L., 1987, Interaction between CD4 and class II MHC molecules mediates cell adhesion, *Nature* (London) **330**:256–259

Figueroa, F., and Klein, J., 1986, The evolution of class II MHC genes, *Immunol. Today* **7**:78–79

Finkelman, F. D., Mond, J. J., and Metcalfe, E. S., 1986, Anti-immunoglobulin antibody induction of B lymphocyte activation and differentiation, in *B-Lymphocyte Differentiation* (J. Cambier, ed.), pp. 41–61, CRC Press, Boca Raton, Fla.

activation I. Inhibition of lipopolysaccharide-induced responses by monoclonal antibodies, *J. Immunol.* **133:**2104–2110

Gay, D., Maddon, P., Sekaly, R., Talle, M. A., Godfrey, M., Long, E., Goldstein, G., Chess, L., Axel, R., Kappler, J., and Marrack, P., 1987, Functional interaction between human T-cell protein CD4 and the major histocompatibility complex HLA-DR antigen, *Nature* (London) **328:**626–629

Goroff, D. K., Stall, A., Mond, J. J., and Finkelman, F. D., 1986, In vitro and in vivo B lymphocyte-activating properties of monoclonal anti-α antibodies. I. Determinants of B lymphocyte-activating properties, *J. Immunol.* **136:**2382–2392

Grupp, S. H., Snow, E. C., and Harmony, A. K., 1987, Phosphatidylinositol response is an early event in physiologically relevant activation of antigen specific B lymphocytes, *Cell. Immunol.* **109:**181–191

Hirohata, S., Jelinek, D. F., and Lipsky, P. E., 1988, T cell-dependent activation of B cell proliferation and differentiation by immobilized monoclonal antibodies to CD3, *J. Immunol.* **140:**3726–3744

Hornbeck, P., and Paul, W. E., 1986, Anti-immunoglobulin and phorbol ester induce phosphorylation of proteins associated with plasma membrane and cytoskeleton in murine B lymphocytes, *J. Biol. Chem.* **261:**14817–14824

Julius, M. H., and Rammensee, H.-G., 1988, T helper cell-dependent induction of resting B cell differentiation need not require cognate cell interactions, *Eur. J. Immunol.* **18:**375–379

Julius, M., Von Boehmer, H., and Sidman, C. L., 1982, Dissociation of two signals required for activation of resting B cells, *Proc. Natl. Acad. Sci. USA* **79:**1989–1993

Julius, M. H., Heusser, C. H., and Hartmann, K.-U., 1984, Induction of resting B cells to DNA synthesis by soluble monoclonal anti-immunoglobulin, *Eur. J. Immunol.* **14:**753–757

Kikkawa, U., Ogita, K., Ono, Y., Asaoka, Y., Shearman, J. S., Tomoko, F., Ase, K., Sekiguchi, K., Igarashi, K., and Nishizuka, K., 1987, The common structure and activities of four subspecies of rat brain protein kinase C family, *FEBS Lett.* **223:**212–216

Klaus, G. G. B., O'Garra, A., Bijsterbosch, M. K., and Holman, M., 1986, Activation and proliferation signals in mouse B cells. VIII. Induction of DNA synthesis in B cells by a combination of calcium ionophores and phorbol myristate acetate, *Eur. J. Immunol.* **16:**92–97

Klemsz, M. J., Palmer, E., and Cambier, J., 1990, B cell activation. IX. Different control mechanisms are operative in anti-Ig and BSF1 induction of increased Ia mRNA and surface expression (submitted for publication)

Klemsz, M. J., Justement, L. B., Palmer, E., and Cambier, J. C., 1989, Induction of c-fos and c-myc expression during B cell activation, *J. Immunol.* **143:**1032–1040

Krusemeier, M., and Snow, C., 1988, Induction of lymphokine responsiveness of hapten-specific B lymphocytes promoted through an antigen-mediated T helper lymphocyte interaction, *J. Immunol.* **140:**367–375

Kung, J. T., Sharrowm, S. O., Sieckman, D. G., Lieberman, R., and Paul, E. W., 1981, A mouse IgM allotypic determinant (Igh 51b) recognized by a monoclonal rat antibody, *J. Immunol.* **127:**873–876

Kupfer, A., and Singer, S. J., 1989, Cell biology of helper T-cell functions, *Annu. Rev. Immunol.* **7:**309–339

Kupfer, A., Swain, S. L., and Singer, S. J., 1987, The specific direct interaction of helper T cells and antigen-presenting B cells. II. Reorientation of the microtubule organizing center and reorientation of the membrane-associated cytoskeleton inside the bound helper T cells, *J. Exp. Med.* **165:**1565–1580

Leclercq, L., Bismuth, G., and Theze, J., 1984, Antigen specific helper T-cell supernatant is sufficient to induce both polyclonal proliferation and differentiation of small resting B lymphocytes, *Proc. Natl. Acad. Sci. USA* **81:**6491–6495

Leclerq, L., Cambier, J. C., Mishel, Z., Julius, M. H., and Theze, J., 1986, Supernatants from a cloned helper T cell stimulates most small resting B cells to undergo increased I-A expression, blastogenesis and progression through cell cycle, *J. Immunol.* **136:**539–545

Leptin, M., 1985, Monoclonal antibodies specific for murine IgM. II. Activation of B lymphocytes by monoclonal antibodies specific for the four constant domains of IgM, *Eur. J. Immunol.* **15:**131–137

MacDougall, S. L., Grinstein, S., and Gelfand, E. W., 1988, Detection of ligand-activated conductive Ca^{2+} channels in human B lymphocytes, *Cell* **54:**229–234

Miles, K., Atweh, S., Otten, G., Arnason, B. G. W., Chelmicka-Schorr, E., 1984, β-Adrenergic receptors on splenic lymphocytes from axotomized mice, *Int. J. Immunopharmacol.* **6:**1171–1177

Mond, J. J., Feuerstein, N., Finkleman, F. D., Huang, F., Huang, K.-P., and Dennis, G., 1987, B-lymphocyte activation mediated by anti-immunoglobulin antibody in the absence of PKC, *Proc. Natl. Acad. Sci. USA* **89:**8588–8592

Monroe, J. G., 1988, Up-regulation of c-fos expression is a component of the mIg signal transduction mechanisms but is not indicative of competence for proliferation, *J. Immunol.* **140:**1454–1460

Mossman, T., 1983, Rapid colorimetric assay for cellular growth and survival: Application to proliferation and cytotoxicity assay, *J. Immunol. Methods* **65:**55–64

Myers, C. D., Kriz, M. K., Sullivan, T. J., and Vitetta, E. S., 1987, Antigen induced changes in phospholipid metabolism in antigen-binding B lymphocytes, *J. Immunol.* **138:**1705–1711

Nel, A. D., Landroth, G. E., Goldschmidt-Clermont, P. J., Tuno, H. E., and Galbraith, R. M., 1984, Enhanced tyrosine phosphorylation in B lymphocytes upon complexing of membrane immunoglobulin, *Biochem. Biophys. Res. Commun.* **125:**859–866

Nel, A. E., Navailles, M., Rosberger, D. F., Landreth, G. E., Goldschmidt-Clermong, P. J., Baldwin, G. J., and Galbraith, R. M., 1985, Phorbol ester induces tyrosine phosphorylation in normal and abnormal human B lymphocytes, *J. Immunol.* **135:**3448–3453

Nel, A. E., Wooten, M. W., Landreth, G. E., Goldschmidt-Clermont, P. J., Stevenson, H. C., Miller, P. J., and Galbraith, R. M., 1986, Translocation of phospholipid/Ca^{++}-dependent protein kinase in B lymphocytes activated by phorbol ester or crosslinking of membrane immunoglobulin, *Biochem. J.* **233:**145–149

Niederhuber, J. E., Frelinger, J. A., Dugan, E., Coutinho, A., and Shreffler, D. C., 1975, Effects of anti-Ia serum on mitogenic responses. I. Inhibition of proliferative response to B cell mitogen, LPS, by specific anti-Ia sera, *J. Immunol.* **115:**1672–1676

Noelle, R. J., Snow, E. C., Uhr, J. W., and Vitetta, E. S., 1983, Activation of antigen-specific B cells: The role of T cells, cytokines and antigens in the induction of growth and differentiation, *Proc. Natl. Acad. Sci. USA* **80:**6628–6631

O'Garra, A., Umland, S., DeFrance, T., and Christiansen, J., 1988, B-cell factors are pleiotropic, *Immunol. Today* **9:**45–54

Oliver, K., Noelle, R., Uhr, J. W., Krammer, P. H., and Vitetta, E. S., 1985, B-cell growth factor (B-cell growth factor I or B-cell-stimulating factor, provisional 1) is a differentiation factor for resting B cells and may not induce cell growth, *Proc. Natl. Acad. Sci. USA* **82:**2465–2467

Owen, T., 1988, A noncognate interaction with anti-receptor antibody-activated helper T cells induces small resting murine B cells to proliferate and to secrete antibody, *Eur. J. Immunol.* **18:**395–401

Palacios, R., Martinez-Maza, O., and Guy, K., 1983, Monoclonal antibodies against HLA-DR antigens replace T helper cells in activation of B lymphocytes, *Proc. Natl. Acad. Sci. USA* **80:**3456–3460

Plaut, M., 1987, Lymphocyte hormone receptors, *Annu. Rev. Immunol.* **5:**621–671

Poo, W.-J. Conrad, L., and Janeway, C. A., Jr., 1988, Receptor-directed focusing of lymphokine release by helper T cells, *Nature* (London) **332:**370–380

Ransom, J. T., Chen, M., Sandoval, V. M., Pasternak, J. A., DiGiusto, D., and Cambier, J. C., 1988, Increased plasma membrane permeability to Ca^{2+} in anti-Ig-stimulated B lymphocytes is dependent on activation of phosphoinositide hydrolysis, *J. Immunol.* **140:**3150–3155

Riedel, C., Owens, T., and Nossal, G. J. V., 1988, A significant proportion of normal resting B cells are induced to secrete immunoglobulin through contact with anti-receptor antibody-activated helper T cells in clonal cultures, *Eur. J. Immunol.* **18:**403–408

Rock, K. L., Benacerraf, B., and Abbas, A. R., 1984, Antigen presentation by hapten-specific B lymphocytes. I. Role of surface immunoglobulin receptors, *J. Exp. Med.* **160:**1102–1113

Rothstein, T. L., Backer, T. R., Miller, R. A., and Kulber, D. L., 1986, Stimulation of murine B cells by the combination of calcium ionophore plus phorbol ester, *Cell. Immunol.* **102:**364–373

Sanders, V. M., Dernandez-Bottran, R., Uhr, J. W., and Vitetta, E. S., 1987, Interleukin 4 enhances the ability of the antigen-specific B cells to form conjugates with T cells, *J. Immunol.* **139:**2349–2354

Schreiner, G. F., and Unanue, E. R., 1976, Membrane and cytoplasmic changes in B lymphocytes induced by ligand-surface immunoglobulin interactions, *Adv. Immunol.* **24:**38–165

Shivdasani, R. A., and Thomas, D. W., 1988, Molecular associations of Ia antigens after T-B cell interactions. I. Identification of new molecular associations, *J. Immunol.* **141:**1252–1260

Singer, A., and Hodes, R. J., 1983, Mechanisms of T cell-B cell interaction, *Annu. Rev. Immunol.* **1:**211–241

Snapper, C. M., and Paul, W. E., 1987, B cell stimulator factor-1 (interleukin 4) prepares resting murine B cells to secrete IgG1 upon stimulation with bacterial lipopolysaccharide, *J. Immunol.* **139:**10–17

Snow, E. C., Fetherston, J. D., and Zimmes, S., 1986, Induction of the cMyc after antigen binding to hapten-specific B cells, *J. Exp. Med.* **164**:944–950

Wade, W. F., Chen, Z. Z., Maki, R., McKercher, S., Palmer, C., Cambier, J. C., and Freed, J. H., 1989, Altered I-A-mediated transmembrane signaling in B cells that express trunscated I-A^k protein, *Proc. Natl. Acad. Sci. USA* **86**:6297–6301

Whalen, B. J., Tony, H.-P., and Parker, D. C., 1988, Characterization of the effector mechanism of help for antigen-presenting and bystander resting B cell growth mediated by Ia-restricted Th2 helper T cell lines, *J. Immunol.* **141**:2230–2239

Yamamoto, K. K., Gonzales, G. A., Biggs, W. H., III, and Montminy, M. R., 1988, Phosphorylation-induced binding and transcriptional efficacy of nuclear factor CREB, *Nature* (London) **334**:494–498

Zitron, I. M., and Clevinger, B. L., 1980, Regulation of murine B cells through surface immunoglobulin. I. Monoclonal anti-δ-antibody that induce allotype-specific proliferation, *J. Exp. Med.* **152**:1135–1146

Chapter 5

IgE Receptors on Lymphocytes and IgE-Binding Factors

Kwang-Myong Kim, Mitsufumi Mayumi, and Haruki Mikawa

1. INTRODUCTION

There are two types of Fc receptors for IgE (FcϵR): the high-affinity FcϵR (FcϵRI) (reviewed in Metzger, 1988) and the low-affinity FcϵR (FcϵRII) (reviewed in Spiegelberg, 1984). The former is expressed on basophils and mast cells, whereas the latter is expressed on lymphocytes (Lawrence *et al.*, 1975), monocytes and macrophages (A. Capron *et al.*, 1975; Joseph *et al.*, 1978), eosinophils (M. Capron *et al.*, 1981a,b), and platelets (Joseph *et al.*, 1983a). Crosslinkage of FcϵRI causes the release of chemical mediators, which is responsible for immediate allergic reactions, from basophils and mast cells. FcϵRIIs on monocytes, eosinophils, and platelets also seem to participate in the activation of cells and the release of biologically active chemical agents (Dessaint *et al.*, 1979b; Rouzer *et al.*, 1982; Joseph *et al.*, 1983b; Rankin *et al.*, 1984; A. Capron *et al.*, 1985; M. Capron *et al.*, 1985; Jouault *et al.*, 1988). Thus, they seem to be related not only to antiparasitic cytotoxic activities (A. Capron *et al.*, 1975; M. Capron *et al.*, 1981b; Haque *et al.*, 1981; Joseph *et al.*, 1983a) but also to the pathogenesis of allergic disorders (A. Capron *et al.*, 1985, 1986, 1987; M. Capron *et al.*, 1985).

In contrast, the crosslinkage of FcϵRIIs on lymphocytes does not cause the release of chemical mediators. However, FcϵRIIs on lymphocytes have attracted attention because of their possible role in the regulation of IgE synthesis, as suggested by the following: (1) the percentage of FcϵRII$^+$ lymphocytes increased in individuals with elevated serum IgE (Spiegelberg *et al.*, 1979; Thompson *et al.*, 1985; Nagai *et al.*, 1985) and (2) FcϵRII$^+$ lymphocytes were shown to secrete an effector molecule with affinity for IgE [IgE-binding factor (IgE-BF)], which enhanced IgE synthesis selectively [IgE-potentiating factor (IgE-PoF)] (reviewed in Ishizaka, 1988; Yodoi *et al.*, 1986).

Kwang-Myong Kim, Mitsufumi Mayumi, and Haruki Mikawa   Department of Pediatrics, Faculty of Medicine, Kyoto University, Sakyo-ku, Kyoto 606, Japan.

The immunoregulatory activity of Ig-BF is not unique to IgE but is common to other Ig classes such as IgG (Fridman and Golstein, 1974), IgA (Yodoi *et al.*, 1983, 1986; Kiyono *et al.*, 1985), and IgD (Adachi and Ishizaka, 1986). Among them, IgG-BF and IgA-BF were described to act class specifically, as does IgE-BF.

In this chapter, we review the activity of FcεRII on lymphocytes and IgE-BF.

2. HISTORICAL OVERVIEW

The history of the study of FcεRII on lymphocytes and IgE-BF can be divided into three periods. The first is from 1975 to 1980, when FcεRIIs were found on human lymphocytes (Lawrence *et al.*, 1975; Gonzalez-Molina and Spiegelberg, 1976) and the difference in function or antigenicity between FcεRI and FcεRII was confirmed (Meinke *et al.*, 1978). The percentage of FcεRII$^+$ lymphocytes is increased and positively correlated with serum IgE levels in atopic patients with elevated serum IgE levels (Spiegelberg *et al.*, 1979). This finding suggests the relationship of FcεRII$^+$ lymphocytes to increased IgE synthesis.

The second period is from 1980 to 1984, when the participation of FcεRII$^+$ lymphocytes in the up regulation of IgE synthesis was demonstrated in rats. The discovery of IgE-BFs was an epoch-making event (Yodoi and Ishizaka, 1980b). IgE-BFs were soluble T-cell factors with affinity for IgE and biologic functions to regulate IgE synthesis class specifically. During this period, the framework of a hypothesis about the IgE-BF-mediated regulation of IgE synthesis, whereby FcεRII$^+$ T lymphocytes play a critical role as a source of IgE-BF, was established.

We are now in the third period. Research is again concentrated on human materials. FcεRII$^+$ B as well as T lymphocytes were demonstrated to produce IgE-BF in humans (Young *et al.*, 1984; Sarfati *et al.*, 1984b). Monoclonal antibodies for human FcεRII were developed (Rector *et al.*, 1985; Suemura *et al.*, 1986; Noro *et al.*, 1986; M. Capron *et al.*, 1986; Bonnefoy *et al.*, 1987). Study with monoclonal anti-FcεRII antibodies revealed that FcεRII has antigenic determinants in common with CD23 antigen (Yukawa *et al.*, 1987; Bonnefoy *et al.*, 1987), which has been studied as an activation marker of B cells (Kintner and Sugden, 1981; Thorley-Lawson and Mann, 1985; Thorley-Lawson *et al.*, 1985). A cDNA clone of human FcεRII was isolated (Kikutani *et al.*, 1986a; Ikuta *et al.*, 1987; Lüdin *et al.*, 1987). Recently, regulation of the expression of FcεRII by cytokines such as interleukin 4 (IL4) and gamma interferon (IFN-γ) has attracted attention (Náray-Fejes-Tóth and Guyre, 1984; Kikutani *et al.*, 1986a,b).

3. FcεRII ON LYMPHOCYTES

3.1. FcεRII-Bearing Cells

FcεRIIs were first demonstrated on human B lymphocytes (Lawrence *et al.*, 1975; Gonzalez-Molina and Spiegelberg, 1976). To date, they have been found on lymphocytes of humans (Gonzalez-Molina and Spiegelberg, 1977), rats (Fritsche and Spiegelberg, 1978), and mice (Chen *et al.*, 1981; Vander-Mallie *et al.*, 1982), on monocytes and macrophages of humans (Joseph *et al.*, 1978; Spiegelberg and Melewicz, 1980; Melewicz and Spiegelberg, 1980), rats (A. Capron *et al.*, 1975, 1977; Dessaint *et al.*, 1979a), and mice (Boltz-Nitulescu *et al.*, 1982; Daeron and Ishizaka, 1986), on eosinophils of humans and rats (M. Capron *et al.*, 1981a,b, 1984), and on platelets of humans and rats (Joseph *et al.*, 1983a).

The percentage of FcεRII$^+$ cells in lymphocyte populations differs with the tissue. In rats, Fritsche and Spiegelberg (1978) reported a higher percentage of FcεRII$^+$ cells in the spleen (18.7%) and peripheral blood lymphocytes (13.3%) than in the lymph nodes (4.1%) or thoracic duct (0%). In humans, Hellström and Spiegelberg (1979) noted that the percentage of FcεRII$^+$ cells was higher in the tonsils (4.2%) and adenoids (5.8%) than in the peripheral blood lymphocytes (1.2%). Kikutani *et al.* (1986b) reported 11.2% in peripheral blood lymphocytes, 28.4% in tonsils, and 0.5% in bone marrow mononuclear cells. In tonsils, there are more FcεRII$^+$ lymphocytes in the germinal center than in the mantle (Kanowith-Klein *et al.*, 1988).

The percentage of FcεRII$^+$ cells in normal human peripheral blood lymphocytes varied considerably among investigators, from approximately 1% to 8%. The low density of FcεRIIs on the nonactivated cell surface may be the main reason for such variations. In our own investigation, the percentage of FcεRII$^+$ cells in normal adults is 4.6%, the same as in normal children, including neonates and infants (Kim *et al.*, 1988). Because the percentage of FcεRII$^+$ cells in cord blood is the same as that in adults (Delespesse *et al.*, 1986a), it is considered to be quite stable in normal individuals during most of their lifetime. However, FcεRII$^+$ lymphocytes are absent in the human fetus at 11–22 weeks of gestation (Kanowith-Klein *et al.*, 1988).

The percentage of FcεRII$^+$ cells was shown to be high in severely atopic patients with extremely high serum IgE levels (Spiegelberg *et al.*, 1979). It also increased in patients with allergic rhinitis during grass pollen season (Spiegelberg and Simon, 1981). The increase in the percentage of FcεRII$^+$ lymphocytes in atopic individuals was detected more easily in younger children than in older children and adults (Kim *et al.*, 1988). The increase in the percentage of FcεRII$^+$ was also shown in patients with hyperimmunoglobulinemia E (Nagai *et al.*, 1985; Thompson *et al.*, 1985).

Lebrun *et al.* (1988) noted that the percentages of FcεRII$^+$ lymphocytes were different among murine strains. However, although infection with the parasite *Nippostrongylus brasiliensis* caused an elevation of serum IgE (Ogilvie and Johns, 1969; Jarret and Bazin, 1974), such differences in the percentage of FcεRII$^+$ lymphocytes did not correlate with the degree of IgE response to *N. brasiliensis* infection (Lebrun *et al.*, 1988).

Most FcεRII$^+$ lymphocytes are thought to be B cells (Gonzalez-Molina and Spiegelberg, 1977; Hellström and Spiegelberg, 1979; Fritsche and Spiegelberg, 1978; Vander-Mallie *et al.*, 1982); there is controversy about the presence of FcεRII$^+$ T cells in humans. FcεRII$^+$ T cells were shown to be present in rats (Yodoi and Ishizaka, 1979a; Yodoi *et al.*, 1979) and in mice (Chen *et al.*, 1981; Marcelletti and Katz, 1984a–d; Mathur *et al.*, 1986). Ishizaka and co-workers claim that FcεRII$^+$ T cells are also present in humans (Yodoi and Ishizaka, 1979a; Ishizaka and Sandberg, 1981). Spiegelberg *et al.* (1979), Young *et al.* (1984), and Delespesse *et al.* (1986a) also noted the presence of FcεRII$^+$ T cells in humans.

On the contrary, Kishimoto and co-workers state that all FcεRII$^+$ lymphocytes are B cells and not T cells except for a human T-cell leukemia virus (HTLV)-transformed T-cell line (Suemura *et al.*, 1986; Kikutani *et al.*, 1986a,b). They found no FcεRII⁺ T cells even in patients with atopic disorders or hyper-IgE syndrome. Moreover, although FcεRIIs could be induced on B cells by some nonviral agents, these researchers failed to induce FcεRII on T cells by such stimuli. They confirmed these results by Northern blot analysis of mRNA of FcεRII, which increased the reliability of their results.

On the basis of these results, FcεRII was postulated to be a differentiation marker expressed specifically on mature B cells (Kikutani *et al.*, 1986b). The authors noted that FcεRIIs were expressed exclusively on surface IgM (sIgM)⁺ sIgD⁺ mature B cells, not on either premature and immature B cells from bone marrow or mature B cells after isotype switching, which bore sIgG or sIgA alone. Waldschmidt *et al.* (1988) reported a similar finding for murine B cells.

It seems valid to conclude that T cells also can express FcεRIIs. However, because the percentage of FcεRII⁺ T cells is very low, a method other than routine immunofluorescent analysis or rosette formation should be devised to clearly demonstrate the presence of FcεRII⁺ T cells in individuals without HTLV infection.

3.2. Function of FcεRII

Little is known about the function of Fcεll on lymphocytes, but the binding of IgE to FcεRII increases with the secretion of IgE-BFs (Yodoi and Ishizaka, 1980b; Ishizaka and Sandberg, 1981). The binding of monoclonal anti-FcεRII antibodies to FcεRII increased the amount of mRNA of FcεRII in an HTLV-I-transformed human T-cell line (Kawabe *et al.*, 1988). Sherr *et al.* (1989) noted that the crosslinkage of FcεRII caused the suppression of ongoing IgE synthesis of a human myeloma cell line, U266/AF10. These findings indicate that FcεRII can transduce some kinds of signals. FcεRII is also likely to be related to the growth promotion of B cells (Gordon *et al.*, 1986a,b) (see Section 7).

Very recently, Kishimoto and co-workers reported that the transfection with cDNA of FcεRII to Epstein-Barr virus nuclear antigen (EBNA)⁻ FcεRII⁻ B-lymphoma cells caused not only the expression of FcεRIIs but also the formation of cellular aggregates, as did infection with Epstein-Barr virus (EBV) (presented at the 18th meeting of the Japanese Society for Immunology, 1988). They postulated that FcεRII was related to cell adhesion.

3.3. Molecular Properties of FcεRII

FcεRII is a different molecule from FcεRI (reviewed in Metzger, 1988). FcεRII consists of a single component (Meinke *et al.*, 1978; Lee and Conrad, 1985), whereas FcεRI contains three components: the α component is expressed on cell surfaces and has an affinity for IgE, whereas the other two, β and γ, are buried in the membrane and have no affinity for IgE (Metzger *et al.*, 1984; Metzger, 1988). The affinity of FcεRII for IgE is about 100-fold lower than that of FcεRI: approximate K_as are 10^7/M for FcεRII (Table I) and 10^{9-10}/M for FcεRI (T. Ishizaka, *et al.*, 1973; Kulczycki and Metzger, 1974; Conrad *et al.*, 1975). Other differences are as follows: antigenic determinant (Meinke *et al.*, 1978), molecular weight (45–50 kDa for FcεRII and 58 kDa for FcεRIα) (Hempstead *et al.*, 1979; Conrad and Peterson, 1984), and isoelectric point (pH 4–5 for FcεRII and pH 6–7 for FcεRIα) (Conrad and Peterson, 1984).

There are no significant differences in molecular weight or isoelectric point between human FcεRII and murine FcεRII (Conrad and Peterson, 1984) (Table I). FcεRIIs expressed on human lymphocytes have the same biochemical characteristics and antigenic determinants as those on human monocytes (Melewicz et al., 1982; Sarfati et al., 1986b), eosinophils, and platelets (Spiegelberg, 1984; A. Capron et al., 1986; Jouault et al., 1988).

The major structure of FcεRII consists of a single polypeptide chain (Meinke et al., 1978). Analysis of human FcεRII by SDS-PAGE usually shows three bands: 86, 47, and 23 kDa (Meinke et al., 1978). Of these, the 47-kDa band is the most prominent and most consistently observed. Proteolytic analysis suggested that the 86-kDa band is a dimer of 47-kDa FcεRII and that the low-molecular-weight band consists of degradation products of 47-kDa FcεRII that retained affinity for IgE (Peterson and Conrad, 1985). The affinity of FcεRII for IgE was lost upon treatment with trypsin (Gonzalez-Molina and Spiegelberg, 1976, 1977). Because tunicamycin, an inhibitor of N-linked glycosylation, inhibited the induction of FcεRII, some oligosaccharide moieties were suggested to be assembled by N-linked glycosylation to the precursor of FcεRII (Yodoi et al., 1981a).

A fine analysis of the biosynthesis of murine FcεRII was performed by Keegan and Conrad (1987). They demonstrated that 49-kDa FcεRII was synthesized from a 36-kDa polypeptide chain. The increase in molecular weight was ascribed mainly to the assembly of N-linked oligosaccharide moieties. Although the biosynthesis of precursor protein was not stopped by tunicamycin treatment, the product (38-kDa FcεRII) could not be expressed on the cell surface.

Letellier et al. (1988) studied the biosynthesis of human FcεRII. They noted that 45-kDa FcεRII contained one N-linked and several O-linked carbohydrate moieties. Although treatment with tunicamycin did not inhibit the biosynthesis of FcεRII, the products were much more susceptible to proteolytic enzymes and were shed rapidly from the cell surface. This finding suggested that the N-linked carbohydrate moiety provided the stability to human FcεRII.

Table I
Molecular Properties of Fcε RII

Cell	Mol. wt. (kDa)	Affinity K_a ($\times 10^7$/M)	Isoelectric point	Binding sites/cell	Reference[a]
Human					
Lymphocytes	47	8.1, 2.8	4.5–5.0	5×10^5, 8×10^5	1
Monocytes	47	7.1, 1.1		5×10^5, 16×10^5	2
Eosinophils	45–50	1.2		5×10^4	3
Platelets	43–45	3.0		600–1000	4
Rat					
Lymphocytes		14		5.9–7.7×10^5	5
Macrophages	54	1–9		4–5×10^4	6
Mouse					
Lymphocytes	45–50	9.5	4.5–5.0	5100	7
Macrophages		0.8		3.3×10^5	8

[a]1, Meinke et al., 1978; Melewicz et al., 1982; Peterson and Conrad, 1985; 2, Melewicz et al., 1982; 3, Jouault et al., 1988; 4, M. Capron et al., 1986; Joseph et al., 1986; A. Capron et al., 1987; Pancré et al., 1988; 5, Manouvriez et al., 1985; 6, Finbloom and Metzger, 1982, 1983; 7, Vander-Mallie et al., 1982; Peterson and Conrad, 1985; 8, Boltz-Nitulescu et al., 1982.

A cDNA clone of human FcεRII was recently isolated (Kikutani *et al.*, 1986a; Ikuta *et al.*, 1987; Lüdin *et al.*, 1987). Human FcεRII contains 321 amino acids, and its NH$_2$-terminal end is oriented toward the cytoplasm. It contains one N-linked glycosylation site. The sequence of amino acids in human FcεRII resembles that of animal lectins, which are involved in pinocytosis. However, it has no similarity to that of rat IgE-BF, (Martens *et al.*, 1985), which was shown to have antigenic determinants common to those of rat FcεRII (Huff *et al.*, 1984).

FcεRII was suggested to be expressed on the cell surface in an associated form with HLA-DR antigen (Bonnefoy *et al.*, 1988b).

Recently, the heterogeneity in the structure of FcεRII was noted by Mathur *et al.* (1988) and Rao *et al.* (1987). Although a polyclonal anti-murine FcεRII antibody reacted to FcεRII on both T and B cells, a monoclonal anti-murine FcεRII antibody specific for FcεRII on B cells failed to react to FcεRII on T cells. This finding suggested that FcεRII on B cells was not identical to that on T cells. Yokota *et al.* (1988) described that there are two types of human FcεRII (FcεRIIa and FcεRIIb) on cells of different intracellular structure. Nonstimulated normal B cells bore FcεRIIa exclusively, whereas nonstimulated B cells of atopic patients expressed FcεRIIb as well as FcεRIIa.

3.4. Regulation of FcεRII Expression on Lymphocytes

3.4.1. Rodents

Infection of rats with the parasite *N. brasiliensis* caused an increase of FcεRII$^+$ lymphocytes (Yodoi *et al.*, 1979). Incubation of normal rat lymphocytes with homologous IgE also caused an increase in the number of FcεRII$^+$ cells (Yodoi and Ishizaka, 1979b; Yodoi *et al.*, 1979). Because *N. brasiliensis*-infection also caused an elevation of serum IgE (Ogilvie and Johns, 1969; Jarret and Bazin, 1974), IgE may play a role in the induction of FcεRII$^+$ lymphocytes in *N. brasiliensis*-infected rats (Yodoi *et al.*, 1979).

Glucocorticoids inhibit the induction of rat FcεRII by infection with *N. brasiliensis* or incubation with IgE (Yodoi *et al.*, 1981b). Glucocorticoids inhibit phospholipase A$_2$ (PLA$_2$) by inducing a PLA$_2$-inhibiting protein, lipomodulin. A monoclonal anti-lipomodulin antibody that removes the PLA$_2$-inhibiting activity from lipomodulin was found to enhance the induction of FcεRII by IgE, as did mellitin, a PLA$_2$ activator (Yodoi *et al.*, 1981c). Thus, the participation of PLA$_2$ in the regulation of FcεRII was suggested.

Glycosylation-enhancing factor (GEF) increased the proportion of FcεRII$^+$ cells (Huff *et al.*, 1983). In contrast, glycosylation-inhibiting factor (GIF) and tunicamycin, an N-linked glycosylation inhibitor, suppressed the induction of FcεRII by IgE (Hirashima *et al.*, 1981c; Yodoi *et al.*, 1981a) (see Section 5).

Chen *et al.* (1981) noted that the level of FcεRII$^+$ lymphocytes increased in *N. brasiliensis*-infected mice. They and Marceletti and Katz (1984a–d) reported that FcεRIIs were also induced on both murine T and B lymphocytes by incubation with IgE. Richards *et al.* (1988) showed that the IgE-antigen complexes were more potent inducers of murine FcεRII than IgE alone. Mathur *et al.* (1988) noted that injection of IgE-secreting hybridoma cells in mice caused not only an increase in serum IgE but also an increase in the percentage of Lyt-2$^+$ FcεRII$^+$ T cells. Treatment of murine B cells with the monoclonal anti-murine FcεRII antibody failed to induce FcεRII (Lee and Conrad, 1987; Rao *et al.*, 1987).

Lee *et al.* (1987) showed that the increase in the number of FcεRII on murine B cells after incubation with IgE was due not to the increase in biosynthesis but to the decrease in decay. This finding indicated that the binding of IgE to FcεRII increased the resistance of FcεRII against proteolytic enzymes.

IL4 and IFN-γ were shown to induce and suppress, respectively, murine FcεRII expression (Hudak *et al.*, 1987; Conrad *et al.*, 1987) (see Section 6).

Recently, the behavior of Ia antigen was shown to be similar to that of FcεRII (Richards *et al.*, 1988).

The agents regulating FcεRII expression on lymphocytes are summarized in Table II.

3.4.2. Humans

In humans, FcεRIIs are not induced on lymphocytes by culture with IgE alone. However, incubation of lymphocytes from ragweed-sensitive allergic patients with IgE after activation with ragweed antigen E increased the number of FcεRII[+] cells (Yodoi and Ishizaka, 1980a). FcεRIIs could also be induced on normal lymphocytes by incubation with IgE after activation with mixed lymphocyte culture (Yodoi and Ishizaka, 1980a).

Kawabe *et al.* (1988) noted that the binding of IgE as well as monoclonal anti-FcεRII antibody to FcεRII caused an increase in the number of FcεRIIs on some FcεRII[+] human T- and B-cell lines.

Lectins such as phytohemagglutinin (PHA)-P, concanavalin A (ConA), pokeweed mitogen (PWM) and lentil lectin, and lymphocytosis-promoting factor (LPF) also induced FcεRII expression on normal human lymphocytes (Delespesse *et al.*, 1986b; Kim *et al.*, 1987). Stimulation of peripheral blood mononuclear cells with lectins and LPF induced FcεRIIs mainly on B cells, aided by the soluble factor secreted by T cells (Delespesse *et al.*, 1986b; Kim *et al.*, 1987). The cell-free supernatant of PHA-stimulated human T cells (PHA-sup) could also induce FcεRIIs on B cells, and IgE enhanced the induction of FcεRIIs by PHA-sup (Suemura *et al.*, 1986; Delespesse *et al.*, 1986b).

Table II
Regulatory Agents of Fcε RII Expression on Lymphocytes[a]

Inducing/enhancing agents
IgE,[1] parasite,[2] virus (EBV,[3] HTLV-I[4]), lectins (ConA, PHA-P, PMW, lentil lectin),[5] LPF,[6] GEF,[7] adjuvant (alum),[8] enzymes (trypsin, plasmin, kallikrein),[9b] mellitin,[10] TPA,[11] lysolectin,[12] bradykinin,[13b] antilipomodulin antibody,[14] IL4,[15] anti-Fcε RII antibody,[16] specific antigen,[17] mixed lymphocyte culture,[17] cyclophosphamide[18b]

Inhibiting/suppressing agents
Glucocorticoids,[19] GIF,[20] adjuvant (CFA),[21] tunicamycin,[22] lipomodulin,[23] IFN-γ,[24] transforming growth factor-β,[25] prostaglandin E$_2$,[26] forskolin,[27] 8-bromo-cyclic AMP,[27] cholera toxin,[27] anti-IL4 antibody[28]

[a]References: 1, Yodoi *et al.*, 1979; Yodoi and Ishizaka, 1980b; Chen *et al.*, 1981; 2, Yodoi *et al.*, 1979b; 3, Sarfati *et al.*, 1984b; Thorley-Lawson *et al.*, 1985; 4, Nutman *et al.*, 1987; 5, Yodoi *et al.*, 1981a; Delespesse *et al.*, 1986b; Kim *et al.*, 1987; 6, Kim *et al.*, 1987; 7, Huff *et al.*, 1983; 8, Uede and Ishizaka, 1982b; 9, Iwata *et al.*, 1983b; 10, Yodoi *et al.*, 1981c; 11, Akasaki *et al.*, 1985; Carlsson *et al.*, 1988; 12, Yodoi *et al.*, 1981c; Huff *et al.*, 1983; 13, Iwata *et al.*, 1983b; 14, Yodoi *et al.*, 1981c; 15, Kikutani *et al.*, 1986a; DeFrance *et al.*, 1986; Conrad *et al.*, 1987; 16, Kawabe *et al.*, 1988; 17, Yodoi and Ishizaka, 1980a; 18, Akasaki and Ishizaka, 1987; 19, Spiegelberg *et al.*, 1979; Yodoi *et al.*, 1981c; Kim *et al.*, 1987; 20, Hirashima *et al.*, 1981c; Uede *et al.*, 1983b; 21, Hirashima *et al.*, 1981c; 22, Yodoi *et al.*, 1981a; 23, Uede *et al.*, 1983b; 24, Kikutani *et al.*, 1986b; DeFrance *et al.*, 1987; Conrad *et al.*, 1987; 25, Petit-Koskas *et al.*, 1988; 26, Rousset *et al.*, 1988; Galizzi *et al.*, 1988; 27, Galizzi *et al.*, 1988; 28, Conrad *et al.*, 1988.
[b]Potential inducers of FCε RII because they are equivalent in activity to GEF or induce the secretion of GEF.

Although Suemura *et al.* (1986) noted that FcεRIIs could not be induced on T cells, Ishizaka and Sandberg (1981) reported that incubation of human T cells that had been precultured in medium containing PHA-sup for longer than 8 days with human IgE caused an increase in the percentage of FcεRII[+] T cells. Delespesse *et al.* (1986b) described the interesting observation that although the percentage of FcεRII[+] B cells reached a peak within 2 days after the initiation of PHA stimulation, the percentage of FcεRII[+] T cells reached the maximum at about 7 days after stimulation. These results suggest a possible difference between T and B cells in the kinetics of FcεRII induction by PHA.

Some viruses were shown to induce FcεRII expression on lymphocytes. Infection of normal B cells with EBV has been shown to cause a marked increase in the number of FcεRIIs (Sarfati *et al.*, 1984b). The relationship of FcεRII (specifically as CD23 antigen) to infection with EBV has been studied precisely (see Section 7). Also, some HTLV-I-transformed T-cell lines were shown to bear FcεRII (Kikutani *et al.*, 1986a; Nutman *et al.*, 1987; Kawabe *et al.*, 1988). Nutman *et al.* (1987) showed that the number of FcεRIIs on T cells increased after infection with HTLV-I.

Recently, some recombinant human cytokines have become available. Among them, IL4 was shown to induce FcεRIIs on B cells (Kikutani *et al.*, 1986a; DeFrance *et al.*, 1987). The activation of B cells by anti-IgM antibody enhanced the induction of FcεRIIs by IL4 (Bonnefoy *et al.*, 1988a; Delespesse *et al.*, 1989). In contrast to IL4, recombinant IFN-γ was shown to inhibit the induction of FcεRIIs on B cells by IL4 (Kikutani *et al.*, 1986b; DeFrance *et al.*, 1987) (see Section 6). Prostaglandin E_2, 8-bromo-cyclic AMP, forskolin, and cholera toxin also inhibited the induction of FcεRIIs by IL4 through an increase of intracellular cyclic AMP (Rousset *et al.*, 1988; Galizzi *et al.*, 1988).

Corticosteroids are thought to suppress the expression of FcεRII because the percentage of FcεRII[+] cells in atopic patients medicated with corticosteroids was significantly lower than in those without such medication (Spiegelberg *et al.*, 1979). Indeed, dexamethasone suppressed the induction of FcεRIIs on human lymphocytes by LPF or PHA (Kim *et al.*, 1987). Dexamethasone inhibits not only the increase of FcεRIIs on B cells but also the production of a T-cell factor that induces FcεRIIs on B cells (Kim *et al.*, 1987).

Recently, Rousset *et al.* (1988) showed that the expression of class II major histocompatibility complex (MHC) antigen (HLA-DQ) was regulated in the same manner as FcεRII.

4. IgE-BINDING FACTORS

4.1. Rat

4.1.1. Characteristics

During the 1970s, many investigators described soluble T-cell factors that regulated IgE synthesis in a class-specific manner (reviewed in Katz, 1980). Ishizaka and coworkers also noted that a rat T-cell factor (IgE-PoF) up regulated IgE synthesis class specifically (Urban *et al.*, 1977; Suemura and Ishizaka, 1979). At the end of the 1970s, they found that this IgE-PoF had an affinity for IgE (IgE-BF), which made it possible to identify the factor (Yodoi and Ishizaka, 1980b). IgE-PoF was demonstrated to be secreted by FcεRII⁺ T cells (Suemura *et al.*, 1980; Hirashima *et al.*, 1981a). This discovery stimulated study of the regulation of rat IgE synthesis, which is now helpful in our studies on human IgE synthesis.

After the discovery of IgE-PoF, another component of IgE-BFs, which was named IgE-suppressing factor (IgE-SuF) because of its inhibiting activity in IgE synthesis, was found (Hirashima *et al.*, 1980a). The source of IgE-SuF is not FcεRII⁺ T cells (Hirashima *et al.*, 1980b).

Two IgE-BFs arise from the same precursor (Yodoi *et al.*, 1981a). They have the same molecular weight of 10–20 kDa (Yodoi and Ishizaka, 1980b; Hirashima *et al.*, 1980a) and common antigenic determinants (Huff *et al.*, 1984) (Table III). They act on B cells to regulate IgE synthesis by binding to surface IgE (Suemura *et al.*, 1980; Uede *et al.*, 1984). IgE-BFs are glycoprotein molecules, as is FcεRII (Yodoi *et al.*, 1980). Because the biologic function of IgE-BFs is lost upon treatment with neuraminidase, carbohydrate moieties are essential for their biologic activity (Yodoi *et al.*, 1980). However, their carbohydrate moieties are different; whereas IgE-PoF binds to lentil lectin and ConA, IgE-SuF binds instead to peanut agglutinin (Yodoi *et al.*, 1980, 1982; Hirashima *et al.*, 1980a). N-linked carbohydrate moieties are essential for IgE-PoF but not for IgE-SuF, such tunicamycin, an inhibtor of N-linked glycosylation, inhibits the formation of IgE-PoF but not IgE-SuF (Yodoi *et al.*, 1981a). The nature of IgE-BFs was determined under physiologic conditions by glycosylation-regulating factors, which regulate N-linked glycosylation during the biosynthesis of IgE-BFs (see Section 5).

The establishment of rat–mouse T-cell hybridomas that secreted rat IgE-BFs facilitated analysis of the molecular properties of IgE-BFs (Huff *et al.*, 1982). The hybridoma secreted IgE-BFs of 60, 30, and 14 kDa in the presence of IgE. Reduction and alkylation of the 60-kDa IgE-BF yielded 30-, 14-, and 10-kDa IgE-BFs. This finding indicated that the 30-, 14-, and 10-kDa IgE-BFs were derived from the 60-kDa IgE-BF (Jardieu *et al.*, 1985b).

The 60- and 30-kDa but not 14- or 10-kDa IgE-BFs were found to react with

Table III
Molecular Properties of IgE-binding factors

Species	Mol. wt. (kDa)	Method	Cell source	N-linked glycosylation	Reference[a]
Human	60, 30, 15	Gel	T	+	1
	30–40, 10–15,	Gel	B	+	2
	25–27, 12–16[b]	SDS-PAGE	T, B	−	3
	14[c]	SDS-PAGE	?	?	4
Rat	60, 30, 10–15	Gel	T	+	5
Mouse	60, 30, 15	Gel	T	+	6
	38[b]	SDS-PAGE	B	?	7

[a]1, Huff *et al.*, 1986; Kisaki *et al.*, 1987, 1988; Young *et al.*, 1986; Stadler *et al.*, 1987; 2, Sarfati *et al.*, 1984b; 3, Peterson and Conrad, 1985; Kikutani *et al.*, 1986a; Ikuta *et al.*, 1987; Lüdin *et al.*, 1987; Nakajima *et al.*, 1987; 4, Sarfati *et al.*, 1986a; 5, Yodoi and Ishizaka, 1980b; Hirashima *et al.*, 19880a; Huff *et al.*, 1982; Jardieu *et al.*, 1985b; Martens *et al.*, 1985; 6, Uede and Ishizaka, 1984; Jardieu *et al.*, 1985a; 7, Lee *et al.*, 1987; Keegan and Conrad, 1987.
[b]sFcε RIIs which are isolated with anti-Fcε RII-affinity column and shown to be degradation products of FcεRII. Other IgE-BFs are isolated with IgE- or lectin-affinity column.
[c]IgE-SuF purified from colostrum.

monoclonal anti-Ia antibody (Jardieu *et al.*, 1985b). The Ia determinant of IgE-BFs is likely associated with an N-linked glycosylation site because IgE-BFs produced in the presence of tunicamycin, an inhibitor of N-linked glycosylation, were found to lack this determinant (Martens *et al.*, 1987).

cDNA clones of rat IgE-BFs were isolated by using these rat–mouse T-cell hybridomas (Martens *et al.*, 1985). cDNA clones encoded two IgE-BFs with molecular weights of 60 and 11 kDa. The 11-kDa IgE-BF was likely a breakdown product of the 60-kDa IgE-BF, since reduction and alkylation of the 60-kDa IgE-BF yielded 11-kDa IgE-BF (Jardieu *et al.*, 1985b). The larger molecule consisted of 556 amino acids and contained two potential N-linked glycosylation sites. The genes were shown to be members of an endogenous retroviruslike intracisternal A-particle gene family (Moore and Martens, 1986; Moore *et al.*, 1986; Kuff *et al.*, 1986; Ymer and Young, 1986; Ymer *et al.*, 1986). Both COS-7 monkey kidney cells and Chinese hamster ovary cells transfected with these cDNAs secreted IgE-BFs. The nature of IgE-BFs could be modulated by glycosylation-regulating factors (Martens *et al.*, 1987), which confirmed that IgE-PoF and IgE-SuF were derived from the same precursor.

4.1.2. Formation

FcεRII[+] T cells collected from rats infected with the nematode *N. brasiliensis* spontaneously released IgE-BFs (Yodoi and Ishizaka, 1980b). Normal rat lymphocytes cultured with IgE also produced IgE-BFs (Yodoi and Ishizaka, 1980b).

Under physiologic conditions, rat IgE-BFs were formed by T cells upon stimulation with IgE-BF-inducing factor, which was secreted by adherent cells or lymphocytes (Yodoi *et al.*, 1981d; Hirashima *et al.*, 1981b,c). IgE-BF-inducing factor had interferonlike properties and was inactivated by anti-mouse type I interferon antibody (Yodoi *et al.*, 1981d).

IgE-BFs induced by IgE-BF-inducing factor were composed of an equal volume of IgE-PoF and IgE-SuF and thus had no biologic activity. Treatment of rats with *Bordetella pertussis* (BP) vaccine cause the spontaneous secretion of three factors: (1) IgE-BF-inducing factor from adherent cells and lymphocytes, (2) GEF from lymphocytes, and (3) IgE-BFs consisting mainly of IgE-PoF from T cells (Hirashima *et al.*, 1981b). The increase in the proportion of IgE-PoF in IgE-BFs was due to the concomitant secretion of GEF. On the contrary, repetitive injection of rats with complete Freund's adjuvant (CFA) alone caused the secretion of GIF as well as IgE-BF-inducing factor and IgE-BFs (Hirashima *et al.*, 1981c). As a result, the major component of IgE-BFs was IgE-SuF in this case.

Different effects of adjuvants on IgE antibody response were shown to be mediated by glycosylation-regulating factors and IgE-BFs with opposite activities: T cells of rats immunized with keyhole limpet hemocyanin (KLH)–alum secreted GEF and IgE-PoF upon antigenic stimulation, whereas T cells immunized with KLH–CFA secreted GIF and IgE-SuF (Uede and Ishizaka, 1982; Uede *et al.*, 1982).

Cell sources of these factors were analyzed. In BCG-primed rats, IgE-BF-inducing factor and GIF were derived from W3/25[−] T cells, whereas IgE-SuF was derived from W3/25[+] T cells (Hirashima *et al.*, 1982). In KLH–alum-primed rats, the IgE-BF-inducing factor was derived from W3/25[+] FcγR[−] T cells, whereas GEF and IgE-BF were derived from W3/25$^+$ FcγR$^+$ T cells (Uede and Ishizaka, 1982). In rats immunized with KLH–CFA, GIF was derived from W3/25$^-$ OX8$^+$ T cells (Uede and Ishizaka, 1982). The reason for the difference in the cell source of IgE-BF-inducing factor is unknown.

4.2. Murine

4.2.1. Characteristics

Murine IgE-BFs also consist of IgE-PoF and IgE-SuF of 15 and 40–60 kDa (Uede *et al.*, 1983a) (Table III). The source of IgE-BFs was FcγR$^+$ Lyt-1$^+$ T cells, and their biologic activities could be modulated by glycosylation-regulating factors, as was that in rats (Uede *et al.*, 1983a; Uede and Ishizaka, 1984).

Jardieu *et al.* (1985a) established antigen-specific murine T-cell hybridomas by the use of antigen-specific T cells collected from mice immunized with ovalbumin (OA). These hybridomas secreted not only IgE-BFs of 60, 30, and 15 kDa but also IgG-BFs, which had not been described in rats, upon stimulation with a specific antigen OA. The 60- and 15-kDa IgE-BFs were IgE-SuFs. IgG-BFs also inhibited IgG synthesis class specifically. The same hybridomas could secret IgE-PoF but not IgG-PoF upon antigenic stimulation in the presence of GEF.

Suemura *et al.* (1981) noted that IgE-TsF, a T-cell-derived IgE-BF, bore the H-2 gene products, which resembled rat IgE-BFs.

Recently, B-cell-derived IgE-BF was described (Table III). B-cell-derived IgE-BFs, which were isolated by using an affinity column coupled with anti-FcεRII antibodies, had a molecular weight of 38 kDa and were shown to be degradation products of FcεRIIs (sFcεRII) (Lee *et al.*, 1987; Keegan and Conrad, 1987). Murine sFcεRII failed to regulate IgE synthesis (Conrad *et al.*, 1988).

4.2.2. Formation

The incubation of murine FcγR$^+$ Lyt-1$^+$ T cells with mouse or rat IgE caused the secretion of IgE-BFs (Uede *et al.*, 1983a). The production of murine IgE-BFs was also induced, as was that in rats, by an interferonlike molecule, IgE-BF-inducing factor (Uede *et al.*, 1983a; Uede and Ishizaka, 1984). IgE-BF-inducing factor was secreted by normal adherent cells upon stimulation with polyinosinic-polycytidylic acid and by Lyt-1$^+$ T cells of OA-primed mice upon stimulation with OA.

Interestingly, the nature of IgF-BFs induced by IgE or IgE-BF-inducing factor varied among murine strains: IgE-PoF was secreted in high-IgE-responder BDF1 mice, IgE-SuF was secreted in low-IgE-responder SJL mice, and a mixture of equal amounts of IgE-PoF and IgE-SuF was secreted in BALB/c mice (Uede and Ishizaka, 1984).

Stimulation of FcγR$^+$ Lyt-1$^+$ T cells of BDF1 mice immunized with alum-absorbed OA with a specific-antigen OA caused the secretion of IgE-PoF, whereas the same subset of T cells of BDF1 mice immunized with OA alone secreted IgE-SuF after the same stimulation (Jardieu *et al.*, 1984). The difference in the nature of IgE-BFs was due to the difference in the nature of glycosylation-regulating factors: GEF was secreted by Lyt-1$^+$ T cells in the former, whereas GIF was secreted by Lyt-2$^+$ I-J$^+$ antigen-specific suppressor T cells in the latter (Uede and Ishizaka, 1984; Jardieu *et al.*, 1984). Thus, the effect of adjuvant was suggested to be mediated by glycosylation-regulating factors, as was that in rats.

Kishimoto and co-workers described that T cells of BALB/c mice primed with dinitrophenyl (DNP)-coupled mycobacterium secreted IgE class-specific suppressor factor (IgE-TsF) upon stimulation with specific antigen (Kishimoto *et al.*, 1976, 1978; Suemura *et al.*, 1977). In 1981, Suemura *et al.* showed that this IgE-TsF had an affinity for IgE, namely, IgE-SuF.

Marcelletti and Katz (1984a–d, 1986) also described murine IgE-PoF and IgE-SuF. They noted that IgE-PoF was derived from FcεRII$^+$ Lyt-2$^+$ T cells, whereas IgE-SuF was derived from FcεRII$^-$ Lyt-1$^+$ T cells.

4.3. Human

4.3.1. Characteristics

Ishizaka and Sandberg (1981) noted the presence of human T-cell-derived IgE-BFs. The IgE-BFs were converted into IgE-PoF and IgE-SuF by GEF and GIF, respectively (Huff and Ishizaka, 1984; Huff *et al.*, 1986). IgE-PoF had an affinity for ConA and lentil lectin, whereas IgE-SuF had an affinity for peanut agglutinin but not for ConA (Huff *et al.*, 1986; Kisaki *et al.*, 1987, 1988). These IgE-BFs have molecular weights of 60, 30, and 15 kDa (Table III). They bind not only to human IgE but also to rat IgE and thus can regulate IgE synthesis in both species (Huff *et al.*, 1986; Kisaki *et al.*, 1988). A monoclonal antibody against these IgE-BFs failed to react with FcεRII (Kisaki *et al.*, 1987). These IgE-BFs are unlikely to be degradation products of FcεRII because IgE-BF-secreting T hybridomas do not bear a detectable amount of FcεRII (Kisaki *et al.*, 1987; Stadler *et al.*, 1987). Other investigators described human IgE-BFs similar to those detected by Ishizaka and co-workers (Young *et al.*, 1984, 1986; Yanagihara *et al.*, 1987; Stadler *et al.*, 1987).

In humans, FcεRII$^+$ B cells were also found to be able to secrete IgE-PoF (Sarfati *et al.*, 1984a,b). Therefore, for the sake of convenience, most studies used B-cell- rather than T-cell-derived IgE-BFs for molecular biologic analysis. Because rat IgE-BFs were noted to have antigenic determinants in common with FcεRII (Huff *et al.*, 1984), monoclonal anti-FcεRII antibodies were used for the isolation of human IgE-BFs rather than IgE or lectins. The human B-cell-derived IgE-BFs were not the same as the IgE-BFs of Ishizaka and co-workers but rather were degradation products of FcεRII [soluble FcεRII (sFcεRII)]. This conclusion was based on the findings that (1) several antigenic determinants were shared by IgE-BFs and FcεRII, (2) the decrease in the amount of cell surface FcεRIIs was associated in a time-related manner with the increase in the amount of IgE-BFs in the cell-free supernatant, (3) surface-radioiodinated B cells released labeled IgE-BFs that had the same antigenic composition and the same migration on SDS-PAGE as purified IgE-BFs (Nakajima *et al.*, 1987), and (4) although FcεRII had one N-glycosylation site, it did not belong to the releasing part (sFcεRII) but to the residual parts (Kikutani *et al.*, 1986a; Ikuta *et al.*, 1987; Lüdin *et al.*, 1987; Letellier *et al.*, 1988).

It seems valid to hypothesize that there are at least two different groups of IgE-BFs: (1) the IgE-BFs isolated by Ishizaka and colleagues and (2) sFcεRII. Although there are some reports suggesting the IgE-potentiating activity of sFcεRII (Sarfati and Delespesse, 198; Pène *et al.*, 1988), the mechanisms of this function remain to be elucidated.

4.3.2. Formation

Human IgE-BFs were first reported by Ishizaka and Sandberg in 1981. They reported that T cells from ragweed-sensitive allergic patients secreted IgE-BF upon stimulation with ragweed antigen E and human IgE (Ishizaka and Sandberg, 1981). Human IgE-BFs were not induced by stimulation with IgE alone. They also showed that normal human T cells secreted IgE-BFs upon incubation with T cell growth factor and IgE (Ishizaka and Sandberg, 1981).

Human IgE-SuF was also produced by purified protein derivative (PPD) stimulation of T cells collected from patients with pulmonary tuberculosis (Deguchi et al., 1983; Suemura and Kishimoto, 1985).

Young et al. (1984) reported that human FcεRII⁺ T cells, which were collected from patients with hyper-IgE syndrome and cultured in IL2-containing medium, secreted IgE-PoF. The IgE-PoF had molecular weights of 15 and 60 kDa and had an affinity for lentil lectin but not for peanut agglutinin. Secretion of IgE-PoF was inhibited by treatment with tunicamycin (Young et al., 1986).

Yanagihara et al. (1987) found that unstimulated FcεRII⁺ T cells collected from asthmatic patients secreted IgE-PoF. The IgE-PoF was converted into IgE-SuF by treatment of FcεRII⁺ T cells with tunicamycin.

Sarfati et al. (1984a) described IgE-BFs derived from FcεRII⁺ B cells. They noted that an FcεRII⁺ EBV-transformed B-cell line, RPMI8866, secreted soluble factors that had an affinity for IgE and could enhance IgE production selectively. These factors were thought to be glycoprotein molecules because of their sensitivity to trypsin and neuraminidase. Their molecular weights were 10–15 and 30–40 kDa. Transformation of lymphocytes of atopic and nonatopic individuals with EBV caused not only increased expression of FcεRII but also production of IgE-PoF (Sarfati et al., 1984b).

Although the cell origin was not defined, Leung et al. (1984) described the presence of IgE-SuF in normal human serum. Sarfati et al. (1986a) also noted the presence of IgE-SuF in human colostrum.

FcεRII⁺ B cells spontaneously released sFcεRII (Peterson and Conrad, 1985; Kikutani et al., 1986a; Ikuta et al., 1987; Lüdin et al., 1987; Delespesse et al., 1987; Nakajima et al., 1987). Recently, several investigators reported that IL4 increased the release of sFcεRII from B cells (see Section 6).

Infection of T cells with HTLV-I induced not only the expression of FcεRII but also the release of sFcεRII, although the biologic activity of the sFcεRII was not examined (Nutman et al., 1987; Sarfati et al., 1987).

5. GLYCOSYLATION-REGULATING FACTORS: GEF AND GIF

5.1. Rat

Results from investigation of the nature of IgE-BFs showed that the assembly of N-linked carbohydrate moieties changed the biologic functions of IgE-BFs. Under physiologic conditions, such modulation is undertaken by the T-cell factors GEF and GIF.

GEF is a protein molecule with a molecular weight of 25 kDa (Iwata et al., 1983a). GEF acted on rat T cells producing unglycosylated IgE-BF (IgE-SuF or inactive IgE-BF) to produce IgE-PoF by the promotion of N-linked glycosylation during the biosynthesis of IgE-BFs (Hirashima *et al.*, 1981a,b; Iwata *et al.*, 1983a). It also increased the expression of FcεRII (Huff *et al.*, 1983). GEF was found to be a kallikreinlike enzyme whose activity was inhibited by serine protease inhibitors and trypsin inhibitors (Iwata *et al.*, 1983b). Bradykinin, a cleavage product of kininogen by kallikrein, has activity similar to that of GEF. GEF had an affinity for D-galactose whereby GEF attached to and acted on IgE-BF-producing cells to secret IgE-PoF (Iwata *et al.*, 1983b). GEF also had an affinity for *p*-aminobenzamidine, by which GEF could be isolated (Iwata *et al.*, 1984a).

PLA$_2$ was suggested to participate in the N-linked glycosylation of IgE-BFs because the inhibition of PLA$_2$ by glucocorticoids suppressed the formation of IgE-PoF (Yodoi *et al.*, 1981b). The biologic activity of GEF was later found to be produced by the activation of PLA$_2$ (Iwata *et al.*, 1984b; Akasaki *et al.*, 1985; T. Ishizaka *et al.*, 1985). GEF first activated phospholipase C and methyltransferase, which increased the formation of di-acylglycerol (DAG). DAG caused the release from cells of lipomodulin, a PLA$_2$-inhibiting protein, which resulted in the relative enhancement of PLA$_2$ activity (Akasaki *et al.*, 1985). DAG also activated protein kinase C (PKC), which phosphorylates various target molecules. Because phosphorylation removed the PLA$_2$-inhibiting activity from lipomodulin, the activation of PKC also seems to contribute to the activation of PLA$_2$ (Akasaki *et al.*, 1985).

GEF was secreted spontaneously by lymphocytes of rats infected with the parasite *N. brasiliensis* (Iwata *et al.*, 1984a) or treated with BP vaccine (Hirashima *et al.*, 1981a,b). LPF, an exotoxin of BP, or ConA also induced normal rat lymphocytes to secrete GEF (Iwata *et al.*, 1983a). T cells from rats primed with alum-absorbed KLH also released GEF upon stimulation with KLH (Uede and Ishizaka, 1982).

GIF is a protein molecule with a molecular weight of 16 kDa (Uede *et al.*, 1983b). GIF inhibited the N-linked glycosylation of IgE-BFs during their biosynthesis (Hirashima *et al.*, 1981c). Therefore, in the presence of GIF, IgE-BF-producing lymphocytes could produce only IgE-SuF, not IgE-PoF. GIF also suppressed the induction of FcεRII (Hirashima *et al.*, 1981c).

GIF reacted with a monoclonal antibody against lipomodulin, a PLA$_2$-inhibiting protein (Uede *et al.*, 1983b). Although GIF did not inhibit the activity of PLA$_2$, it did so in dephosphorylated form (Uede *et al.*, 1983b). Lipomodulin, which has a larger molecular weight in intact form than GIF (Hirata *et al.*, 1980), was demonstrated to have a glycosylation-inhibiting activity (Uede *et al.*, 1983b). These results indicated that GIF is a phosphorylated fragment of lipomodulin.

GIF was secreted by BCG-primed rat T cells after stimulation with PPD (Hirashima *et al.*, 1982) or by T cells from rats primed with CFA-absorbed KLH after stimulation with KLH (Uede and Ishizaka, 1982). It was also secreted spontaneously by T cells from rats receiving repetitive injections with CFA alone (Hirashima *et al.*, 1981c).

5.2. Murine

Glycosylation-regulating factors have also been studied in mice. However, the murine glycosylation-regulating factors were considerably different from those of rats. Major characteristics of murine glycosylation-regulating factors, which are unique to mice and not observed in rats, are (1) antigen specificity and (2) antigenic determinants common to H-2 subregion products. Murine glycosylation-regulating factors resemble antigen-specific suppressor factor (TsF) (Tada *et al.*, 1976; Kapp *et al.*, 1976) and antigen-specific augmenting factor (TaF) (Tokuhisa *et al.*, 1978).

Murine antigen-specific GEF were formed by Lyt-1$^+$ T cells of BDF1 mice immunized with alum-absorbed OA upon stimulation with specific antigen (Uede and Ishizaka, 1984). This GEF was found to have an affinity for OA and enhanced the anti-hapten IgE antibody response to DNP–OA but not to DNP–KLH (Iwata *et al.*, 1987). Antigen-specific GEF consisted of 65–85- and 40–55-kDa molecules, which were larger than those of nonspecific GEF (50–70 and 20–30 kDa) produced upon stimulation of normal spleen cells with BP toxin (Iwata *et al.*, 1987). Antigen-specific GEF was shown to have antigenic determinants common to the product of the I-A subregion of MHC (Iwata *et al.*, 1988). Reduction and alkylation of antigen-specific GEF yielded nonspecific GEF. This finding indicated that antigen-specific GEF is composed of nonspecific GEF and an antigen-specific polypeptide (Iwata *et al.*, 1988). Murine GEF had characteristics of serine proteases, as did rat GEF (Iwata *et al.*, 1988).

Jardieu *et al.* (1984) found that immunization of BDF1 mice with OA alone induced Lyt-2$^+$ I-J$^+$ antigen-specific suppressor T cells. Murine GIF of 25–30 kDa was produced by this subset of T cells upon stimulation with specific-antigen OA. This GIF was shown to have an affinity for OA and to react with anti-I-J alloantibodies and an anti-lipomodulin antibody (Jardieu *et al.*, 1984).

The characteristics of antigen-specific GIF were analyzed further by the construction of antigen-specific T-cell hybridomas (Jardieu *et al.*, 1985a). Although these hybridomas spontaneously secreted nonspecific GIF, they could secrete antigen-specific GIF in the presence of antigen-presenting cells (Jardieu *et al.*, 1987). The OA-specific GIF consisted of 80- and 30–40-kDa molecules, which were larger than nonspecific GIF (15 kDa). Although both nonspecific and antigen-specific GIF suppressed anti-hapten IgE and IgG1 antibody responses in BDF1 mice immunized with alum-absorbed DNP–OA, antigen-specific GIF was much more potent than nonspecific GIF (Jardieu *et al.*, 1987). OA-specific GIF suppressed anti-DNP antibody response of DNP–KLH-immunized mice to the same degree as did nonspecific GIF (Jardieu *et al.*, 1987). Thus, antigen-specific GIF suppressed the antibody response in a carrier-specific manner. Since such carrier-specific suppressing activity of antigen-specific GIF could not be carried over to other strains, the effects of antigen-specific GIF were indicated to be MHC restricted (Iwata *et al.*, 1988).

6. INTERLEUKIN 4 AND GAMMA INTERFERON

IL4 is a T-cell-derived lymphokine. It has multiple effects on various blood cells such as B cells, T cells, and monocytes, although its growth-promoting effects on B cells was detected first. In 1986, Coffman and Carty, Coffmann *et al.*, and Lee *et al.* reported that mouse IL4 enhanced not only IgG1 but also IgE production by lipopolysaccharide-stimulated mouse B cells. At the same time, Finkelman *et al.* (1986) reported that the injection with monoclonal anti-IL4 antibody to mice infected with the parasite *N. brasiliensis* suppressed IgE synthesis. Very recently, human recombinant IL4 (rIL4) was also demonstrated to enhance human IgE synthesis (Del Prete *et al.*, 1988; Sarfati and Delespesse, 1988).

Human rIL4 was shown to be able to induce FcεRIIs on B cells (Kikutani *et al.*, 1986a; DeFrance *et al.*, 1987; Hudak *et al.*, 1987). The induction of FcεRIIs was accompanied by an increased release of sFcεRII (Kawabe *et al.*, 1988; Bonnefoy *et al.*, 1988a; Delespesse *et al.*, 1989). In contrast to IgE, IL4 did not affect the degradation of FcεRIIs but increased the biosynthesis of FcεRIIs (Conrad *et al.*, 1987; Kawabe *et al.*, 1988).

These results raise the question of whether the increase in IgE synthesis upon treatment with rIL4 is a direct effect of rIL4 itself or is an indirect effect caused by the increased release of sFcεRII. Conrad *et al.* (1988) studied the relationship between the induction of FcεRII, which was accompanied by the increased release of sFcεRII, and the induction of IgE synthesis by rIL4 in mice. They reported that there was no relationship between them and that the addition of purified sFcεRII did not enhance IgE synthesis. In contrast, Sarfati and Delespesse (1988) noted that the addition of monoclonal anti-human FcεRII antibody suppressed IL4-induced IgE synthesis class specifically. Pène *et al.* (1988) found that sFcεRII (soluble CD23 antigen) enhanced IL4-induced human IgE synthesis. Therefore, they concluded that FcεRII as well as sFcεRII plays an essential role in IgE synthesis.

Although IFN-γ induced FcεRIIs on monocytes and macrophages (Náray-Fejes-Tóth and Guyre, 1984; Mayumi *et al.*, 1988; Vercelli *et al.*, 1988) and platelets (Pancré *et al.*, 1988), it did not induce them on lymphocytes (Kim *et al.*, 1987). By contrast, it inhibited the induction of FcεRIIs (Kikutani *et al.*, 1986b; DeFrance *et al.*, 1987). It also inhibited the induction of IgE synthesis and sFcεRII (Coffman and Carty, 1986; Bonnefoy *et al.*, 1988a). Rousset *et al.* (1988) reported that IFN-γ blocked the IL4-induced increase in the transcription of mRNA of FcεRII. However, Kawabe *et al.* (1988) postulated a different mechanism for the inhibitory activity of IFN-γ because it did not seem to suppress the IL4-induced increase in the transcription of mRNA.

7. CD23 ANTIGEN

CD23 was shown to have antigenic determinants common to those of FcεRII (Yukawa *et al.*, 1987; Bonnefoy *et al.*, 1987). Indeed, CD23 has an affinity for IgE. Other biologic and biochemical properties of CD23 are also similar to those of FcεRII. Therefore, CD23 is now accepted as the same molecule as FcεRII.

CD23 has been studied as a B-cell-specific surface antigen. It was induced on B cells by stimulation with PWM, protein A, anti-IgM, or EBV, among which EBV was most potent (Thorley-Lawson *et al.*, 1985). CD23 appeared within 24 hr after stimulation (Thorley-Lawson and Mann, 1985). It was degraded immediately (half-life of 1–2 hr) and shed into the medium as soluble CD23 (Thorley-Lawson *et al.*, 1986). CD23 was also expressed on chronic B lymphocytic leukemia (BCLL) cells (Thorley-Lawson *et al.*, 1985), and the serum level of soluble CD23 was increased in patients with BCLL (Sarfati *et al.*, 1988).

CD23 is a glycoprotein molecule with a molecular weight of 47 kDa in intact form and 25 and 12 kDa in soluble form (Thorley-Lawson *et al.*, 1986; Sarfati *et al.*, 1988). A 47-kDa molecule was formed by the assembly of mannose-rich oligosaccharides to the 43-kDa precursor polypeptide by N-linked glycosylation during biosynthesis, which was inhibited by tunicamycin (Thorley-Lawson *et al.*, 1986). The isoelectric point of both the 47- and 25-kDa forms of CD23 was 4.5–5.0 (Thorley-Lawson *et al.*, 1986; Sarfati *et al.*, 1988).

The biologic functions of CD23 are now being investigated. Gordon *et al.* (1986a) reported that the ligation of CD23 by monoclonal anti-CD23 antibody promoted DNA synthesis of tonsillar B cells preactivated with phorbol ester. Such activity resembled that of low-molecular-weight B-cell growth factor (BCGF). Indeed, the absorption of low-molecular-weight BCGF by target B cells was inhibited by pretreatment with some monoclonal anti-CD23 antibodies. Moreover, the ligation of Fab fragments of monoclonal anti-CD23 antibody to CD23, which mimics the ligation of hormonal factors to their receptors, was enough to deliver growth-promoting signals (Gordon *et al.*, 1987). Thus, the authors concluded that CD23 was a receptor for low-molecular-weight BCGF (Gordon *et al.*, 1986b). IgE was suggested to coordinate with BCGF in up regulation of B-cell growth (Guy and Gordon, 1987).

Swendeman and Thorley-Lawson (1987) showed that soluble CD23 enhanced DNA synthesis of normal B cells preactivated with anti-IgM antibody. They concluded that soluble CD23 was an autocrine BCGF (Gordon *et al.*, 1984; Cairns *et al.*, 1988). Gordon *et al.* (1988) and Armitage and Goff (1988) also reported that soluble CD23 increased the proliferative responses of B cells. However, Uchibayashi *et al.* (1989) reported that recombinant sFcɛRII did not enhance DNA synthesis of preactivated B cells.

8. CONCLUSION

IgE-BFs were shown by Ishizaka and co-workers to have regulatory activities in IgE synthesis. FcɛRII$^+$ T lymphocytes were concluded to play an important role in this IgE-BF-mediated regulation of IgE synthesis as a source of IgE-PoF. Because the incubation of IgE-BF-secreting lymphocytes with IgE caused an increase in the secretion of IgE-BFs, FcɛRII on lymphocytes is also likely to play a role in this system.

A remarkable advance in the study of human FcɛRII was achieved during the last several years, which led to the isolation of cDNA for FcɛRII. However, this effort failed to encompass the IgE-BFs of Ishizaka and colleagues. Although the molecular structure of a human IgE-BF was clarified, it was revealed to be a degradation product of FcɛRII (sFcɛRII) and to have molecular properties different from those of the IgE-BFs detected by Ishizaka and co-workers. The ability of sFcɛRII to regulate IgE synthesis is still controversial, although it seems to be related in some way to the regulation of IgE synthesis. On the other hand, although their function is evident, the molecular characteristics of the human IgE-BFs of Ishizaka and colleagues remains less clear than that of sFcɛRII. Further efforts to clarify the molecular structure of the IgE-BFs of Ishizaka and colleagues, including isolation of cDNA, are needed.

The source of rat IgE-PoF was shown to be FcɛRII$^+$ T cells. However, in humans, the IgE-BFs of Ishizaka and co-workers, including IgE-PoF, could be secreted by T cells bearing an undetectable amount of FcɛRIIs. By contrast, sFcɛRII was demonstrated to be released from FcɛRII$^+$ T as well as B cells. The relationship between FcɛRII$^+$ lymphocytes and the secretion of IgE-PoF should be reexamined in humans.

Recently, IgE-BFs other than those of Ishizaka and colleagues and sFcϵRII have been described (Liu *et al.*, 1986; Haak-Frendscho *et al.*, 1988). To avoid confusion, agreement as to the use of the term "IgE-BFs" is needed.

Much remains to be learned about the function of FcϵRII itself. Recently, however, FcϵRII was found to be the same molecule as CD23. This observation suggested that FcϵRII plays an important role in delivering a growth-promoting signal in B cells and that sFcϵRII might be an autocrine BCGF. This finding may increase the biologic significance of FcϵRII and sFcϵRII such that more attention will be directed to them.

9. REFERENCES

Adachi, M., and Ishizaka, K., 1986, IgD-binding factors from mouse T lymphocytes, *Proc. Natl. Acad. Sci. USA* **83**:7003–7007

Akasaki, M., and Ishizaka, K., 1987, Effects of cyclophosphamide treatment and gamma irradiation of SJL mice on the IgE antibody response and the nature of IgE-binding factors, *Int. Arch. Allergy Appl. Immunol.* **82**:417–418

Akasaki, M., Iwata, M., and Ishizaka, K., 1985, Modulation of the biologic activities of IgE-binding factors. VII. Biochemical mechanisms by which glycosylation-enhancing factor activates phospholipase in lymphocytes, *J. Immunol.* **135**:4069–4077

Armitage, R. J. and Goff, L. K., 1988, Functional interaction between B cell subpopulations defined by CD23 expression, *Eur. J. Immunol.* **18**:1753–1760

Boltz-Nitulescu, G., Plummer, J. M., and Spiegelberg, H. L., 1982, Fc receptor for IgE on mouse macrophages and macrophage-like cell lines, *J. Immunol.* **128**:2265–2268

Bonnefoy, J. Y., Aubry, J.-P., Peronne, C., Wijdenes, J., and Banchereau, J., 1987, Production and characterization of a monoclonal antibody specific for the human lymphocyte low affinity receptor for IgE: CD 23 is a low affinity receptor for IgE, *J. Immunol.* **138**:2970–2978

Bonnefoy, J. Y., Defrance, T., Perrone, C., Rousset, F., Pène, J., De Vries, J. E., and Banchereau, J., 1988a, Human recombinant interleukin 4 induces normal B cells to produce soluble CD23/IgE-binding factor analogous to that spontaneously released by lymphoblastoid B cell lines, *Eur. J. Immunol.* **18**:117–122

Bonnefoy, J. Y., Guillot, O., Spits, H., Blanchard, D., Ishizaka, K., and Banchereau, J., 1988b, The low-affinity receptor for IgE (CD23) on B lymphocytes is spatially associated with HLA-DR antigens, *J. Exp. Med.* **167**:57–72

Cairns, J., Flores-Romo, L., Millsum, M. J., Guy, G. R., Gillis, S., Ledbetter, J. A., and Gordon, J., 1988, Soluble CD23 is released by B lymphocytes cycling in response to interleukin 4 and anti-Bp50 (CDw40), *Eur. J. Immunol.* **18**:349–353

Capron, A., Dessaint, J.-P., Capron, M., and Bazin, H., 1975, Specific IgE antibodies in immune adherence of normal macrophages to *Schistosoma mansoni* schistosomules, *Nature* (London) **253**:474–475

Capron, A., Dessaint, J.-P., Joseph, M., Rousseaux, R., Capron, M., and Bazin, H., 1977, Interaction between IgE complexes and macrophages in the rats: A new mechanism of macrophage activation, *Eur. J. Immunol.* **7**:315–322

Capron, A., Ameisen, J. C., Joseph, M., Auriault, C., Tonnel, A. B., and Caen, J., 1985, New functions for platelets and their pathological implications, *Int. Arch. Allergy Appl. Immunol.* **77**:107–114

Capron, A., Dessaint, J. P., Capron, M., Joseph, M., Ameisen, J. C., and Tonnel, A. B., 1986, From parasites to allergy: A second receptor for IgE, *Immunol. Today* **7**:15–18

Capron, A., Joseph, M., Ameisen, J. C., Capron, M., Pancré, V., and Auriault, C., 1987, Platelets as effectors in immune and hepersensitivity reactions, *Int. Arch. Allergy Appl. Immunol.* **82**:307–312

Capron, M., Capron, A., Dessaint, J.-P., Torpier, G., Gunner, S., Johansson, O., and Prin, L., 1981a, Fc receptors for IgE on human and rat eosinophils, *J. Immunol.* **126**:2087–2092

Capron, M., Bazin, H., Joseph, M., and Capron, A., 1981b, Evidence for IgE-dependent cytotoxicity by rat eosinophils, *J. Immunol.* **126**:1764–1768

Capron, M., Spiegelberg, H. L., Prin, L., Bennich, H., Butterworth, A. E., Pierce, R. J., AliOuaissi, M., and

Capron, A., 1984, Role of IgE receptors in effector function of human eosinophils, *J. Immunol.* **132**:462–468

Capron, M., Kusnierz, J. P., Prin, L., Spiegelberg, H. L., Khalife, J., Tonnel, A. B., and Capron, A., 1985, Cytophilic IgE on human blood and tissue eosinophils, *Int. Arch. Allergy. Appl. Immunol.* **77**:246–248

Capron, M., Jouault, T., Prin, L., Joseph, M., Ameisen, J. C., Butterworth, A. E., Papin, J. P., Kusnierz, J. P., and Capron, A., 1986, Functional study of a monoclonal antibody to IgE Fc receptor (FcεR2) of eosinophils, platelets, and macrophages, *J. Exp. Med.* **164**:72–89

Carlsson, M., Totterman, T. H., Matsson, P., and Nilsson, K., 1988, Cell cycle progression of B-chronic lymphocytic leukemia cells induced to differentiate by TPA, *Blood* **71**:415–421

Chen, S.-S., Bohn, J. W., Liu, F.-T., and Katz, D. H., 1981, Murine lymphocytes expressing Fc receptor for IgE (FcεR). I. Conditions for inducing FcεR+ lymphocytes and inhibition of the inductive events by suppressive factor of allergy (SFA), *J. Immunol.* **127**:166–173

Coffman, R. L., and Carty, J., 1986, A T cell activity that enhances polyclonal IgE production and its inhibition by interferon-γ, *J. Immunol.* **136**:949–954

Coffman, R. L., Ohara, J., Bond, M. W., Carty, J., Zlotnik, A., and Paul, W. E., 1986, B cell stimulatory factor-1 enhances the IgE response of lipopolysaccharide-activated B cells, *J. Immunol.* **136**:4538–4541

Conrad, D. H., and Peterson, L. H., 1984, The murine lymphocytes receptor for IgE. I. Isolation and characterization of the murine B cell Fcε receptor and comprison with Fcε receptors from rat and human, *J. Immunol.* **132**:796–803

Conrad, D. H., Bazin, H., Sehon, A. H., and Froese, A., 1975, Binding parameters of the interaction between rat IgE and rat mast cell receptors, *J. Immunol.* **114**:1688–1691

Conrad, D. H., Waldschmidt, T. J., Lee, W. T., Rao, M., Keegan, A. D., Noelle, J., Lynch, R. G., and Kehry, M. R., 1987, Effect of B cell stimulatory factor-1 (interleukin 4) on Fcε and Fcγ receptor expression on murine B lymphocytes and B cell lines, *J. Immunol.* **139**:2290–2296

Conrad, D. H., Keegan, A. D., Kalli, K. R., Van Dusen, R., Rao, M., and Levine, A. D., 1988, Superinduction of low affinity IgE-receptors on murine B lymphocytes by lipopolysaccharide and IL-4, *J. Immunol.* **141**:1091–1097

Daeron, M., and Ishizaka, K., 1986, Induction of Fcε receptors on mouse macrophages and lymphocytes by homologous IgE, *J. Immunol.* **136**:1612–1619

DeFrance, T., Aubry, J. P., Rousset, F., Vanbervliet, B., Bonnefoy, J. Y., Arai, N., Takebe, Y., Yokota, T., Lee, F., Arai, K., De Vries, J., and Banchereau, 1987, Human recombinant interleukin 4 induces Fcε receptors (CD23) on normal human B lymphocytes, *J. Exp. Med.* **165**:1459–1467

Deguchi, H., Suemura, M., Ishizaka, A., Ozaki, Y., Kishimoto, S., Yamamura, Y., and Kishimoto, T., 1983, IgE class-specific suppressor T cells and factors in humans, *J. Immunol.* **131**:2751–2756

Delespesse, G., Sarfati, M., Rubio-Trujilo, M., and Wolowiec, T., 1986a, IgE receptors on human lymphocytes. II. Detection of cells bearing IgE receptors in unstimulated mononuclear cells by means of a monoclonal antibody, *Eur. J. Immunol.* **16**:815–821

Delespesse, G., Sarfati, M., Rubio-Trujilo, M., and Wolowiec, T., 1986b, IgE receptors on human lymphocytes. III. Expression of IgE receptors on mitogen-stimulated human mononuclear cells, *Eur. J. Immunol.* **16**:1043–1047

Delespesse, G., Sarfati, M., and Rubio-Trujilo, M., 1987, *In vitro* production of IgE-binding factors by human mononuclear cells, *Immunology* **60**:103–110

Delespesse, G., Sarfati, M., and Peleman, R., 1989, Influence recombinant IL-4, IFN-α, and IFN-γ on the production of human IgE-binding factor (soluble CD23), *J. Immunol.* **142**:134–138

Del Prete, G., Maggi, E., Parronchi, P., Chrétien, I., Tiri, A., Macchia, D., Ricci, M., Banchereau, J., De Vries, J., and Romagnani, S., 1988, IL-4 is an essential factor for the IgE synthesis induced in vitro by human T cell clones and their supernatants, *J. Immunol.* **140**:4193–4198

Dessaint, J. P., Torpier, G., Capron, M., Bazin, H., and Capron, A., 1979a, Cytophilic binding of IgE to the macrophage. I. Binding characteristics of IgE on the surface of macrophages in the rat, *Cell. Immunol.* **46**:12–23

Dessaint, J. P., Capron, A., Joseph, M., and Bazin, H., 1979b, Cytophilic binding of IgE to the macrophage. II. Immunologic release of lysosomal enzyme from macrophages by IgE and anti-IgE in the rat: A new mechanism of macrophage activation, *Cell. Immunol.* **46**:24–34

Finbloom, D. S., and Metzger, H., 1982, Binding of immunoglobulin E to the receptor on rat peritoneal macrophage, *J. Immunol.* **129**:2004–2008

Finbloom, D. S., and Metzger, H., 1983, Isolation of cross-linked IgE-receptor complexes from rat macrophages, *J. Immunol.* **130:**1489–1491

Finkelman, F. D., Kotona, I. M., Urban, J. F., Snapper, C. M., Ohara, J., and Paul, W. E., 1986, Suppression of *in vivo* polyclonal IgE responses by monoclonal antibody to the lymphokine B-cell stimulatory factor 1, *Proc. Natl. Acad. Sci. USA* **83:**9675–9678

Fridman, W. H., and Golstein, P., 1974, Immunoglobulin-binding factor present on and produced by thymus-processed lymphocytes (T cells), *Cell. Immunol.* **11:**442–455

Fritsche, R., and Spiegelberg, H. L., 1978, Fc receptors for IgE on normal rat lymphocytes, *J. Immunol.* **121:**471–478.

Galizzi, J.-P., Cabrillat, H., Rousset, F., Ménétrier, C., De Vries, J. E., and Banchereau, J., 1988, IFN-γ and prostaglandin E2 inhibit IL4-induced expression of FcεR2/CD23 on B lymphocytes through different mechanisms without altering binding of IL-4 to its receptor, *J. Immunol.* **141:**1982–1988

Gonzalez-Molina, A., and Spiegelberg, H. L., 1976, Binding of IgE myeloma proteins to human cultured lymphoblastoid cells, *J. Immunol.* **117:**1838–1845

Gonzalez-Molina, A., and Spiegelberg, H. L., 1977, A subpopulation of normal human peripheral blood B lymphocytes that bind IgE, *J. Clin. Invest.* **59:**616–624

Gordon, J., Ley, S. C., Melamed, M. D., English, L. S., and Hughes-Jones, N. C., 1984, Immortalized B lymphocytes produce B-cell growth factor, *Nature* (London) **310:**145–147

Gordon, J., Rowe, M., Walker, L., and Guy, G., 1986a, Ligation of the CD23, p45 (BLAST-2, EBVCS) antigen triggers the cell-cycle progression of activated B lymphocytes, *Eur. J. Immunol.* **16:**1075–1080

Gordon, J., Webb, A. J., Walker, L., Guy, G. R., and Rowe, M., 1986b, Evidence for an association between CD23 and the receptor for a low molecular weight B cell growth factor, *Eur. J. Immunol.* **16:**1627–1630

Gordon, J., Webb, A. J., Guy, G. R., and Walker, L., 1987, Triggering of B lymphocytes through CD23: Epitope mapping and studies using antibody derivatives indicate an allosteric mechanism of signalling, *Immunology* **60:**517–521

Gordon, J., Cairns, J. A., Millsum, M. J., Gillis, S., and Guy, G. R., 1988, Interleukin 4 and soluble CD23 as progression factors for human B lymphocytes: Analysis of their interactions with agonists of the phosphoinositide "dual pathway" of signalling, *Eur. J. Immunol.* **18:**1561–1565

Guy, G. R., and Gordon, J., 1987, Coordinated action of IgE and a B-cell-stimulatory factor on the CD23 receptor molecule up-regulates B-lymphocyte growth, *Proc. Natl. Acad. Sci. USA* **84:**6239–6243

Haak-Frendscho, M., Sarfati, M., Delespesse, G., and Kaplan, A. P., 1988, Comparison of mononuclear cell and B-lymphoblastoid histamine-releasing factor and their distinction from an IgE-binding factor, *Clin. Immunol. Immunopathol.* **49:**72–82

Haque, A., Ouaiss, A., Joseph, M., Capron, M., and Capron, A., 1981, IgE antibody in eosinophil- and macrophage-mediated *in vitro* killing of *Dipetalonema viteae* microfilariae, *J. Immunol.* **127:**716–725

Hellström, U., and Spiegelberg, H. L., 1979, Characterization of human lymphocytes bearing Fc receptors for IgE isolated from blood and lymphoid organs, *Scand. J. Immunol.* **9:**75–86

Hempstead, B. L., Parker, C. W., and Kulczycki, A., 1979, Characterization of the IgE receptor isolated from human basophils, *J. Immunol.* **123:**2283–2291

Hirashima, M., Yodoi, J., and Ishizaka, K., 1980a, Regulatory role of IgE-binding factors from rat T lymphocytes. III. IgE-specific suppressive factor with IgE-binding activity, *J. Immunol.* **125:**1442–1448

Hirashima, M., Yodoi, J., and Ishizaka, K., 1980b, Regulatory role of IgE-binding factors from rat T lymphocytes. IV. Formation of IgE-binding factors in rats treated with complete Freund's adjuvant, *J. Immunol.* **125:**2154–2160

Hirashima, M., Yodoi, J., and Ishizaka, K., 1981a, Regulatory role of IgE-binding factors from rat T lymphocytes. V. Formation of IgE-potentiating factor by T lymphocytes from rats treated with *Bordetella pertussis* vaccine, *J. Immunol.* **126:**838–842

Hirashima, M., Yodoi, J., and Ishizaka, K., 1981b, Formation of IgE-binding factors by rat T lymphocytes. II. Mechanism of selective formation of IgE-potentiating factor by treatment with *Bordetella pertussis* vaccine, *J. Immunol.* **127:**1804–1810

Hirashima, M., Yodoi, J., Huff, T. F., and Ishizaka, K., 1981c, Formation of IgE-binding factors by rat T lymphocytes. III. Mechanism of selective formation of IgE-suppressive factor by treatment with complete Freund's adjuvant, *J. Immunol.* **127:**1810–1816

Hirashima, M., Uede, T., Huff, T., and Ishizaka, K., 1982, Formation of IgE-binding factors by rat T lymphocytes. IV. Mechanism for the formation of IgE-suppressive factors by antigen stimulation of BCG-primed spleen cells, *J. Immunol.* **128**:1909–1916

Hirata, F., Schiffman, E., Venkatasubramamian, K., Solomon, D., and Axelord, J., 1980, A phospholipase A₂ inhibitory protein in rabbit neutrophils induced by glucocorticoids, *Proc. Natl. Acad. Sci. USA* **77**:2533–2536

Hudak, S. A., Gollnick, S. O., Conrad, D. H., and Kehry, M. R., 1987, Murine B cell stimulatory factor 1 (interleukin 4) increases expression of the Fc receptor for IgE on mouse B cells, *Proc. Natl. Acad. Sci. USA* **84**:4606–4610

Huff, T. F., and Ishizaka, K., 1984, Formation of IgE-binding factors by human T-cell hybridomas, *Proc. Natl. Acad. Sci. USA* **81**:1514–1518

Huff, T. F., Uede, T., and Ishizaka, K., 1982, Formation of rat IgE-binding factors by rat-mouse T cell hybridomas, *J. Immunol.* **129**:509–514

Huff, T. F., Uede, T., Iwata, M., and Ishizaka, K., 1983, Modulation of the biologic activities of IgE-binding factors. III. Switching of a T cell hybrid clone from the formation of IgE-suppressive factor to the formation of IgE-potentiating factor, *J. Immunol.* **131**:1090–1095

Huff, T. F., Yodoi, J., Uede, T., and Ishizaka, K., 1984, Presence of an antigenic determinant common to rat IgE-potentiating factor, IgE-suppressive factor, and Fcε receptors on T and B lymphocytes, *J. Immunol.* **132**:406–412

Huff, T. F., Jardieu, P., and Ishizaka, K., 1986, Regulatory effects of human IgE-binding factors on the IgE response of rat lymphocytes, *J. Immunol.* **136**:955–962

Ikuta, K., Takami, M., Kim, C. W., Honjo, T., Miyoshi, T., Tagaya, Y., Kawabe, T., and Yodoi, J., 1987, Human lymphocyte Fc receptor for IgE: Sequence homology of its cloned cDNA with animal lectins, *Proc. Natl. Acad. Sci. USA* **84**:819–823

Ishizaka, K., 1988, IgE-binding factors and regulation of the IgE antibody response, *Annu. Rev. Immunol.* **6**:513–534

Ishizaka, K., and Sandberg, K., 1981, Formation of IgE binding factors by human T lymphocytes, *J. Immunol.* **126**:1692–1696

Ishizaka, T., Soto, C. S., and Ishizaka, K., 1973, Mechanisms of passive sensitization. III. Number of IgE molecules and their receptor sites on human basophil granulocytes, *J. Immunol.* **111**:500–511

Ishizaka, T., Iwata, M., and Ishizaka, K., 1985, Release of histamine and arachidonate from mouse mast cells induced by glycosylation-enhancing factor and bradykinin, *J. Immunol.* **134**:1880–1887

Iwata, M., and Ishizaka, K., 1988, Construction of antigen-specific suppressor T cell hybridomas from spleen cells of mice primed for the persistent IgE antibody formation, *J. Immunol.* **141**:3270–3277

Iwata, M., Huff, T. F., Uede, T., Munoz, J. J., and Ishizaka, K., 1983a, Modulation of the biologic activities of IgE-binding factor. II. Physicochemichal properties and cell sources of glycosylation-enhancing factor, *J. Immunol.* **130**:1802–1808

Iwata, M., Munoz, J. J., and Ishizaka, K., 1983b, Modulation of the biologic activities of IgE-binding factor. IV. Identification of glycosylation-enhancing factor as a kallikrein-like enzyme, *J. Immunol.* **131**:1954–1960

Iwata, M., Huff, T. F., and Ishizaka, K., 1984a, Modulation of the biologic activities of IgE-binding factor. V. The role of glycosylation-enhancing factor and glycosylation-inhibiting factor in determining the nature of IgE-binding factors, *J. Immunol.* **132**:1286–1293

Iwata, M., Akasaki, M., and Ishizaka, K., 1984b, Modulation of the biologic activities of IgE-binding factor. VI. The activation of phospholipase by glycosylation enhancing factor, *J. Immunol.* **133**:1505–1512

Iwata, M., Fukutomi, Y., Hashimoto, T., Sato, Y., Sato, H., and Ishizaka, K., 1987, Augmentation of the antibody response by antigen-specific glycosylation-enhancing factor, *J. Immunol.* **138**:2561–2567

Iwata, M., Adachi, M., and Ishizaka, K., 1988, Antigen-specific T cells that form IgE-potentiating factor, IgG-potentiating factor, and antigen-specific glycosylation-enhancing factor on antigenic stimulation, *J. Immunol.* **140**:2534–2542

Jardieu, P., Uede, T., and Ishizaka, K., 1984, IgE-binding factors from mouse T lymphocytes. III. Role of antigen-specific suppressor T cells in the formation of IgE-suppressive factor, *J. Immunol.* **133**:3266–3273

Jardieu, P., Uede, T., and Ishizaka, K., 1985a, Presence of an antigen-specific T cell subset that forms IgE-suppressive factor and IgG-suppressive factor on antigenic stimulation, *J. Immunol.* **135**:922–929

Jardieu, P., Moore, K., Martens, C., and Ishizaka, K., 1985b, Relationship among IgE-binding factors with various molecular weights, *J. Immunol.* **135**:2727–2734

Jardieu, P., Akasaki, M., and Ishizaka, K., 1987, Carrier-specific suppression of antibody responses by antigen-specific glycosylation-inhibiting factor, *J. Immunol.* **138**:1494–1501

Jarret, E., and Bazin, H., 1974, Elevation of total serum IgE in rats following helminth parasite infection, *Nature* (London) **251**:613–614

Joseph, M., Capron, A., Butterworth, A. E., Sturrock, R. F., and Houba, V., 1978, Cytotoxity of human and baboon mononuclear phagocytes against Shistosomula *in vitro:* Induction of immune complexes containing IgE and *Schistosoma mansoni* antigens, *Clin. Exp. Immunol.* **33**:48–56

Joseph, M., Auriault, C., Capron, A., Vorng, H., and Viens, P., 1983a, A new function for platelets: IgE-dependent killing of schistosomes, *Nature* (London) **303**:810–812

Joseph, M., Tonnel, A. B., Torpier, G., Capron, A., Arnoux, B., and Benveniste, J., 1983b, Involvement of immunoglobulin E in the secretory processes of alveolar macrophages from asthmatic patients, *J. Clin. Invest.* **71**:221–230

Joseph, M., Capron, A., Ameisen, J. C., Capron, M., Vorng, H., Pancre, V., Kusnierz, J. P., and Auriault, C., 1986, The receptor for IgE on blood platelets, *Eur. J. Immunol.* **16**:306–312

Jouault, T., Capron, M., Balloul, J.-M., Ameisen, J.-C., and Capron, A., 1988, Quantitative and qualitative analysis of the Fc receptor for IgE (FcεRII) on human eosinophils, *Eur. J. Immunol.* **18**:237–241

Kanowith-Klein, S., Hofman, F., and Saxon, A., 1988, Expression of Fcε receptors and surface and cytoplasmic IgE on human fetal and adult lymphopoietic tissue, *Clin. Immunol. Immunopathol.* **48**:214–224

Kapp, J. A., Pierce, C. W., dela Croix, F., and Benacerraf, B., 1976, Immunosuppressive factor(s) extracted from lymphoid cells of nonresponder mice primed with L-glutamic acid^{60}-L-alanine^{30}-L-tyrosine^{10}(GAT). I. Activity and antigenic specificity, *J. Immunol.* **116**:305–309.

Katz, D. H., 1980, Recent studies on the regulation of IgE antibody synthesis in experimental animals and man, *Immunology* **41**:1–24

Kawabe, T., Takami, M., Hosoda, M., Maeda, Y., Sato, S., Mayumi, M., Mikawa, H., Arai, K., and Yodoi, J., 1988, Regulation of FcεR$_2$/CD23 gene expression by cytokines and specific ligands (IgE and anti-FcεR$_2$ monoclonal antibody), *J. Immunol.* **141**:1376–1382

Keegan, A. D., and Conrad, D. H., 1987, The murine lymphocyte receptor for IgE. V. Biosynthesis, transport, and maturation of the B cell Fcε receptor, *J. Immunol.* **139**:1199–1205

Kikutani, H., Inui, S., Sato, R. Barsumian, E. L., Owaki, H., Yamasaki, K., Kaisho, T., Uchibayashi, N., Hardy, R. R., Hirano, T., Tsunasawa, S., Sakiyama, F., Suemura, M., and Kishimoto, T., 1986a, Molecular structure of human lymphocyte receptor for immunoglobulin E, *Cell* **47**:657–665

Kikutani, H., Suemura, M., Owaki, H., Nakamura, H., Sato, R., Yamasaki, K., Barsumian, E. L., Hardy, R. R., and Kishimoto, T., 1986b, Fcε receptor, a specific differentiation marker transiently expressed on mature B cells before isotype switching, *J. Exp. Med.* **164**:1455–1469

Kim, K.-M., Tanaka, M., Yoshimura, T., Katamura, K., Mayumi, M., and Mikawa, H., 1987, Regulation of IgE receptor expression on human peripheral blood lymphocytes by lymphocytosis promoting factor (LPF), lectins and dexamethasone, *Clin. Exp. Immunol.* **68**:418–426

Kim, K.-M., Mayumi, M., Iwai, Y., Tanaka, M., Ito, S., Shinomiya, K., and Mikawa, H., 1988, IgE receptor-bearing lymphocytes in allergic and nonallergic children, *Pediatr. Res.* **24**:254–257

Kintner, C., and Sugden, B., 1981, Identification of antigenic determinants unique to the surfaces of cells transformed by Epstein-Barr virus, *Nature* (London) **294**:458–460

Kisaki, T., Huff, T. F., Conrad, D. H., Yodoi, J., and Ishizaka, K., 1987, Monoclonal antibody specific for T cell-derived human IgE binding factors, *J. Immunol.* **138**:3345–3351

Kisaki, T., Leung, D. Y. M., Jardieu, P., Geha, R. S., and Ishizaka, K., 1988, Regulatory effects of human IgE-binding factors in the IgE synthesis by human and rat lymphocytes, *Eur. J. Immunol.* **18**:1663–1670

Kishimoto, T., Hirai, Y., Suemura, M., and Yamamura, Y., 1976, Regulation of antibody response in different immunoglobulin classes. I. Selective suppression of anti-DNP IgE antibody response by preadministration of DNP-coupled Mycobacterium, *J. Immunol.* **117**:396–404

Kishimoto, T., Hirai, Y., Suemura, M., Nakanishi, K., and Yamamura, Y., 1978, Regulation of antibody responses in different immunoglobulin classes. IV. Properties and functions of "IgE class-specific" suppressor factor(s) released from DNP-Mycobacterium-primed T cells, *J. Immunol.* **121**:2106–2111

Kiyono, H., Mosteller-Barnum, L. M., Pitts, A. M., Williamson, S. I., Michalek, S. M., and McGhee, J. R., 1985, Isotype-specific immunoregulation. IgA-binding factors produced by Fcα receptor positive T cell hybridomas regulate IgE responses, *J. Exp. Med.* **161**:731–747

Kuff, E. L., Mietz, J. A., Trounstine, M. L., Moore, K. W., and Martens, C. L., 1986, cDNA clones encoding murine IgE-binding factors represent multiple structural variants of intracisternal A-particle genes, *Proc. Natl. Acad. Sci. USA* **83**:6583–6587

Kulczycki, A., Jr., and Metzger, H., 1974, The interaction of IgE with rat basophilic leukemia cells, *J. Exp. Med.* **140**:1676–1695

Lawrence, D. A., Weigle, W. O., and Spiegelberg, H. L., 1975, Immunoglobulins cytophilic for human lymphocytes, monocytes, and neutrophils, *J. Clin. Invest.* **55**:368–376

Lebrun, P., Sidman, C. L., and Spiegelberg, H. L., 1988, IgE formation and Fcε receptor-positive lymphocytes in normal, immuno-deficient, and auto-immune mice injected with *Nippostrongylus brasiliensis*, *J. Immunol.* **141**:249–257

Lee, F., Yokota, T., Otsuka, T., Meyerson, P., Villaret, D., Coffman, R., Mosmann, T., Rennick, D., Roehm, N., Smith, C., Zlotnik, A., and Arai, K., 1986, Isolation and characterization of a mouse interleukin cDNA clone that expresses B-cell stimulatory factor 1 activities and T-cell and mast-cell-stimulating activities, *Proc. Natl. Acad. Sci. USA* **83**:2061–2065

Lee, W. T., and Conrad, D. H., 1985, The murine lymphocyte receptor for IgE. III. Use of chemical cross-linking reagents to further characterize the B lymphocyte Fcε receptor, *J. Immunol.* **134**:518–524

Lee, W. T., and Conrad, D. H., 1987, Murine B cell hybridomas bearing ligand-inducible Fc receptors for IgE, *J. Immunol.* **136**:4573–4580

Lee, W. T., Rao, M., and Conrad, D. H., 1987, The murine lymphocyte receptor for IgE. IV. The mechanism of ligand-specific receptor upregulation on B cells, *J. Immunol.* **139**:1191–1198

Letellier, M., Nakajima, T., and Delespesse, G., 1988, IgE receptor on human lymphocytes. IV. Further analysis of its structure and of the role of N-linked carbohydrates, *J. Immunol.* **141**:2374–2381

Leung, D. Y. M., Brozek, C., Frankel, R., and Geha, R. S., 1984, IgE-specific suppressor factors in normal human serum, *Clin. Immunol. Immunopathol.* **32**:339–350

Liu, M. C., Proud, D., Lichtenstein, L. M., MacGlashan, D. W., Schleimer, R. P., Adkinson, N. F., Kagey-Sobotka, A., Schulman, E. S., and Plaut, M., 1986, Human lung macrophage-derived histamine-releasing activity is due to IgE-dependent factors, *J. Immunol.* **136**:2588–2595

Lüdin, C., Hofstetter, H., Sarfati, M., Levy, C. A., Suter, U., Alaimo, D., Kilchherr, E., Frost, H., Delespesse, G., 1987, Cloning and expression of the cDNA coding for a human lymphocyte IgE-receptor, *EMBO J.* **6**:109–114

Manouvriez, P., Ravoet, A. M., and Bazin, H., 1985, Fc epsilon receptors on rat B-lymphocytes: Specificity and binding kinetics, *Mol. Immunol.* **22**:1201–1208

Marcelletti, J. F., and Katz, D. H., 1984a, FcRε+ lymphocytes and regulation of the IgE antibody system. I. A new class of molecules, termed IgE-induced regulants (EIR), which modulate FcRε expression by lymphocytes, *J. Immunol.* **133**:2821–2828

Marcelletti, J. F., and Katz, D. H., 1984b, FcRε+ lymphocytes and regulation of the IgE antibody system. II. FcRε+ B lymphocytes initiate a cascade of cellular and molecular interactions that control FcRε expression and IgE production, *J. Immunol.* **133**:2829–2837

Marcelletti, J. F., and Katz, D. H., 1984c, FcRε+ lymphocytes and regulation of the IgE antibody system. III. Suppressive factor of allergy (SFA) is produced during the in vitro FcRε expression cascade and displays corollary physiologic activity in vivo, *J. Immunol.* **133**:2837–2844

Marcelletti, J. F., and Katz, D. H., 1984d, FcRε+ lymphocytes and regulation of the IgE antibody system. IV. Delineation of target cells and mechanism of action of SFA and EFA in inhibiting in vitro induction of FcRε expression, *J. Immunol.* **133**:2845–2851

Marcelletti, J. F., and Katz, D. H., 1986, FcRε+ lymphocytes and regulation of the IgE antibody system. V. Preliminary physicochemical characterization of the T cell-selective IgE-induced regulant EIR$_T$, *J. Immunol.* **137**:2599–2610

Martens, C. L., Huff, T. F., Jardieu, P., Trounstine, M. L., Coffman, R. L., Ishizaka, K., and Moore, K. W., 1985, cDNA clones encoding IgE-binding factors from a rat-mouse T-cell hybridoma, *Proc. Natl. Acad. Sci. USA* **82**:2460–2464

Martens, C. L., Jardieu, P., Trounstine, M. L., Stuart, S. G., Ishizaka, K., and Moore, K. W., 1987, Potentiating and suppressive IgE-binding factors are expressed by a single cloned gene, *Proc. Natl. Acad. Sci. USA* **84**:809–813

Mathur, A., Maekawa, S., Ovary, Z., and Lynch, R. G., 1986, Increased T cells in Balb/c mice with an IgE-secreting hybridoma, *Mol. Immunol.* **23**:1193–1201

Mathur, A., Conrad, D. H., and Lynch, R. G., 1988, Characterization of the murine T cell receptor for IgE (FcεRII). Demonstration of shared and unshared epitopes with the B cell FcεRII, *J. Immunol.* **141:**2661–2667

Mayumi, M., Kawabe, T., Kim, K.-M., Heike, T., Katamura, K., Yodoi, J., and Mikawa, H., 1988, Regulation of Fcε receptor expression on a human monoblast cell line U937, *Clin. Exp. Immunol.* **71:**202–206

Mazingue, C., Carrière, V., Dessaint, J.-P., Detoeut, F., Turz, T., Auriault, C., and Capron, A., 1987, Regulation of IgE synthesis by macrophage expressing FcE-receptors: Role of interleukin 1, *Clin. Exp. Immunol.* **67:**587–593

Meinke, G. C., Margo, A. M., Lawrence, D. A., and Spiegelberg, H. L., 1978, Characterization of an IgE receptor isolated from cultured B-type lymphoblastoid cells, *J. Immunol.* **121:**1321–1328

Melewicz, F. M., and Spiegelberg, H. L., 1980, Fc receptors for IgE on a subpopulation of human peripheral blood monocytes, *J. Immunol.* **125:**1026–1031

Melewicz, F. M., Plummer, J. M., and Spiegelberg, H. L., 1982, Comparison of the Fc receptors for IgE on human lymphocytes and monocytes, *J. Immunol.* **129:**563–569

Metzger, H., 1988, Molecular aspects of receptors and binding factors for IgE, *Adv. Immunol.* **43:**277–312

Metzger, H., Rivnay, B., Henkart, M., Kanner, B., Kinet, J.-P., and Perez-Montfort, 1984, Analysis of the fine structure and function of the receptor for immunoglobulin E, *Mol. Immunol.* **21:**1167–1173

Moore, K. W., and Martens, C. L., 1986, Homology between IgE-binding factor and retrovirus genes, *Nature* (London) **322:**484

Moore, K. W., Jardieu, P., Mietz, J. A., Trounstine, M. L., Kuff, E. L., Ishizaka, K., and Martens, C. L., 1986, Rodent IgE-binding factor genes are members of an endogenous retrovirus-like gene family, *J. Immunol.* **136:**4283–4290

Nagai, T., Adachi, M., Noro, N., Yodoi, J., and Uchino, H., 1985, T and B lymphocytes with immunoglobulin E Fc receptors (FcεR) in patients with non allergic hyperimmunoglobulinemnia E: Demonstration using a monoclonal antibody against FcεR-associated antigen, *Clin. Immunol. Immunopathol.* **35:**261–275

Nakajima, T., Sarfati, M., and Delespesse, G., 1987, Relationship between human IgE-binding factors (IgE-BF) and lymphocyte receptor for IgE, *J. Immunol.* **139:**848–854

Náray-Fejes-Tóth, A., and Guyre, P. M., 1984, Recombinant human immune interferon induces increased IgE receptor expression on the human monocyte cell line U-937, *J. Immunol.* **133:**1914–1919

Noro, N., Yoshioka, A., Adachi, M., Yasuda, K., Masuda, T., and Yodoi, J., 1986, Monoclonal antibody (H107) inhibiting IgE binding to FcεR(+) human lymphocytes, *J. Immunol.* **137:**1258–1263

Nutman, T. B., Delespesse, G., Sarfati, M., and Volkman, D. J., 1987, T cell-derived IgE-binding factors. I. Cloned and transformed T cells producing IgE-binding factors, *J. Immunol.* **139:**4049–4054

Ogilvie, B. M., and Johns, V. E., 1969, Reaginic antibodies and helminth infections, in *Cellular and Humoral Mechanisms in Anaphylaxis and Allergy* (H. Z. Movot, ed.), pp. 13–22, S. Karger, Basel

Pancré, V., Joseph, M., Capron, A., Wietzerbin, J., Kusnierz, J.-P., Vorng, H., and Auriault, C., 1988, Recombinant human interferon-γ induces increased IgE receptor expression on human platelets, *Eur. J. Immunol.* **18:**829–832

Pène, J., Rousset, F., Brière, F., Chrétien, I., Wideman, J., Bonnefoy, J. Y., and De Vries, J. E., 1988, Interleukin 5 enhances interleukin 4-induced IgE production by normal human B cells. The role of soluble CD23 antigen, *Eur. J. Immunol.* **18:**929–935

Peterson, L. H., and Conrad, D. H., 1985, Fine specificity, structure, and proteolytic susceptibility of the human lymphocyte receptor for IgE, *J. Immunol.* **135:**2654–2660

Petit-Koskas, E., Genot, E., Lawrence, D., and Kolb, J.-P., 1988, Inhibition of the proliferative response of human B lymphocytes to B cell growth factor by transforming growth factor-beta, *Eur. J. Immunol.* **18:**111–116

Rankin, J. A., Hitchcock, M., Merrill, W. W., Huang, S. S., Brashler, J. R., Bach, M. K., and Askenase, P. W., 1984, IgE immune complexes induce immediate and prolonged release of leukotriene C_4 (LTC_4) from rat alveolar macrophages, *J. Immunol.* **132:**1993–1999

Rao, M., Lee, W. T., and Conrad, D., 1987, Characterization of a monoclonal antibody directed against the murine B lymphocyte receptor for IgE, *J. Immunol.* **138:**1845–1851

Rector, E., Nakajima, T., Rocha, C., Duncan, D., Lestourgeon, D., Mitchell, R. S., Fisher, J., Sehon, A. H., and Delespesse, G., 1985, Detection and characterization of monoclonal antibodies specific to IgE receptors on human lymphocytes by flow cytometry, *Immunology* **55:**481–488

Richards, M. L., Marcelletti, J. F., and Katz, D. H., 1988, IgE-antigen complexes enhance FcεR and Ia expression by murine B lymphocytes, *J. Exp. Med.* **168:**571–587

Rousset, F., Malefijt, R. W., Slierendregt, B., Aubry, J.-P., Bonnefoy, J. Y., Defrance, T., Banchereau, J., and De Vries, J. E., 1988, Regulation of Fc receptor for IgE (CD23) and class II MHC antigen expression on Burkitt's lymphoma cell lines by human IL-4 and IFN-γ, *J. Immunol.* **140:**2625–2632

Rouzer, C. A., Scott, W. A., Hamil, A. L., Liu, F. T., Katz, D. H., and Cohn, Z. A., 1982, Secretion of leukotriene C and other arachidonic acid metabolites by macrophages challenged with immunoglobulin E immune complexes, *J. Exp. Med.* **156:**1077–1086

Sarfati, M., and Delespesse, G., 1988, Possible role of human lymphocyte receptor for IgE (CD23) or its soluble fragments in the *in vitro* synthesis of human IgE, *J. Immunol.* **141:**2195–2199

Sarfati, M., Rector, E., Wong, K., Rubio-Trujillo, M., Sehon, A. H., and Delespesse, G., 1984a, In vitro synthesis of IgE by human lymphocytes. II. Enhancement of the spontaneous IgE synthesis by IgE-binding factors secreted by RPMI 8866 lymphoblastoid B cells, *Immunology* **53:**197–205

Sarfati, M., Rector, E., Rubio-Trujillo, M., Wong, K., Sehon, A. H., and Delespesse, G., 1984b, In vitro synthesis of IgE by human lymphocytes. III. IgE-potentiating activity of culture supernatants from Epstein-Barr virus (EBV) transformed B cells, *Immunology* **53:**207–214

Sarfati, M., Vanderbeeken, Y., Rubio-Trujillo, M., Duncan, D., Delespesse, G., 1986a, Presence of IgE suppressor factors in human colostrum, *Eur. J. Immunol.* **16:**1005–1008

Sarfati, M., Nutman, T., Fonteyn, C., and Delespesse, G., 1986b, Presence of antigenic determinants common to FcIgE receptors on human macrophages, T and B lymphocytes and IgE-binding factors, *Immunology* **59:**569–575

Sarfati, M., Nutman, T. B., Suter, U., Hofstetter, H., and Delespesse, G., 1987, T cell-derived IgE-binding factors. II. Purification and characterization of IgE-binding factors produced by human T cell leukemia/lymphoma virus-1-transformed T lymphocytes, *J. Immunol.* **139:**4055–4060

Sarfati, M., Bron, D., Lagneaux, L., Fonteyn, C., Frost, H., and Delespesse, G., 1988, Elevation of IgE-binding factors in serum of patients with B cell-derived chronic lymphocytic leukemia, *Blood* **71:**94–98

Sherr, E., Macy, E., Kimata, H., Gilly, M., and Saxon, A., 1989, Binding the low affinity FcεR on B cells suppresses ongoing human IgE synthesis, *J. Immunol.* **142:**481–489

Spiegelberg, H. L., 1984, Structure and function of Fc receptors for IgE on lymphocytes, monocytes and macrophages, *Adv. Immunol.* **35:**61–88

Spiegelberg, H. L., and Melewicz, F. M., 1980, Fc receptors specific for IgE on subpopulations of human lymphocytes and monocytes, *Clin. Immunol. Immunopathol.* **15:**424–433

Spiegelberg, H. L., and Simon, R. A., 1981, Increase of lymphocytes with Fc receptors for IgE in patients with allergic rhinitis during the grass pollen season, *J. Clin. Invest.* **68:**845–852

Spiegelberg, H. L., O'Connor, R. D., Simon, R. A., and Mathison, D. A., 1979, Lymphocytes with immunoglobulin E Fc receptors in patients with atopic disorders, *J. Clin. Invest.* **64:**714–720

Stadler, B. M., Gauchat, D., Hildbrand, M.-L., Yang, X., and de Weck, A. L., 1987, Human T hybridoma-derived immunoglobulin-binding factors, *Int. Arch. Allergy Appl. Immunol.* **82:**405–407

Suemura, M., and Ishizaka, K., 1979, Potentiation of IgE response in vitro by T cells from rats infected with *Nippostrongylus brasiliensis*, *J. Immunol.* **123:**918–924

Suemura, M., and Kishimoto, T., 1985, Regulation of human IgE responses by T cells and their products, *Int. Arch. Allergy Appl. Immunol.* **77:**26–31

Suemura, M., Kishimoto, T., Hirai, Y., and Yamamura, Y., 1977, Regulation of antibody response in different immunoglobulin classes. III. In vitro demonstration of "IgE class-specific" suppressor functions of DNP-Mycobacterium-primed T cells and the soluble factor released from these cells, *J. Immunol.* **119:**149–155

Suemura, M., Yodoi, J., Hirashima, M., and Ishizaka, K., 1980, Regulatory role of IgE-binding factors from rat T lymphocytes. I. Mechanism of enhancement of IgE response by IgE-potentiating factor, *J. Immunol.* **125:**148–154

Suemura, M., Shiho, O., Deguchi, H., Yamamura, Y., Bottcher, I., and Kishimoto, T., 1981, Characterization and isolation of IgE class-specific suppressor factor (IgE-TsF). I. The presence of the binding site(s) for IgE and H-2 gene products in IgE-TsF, *J. Immunol.* **127:**465–470

Suemura, M., Kikutani, H., Barsumian, E. L., Hattori, Y., Kishimoto, S., Sato, R., Maeda, A., Nakamura, H., Owaki, H., Hardy, R. R., and Kishimoto, T., 1986, Monoclonal anti-Fcε receptor antibodies with different specificities and studies on the expression of Fcε receptors on human B and T cells, *J. Immunol.* **137:**1214–1220

Swendeman, S., and Thorley-Lawson, D. A., 1987, The activation antigen BLAST-2, when shed, is an autocrine BCGF for normal and transformed B cells, *EMBO J.* **6:**1637–1642

Tada, T., Taniguchi, M., and David, C. S., 1976, Properties of antigen-specific suppressive T cell factor in the regulation of antibody response of the mouse. IV. Special subregion assignment of the gene that codes for the suppressive T cell factor in the H-2 histocompatibility complex, *J. Exp. Med.* **144:**713–725

Thompson, L. F., Spiegelberg, H. L., and Buckley, R. H., 1985, IgE Fc receptor positive T and B lymphocytes in patients with the hyper IgE syndrome, *Clin. Exp. Immunol.* **59:**77–84

Thorley-Lawson, D., and Mann, K. P., 1985, Early events in Epstein-Barr virus infection provide a model for B cell activation, *J. Exp. Med.* **162:**45–59

Thorley-Lawson, D., Nadler, L. M., Bhan, A. K., and Schooley, R. T., 1985, BLAST-2 (EBVCS), an early cell surface marker of human B cell activation, is superinduced by Epstein-Barr virus, *J. Immunol.* **134:**3007–3012

Thorley-Lawson, D., Swendeman, S. L., and Edson, C. M., 1986, Biochemical analysis suggests distinct functional roles for the BLAST-1 and BLAST-2 antigens, *J. Immunol.* **136:**1745–1751

Tokuhisa, T., Taniguchi, M., Okumura, K., and Tada, T., 1978, An antigen-specific I region gene product that augments the antibody response, *J. Immunol.* **120:**414–421

Uchibayashi, N., Kikutani, H., Barsumian, E. L., Hauptmann, R., Schneider, F.-J., Schwendenwein, R., Sommergruber, W., Spevak, W., Maurer-Fogy, I., Suemura, M., and Kishimoto, T., 1989, Recombinant soluble Fcε receptor II (FcεRII/CD23) has IgE binding activity but no B cell growth promoting activity, *J. Immunol.* **142:**3901–3908

Uede, T., and Ishizaka, K., 1982, Formation of IgE-binding factors by rat T lymphocytes. VI. Cellular mechanism for the formation of IgE-potentiating factor and IgE-suppressive factor by antigenic stimulation of antigen-primed spleen cells, *J. Immunol.* **129:**1391–1397

Uede, T., and Ishizaka, K., 1984, IgE-binding factors from mouse T lymphocytes. II. Strain differences in the nature of IgE-binding factor, *J. Immunol.* **133:**359–367

Uede, T., Huff, T. F., and Ishizaka, K., 1982, Formation of IgE-binding factors by rat T lymphocytes. V. Effect of adjuvant for the priming immunization on the nature of IgE binding factors formed by antigenic stimulation, *J. Immunol.* **129:**1384–1390

Uede, T., Sandberg, K., Bloom, B. R., and Ishizaka, K., 1983a, IgE-binding factors from mouse T lymphocytes. I. Formation of IgE-binding factors by stimulation with homologous IgE and interferon, *J. Immunol.* **130:**649–654

Uede, T., Hirata, F., Hirashima, M., and Ishizaka, K., 1983b, Modulation of the biologic activities of IgE-binding factors. I. Identification of glycosylation-inhibiting factor as a fragment of lipomodulin, *J. Immunol.* **130:**878–884

Uede, T., Huff, T. F., and Ishizaka, K., 1984, Suppression of IgE synthesis in mouse plasma cells and B cells by rat IgE-suppressive factor, *J. Immunol.* **133:**803–808

Urban Jr., J. F., Ishizaka, T., and Ishizaka, K., 1977, IgE formation in the rat following infection with *Nippostrongylus brasiliensis*. III. Soluble factor for the generation of IgE-bearing lymphocytes, *J. Immunol.* **119:**583–590

Vander-Mallie, R., Ishizaka, T., and Ishizaka, K., 1982, Lymphocytes bearing Fc receptors for IgE. VIII. Affinity of mouse IgE for FcεR on mouse B lymphocytes, *J. Immunol.* **128:**2306–2312

Vercelli, D., Jabara, H. H., Lee, B.-W., Woodland, N., Geha, R. S., and Leung, D. Y. M., 1988, Human recombinant interleukin 4 induces FcεR2/CD23 on normal human monocytes, *J. Exp. Med.* **167:**1406–1416

Waldschmidt, T. J., Conrad, D. H., and Lynch, R. G., 1988, The expression of B cell surface receptors. I. The ontogeny and distribution of the murine B cell IgE Fc receptor, *J. Immunol.* **140:**2148–2154

Yanagihara, Y., Kajiwara, K., Kiniwa, M., Kamisaki, T., Yui, Y., Shida, T., and Delespesse, G., 1987, Modulation of IgE-synthesis by IgE-binding factors released by T cells of asthmatic patients with elevated serum IgE, *Microbiol. Immunol.* **31:**261–274

Ymer, S., and Young, I. G., 1986, Homology between IgE-binding factor gene and endogenous retroviruses, *Nature* (London) **323:**186

Ymer, S., Tucker, W. Q. J., Campbell, H. D., and Young, I. G., 1986, Nucleotide sequence of the intracisternal A-particle genome inserted 5′ to the interleukin-3 gene of the leukemia cell line WEHI-3B, *Nucleic Acids Res.* **14:**5901–5918

Yodoi, J., and Ishizaka, K., 1979a, Lymphocytes bearing Fc receptor for IgE. I. Presence of human and rat T lymphocytes with Fcε receptors, *J. Immunol.* **122:**2577–2583

Yodoi, J., and Ishizaka, K., 1979b, Lymphocytes bearing Fc receptors for IgE. III. Transition of FcγR(+) cells to FcεR(+) cells by IgE, *J. Immunol.* **123:**2004–2010

Yodoi, J., and Ishizaka, K., 1980a, Induction of Fcϵ-receptor bearing cells in vitro in human peripheral lymphocytes, *J. Immunol.* **124:**934–938

Yodoi, J., and Ishizaka, K., 1980b, Lymphocytes bearing Fc receptors for IgE. IV. Formation of IgE-binding factor by rat T lymphocytes, *J. Immunol.* **124:**1322–1328

Yodoi, J., Ishizaka, T., and Ishizaka, K., 1979, Lymphocytes bearing Fc receptor for IgE. II. Induction of Fcϵ-receptor bearing rat lymphocytes by IgE, *J. Immunol.* **123:**455–462

Yodoi, J., Hirashima, M., and Ishizaka, K., 1980, Regulatory role of IgE-binding factors from rat T lymphocytes. II. Glycoprotein nature and source of IgE-potentiating factor, *J. Immunol.* **125:**1436–1441

Yodoi, J., Hirashima, M., and Ishizaka, K., 1981a, Lymphocytes bearing Fc receptors for IgE. V. Effect of tunicamycin on the formation of IgE-potentiating factor and IgE-suppressive factor by Con A-activated lymphocytes, *J. Immunol.* **126:**877–882

Yodoi, J., Hirashima, M., and Ishizaka, K., 1981b, Lymphocytes bearing Fc receptors of IgE. VI. Suppressive effect of glucocorticoids on the expression of Fcϵ receptors and glycosylation of IgE-binding factors, *J. Immunol.* **127:**471–476

Yodoi, J., Hirashima, M., Hirata, F., DeBlas, A. L., and Ishizaka, K., 1981c, Lymphocytes bearing Fc receptors for IgE. VII. Possible participation of phospholipase A2 in the glycosylation of IgE-binding factors, *J. Immunol.* **127:**476–482

Yodoi, J., Hirashima, M., Bloom, B. R., and Ishizaka, K., 1981d, Formation of IgE-binding factors by rat T lymphocytes. I. Induction of IgE-binding factors by poly I:C and interferon, *J. Immunol.* **127:**1579–1585

Yodoi, J., Hirashima, M., and Ishizaka, K., 1982, Regulatory role of IgE-binding factors from rat T lymphocytes. V. The carbohydrate moieties in IgE-potentiating and IgE-suppressive factors, *J. Immunol.* **128:**289–295

Yodoi, J., Adachi, M., Teshigawara, K., Miyama-Inaba, M., Masuda, T., and Fridman, W. H., 1983, T cell hybridomas coexpressing Fc receptors (FcR) for different isotype. II. IgA-induced formation of suppressive IgA-binding factor(s) by a murine T hybridoma bearing FcγR and FcαR, *J. Immunol.* **131:**303–310.

Yodoi, J., Adachi, M., and Noro, N., 1986, IgA binding factors and Fc receptors for IgA: Comparative studies between IgA and IgE Fc receptor systems, *Int. Rev. Immunol.* **2:**73–97

Yokota, A., Kikutani, H., Tanaka, T., Sato, R., Barsumian, E. L., Suemura, M., and Kishimoto, T., 1988, Two species of human Fcϵ receptor II (FcϵRII/CD23): Tissue-specific and IL-4-specific regulation of gene expression, *Cell.* **55:**611–618

Young, M. C., Leung, D. Y. M., and Geha, R. S., 1984, Production of IgE-potentiating factor in man by T cell lines bearing Fc receptors for IgE, *Eur. J. Immunol.* **14:**871–878

Young, M. C., Geha, R. S., Maksad, K. N., and Leung, D. Y. M., 1986, Characterization of human T cell-derived IgE-potentiating factor, *Eur. J. Immunol.* **16:**985–991

Yukawa, K., Kikutani, H., Owaki, H., Yamasaki, K., Yokota, A., Nakamura, H., Barsumian, E. L., Hardy, R. R., Suemura, M., and Kishimoto, T., 1987, A B cell-specific differentiation antigen, CD23, is a receptor for IgE (FcϵR) on lymphocytes, *J. Immunol.* **138:**2576–2580

Chapter 6

Lymphocyte-Mediated Cytolysis
Role of Granule Mediators

John Ding-E Young, Byoung S. Kwon, Joseph A. Trapani,
Chau-Ching Liu, and Lucy H. Y. Young

Cytotoxic T lymphocytes (CTL) and natural killer (NK) cells represent the two main effector lymphocyte populations thought to be involved in killing virus-infected and transformed cells (reviewed in Berke, 1983; Herberman *et al.*, 1986; Trinchieri and Perussia, 1984; Goldfarb, 1986). Killing by lymphocytes is known to require a contact-dependent mechanism. Although the precise mechanisms by which lymphocytes kill their targets are still unclear, a large body of information has recently accumulated on various mediators that may play a direct role in lymphocyte-mediated killing. This information will be reviewed briefly here.

1. ROLE OF GRANULES AND PERFORIN IN LYMPHOCYTE-MEDIATED KILLING

1.1. The Granule Exocytosis or Secretion Model for Cell Killing

With the advent of a better understanding of the growth requirements of lymphocytes, numerous CTL and NK cell clones have been derived *in vitro* and propagated in long-term cultures containing the T-cell growth hormone interleukin 2 (IL2) (reviewed in Brooks *et al.*, 1983). The availability, for the first time, of homogeneous populations of

John Ding-E Young and Chau-Ching Liu   Laboratory of Cellular Physiology and Immunology, The Rockefeller University, New York, New York 10021, USA.   Byoung S. Kwon   Department of Microbiology and Immunology and Walther Oncology Center, Indiana University School of Medicine, Indianapolis, Indiana 46223, USA.   Joseph A. Trapani   Laboratory of Human Immunogenetics, Memorial Sloan-Kettering Cancer Center, New York, New York 10021, USA.   Lucy H. Y. Young   The Massachusetts Eye and Ear Infirmary, Harvard Medical School, Boston, Massachusetts 02114, USA.

effector lymphocytes that could be grown to large numbers was instrumental in allowing subsequent biochemical analysis of the various cytolytic mediators produced by this cell type.

One of the more prominent features of CTL and NK cell clones that have been maintained *in vitro* is the presence of large cytoplasmic granules. These granules are azurophilic, containing electron-dense cores bounded by vesicular material. A role of these cytoplasmic granules in cell-mediated killing has been suggested by morphologic studies demonstrating accumulation of granules at the site of contact with target cells (Henkart and Henkart, 1982; Kupfer and Dennert, 1984; Yannelli *et al.*, 1986). Following cell-to-cell contact, this displacement of granules occurs by means of an active reorientation of granules and the microtubular network in the direction of the target cells (Kupfer and Dennert, 1984; Dennert *et al.*, 1985). The involvement of granules in lymphocyte-mediated killing was largely supported by studies in which granules isolated from CTL and NK cell clones were shown to be cytolytic (reviewed in Henkart, 1985; Podack, 1985; Young and Cohn, 1987; Tschopp and Jonheneel, 1988). The cytoplasmic granules of lymphocytes have been isolated by several subcellular fractionation procedures, all of which make use of differential equilibration of granules in density gradients. Isolated granules are cytolytic to a variety of target cells and show little specificity for the species or the nature of the target cells used.

Numerous studies in the past have attempted to identify morphologic lesions that may putatively be associated with lymphocyte-mediated killing. Although several earlier studies have suggested that lymphocytes may produce discrete lesions on target cell membranes (Henney, 1973, 1974; Martz, 1976), this inference was based on functional studies that determined an increased diffusion of macromolecular markers through plasma membranes of target cells which had been exposed to lymphocytes. A direct demonstration of membrane lesions came only with a study by Dourmashkin and colleagues (1980) in which peripheral blood lymphocytes were shown to assemble tubular lesions with internal diameters of approximately 15 nm. This study was later confirmed and significantly extended to CTL and NK cell clones (Podack and Dennert, 1983; Dennert and Podack, 1983). Both of these effector cell populations were shown to assemble ringlike lesions of two distinct diameters: 16 nm and 5 nm. These lesions resemble the tubular lesions produced by complement (C), which had previously been shown to assemble lesions with internal diameters of 10 nm (Mayer *et al.*, 1981; Bhakdi and Tranum-Jensen, 1983). The two distinct tubular structures of 16 and 5 nm were named polyperforins 1 and 2, respectively, and their putative mediators were designated perforins 1 and 2. These putative mediators have also been called cytolysin (Henkart, 1985), lymphocyte pore-forming protein [PFP (Young and Cohn, 1987)], and C9-related polypeptide [C9RP (Müller-Eberhard, 1988)]. Collectively, this type of cell killing has been known as the granule exocytosis or the secretion model of cytotoxicity. Accordingly, cell death by this mechanism is thought to occur as a direct result of membrane perturbation and increased leakiness, with target cell rupture occurring as a consequence of colloid-osmotic swelling. This proposed mechanism closely resembles killing mediated by assembly of C lesions.

1.2. Cytoplasmic Granules and Perforin as Mediators of Cytotoxicity

Subsequent work from several laboratories has identified the cytoplasmic granules as the organelles responsible for assembly of tubular lesions (reviewed extensively in

Henkart, 1985; Podack, 1985; Young and Cohn, 1987; Tschopp and Jonheneel, 1988). Proteins of isolated granules are also capable of producing this type of lesion (Figure 1). The PFP/perforin monomer has been identified as a protein with a single subunit that migrates in SDS-gels under reducing conditions with a molecular masses of 66–68 kDa (Masson and Tschopp, 1985) and 70 kDa (Podack *et al.*, 1985; Young *et al.*, 1986a; Henkart *et al.*, 1986). To date, only one perforin species has been identified. In our hands, isolated perforin gives rise to lesions of multiple sizes, accounting probably for both polyperforin 1 and polyperforin 2 lesions. The isolated protein is cytolytic to a variety of target cells tested. Perforin-mediated lysis is inextricably dependent on the presence of submillimolar concentrations of calcium. Perforin has been incorporated successfully into high-resistance planar lipid bilayers and shown to produce ion channels of large unitary conductances (in the order of nanosiemens in physiological salt concentrations), which are comparable to the pores produced by C (Young *et al.*, 1986b). In the presence of calcium, the perforin monomer polymerizes into supramolecular structures with molecular masses exceeding 1 million Da that resist dissociation by SDS, disulfide bond-reducing reagents, and boiling (Young *et al.*, 1986b).

1.2.1. Structure and Function of Perforin

The perforin monomer has been shown to be immunologically related to various C components, namely, the terminal membrane attack complex components C5b-6, C7, C8, and C9 (Young *et al.*, 1986a; Tschopp *et al.*, 1986; Young *et al.*, 1986c; Stanley and Luzio, 1988). In particular, perforin bears close structural and functional resemblance to C9, which had previously been shown to be capable of assembling by itself, under certain conditions, into polymeric structures (poly-C9 lesions) (Podack and Tschopp, 1982).

The homology between perforin and the terminal components was subsequently confirmed by cDNA cloning, which revealed a 25–30% amino acid identity between perforin and C9 (Shinkai *et al.*, 1988a,b; Lichtenheld *et al.*, 1988; Lowrey *et al.*, 1989; Kwon *et al.*, 1989). The primary sequences of the mouse and human forms are about 70% identical, and various segments with putative secondary alpha-helical or beta-coil structures are conserved in the two species (Kwon *et al.*, 1989) (Figure 2). These segments may represent putative membrane-spanning domains. With respect to domain-specific functions, perforin appears to partially resemble the terminal C components (Stanley and Luzio, 1988). One important difference between C and perforin should be noted. Unlike C-mediated lesions, the assembly of perforin lesions does not appear to require any other intermediate proteins. Perforin binds directly to lipid molecules, which may function as perforin receptors in the membrane (Yue *et al.*, 1987; Young *et al.*, 1987b; Tschopp *et al.*, 1989). The C components, on the other hand, produce heteropolymeric tubules that require successive binding of a cascade of intermediate proteins. The C pore interior is currently thought to be lined up simultaneously by various C components (that is, by at least one C5b-6, one C7, one C8, and one or multiple molecules of C9; see Bhakdi and Tranum-Jensen, 1987). Our current working model for perforin assembly proposes the formation of nucleation sites by membrane-inserted perforin monomers, in which form they may bind to additional perforin molecules to create polymeric lesions. Accordingly, the perforin lesion would grow in size in proportion to the number of recruited monomers. This "barrel stave" model would account for the marked heterogeneity of sizes associated with perforin lesions observed. This model also predicts that functionally active channels

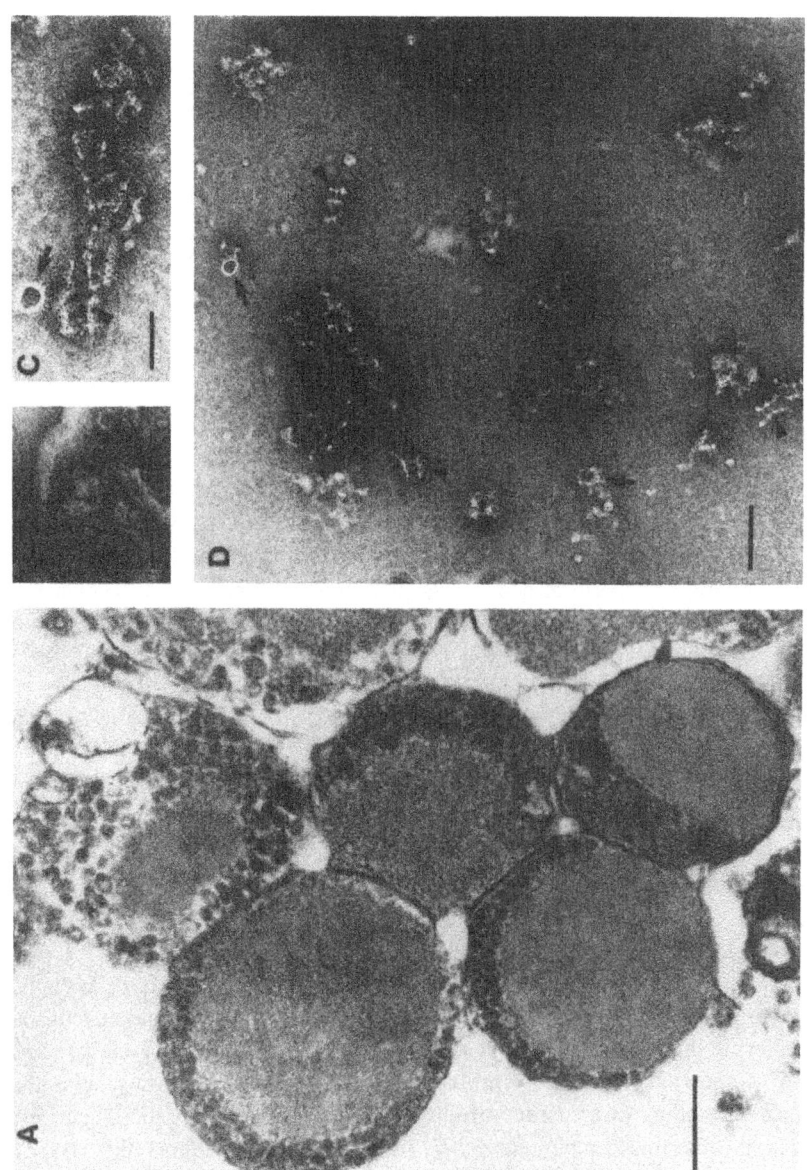

FIGURE 1. Morphology of isolated granules and granule-derived lesions. (A) Granules isolated by centrifugation through Percoll gradients. (B–D) Membrane lesions produced by granules on sheep erythrocyte membranes. Arrows point to top views of circular lesions. Arrowheads correspond to longitudinal views of the tubular lesions. Scale bars = 270 nm (A), 57 nm (B), 38 nm (C), and 87 nm (D). [Reproduced from Young *et al.* (1986b), with permission.]

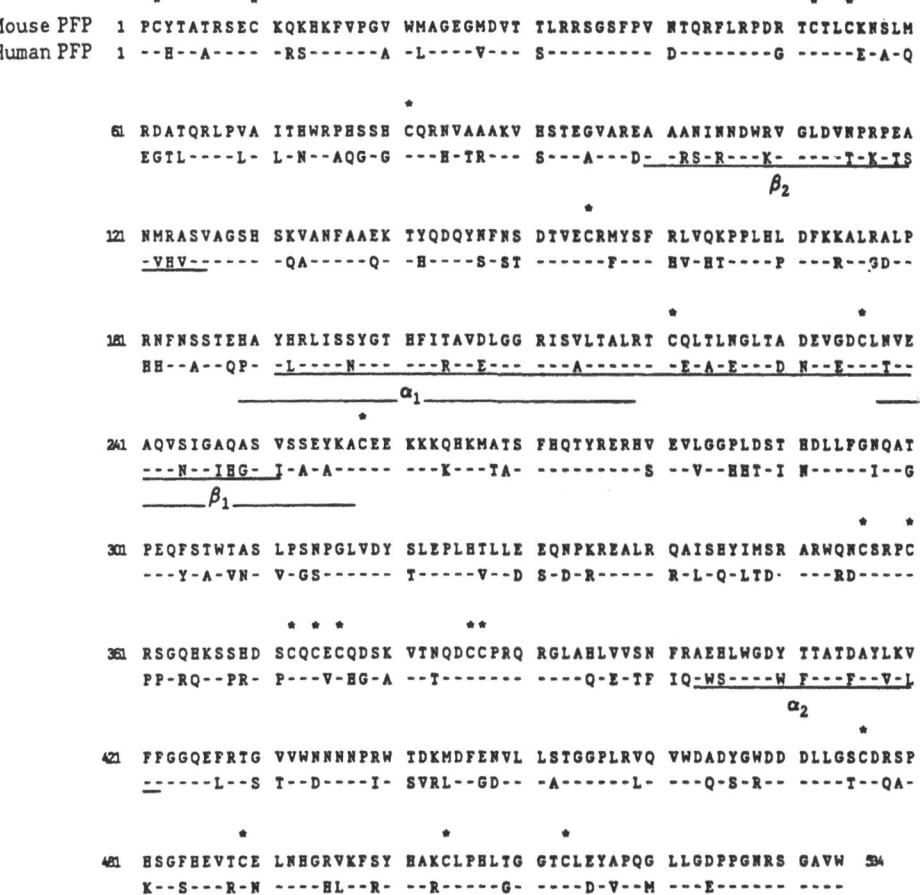

FIGURE 2. Alignment of amino acid sequences of mouse and human PFP/perforin. Identical amino acids in the human sequence are not displayed (dashes). Cysteines are indicated by asterisks. The region of high hydrophobicity (191–251) and the candidate amphiphilic α or β domains are underlined. [Reproduced from Kwon *et al.* (1989), with permission.]

may form even prior to the stage during which lesions become visible as tubular structures when observed by electron microscopy. It is possible that target cells may become leaky or damaged even in the absence of ultrastructurally visible membrane lesions; these lesions would be expected to form only when high concentrations of perforin monomer are optimally concentrated near the target membrane. Although it is likely that binding to calcium ions produces a profound conformational or structural change in perforin monomers, enabling them to bind to and to insert into lipid bilayers (Young *et al.*, 1987b), the putative calcium-binding sites have not been identified, and they are not obvious from analysis of the primary sequence of perforin (Kwon *et al.*, 1989).

The perforin mRNA is approximately 2.9 kb in size. The coding region of perforin is localized to a single locus spanning less than 10 kb of genomic sequence (Lichtenheld *et al.*, 1988; Kwon *et al.*, 1989). The mouse perforin gene has been mapped to chromosome 10 (Trapani *et al.*, 1990), and the genomic sequence of mouse perforin has been obtained

(J. A. Trapani *et al.*, 1990). The coding region of the mouse perforin is interrupted by a single intron of approximately 2.3 kb. This is in marked contrast to the gene encoding C9, which spans about 80 kb and contains at least 12 introns (Stanley, 1988).

1.2.2. Distribution of Perforin

Both mouse and human forms of perforin are found predominantly in lymphocytes bearing surface CD8 (CTL marker) or CD16 (Fc receptor, an NK cell marker). It is generally absent from nonlymphocyte populations or noncytolytic T-cell subsets (Figure 3). However, it should be noted that a weak perforin mRNA signal is present in HuT 78, a human T-cell leukemia virus type I-transformed T-cell line of CD4 (T helper) phenotype (Lichtenheld *et al.*, 1988; see also Figure 3). Resting lymphocyte populations do not bear perforin mRNA or product. However, upon stimulation with lectins, phorbol esters, IL2, and anti-CD3 antibodies, lymphocytes are activated to produce perforin (unpublished). The most potent signal for perforin production appears to be IL2, which also drives CTL and NK cells to proliferate. The production of perforin after IL2 stimulation does not appear to require cell proliferation, since an IL2-independent CTL line has been found to produce perforin mRNA and protein but not to proliferate in the presence of exogenous IL2 (Liu *et al.*, 1990). The same CTL line also responds directly to IL3, IL4, and IL6 stimulation with enhanced production of perforin (unpublished).

All NK cell populations examined to date have been shown to contain perforin mRNA and protein, including primary cell populations that have been stimulated in short-term cultures in the presence of IL2 (Zalman *et al.*, 1986a; Liu *et al.*, 1986; Henkart *et al.*, 1986; Zychlinsky *et al.*, 1990). Antibodies produced against NK cell granules have been shown to effectively block NK-mediated lysis (Reynolds *et al.*, 1987).

The distribution and role of perforin in CTL have been somewhat more controversial. A panel of CTL lines was assessed for production of perforin (Young *et al.*, 1987a; Allbritton *et al.*, 1988b) and its mRNA (Joag *et al.*, 1990). Only some of the CTL lines tested produced measurable amounts of perforin. This result is puzzling in view of the putative role of perforin as a mediator of CTL-mediated cytotoxicity. Moreover, Berke and Rosen (1987, 1988) and Dennert *et al.* (1987) have shown that certain CTL popula-

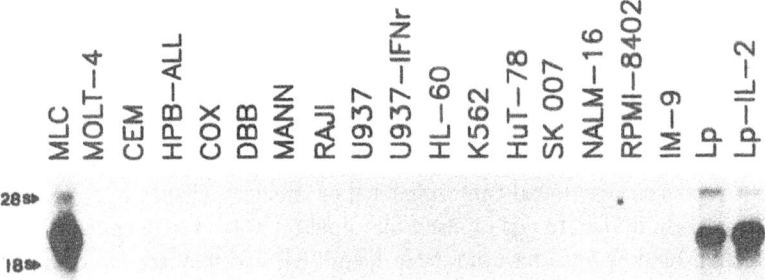

FIGURE 3. Northern blot analysis of human PFP/perforin. Expression of perforin message was assessed in various cell lines. Lanes contain poly(A)+ RNA (1 μg). MLC represents a bulk culture of allogenically stimulated human CTL and NK cells. Lp represents the cells of a patient with a marked leukocytosis involving large granular lymphocytes. Lp-IL2 RNA was derived from the same cells cultured with recombinant IL2 for 20 days. The autoradiogram was exposed for 5 days.

tions that have been primed *in vivo* do not produce any measurable levels of perforin activity. However, upon *in vitro* stimulation with IL2, these CTL acquire granules and perforin (Berke and Rosen, 1987, 1988). On the basis of these results, the two groups have argued against a role for perforin in mediating the cytotoxicity of CTL primed *in vivo*, raising the possibility that perforin may be inextricably associated with CTL that have been propagated in long-term cultures in the presence of IL2.

To address this issue and to assess the physiological role of perforin, we have recently initiated immunolocalization studies to determine the tissue distribution of perforin. Preliminary studies have shown that the perforin antigen is not found in murine resident tissues. We then sought to determine the presence of perforin in tissues afflicted with certain pathological conditions, such as acute viral infections and cell-mediated autoimmune disorders. We have studied experimental murine infections produced by lymphocytic choriomeningitis virus (LCMV) and coxsackievirus B type 3. The hepatotropic and neurotropic strains of LCMV and the myocardiotropic strain of coxsackievirus B were used in our studies. In tissues infected with LCMV, we detected a marked infiltration of CTL blasts bearing the perforin antigen (Young *et al.*, 1989a) (Figure 4). Very few NK cells were found in LCMV-infected tissues. In contrast, the myocardia of mice infected with coxsackievirus B contained a greater proportion of NK cells which were also perforin positive (unpublished).

We also studied the presence of perforin in the pancreas of nonobese diabetic (NOD) mice. NOD mice develop spontaneous, insulin-dependent (or type I) diabetes mellitus. Furthermore, young NOD mice can be induced to become diabetic upon adoptive transfer of lymphocytes obtained from adult, spontaneously diabetic NOD mice. Both in adult, spontaneously diabetic NOD mice and in adoptively transferred recipients, we observed the presence of a small subset of CD8-positive lymphocytes (presumably of the CTL phenotype) bearing the perforin antigen (Young *et al.*, 1989b). In all three pathologies examined, the perforin antigen was detected only in lymphocytes actively engaged in proliferation or blastogenesis. This observation is consistent with the fact that all previous studies on perforin were performed with CTL or NK cell populations actively engaged in proliferation. Thus, these studies demonstrate that perforin is in fact found associated with CTL and NK cells primed *in vivo* and may thereby play an important role in limiting viral infections and in producing cell-mediated, autoimmune disorders by mediating lysis of virus-infected and other targeted host cells. Our observation that perforin is found associated with effector lymphocytes actively engaged in blastogenesis suggests that the key issue is not whether perforin is found *in vivo* but whether CTL and NK cell blasts play an important role in cell-mediated cytotoxic reactions. Perforin may be produced *in vivo* only under intense stimulatory conditions (as in a viral focus) that lead to local tissue increase of IL2 levels, which in turn may drive lymphocytes to undergo blastogenesis. The development of specific antibody and cDNA probes for perforin should permit a more detailed study on the physiological role of perforin in the near future.

1.3. A Family of Serine Esterases Localized in Lymphocyte Granules

In addition to perforin, the lymphocyte granules also contain several serine esterases (Pasternack and Eisen, 1985; Pasternack *et al.*, 1986; Masson *et al.*, 1986; Young *et al.*, 1986c; Simon *et al.*, 1986; Masson and Tschopp, 1987). Serine esterase I [also designated

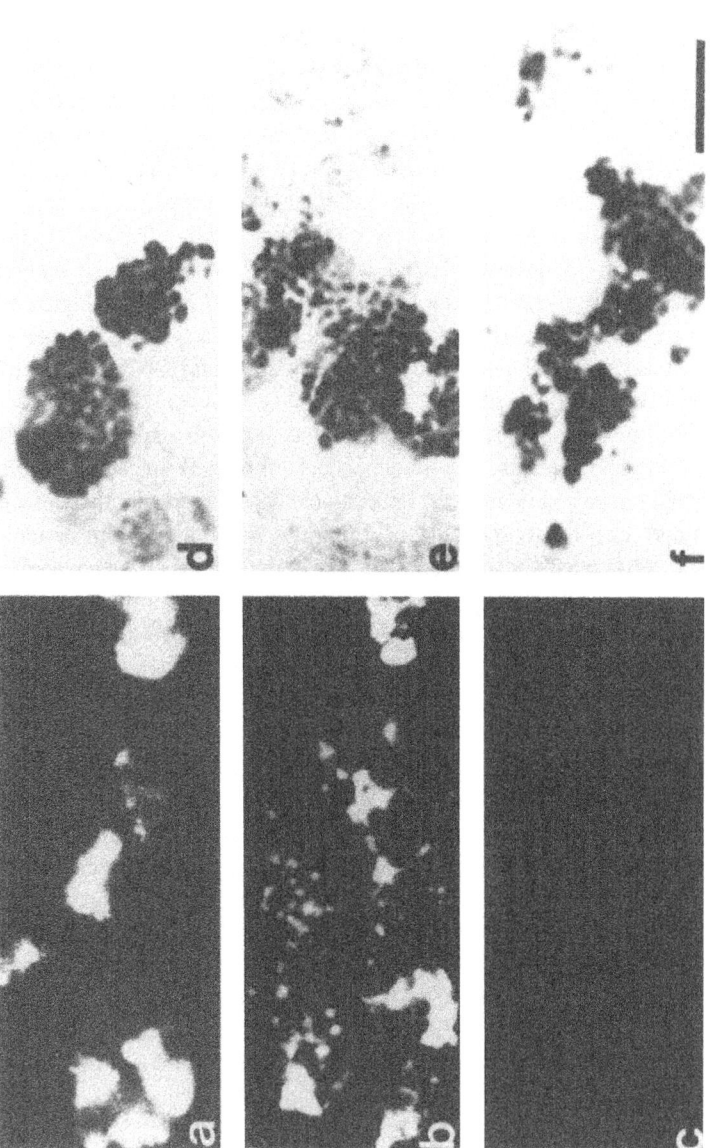

FIGURE 4. Immunofluorescent staining (a–c) and Avidin-Biotin complex (ABC)-immunoperoxidase staining (d–f) of frozen sections of LCMV (Armstrong strain)-infected brains with rabbit anti-PFP/perforin antiserum (a, b, and d–f) and preimmune serum (c). Shown are selected images of frozen sections obtained from SWR/J mice at post mortem day 5 postinfection. Note the intense granularity of stain seen in some cells (b and d–f). Bar = 12 μm (a–c) and 9 μm (d–f). [Reproduced from Young *et al.* (1989a), with permission.]

granzyme A by Masson *et al.* (1986) and T-cell-specific protease I by Simon *et al.* (1986)] is a 62-kDa disulfide-linked homodimer of two 35-kDa subunits that displays trypsinlike activity. To date, this enzyme has been studied in most detail (for reviews, see Kramer and Simon, 1987, and Jenne and Tschopp, 1988). Several synthetic substrates have been used to define its specificity. However, the precise function of this enzyme and of six other closely related enzymes (Masson and Tschopp, 1987) is still unclear.

An alternative approach has also been used to study the lymphocyte serine esterases. Several laboratories have independently derived CTL-specific cDNA libraries by using the subtractive hybridization method (Lobe *et al.*, 1986; Gershenfeld and Weissman, 1986; Brunet *et al.*, 1986; Kwon *et al.*, 1987). The CTL-associated cDNA clones obtained by this strategy have turned out to consist predominantly of transcripts coding for serine esterases. This is not surprising in view of the abundance of these transcripts in CTL. To date, several of these enzymes have been cloned and sequenced (see also Gershenfeld *et al.*, 1988; Trapani *et al.*, 1988; Jenne *et al.*, 1988; Kwon *et al.*, 1988). The deduced primary sequences of these transcripts reveal a high degree of homology among the various enzymes (Figure 5 shows alignment of some of these enzymes). These enzymes are not exclusively associated with CTL, as originally presumed. Some non-CTL populations, such as mast cells and T-helper subsets, may also produce mRNAs specific for some of these enzymes (Brunet *et al.*, 1987; Velotti *et al.*, 1987; our unpublished observations).

Although the specificity of the cell type distribution of these enzymes must still be carefully evaluated, it is evident that the combined use of perforin and serine esterase probes will undoubtedly provide powerful tools for *in situ* hybridization studies of tissues containing lymphocytic infiltrates. Preliminary *in situ* hybridization studies using a serine esterase probe alone have already revealed the presence of serine esterase-containing lymphocytes in cardiac allografts undergoing rejection (Mueller *et al.*, 1988) and in a number of lymphocyte-associated dermatoses (Wood *et al.*, 1988). It will be exciting to determine whether the same T-cell populations also produce perforin.

1.4. Proteoglycans

The lymphocyte granules are also enriched in acidic proteoglycans of the chondroitin sulfate type (MacDermott *et al.*, 1985; Schmidt *et al.*, 1985). The peptide core of these proteoglycans is identical to the peptide repeat found in mast cells and in HL-60 promyelocytic cells (Stevens *et al.*, 1987, 1988). Although the function of these proteoglycans is still unclear, they have been speculated to play a perforin-inhibitory function in restricting self-lysis mediated by perforin (Tschopp and Conzelmann, 1986). It should be noted that in this regard, perforin itself has an acidic pI (Persechini and Young, 1988), making it less likely that two acidic molecules may be strongly associated in the same compartment. However, a protective role for proteoglycans cannot be excluded at the moment.

1.5. Leukalexins and Cytokines Related to Tumor Necrosis Factor and Lymphotoxin

The lymphocyte granules have also been shown to contain a slowly acting cytotoxic lymphokine that mediates killing of a more selected group of target cells (Liu *et al.*, 1987;

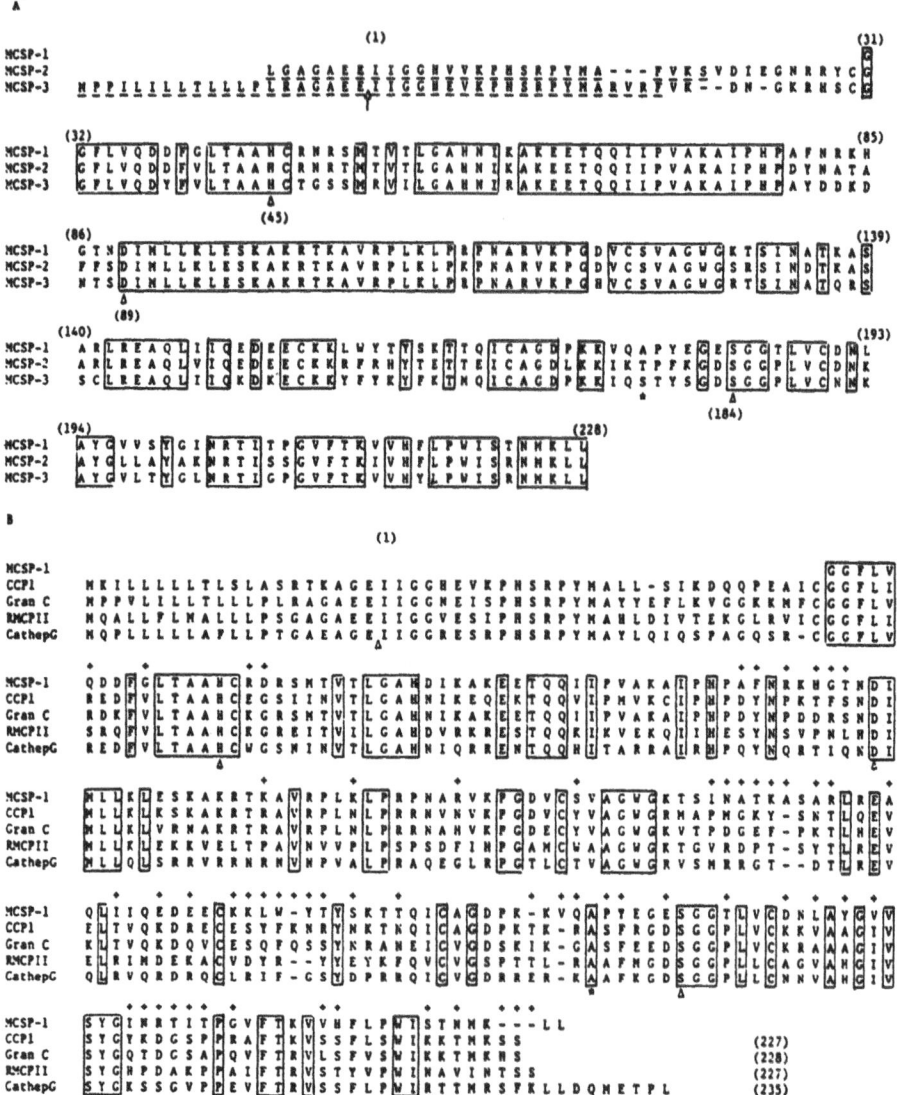

FIGURE 5. Alignment of predicted amino acid sequences of T-cell serine proteases. Comparisons were made among mouse CTL serine protease 1 (MCSP-1), MCSP-2, and MCSP-3 deduced amino acid sequences (A), and MCSP-1 amino acids were aligned with four other predicted serine proteases whose active-site pocket residue is alanine (B). Symbols: ⇑, putative cleavage site to generate an active enzyme; Δ, the "charge relay" system of the active site of serine proteases; ∗, the potential amino acid residue that participates in the primary substrate binding site in serine proteases. Numbering begins at the NH₂ terminus of the predicted active enzyme. Gaps were introduced to optimize the alignment. Amino acids that are identical among the sequences are boxed. Numbers in parentheses indicate the positions of amino acids. (A) The underlined NH₂-terminal amino acids of MCSP-2 and MCSP-3 are identical to those of granzymes E and F, respectively. (B) Positions at which the amino acids of MCSP-1 are different from those of the other four proteins (+). CCCP1, cytotoxic cell protease I; Gran C, granzyme C; RMCPII, rat mast cell protease II; Cathep G, cathepsin G. [Reproduced from Kwon *et al.* (1988), with permission.]

Konigsberg and Podack, 1986; Zychlinsky *et al.*, 1990). This lymphokine, also termed leukalexin, has a molecular mass of 50–60 kDa and is immunologically related to two other well-known cytokines, tumor necrosis factor (TNF) and lymphotoxin (LT), produced by activated macrophages and lymphocytes, respectively (Liu *et al.*, 1987). Leukalexin is distinct from TNF and LT, since cDNA probes specific for either TNF or LT do not hybridize with mRNA of cells that produce leukalexin. Leukalexin and perforin are colocalized in the same cytoplasmic granules, but in addition to this granular compartment, leukalexin is also found in the cytosol of effector lymphocytes (Figure 6). The target cell lysis mediated by leukalexin becomes detectable only after several hours of incubation, in contrast to perforin, which lyses target cells within 30–60 min. Leukalexin also produces DNA fragmentation of susceptible target cells, but this phenomenon can also be detected only after 12–18 hr of incubation (Liu *et al.*, 1987). In marked contrast, it is still unclear whether perforin produces any DNA fragmentation in target cells. Thus, while

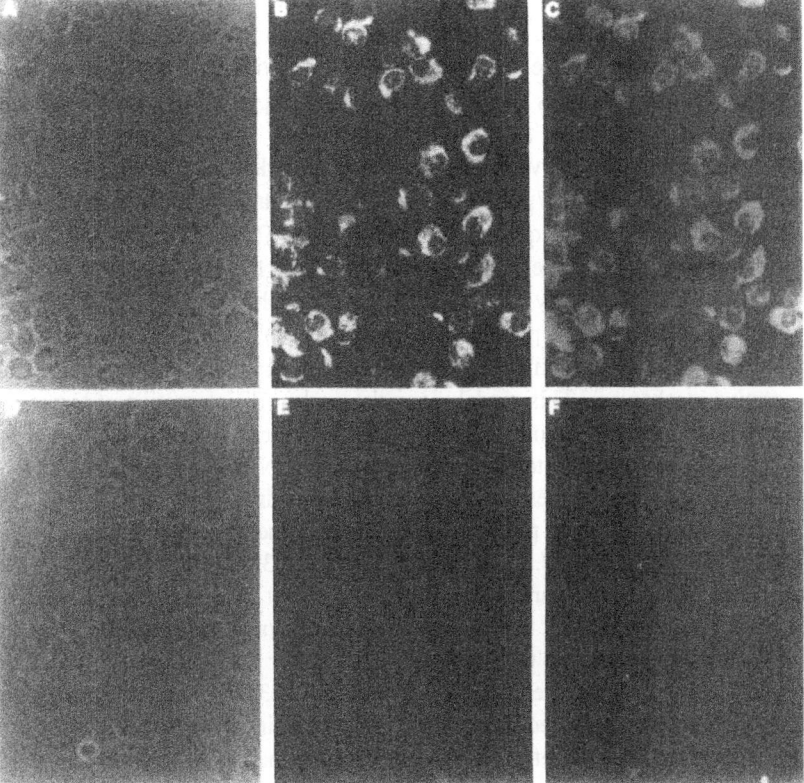

FIGURE 6. Double-labeling immunofluorescence using anti-PFP/perforin and anti-TNF antibodies. CTLL-R8 cells, sedimented in the cytocentrifuge, were stained with antisera specific for PFP/perforin (B, fluorescein channel) and murine recombinant TNF (C, rhodamine channel) or with preimmune sera (E, control for PFP/perforin, fluorescein channel; F, control for murine recombinant TNF, rhodamine channel). Corresponding phase contrast micrographs are shown in panels A and D. [Reproduced from Liu *et al.* (1987), with permission.]

some groups have shown pronounced DNA fragmentation of target cells produced by isolated granules (Allbritton *et al.*, 1988a) and purified perforin (Hameed *et al.*, 1989), other investigators have not been able to detect any measurable amount of DNA fragmentation in target cells after incubation (for up to 8 hr) with granules (Duke *et al.*, 1988; Gromkowski *et al.*, 1988) or highly purified preparations of perforin (Duke *et al.*, 1989).

Other TNF- and LT-like cytokines have been detected in cytotoxic T-lymphocyte populations (Kobayashi *et al.*, 1986; Green *et al.*, 1985, 1986) and in NK cells (Wright and Bonavida, 1983, 1987). Their precise subcellular localization and cell-type distribution are still unclear. Since they all mediate slow cytotoxicity, generally taking several hours for detection, their role in cytotoxicity must be viewed with caution. The elucidation of their primary amino acid structures will provide clearer insight into their biology and function.

2. OTHER CANDIDATE MECHANISMS OF LYMPHOCYTE-MEDIATED KILLING

As mentioned earlier, some CTL populations were found recently to lack measurable levels of perforin activity while displaying potent cytolytic activity. Other studies have revealed that certain CTL lines are capable of lysing targets in the absence of calcium (Trenn *et al.*, 1987; Ostergaard *et al.*, 1987; Young *et al.*, 1987a). To the extent that the release of a serine esterase activity may be used as a marker for granule exocytosis, the same studies (Trenn *et al.*, 1987; Ostergaard *et al.*, 1987) also showed that for some CTL lines studied, killing may not require granule exocytosis. These studies support the notion introduced by several other investigators (Dennert *et al.*, 1987; Berke and Rosen, 1987) that argue against an obligatory role for perforin in lymphocyte-mediated killing.

Among other mechanisms previously proposed to explain lymphocyte-mediated killing, the internal disintegration model proposed by Russell (1983) stands out as an attractive alternative mechanism. This model views the nuclear disintegration of target cells as the primary event of cell killing. Accordingly, lymphocytes are thought to trigger a program of self-death within target cells, involving DNA fragmentation and nuclear dissolution. Some earlier studies have shown that following attack by lymphocytes, the DNA fragmentation may even precede the increase in membrane permeability (Russell and Dobes, 1980; Russell *et al.*, 1982; Duke *et al.*, 1983; see also Gromkowski *et al.*, 1986). Since the DNA strands are broken down into repeat units of about 150–180 bp, an endonuclease activity is thought to be activated within the target cell nucleus. This activity is presumed to be derived from the target cell itself rather than being actively introduced by the effector lymphocyte (see also Duke *et al.*, 1983). Proponents of the granule exocytosis model have suggested that granules may contain an endonuclease activity that could be introduced into target cells via the perforin-mediated pores (Munger *et al.*, 1988). However, these results have not been confirmed by other investigators.

The internal disintegration or induced-suicide model [as termed by Golstein (1987)] was corroborated by a study showing that a mutant cell line selected for its resistance to both glucocorticoid- and lymphocyte-mediated cytolysis and DNA fragmentation could be rendered susceptible to both forms of treatment by a one-step genetic reversion (Ucker, 1987). However, subsequent studies by a number of other investigators (Dennert *et al.*,

1988; E. Martz and R. C. Duke, personal communication) have failed to confirm these findings.

The induced-suicide model is attractive in that it suggests an analogy between lymphocyte-mediated killing and programmed cell death observed during embryologic morphogenesis. If this interpretation is correct, then it would be relevant to determine the nature of the signal(s) turned on by lymphocytes that presumably initiates the program of cell death. However, to date, this hypothetical model is still in search of a putative mediator(s) that could account for the rapid onset of DNA fragmentation observed during lymphocyte-mediated killing.

3. RESISTANCE OF LYMPHOCYTES TO SELF-MEDIATED KILLING

While lymphocytes vigorously kill their target cells, they remain refractory to self-mediated killing. Thus, recent studies from several laboratories have shown that CTL populations are much more resistant to killing mediated by other effector CTL than are non-CTL target cells (Kranz and Eisen, 1987; Blakely et al., 1987; Skinner and Marbrook, 1987; for a review, see Young and Liu, 1988). CTL and NK cells are also resistant to lysis mediated by granules and purified perforin (Verret et al., 1987; Blakely et al., 1987; Shinkai et al., 1988a; Jiang et al., 1988). The nature of this resistance is still unclear. Some recent observations have suggested that a membrane factor, designated homologous restriction factor (HRF) or C8-binding protein (C8BP), may be involved in conferring perforin resistance (reviewed in Muller-Eberhard, 1988). HRF/C8BP had previously been shown to mediate protection of erythrocytes of a given species against lysis mediated by the terminal C components of the same or homologous species (Zalman et al., 1986b; Schonermark et al., 1986; Shin et al., 1986). HRF binds to both C8 and C9 and is thought to protect cells by interrupting the binding and polymerization of C9 in the plane of a target bilayer. HRF/C8BP is thought to be phosphoinositol (PI) anchored to the surface of cells, since (1) it is deficient in type III erythrocytes of patients with paroxysmal nocturnal hemoglobinuria (PNH), a condition now known to be associated with a deficiency in PI anchoring, and (2) it can be cleaved from the cell surface by PI-specific phospholipase C (Hansch et al., 1987; Zalman et al., 1987b). An involvement of HRF in mediating lymphocyte resistance to perforin has been suggested by the following evidence: (1) HRF that has been reconstituted into HRF-deficient erythrocytes of PNH patients is capable of conferring to these cells protection against perforin-mediated lysis (Zalman et al., 1987a); (2) peripheral blood lymphocytes become more resistant to perforin and acquire enhanced expression of surface HRF upon cell stimulation with anti-CD3 antibodies, and this increase in perforin resistance can be abrogated by $F(ab)'_2$ specific for HRF (Martin et al., 1988); and (3) a soluble form of HRF has been detected in granules of peripheral blood-derived large granular lymphocytes (consisting predominantly of NK cells) that confers protection against perforin-mediated lysis when present in the fluid phase (Zalman et al., 1988). In our own studies, we have not been able to confirm these findings. Thus, while PNH type III erythrocytes are indeed deficient in HRF and are more susceptible to lysis mediated by homologous (human) C than are normal erythrocytes, they are equally susceptible to lysis mediated by perforin of both homologous and heterologous species (Jiang et al., 1989a). These results are also supported by a study in which perforin obtained from three species (human, mouse, and rat) was tested against erythrocytes of 10 different species (Jiang *et al.*, 1988). No correlation was observed between the species of perforin and target cells used, whereas erythrocytes displayed clear homologous species restriction toward C-mediated lysis. Moreover, we have not been able to detect any soluble HRF activity in lymphocyte granules or cytosol (Jiang *et al.*, 1989b). Our studies indicate that the role of HRF in this type of lymphocyte resistance must be carefully reevaluated. Other candidate mediators of protection (protectins?) must be considered. Since cytolytic factors other than perforin may be involved in mediating cytotoxicity, it is likely that the resistance of lymphocytes to self-mediated lysis may also involve multiple mechanisms and mediators.

4. CONCLUSION

The recent reductionist approach used to define lymphocyte-mediated killing as an interplay of discrete events and mediators has already resulted in the accumulation of a wealth of biochemical information on this cellular reaction. Recent biochemical studies have heavily emphasized the role of granule mediators in this type of killing. Although we still do not have a consensus on the role of these mediators in lymphocyte-mediated killing, the identification and characterization of granule mediators have helped define new biochemical entities that now can be studied in detail. Several such candidate cytolytic mediators have been cloned and sequenced, and their roles in cytotoxicity are now being evaluated by using specific cDNA and antibody probes. We hope that this approach will eventually clarify a number of problems and issues related to cell-mediated cytotoxicity that presently remain unresolved.

ACKNOWLEDGMENTS. We thank Zanvil A. Cohn and Bo Dupont for continuous advice and support. The work described here was supported in part by U.S. Public Health Service grants CA-47307, CA-22507, AI-07012, and AI-28175, by a special grant for research from the American Cancer Society, New York City Division, by grants from the Schering Corporation, the American Diabetes Association, the Cancer Research Institute/Frances L. & Edwin L. Cummings Memorial Fund Investigator award, the American Heart Association, New York City affiliate, and the Lucille P. Markey Charitable Trust, and by a fellowship from the Irvington Institute for Medical Research (C.C.L.). J.A.T. holds a C. J. Martin traveling fellowship, NH and MRC, Australia. J.D.-EY. is a Lucille P. Markey Scholar. L.H.Y.Y. is a Heed Fellow and is also supported by a National Research Service Award fellowship.

5. REFERENCES

Allbritton, N. L., Verret, C. R., Wolley, R. C., and Eisen, H. N., 1988a, Calcium ion concentrations and DNA fragmentation in target cell destruction by murine cloned cytotoxic T lymphocytes, *J. Exp. Med.* **167**:514–527

Allbritton, N. L., Nagler-Anderson, C., Elliott, T. J., Verret, C. R., and Eisen, H. N., 1988b, Target cell lysis by cytotoxic T lymphocytes that lack detectable hemolytic perforin activity, *J. Immunol.* **141**:3243–3248

Berke, G., 1983, Cytotoxic T-lymphocytes. How do they function, *Immunol. Rev.* **72**:5–42

Berke, G., and Rosen, D., 1987, Are lytic granules and perforin 1 involved in lysis induced by in vivo-primed peritoneal exudate cytolytic T lymphocytes, *Transplant. Proc.* **19**:412–416

Berke, G., and Rosen, D., 1988, Highly lytic in vivo primed cytolytic T lymphocytes devoid of lytic granules and BLT-esterase activity acquire these constituents in the presence of T cell growth factors upon blast transformation in vitro, *J. Immunol.* **141**:1429–1436

Bhakdi, S., and Tranum-Jensen, J., 1983, Membrane damage by complement, *Biochim. Biophys. Acta* **737**:343–372

Bhakdi, S., and Tranum-Jensen, J., 1987, Damage to mammalian cells by proteins that form transmembrane pores, *Rev. Physiol. Biochem. Pharmacol.* **107**:147–233

Blakely, A., Gorman, K., Ostergaard, H., Svoboda, K., Liu, C.-C., Young, J. D.-E, and Clark, W. R., 1987, Resistance of cloned cytotoxic T lymphocytes to cell-mediated cytotoxicity, *J. Exp. Med.* **166**:1070–1083

Brooks, C. G., Urdal, D. L., and Henney, C. S., 1983, Lymphokine-driven "differentiation" of cytotoxic T-cell clones into cells with NK-like specificity: Correlations with display of membrane macromolecules, *Immunol. Rev.* **72**:44–72

Brunet, J.-F., Dosseto, M., Denizot, F., Mattei, M. G., Clark, W. R., Haqqi, T. M., Ferrier, P., Nabholz, M., • Schmitt-Verhulst, A.-M., Luciani, M.-F., and Golstein, P., 1986, The inducible cytotoxic T-lymphocyte-associated gene transcript CTLA-1 sequence and gene localization to mouse chromosome 14, *Nature* (London) **322**:268–271

Brunet, J.-F., Denizot, F., Suzan, M., Haas, W., Mencia-Huerta, J.-M., Berke, G., Luciani, M.-F., and Golstein, P., 1987, CTLA-1 and CTLA-3 serine esterase transcripts are detected mostly in cytotoxic T cells, but not only and not always, *J. Immunol.* **138**:4102–4105

Dennert, G., and Podack, E. R., 1983, Cytolysis by H-2-specific T killer cells. Assembly of tubular complexes on target membranes, *J. Exp. Med.* **157**:1483–1495

Dennert, G., Kupfer, A., Anderson, C. G., and Singer, S. J., 1985, Reorientation of the Golgi apparatus and the microtubule organizing center: Is it a means to polarize cell-mediated cytotoxicity?, *Adv. Exp. Med. Biol.* **184**:83–97

Dennert, G., Anderson, C. G., and Prochazka, G., 1987, High activity of N-α-benzyloxycarbonyl-L-lysine thiobenzyl ester serin esterase and cytolytic perforin in cloned cell lines is not demonstrable in in vivo-induced cytotoxic effector cells, *Proc. Natl. Acad. Sci. USA* **84**:5004–5008

Dennert, G., Landon, C., and Nowicki, M., 1988, Cell-mediated and glucocorticoid-mediated target cell lysis do not appear to share common pathways, *J. Immunol.* **141**:785–791

Dourmashkin, R. R., Deteix, P., Simone, C. B., and Henkart, P., 1980, Electron microscopic demonstration of lesions in target cell membranes associated with antibody-dependent cellular cytotoxicity, *Clin. Exp. Immunol.* **42**:554–560

Duke, R. C., Chervenak, R., and Cohen, J. J., 1983, Endogenous endonuclease-induced DNA fragmentation: An early event in cell-mediated cytolysis, *Proc. Natl. Acad. Sci. USA* **80**:6361–6365

Duke, R. C., Sellins, K. S., and Cohen, J. J., 1988, Cytotoxic lymphocyte-derived lytic granules do not induce DNA fragmentation in target cells, *J. Immunol.* **141**:2191–2194

Duke, R. C., Persechini, P. M., Chang, S., Liu, C.-C., Cohen, J. J., and Young, J. D.-E 1989, Purified perforin induces target cell lysis but not DNA fragmentation, *J. Exp. Med.* **170**:1451–1456

Gershenfeld, H. K., and Weissman, I. L., 1986, Cloning of a cDNA for a T cell-specific serine protease from a cytotoxic T lymphocyte, *Science* **232**:854–858

Gershenfeld, H. K., Hershberger, R. J., Shows, T. B., and Weissman, I. L., 1988, Cloning and chromosomal assignment of a human cDNA encoding a T cell- and natural killer cell-specific trypsin-like serine protease, *Proc. Natl. Acad. Sci. USA* **85**:1184–1188

Goldfarb, R. H., 1986, Cell-mediated cytotoxic reactions, *Hum. Pathol.* **17**:138–145

Golstein, P., 1987, Cytolytic T-cell melodrama, *Nature* (London) **327**:12

Green, L. M., Stern, M. L., Haviland, D. L., Mills, B. J., and Ware, C. F., 1985, Cytotoxic lymphokines produced by cloned human cytotoxic T lymphocytes. I. Cytotoxins produced by antigen-specific and natural killer-like CTL are dissimilar to classical lymphotoxins, *J. Immunol.* **135**:4034–4043

Green, L. M., Reade, J. L., Ware, C. F., Devlin, P. E., Liang, C.-M., and Devlin, J. J., 1986, Cytotoxic lymphokines produced by cloned human cytotoxic T lymphocytes. II. A novel CTL-produced cytotoxin that is antigenically distinct from tumor necrosis factor and alpha-lymphotoxin, *J. Immunol.* **137**:3488–3493

Gromkowski, S. H., Brown, T. C., Cerutti, P. A., and Cerottini, J.-C., 1986, DNA of human Raji target cells is damaged upon lymphocyte-mediated lysis, *J. Immunol.* **136**:752–756

Gromkowski, S. H., Brown, T. C., Masson, D., and Tschopp, J., 1988, Lack of DNA degradation in target cells lysed by granules derived from cytolytic T lymphocytes, *J. Immunol.* **141**:774–778

Hameed, A., Olsen, K. J., Lee, M.-K., Lichtenheld, M. G., and Podack, E. R., 1989, Cytolysis by Ca-permeable transmembrane channels. Pore formation causes extensive DNA degradation and cell lysis, *J. Exp. Med.* **169**:765–777

Hansch, G. M., Schonermark, S., and Roelcks, D., 1987, Paroxysmal nocturnal hemoglobinuria type III: Lack of an erythrocyte membrane protein restricting the lysis by C5b-9, *J. Clin. Invest.* **80**:7–12

Henkart, P. A., 1985, Mechanism of lymphocyte-mediated cytotoxicity, *Ann. Rev. Immunol.* **3**:31–58

Henkart, M. P., and Henkart, P. A., 1982, Lymphocyte mediated cytolysis as a secretory phenomenon, *Adv. Exp. Med. Biol.* **146**:227–247

Henkart, P. A., Yue, C. C., Yang, J., and Rosenberg, S. A., 1986, Cytolytic and biochemical properties of cytoplasmic granules of murine lymphokine-activated killer cells, *J. Immunol.* **137**:2611–2617

Henney, C. S., 1973, Studies on the mechanism of T-cell mediated cytolysis, *Transplant. Rev.* **17**:37–70

Henney, C. S., 1974, Estimation of the size of a T-cell-induced lytic lesion, *Nature* (London) **249**:456–458

Herberman, R. B., Reynolds, C. W., and Ortaldo, J. R., 1986, Mechanism of cytotoxicity by natural killer (NK) cells, *Annu. Rev. Immunol.* **4**:651–680

Jenne, D. E., and Tschopp, J., 1988, Granzymes, a family of serine proteases released from granules of cytolytic T lymphocytes upon T cell receptor stimulation, *Immunol. Rev.* **103**:53–72

Jenne, D., Rey, C., Haefliger, J.-A., Qiao, B.-Y., Groscurth, P., and Tschopp, J., 1988, Identification and sequencing of cDNA clones encoding the granule-associated serine proteases granzymes D, E, and F of cytolytic T lymphocytes, *Proc. Natl. Acad. Sci. USA* **85**:4814–4818

Jiang, S., Persechini, P. M., Zychlinsky, A., Liu, C.-C., Perussia, B., and Young, J. D.-E, 1988, Resistance of cytolytic lymphocytes to perforin-mediated killing. Lack of correlation with complement-associated homologous species restriction, *J. Exp. Med.* **168**:2207–2219

Jiang, S., Persechini, P. M., Rosse, W. F., Perussia, B., and Young, J. D.-E, 1989a, Differential susceptibility of type III erythrocytes of paroxysmal nocturnal hemoglobinuria to lysis mediated by complement and perforin, *Biochem. Biophys. Res. Commun.* **162**: 316–325.

Jiang, S., Persechini, P. M., Perussia, B., and Young, J. D.-E, 1989b, Resistance of cytolytic lymphocytes to perforin-mediated killing: Murine CTLs and human NK cells do not contain functional soluble homologous restriction factor or other specific soluble protective factors, *J. Immunol.* **143**:1453–1460.

Joag, S. V., Liu, C.-C., Kwon, B. S., Clark, W. R., and Young, J. D.-E 1990, The expression of mRNAs for pore-forming protein and two serine esterases in murine primary and cloned effector lymphocytes, *J. Cell. Biochemis.* (in press)

Kobayashi, M., Plunkett, J. M., Masunaka, I. K., Yamamoto, R. S., and Granger, G. A., 1986, The human LT system. XII. Purification and functional studies of LT and "TNF-like" LT forms from a continuous human T cell line, *J. Immunol.* **137**:1885–1892

Konigsberg, P. J., and Podack, E. R., 1984, DNA damage of target cells by cytolytic T-cell granules, *J. Cell. Biochem.* (Suppl.) **10B**:85

Kramer, M. D., and Simon, M. M., 1987, Are proteinases functional molecules of T lymphocytes, *Immunol. Today* **8**:140–142

Kranz, D. M., and Eisen, H. N., 1987, Resistance of cytotoxic T lymphocytes to lysis by a clone of cytotoxic T lymphocytes, *Proc. Natl. Acad. Sci USA* **84**:3375–3379

Kupfer, A., and Dennert, G., 1984, Reorientation of the microtubule-organizing center and the Golgi apparatus in cloned cytotoxic lymphocytes triggered by binding to lysable target cells, *J. Immunol.* **133**:2762–2766

Kwon, B. S., Kim, G. S., Prystowsky, M. B. Lancki, D. W., Sabath, D. E., Pan, J., and Weisman, S. M., 1987, Isolation and initial characterization of multiple species of T-lymphocyte subset cDNA clones, *Proc. Natl. Acad. Sci. USA* **84**:2896–2900

Kwon, B. S., Kestler, D., Lee, E., Wakulchik, M., and Young, J. D.-E, 1988, Isolation and sequence analysis of serine protease cDNAs from mouse cytolytic T lymphocytes, *J. Exp. Med.* **168**:1839–1854

Kwon, B. S., Wakulchik, M., Liu, C.-C., Persechini, P. M., Trapani, J. A., Haq, A. K., Kim, Y., and Young, J. D.-E, 1989, The structure of the lymphocyte pore-forming protein perforin, *Biochem. Biophys. Res. Commun.* **158**:1–10

Lichtenheld, M. G., Olsen, K., Lu, P., Lowrey, D. M., Hameed, A., Hengartner, H., and Podack, E. R., 1988, Structure and function of human perforin, *Nature* (London) **335**:448–451

Liu, C.-C., Perussia, B., Cohn, Z. A., and Young, J. D.-E, 1986, Identification and characterization of a pore-forming protein of human peripheral blood natural killer cells, *J. Exp. Med.* **164**:2061–2076

Liu, C.-C., Steffen, M., King, F., and Young, J. D.-E, 1987, Identification, isolation, and characterization of a novel cytotoxin in murine cytolytic lymphocytes, *Cell* **51**:393–403

Liu, C.-C., Joag, S. V., Kwon, B. S., and Young, J. D.-E, 1990, Induction of perforin and serine esterases in a murine cytotoxic T lymphocyte clone, *J. Immunol.* **144**:1196–1201

Lobe, C. G., Finlay, B. B., Paranchych, W., Paetkau, V. H., and Bleackley, R. C., 1986, Novel serine proteases encoded by two cytotoxic T lymphocyte-specific genes, *Science* **232**:858–861

Lowrey, D. M., Aebischer, T., Olsen, K., Lichtenheld, M., Rupp, F., Hengartner, H., and Podack, E. R., 1989, Cloning, analysis, and expression of murine perforin 1 cDNA, a component of cytolytic T-cell granules with homology to complement component C9, *Proc. Natl. Acad. Sci. USA* **86**:247–251

MacDermott, R. P., Schmidt, R. E., Caulfield, J. P., Hein, A., Bartley, G. T., Ritz, J., Schlossman, S. F., Austen, K. F., and Stevens, R., 1985, Proteoglycans in cell mediated cytotoxicity. Identification, localisation, and exocytosis of a chondroitin sulfate proteoglycan from human cloned natural killer cells during target cell lysis, *J. Exp. Med.* **162**:1771–1787

Martin, D. E., Zalman, L. S., and Muller-Eberhard, H. J., 1988, Induction of expression of cell-surface homologous restriction factor upon anti-CD3 stimulation of human peripheral lymphocytes, *Proc. Natl. Acad. Sci. USA* **85**:213–217

Martz, E., 1976, Multiple target cell killing by the cytolytic T lymphocytes and the mechanism of cytotoxicity, *Transplantation* **21**:5–11

Masson, D., and Tschopp, J., 1985, Isolation of a lytic, pore-forming protein (perforin) from cytolytic T-lymphocytes, *J. Biol. Chem.* **260**:9069–9072

Masson, D., and Tschopp, J., 1987, A family of serine esterases in lytic granules of cytolytic T lymphocytes, *Cell* **49**:679–685

Masson, D., Nabholz, M., Estrade, C., and Tschopp, J., 1986, Granules of cytolytic T-lymphocytes contain two serine esterases, *EMBO J.* **5**:1595–1600

Mayer, M. M., Michaels, D. W., Ramm, L. E., Whitlow, M. B., Willoughby, J. B., and Shin, M. L., 1981, Membrane damage by complement, *Crit. Rev. Immunol.* **2**:133–165

Mueller, C., Gershenfeld, H. K., Lobe, C. G., Okada, C. Y., Bleackley, R. C., and Weissman, I. L., 1988, A high proportion of T lymphocytes that infiltrate H-2-incompatible heart allografts in vivo express genes encoding cytotoxic cell-specific serine proteases, but do not express the MEL-14-defined lymph node homing receptor, *J. Exp. Med.* **167**:1124–1136

Muller-Eberhard, H. J., 1988, The molecular basis of target cell killing by human lymphocytes and of killer cell self-protection, *Immunol. Rev.* **103**:87–98

Munger, W. E., Berrebi, G. A., and Henkart, P. A., 1988, Possible involvement of CTL granule proteases in target cell DNA breakdown, *Immunol. Rev.* **103**:99–109

Ostergaard, H. L., Kane, K. P., Mescher, M. F., and Clark, W. R., 1987, Cytotoxic T lymphocyte mediated lysis without release of serine esterase, *Nature* (London) **330**:71–72

Pasternack, M. S., and Eisen, H. N., 1985, A novel serine esterase expressed by cytotoxic T lymphocytes, *Nature* (London) **314**:743–745

Pasternack, M. S., Verret, C. R., Liu, M. A., and Eisen, H. N., 1986, Serine esterase in cytolytic T lymphocytes, *Nature* (London) **322**:740–743

Persechini, P. M., and Young, J. D.-E, 1988, The primary structure of the lymphocyte pore-forming protein perforin: Partial amino acid sequencing and determination of isoelectric point, *Biochem. Biophys. Res. Commun.* **156**:740–745

Podack, E. R., 1985, The molecular mechanism of lymphocyte-mediated tumor cell lysis, *Immunol. Today* **6**:21–27

Podack, E. R., and Dennert, G., 1983, Assembly of two types of tubules with putative cytolytic function by cloned natural killer cells, *Nature* (London) **302**:442–445

Podack, E. R., and Tschopp, J., 1982, Circular polymerization of the ninth component of complement. Ring closure of the tubular complex confers resistance to detergent dissociation and to proteolytic degradation, *J. Biol. Chem.* **257**:15204–15212

Podack, E. R., Young, J. D.-E, and Cohn, Z. A., 1985, Isolation and biochemical and functional characterization of perforin 1 from cytolytic T-cell granules, *Proc. Natl. Acad. Sci. USA* **82**:8629–8633

Reynolds, C. W., Reichardt, D., Henkart, M., Millard, P., and Henkart, P., 1987, Inhibition of NK and ADCC activity by antibodies against purified cytoplasmic granules from rat LGL tumors, *J. Leukocyte Biol.* **42**:642–652

Russell, J. H., 1983, Internal disintegration model of cytotoxic lymphocyte-induced target damage, *Immunol. Rev.* **72**:97–118

Russell, J. H., and Dobes, C. B., 1980, Mechanisms of immune lysis. II. CTL-induced nuclear disintegration of the target begins within minutes of cell contact, *J. Immunol.* **125**:1256–1261

Russell, J. H., Masakowski, V., Rucinsky, T., and Phillips, G., 1982, Mechanisms of immune lysis. III. Characterization of the nature and kinetics of the cytotoxic T lymphocyte-induced nuclear lesion in the target, *J. Immunol.* **128**:2087–2094

Schmidt, R. E., MacDermott, R. P., Bartley, G. T., Bertovich, M., Amato, D. A. Austen, K. F., Schlossman, S. F., Stevens, R. L., and Ritz, J., 1985, Specific release of proteoglycans from human natural killer cells during target lysis, *Nature* (London) **318**:289–291

Schonermark, S., Rauterberg, E. W., Shin, M. L., Loke, S., Roelcke, D., and Hansch, G. M., 1986, Homologous species restriction in lysis of human erythrocytes: A membrane-derived protein with C8-binding capacity functions as an inhibitor, *J. Immunol.* **136**:1772–1776

Shin, M. L., Hansch, G., Hu, V. W., and Nicholson-Weller, A., 1986, Membrane factors responsible for homologous species restriction of complement-mediated lysis: Evidence for a factor other than DAF operating at the stage of C8 and C9, *J. Immunol.* **136**:1777–1782

Shinkai, Y., Ishikawa, H., Hattori, M., and Okumura, K., 1988a, Resistance of mouse cytolytic cells to pore-forming protein-mediated cytolysis, *Eur. J. Immunol.* **18**:29–20

Shinkai, Y., Takio, K., and Okumura, K., 1988b, Homology of perforin to the ninth component of complement (C9), *Nature* (London) **334**:525–527

Simon, M. M., Hoschutzky, H., Fruth, U., Simon, H.-G., and Kramer, M. D., 1986, Purification and characterization of a T cell specific serine proteinase (TSP-1) from cloned cytolytic T lymphocytes, *EMBO J.* **5**:3267–3274

Skinner, M., and Marbrook, J., 1987, The most efficient cytotoxic T lymphocytes are the least susceptible to lysis, *J. Immunol.* **139**:985–987

Stanley, K. K., 1988, The molecular mechanism of complement C9 insertion and polymerzation in biological membranes, *Curr. Top. Microbiol. Immunol.* **140**:49–65

Stanley, K. K., and Luzio, P., 1988, A family of killer proteins, *Nature* (London) **334**:475–476

Stevens, R. L., Otsu, K., Weis, J. H., Tantravahi, R. V., Austen, K. F., Henkart, P. A., Galli, M. C., and Reynolds, C. W., 1987, Co-sedimentation of chondroitin sulfate A glycosaminoglycans and proteoglycans with the cytolytic secretory granules of rat large granular lymphocyte (LGL) tumor cells, and identification of a mRNA in normal and transformed LGL that encodes proteoglycans, *J. Immunol.* **139**:863–868

Stevens, R. L., Kamada, M. N., and Serafin, W. E., 1988, Structure and functions of the family of proteoglycans that reside in the secretory granules of natural killer cells and other effector cells of the immune response, *Curr. Top. Microbiol. Immunol.* **140**:93–108

Trapani, J. A., Klein, J. L., White, B. C., and Dupont, B., 1988, Molecular cloning of an inducible serine esterase gene from human cytotoxic lymphocytes, *Proc. Natl. Acad. Sci. USA* **85**:6924–6928

Trapani, J. A., Kwon, B. S., Kozak, C. A., Chintamaneni, C., Young, J. D.-E, and Dupont, B., 1990, Genomic organization of the pore-forming protein (perforin) gene and localization to chromosome 10; similarities to and differences from C9. *J. Exp. Med.* **171**:545–557

Trenn, G., Takayama, H., and Sitkovsky, M. V., 1987, Exocytosis of cytolytic granules may not be required for target cell lysis by cytotoxic T-lymphocytes, *Nature* (London) **330**:72–74

Trinchieri, G., and Perussia, B., 1984, Human natural killer cells: Biologic and pathologic aspects, *Lab. Invest.* **50**:489–513

Tschopp, J., and Conzelmann, A., 1986, Proteoglycans in secretory granules of NK cells, *Immunol. Today* **7**:135–136

Tschopp, J., and Jonheneel, C. V., 1988, Cytotoxic T lymphocyte mediated cytolysis, *Biochemistry* **27**:2641–2646

Tschopp, J., Masson, D., and Stanley, K. K., 1986, Structural/functional similarity between proteins involved in complement and cytotoxic T-lymphocyte-mediated cytolysis, *Nature* (London) **322**:831–834

Tschopp, J., Schafer, S., Masson, D., Peitsch, M. C., and Heusser, C., 1989, Phosphorylcholine acts as a Ca^{2+}-dependent receptor molecule for lymphocyte perforin, *Nature* (London) **337**:272–274

Ucker, D. S., 1987, Cytotoxic T lymphocytes and glucocorticoids activate an endogenous suicide process in target cells, *Nature* (London) **327**:62–64

Velotti, F., MacDonald, H. R., and Nabholz, M., 1987, Granzyme A secretion by normal activated Lyt-2$^+$ and L3T4$^+$ T cells in response to antigenic stimulation, *Eur. J. Immunol.* **17**:1095–1099

Verret, C. R., Firmenich, A. A., Kranz, D. M., and Eisen, H. N., 1987, Resistance of cytotoxic T lymphocytes to the lytic effects of their toxic granules, *J. Exp. Med.* **166**:1536–1547

Wood, G. S., Mueller, C., Warnke, R. A., and Weissman, I. L., 1988, *In situ* localization of HuHF serine protease mRNA and cytotoxic cell-associated antigens in human dermatoses. A novel method for the detection of cytotoxic cells in human tissues, *Am. J. Pathol.* **133**:218–225

Wright, S. C., and Bonavida, B., 1983, Studies on the mechanism of natural killer cell-mediated cytotoxicity. IV. Interferon-induced inhibition of NK target cell susceptibility to lysis is due to a defect in their ability to stimulate release of natural killer cytotoxic factors (NKCF), *J. Immunol.* **130**:2965–2968

Wright, S. C., and Bonavida, B., 1987, Studies on the mechanism of natural killer cell-mediated cytotoxicity. VII. Functional comparison of human natural killer cytotoxic factors with recombinant lymphotoxin and tumor necrosis factor, *J. Immunol.* **138**:1791–1798

Yannelli, J. R., Sullivan, J. A., Mandell, G. L., and Engelhard, V. H., 1986, Reorientation and fusion of cytotoxic T lymphocyte granules after interaction with target cells as determined by high resolution cinemicrography, *J. Immunol.* **136**:377–382

Young, J. D.-E, and Cohn, Z. A., 1987, Cellular and humoral mechanisms of cytotoxicity: Structural and functional analogies, *Adv. Immunol.* **41**:269–331

Young, J. D.-E, and Liu, C.-C., 1988, How do cytotoxic T lymphocytes avoid self-lysis?, *Immunol. Today* **9**:14–15

Young, J. D.-E, Cohn, Z. A., and Podack, E. R., 1986a, The ninth component of complement and the pore-forming protein (perforin 1) from cytotoxic T cells: Structural, immunological, and functional similarities, *Science* **233**:184–190

Young, J. D.-E, Hengartner, H., Podack, E. R., and Cohn, Z. A., 1986b, Purification and characterization of a cytolytic pore-forming protein from granules of cloned lymphocytes with natural killer activity, *Cell* **44**:849–859

Young, J. D.-E, Leong, L. G., Liu, C.-C., Damiano, A., Wall, D. A., and Cohn, Z. A., 1986c, Isolation and characterization of a serine esterase from cytolytic T cell granules, *Cell* **47**:183–194

Young, J. D.-E, Liu, C.-C., Leong, L. G., and Cohn, Z. A., 1986d, The pore-forming protein (perforin) of cytolytic T lymphocytes is immunologically related to the components of membrane attack complex of complement through cysteine-rich domains, *J. Exp. Med.* **164**:2077–2082

Young, J. D.-E, Clark, W. R., Liu, C.-C., and Cohn, Z. A., 1987a, A calcium- and perforin-independent pathway of killing mediated by murine cytolytic lymphocytes, *J. Exp. Med.* **166**:1894–1899

Young, J. D.-E, Damiano, A., DiNome, M. A., Leong, L. G., and Cohn, Z. A., 1987b, Dissociation of membrane binding and lytic activities of the lymphocyte pore-forming protein (perforin), *J. Exp. Med.* **165**:1371–1382

Young, L. H. Y., Klavinskis, L. S., Oldstone, M. B. A., and Young, J. D.-E, 1989a, In vivo expression of perforin by CD8$^+$ lymphocytes during an acute viral infection, *J. Exp. Med.*, **169**:2159–2172

Young, L. H. Y., Peterson, L. B., Wicker, L. S., Persechini, P. M., and Young, J. D.-E, 1989b, In vivo expression of perforin by CD8$^+$ lymphocytes in autoimmune disease: studies on spontaneous and adoptively transferred diabetes in nonobese diabetic mice. *J. Immunol.* **143**:3994–3999

Yue, C. C., Reynolds, C. W., and Henkart, P. A., 1987, Inhibition of cytolysin activity in large granular lymphocyte granules by lipids: Evidence for a membrane insertion mechanism of lysis, *Mol. Immunol.* **24**:647–653

Zalman, L. S., Brothers, M. A., Chiu, F. J., and Müller-Eberhard, H. J., 1986a, Mechanism of cytotoxicity of human large granular lymphocytes: Relationship of the cytotoxic lymphocyte protein to the ninth component (C9) of human complement, *Proc. Natl. Acad. Sci. USA* **83**:5262–5266

Zalman, L. S., Wood, L. M., and Müller-Eberhard, H. J., 1986b, Isolation of a human erythrocyte membrane protein capable of inhibiting expression of homologous complement transmembrane channels, *Proc. Natl. Acad. Sci. USA* **83**:6975–6979

Zalman, L. S., Wood, L. M., and Müller-Eberhard, H. J., 1987a, Inhibition of antibody-dependent lymphocyte cytotoxicity by homologous restriction factor incorporated into target cell membranes, *J. Exp. Med.* **166:**947–955

Zalman, L. S., Wood, L. M., Frank, M. M., and Müller-Eberhard, H. J., 1987b, Deficiency of the homologous restriction factor in paroxysmal nocturnal hemoglobinuria, *J. Exp. Med.* **165:**572–577

Zalman, L. S., Brothers, M. A., and Müller-Eberhard, H. J., 1988, Self-protection of cytotoxic lymphocytes: A soluble form of homologous restriction factor in cytoplasmic granules, *Proc. Natl. Acad. Sci. USA* **85:**4827–4831

Zychlinsky, A., Joag, S. V., Liu, C.-C., and Young, J. D.-E, 1990, Cytotoxic mechanisms of murine lymphokine-activated killer cells: Functional and biochemical characterization of homogeneous populations of spleen LAK cells, *Cell. Immunol.* **126:**377–390

Chapter 7

CR1–Cytoskeleton Interactions in Neutrophils

James D. Katz

1. INTRODUCTION

The C3b receptor (CR1) is a polymorphic membrane glycoprotein (Dykman *et al.*, 1983a,b, 1984, 1985; Wong *et al.*, 1983). It is found in a variety of cell types and has several important immunologic functions (see reviews in Fearon, 1985, Schreiber, 1984, and Arnaout and Colten, 1984). For example, aside from mediating attachment of phagocytic cells to opsonized targets, erythrocyte CR1 may function in systemic clearance of complement-containing immune complexes (Cornacoff *et al.*, 1983) or in local inhibition of complement-mediated damage to autologous tissues (Iida and Nussenzweig, 1983). It has been proposed that this function inadvertently plays a roll in the pathogenesis of CR1 deficiency found in patients with systemic lupus erythematosus (Iida *et al.*, 1982). In addition, the human neutrophil CR1 mediates endocytosis of soluble multivalent ligand or soluble C3b-bearing complexes (Fearon *et al.*, 1981; Abrahamson and Fearon, 1983). How this is effected has prompted researchers to investigate the relationship between CR1 and the cytoskeletal apparatus. Among the tools available to accomplish this is nonionic detergent lysis (or extraction) of cells.

2. DETERGENT EXTRACTION OF CELLS

It is well documented that mild nonionic detergents such as Triton X-100 and Nonidet P-40 minimize interference with either the biologic activity of proteins or protein–protein interactions (Helenius and Simons, 1975). Such detergents have gained popularity

James D. Katz   Department of Medicine, The Western Pennsylvania Hospital, Pittsburgh, Pennsylvania 15224, USA.   *Present Address*: Division of Rheumatic Diseases, Department of Medicine, The University of Connecticut Health Center, Farmington, Connecticut 06030, USA.

since they enable the immunochemical (and morphologic) characterization of the cell nucleus as well as the cytoskeleton (Osborn and Weber, 1977). Further research with this technique showed not only that the detergent-insoluble portion of cells consists of the nucleus and cytoskeleton but that certain plasma membrane proteins are retained as well during detergent extraction (Yu *et al.*, 1973). An important extension of this work was accomplished by Ben Ze'ev *et al.* (1979). They examined the surface lamina covering the cytoskeleton after detergent extraction of over 90% of cellular lipids. By prior radio-iodination of surface proteins, they demonstrated that most (80%) of the cell surface protein remained with the extracted cytoskeleton while three-fourths of the cytoplasmic protein was removed. Subsequently, Sheterline and Hopkins (1981) argued that specific linkages exist between surface proteins and the peripheral filamentous network of the cytoskeleton rather than such proteins being trapped within the network. They went beyond Ben Ze'ev *et al.* to note a selectivity of association between surface proteins and the neutrophil detergent-extracted "ghosts." These authors also presented evidence that proteolysis occurs during lysis and therefore added protease inhibitors to their lytic solution. This is noteworthy because it is an indication that the lysis procedure itself encompasses a biochemically active time.

3. RECEPTOR–CYTOSKELETON INTERACTIONS

A significant precedent for investigating cytoskeleton–surface protein interactions can be found in the work of Flanagan and Koch (1978). They labeled cell surface immunoglobulin with radioiodinated anti-immunoglobulin antibody fragments. Then, using the myosin affinity technique on cell lysates, they demonstrated that divalent labeling reagent (which induces patching but not necessarily capping) resulted in greater attachment of surface immunoglobulin to cellular actin than did monovalent labeling reagent. Presumably, crosslinking is a transmembrane signal for surface component attachment to the cytoskeleton. In this research, the authors addressed the issue of postlysis binding of crosslinked surface immunoglobulin to actin by carrying out the lysis and myosin extraction over a wide range of dilution of cells. They found that the association was not dependent on the concentration of cellular "reagents." Braun *et al.* (1982) confirmed these results directly by demonstrating that ligand interaction with cell surface immunoglobulin converts surface immunoglobulin from detergent-soluble to the cytoskeleton-associated detergent-insoluble state. Again, this conversion required divalent ligand interaction and hence is taken to imply crosslinking. As found by Flanagan and Koch (1978), capping was not necessary to induce surface immunoglobulin detergent insolubility.

Another approach to investigating receptor–cytoskeleton interactions was used by Prives *et al.* (1982). They explored detergent extractability of receptors with respect to maturation and development of embryonic cells in culture. In particular, they found that the proportion of acetylcholine receptors retained by the cytoskeleton after detergent extraction increased as the embryonic muscle cell matured. They further proposed that the previously noted lateral (within the plane of the plasma membrane) immobility of patches of acetylcholine receptors is therefore due to cytoskeletal attachment. This latter reference is to a technique that involves measurement of diffusion coefficients of cell surface receptors. Alterations in diffusion coefficients reflect changes in mobility states. Hence, manipulation of factors influencing the lateral mobility of receptors enables inference about receptor–cytoskeleton interactions. For example, Schlessinger et al. (1977), using the fluorescence photobleaching recovery method, demonstrated that the diffusion coefficient of cell surface receptors is altered by microtubular disrupting agents. In particular, surface receptor modulation by concanavalin A was partially reversed by agents such as vinblastin and podophyllotoxin. Subsequently, Tank et al. (1982) found that receptors (including acetylcholine receptors) were freely diffusing within membrane blebs separated from the cytoskeleton of isolated muscle cells and myoblasts. Hence, indirect evidence arguing for cell surface receptor association and the cytoskeleton has accumulated. Through the work of researchers such as Kordeli et al. (1987), more direct evidence is emerging. In the torpedo electrocyte model, they used immunohistochemical methods to identify a candidate protein that may serve to anchor surface receptors to intermediate filaments.

4. THE C3b RECEPTOR (CR1)

All of the previously mentioned techniques for investigating surface protein–cytoskeleton interactions have been applied to the C3b receptor. Not surprisingly, many aspects of surface CR1 behavior parallel the findings for other surface proteins. Early work by Atkinson et al. (1977) showed that CR1 function (as measured by a quantitative rosette assay) was modulated by cytochalasin-, vinblastin-, and colchicine-sensitive structures. This was important in implicating the central role that microfilaments and/or microtubules play in CR1 functioning. If cytoskeletal elements influence CR1 behavior, then one might expect to find both anchoring and free diffusibility with lateral mobility studies reminiscent of what Schlessinger et al. (1977) had found with surface receptors on mouse fibroblasts. Indirect yet detailed studies with the mouse macrophage (Griffin and Mullinax, 1981) revealed that C3b receptors migrate within the plane of the plasma membrane. The interpretation by Griffin and Mullinax was that such free lateral mobility is essential to complement receptor-mediated phagocytosis. Furthermore, it appears that receptor diffusion is necessary for high affinity binding of soluble C3b oligomers to CR1 (Brown, 1989). However, direct measurement of CR1 mobility by Hafeman et al. (1982) using epifluorecence and photobleaching techniques showed that while certain solid supports induce receptor patching, neutrophils resting on lipid monolayer-coated glass maintain, a uniform and freely diffusing population of CR1.

At this point, two separate mechanisms of CR1 behavior have been implicated. One is aggregation of passively diffusing receptors through ligand crosslinking, and the other is receptor modulation by the cytoskeletal apparatus as triggered by transmembrane signaling (consequent to ligand–receptor interaction). Strong evidence for the latter was provided by Jack and Fearon (1984). Using monospecific anti-CR1 antibody fragments, these researchers demonstrated that bivalent (but not monovalent) ligand binding to CR1 induced patches and capping in neutrophils. Morphologic analysis revealed that such capping was associated with underlying accumulation of myosin. They further demonstrated that receptor redistribution could be inhibited by pretreatment of cells with either cytochalasin D or chlorpromazine, suggesting that microfilaments or calcium-dependent reactions modulate cell surface CR1. An additional tantalizing observation derived from this research was that induction of capping of CR1 resulted in similar rearrangement of Fc receptors and vice versa. This co-redistribution of receptors did not occur with all cell surface proteins but could be inhibited with cytochalasin D and chlorpromazine. Nevertheless, it is important to note that crosslinking of CR1 is probably not a sufficient stimulus to induce ligand internalization (Brown, 1989); which brings us to a third mechanism of CR1 regulation. Changelian *et al.* (1985) found that phorbol myristate acetate induces not only externalization but also internalization of cellular CR1 which, at appropriate doses, can result in a net decrease in expression of cell surface CR1. They concluded that since no C3b-coated particles were present, phorbol myristate acetate induces ligand-independent down regulation of CR1 expression. This was confirmed by O'Shea *et al.* (1985b), who provided a deeper level of understanding. They found that phorbol ester-induced internalization of CR1 could be reversibly inhibited by cytochalasin B. They also demonstrated that phorbol esters increased the detergent insolubility of CR1. Because phorbol esters do not cause crosslinking of CR1, one may speculate that internalization is regulated by protein kinase C and that microfilaments not only function in surface receptor redistribution but are involved in receptor internalization as well.

Since phorbol myristate acetate causes phosphorylation, it is not surprising to learn from Wright and Meyer (1986) that phosphorylation plays a role in CR1 function. Specifically, they reported that phorbol esters caused activation and then deactivation of CR1, as measured by the phagocytosis-promoting capacity of CR1 in human neutrophils. Incorporation of the phosphate analog thiophosphate into neutrophils both enhanced receptor activity and blocked the receptor deactivation phase. Additional evidence for CR1 phosphorylation was provided by Changelian and Fearon (1986). They showed that phorbol myristate acetate induced CR1 phosphorylation in phagocytic cells but not in non-myelomonocytic cells. Whereas phorbol esters promote phagocytic function, NaF has been shown to inhibit this aspect of CR1 behavior (Okada and Brown, 1988). This effect does not appear to be due to NaF-induced ATP depletion because phagocytic capability returned before ATP repletion. The fact that NaF did not, however, inhibit CR1-mediated endocytosis implies that different mechanisms control CR1 function during phagocytosis versus endocytosis. At the very least, phagocytosis appears to be dependent on an intact cytoskeleton. Finally, the fate of internalized CR1 has recently been investigated. Through the use of flow cytometry with fixed and permeabilized neutrophils, Turner *et al.* (1988) quantified total cellular CR1 in the presence and absence of a protease inhibitor or lysosomal inhibitor and therefore deduced the existence of intracellular degradation of these receptors under conditions that cause ligand-independent internalization. Apparently, a pathway exists for receptors to be translocated to lysosomes for localized intracellular digestion. This confirms the earlier observation by Abrahamson and Fearon (1983), using electron microscopy, that much of the probe for CR1 could be found within lysosomes 20 min after neutrophils and monocytes exhibited CR1-mediated endocytosis. On the other hand, Malbran *et al.* (1988) demonstrated that some of the internalized receptor–CR1 complexes are recycled to the cell surface. By eluting cell surface C3b or monoclonal monovalent probe, they could follow the fate of internalized ligand and therefore suggested that recycling occurs via a prelysosomal predegradative compartment.

Direct evidence for association between crosslinked CR1 and the cytoskeleton was provided by Jack and Fearon (1983), who showed that such conditions induced increased detergent insolubility of bivalent anti-CR1 antibody fragments. They went on to define a significant relationship between detergent-insoluble CR1 and actin (Jack *et al.*, 1986). Using low-salt versus high-salt detergent as well as direct depolymerization of F-actin, they demonstrated that detergent insolubility parallels the presence of polymerized actin. Whether this CR1–microfilament interaction requires an intermediary protein has not yet been determined. However, the fact that this finding was not reproduced with another membrane protein, major histocompatibility complex class I, suggests that association between CR1 and the microfilamentary component of the cytoskeleton is specific.

Although nonionic detergent lysis appears to be a direct method for evaluating cytoskeleton–receptor interaction, the validity of this procedure must be examined more closely. We have already seen how biochemical reactions (e.g., proteolysis) can occur during detergent extraction. More germane to the interests of this chapter is the question of whether a steady state exists or develops between detergent-solubilized and bound CR1. Additional detail about the detergent extraction procedure will be useful here.

The procedure as it is typically used involves cell surface receptor labeling with receptor-specific probe (e.g., C3b-coated particles or anti-CR1 antibody fragments). The probe must be well characterized with regard to its dissociation constant. If, for example, the probe is fluorescent or radioactive, its behavior can be monitored. Lipid-soluble components can be solubilized in one of two ways, each using mild nonionic detergent. Either (1) detergent can be added to a suspension of cells or (2) cells pelleted by centrifugation can be resuspended in detergent. Next, the soluble and insoluble components can be separated by centrifugation. Aspiration of the supernatant allows resuspension of the cytoskeleton-containing pellet in buffered solution.

In our laboratory (Katz *et al.*, 1987), we found that the time between exposure to detergent and centrifugation of the lysate influences the measured quantity of probe associated with the deterget-insoluble cell residue. In particular, the association progresses with the length of time that the lysate is allowed to incubate before centrifugation. This indicates solubilized CR1 association with the detergent-insoluble cellular residue. By extrapolation to the zero time point of lysis, we concluded that less than 10% of cell surface CR1 in unactivated neutrophils was associated in a detergent-stable fashion with the cytoskeleton. This conclusion was consistent with the previously cited research of Hafeman *et al.* (1982), who found that "unperturbed" neutrophils exhibited freely diffusing surface CR1 during lateral mobility experiments.

Using an independent method, we confirmed postlysis association of CR1 with the detergent-insoluble residue. In a method similar to that used by Flanagan and Koch (1978), we varied the volume of lysis while maintaining the lysate incubation time constant. In this way, it was observed that the higher the concentration of solubilized CR1, the greater the extent of specific binding. Clearly, this phenomenon of postlysis binding presents a confounding variable that must be accounted for when interpreting CR1 detergent insolubility experiments.

A much more difficult question remains: Is the observed postlysis association of receptors to the detergent-insoluble residue representative of *in vivo* behavior? In other words, is this interaction specific for the cytoskeleton, or does it merely represent nonspecific cytoskeletal adsorption? One indicator that this interaction was not nonspecific was the presence of salt-poor bovine serum albumin in the lytic solution. Another approach to this question is to evaluate whether the affinity of CR1 for the cytoskeleton-containing cell residue can be altered by physiologic interventions. For example, it might be expected that calcium mobilization or exposure to synthetic chemotactic peptide would regulate or influence any CR1–cytoskeleton association. From research by O'Shea et al. (1985c), we know that calcium plays a role in CR1 expression and function. They found that antagonizing intracellular calcium inhibits phorbol ester-induced CR1 internalization. Inhibiting extracellular calcium, however, does not influence CR1 expression. Our research complements these data: we found that chelating extracellular calcium does not alter the percentage of surface CR1 associated with the nonionic detergent-insoluble residue.

While an increase in cytosolic free calcium (using ionophore A23187) leads to increased receptor expression, this effect was blocked by EDTA (Berger et al., 1985). That the calcium–calmodulin complex is necessary for receptor expression is suggested by the ability of phenothiazines to block increased receptor expression (Berger et al., 1985). The fact that TMB-8, an antagonist of calcium release from intracellular stores, prevents increased CR1 expression caused Berger et al. to conclude that calcium release from intracellular stores is necessary and sufficient for translocation of CR1 to the surface. Since receptor occupancy alone is not sufficient to increase surface CR1 expression (Brown, 1989) an additional signal must be postulated.

There is strong documentation that formylmethionyl-leucyl-phenylalanine (fMLP) affects both CR1 expression and the polymerization and depolymerization of actin. Several researchers, such as Kay et al. (1979), Fearon and Collins (1983), and Berger et al. (1984), have noted that fMLP can augment CR1 expression on the cell surface. In addition, Berger's group demonstrated that surface CR1 up regulation is not blocked by puromycin or cycloheximide. Since protein synthesis is not involved, this finding suggests that rather than being synthesized de novo, receptors translocate from intracellular sites. This evidence for the existence of a latent pool of intracellular CR1 has been elegantly illuminated by O'Shea et al. (1985a). Subsequently, Fyfe et al. (1987) also delineated distinct pools of neutrophil CR1 by using biochemical means. They reported that energy inhibitors such as dinitrophenol, rotenone, and antimycin A can block the increased expression of CR1 induced by fMLP but not the increase caused by warming. Similarly, cytochalasin B and colchicine blocked the increased expression of CR1 seen with fMLP but did not do so with warming alone. Hence, at least one pool of intracellular CR1 requires energy and an intact cytoskeleton to translocate to the cell surface. Other studies by Rao and Varani (1982), Howard and Oresajo (1985), and Sklar et al. (1985), have investigated the time course and regulating factors involved in fMLP-induced changes in cellular actin content and distribution. In this way, they have provided evidence for receptor-induced cytoskeleton changes that supports nicely the aforementioned research by Jack et al. 1986). Taken together, the data suggest that fMLP-induced changes in actin may be an intermediate step in fMLP-induced modulation of CR1.

In our laboratory, we found that conditions that lead to capping and CR1-mediated endocytosis [bivalent ligand binding with warming (Jack and Fearon, 1984; Abrahamson and Fearon, 1983)] lead to a decrease in the measured detergent insolubility of CR1. This reflects either a decrease in the quantity of CR1 associated with the cytoskeleton at the time of lysis or a decrease in the affinity of solubilized CR1 for the detergent-insoluble residue after lysis. Since the data were not reproducible with a monovalent probe, it is most likely that bivalent binding was responsible for this observation. In addition, we found that the decrease in detergent insolubility is evident for fMLP-stimulated cells regardless of the valence of the probe bound to CR1, indicating that the presence of crosslinked receptors is not a necessary condition for this effect. It may be that other researchers who showed enhanced CR1-cytoskeleton association had results confounded by postlysis association. Alternatively, our results may reflect transient actin polymerization (decay of cytoskeletal activation) due to removal of the chemotactic peptide stimulus (Sklar *et al.*, 1985).

If we attempt to shed a teleologic light on fMLP-induced changes in CR1, we must take into account the fact that neutrophils stimulated by fMLP show an increase in cell surface CR1 without a concommitant increase in the fraction of surface receptors associated with the cytoskeleton. This would suggest that fMLP serves to enhance the probability of C3b-mediated adherence but that some additional stimulus, leading to phosphorylation, is necessary to induce endocytosis. In a different cell system, Boxer *et al.* (1979) made a similar inference about fMLP. They reported that fMLP enhances neutrophil adherence to nylon without affecting the rate of phagocytosis of C3-coated particles. However, it should be pointed out that such an effect probably depends on the nature, concentration, and duration of exposure to chemotactic peptides. That complement receptor enhancement by fMLP is not sufficient to account for changes in CR1 function has also been proposed by Richerson *et al.* (1985). They reported that CR1 adherence, in the setting of fMLP-induced augmentation of cell surface CR1, is influenced by the nature of ligand binding.

5. REFERENCES

Abrahamson, D. R., and Fearon, D. T., 1983, Endocytosis of the C3b receptor of complement within coated pits in human polymorphonuclear leukocytes and monocytes, *Lab. Invest.* **48**:162–168

Arnaout, M. A., and Colten, H. R., 1984, Complement C3 receptors: Structure and function, *Mol. Immunol.* **21**:1191–1199

Atkinson, J. P., Michael, J. M., Chaplin, H., Jr., and Parker, C. W., 1977, Modulation of macrophage C3b receptor function by cytochalasin-sensitive structures, *J. Immunol.* **118**:1292–1299

Ben-Ze'ev, A., Duerr, A., Solomon, F., and Penman, S., 1979, The outer boundary of the cytoskeleton: A lamina derived from plasma membrane proteins, *Cell* **17**:859–865

Berger, M., O'Shea, J., Cross, A. S., Folks, T. M., Chused, T. M., Brown, E. J., and Frank, M. M., 1984, Human neutrophils increase expression of C3bi as well as C3b receptors upon activation, *J. Clin. Invest.* **74**:1566–1571

Berger, M., Birx, D. L., Wetzler, E. M., O'Shea, J. J., Brown, E. J., and Cross, A. S., 1985, Calcium requirements for increased complement receptor expression during neutrophil activation, *J. Immunol.* **135**:1342–1348

Boxer, L. A., Yoder, M., Bonsib, S., Schmidt, M., Ho, P., Jersild, R., and Baehner, R. L., 1979, Effects of a chemotactic factor, N-formylmethionyl peptide, on adherence, superoxide anion generation, phagocytosis, and microtubule assembly of human polymorphonuclear leukocytes. *J. Lab. Clin. Med.* **93**:506–514

Braun, J., Hochman, P. S., and Unanue, E. R., 1982, Ligand-induced association of surface immunoglobulin with the detergent-insoluble cytoskeletal matrix of the B lymphocyte, *J. Immunol.* **128**:1198–1203

Brown, E. J., 1989, The interaction of small oligomers of complement 3B (C3B) with phagocytes, *J. Biol. Chem.* **264(11)**:6196–6201

Changelian, P. S., and Fearon, D. T., 1986, Tissue-specific phosphorylation of complement receptors CR1 and CR2, *J. Exp. Med.* **163**:101–115

Changelian, P. S., Jack, R. M., Collins, L. A., and Fearson, D. T., 1985, PMA induces ligand-independent internalization of CR1 on human neutrophils, *J. Immunol.* **134**:1851–1858

Comacoff, J. B., Hebert, L. A., Smead, W. L., Van Aman, M. E., Birmingham, D. J., and Waxman, F. J., 1983, Primate erythrocyte-immune complex-clearing mechanism, *J. Clin. Invest.* **71**:236–247

Dykman, T. R., Cole, J. L., Iida, K., and Atkinson, J. P., 1983a, Structural heterogeneity of the C3b/C4b receptor (CR1) on human peripheral blood cells, *J. Exp. Med.* **157**:2160–2165

Dykman, T. R., Cole, J. L., Iida, K., and Atkinson, J. P., 1983b, Polymorphism of human erythrocyte C3b/C4b receptor, *Proc. Natl. Acad. Sci. USA* **80**:1698–1702

Dykman, T. R., Hatch, J. A., and Atkinson, J. P., 1984, Polymorphism of the human C3b/C4b receptor: Identification of a third allele and analysis of receptor phenotypes in families and patients with systemic lupus erythematosus, *J. Exp. Med.* **159**:691–705

Dykman, T. R., Hatch, J. A., Aqua M. S., and Atkinson, J. P., 1985, Polymorphism of the C3b/C4b receptor (CR1): Characterization of a fourth allele, *J. Immunol.* **134**:1787–1789

Fearon, D. T., 1985, Human complement receptors for C3b (CR1) and C3d (CR2), *J. Invest. Dermatol.* **85**:55s–57s

Fearon, D. T., and Collins, L. A., 1983, Increased expression of C3b receptors on polymorphonuclear leukocytes induced by chemotactic factors and by purification procedures, *J. Immunol.* **130**:370–375

Fearon, D. T., Kaneko, I., and Thomson, G. G., 1981, Membrane distribution and adsorptive endocytosis by C3b receptors on human polymorphonuclear leukocytes, *J. Exp. Med.* **153**:1615–1628

Flanagan, J., and Koch, G. L. E., 1978, Cross-linked surface Ig attaches to actin, *Nature* (London) **273**:278–281

Fyfe, A., Holme, E. R., Zoma, A., and Whaley, K., 1987, C3b receptor (CR1) expression on the polymorphonuclear leukocytes from patients with systemic lupus erythematosus, *Clin. Exp. Immunol.* **67**:300–308

Griffin, F. M., Jr., and Mullinax, P. J., 1981, Augmentation of macrophage complement receptor function in vitro, *J. Exp. Med.* **154**:291–305

Hafeman, D. G., Smith, L. M., Fearon, D. T., and McConnell, H. M., 1982, Lipid monolayer-coated solid surfaces do not perturb the lateral motion and distribution of C3b receptors on neutrophils, *J. Cell Biol.* **94**:224–227

Helenius, A., and Simons, K., 1975, Solubilization of membranes by detergents, *Biochim. Biophys. Acta* **415**:29–79

Howard, T. H., and Oresajo, C. O., 1985, The kinetics of chemotactic peptide-induced change in F-actin content, F-actin distribution and the shape of neutrophils, *J. Cell Biol.* **101**:1078–1085

Iida, K., and Nussenzweig, V., 1983, Functional properties of membrane-associated complement receptor CR1, *J. Immunol.* **130**:1876–1880

Iida, K., Mornaghni R., and Nussenzweig, V., 1982, Complement receptor (CR1) deficiency in erythrocytes from patients with systemic lupus erythematosis, *J. Exp. Med.* **155**:1427–1438

Jack, R. M., and Fearon, D. T., 1983, Cytoskeletal attachment of C3b receptor (C3bR) and reciprocal co-redistribution with Fc receptor (FcR) on human polymorphonuclear leukocytes (PMN), *Fed. Proc.* **42**:1235

Jack, R. M., and Fearon, D. T., 1984, Altered surface distribution of both C3b receptors and Fc receptors on neutrophils induced by anti-C3b receptor or aggregated IgG, *J. Immunol.* **132**:3028–3033

Jack, R. M., Ezzell, R. M., Hartwig, I., and Fearon, D. T., 1986, Differential interaction of the C3b/C4b receptor and MHC class I with the cytoskeleton of human neutrophils, *J. Immunol.* **137**:3996–4003

Katz, J. D., Rimmerman, C. M., Berrettoni, C. M., and Hafeman, D. G., 1987, Receptors for C3b on the neutrophil surface associate with the cytoskeleton-containing cell residue subsequent to cell lysis with nonionic detergent, *J. Immunol. Methods* **99**:83–93

Kay, A. B., Glass, E. J., and Salter, D. McG., 1979, Leukoattractants enhance complement receptors on human phagocytic cells, *Clin. Exp. Immunol.* **38**:294–299

Kordeli, E., Cartaud, J., Nghiem, H. O., and Changeux, J. P., 1987, The torpedo electrocyte: A model system for the study of receptor-cytoskeleton interactions, *J. Receptor Res.* **7**:71–88

Malbran, A., Siwik, S., Frank, M. M., and Fries, L. F., 1988, Cr1-receptor recycling in phorbol ester-activated polymorphonuclear leukocytes, *Immunology* **63**:325–330

Okada, K., and Brown, E. J., 1988, Sodium fluoride reveals multiple pathways for regulation of surface expression of the C3b/C4b receptor (CR1) on human polymorphonuclear leukocytes, *J. Immunol.* **140**:878–884

Osborn, M., and Weber, K., 1977, The detergent-resistant cytoskeleton of tissue culture cells includes the nucleus and the microfilament bundles, *Exp. Cell Res.* **106**:339–349

O'Shea, J. J., Brown, E. J., Seligmann, B. E., Metcalf, J. A., Frank, M. M., and Gallin, J. I., 1985a, Evidence for distinct intracellular pools of receptors for C3b and C3bi in human neutrophils, *J. Immunol.* **134:**2580–2587

O'Shea, J. J., Brown, E. J., Gaither, T. A., Takahashi, T., and Frank, M. M., 1985b, Tumor-promoting phorbol esters induce rapid internalization of the C3b receptor via a cytoskeleton-dependent mechanism, *J. Immunol.* **135:**1325–1330

O'Shea, J. J., Siwik, S. A., Gaither, T. A., and Frank, M. M., 1985c, Activation of the C3b receptor: Effect of diacylglyerols and calcium mobilization, *J. Immunol.* **135:**3381–3387

Prives, J., Fulton, A. B., Penman, S., Daniels, M. P., and Christian, C. N., 1982, Interaction of the cytoskeletal framework with acetylcholine receptor on the surface of embryonic muscle cells in culture, *J. Cell Biol.* **92:**231–236

Rao, K. M. K., and Varani, J., 1982, Actin polymerization induced by chemotactic peptide and concanavalin A in rat neutrophils, *J. Immunol.* **129:**1605–1607

Richerson, H. B., Walsh, G. M., Walport, M. J., Moqbel, R., and Kay, A. B., 1985, Enhancement of human neutrophil complement receptors: A comparison of the rosette technique with the uptake of radio-labelled anti-CR1 monoclonal antibody, *Clin. Exp. Immunol.* **62:**442–448

Schlessinger, J., Elson, E. L., Webb, W. W., Yahara, I., Rutishauser, U., and Edelman, G. M., 1977, Receptor Diffusion on cell surfaces modulated by locally bound concanavalin A, *Proc. Natl. Acad. Sci. USA* **74:**1110–1114

Schreiber, R. D., 1984, The chemistry and biology of complement receptors, *Springer Semin. Immunopathol.* **7:**221–249

Sheterline, P., and Hopkins, C. R., 1981, Transmembrane linkage between surface glycoproteins and components of the cytoplasm in neutrophil leukocytes, *J. Cell Biol.* **90:**743–754

Sklar, L. A., Omann, G. M., and Painter, R. G., 1985, Relationship of actin polymerization and depolymerization to light scattering in human neutrophils: Dependence on receptor occupancy and intracellular Ca^{++}, *J. Cell Biol.* **101:**1161–1166

Tank, D. W., Wu, E., and Webb, W. W., 1982, Enhanced molecular diffusibility in muscle membrane blebs: Release of lateral constraints, *J. Cell Biol.* **92:**207–212

Turner, J. R., Tartakoff, A. M., and Berger, M., 1988, Intracellular degradation of the complement C3b/C4b receptor in the absence of ligand, *J. Biol. Chem.* **263:**4914–4920

Wong, W. W., Wilson, J. G., and Fearon, D. T., 1983, Genetic regulation of a structural polymorphism of human C3b receptor, *J. Clin. Invest.* **72:**685–693

Wright, S. D., and Meyer, B. C., 1986, Phorbol esters cause a sequential activation and deactivation of complement receptors on polymorphonuclear leukocytes, *J. Immunol.* **136:**1759–1764

Yu, J., Fischman, D. A., and Steck, T. L., 1973, Selective solubilization of proteins and phospholipids from red blood cell membranes by nonionic detergents, *J. Supramol. Struct.* **1:**233–248

Chapter 8

The Flow of Granular Organelles in Leukocyte Differentiation

Arthur K. Sullivan

1. INTRODUCTION

In following the milestones *en route* to our current image of life inside the cell, one is reminded of the aphorism that medical science begins and ends at the bedside. Those careful and insightful clinical observations, unfolding with the suspense of a great mystery novel, have stimulated many of the basic advances in the field. From their studies on families with premature heart disease and abnormal lipoproteins, Brown, Goldstein, and co-workers expanded the concept of intracellular vesicular traffic (Goldstein *et al.*, 1985). From observations on children with skeletal deformities and psychomotor retardation came a new model for how enzymes are directed to lysosomes (Kornfeld, 1986). From experiments on abnormal blood cells of children with recurrent severe infections were defined new mechanisms of cell adhesion (Springer *et al.*, 1987).

Just as the morphogenesis of a whole organism proceeds according to finely timed and regulated events, so does the synthesis of each organelle within a single cell. According to the dictates of its genetic design, a developing blood cell forms granules to compartmentalize the elements of its secretory apparatus. In a mature neutrophil, macrophage, platelet, eosinophil, or basophil, these specialized organelles constitute a major determinant of function, and if the process fails at any step along the way, disease soon will follow. As an example, consider the transition of a myeloblast into a polymorphonuclear neutrophil. The first stage, the promyelocyte, is defined by the presence of specialized lysosomelike structures called azurophilic granules. Later, the emergence of secondary or specific granules identifies the myelocyte. Both of these organelles are formed in the Golgi apparatus by a series of synchronized reactions that assemble the

Arthur K. Sullivan   McGill Cancer Centre, and Division of Hematology, Royal Victoria Hospital, Montreal, Quebec H3G 1Y6, Canada.

constituent proteins into a precursor vesicle. When myeloid precursors become neoplastic, they often deviate from their normal program of maturation and fail to produce granules of sufficient number or quality. This, along with deficiencies in both the phagocytic apparatus and essential plasma membrane components, results in a predominance of cells that lack adequate microbicidal competence to prevent recurrent infection, sepsis, and eventual death of the host.

Throughout this review, intended more to whet the appetite than to be comprehensive, I will try to connect different areas of research pertinent to leukocyte differentiation, summarize the current view of how membrane organelles are formed, discuss a few illustrative clinical conditions, and finally ponder some of the questions that await an answer. Because justice to these different topics would require a separate review for each, I have had to be selective, omitting individual papers and whole areas that might be significant, and have referred the reader to other publications where details can be found.

2. BIOGENESIS OF MEMBRANE-BOUND ORGANELLES

2.1. Introduction

Although other tissues (e.g., pancreas) can differentiate to form an organ containing a stable population of secretory cells, leukocytes are exceptional because they are in a state of constant turnover—each species with its unique granules is replenished from a common progenitor. From the experimentalist's point of view, however, this complexity presents problems in trying to dissect how the intracellular machinery is formed. For this reason, much of the background information has been derived from simpler models, and at this point I will review some of the present concepts of how the internal membrane compartments are made.

2.2. Endoplasmic Reticulum

Any protein destined to reside in a membranous organelle, either on the surface or within the cell, is synthesized, processed, and transported as part of a smaller vesicular package. It has been one of the major challenges in cell biology to understand the logic used by these wandering bodies along the way to where they should go. The importance of this process already has been emphasized by some of the catastrophic diseases that result when the system does not work.

2.2.1. How Does a Protein Enter the Membrane Compartment?

The first problem was to understand how a nascent protein in a ribosomal complex on the cytoplasmic side of the endoplasmic reticulum (ER) passes through the membrane into the lumen, and after it is assembled how it is incorporated into a vesicular carrier bound for other loci. Stated another way, the question was: What are the signals that direct the RNA complex to the right place on the ER membrane, and how can a chain of polar amino acids pass through the lipid bilayer while obeying the rules of thermodynamics? Out of this evolved the model of the signal recognition particle cycle. This process has been reviewed in detail (Rapoport and Wiedmann, 1985; Duffaud *et al.*, 1985; Rapoport, 1986; Walter and Lingappa, 1986; Von Heijne, 1988; Lingappa, 1989) and will be summarized here.

Usually, translation begins at a methionine codon near the 5' end of the mRNA and results in a 15–30-amino-acid sequence that becomes attached to a cytosolic complex composed of a small RNA chain and six proteins, which together are called the signal recognition particle (SRP). After the SRP–ribosomal complex has been formed, the rate of translation slows until the complex binds to a receptor on the cytoplasmic side of the ER membrane called the docking protein. This interaction results in release of the complex from the receptor, further elongation of the developing protein, and the beginning of its intrusion across the membrane. Two predominant hypotheses compete to explain the mode of vectorial translocation: one proposes that the protein assumes a conformation that energetically is favorable for interaction with the lipid and that allows it to move in one direction into the lumen; the other suggests that there is a chain of receptor-mediated reactions that directs the flow of energy in the appropriate direction. After being translocated across the lipid layer, the polypeptide encounters a signal peptidase on the inner ER membrane, and the signal sequence is cleaved, allowing the remainder of the protein to be deposited inside the ER lumen. While on the way through the ER membrane, a decision must be made as to whether the protein will be cleaved to a free form (as for secretory proteins) or whether it will remain embedded in the membrane (as for integral membrane proteins), and if fixed in the membrane, how many passes it will weave itself through. Part of this process may be controlled by a stop-transfer codon, but other evidence from studies on hybrid proteins, produced by recombining DNA from parts of genes encoding different soluble and membrane-bound structures, indicates that the necessary information may reside in the primary polypeptide sequence itself.

While still in the ER, a new protein acquires its *N*-glycosyl groups by transfer from a dolichol-oligosaccharide intermediate, and then its mannose-rich core sugars undergo trimming of their terminal triglucose (Hubbard and Ivatt, 1981). Other modifications begin here also, including addition of *N*-acetylgalactosamine to form O linkages (Tooze *et al.*, 1988), as well as acyl groups used for a selective kind of anchoring to the membrane (Hu *et al.*, 1988). At present there is not sufficient information to predict which of the proteins that exit from the ER are destined to be soluble secretory products or integral membrane components. There is equally little indication of how a single cell at different stages in its differentiation can produce many kinds of granular organelles.

2.2.2. What Regulates Exit from the Endoplasmic Reticulum?

After its synthesis is complete, the polypeptide chain is transferred from the ER to the Golgi apparatus by a process that appears to be the limiting step. Although the *passe-partout* contained in the primary sequence, or the conformation that permits passage and determines the rate of flow, has not been identified, it is evident that different proteins proceed with different kinetics. From studies on three structures that remain in the ER, it has been suggested that there may be a "gatekeeper" conformation possessing a common C-terminal tetrapeptide sequence (reviewed in Lodish, 1988). Other work with a mutant low-density lipoprotein receptor from the WHHL rabbit has shown that the abnormal precursor protein is of normal size but is degraded before it reaches the plasma membrane.

Analysis of its encoding gene showed that there is a four-amino-acid deletion in the region of the membrane junction in a cysteine-rich domain that may prevent exit from the ER. The authors postulate that this gatekeeper in some way is mediated by formation of disulfide bonds (Yamamoto, 1986). In another example, a comparison between homologous H2-D and -K murine major histocompatibility complex antigens revealed that the rate of transfer from ER to Golgi was different for each of them. Such a divergence between two so closely related structures implies that even more subtle conformational differences may be the controlling parameters (Williams *et al.*, 1985). In a simpler system, the slime mold *Dictyostelium*, mutants have been isolated in which both the mannosidase and glucosidase contained the correct primary sequence, but neither was proteolytically processed in a normal manner. These enzymes accumulated in the ER and did not reach the Golgi, again suggesting that unless a protein can attain the proper conformation, the localization mechanism cannot function as usual (Woychik and Dimond, 1987).

The carrier that passes proteins on to the *cis* Golgi appears to be a membrane vesicle. It has been identified on density gradients in the very light fraction and enters the stack at a rate proportional to the rate of secretion of the protein (Lodish, 1988). Distinct from endosomes, these particles do not contain clathrin but bear another type of coating that has not yet been identified (Orci *et al.*, 1986). Little is known in mammals of how this intermediate organelle finds the Golgi, but work with the *sec-23* yeast mutant has indicated that there is an acceptor protein essential for the process of transfer from the ER (Ruohla *et al.*, 1988).

2.3. Golgi Complex

2.3.1. Pattern of Flow Through the Golgi Apparatus

After a protein has entered one of the cisternae along the *cis* face of the Golgi apparatus, its sugars undergo extensive modification before it finally is released from the distal *trans* side (reviewed in Krag, 1985; Farquhar, 1985; Pfeffer and Rothman, 1987). Early in the course of transit, further mannoses are trimmed from the core before it is rebuilt into complex and hybrid chains by a host of glycosyl transferases located in an assembly line fashion along the Golgi membranes. It is not difficult to imagine many levels at which the character of oligosaccharides can be influenced, depending on the type and conformation of substrate encountered by the enzyme, how their activities are modulated, and how the accessibility and function of sugar nucleotide transporters are regulated (reviewed in Schacter, 1986; Deutscher *et al.*, 1984; Deutscher and Hirschberg, 1986; Hirschberg and Snider, 1987). As demonstrated by Pfeffer and Rothman (1987) in cell-free systems, transfer of product between the *cis*, medial, and *trans* compartments is mediated by vesicles. Although specific signals or "routing labels" have not been identified for this process, there is evidence that transport from one Golgi subcompartment to the next is not random but follows the direction of the recognized compartmental boundaries. After emerging from the *trans* face of the stacks, the particle must undergo homo- or heterotypic fusion events or migrate to another site to finally become the mature organelle it is destined to be. This implies the existence of a further unidentified signal molecule(s), especially since the role of clathrin as an element of intermediate vectorial flow remains controversial.

Quite relevant to differentiation and cell function is the possibility that this complex interplay of signals may not be fixed in a cell but may be subject to further regulation. For example, in rat hepatoma cells infected with mouse mammary tumor virus, a major viral protein is synthesized in the ER as a 74-kDa precursor and, after proteolytic processing in the Golgi, finally appears on the plasma membrane as a 50-kDa form. However, after the cells have been exposed to dexamethasone, the same precursor gives rise to three plasma membrane species of 78, 70, and 32 kDa. Because the effect is inhibited by monensin, and further sugars have been added to the mannose core, it is possible that the proteolytic processing occurred in the medial or early *trans* Golgi elements (Haffar *et al.*, 1988).

2.3.2. Where Does the Sorting Mechanism for Membranes Reside?

Evidence is beginning to emerge that the place for actual sorting of products destined for membranes of cytoplasmic granules is in the *trans* Golgi stack or in the more distal *trans* Golgi network (reviewed in Griffiths and Simons, 1986). After a protein has been fully processed and is ready to migrate to its final destination, what are the determinants that dictate either secretion or transport to a membrane organelle? Kelly (1985) has proposed a model in which the process of protein migration, regardless of final destination, can be either constitutive (implying a constant low-level release or localization) or regulated (implying sudden discharges of activity following an appropriate stimulus) (Burgess and Kelly, 1987). Unless the proper signals are supplied, the product follows a process of passive bulk membrane flow by which the system is programmed to move vesicles constantly to the plasma membrane.

To identify the point of divergence between the regulated and constitutive pathways, experiments were designed with the AtT20 pituitary cell line, which has retained its capacity to make secretory granules. A transfected influenza virus hemagglutinin (HA) protein was the model constitutive protein, and insulin inserted into the pathway via its cDNA was the regulated species. Both of these proteins were randomly distributed throughout the Golgi complex, but outside of the *trans* face HA was found only in 100–300-nm vesicles and insulin only in typical dense granules. Applying this technique to pancreatic beta cells, the investigators showed that both naturally secreted insulin and HA were found in the *trans*-most cisternae, but coated granules contained only insulin. Because it already had been shown that immature vesicles are clathrin coated immediately after their exit from the Golgi and mature secretory granules no longer are coated, they concluded that the sorting of HA and insulin preceded formation of the coated particle. Presumably, this occurs somewhere in the *trans*-most compartment, and other specific signals are required at that point for entry into the regulated pathway (Orci *et al.*, 1987). Similarly, it has been shown in AtT20 cells infected with hepatitis virus that both adrenocorticotropin and viral proteins travel the same course through the Golgi but diverge into separate units after they exit from the *trans* compartment (Tooze *et al.*, 1987). Because electron microscopic (EM) studies have shown that it can contain both recycling transferrin receptor and newly formed alpha-1-antitrypsin at the same time, this organelle may contain also the apparatus for redirecting the endosomal/endocytic and exocytic pathways (Stoorvogel *et al.*, 1988).

Other evidence does not easily fit into a model of simple passive flow through the proximal Golgi, with active sorting at the distal end. In studies comparing the maturation of a constitutively secreted *Drosophila* protein (YP2) that contains a single tyrosine sulfation site with that of a mutant that lacks the site, it was noted that the defective form resided in the *trans* Golgi for a longer time. The authors interpreted this to imply that there may be a conformational or S-mediated controlling element (Friederich *et al.*, 1988). Furthermore, detailed ultrastructural observations and reconstruction of the entire three-dimensional topography of the Golgi apparatus of pancreatic acinar cells have shown that prosecretory granules appear initially as dilatations on the fenestrated saccules of the *cis* (!) side and continue to enlarge as they migrate toward the *trans* elements prior to budding (Rambourg *et al.*, 1988).

2.4. Lysosomes

Whatever the mechanism that determines the route of a glycoprotein through the Golgi apparatus, we still are left with the problem of how the soluble contents and membranes come together in a coordinated manner to form organelles such as lysosomes or leukocyte granules. At present, one subset of lysosomal enzymes and a few lysosomal membrane proteins have been studied in detail.

2.4.1. The Mannose-6-Phosphate Pathway for Localization of Lysosomal Enzymes

In the course of studies on fibroblasts from children with a rare syndrome of mental retardation, skeletal deformities, and hepatosplenomegaly (I-cell disease or mucolipidosis II), it was noticed that abnormal products remained stored in the lysosomes. Further experiments tracing the pathway of glycosylation of lysosomal proteins revealed that many acid hydrolases were not phosphorylated at the 6 position of mannose as they are in normal fibroblasts because the *N*-acetylglucosaminyl-phosphotransferase enzyme was either absent or defective. This enzyme mediates transfer of P-GlcNAc to protein from UDP-GlcNAc, forming a substrate for phosphoglycosidase, which in turn cleaves the sugar and leaves the phosphomannose-protein (reviewed in Kornfeld and Sly, 1985). By combining with a receptor for mannose-6-phosphate (M6PR) in the Golgi apparatus, the product is concentrated and targeted to a prelysosomal vesicle. More recently, two M6PRs have been identified, one a 46-kDa protein that binds its ligand by a Ca^{2+}-dependent process and the other a 215-kDa, Ca^{2+}-independent structure. Both species have been found also on the cell surface, but the smaller form appears not bind to exogenous ligands (Stein *et al.*, 1987). The pH dependence of the receptor–ligand interaction enables the complex to dissociate as the maturing lysosome becomes progressively acidified, freeing the active enzyme. A special adaptation of this mechanism has been made by osteoclasts, which appear to use the M6PR to target hydrolytic enzymes to their apical ruffled border for secretion at sites of bone resorption (Baron *et al.*, 1988).

Despite the elegance of this work, some very important gaps remain to be filled. Because not all proteins that bear the requisite core sugars are phosphorylated and sugars themselves are poor substrates, there must be other control mechanisms that determine the selectivity of the phosphorylation reaction. Also, because children with I-cell disease do not suffer from recurrent infections and their neutrophils appear to have normal granule contents, there must be parallel but different targeting mechanisms operating in other cells

(Varki *et al.*, 1982). Evidence for an alternate pathway has been suggested by the observation that beta-galactosidase of epididymal cells binds to membranes through a mechanism that is inhibited much better by fructose-1,6-diphosphate than by mannose-6-phosphate (Sosa *et al.*, 1987).

2.4.2. Lysosomal Membranes Contain Specific Proteins

To form a complete organelle, the lysosomal enzymes must be encased in a proper membrane package. A few proteins have been described that are restricted primarily to lysosomal membranes, but the recognition code that determines why they localize there remains a mystery. Two lysosome-associated membrane proteins (LAMPs) of murine cells have been studied in detail. LAMP-1 and LAMP-2 are highly glycosylated, highly sialylated proteins of 105–115 and 100–110 kDa, respectively, originating from a core protein of 45 kDa. Their tissue distribution is quite interesting in that the major sites of expression include macrophages as well as epithelioid cells such as kidney tubules, islets of Langerhans, and hepatocytes but not most stromal cell types. Functionally, they may be involved in proton transport, as indicated by the fact that polyclonal antibodies to them also reacted with the H^+/K^+-ATPase. During their synthesis they are concentrated in the ER and processed by the Golgi, but they appear not to be phosphorylated on glycosyl groups (Chen *et al.*, 1986; D'Souza and August, 1986). More recently, a cDNA for "LAMP-A" has been cloned from human tissue; it was found to have homology to immunoglobulin A in the hinge region but did not show any similarity to the M6PR. This was accomplished after immunopurifying a major sialylated glycoprotein called leukosialin from an extract of chronic myelogenous leukemia (CML) cells and later demonstrating immunological cross-reactivity with murine LAMP-1 and -2. The high sialic acid content suggests that it is sorted to the lysosome after passage through the *trans* Golgi, because that is where most of the enzymes reside for sialylation and fucosylation reactions (Viitala *et al.*, 1988).

A second group of lysosomal glycoproteins (lgp's) has been purified from rat liver; this group includes species of 120, 100, and 80 kDa that also are heavily glycosylated and rich in sialic acid. The distribution of gp120 appears to be quite restricted, not detected even in endosomes (Lewis *et al.*, 1985). Later cloning of a cDNA showed that the gp120 protein contains five domains similar to that of LEP 100 of the chicken (see below) and lgp 110. Previous speculation that the rich glycosylation of these proteins protected them from lysosomal proteases did not prove true, because they were shown to be as sensitive to trypsin as other proteins that bore fewer oligosaccharides (Howe *et al.*, 1988).

A third group of lysosome/endosome/plasma membrane (LEP) proteins related to the lgp's was described in chicken embryo fibroblasts. LEP 100, another highly glycosylated species derived from a core polypeptide of 45 kDa, was different from other similar structures in that 90% of it was found in the lysosome, 5–8% was in the endosomal compartment, and 2–3% was in the plasma membrane. The endosomal fraction appeared to be localized by a cloroquine-sensitive process, indicative of a unique recycling pathway between the lysosome, endosome, and cell surface (Lippincott-Schwartz and Fambrough, 1987). A cDNA of avian origin has been expressed in mouse cells, and the protein product has been colocalized to lysosomes with LAMP-1 and -2. This implies that the necessary information for lysosomal targeting lies within the protein itself and does not depend on modifications restricted to a particular cell (Fambrough *et al.*, 1988).

2.4.3. How Are Lysosomal Membrane Proteins Processed in the Golgi Complex?

From what has been described above, it is clear that many lysosomal membrane proteins are glycosylated and probably continue in the Golgi complex through to the *trans* elements. The M6PR has been used as a model to follow this process in a number of studies because it is an integral protein of the Golgi and plasma membranes and carries enzymes to the prelysosomal compartment but itself does not go into the lysosome. Thus, it is a useful marker in pinpointing where the final sorting events might occur. In rat liver, M6PR is found in the *trans* Golgi as well as in the coated vesicles that are thought to shuttle the receptor on to distinct endosomes. Also, the Ca^{2+}-independent form has been shown to be the receptor for type II insulin-like growth factor, to be phosphorylated on tyrosine and serine, and to be concentrated in a detergent-insoluble, clathrin-coated subcellular fraction (Corvera *et al.*, 1988). An indication of where the M6PR and lysosomal components diverge has come from experiments showing that after they pass through the *trans* Golgi, lgp's are found in high concentration in the lysosomes and low concentrations in the Golgi, plasma membrane, and endosomes, but M6PR is not detected in the lysosomes at all. Thus, there must exist a post-Golgi, prelysosomal sorting organelle. Trying to identify such a structure by EM, Geuze *et al.* (1988) found the M6PR of rat hepatoma cells in endosomal tubules and vacuoles but lgp's only in vacuoles. Monitoring the course of an endocytic marker (horseradish peroxidase–ferritin), they observed that it appeared sequentially through $M6PR^+ lgp^- \rightarrow M6PR^- lgp^+$ vesicles and concluded that the sorting compartment is the $M6PR^+ lgp^+$ endosome. A similar conclusion was reached by Griffiths *et al.* (1988) from studies on rat kidney cells. They identified the M6PR in the perinuclear, para-Golgi, multilamellar organelle that contained lgp 120 as well as cathepsin D and beta-glucuronidase. This entity, which they proposed to be the "intermediate compartment binding the lysosomal enzyme biosynthetic and endocytic pathways to lysosomes," was in the process of becoming acidified and was distinct from the *trans* Golgi network.

Using methods whereby coated vesicles from endocytic and exocytic pathways could be separated and immature and mature forms of representative enzymes (cathepsins C and D) could be distinguished, Lemansky *et al.* (1987) showed that the exocytic coated vesicles contained only immature enzymes and that lysosomes contained only mature enzymes. These observations led them to conclude that the process of sorting to lysosomes must occur distal to the *trans* Golgi and that coated vesicles are involved in transport of newly synthesized lysosomal enzymes. The conclusion is supported by other studies investigating the pathway of lgp 110 and 120 to the lysosome and of Fc receptor and HA to the plasma membrane. Although the two classes of proteins flowed through the Golgi complex at the same rate, the lysosomal components reached their destination too quickly to be explained by an indirect route through the plasma membrane and back to the lysosome. Thus, they go directly to the lysosome, probably after being sorted in the *trans* Golgi network (Green *et al.*, 1987).

On the basis of this descriptive foundation, several laboratories are trying to elucidate the mechanisms. Immunolocalization studies have shown that three typical highly glycosylated and sialylated lysosomal membrane proteins of rat liver (LIMPs) pass through the Golgi, distribute to both coated and uncoated vesicles, and pass on into the same lysosomal granule. Pulse-chase experiments of the three different structures revealed that they flowed through the Golgi at different rates, suggesting that all proteins bound for the lysosomal membrane do not have a common carrier. Incubation of cells with tunicamycin actually increased the rate of incorporation into the lysosomes, supporting the notion that N-linked oligosaccharides are not essential for targeting (Barriocanal *et al.*, 1986). It still is not known what triggers the movement of M6PR from binding sites in the Golgi to the site of ligand dissociation. Although it has been proposed that the binding process itself was the activator of forward movement and that dissociation signaled the return of free receptors back to the site of binding, Braulke *et al.* (1987) showed that the kinetics of the endocytic and sorting pathways follows a constitutive mode not altered by ligand binding.

The lysosomes, and presumably other granular organelles, are not static bags of enzymes. After or slightly before they exit from the *trans* Golgi network, the precursor forms become progressively more acidic (Anderson and Orci, 1988). Other studies have shown that after interspecies hybrids have been produced, the lysosomes intermingle and intercommunicate, possibly through a distinct set of endosomes (Deng and Storrie, 1988). Finally, in order to mediate their ultimate role of fusing with phagocytic vacuoles, they must possess elements that enable them to distinguish their microtubule tracks from the microfilaments (Matteoni and Kreis, 1987).

2.5. Communication between Compartments: Endosomes

Throughout the cytoplasm, heterogeneous membrane microvesicles, collectively called the endosomes, mediate regeneration of the plasma membrane and communication between compartments. There is an extensive literature on the movement and function of these organelles in specific systems, and only selected points will be discussed here (reviewed in Willingham and Pastan, 1984; Goldstein *et al.*, 1985; Bergeron *et al.*, 1985, 1988).

2.5.1. Generation of Plasma Membrane Heterogeneity: Polarity

Another challenging aspect of the localization problem is to unravel how the polarity of selected plasma membrane proteins is generated in different types of cells (reviewed in Simons and Fuller, 1985; Simons, 1987). When grown under appropriate conditions in culture, the colon carcinoma line CaCo-2 establishes a functional polarity for solute transport and the topographical specialization in structure to support it. On the basis of the disruption caused by tunicamycin or lysosomotropic agents, Rindler and Traber (1988) proposed that the constitutive pathway may lead to genesis of membrane at the basolateral side, but positive signals are needed for subsequent apical localization. Indeed, microtubule poisons allow apical proteins to sort erroneously in a basolateral direction and block apical secretion of lysosomal enzymes (Eilers *et al.*, 1989). In the Madin-Darby canine kidney cell model, in which renal tubular epithelial cells form an up–down orientation in culture, Vega-Salas *et al.* (1988) showed that intercellular contacts fix a first stage of weak polarity during which most of the marker proteins associated with the apical surface are stored in a vacuolar organelle. This structure then fuses ("degranulates") at an area of cell–cell contact at the basolateral side, after which the apical proteins migrate to the free surface. From detailed kinetic studies, Bartles *et al.* (1987) showed that proteins are first transported to the basolateral side and only later are brought to the apex by another retrieval process.

2.5.2. Communication between Plasma Membranes and Lysosomes

Assuming that the endosomes are a definite functional compartment with a distinct set of rules to follow, what evidence is there for traffic between the organelles? Clearly there is intercommunication between the plasma membrane and the lysosome, but it is not a process of simple melding of one into the other. It has been estimated that only approximately 1% of the internalized plasma membrane is recycled by way of the lysosome, and the bulk is returned by an independent route (reviewed in Haylett and Thilo, 1986). However, the mechanisms governing this course are not evident. Except for those structures that bind their ligands in clathrin-coated pits and follow a distinct pathway, it appears that there is a similar rate of turnover for most of the plasma membrane proteins, of which only a small fraction ever becomes associated with the lysosomal membrane (Draye *et al.*, 1987).

If the bulk of membrane proteins do not pass through the lysosome, where do they go after endocytosis? On the basis of the fact that the Golgi apparatus is organized to expose proteins to glysosyl transferases in a sequential fashion, the point of entry and route of a protein through the stacks can be inferred by modifications of known model oligosaccharide carriers. Results from such studies have led to the conclusion that internalized plasma membrane does not recirculate through the *cis* or *medial* Golgi compartments but passes through the *trans* cisterns (Duncan and Kornfeld, 1988; Neefjes *et al.*, 1988). There can be exceptions to this rule, however, in specialized sites such as the synaptic vesicule membrane. After fusing with the plasma membrane, certain granule proteins do not recirculate all the way back to the Golgi but are recovered by a local recycling process, presumably to keep the material closer to the synaptic terminal for reuse (Lowe *et al.*, 1988).

2.5.3. Are All Endosomes the Same?

Much of what is known about the activity of these particles has come from studies on the routes followed by receptor–ligand complexes after they have been enclosed in an endocytic vesicle pinched off from the cell surface. Depending on the specific molecules involved, different pathways result: the receptor–ligand complex can return to the plasma membrane intact (e.g., transferrin), it can be brought to the lysosome and degraded (e.g., Fc receptors), or it can dissociate, with only the receptor returning intact to the surface. To address the question of whether there might be subpopulations of endosomes that mediate flow along different routes, Bergeron *et al.* (1988) showed that there are two fractions (early and late), distinguishable by density, degree of chloroquine binding (acidification), and pattern of processing of insulin and prolactin, which were named early and late because of the time sequence noted for the passage of internalized ligand–receptor complexes. Possibly, the late fraction may bear unique components, as determined by immunological markers (Ahluwalia *et al.*, 1988). In agreement with these previous results, Schmid *et al.* (1988) used free-flow electrophoresis to separate endosomes into early and late fractions and found that their markers could be traced sequentially from one to the other. On the basis of the patterns obtained, they concluded that the early group is a major vehicle for recycling of plasma membrane receptors but the late form is involved in delivery to the lysosomes. However, each type contained distinct proteins and showed quantal differences in surface charge, consistent with the hypothesis that one is qualitatively different from the other, of independent origin, and not wholly derived from the plasma membrane.

These intermediates introduce yet another locus at which specialization can develop in a differentiated tissue and, in a regulated fashion, respond to the cell's programming. For example, in the basal state, the glucose transporters of adipocytes are stored in tubulovesicular structures of the *trans* Golgi and in small cytoplasmic vesicles, but after the cell has been stimulated by insulin, they rapidly are translocated to the plasma membrane (Blok *et al.*, 1988).

2.5.4. What Might Be the Regulators?

As one can see in this short survey of membrane protein synthesis, there are many points from whence a minor deviation could be amplified into severe pathology for the organism. In view of the links that have been made between the *ras* family of oncogenes and the complex of GTP-binding structures (the G proteins), it may be opportune at this time to review some of the evidence for their role in regulating intracellular organelles (reviewed in Barbacid, 1987; Gilman, 1987; Stryer and Bourne, 1986). Recent reports have implicated GTP at several controlling points in the process of protein flow through the synthetic apparatus. It has been shown that for some polypeptides to be inserted into the ER membrane during formation of the SRP–receptor complex, a GTP-binding protein is required (Wilson *et al.*, 1988). In yeast cells, excess GTP can inhibit transfer from the ER to the Golgi; in a cell-free system designed to study transport between compartments of the Golgi stacks, addition of the hydrolysis-resistant GTP analog and G-protein activator GTPγS resulted in a block at the site of the acceptor membrane, possibly at the step of vesicle attachment prior to its fusion (Melançon *et al.*, 1987). Further work with yeast cells has revealed yet another site in the temperature-sensitive *sec4* mutant, in which transport of vesicles to the plasma membrane is blocked at the restrictive temperature. In these cells the *sec4* protein, a *ras*-like GTP binder normally located on the cytoplasmic face of the plasma membrane and present in some secretory vesicles, does not function. The authors propose that there may be a group of GTP-binding proteins that can regulate different steps in exocytosis and that vesicular carriers of different classes might even have their own form of GTP binders (Goud *et al.*, 1988). Some of these steps may rely on fusion reactions of microsomes and endosomes that have been shown to depend on GTP-dependent-glycosylation steps (Paiement and Bergeron, 1983). In addition to regulating the outward flow, GTP is required also for M6PR to recycle to the *trans* Golgi network (Goda and Pfeffer, 1988). In an effort to unify all of this information into a conceptual rationale, Bourne (1988) has proposed the "G-protein cycle" that regulates the shuttle between donor and acceptor membranes. After the granular organelles have matured, these molecules continue to exhibit their effects. Indeed, Bar-Sagi and Gomperts (1988) have shown that microinjection of human H-*ras* proteins into rat mast cells induces degranulation, extending other studies with neutrophils that demonstrate a role for guanine nucleotides at multiple steps in the secretion pathway (reviewed in Boxer and Smolen, 1988). In addition to these observations involving G proteins, the p60 of the c-*src* gene family of kinases has been found in high concentrations in adrenal chromaffin granules and has been implicated in control of the exocytic process (Grandori and Hana-fusa, 1988).

2.6. Secretory Granules

2.6.1. Granule Genesis

Bringing us another step closer to what we encounter in leukocytes are the mecha-nisms of regulated secretion in endocrine cells. Most hormone-containing granules rest in the cytoplasm until they receive a proper stimulus to migrate, fuse with the plasma membrane, and release their contents. In the AtT20 pituitary tumor cell model, the proteins of the granule contents, those of the granule membrane, and those of the plasma membrane migrate together through the Golgi complex, and separate only later in a post-Golgi fraction. Evidence from experiments using fused genes of vesicular stomatitis virus G protein (constitutive plasma membrane localization) and growth hormone (regulated secretion), the products of which can be routed differentially depending on how the construct is made, indicates that the process depends on information carried by the protein. Although the code has not been deciphered, the implication of this experiment is that there are signals for routing to the regulated pathway that appear not to depend on glycosylation or sulfation (reviewed in Moore, 1987). Even greater complexity is evident from observations on bovine anterior pituitary cells that can secrete growth hormone, prolactin, and secretogranin II. By EM, all of these hormones were found together in the Golgi, but prolactin and growth hormone separated at the stage of budding of the pre-secretory vesicle and usually were distributed into separate secretory granules. Because the separate granules were not the result of asynchronous synthesis, the implication is that even within the regulated pathway there may be different controls operating in parallel (Hashimoto *et al.*, 1987).

2.6.2. Synthesis of Granule Membranes

To study the membrane components, Grimaldi *et al.* (1987a) prepared monoclonal antibodies to rat insulinoma secretory granules and identified three main classes of inte-gral membrane proteins. One, called gp100-110, selected for further examination, was found also in the granules of adrenal chromaffin cells, anterior pituitary, and the liver. They showed that glucose increased the rate of production of gp110 and proinsulin at the same time, but induction of degranulation by tolbutamide did not turn on its production; they concluded that although synthesis and secretion appear not to be coupled, regulated synthesis of the membrane components is coordinated with that of the granule contents (Grimaldi *et al.*, 1987b).

In summary, at this juncture we can say that the understanding of how a protein component of a membrane organelle finds its destination remains quite rudimentary. It is reasonable to suspect that subtle changes in the regulation of these events may play a significant role in controlling a cell's interaction with its internal and external environments. Now that the basic concepts have been formed and the tools are becoming available, the field is ready to focus on model systems that examine the flow of molecules as a cell evolves in its differentiation or is altered by disease.

3. LEUKOCYTE GRANULES ARE MAJOR DETERMINANTS OF FUNCTION

3.1. Neutrophils

3.1.1. Granules Reflect the Plan of Differentiation

Common to all nonerythroid blood cells is their ability to secrete products related to their specialized function. Unique to each of them are their granules, evolved to another level of complexity during differentiation of each respective lineage. Because these cells have attained such refinement in their process of organelle synthesis, there is much to learn from them about the dynamics of the internal membrane system.

For several years most of the interest focused on the granules of neutrophil precursors because of their easy visualization by routine cytochemical staining as well as their relevance to bactericidal function. Later, after the platelet release reaction became recognized as a significant mechanism in blood clotting, its granule components also came under more intense investigation. Most recently, with the development of assays to measure lymphocyte-mediated cellular cytotoxicity, the importance of lymphoid granules has been acknowledged. In the following section, I will survey the major granule-containing leukocyte populations, concentrating on neutrophils, briefly including platelets and cytotoxic lymphocytes, but leaving eosinophils and mast cells to others. (See Chapter and Volume by A. M. Dvorak).

3.1.1a. Subpopulations of Neutrophil Granules: Morphology. How many different types of granule are there in a human polymorphonuclear neutrophilic leukocyte (PMN)? This question is not just a rhetorical query on how many angels dance on the head of a pin, because for each structure identified there must be a routing and sorting mechanism to coordinate protein synthesis for the granule membrane and its contents. Early histochemical observations on rabbit bone marrow showed that different membrane-bound structures appeared in sequence during differentiation of myeloid cells. The seminal work of Bainton and colleagues established the paradigm of two independent granule populations that provided the temporal and morphological landmarks to begin further classification. According to this model, after a committed precursor cell (myeloblast) has developed to the point that it produces prominent azurophilic (primary) granules and myeloperoxidase (MPO), it is called a promyelocyte. After its proliferative capacity begins to diminish and the chromatin pattern matures, it is defined as a myelocyte, which is characterized by another wave of synthesis that generates peroxidase-negative specific (secondary) granules. For the most part this distinction remains true, although as we shall see, the borders are becoming blurred and some significant exceptions have been recognized (reviewed in Bainton, 1977; Rice et al., 1986).

Although one can identify two types of granule that fulfill these classical criteria, in human cells the distinction between them is not as clear. Furthermore, conflicting terminologies and variations in EM technique have led to controversy over a strict morphological classification. What is called primary, secondary, or tertiary depends on the starting point; what may be described as dense can depend on the method of fixation; what may be considered as variation in shape may depend on the plane of section. Other qualitative changes occur during maturation of a single structure, as demonstrated by the diminution of granule staining by cationic dyes between the promyelocyte and PMN stages as a result of masking of ionic groups on the sulfated glycosaminoglycans (Parmley *et al.*, 1986). These problems have led to much reevaluation.

Observing the kinds of granules that emerge in myeloid precursors of human bone marrow, Brederoo *et al.* (1983, 1986) identified three stages relating to the presence of MPO. The earliest cell, which they called the eomyelocyte, was recognized by the presence of nucleated granules, so called because of the central condensation in a round, oval, or elongated structure. Although MPO was prominent in the rough ER and Golgi, it was not present in the granule itself. This could imply that before the genes are activated to make the enzymes of the granule contents, the apparatus must be in place to synthesize the membrane bags. The second stage, the promyelocyte, showed MPO developing in small para-Golgi vesicles, then in the nucleated granules, and eventually in azurophilic granules with their homogeneous matrix and round or oval shape. Although fusions were not seen, the proximity of vesicles to the granules suggests that they might occur; however, whether MPO is shuttled from the Golgi to a preformed granule remains to be demonstrated formally. Later, as the size of the Golgi apparatus increases, the activity of MPO decreases both there and in the ER. Finally, after MPO has disappeared from the ER and Golgi, a population of elongated, round, or dumbbell-shaped structures begins to emerge, defining the myelocyte. The question of whether other discrete types, such as tertiary granules, may develop *de novo* after this stage remains controversial. A more recent reevaluation by Tang and Clermont (1989) has demonstrated what appear to be nucleated primary progranules forming as bulging dilatations out of the *trans* Golgi of promyelocytes. In addition, fusions between granules were clearly shown, some of the same size and others possibly smaller. Secondary granules also appeared to come from the myelocyte *trans* Golgi in a "peeling-off" pattern. Of special interest in this study was their demonstration that although the *cis/trans* biochemical polarity of the Golgi apparatus was preserved throughout granulocyte maturation, the concave/convex configuration changed.

Using a modified technique for MPO visualization, a very strongly staining smaller (100–200 nm) structure was identified in the late promyelocyte, after formation of the usual primary granules (spherical, 500 nm; elongated, 200–600 nm) was complete. Often they appeared in chains attached to trabeculae and tended to accumulate in the trailing end of polarized polymorphs or at the periphery of spread cells. Appropriate controls showed that they were not peroxisomes staining by catalase activity (Pryzwansky and Breton-Gorius, 1985). In another careful morphometric study correlating granule formation and maturation of heterochromatin in feline PMN, it was concluded that some MPO-positive granules do form at this stage, possibly by fragmentation of existing structures as well as the addition of new material. How this might relate to the 100–200-nm particles and concur with present models of vesicle traffic remains to be seen (Fittschen *et al.*, 1988). As is evident by now, much of the current view is based on the distribution of MPO.

Evidence for heterogeneity within the primary granule population has come from colocalization of MPO activity with immunoreactive elastase, showing in mature PMN that there tends to be one subgroup with strong MPO and weak elastase and another with weak MPO and strong elastase (Damiano *et al.*, 1988).

Other proteins, such as lactoferrin (LF), have been found to segregate into the MPO-nonreactive granule population (reviewed in Boxer and Smolen, 1988). Earlier ultrastructural studies had suggested that iron-binding proteins and immunoreactive LF colocalized with peroxidase and were not in secondary granules (Parmley *et al.*, 1982). However, more recent electron microscopic immunolocalization (immuno-EM) of LF in mature PMN has demonstrated it to be concentrated in round or dumbbell-shaped structures of approximately 160-nm mean diameter (calculated from the area published, assuming a circle). In cells at this stage the Golgi and ER did not stain, but in myelocytes from the bone marrow LF was detected in the Golgi, slightly in the ER, and in more spherical particles varying between 160 and 225 nm in mean diameter. In promyelocytes, all organelles lacked LF. Morphometric analysis revealed that the frequency distribution curves of the sizes of the LF^+ and MPO^- granules were congruent and that these particles accounted for nearly 100% of granules, indicating that if other populations exist, they are a small minority (Miyauchi and Watanabe, 1987). Another interesting family of molecules, localized to the membranes of specific granules and synthesized during the myelocyte stage, are the leukocyte adhesion receptors (to be discussed in greater detail later). By immuno-EM they have been visualized in only MPO^- granules, and by simultaneous staining they were found to be in a subpopulation of LF^+ structures (Bainton *et al.*, 1987).

The granules of other leukocytes have not been subjected to such intense scrutiny. However, in monocytes the peroxidase is encased in granules of a more complex morphology. These include round particles of variable size as well as other elongated structures not found in PMN, which by computer-assisted reconstruction of serial EM sections were shown to form a network. Because they contain other acid hydrolases and can be released by agents such as phorbol esters, they are considered to be another type of lysosome (Deimann and Teckhous, 1985; Swanson *et al.*, 1987).

3.1.1b. Subpopulations of Neutrophil Granules: Degranulation. Another approach to the study of granule heterogeneity is by a parameter related to the cell's function. The release reaction of granulocytes is a complex process of excitation–response coupling through which the contents of granule subpopulations are selectively discharged, depending on the agonist used. Particulate activators, such as immune complexes, and the chemotactic peptide formylmethionyl-leucyl-phenylalanine (fMLP) tend to stimulate exocytosis of both primary and secondary types, whereas concanavalin A, phorbol ester, lithium, and ionophore A23187 induce release of predominantly secondary granule contents. However, there are subtle concentration effects that are not well understood, as in the example of synergy between cytochalasin D and fMLP to provoke the preferential release of primary granule products. Thus, there is further evidence that these two types of particle are under independent control (reviewed in Baggiolini and Dewald, 1984; Boxer and Smolen, 1988; Murphy and Hart, 1987).

Even within the two major classes of granules there appear to be functional subtypes. Reflecting a possible subpopulation of primary granules is the observation that only two of the three chromatographic variants of MPO were released by fMLP (Pember and Kincade,

1983). The small (150-nm) MPO-containing granules that aggregate near the Golgi and were found in the lightest-density fraction on the gradients described by Parmley *et al.* (1987a) were not released by A23187. Using the probe fura-2 to monitor changes in the concentration of intracellular free Ca^{2+} during fMLP-induced exocytosis of secondary granules, Perez *et al.* (1987) demonstrated that less was required for release of vitamin B_{12}-binding protein than for LF, implying that there are functional subpopulations of morphologically similar granules.

Further interest has focused on the behavior of the leukocyte adhesion receptors stored in secondary granule membranes. After simulation of PMN with phorbol ester or low concentrations of fMLP, they are externalized onto the surface of the plasma membrane. Another particle that may be a distinct subclass of specific granules is the one that contains the enzyme gelatinase. This enzyme is localized to a subpopulation of the LF^+ structures, but other reports have indicated that it can be released by low concentrations fMLP, concurrently with externalization of adhesion receptors (Mo1 antibody to the beta chain) but independently of other primary or secondary markers (Bainton *et al.*, 1987; Petrequin *et al.*, 1987; Stevenson *et al.*, 1987). The three known members of this group of adhesion proteins share a common light (beta) chain, but each has a different heavy (alpha) chain, which according to the markers used by the International Workshop on Human Leukocyte Differentiation Antigens have the designations CD18, and CD11a, -11b, and -11c, respectively. When PMN were induced to degranulate by fMLP or by A23187 and cytochalasin B, there was increased surface expression of CD18, -11b, and -11c but not CD11a. Further experiments showed that cytochalasin B provoked more release of gelatinase than of either lysozyme or beta-glucuronidase and that after density separation of granules, CD11b and -11c appeared in the lightest fraction. The conclusion was that molecules carrying different alpha subunits of the adherence glycoprotein family are processed differently, with CD11a being directed mostly to the cell surface but DC11b and -11c to a tertiary granule (Lacal *et al.*, 1988).

3.1.1c. Subpopulations of Neutrophil Granules: Density. As discussed above, there is evidence for morphological and functional heterogeneity among subpopulations of MPO- and LF-bearing granules of neutrophils. Also, density gradient fractionation has been applied to examine further differences in their physical composition. Using high-resolution isopycnic Percoll gradients, Rice *et al.* (1986) defined five heavy and eight light granule subpopulations from mature PMN. Overall, the heavy corresponded to the primary granules in that most of the MPO activity was concentrated in this fraction and the light contained most of the lactoferrin, implying the presence of specific granules. In general agreement with other morphological observations, ultrastructural examination of samples from the gradients confirmed that those from the heavy fractions were larger, the "typical" primary granules being 290 nm in mean diameter and the light fraction approximately 210 nm. Furthermore, in addition to the presence of a significant amount of MPO in the lower densities and skewing of LF into the higher densities, there was not complete congruence of the density profiles for MPO and another lysosomal marker, beta-glucuronidase. This heterogeneity is in agreement with what had been reported earlier (discussed in Shannon and Zellmer, 1982). Indeed, Bretz and Baggiolini as early as 1974 had claimed that "azurophil granules must be considered a special kind of primary lysosome which contains microbicidal agents in addition to the characteristic digestive enzymes" (Brets and Baggiolini, 1974). Even in the lightest fraction called the micro-granules, which had a mean diameter of 150 nm, there was a significant quantity of MPO, comprising 3% of the cell's total. The smaller structures stained strongly for carbohydrate, were formed during the promyelocyte stage in clusters near the Golgi complex, and were not induced to exocytose by ionophore A23187. It remains to be seen whether they are related to the small MPO-containing particles described previously (Pryzwansky and Breton-Gorius, 1985; Parmley et al., 1987a). Although the usual primary granule markers (MPO, beta-glucuronidase, and elastase) were distributed throughout the five heavy fractions, they were more concentrated in fractions 1 through 4. The heavy fraction 5, consisting of large (320-nm mean diameter) electron-dense particles present in promyelo-cytes, was unique in that it contained most of the defensin proteins (Ganz et al., 1985; Rice et al., 1987).

By relating a histochemical reaction for complex glycoconjugates to different density particles, four predominant classes of granules have been defined: type I, mature second-ary with intense matrix but not membrane staining; type II, immature secondary with intense matrix and membrane staining; type III, mature azurophilic with weak matrix but not membrane staining; and type IV, immature primary, some membrane, and weak matrix staining. The major conclusion from all of these studies was that there clearly is heterogeneity among the classic primary/secondary designations and that the classic MPO and LF markers do not strictly follow a classification based on either size or density alone (Rice et al., 1986). How much of this heterogeneity reflects intermediates in the synthetic pathways and vesicular traffic remains to be shown. Now that antibody markers are becoming available to identify granule membranes, some of these apparent discrepancies soon may be resolved.

3.1.2. Biosynthesis of Granule Contents

Beyond these morphological and functional descriptions, there has not been the same detailed characterization of the synthesis of leukocyte granules as there has been for the cytoplasmic structures of some other tissues. Nonetheless, recent attention has focused upon MPO and LF as model proteins for studying the contents of primary and secondary granules. Both molecules have been purified, their encoding DNAs have been cloned, and it has been established that their biosynthesis and processing are temporally linked to the stages of granulocyte development.

The gene for MPO, located on human chromosome 17, is activated early in the promyelocyte stage. As the common single-chain precursor protein (80 kDa) matures through the ER and Golgi apparatus, it is glycosylated, acquires a heme group, and finally is processed into a homodimer of two heavy/light-chain (60 and 12 kDa) complexes. Although one oligosaccharide chain is phosphorylated, it appears not to be essential for targeting and thus not dependent on the mannose-6-phosphate mechanism. Because many proteins contained in the primary granule are cationic, the speculation has been tendered that proteoglycans might be involved as a carrier. Pulse-chase experiments with radioac-tive amino acid precursors localize enzyme activity to the heavy granule fraction as well as to lighter vesicles. Although the incompletely processed enzyme is found in these smaller particles, the mature primary granules contain only the fully processed species. Furthermore, correct processing requires an acidic prelysosomal compartment, suggesting an acid-dependent targeting mechanism (reviewed in Nauseef et al., 1988; Akin and Kinkade, 1987).

Attempts to understand why MPO synthesis stops at the late promyelocyte stage have used the human promyelocytic leukemia line HL60. These cells can be induced to acquire some of the characteristics of mature polymorphic neutrophils but conveniently do not make secondary granules or LF. During the course of their maturation, MPO-specific RNA decreases in proportion to the overall RNA level, suggesting that posttranslational processing may lead to its degradation before it reaches the stable pool and that there might be an intact negative regulator for the gene itself (Tobler et al., 1988).

In normal marrow cells, biosynthesis of the LF protein begins at the myelocyte and ends before the band stage (Rado et al., 1984). In both granulocytes from normal PMN and those from patients with chronic granulocytic leukemia (CGL), an immature precursor form of LF could not be detected; its oligosaccharide groups were processed more rapidly than those of MPO, but its appearance in the granules, unlike that of MPO, was blocked by monensin and NH_4^+. Furthermore, the oligosaccharides of LF became resistant to endoglycosidase H within 30 min, but MPO remained sensitive even in its mature form. This has led to the speculation that MPO might be released from the *cis* Golgi and not be transported through the *medial/trans* elements (Olsson et al., 1988). Should this prove true, in view of the observations of Tang and Clermont (1989) that primary granules come directly from the *trans* Golgi, then one might expect to find another particle with a high concentration of MPO fusing into a budding progranule that lacks MPO. From these data taken together, it is quite clear that the two products are processed by different pathways. Using a cDNA probe, Rado et al. (1987) were not able to detect any LF-specific mRNA in HL60 promyelocytes; these leukemic cells, however, may not accurately reflect the timing of normal marrow precursors. To understand how granulocyte maturation is regulated, it is of utmost importance to know whether there is any overlap in expression of the MPO and LF genes, now that the necessary tools are available.

3.1.3. Granule Membranes

Very little work has been done on the composition and genesis of the membrane moiety of leukocyte granules. Antibodies to the LIMP structures of hepatocyte lysosomal membranes (discussed above) localize to serotonin-containing granules in rat basophils, indicating some proteins in common in the organelles of these two tissues. Of the four proteins studied, LIMP-1 was on the plasma membrane also, and its concentration increased after degranulation was stimulated by A23187, whereas that of LIMP-2 and -3 did not change (Suarez-Quian, 1987). Monoclonal antibodies D46 and K101 have been reported to show a granular pattern by immunoperoxidase staining and light microscopy of HL60 promyelocytes and normal marrow cells and appear not to react with the plasma membranes of normal human granulated cells, within the limits of detection by flow cytometry (Fitz-Gibbon et al., 1985; Sullivan et al., 1986).

To attack the problem of granule genesis, the HL60 model can be a useful reference for examining the process in a more controlled fashion than is possible with whole bone marrow, as long as one does not forget its limitations. These cells, established from a patient with acute myeloblastic leukemia (AML), continue to make peroxidase-containing granules but few, if any, secondary granules (Gallagher *et al.*, 1979; Fontana *et al.*, 1980; Dalton *et al.*, 1988). Ultrastructural studies have shown that the MPO$^+$ structures are of two sizes, the relative frequencies of which can change with time of passage in culture (Parmley *et al.*, 1987b). Further work with the D46 antigen revealed that it is expressed on the surface of a blastic (agranular) variant of HL60 (HL60-D) and on some leukemic T-lymphoblastic lines that do not make any detectable cytoplasmic granules (Fitz-Gibbon *et al.*, 1983). Immunoprecipitation studies on biosynthetically labeled cells showed that the determinant bearing the epitope on the plasma membrane of HL60-D was of a lower apparent molecular mass (88 kDa) than the integral membrane protein of the parental HL60 granules (100–130 kDa) (Peyman and Sullivan, 1987). One could speculate that the localization of this structure on the plasma membrane is a reflection of constitutive expression in cells that are not able to form granules. However, in cells that have the mechanisms activated to produce a complete primary granule, the protein localization comes under the influence of the regulated pathway. Because the HL60-D cells lack any detectable MPO mRNA (unpublished observations), the results suggest that the genes for granule membranes might be under control mechanisms independent of those encoding granule contents.

As determined by their mobility on sucrose density gradients, two major populations of granules have been identified from HL60 cells. The heavier fraction contained the MPO and cathepsin G activities, whereas the lighter contained acid phosphatase and beta-glucuronidase, in agreement with studies discussed earlier showing that in normal PMN the glucuronidase activity tended to be skewed to lighter fractions compared with MPO. Further work with another HL60 variant line (HL60-A7), containing large aberrant granules deficient in both MPO protein and mRNA, showed that the D46 marker was concentrated in the lighter fraction. This finding implies that some of the membrane components of primary granule subpopulations might be unique (Sullivan *et al.*, 1986; Peyman and Sullivan, 1987).

3.2. Platelets

The presence of platelet granules and their role in the release reaction are well described. Although very little has been reported on the control of granule production, there is a voluminous literature on the plasma membrane and its role in platelet activation (reviewed in Colman, 1986; Steen and Holmsen, 1987; Clementson and Luscher, 1988; Phillips *et al.*, 1988). Here, I will briefly discuss some of the recent findings that are pertinent to our inquiry into leukocyte granules.

At least two subpopulations of cytoplasmic particles, called the alpha and dense granules, have been recognized, each containing some components that are unique. In the disease referred to as the gray platelet syndrome, so called because the platelets appear faint on routine stains, there is a mild bleeding disorder and the alpha granules are depleted of all seven of the associated proteins (beta-thromboglobulin, platelet factor 4, thrombospondin, fibrinogen, fibronectin, platelet-derived growth factor, and von Willebrand factor). Some of the lysosomal enzymes (e.g., glucuronidase) are present in a normal concentration, whereas some others are decreased; aggregation of platelets and release of the dense bodies appear to be normal in response to all stimulants tested except thrombin. The authors suggest that there may be a packaging defect to account for a lesion involving so many different pathways (Srivastava *et al.*, 1987). Using an antibody (S12) to a 140-kDa protein of the alpha granule, others have found that the abnormal large vacuoles seen in gray platelets are reactive, implying that the megakaryocyte can produce the organelle membrane but is not able to incorporate the content proteins normally (Rosa *et al.*, 1987). Another group of conditions has been described under the classification of platelet storage pool diseases, some of which are associated with abnormalities of melanin granules and abnormal secretion of lysosomal enzymes from the kidney, as well as the defect in contents of the dense bodies. In the murine analog of this disease, the dense bodies can be visualized with fluorescent dyes, but the chromogen is released prematurely in both the mature platelet and the megakaryocyte, again suggesting a defect in the ability of the structure to properly concentrate its usual contents (Reddington *et al.*, 1987). Morphological analysis of the release reaction has shown that some of the granules migrate to and appear to fuse with the plasma membrane. Further confirmation of this has come from studies with another antibody reactive with a 53-kDa protein present in cathepsin D-containing granules of both megakaryocytes and endothelial cells. After stimulation of platelets by thrombin, this structure is externalized to the plasma membrane, a phenomenon that has been used clinically to detect platelet activation in patients with venous thrombosis and after extracorporeal circulation (Nieuwenhuis *et al.*, 1987, and references therein). These studies emphasize again in another type of cell the principle of granule specialization in preparing a leukocyte for its task. Because some of the granule constituents are synthesized by the developing megakaryocytes and others are incorporated from the plasma, there must be further mechanisms for routing both kinds of products to the right place in the cell (Handagama *et al.*, 1987).

3.3. Cytotoxic Lymphocytes

Much in the news for their potential antitumor activity are the large lymphoid cells that contain azurophilic granules (LGL). Such cells include lymphokine-activated killers (LAK cells), cytotoxic T lymphocytes (CTL), effectors of antibody-dependent cytotoxicity (ADCC or K cells), and those known as natural killer (NK) cells, which lack the CD3 surface antigen or rearranged T-cell receptors. Whatever the recognition mechanism used, which remains controversial for LAK and NK cells, an essential step in the killing process involves polarization of granules between the Golgi complex and the nucleus and eventual release at the site of contact with its target. Further proof that granule components are direct participants in cytolysis comes from the observation that granules purified from both a rat NK leukemia cell line and preparations of human LAK cells are able to kill targets that are resistant to intact effectors. During their genesis, these organelles accumulate the lytic effector molecules, which appear to include serine esterases, perforins related to the ninth complement component (C′9), and another protein called NK cytotoxic factor (reviewed in Henkart, 1985, Herberman *et al.*, 1986, Tschopp and Nabholz, 1987, Ortaldo and Longo, 1988, and Henkart and Yue, 1988; Lowrey *et al.*, 1988).

In trying to understand this process from the point of view of leukocyte differentiation, two questions arise: Are LAK and NK granules different, and how are they prevented from leading to suicide of the host cell? Reynolds *et al.* (1987) have demonstrated with rabbit polyclonal antibodies to rat LGL granules that cytolysis by NK cells, but not by CTL or macrophages, could be blocked by F(ab)′₂ fragments, concluding that either there were different lytic mechanisms operating in the two types of cytotoxic cells or the CTL junctions could not be penetrated. Conversely, Henkart *et al.* (1987) found that antibodies to murine LGL granules neutralized the activity of both, concluding that the two processes were similar. Assuming that both conclusions are correct, these data might be evidence that the lytic granules of the two cell types bear both common and distinct proteins. After being released into the effector–target junction, the lytic molecules must be prevented from destroying the parent cell, because previous studies have shown that NK cells are able to recycle and kill several times. In addition to possible specificity in the form of membrane receptors for the toxins, other factors have been identified. A 65-kDa granule protein, called homologous restriction factor, has been purified that can interfere with channel formation induced by the perforin-type of membrane attack complex of CTL as well as protect from killing by NK, ADCC, and C'9. This molecule showed immunological cross-reactivity to a structure previously found on erythrocyte plasma membranes (Zalman *et al.*, 1988). Concurrent with granule exocytosis from rat LGL are proteoglycans of chondroitin sulfate type A, which may be able to bind and restrict the activity of granule serine esterases that have been implicated in the killing process, either directly by their proteolytic action or indirectly as activators of other lytic molecules (Stevens *et al.*, 1987).

4. GRANULES INTERACT WITH THE PLASMA MEMBRANE

4.1. Neutrophils

4.1.1. Interactions That Promote Cell–Cell Adhesion: Integrins

In addition to being containers of stored soluble products, granule membranes also may be repositories for preassembled functional complexes that can be translocated to the cell surface on demand, as has been shown for glucose transporters in adipocytes (Blok *et al.*, 1988). One group of molecules with this capability in blood cells are the leukocyte cell adhesion molecules (Leu-CAM) which, with the very late antigens (VLA) and cytoadhesins, constitute the integrin superfamily. These protein complexes all consist of a group-specific light (beta) chain of 95–130 kDa and an individually specific heavy (alpha) chain of 130–210 kDa. Some of the ligands that have been found to interact with the integrin proteins include fibronectin, laminin, and vitronectin, all of which contain a common tripeptide Arg-Gly-Asp (RGD) sequence involved in the binding site (reviewed in Hynes, 1987; Ruoslahti and Pierschbacher, 1987; Ginsberg *et al.*, 1988).

Within the Leu-CAM group are three separate bimolecular complexes with a common beta chain recognized by antibodies to the CD18 determinants: LFA-1, present on T lymphocytes, monocytes, and neutrophils (alpha chain reactive with CD11a antibodies); Mac-1, present on monocytes, NK cells, and neutrophils (CD11b, also a receptor for complement fragment C3bi); and p150,95, present on neutrophils (CD11c). Although the presence of CD11b and -11c on a subpopulation of neutrophil secondary granules indicates that these granules are synthesized during the myelocyte stage, studies on HL60 promyelocytes have demonstrated that CD11a and -11c, but not CD11b, are present on the plasma membrane and that with induction of maturation to either polymorphs or macrophages the surface representation of all three increases. A similar maturation-related pattern was observed with spontaneously maturing CGL cells. Because all of these proteins are thought to mediate adhesive reactions that are essential for neutrophil function (see Section 5.1), it has been suggested that the granule reservoir serves to augment the ability of the cell to invade inflammatory sites (Hickstein *et al.*, 1987).

4.1.2. Other Candidate Cell Adhesion Molecules of Neutrophils

Another fascinating set of membrane glycoproteins are those bearing a fucosyl lactosamine determinant [Gal b1 $\rightarrow$ 4(Fuc a1 $\rightarrow$ 3)GlcNAc], recognized by CD15 monoclonal antibodies. This antigen originally was demonstrated as the murine glycolipid stage-specific embryonic antigen (SSEA-1), thought to be involved in adhesion reactions during the early stages of rodent development (Solter and Knowles, 1978; Fox *et al.*, 1981). It was first detected on human granulocytes and other tissues by Beverly *et al.* (1980) with the TG-1 antibody and later by Civen *et al.* (1981) as the "granulocyte-specific" My1 antigen. Light microscopy has shown that it is present both on the plasma membranes and granules of HL60 promyelocytes and on normal granulocytic precursors from promyelocyte to neutrophil (Fitz-Gibbon *et al.*, 1985); EM analysis has confirmed its presence on the plasma membrane and localized it to both the matrix and membrane of lysosomal granules of normal neutrophils (Warhol *et al.*, 1987). Further biochemical analysis has indicated that it is on the oligosaccharides of LF and a subpopulation of the Leu-CAM family (Prieels *et al.*, 1978; Skubitz and Snook, 1987). Although this highly immunogenic structure is present in places where one might expect to find cell adhesion molecules, its role in granulocyte function remains to be demonstrated.

Another family of molecules that recently has been implicated in cell–cell adhesion reactions are the carcinoembryonic antigens (CEA) of epithelial tissues (Benchimol *et al.*, 1989). Structures identified by immunological cross-reactivity (NCA) have been found on the plasma membranes and in the granules of developing marrow cells, but different authors have reported variable results as to whether it first appears at the promyelocyte (implying primary granules) or myelocyte (implying secondary granules) stage. Because the studies were performed with polyclonal antisera, reagents directed at different epitopes might be an explanation. To date there have not been any reports indicating whether these highly glycosylated proteins are exocytosed during degranulation or mediate adhesion events of neutrophils (Burtin *et al.*, 1980; Noworolska *et al.*, 1985; Wahren *et al.*, 1983; Audette *et al.*, 1987). Because the CEA family is derived from at least 10 genes and the NCA species that predominates in mature granulocytes (2.3-kb mRNA) appears to be unique, very selective monoclonal antibodies may be required before the precise pattern of expression is understood in relation to myeloid maturation and function (Thompson and Zimmerman, 1988; Cournoyer *et al.*, 1988).

4.1.3. The Regulation of Granules and the Regulation of Plasma Membranes Are Linked.

During maturation of neutrophils, there appears to be close coordination between the timing of granule development and that of surface adhesion molecules. Also, the expression of receptors for the chemotactic peptide fMLP increases from the promyelocyte through to the polymorph stage (Sullivan *et al.*, 1987). Recently, the CALLA (common acute lymphoblastic leukemia antigen; CD10) structure has been implicated in this regulatory network (reviewed in Le Bien and McCormack, 1989). This antigen was so named because of where it was first detected, but later it was shown to be present in a wide range of cells, including fetal liver, proximal renal tubules, glomerular epithelium, and brain. CALLA is not present on HL60 promyelocytes, and in normal bone marrow it is either absent or in very low concentration on promyelocytes and myelocytes; it begins to be expressed on bands and is maximal on mature polymorphs (Braun *et al.*, 1983). From information available comparing the DNA sequence with others in the data bank, it is now proven that CALLA is a neutral endopeptidase, implicated in the degradation of peptide hormones such as angiotensins, encephalins, and fMLP (Letarte *et al.*, 1988; Jongineel *et al.*, 1989). Such a physiological role is supported by an experiment in which the majority (>95%) of CALLA$^+$ neutrophils were separated from the negative minority population by cell sorting. It was shown that the negative population had an increased chemotactic response to complement. However, in view of the hormone degradation hypothesis, it remains to be explained why the CALLA$^-$ cells did not respond differently to fMLP (McCormack *et al.*, 1987).

As part of the amplification process, when C3bi binds to its receptor on the neutrophil plasma membrane, there is exocytosis of further CD11b protein in a large cluster pattern, as detected by fluorescent staining with the Mo1 antibody. Further evidence for this being a granule-mediated event is the observation that patients with specific granule deficiency do not form clusters. These results have been interpreted as supporting the idea that microdomains are formed at local sites of fusion (Petty *et al.*, 1987).

What controls the fusion events? As discussed above, the process of degranulation proceeds from an initial ligand–receptor interaction at the plasma membrane, through the phospholipase C/diacylglycerol/inositol phosphate mechanisms, to eventual elevation of intracellular free Ca^{2+} (reviewed in Boxer and Smolen, 1988). Some insights into how this is coupled to membrane fusion events have come from studies on degranulation of bovine adrenal medullary cells identifying a 47-kDa protein called synexin. Three immunologically cross-reactive proteins of 67, 47, and 28 kDa were found in the cytosol of human polymorphs that promoted aggregation of isolated secondary granules in a Ca^{2+}-dependent manner. An intermediary such as this may be an essential step preceding fusion of the granules with a phagocytic vacuole or the plasma membrane (Meers *et al.*, 1987).

4.2. Platelets

Cell interaction molecules of the integrin family are major regulators of platelet aggregation and granule release. Although the Leu-CAMs have not been demonstrated unequivocally on human platelet membranes, the gpIIb-IIIa complex is a member of the cytoadhesin group, gpIa-IIa is VLA-2, and gpIc may be the VLA-3 or VLA-5 alpha subunit. As would be expected of these proteins, the gpIIb-IIIa complex binds the RGD peptide present in fibrinogen, fibronectin, and von Willebrand factor, but it appears that other sites are necessary for optimal interaction (reviewed in Ginsberg *et al.*, 1988). Also, other reactions are being recognized that implicate similar molecules in different ways. Altieri *et al.* (1988a) have shown that the CD11b (Mac-1) structure can bind fibrinogen independent of the Arg-Gly-Asp site, and after being stimulated with ADP, monocytes can bind coagulation factor X through the CD11–CD18 complex and serve as a surface for its activation (Altieri *et al.,* 1988b).

Although essential for normal hemostasis, as evident in the bleeding disorder of patients with Glanzmann's thrombasthenia who lack GrpIIb-IIIa, these molecules also may be mediators for other kinds of disease. It has been proposed from animal models that platelet aggregation around a tumor can mediate metastasis; indeed, the murine sarcoma (PAK 17.15) that requires platelets for metastasis to lungs induces platelet aggregation in an ADP-dependent fashion. *In vitro,* this aggregation is inhibited by the RGD peptide, and injection of the peptide into animals inhibits colonization of the lungs (Ugen *et al.,* 1988).

Another antibody-defined molecule (D51) has been described as a panmyeloid antigen because of its presence on the surface of neutrophils, granulocytic precursors (blasts through PMN), monocytes, a minor subpopulation of lymphoid cells, and, most strongly, platelets. However, staining of granules was not noted in normal marrow cells, HL60 promyelocytes, or U937 monoblasts. *In vitro* studies have shown that the presence of purified D51 antibody in the reaction mixture inhibits the second wave of platelet aggregation induced by the weak agonists (epinephrin, ADP, and platelet-activating factor) but not by the more potent stimulators (collagen, thrombin, and ionophore A23187). Although its distribution on other types of cells is similar to that of the CD11–CD18 (Leu-CAM) molecules, the fact that it is strongly expressed on platelets attests to it not being part of this group; however, it has not been disproven that the antibody reacts to an epitope common to all integrin molecules (R. Gareau and A. K. Sullivan, unpublished observations).

5. PATHOLOGY OF MYELOID GRANULES AND PLASMA MEMBRANES

5.1. Nonneoplastic

5.1.1. Leukocyte Adhesion Protein Deficiency

If proper behavior of granular organelles is important for blood cell function, then its failure should cause disease. A forceful illustration of this, in which the transition of applied technology has moved very rapidly from thermometers to Southern blots, is the leukocyte adhesion deficiency syndrome. In this rare group of patients with impaired wound healing and recurrent severe pyogenic and fungal infections, it was observed that the response to complement-independent activators of the respiratory burst and degranulation were normal, but many parameters of neutrophil function were defective, including impairment of adherence, aggregation, chemotaxis, and C'3-dependent phagocytosis.

Further examination of leukocytes from these patients revealed that none of the anti-CD11 or -CD18 antibodies (OKM1, Mo1, or Leu-M5) reacted, implying that the initiating lesion was linked to production of the beta chain. Later studies on a larger group demonstrated heterogeneity in the clinical behavior, with at least five patterns, ranging from very severe to moderate, that reflect quantitative deficits in beta-chain production and qualitative defects in mRNA or protein processing. In addition to abnormal granulocytes, they show diminished lymphocyte-mediated immunity assessed by NK/ADCC and helper T-cell function, to the extent that allografts are poorly rejected. This, however, has worked to the advantage of some in that a bone marrow transplant, which cures the disease, can be sustained with minimal need for immunosuppressive drugs. On the basis of these observations, clinical trials are in progress testing antibodies to LFA-1 to assess their possible value in blocking rejection of HLA-mismatched marrow transplanted for other conditions (clinical syndrome reviewed in Anderson *et al.*, 1985; reviewed in Springer *et al.*, 1987, and Todd and Freyer, 1988; Kishimoto *et al.*, 1987). In agreement with the lack of Leu-CAM expression on platelets, coagulopathies are not part of the disease complex.

The receptor for LFA-1 has been identified as the intracellular adhesion molecule 1 of the immunoglobulin gene superfamily. From evidence that this structure is used by lymphocytes to bind to endothelial cells and the observation that it can be up regulated by interleukin-1, tumor necrosis factor, and gamma interferon, it appears that it may be part of a significant regulatory mechanism for leukocyte traffic through tissues (Dustin and Springer, 1988). On the other hand, because the known Leu-CAM molecules appear not to be expressed on myeloblasts and patients with deficiency states have normal blood cell counts, these structures probably are not part of the elusive regulatory system mediating interaction between hemopoietic precursors and the bone marrow stroma.

5.1.2. Primary Granule Abnormalities

Another disorder associated with recurrent bacterial and fungal infections is Chédiak-Higashi disease (CHD). In this disease, the neutrophils are remarkable for their large lysosomelike granules, the origin of which remains controversial (reviewed in Rotrosen and Gallin, 1987; Lehrer *et al.*, 1988). In two cases studied by White and Krumweide (1987), the granules varied in size, shape, and internal structure, and the peroxidase reaction showed that there was a mixture of normal- and abnormal-size structures, some of which appeared in a chaotic array of aggregations. Surveying the pattern of granule development in myeloid precursors at different stages of maturation, they found that the abnormal coalescing structures were more common in circulating PMN than in earlier forms and proposed that the huge granules are secondary lysosomes formed by an ongoing process of inappropriate fusion. Ganz *et al.* (1988) purified granules from the neutrophils of another three patients, all of which showed decreased levels of elastase and cathepsin G but a near-normal concentration of defensins. On the basis of this and other information from previous reports that MPO and glucuronidase activities are normal in CHD, the authors proposed that this "may be a disorder of regulation of protein synthesis, protein processing, or granule assembly rather than a primary lesion in elastase and cathepsin G genes." This is supported by evidence from the beige mouse model showing that elastase and cathepsin G concentrations are normal in late marrow precursors but absent in circulating PMN (Takeuchi *et al.*, 1987). The fact that these patients are partially albino and their NK cells also exhibit the morphological abnormality supports the idea that CHD may be the result of a more generalized granulopathy.

Isolated enzyme deficiency can occur without any conspicuous aberration in granule structure, as in one of the most common human genetic diseases, MPO deficiency. Patients with this condition do not have as severe septic episodes as do those with some of the other phagocyte disorders, but they may have an increased susceptibility to fungal infections. mRNA has been studied from some of these patients, and there appears to be a variety of posttranscriptional defects (reviewed in Nauseef, 1988, 1989). Another HL60 variant (HL60-A7) has been reported that has abnormal fused primary granules and less than 5% MPO activity or RNA (by cell-free translation or Northern blotting) but by immunohistological staining has normal elastase and cathepsin G proteins. After incubation of HL60 and HL60-A7 cells with radioactive mannose, purification of the granule membranes, and analysis of the proteins by gel electrophoresis, it was shown that the major band seen in extracts of parental HL60 cells (gp110-170) was not found in HL60-A7 cells (Sullivan *et al.*, 1986; Peyman and Sullivan, 1987; unpublished results). Although multiple mutations are possible, an alternative possibility could be that activation of the MPO gene is linked to the genesis of an essential component of the granule membrane, as implied by Ganz *et al.* (1988) to explain the multiple defects of CHD.

5.1.3. Secondary Granule Abnormalities

Patients with a quantitative defect in secondary granules (SGD) often suffer from recurrent bacterial infections (reviewed in Gallin, 1985; Boxer and Smolen, 1988; Lehrer *et al.*, 1988). Investigating the nature of the morphological abnormality of a single patient, Parmley *et al.* (1983) showed that the PMN contained MPO$^+$ primary granules, as well as small aberrant MPO$^-$ elongated particles (100–200 nm) that were more numerous in bands and PMN than in less mature marrow cells. The primary granules for these patients, however, continued to stain heavily for sugars, unlike those that develop in normal cells. Detailed studies on another two patients confirmed that the concentrations of the primary granule proteins MPO, elastase, and cathepsin G were normal, but LF was not detected by immunoprecipitation. Analysis of RNA from marrow cells revealed that the LF message was markedly decreased, but the small amount present was of normal size. Because LF is produced by several other tissues, it was sought and shown to be released normally into nasal secretions; Southern blotting of DNA from nasal tissue showed that the gene was present in a single copy and yielded restriction fragments of normal size. However, not compatible with a strict interpretation of the two-granule model are two other observations: the defensins, previously shown to be localized to the largest and heaviest of the putative primary granules, were at only 10% of the normal level, and gelatinase, a tertiary granule product, was not released after induction of degranulation (Petty *et al.*, 1987; Ganz *et al.*, 1988; Lomax *et al.*, 1989). Recently, the restricted nature of the lesion in SGD has been called into question with the observation that the MPO$^+$ granules are of a much smaller diameter than normal (mean of 160 versus 240 nm), although they are not decreased in number. Thus, the marked decreases in both typical secondary granules and the large defensin-rich structures, and now abnormal primary granules, all suggest that there is a more generalized synthetic defect (Parmley *et al.*, 1989).

On the basis of these patterns of coupled deficiencies, it would be reasonable to search for targeting mechanisms common to the proteins of specific progranule membranes and their contents. Possible independent segregating units in subpopulations of primary and secondary granules might include MPO to one class (evidence from HL60-

A7), elastase and cathepsin to another (from CHD lesion), and LF, defensins, and gelatinase to others (from SGD).

5.2. Leukemic

5.2.1. Granule Failure in Myeloid Leukemia

During normal hemopoiesis, the precursors of granulocytes, macrophages, and platelets concentrate a significant measure of their activity on forming granular organelles. According to the model proposed by Sachs (1982), an essential step in leukemogenesis is the uncoupling of signals that regulate a cell's decision to self-renew or to mature. As a consequence of this loosened control, one would predict that in myeloid leukemia a variety of granule defects would occur in one or more lineages. Indeed, this is what happens. Although solid tumors expand to compress a vital structure, leukemia is fatal most often because the bone marrow does not produce leukocytes and platelets that are sufficiently competent to prevent sepsis and hemorrhage.

5.2.1a. Leukemia-Related Granulopathies: Acute Leukemia. One can see quite readily that the blastic cells in myeloid leukemia are too immature to produce a normal complement of granules, but it is not so obvious that the residual mature cells also are deficient in granule-related enzymes. In a series of papers, Bendix-Hansen and collaborators (1987) reported their results on a large number of patients with different types of leukemia monitored over an extended period of time. They found that the few circulating mature neutrophils were deficient in MPO in 40–50% of cases with AML of the differentiating myeloblastic and myelomonocytic types (M2 and M4 by the French-American-British classification), in 20% of cases with chronic-phase CGL, but not in any with acute or chronic lymphoid leukemia. In another series, deficiency of elastase was reported to be even more frequent than that of MPO (Havemann *et al.*, 1983). Although this parameter was not useful in predicting who might attain a complete remission, of those who had deficient PMN detected at presentation and went on to relapse, most continued either to produce the deficient PMN throughout their course or to show an increase prior to the recurrence of overt disease. In two patients, this preceded other clinical signs by 2 and 8 months (Bendix-Hansen, 1987, and references cited therein). This observation indicates that there is a subgroup of AML in which the neoplastic cells can escape the drive to replicate but are not able to activate the mechanism(s) necessary for synthesis of normal granules.

In another series of leukemic patients, Davey *et al.* (1988) confirmed that 59% of patients with AML [type M1 (nondifferentiating myeloblastic) and types M2, M4, and M5 (monoblastic)] produced mature PMN that lacked both MPO enzymatic activity and immunoreactivity; these authors extended their previous results to show that elastase and LF often were deficient as well. Of further significance in this study was the observation that a number of patients appeared to have a defect in genesis of primary granule enzymes, without a measurable decrease in LF content. However, it is difficult to discern from such qualitative histological methods whether this merely reflects differences in the relative sensitivities of the assays and the larger proportion of secondary granules in PMN. Because there has not been described a congenital disease of primary granule membrane deficiency, and because it is not known whether the signals needed to initiate production of specific granules depend on information generated during synthesis of the primary granules, it will be important to study these "experiments of nature" with antibodies to granule membranes in order to ascertain whether a cell can make the second type independently of the first.

A classic morphological feature used in the diagnosis of AML is the Auer rod phenomenon. These cytoplasmic structures, found in various forms in AML types M2, M3, M4, and M5, contain MPO and elastase and are thought to have evolved from primary granules. Although large bizarre-shaped granules (pseudo-Chédiak-Higashi) occur often in leukemic cells, their relationship to the process of Auer rod formation is not certain (Bessis, 1973; Tulliez and Breton-Gorius, 1979; Payne and Harrow, 1982; Havemann et al., 1983; Jain et al., 1987). On the basis of their EM observations using tilted specimens, Dixon et al. (1984) have proposed that after the granules fuse, their membranes form tightly rolled lamellae that eventually break apart to form multiple small tubules. Of interest in this regard is one patient with acute promyelocytic leukemia (M3) in whom a large percentage of blast cells contained Auer rods, none of which stained with the D46 anti-granule membrane antibody, although the primary granules themselves reacted well (A. K. Sullivan, unpublished observations).

5.2.1b. Leukemia-Related Granulopathies: Myelodysplasia. At the other end of the leukemia spectrum are the myelodysplastic syndromes (MDS), in which most of the polymorphs appear mature superficially and blastic cells are present in a minority. Patients with the pattern referred to as refractory anemia and excess myeloblasts (RAEB) suffer from frequent infections due to both quantitative and qualitative granulocyte defects, as well as occasional bleeding due to similar deficiencies in their platelets. If they do not succumb to these problems, they usually progress to a form of acute leukemia that seldom responds to any treatment other than bone marrow transplant (reviewed in Jacobs and Clark, 1986, and Mufti and Galton, 1986). Likewise, in some of the clonal myeloproliferative disorders (CGL, polycythemia rubra vera, and myeloid metaplasia) there is a tendency for both bleeding and thrombosis, and often the platelets show abnormal structure, aggregation, and granulation (reviewed in Schafer, 1984).

Similar to what they observed in AML, Bendix-Hansen and collaborators (1987) found that in 26% of MDS cases the polymorphs were deficient in MPO. Similarly, of the patients reported by Davey et al. (1988), 43% showed a granulocyte dyscrasia, including decreased MPO, elastase, and LF. Although difficult to distinguish from MDS, a single patient with aplastic anemia has been reported to have residual neutrophils deficient in MPO (Bizzaro et al., 1988). When the dysplasia began to transform into leukemia or when peroxidase-deficient PMN arose de novo, chromosomal aberrations usually developed concurrently but did not reflect any of the known cytogenetic abnormalities associated with MDS. From a different direction, Friedman et al. (1985) used a CD15-related antibody (H36/71) to stain polymorphs from patients with RAEB and found that in 8 of 13 the cells were deficient. Again, this finding supports the notion that the leukemic granulopathy is not restricted to enzyme deficiencies but consists of a more general defect in granule formation. However, these studies did not determine whether the lack of staining was due to a masked or absent antigen or to failure to produce the entire granule.

Even in some leukemic myeloblasts and polymorphs in which MPO can be detected, the predominant type of primary granule is the early micro form, affirming that even when routine histochemical stains suggest there is normal function, more subtle defects must be sought (Parmley *et al.*, 1987a). It remains to be seen where these overt patterns of granule failure originate and what steps are disrupted in the normal flow of vesicular intermediates. As more is understood of how G proteins regulate reactions along the granule synthetic pathway, it will be reasonable to ask whether the leukemic granulopathies are a consequence of the *ras* gene mutations at codons 12 and 13 reported in AML and myelodysplasia (Lyons *et al.*, 1988; Bar-Eli *et al.*, 1989). It appears that expression of this oncogene is not essential for the proliferative component of the neoplastic phenotype, but it becomes more pronounced in both MDS and CML as they progress to the agranular, blastic stages (Hirai *et al.*, 1988; Le Maistre *et al.*, 1989).

5.2.1c. Leukemia-Related Granulopathies: Lessons from HL60. As summarized previously, the HL60 cell line produces prominent azurophilic granules that contain peroxidase, elastase, cathepsin G, and other primary granule-associated products. In many ways, the behavior of these cells mimics different events seen in primary leukemic myeloid cells, although not necessarily as the result of identical mechanisms. Although they replicate as promyelocytes, 5–10% of them mature spontaneously, and almost all of them can be induced to stop dividing and progress to metamyelocytes or macrophages by many different reagents (reviewed in Collins, 1987). Within a culture of HL60 cells there is a small fraction (less than 0.1%) that have become completely blastic, agranular, and nonmaturing and can be selected by repeated exposure to inducing agents (e.g., dimethyl sulfoxide) or by flow cytometric sorting for cells that lack the CD15 surface antigen (Fitz-Gibbon *et al.*, 1983; Sullivan, unpublished). In one sense, this mimics what occurs in CML during progression to "blastic crisis," when the stem cells no longer give rise to progeny that are capable of maturing and produce only agranular forms.

Although the MPO-rich HL60 cells appear to be promyelocytic, their azurophilic granules are not completely normal. On density gradients they are lighter than those from mature PMN, and by EM the predominant type is similar to the immature microgranular variety that is apparent with stains for sulfated glycoconjugates (Rice *et al.*, 1986); Parmley *et al.*, 1987a). Overall, the MPO enzyme appears to mature in a normal sequence, with the 89-kDa precursor being associated with the lighter fraction on density gradients (Golgi, ER, and plasma membranes) and the 59-kDa mature form localizing to the denser mature granule fraction (Nauseef and Clark, 1986). Ultrastructural localization of MPO has shown that most cells stain in the perinuclear cistern and ER, indicating that they are actively making the enzyme, but in only 5–10% is MPO visible in the Golgi. Furthermore, the granule membranes themselves are thinner than normal and closer in thickness to the ER membrane. This deviation from normal led Bainton (1988) to arrive at a conclusion similar to that of Olsson *et al.* (1988) that in most of these cells transfer of vesicles and/or fusion with the Golgi may be blocked during most of the cell cycle, resulting in budding of MPO-containing vesicles directly from the ER or an early Golgi compartment.

If HL60 cells are leukemic promyelocytes and they produce primary granules, why do they not make Auer rods? Also, why do these structures not form in many leukemic cells that appear to possess the prerequisite granules? It is reasonable to anticipate that the answers to such questions might point to a step in the pathway of vesicular traffic that is altered in leukemic precursor cells. The observations that HL60 granule membranes appeared less mature than normal and that the originating patient (whose cells did not contain Auer rods) may have had acute myeloblastic (M2) and not acute promyelocytic (M3) leukemia introduce the possibility that the primary granules of leukemic patients must attain a minimal level of maturity before they become fusogenic (Dalton et al., 1988). Furthermore, a variant HL60 line (HL60-A7) has been described by Sullivan et al. (1986) in which most of the granules are large and fused, similar to the pseudo-Chédiak-Higashi structures suggested to be possible precursors of Auer rods (Bessis, 1973; Payne and Harrow, 1983). Analysis of the granule membranes has shown that the major glycoprotein that was identified by biosynthetic incorporation of tritiated mannose into parental HL60 was not evident in HL60-A7 (Peyman and Sullivan, 1987). Although this clearly points to a processing defect in the membrane proteins of the aberrant granules, it is premature to extend this observation to suggest that a similar lesion necessarily would be present in primary leukemic cells or that there is any link to the observed N-*ras* codon 61 mutations reported in HL60 (reviewed in Collins, 1987).

5.2.2. Plasma Membrane Antigens in Diagnosis and Prognosis

By focusing on individual molecular entities, one hopes to understand mechanisms of disease more precisely than is possible by using classical morphology and enzyme cytochemistry. Over the past few years monoclonal antibody technology, in addition to its expanding commercial appeal, has brought a new dimension to the diagnosis and classification of leukemia. This has resulted in a proliferation of market reagents directed to different plasma membrane antigens and a profusion of publications too voluminous to catalog in this review. Recently, a very comprehensive summary of leukocyte cell surface proteins has been published (Horejsi and Bazil, 1988).

Many of the target antigens selected for study were those shown to be expressed at different stages of granulocytic maturation, and most of them have been detected on different subclasses of leukemic cells. Some have proven quite useful for distinguishing myeloid from lymphoid leukemias, especially those cells that do not exhibit sufficient morphological features to enable absolute distinction by standard criteria (Linch et al., 1984; Strauss et al., 1984; Pessano et al., 1984; Lange et al., 1984; Lowenberg and Bauman, 1985; Hanson et al., 1987; Davey et al., 1987). As shown in the large Cancer and Acute Leukemia Group B study, it appears that there is no single marker that can do this definitively, but by selecting an appropriate panel one may ascertain lineage in over 95% of cases. Interestingly, the resulting clusters of surface antigen phenotypes exhibited by leukemic cells from different patients do not correlate strongly with the French-American-British classification except for types M6 (erythroleukemia) and M7 (mega-karyoblastic).

The potential for using a more objective (albeit expensive) system of classification now is pointing the way to new therapeutic possibilities. For example, in acute non-lymphoblastic leukemia, the expression of My4 (CD14) and My7 (CD13) antigens predicted a lower rate of complete remission, and expression of HLA-DR, My8, and Mo1 (CD11) predicted a sooner relapse. Using these reagents, a new type of leukemia—biphenotypic—has been identified, which usually has lymphoblastoid morphology but bears a mixture of lymphoid- and myeloid-associated antigens. Such patients have a poorer prognosis and enter remission only 35% of the time, instead of the 76% observed for the comparison group (Griffin, 1987; Merle-Béral *et al.*, 1988). If these exemplary observations continue to be confirmed, it is realistic to anticipate that improved therapeutic strategies will follow; patients in the higher-risk groups would be selected to receive more aggressive treatments, while others would have their treatment tailored more directly to the biological characteristics of their leukemic cells.

Further attempts are being made to correlate the cell surface antigenic phenotype with other functional parameters. Gallin *et al.* (1986) have identified a structure (antibody 31D8) that is present on >95% of normal PMN but is decreased on the neutrophils of a minor subgroup of patients with CML. These cells generate only a low level of superoxide in response to fMLP, and the patients from whom they were obtained all progressed to blast crisis during the 10-month period of study. In the same disease, Todd *et al.* (1987) used three antibodies, one of which was to CD11b, to examine the patterns of staining on sections of bone marrow. Again, loss of antigen expression predicted progression to blastic crisis.

6. CONCLUSION

Over the past two decades, much has been learned about life inside the cell, how the organelles are formed, and how the internal dynamics flow with the endosomal currents. In Virchow's 1860 English edition of *Cellular Pathology,* he noted that the "colourless corpuscles" contained granules, but did not give any indication that he thought they were relevant to function. About 50 years later, in 1912, Gulland and Goodall of Edinburgh, in *The Blood: A Guide to Its Examination and to the Diagnosis and Treatment of Its Diseases,* mentioned in passing that the "view that the neutrophils break down to form antitoxins is plausible . . . ," but they did not consider the possibility that the red cytoplasmic dots in their drawings were pertinent. It was only after another 50 years, with the advent of the electron microscope, that one was able to appreciate the complexity of granular organelles and their contribution to our survival among the microbes. Today, with new immunological probes arriving almost daily, we are in the middle of a very exciting period as many laboratories are reconstructing the architecture of blood cells as they differentiate and perform or become distorted by disease.

Questions that one can hope will be answered soon include the following: What are the mechanisms that enable a myeloid cell to make so many different granules at different times? What are the signals that enable the different granule proteins to find the right membrane at the right time? What are the relationships between the known regulators of internal vesicle traffic, such as the G proteins, and the synthesis of neutrophil granules? Is the granulopathy of leukemia merely an epiphenomenon to the neoplastic cell's prolonged replication, or are both attributes the consequence of a common lesion?

As a final aside, these vignettes in the history of medicine might serve to give perspective and guidance to reviewers of research grants who may have drifted so far from the guiding shores of clinical perspective that they rebuff endeavors they view as "too descriptive." It should be evident that very little has been accomplished in medical basic science that does not rest on the shoulders of clinical observation or follow the road map drawn by the careful cartographers of descriptive pathology.

7. REFERENCES

Ahluwalia, J. P., Doherty, J. J., Troulis, M., Posner, B. I., and Bergeron, J. J. M., 1988, Identification of antigen(s) specific to a "late" endosome fraction, *Prog. Clin. Biol. Res.* **270**:411–415

Akin, D. T., and Kinkade, J. M., 1987, Evidence for the involvement of an acidic compartment in the processing of myeloperoxidase in human promyelocytic leukemia HL60 cells, *Arch. Biochem. Biophys.* **255**:428–436

Altieri, D. C., Bader, R., Mannucci, P. M., and Edgington, T. S., 1988a, Oligospecificity of the cellular adhesion receptor MAC-1 encompasses an inducible recognition specificity for fibronogen, *J. Cell Biol.* **107**:1893–1900

Altieri, D. C., Morrissey, J. H., and Edgington, T. S., 1988b, Adhesive receptor Mac-1 co-ordinates the activation of factor X on stimulated cells of monocytic and myeloid differentiation: An alternative initiation of the coagulation protease cascade, *Proc. Natl. Acad. Sci. USA* **85**:7462–7466

Anderson, D. C., Schmalstieg, F. C., Finegold, M. J., Hughes, B. J., Rothlein, R., Miller, L. J., Kohl, S., Tosi, M. F., Jacobs, R. L., Waldrop, T. C., Goldman, A. S., Shearer, W. T., and Springer, T. A., 1985, The severe and moderate phenotypes of heritable Mac-1, LFA-1 deficiency: Their quantitative definition and relation to leukocyte dysfunction and clinical features, *J. Infect. Dis.* **152**:668–689

Anderson, R. G. W., and Orci, L., 1988, A view of acidic intracellular compartments, *J. Cell Biol.* **106**:539–543

Audette, M., Buchegger, F., Schreyer, M., and Mach, J.-P., 1987, Monoclonal antibody against carcinoembryonic antigen (CEA) identifies two new forms of crossreacting antigens of molecular weight 90,000 and 160,000 in normal granulocytes, *Mol. Immunol.* **24**:1177–1186

Baggiolini, M., and Dewald, B., 1984, Exocytosis by neutrophils, in *Contemporary Topics in Immunobiology* (R. Snyderman, ed.), pp. 221–246, Plenum Press, New York

Bainton, D. F., 1977, Differentiation of human neutrophilic granulocytes: Normal and abnormal, in *The Granulocyte: Function and Clinical Utilization* (T. Greenwalt and G. A. Jamieson, eds.), pp. 1–27, Alan R. Liss, New York

Bainton, D. F., 1988, HL-60 cells have abnormal myeloperoxidase transport and packaging, *Exp. Hematol.* **16**:150–158

Bainton, D. F., Miller, L. J., Kishimoto, T. K., and Springer, T. A., 1987, Leukocyte adhesion receptors are stored in peroxidase-negative granules of human neutrophils, *J. Exp. Med.* **166**:1641–1653

Barbacid, J., 1987, *ras* genes, *Annu. Rev. Biochem.* **56**:779–827

Bar-Eli, M., Ahuja, H., Gonzalez-Cadavid, N., Foti, A., and Cline, M. J., 1989, Analysis of N-*ras* exon-1 mutations in myelodysplastic syndromes by polymerase chain reaction and direct sequencing, *Blood* **73**:281–283

Baron, R., Neff, L., Brown, W., Courtoy, P. J., Louvard, D., and Farquhar, M. G., 1988, Polarized secretion of lysosomal enzymes: Co-distribution of cation-independent mannose 6-phosphate receptors and lysosomal enzymes along the osteoclast exocytic pathway, *J. Cell Biol.* **106**:1863–1872

Barriocanal, J. G., Bonifacino, J. S., Yuan, L., and Sandoval, I. V., 1986, Biosynthesis, glycosylation, movement through the Golgi system, and transport to lysosomes by an N-linked carbohydrate-independent mechanism of three lysosomal internal membrane proteins, *J. Biol. Chem.* **261**:16755–16763

Bar-Sagi, D., and Gomperts, B. D., 1988, Stimulation of exocytic degranulation by microinjection of the *ras* oncogene protein into rat mast cells, *Oncogene* **3**:463–469

Bartles, J. R., Feracci, H. M., Stieger, B., and Hubbard, A. L., 1987, Biogenesis of the rat hepatocyte plasma membrane in vivo: Comparison of the pathways taken by apical and vasolateral proteins using subcellular fractionation, *J. Cell Biol.* **105**:1241–1251

Benchimol, S., Fuks, A., Jothy, S., Beauchemin, N., Shirota, K., and Stanners, C. P., 1989, Carcinoembryonic antigen, a human tumor marker, functions as an intercellular adhesion molecule, *Cell* **57**:327–334

Bendix-Hansen, K., 1987, Annotation: Enzyme cytochemistry of neutrophil granulocytes, *Br. J. Haematol.* **65**:127–129

Bergeron, J. J. M., Cruz, J., Khan, M. N., and Posner, B. I., 1985, Uptake of insulin and other ligands into receptor rich endocytic components of target cells: The endosomal apparatus, *Annu. Rev. Physiol.* **47:**383–403

Bergeron, J. J. M., Kay, D. G., Lai, W. H., Doherty, J. J., Smith, C. E., Khan, M. N., and Posner, B. I., 1988, Functional characteristics of the endosomal apparatus of rat liver parenchyma, *Prog. Clin. Biol. Res.* **270:**391–409

Bessis, M., 1973, *Living Blood Cells and Their Ultrastructure,* Springer-Verlag, New York

Beverly, P. C. L., Linch, D., and Delia, D., 1980, Isolation of human haematopoietic progenitor cells using monoclonal antibodies, *Nature* (London) **287:**332–333

Bizzaro, N., Briani, G., and Boccato, P., 1988, Acquired myeloperoxidase deficiency of neutrophils in a patient with aplastic anemia (idiopathic marrow aplasia), *Acta Haematol.* **80:**71–73

Blok, J., Gibbs, M. E., Lienhard, G. E., and Geuze, H. J., 1988, Insulin-induced translocation of glucose transporters from post-Golgi compartments to the plasma membrane of 3T3-L1 adipocytes, *J. Biol. Chem.* **106:**69–76

Bourne, H. R., 1988, Do GTP'ases direct membrane traffic in secretion, *Cell* **53:**669–671

Boxer, L. A., and Smolen, J. E., 1988, Neutrophil granule constituents and their release in health and disease, *Hematol. Oncol. Clin. North Am.* **2:**101–134

Braulke, T., Gartung, C., Hasilik, A., and Von Figura, K., 1987, Is movement of mannose 6-phosphate-specific receptor triggered by binding of lysosomal enzymes, *J. Cell. Biol.* **104:**1735–1742

Braun, M. P., Martin, P. J., Ledbetter, J. A., and Hansen, J. A., 1983, Granulocytes and cultured human fibroblasts express common acute lymphoblastic leukemia-associated antigens, *Blood* **61:**718–725

Brederoo, P., Van der Meulen, J., and Mommaas-Kienhuis, A. M., 1983, Development of the granule population in neutrophil granulocytes from human bone marrow, *Cell Tissue Res.* **234:**469–496

Brederoo, P., Van der Meulen, J., and Daems, W. T., 1986, Ultrastructural localization of peroxidase activity in developing neutrophil granulocytes from human bone marrow, *Histochemistry* **84:**445–453

Bretz, U., and Baggiolini, M., 1974, Biochemical and morphological characterization of azurophil and specific granules of human neutrophilic polymorphonuclear leukocytes, *J. Cell Biol.* **63:**251–269

Burgess, T. L., and Kelly, R. B., 1987, Constitutive and regulated secretion of proteins, *Annu. Rev. Cell Biol.* **3:**243–293

Burtin, P., Flandrin, G., and Fondaneche, M. C., 1980, Presence of NCA (nonspecific cross-reacting antigen) in the cells of the human granulocyte series, *Blood Cells* **6:**263–273

Chen, J. W., Chen, G. L., D'Souza, M. P., Murphy, T. L., and August, J. T., 1986, Lysosomal membrane glycoproteins: Properties of LAMP-1 and LAMP-2, *Biochem. Soc. Symp.* **51:**97–112

Civin, C. J., Mirro, J., and Banquerigo, M. L., 1981, My-1, a new myeloid-specific antigen identified by a mouse monoclonal antibody, *Blood* **57:**842–845

Clementson, K. J., and Luscher, E. F., 1988, Membrane glycoprotein abnormalities in pathological platelets, *Biochim. Biophys. Acta* **947:**53–73

Collins, S. J., 1987, The HL-60 promyelocytic leukemia cell line: Proliferation, differentiation, and cellular oncogene expression, *Blood* **70:**1233–1244

Colman, R. W., 1986, Platelet activation: Role of an ADP receptor, *Semin. Hematol.* **23:**119–128

Corvera, S., Folander, K., Clairmont, K. B., and Czech, M. P., 1988, A highly phosphorylated subpopulation of insulin-like growth factor II/mannose 6-phosphate receptors is concentrated in a clathrin-enriched plasma membrane fraction, *Proc. Natl. Acad. Sci.USA* **85:**7567–7571

Cournoyer, D., Beauchemin, N., Boucher, D., Benchimol, S., Fuks, A., and Stanners, C. P., 1988, Transcription of genes of the carcinoembryonic antigen family in malignant and non-malignant human tissues, *Cancer Res.* **48:**3153–3157

Dalton, W. T., Ahearn, M. J., McCredie, K. B., Freireich, E. J., Stass, S. A., and Trujillo, J. M., 1988, HL-60 cell line was derived from a patient with FAB-M2 and not FAB-M3, *Blood* **71:**242–247

Damiano, V. V., Kucich, U., Murer, E., Laudenslager, N., and Weinbaum, G., 1988, Ultrastructural quantitation of peroxidase- and elastase-containing granules in human neutrophils, *Am. J. Pathol.* **131:**235–245

Davey, F. R., Erber, W. N., Gatter, K. C., and Mason, D. Y., 1987, Immunophenotyping of acute myeloid leukemia by immunoalkaline phosphatase (APAAP) labeling with a panel of antibodies, *Am. J. Hematol.* **26:**157–166

Davey, F. R., Erber, W. N., Gatter, K. C., and Mason, D. Y., 1988, Abnormal neutrophils in acute myeloid leukemia and myelodysplastic syndrome, *Hum. Pathol.* **19:**454–459

Deimann, W., and Teckhous, L., 1985, Formation of a morphologically complex system of peroxidase-positive lysosomal elements in human monocytes, *Blood* **66**:514–521

Deng, Y., and Storrie, B., 1988, Animal cell lysosomes rapidly exchange membrane proteins, *Proc. Natl. Acad. Sci. USA* **85**:3860–3864

Deutscher, S. L., and Hirschberg, C. B., 1986, Mechanism of galactosylation in the Golgi apparatus, *J. Biol. Chem.* **261**:96–100

Deutscher, S. L., Nuwayhid, N., Stanley, P., Briles, E. I. B., and Hirschberg, C. B., 1984, Translocation across Golgi vesicle membranes: A CHO glycosylation mutant deficient in CMP-sialic acid transport, *Cell* **39**:295–299

Dixon, B. R., Mukherjee, T. M., and Ho, J. Q. K., 1984, The ultrastructural identification of Auer body precursors in a case of acute promyelocytic leukemia using high-angle specimen tilt, *Am. J. Clin. Pathol.* **81**:132–137

Draye, J.-P., Quintart, J., Courtoy, P. J., and Bandhuin, P., 1987, Relations between plasma membrane and lysosomal membrane. I. Fate of covalently labelled plasma membrane protein, *Eur. J. Biochem.* **170**:395–403

D'Souza, M. P., and August, J. T., 1986, A kinetic analysis of biosynthesis and localization of a lysosome-associated membrane glycoprotein, *Arch. Biochem. Biophys.* **249**:522–532.

Duffaud, G. D., Lehnhardt, S. K., March, P. E., and Inouye, M., 1985, Structure and function of the signal peptide, *Curr. Top. Membr. Transp.* **24**:65–104

Duncan, J. R., and Kornfeld, S., 1988, Intracellular movement of two mannose 6-phosphate receptors: Return to the Golgi apparatus, *J. Cell Biol.* **106**:617–628

Dustin, M. L., and Springer, T. A., 1988, Lymphocyte function-associated antigen-1 (LFA-1) interaction with intracellular adhesion molecule-1 (ICAM-1) is one of at least three mechanisms for lymphocyte adhesion to cultured endothelial cells, *J. Cell Biol.* **107**:321–331

Eilers, U., Klumperman, J., and Hauri, H.-P., 1989, Nocodazole, a microtubule-active drug, interferes with apical protein delivery in cultured intestinal epithelial cells (CaCo-2), *J. Cell Biol.* **108**:13–22

Fambrough, D. M., Takeyasu, K., Lippincott-Schwartz, J., and Siegel, N. R., 1988, Structure of LEP 100, a glycoprotein that shuttles between lysosomes and the plasma membrane, deduced from the nucleotide sequence of the encoding DNA, *J. Cell Biol.* **106**:61–67

Farquhar, M. G., 1985, Progress in unraveling pathways of Golgi traffic, *Annu. Rev. Cell Biol.* **1**:447–488

Fittschen, C., Parmley, R. T., Bishop, S. P., and Williams, J. C., 1988, Morphometry of feline neutrophil granule genesis, *Am. J. Anat.* **181**:195–202

Fitz-Gibbon, L., Price, G. B., and Sullivan, A. K., 1983, Loss of promyelocytic maturation in HL60 sublines: A potential model for leukemia progression, *Br. J. Haematol.* **55**:311–318

Fitz-Gibbon, L., Shematek, G., and Sullivan, A. K., 1985, Antigens differentially expressed on surface and cytoplasmic structures of human myeloid cells, *Leuk. Res.* **9**:123–134

Fontana, J. A., Wright, D. G., Schiffman, E., Corcoran, B. A., and Deisseroth, A. B., 1980, Development of chemotactic responsiveness in myeloid precursor cells: Studies with a human leukemia cell line, *Proc. Natl. Acad. Sci. USA* **77**:3664–3668

Fox, N., Damjanov, I., Martinez-Hernandez, A., Knowles, B., and Solter, D., 1981, Immunohistochemical localization of the early embryonic antigen (SSEA-1) in postimplantation mouse embryos and fetal and adult tissues, *Dev. Biol.* **83**:391–398

Friederich, E., Fritz, H.-J., and Huttner, W. B., 1988, Inhibition of tyrosine sulfation in the *trans*-Golgi retards transport of a constitutively secreted protein to the cell surface, *J. Cell. Biol.* **107**:1655–1667

Friedman, R. M., Lister, J., Fitz-Gibbon, L., and Sullivan, A. K., 1985, Deficiency of a major myeloid antigen in neutrophils of patients with refractory anemia and excess myeloblasts, *Leuk. Res.* **9**:435–440

Gallagher, R., Collins, S., Trujillo, J., McCredie, K., Ahearn, M., Tsai, S., Metzar, R., Aulakh, G., Ting, R., Ruscetti, F., and Gallo, R., 1979, Characterization of the continuous differentiating myeloid cell line (HL60) from a patient with acute promyelocytic leukemia, *Blood* **54**:713–733

Gallin, J. I., 1985, Neutrophil specific granule deficiency, *Annu. Rev. Med.* **36**:263–274

Gallin, J. I., Jacobson, R. J., Seligman, B. E., Metcalf, J. A., McKay, J. H., Sacher, R. A., and Malech, H. L., 1986, A neutrophil membrane marker reveals two groups of chronic myelogeneous leukemia and its absence may be a marker for disease progression, *Blood* **68**:343–346

Ganz, T., Selsted, M. E., Szklarek, D., Harwig, S. S. L., Daher, K., Bainton, D. F., and Lehrer, R. I., 1985, Defensins—natural antibiotics of human neutrophils, *J. Clin. Invest.* **76**:1427–1435

Ganz, T., Metcalf, J. A., Gallin, J. I., Boxer, L. A., and Lehrer, R. I., 1988, Microbicidal/cytotoxic proteins of neutrophils are deficient in two disorders: Chediak-Higashi syndrome and "specific" granule deficiency, *J. Clin. Invest.* **82**:552–556

Geuze, H., Stoorvogel, W., Strous, G. J., Slot, J. W., Bleekemden, J. E., and Mellman, I., 1988, Sorting of mannose 6-phosphate receptors and lysosomal membrane proteins in endocytic vesicles, *J. Cell Biol.* **107**:2491–2501

Gilman, A. G., 1987, G proteins: Transducers of receptor-generated signals, *Annu. Rev. Biochem.* **56**:615–649

Ginsberg, M. H., Loftus, J. C., and Plow, E. F., 1988, Cytoadhesins, integrins and platelets, *Thromb. Haemostasis* **59**:1–6

Goda, Y., and Pfeffer, S. R., 1988, Selective recycling of the mannose 6-phosphate/IGF-II receptor to the *trans*-Golgi network *in vitro, Cell* **55**:309–320

Goldstein, J. L., Brown, M. S., Anderson, R. G. W., Russell, D. W., and Schneider, W. J., 1985, Receptor-mediated endocytosis, *Annu. Rev. Cell Biol.* **1**:1–40

Goud, B., Salminen, A., Walworth, N. C., and Novick, P. J., 1988, A GTP-binding protein required for secretion rapidly associated with secretory vesicles and the plasma membrane in yeast, *Cell* **53**:753–768

Grandori, C., and Hanafusa, H., 1988, p60[c-src] is complexed with a cellular protein in subcellular compartments involved in exocytosis, *J. Cell Biol.* **107**:2125–2135

Green, S. A., Zimmer, K.-P., Griffiths, G., and Mellman, I., 1987, Kinetics of intracellular transport and sorting of lysosomal membrane and plasma membrane proteins, *J. Cell Biol.* **105**:1227–1240

Griffin, J. D., 1987, The use of monoclonal antibodies in the characterization of myeloid leukemias, *Hematol. Pathol.* **1**:81–91

Griffiths, G., and Simons, K., 1986, The *trans* Golgi network: Sorting at the exit site of the Golgi complex, *Science* **234**:438–443

Griffiths, G., Hoflack, B., Simons, K., Mellmann, I., and Kornfeld, S., 1988, The mannose 6-phosphate receptor and the biogenesis of lysosomes, *Cell* **53**:329–341

Grimaldi, K. A., Hutton, J. C., and Siddle, K., 1987a, Production and characterization of monoclonal antibodies to insulin secretory granule membranes, *Biochem. J.* **245**:557–566

Grimaldi, K. A., Siddle, K., and Hutton, J. C., 1987b, Biosynthesis of insulin secretory granule membrane proteins, *Biochem. J.* **245**:567–573

Haffar, O. K., Aponte, G. W., Bravo, D. A., John, N. J., Hess, R. T., and Firestone, G. L., 1988, Glucocorticoid-regulated localization of cell surface glycoproteins in rat hepatoma cells is mediated within the Golgi complex, *J. Cell Biol.* **106**:1463–1474

Handagama, P. J., George, J. N., Shuman, M. A., McEver, R. P., and Bainton, D. F., 1987, Incorporation of a circulating protein into megakaryocyte and platelet granules, *Proc. Natl. Acad. Sci. USA* **84**:861–865

Hanson, C. A., Gajl-Peczalska, K. J., Parkin, J. L., and Brunning, R. D., 1987, Immunophenotyping of acute myeloid leukemia using monoclonal antibodies and the alkaline phosphatase-anti-alkaline phosphatase technique, *Blood* **70**:83–89

Hashimoto, S., Fumagalli, G., Zanini, A., and Meldolesi, J., 1987, Sorting of three secretory proteins to distinct secretory granules in acidophilic cells of cow anterior pituitary, *J. Cell Biol.* **105**:1579–1586

Havemann, K., Gramse, M., and Gassel, W. D., 1983, Cytochemical determination of granulocyte elastase and chymotrypsin in human myeloid cells and its application in acquired deficiency states and diagnosis of myeloid leukemia, *Klin. Wochenschr.* **61**:49–56

Haylett, T., and Thilo, L., 1986, Limited and selective transfer of plasma membrane glycoproteins to membrane of secondary lysosomes, *J. Cell Biol.* **103**:1249–1256

Henkart, P. A., 1985, Mechanism of lymphocyte-mediated cytoxicity, *Annu. Rev. Immunol.* **3**:31–58

Henkart, P., and Yue, C. C., 1988, The role of cytoplasmic granules in lymphocyte cytotoxicity, *Prog. Allergy* **40**:82–110

Henkart, P. A., Yue, C. C., and Yang, J., and Rosenberg, S. A., 1987, Cytolytic and biochemical properties of cytoplasmic granules of murine lymphokine-activated killer cells, *J. Immunol.* **137**:2611–2617

Herberman, R. B., Reynolds, C. W., and Ortaldo, J., 1986, Mechanisms of cytoxicity by natural killer cells, *Annu. Rev. Immunol.* **4**:651–680

Hickstein, D. D., Smith, A., Fisher, W., Beatty, P. G., Schwartz, B. R., Harlan, J. M., Root, R. K., and Locksley, R. M., 1987, Expression of leukocyte adherence-related glycoproteins during differentiation of HL-60 promyelocytic leukemia cells, *J. Immunol.* **138**:513–519

Hirai, H., Okada, M., Mizoguchi, H., Mano, H., Kobayashi, Y., Nishida, J., and Takaku, F., 1988, Relationship between and activated N-*ras* oncogene and chromosomal abnormality during leukemic progression from myelodysplastic syndrome, *Blood* **71**:256–258

Hirschberg, C. B., and Snider, M. D., 1987, Topography of glycosylation in the rough endoplasmic reticulum and Golgi apparatus, *Annu. Rev. Biochem.* **56**:63–87

Horejsi, V., and Bazil, V., 1988, Surface proteins and glycoproteins of human leukocytes, *Biochem. J.* **253**:1–26

Howe, C. L., Granger, B. L., Hull, M., Green, S. A., Gabel, C. A., Helenius, A., and Mellman, I., 1988, Derived protein sequence, oligosaccharides, and membrane insertion of the 120 kDa lysosomal membrane glycoprotein (lgp 120): Identification of a highly conserved family of lysosomal membrane glycoproteins, *Proc. Natl. Acad. Sci USA* **85**:7577–7581

Hu, J.-S., James, G., and Olson, E. N., 1988, Protein fatty acylation: A novel mechanism for association of proteins with membranes and its role in transmembrane regulatory pathways, *Bio Factors* **1**:219–226

Hubbard, S. C., and Ivatt, R. J., 1981, Synthesis and processing of asparagine-linked oligosaccharides, *Annu. Rev. Biochem.* **50**:555–583

Hynes, R. O., 1987, A family of cell surface receptors, *Cell* **48**:549–554

Jacobs, A., and Clark, R. E., 1986, Pathogenesis and clinical variations in the myelodysplastic syndromes, *Clin. Hematol.* **15**:925–951

Jain, N. C., Cox, C., and Bennett, J. M., 1987, Auer rods in the acute myeloid leukemias: Frequency and methods of demonstration, *Hematol. Oncol.* **5**:197–202

Jongineel, C. V., Quackenbush, E. J., Verroust, P., Carrel, S., and Letarte, M., 1989, Common acute lymphoblastic leukemia antigen expressed on leukemia and melanoma cell lines has neutral endopeptidase activity, *J. Clin. Invest.* **83**:713–717

Kelly, R. B., 1985, Pathways of protein secretion in eukaryotes, *Science* **230**:25–32

Kishimoto, T. K., Hollander, N., Roberts, T. M., Anderson, D. C., and Springer, T. A., 1987, Heterogeneous mutations in the β-subunit common to the LFA-1, Mac-1, and p150,95 glycoproteins cause leukocyte adhesion deficiency. *Cell* **50**:193–202

Kornfeld, S., 1986, Trafficking of lysosomal enzymes in normal and disease states, *J. Clin. Invest.* **77**:1–6

Kornfeld, S., and Sly, W. S., 1985, Lysosomal storage defects, *Hosp. Pract.*: 71–82

Krag, S. S., 1985, Mechanisms and functional role of glycosylation on membrane protein synthesis, *Curr. Top. Membr. Transp.* **24**:181–249

Lacal, P., Pulido, R., Sanchez-Madrid, F., Cabanas, C., and Mollinedo, F., 1988, Intracellular localization of a leukocyte adhesion glycoprotein family in the tertiary granules of human neutrophils, *Biochem. Biophys. Res. Commun.* **154**:641–647

Lange, B., Ferrero, D., Pessandro, S., Palumbo, A., Faust, J., Meo, P., and Rovero, G., 1984, Surface phenotype of clonogenic cells in acute myeloid leukemia defined by monoclonal antibodies, *Blood* **64**:693–700

Le Bien, T. W., and McCormack, R. T., 1989, The common acute lymphoblastic leukemia antigen (CD10)—emancipation from a functional enigma, *Blood* **73**:625–635

Lehrer, R. I., Ganz, T., and Selsted, M. E., 1988, Oxygen-independent bactericidal systems: Mechanisms and disorders, *Hematol. Oncol. Clin. North Am.* **2**:159–170

Le Maistre, A., Lee, M.-S., Talpaz, M., Kantarjian, H. M., Freireich, E. J., Deisseroth, A. B., Trujillo, J. M., and Stass, S. A., 1989, *Ras* oncogene mutations are rare late stage events in chronic myelogenous leukemia, *Blood* **73**:889–891

Lemansky, P., Hasilik, A., Von Figura, K., Helmy, S., Fishman, J., Fine, R. E., Kodersha, N. L., and Rome, L. H., 1987, Lysosomal enzyme precursors in coated vesicles derived from the exocytic and endocytic pathways, *J. Cell Biol.* **104**:1743–1748

Letarte, M., Vera, S., Tran, R., Addis, J. B. L., Onizuka, R. J., Quackenbush, E. J., Jongeneel, C. V., and McInnes, R. R., 1988, Common acute lymphocytic leukemia antigen is identical to neutral endopeptidase, *J. Exp. Med.* **168**:1247–1253

Lewis, V., Green, S. A., March, M., Viliko, P., Helenius, A., and Mellman, I., 1985, Glycoproteins of the lysosomal membrane, *J. Cell Biol.* **100**:1839–1847

Linch, D. C., Allen, C., Beverly, P. C. L., Bynoe, A. G., Scott, C. S., and Hogg, N., 1984, Monoclonal antibodies differentiating between monocytic and non-monocytic variants of AML, *Blood* **63**:566–573

Lingappa, V. R., 1989, Intracellular traffic of newly synthesized proteins, *J. Clin. Invest.* **83**:739–751

Lippincott-Schwartz, J., and Fambrough, D. M., 1987, Cycling of the integral membrane and lysosomes: Kinetic and morphological analysis, *Cell* **49**:669–677

Lodish, H. F., 1988, Transplant of secretory and membrane glycoproteins from the rough endoplasmic reticulum to the Golgi, *J. Biol. Chem.* **263**:2107–2110

Lomax, K. J., Gallin, J. I., Rotrosen, D., Raphael, G. D., Kaliner, M. A., Benz, E. J., Boxer, L. A., and Malech, H. L., 1989, Selective defect in myeloid cell lactoferrin gene expression in neutrophil specific granule deficiency, *J. Clin. Invest.* **83**:514–519

Lowe, A. W., Madeddu, L., and Kelly, R. B., 1988, Endocrine secretory granules and neuronal synaptic vesicles have three integral membrane proteins in common, *J. Cell Biol.* **106**:51–59

Lowenberg, B., and Bauman, J. G. J., 1985, Further results in understanding the subpopulation structure of AML: Clonogenic cells and their progeny identified by differentiation markers, *Blood* **66**:1225–1232

Lowrey, D. M., Hameed, A., Lichtenheld, M., and Podack, E. R., 1988, Isolation and characterization of cytotoxic granules from human lymphokine (interleukin 2) activated killer cells, *Cancer Res.* **48**:4681–4688

Lyons, J. J., Janssen, J. W. G., Bartram, C., Layton, M., and Mufti, G. J., 1988, Mutation of Ki-*ras* and N-*ras* oncogenes in myelodysplastic syndromes, *Blood* **71**:1707–1712

Matteoni, R., and Kreis, T. E., 1987, Translocation and clustering of endosomes and lysosomes depends on microtubules, *J. Cell Biol.* **105**:1253–1265

McCormack, R. T., Nelson, R. D., Chenoweth, D. E., and LeBien, T. W., 1987, Identification and characterization of a unique subpopulation (CALLA/CD10-negative) of human neutrophils manifesting a heightened chemotactic response to activated complement, *Blood* **70**:1624–1629

Meers, P., Ernst, J. D., Düzgünes, N., Hong, K., Fedor, J., Goldstein, I. M., and Papahadjopoulos, D., 1987, Synexin-like proteins from human polymorphonuclear leukocytes, *J. Biol. Chem.* **262**:7850–7858

Melançon, B. S., Malhotra, V., Weidman, P. J., Serafini, T., Gleason, M. L., Orci, L., and Rothman, J. E., 1987, Involvement of GTP-binding "G" proteins in transport through the Golgi stack, *Cell* **51**:1053–1062

Merle-Béral, H., Laabid, M., and Debré, P., 1988, Antigène de différenciation myéloïde: Caractérisation et expression dans les leucémies aiguës myéloblastiques, *Nouv. Rev. Fr. Hematol.* **29**:327–331

Miyauchi, J., and Watanabe, Y., 1987, Immunocytochemical localization of lactoferrin in human neutrophils. An ultrastructural and morphometrical study, *Cell Tissue Res.* **247**:249–258

Moore, H.-P. H., 1987, Factors controlling packaging of peptide hormones into secretory granules, *Annu. N.Y. Acad. Sci.* **493**:50–61

Mufti, G. J., and Galton, D. A. G., 1986, Myelodysplastic syndromes: Natural history and features of prognostic importance, *Clin. Hematol.* **15**:953–971

Murphy, P., and Hart, D. A., 1987, Regulation of enzyme release from human polymorphonuclear leukocytes: Further evidence for the independent regulation of granule subpopulations, *Biochem. Cell Biol.* **65**:1007–1015

Nauseef, W. M., 1988, Myeloperoxidase deficiency, *Hematol. Oncol. Clin. North Am.* **2**:135–158

Nauseef, W. M., 1989, Aberrant restriction endonuclease digests of DNA from subjects with hereditary myeloperoxidase deficiency, *Blood* **73**:290–295

Nauseef, W. M., and Clark, R. A., 1986, Separation and analysis of subcellular organelles in a human promyelocytic leukemia cell line HL-60: Application to the study of myeloid lysosomal enzyme synthesis and processing, *Blood* **68**:442–449

Nauseef, W. M., Olsson, I., and Arnljots, K., 1988, Biosynthesis and processing of myeloperoxidase-A marker for myeloid cell differentiation, *Eur. J. Hematol.* **40**:97–110

Neefjes, J. J., Verkerk, J. M. H., Broxterman, H. J. G., Van der Marel, G. A., Von Boom, J. H., and Ploegh, H. L., 1988, Recycling glycoproteins do not return to the *cis*-Golgi, *J. Cell Biol.* **107**:79–87

Nieuwenhuis, H. K., Van Oosterhout, J. J. G., Rozemuller, E., Van Iwaarden, F., and Sixma, J. J., 1987, Studies with a monoclonal antibody against activated platelets: Evidence that a secreted 53,000 molecular weight lysosome-like granule protein is exposed on the surface of activated platelets in the circulation, *Blood* **70**:838–845

Noworolska, A., Harlozinska, A., Richter, R., and Brodzka, W., 1985, Nonspecific cross-reactive antigen (NCA) in individual maturation stages of myelocytic cell series, *Br. J. Cancer* **51**:371–377

Olsson, I., Lantz, M., Persson, A.-M., and Arnljots, K., 1988, Biosynthesis and processing of lactoferrin in bone marrow cells, a comparison with processing of myeloperoxidase, *Blood* **71**:441–447

Orci, L., Glick, B. S., and Rothman, J. E., 1986, A new type of coated vesicular carrier that appears not to contain clathrin: Its possible role in protein transport within the Golgi stack, *Cell* **46**:171–184

Orci, L., Ravazzola, M., Amherst, M., Perrelet, A., Powell, S. K., Quinn, D. L., and Moore, H. H., 1987, The *trans*-most cisternae of the Golgi complex: A compartment for sorting of secretory and plasma membrane proteins, *Cell* **51**:1039–1051

Ortaldo, J. R., and Longo, D. L., 1988, Human natural lymphocyte effector cells: Definition, analysis of activity and clinical effectiveness, *J. Natl. Cancer Inst.* **8**:999–1009

Paiement, J., and Bergeron, J. J. M., 1983, Localization of GTP-stimulated core glycosylation to fused membranes, *J. Cell Biol.* **96**:1791–1796

Parmley, R. T., Takagi, M., Barton, J. C., Boxer, L. A., and Austin, R. L., 1982, Ultrastructural localization of lactoferrin and iron-binding protein in human neutrophils and rabbit heterophils, *Am. J. Pathol.* **109**:343–358

Parmley, R. T., Tzeng, D. Y., Baehner, R. L., and Boxer, L. A., 1983, Abnormal distribution of complex carbohydrates in neutrophils of a patient with lactoferrin deficiency, *Blood* **62**:538–548

Parmley, R. T., Doran, T., Boyd, R. L., and Gilbert, C., 1986, Unmasking and redistribution of lysosomal sulfated glycoconjugates in phagocytic polymorphonuclear leukocytes, *J. Histochem. Cytochem.* **34**:1701–1707

Parmley, R. T., Rice, W. G., Kinkade, J. M., Gilbert, C., and Barton, J. C., 1987a, Peroxidase-containing microgranules in human neutrophils: Physical, morphological, cytochemical, and secretory properties, *Blood* **70**:1630–1638

Parmley, R. T., Akin, D. T., Barton, J. C., Gilbert, C. S., and Kinkade, J. M., 1987b, Cytochemistry and ultrastructural morphometry of cultured HL60 myeloid leukemia cells, *Cancer Res.* **47**:4932–4940

Parmley, R. T., Gilbert, C. S., and Boxer, L. A., 1989, Abnormal peroxidase-positive granules in "specific granule" deficiency, *Blood* **73**:838–844

Payne, C. M., and Harrow, E. J., 1983, A cytochemical and ultrastructural study of acute myelomonocytic leukemia exhibiting the pseudo-Chediak-Higashi anomaly of leukemia and "splinter-type" Auer rods, *Am. J. Clin. Pathol.* **80**:216–223

Pember, S. O., and Kinkade, J. M., 1983, Differences in myeloperoxidase activity from neutrophilic poly-morphonuclear leukocytes of differing density: Relationship to selective exocytosis of distinct forms of the enzyme, *Blood* **61**:1116–1124

Perez, H. D., Marder, S., Elfman, F., and Ives, H. E., 1987, Human neutrophils contain subpopulations of specific granules exhibiting different sensitivities to changes in cytosolic free calcium, *Biochem. Biophys. Res. Commun.* **145**:976–981

Pessano, S., Palumbo, A., Ferrero, D., Pagliardi, G. L., Bottero, L., Lai, S. K., Meo, P., Carter, C., Hubbell, H., Lange, B., and Rovera, G., 1984, Subpopulation heterogeneity in human acute myeloid leukemia determined by monoclonal antibodies, *Blood* **64**:275–281

Petrequin, P. R., Todd, R. F., Devall, L. J., Boxer, L. A., and Curnutte, J. T., 1987, Association between gelatinase release and increased plasma membrane expression of the Mo1 glycoprotein, *Blood* **69**:605–610

Petty, H. R., Francis, J. W., Todd, R. F., Petrequin, P., and Boxer, L. A., 1987, Neutrophil C3bi receptors: Formation of membrane clusters during cell triggering requires intracellular granules, *J. Cell Physiol.* **133**:235–242

Peyman, J. A., and Sullivan, A. K., 1987, Different molecular forms of a glycoprotein antigen found on azurophilic granule membranes of cultured human HL60 promyelocytes and on the plasma membrane of a myeloblastoid variant line, *Leuk. Res.* **11**:385–396

Pfeffer, S. R., and Rothman, J. E., 1987, Biosynthetic protein transport and sorting by the endoplasmic reticulum and Golgi, *Annu. Rev. Biochem.* **56**:829–852

Phillips, D. R., Charo, I. F., Parise, L. V., and Fitzgerald, L. A., 1988, The platelet membrane glycoprotein IIb-IIIa complex, *Blood* **71**:831–843

Prieels, J. P., Pizzo, S. V., Glasgow, L. R., Paulson, J. C., and Hill, R. L., 1978, Hepatic receptor that specifically binds oligosaccharides containing fucosyl-$\alpha 1 \rightarrow 3$ N-acetylglucosamine linkages, *Proc. Natl. Acad. Sci. USA* **75**:2215–2219

Pryzwansky, K. B., and Breton-Gorius, J., 1985, Identification of a subpopulation of primary granules in human neutrophils based upon maturation and distribution, *Lab. Invest.* **53**:664–671

Rado, T. A., Bollekens, J., St. Laurent, G., Parker, L., and Benz, E. J., 1984, Lactoferrin biosynthesis during granulopoiesis, *Blood* **64**:1103–1109

Rado, T. A., Wei, X., and Benz, E. J., 1987, Isolation of lactoferrin cDNA from a human myeloid library and expression of mRNA during normal and leukemic myelopoiesis, *Blood* **70**:989–993

Rambourg, A., Clermont, Y., and Hermo, L., 1988, Formation of secretion granules in the Golgi apparatus of pancreatic acinar cells of the rat, *Am. J. Anat.* **183**:187–199

Rapoport, T. A., 1986, Protein translocation across and integration into membranes, *Crit. Rev. Biochem.* **20**:73–137

Rapoport, T. A., and Wiedmann, M., 1985, Application of the signal hypothesis to the incorporation of integral membrane proteins, *Curr. Top. Membr. Transp.* **24**:1–63

Reddington, M., Novak, E. K., Hurley, E., Medda, C., McGarry, M. P., and Swank, R. T., 1987, Immature dense granules in platelets from mice with platelet storage pool disease, *Blood* **69**:1300–1306

Reynolds, C. W., Reichardt, D., Henkart, M., Millard, P., and Henkart, P., 1987, Inhibition of NK and ADCC activity by antibodies against purified cytoplasmic granules from rat LGL tumors, *J. Leukocyte Biol.* **42**:642–652

Rice, W. G., Kinkade, J. M., and Parmley, R. T., 1986, High resolution heterogeneity among human neutrophil granules: Physical, biochemical and ultrastructural properties of isolated fractions, *Blood* **68**:541–555

Rice, W. G., Ganz, T., Kinkade, J. M. Selsted, M. E., Lehrer, R. I., and Parmley, R. T., 1987, Defensin-rich dense granules of human neutrophils, *Blood* **70**:757–765

Rindler, M. J., and Traber, M. G., 1988, A specific sorting signal is not required for the polarized secretion of newly synthesized proteins from cultured intestinal epithelial cells, *J. Cell Biol.* **107**:471–479

Rosa, J.-P., George, J. N., Bainton, D. F., Nurden, A. T., Caen, J. P., and McEver, R. P., 1987, Grey platelet syndrome. Demonstration of alpha granule membranes that can fuse with the cell surface, *J. Clin. Invest.* **80**:1138–1146

Rotrosen, D., and Gallin, J. I., 1987, Disorders of phagocyte function, *Annu. Rev. Immunol.* **5**:127–150

Ruohla, H., Kabcenell, A. K., and Ferro-Novick, S., 1988, Reconstitution of protein transport from the endoplasmic reticulum to the Golgi complex in yeast: The acceptor Golgi compartment is defective in the *sec 23* mutant, *J. Cell Biol.* **107**:1465–1476

Ruoslahti, E., and Pierschbacher, M. D., 1987, New perspectives in cell adhesion: RGD and integrins, *Science* **238**:491–497

Sachs, L., 1982, Control of growth and normal differentiation in leukemic cells: Regulation of the developmental program and restoration of the normal phenotype in myeloid leukemia, *J. Cell Physiol. Suppl.* **1**:151–164

Schacter, H., 1986, Biosynthetic controls that determine the branching and microheterogeneity of protein-bound oligosaccharides, *Biochem. Cell. Biol.* **64**:163–181

Schmid, S. L., Fuchs, R., Male, P., and Mellman, I., 1988, Two distinct subpopulations of endosomes involved in membrane recycling and transport to lysosomes, *Cell* **52**:73–83

Shafer, A. I., 1984, Bleeding and thrombosis in the myeloproliferative disorders, *Blood* **64**:1–12

Shannon, W. A., and Zellmer, D. M., 1982, Heterogeneity in polymorphonuclear leukocyte neutrophil granules, *Histochem. J.* **14**:847–850

Simons, K., 1987, Membrane traffic in an epithelial cell line derived from the dog kidney, *Kidney Int. Suppl.* **23**:S201–210

Simons, K., and Fuller, S. D., 1985, Cell surface polarity in epithelia, *Annu. Rev. Cell Biol.* **1**:243–288

Skubitz, K. M., and Snook, R. W., 1987, Monoclonal antibodies that recognize lacto-N-fucopentaose III (CD15) react with the adhesion-promoting glycoprotein family (LFA-1/HMAC-1/GP 150,95) and CR1 on human neutrophils, *J. Immunol.* **139**:1631–1639

Solter, D., and Knowles, B. B., 1978, Monoclonal antibody defining a stage-specific mouse embryonic antigen (SSEA-1), *Proc. Natl. Acad. Sci USA* **75**:5565–5569

Sosa, M. A., Mayorga, L. S., and Bertini, F., 1987, β-Galactosidase from rat epididymal fluid is bound by a recognition site attached to membranes of the epididymis different from the phosphomannosyl receptor, *Biochem. Biophys. Res. Commun.* **143**:799–807

Springer, T. A., Dustin, M. L., Kishimoto, and Marlin, S. D., 1987, The lymphocyte function-associated LFA-1, CD2, and LFA-3 molecules: Cell adhesion receptors of the immune system, *Annu. Rev. Immunol.* **5**:223–252

Srivastava, P. C., Powling, M. J., Nokes, T. J. C., Patrick, A. D., Dawes, J., and Hardisty, R. M., 1987, Grey platelet syndrome: Studies on platelet alpha-granules, lysosomes and defective response to thrombin, *Br. J. Haematol.* **65**:441–446

Steen, V. M., and Holmsen, H., 1987, Current aspects on human platelet activation and responses, *Eur. J. Haematol.* **38**:383–399

Stein, M., Zijderhand-Bleekemden, J. E., Geuze, H., Hasilik, A., and Von Figura, K., 1987, Mr 46,000 mannose 6-phosphate receptor: Its role in targeting of lysosomal enzymes, *EMBO J.* **6**:2677–2681

Stevens, R. L., Otsu, K., Weis, J. H., Tantravahi, R. V., Austen, K. F., Henkart, P. A., Galli, M. C., and Reynolds, C. W., 1987, Co-sedimentation of chondroitin sulfate A glycosaminoglycans and proteoglycans with the cytolytic secretory granules of rat large granular lymphocyte (LGL) tumor cells, and transformed LGL that encodes proteoglycans, *J. Immunol.* **139**:863–868

Stevenson, K. B., Nauseef, W. M., and Clark, R. A., 1987, The neutrophil glycoprotein Mo 1 is an integral membrane protein of plasma membranes and specific granules, *J. Immunol.* **139**:3759–3763

Stoorvogel, W., Geuze, H. J., Griffith, J. M., and Strous, G. J., 1988, The pathways of endocytosed transferrin and secretory protein are connected in the *trans*-Golgi reticulum, *J. Cell Biol.* **106**:1821–1829

Strauss, L. C., Skubitz, K. M., August, J. T., and Civin, C. I., 1984, Antigenic analysis of hematopoiesis. II. Expression of human neutrophil antigens on normal and leukemic marrow cells, *Blood* **63**:574–578

Stryer, L., and Bourne, H. R., 1986, G proteins: A family of signal transducers, *Annu. Rev. Cell Biol.* **2**:391–419

Suarez-Quian, C. A., 1987, The distribution of four lysosomal integral membrane proteins (LIMPs) in rat basophilic leukemia cells, *Tissue Cell* **19**:495–504

Sullivan, A. K., Amatruda, T. T., Fitz-Gibbon, L., Koeffler, H. P., Peyman, J., Rowden, G., Shematek, G., and Shihab-El-Deen, A., 1986, Deficiency of myeloperoxidase and abnormal chromosome 1 occurs in variant (HL60) promyelocytes, *Leuk. Res.* **10**:501–513

Sullivan, R., Griffin, I. D., and Malech, H. L., 1987, Acquisition of formyl peptide receptors during normal human myeloid differentiation, *Blood* **70**:1222–1224

Swanson, J., Bushnell, A., and Silverstein, S. C., 1987, Tubular lysosome morphology and distribution within macrophages depend on the integrity of cytoplasmic microtubules, *Proc. Natl. Acad. Sci. USA* **84**:1921–1925

Takeuchi, K. H., McGarry, M. P., and Swank, R. T., 1987, Elastase and cathepsin G activities are present in immature bone marrow neutrophils and absent in late marrow and circulating neutrophils of beige (Chediak-Higashi) mice, *J. Exp. Med.* **166**:1362–1376

Tang, X. M., and Clermont, Y., 1989, Granule formation and polarity of the Golgi apparatus in neutrophil granulocytes of the rat, *Anat. Rec.* **223**:128–138

Thompson, J., and Zimmerman, W., 1988, The carcinoembryonic antigen gene family: Structure, expression and evolution, *Tumour Biol.* **9**:63–83

Tobler, A., Miller, C. W., Johnson, K. R., Selsted, M. E., Rovera, G., and Koeffler, H. P., 1988, Regulation of gene expression of myeloperoxidase during myeloid differentiation, *J. Cell Physiol.* **136**:215–225

Todd, M. B., Waldron, J. A., Jennings, T. A., Rome, L. S., Markowitz, S. D., Holford, T. R., Gardner, J. P., Wolak, J. P., and Malech, H. L., 1987, Loss of myeloid differentiation antigens precedes blastic transformation in chronic myelogenous leukemia, *Blood* **70**:122–131

Todd, R. F., and Freyer, D. R., 1988, The CD11/CD18 leukocyte glycoprotein deficiency, *Hematol. Oncol. Clin. North Am.* **2**:13–31

Tooze, J., Tooze, S. A., and Fuller, S. D., 1987, Sorting of progeny corona virus from condensed secretory proteins at the exist from the *trans*-Golgi network of AtT20 cells, *J. Cell Biol.* **105**:1215–1226

Tooze, S. A., Tooze, J., and Warren, G., 1988, Site of addition of N-acetyl-galactosamine to the E1 glycoprotein of mouse hepatitis virus-A59, *J. Cell Biol.* **106**:1475–1487

Tschopp, J., and Nabholz, M., 1987, The role of cytoplasmic granule components in cytolytic lymphocyte-mediated cytolysins, *Annu. Inst. Pasteur Immunol.* **138**:290–295

Tulliez, M., and Breton-Gorius, J., 1979, Three types of Auer bodies in acute leukemia—visualization of their protein by negative contrast after peroxidase cytochemistry, *Lab.Invest.* **41**:419–426

Ugen, K. E., Mahalingam, M., Klein, P. A., and Kao, K.-J., 1988, Inhibition of tumor cell-induced platelet aggregation and experimental tumor metastasis by the synthetic gly-arg-gly-asp-ser peptide, *J. Natl. Cancer Inst.* **80**:1461–1466

Varki, A., Reitman, M. L., Vannier, A., Kornfeld, S., Grugg, J. H., and Sly, W. S., 1982. Demonstration of the heterozygous state for I-cell disease and pseudo-Hurler polydystrophy by assay of N-acetylglucosaminylphosphotransferase in white blood cells and fibroblasts, *Am. J. Hum. Genet.* **34**:717–729

Vega-Salas, D. E., Salas, P. J. I., and Rodriguez-Boulau, E., 1988, Exocytosis of vacuolar opical compartment

(VAC): A cell-cell contact controlled mechanism for the establishment of the apical plasma membrane domain in epithelial cells, *J. Cell Biol.* **107**:1717–1728

Viitala, J., Carlsson, S. R., Siebert, P. D., and Fukuda, M., 1988, Molecular cloning of cDNAs encoding lamp A, a human lysosomal membrane glycoprotein with apparent Mr 120,000, *Proc. Natl. Acad. Sci. USA* **85**:3743–3747

Von Heijne, G., 1988, Transcending the impenetrable: How proteins come to terms with membranes, *Biochim. Biophys. Acta* **947**:307–333

Wahren, B., Gadler, F., Gahrton, G., Hammarstrom, S., Hareland, Y., Hyden, N., Ljungdahn, E., Mahlen, A., Ruden, U., and Wilkund, M., 1983, NCA: A differentiation antigen of myelopoietic cells in humans and hominoid monkeys, *Annu. N.Y. Acad. Sci.* **417**:344–358

Walter, P., and Lingappa, V. R., 1986, Mechanism of protein translocation across the endoplasmic reticulum membrane, *Annu. Rev. Cell Biol.* **2**:499–516

Warhol, M. J., Pinkis, G. S., and Said, J. W., 1987, Ultrastructural localization of Leu M1 in Reed-Sternberg cells and normal myeloid cells, *Hum. Pathol.* **18**:824–829

White, J. G., and Krumweide, M., 1987, Normal-sized primary lysosomes are present in Chediak-Higashi syndrome neutrophils, *Pediatr. Res.* **22**:208–215

Williams, D. B., Swiedler, S. J., and Hart, G. W., 1985, Intracellular transport of membrane glycoproteins: Two closely related histocompatibility antigens differ in their rates of transmit to the cell surface, *J. Cell Biol.* **101**:725–734

Willingham, M. C., and Pastan, I., 1984, Endocytosis and exocytosis: Current concepts of vesicle traffic in animal cells, *Int. Rev. Cytol.* **92**:51–92

Wilson, C., Connolly, T., Morrison, T., and Gilmore, R., 1988, Integration of membrane proteins into the endoplasmic reticulum requires GTP, *J. Cell Biol.* **107**:69–77

Woychick, N. A., and Dimond, R. L., 1987, A single mutation prevents the normal intracellular transport of multiple lysosomal proteins from the rough endoplasmic reticulum, *J. Biol. Chem.* **262**:10008–10014

Yamamoto, T., 1986, Deletion of cysteine-rich region of LDL receptor impedes transport to cell surface in WHHL rabbit, *Science* **232**:1230–1237

Zalman, L. S., Brothers, M. A., and Müller-Eberhard, H. J., 1988, Self-protection of cytotoxic lymphocytes: A soluble form of homologous restriction factor in cytoplasmic granules, *Proc. Natl. Acad. Sci. USA* **85**:4827–4831

Chapter 9

The Elusive Oxidase
The Respiratory Burst Oxidase
of Human Phagocytes

Robert M. Smith, Richard C. Woodman, and Bernard M. Babior

1. INTRODUCTION

The blood's phagocytic cells (neutrophils, eosinophils, and mononuclear phagocytes) are the body's elite infantry troops. Capable of utilizing a wide range of weapons in the name of host defense, they circulate vigilantly, waiting at short notice to swarm into infected tissues to assault and eliminate invading microorganisms. For over three decades, the generation of reactive oxidants by stimulated phagocytic cells has been recognized as one of the essential armaments used by these cells. Upon ingesting an invading microorganism *in vivo* or upon contact with a wide variety of soluble or particulate stimuli *in vitro*, the phagocytic cell undergoes a dramatic metabolic event termed the respiratory burst. The hallmark of this event is a several hundredfold increase in the oxygen consumption of the cell—hence the name—and the subsequent conversion of this metabolized oxygen to highly reactive reduced oxygen species such as superoxide anion (O_2^-), hydrogen peroxide (H_2O_2), hydroxyl radical ($OH\cdot$), and oxidized halogens.

Unfortunately, the battlefield on which the phagocytes loose their weapons is human tissue, and the damage done to these tissues may, in some diseases, overshadow the clear benefits usually provided by these protective infantry. Thus, although initially recognized for their microbicidal properties, the oxidants produced in the respiratory burst have more recently been implicated in a number of inflammatory diseases, including the adult respiratory distress syndrome (McGuire *et al.*, 1982), ischemic tissue injury (Jolly *et al.*,

Robert M. Smith   Division of Pulmonary and Critical Care Medicine, Department of Internal Medicine, University of California Medical Center, San Diego, California 92103, USA.   **Richard C. Woodman** Department of Medicine, University of Calgary, Calgary, Alberta T2N 2T9 Canada.   **Bernard M. Babior** Department of Molecular and Experimental Medicine, Research Institute of Scripps Clinic, La Jolla, California 92037, USA.

1984; Arroyo *et al.*, 1987), arthritis (Halliwell *et al.*, 1985; Halliwell, 1987), and glomerulonephritis (Johnson *et al.*, 1987); they may also play an important role in carcinogenesis (Trush *et al.*, 1985; Weitzman *et al.*, 1985; Cerutti, 1985; Frenkel *et al.*, 1986). Despite the possibility that the products of the respiratory burst may be pathogenic, the capacity of phagocytic cells to undergo the respiratory burst is vital for host defense. The inability of the phagocyte to express the respiratory burst is the underlying defect in chronic granulomatous disease (CGD), a rare disorder of heterogeneous inheritance characterized by recurrent systemic infections usually resulting in significant morbidity and in early mortality. Although the incidence of this latter disorder is quite rare in the general populace, the inability of CGD cells to produce oxidants during the respiratory burst has made them an invaluable tool for probing the role of oxidants in disease states and for elucidating the nature of the enzyme system responsible for oxidant production in normal cells.

A complete understanding of the biochemical mechanisms underlying the respiratory burst has proven elusive. However, over the last decade, extensive investigation of the respiratory burst oxidase, the enzyme system responsible for the production of oxidants, has begun to clarify many of the areas of uncertainty. This chapter will review the current biochemical and genetic knowledge of the respiratory burst oxidase and will examine some of the recent investigations into the genetics of CGD as well as the directions that ongoing research is presently heading.

2. THE RESPIRATORY BURST

2.1. General Description

In an unstimulated state, phagocytic cells, neutrophils in particular, are relatively inactive metabolically, with rates of oxygen uptake of $0.1–1.0$ nmol $O_2 \cdot min^{-1} \cdot 10^7$ cells^{-1}. However, if the cell is allowed to contact an agent that stimulates the respiratory burst, a complex series of events begins to unfold. After a delay of seconds to a minute, the rate of oxygen uptake by the cell increases markedly, sometimes by two to three orders of magnitude ($100–200$ nmol $O_2 \cdot min^{-1} \cdot 10^7$ cells^{-1} for a neutrophil; Figure 1). Concurrent with this increase in oxygen uptake, there is production of reduced oxygen intermediates and a $10–100$-fold increase in movement of substrate through the hexose monophosphate (HMP) shunt. Depending on the stimulus, the cells may exhibit a spectrum of additional responses, including chemotaxis, adhesion, degranulation, mobilization of intracellular calcium stores, and membrane depolarization.

A variety of stimuli are able to elicit the respiratory burst. Particulate stimuli such as opsonized bacteria or yeast cell wall fragments (zymosan) appear to induce the respiratory burst only on the portion of the cell with which they come in contact (Bellavite *et al.*, 1982); with phagocytosis, the products of the respiratory burst are contained within the phagosome containing the ingested particle. In contrast, soluble stimuli appear to elicit the respiratory burst globally over the entire membrane surface of the cell either by direct perturbation of the cell surface (e.g., by certain fatty acids and detergents) or by receptor-mediated events (e.g., the binding of formylated bacterial tripeptides, complement fragments, or phorbol esters). With the use of any of these stimuli, there is an initial lag period

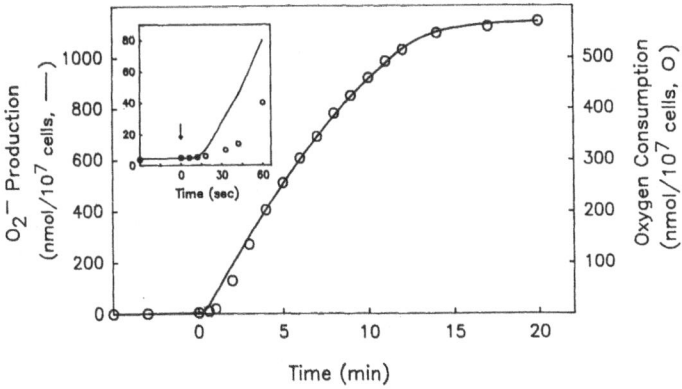

FIGURE 1. Representative profiles of O_2^- production ($-$) and oxygen consumption ($\bigcirc$), by human neutrophils are shown. At time zero (indicated by the arrow), the respiratory burst was initiated by the addition of opsonized zymosan. After a brief lag, shown best in the inset, both oxygen consumption and O_2^- production commence and then closely parallel each other in a 1 : 2 stoichiometry.

(the length varies with the stimulus) before any respiratory burst activity is detected, followed by an acceleration in oxygen uptake until a maximum velocity is attained over seconds to minutes. As discussed in other chapters of this volume, a subthreshold application of one stimulus appears to cause the respiratory burst oxidase to enter a primed state; subsequent stimuli cause a much greater level of oxygen uptake than would have otherwise resulted. Furthermore, the respiratory burst appears to be a reversible phenomenon; removal of the stimulus typically results in a rapid reversal of the cell to the unstimulated state, with a concomitant drop in oxygen utilization (Curnutte *et al.*, 1979; Cohen *et al.*, 1982; Badwey *et al.*, 1984). Repeat stimulation results in the reappearance of the respiratory burst at previous levels of activity. However, repeated application of a stimulus can also induce apparent deactivation of the respiratory burst enzyme or of its activation system (Kitagawa and Johnston, 1986), yielding a state that may mimic CGD (Matzner *et al.*, 1982). The diverse mechanisms of activation of the respiratory burst, although not the subject of this chapter, remain one of the most complex and least understood phenomena of the neutrophil.

Although the respiratory burst had been described as early as 1933, it was initially thought to fuel mitochondrial oxidative metabolism in order to provide energy for phagocytosis. However, since the seminal work by Sbarra and Karnovsky (1959), it has been recognized that the increased oxygen consumption is insensitive to cyanide and azide and thus is not used in mitochondrial oxidative metabolism. As discussed below, more recent studies have shown that the oxygen consumed in the burst is used almost exclusively for the production of oxygen radical species.

A number of different reactive oxidants can be found after stimulation of the respiratory burst in living systems. The key element in the generation of all of these radical species is an oxidase, the NADPH oxidase, which catalyzes the one-electron reduction of molecular oxygen to superoxide anion (O_2^-) according to the equation

$$\tfrac{1}{2}NADPH + O_2 \rightarrow \tfrac{1}{2}NADP^+ + \tfrac{1}{2}H^+ + O_2^- \tag{1}$$

When activated, this enzyme system is confined to the plasma membrane and preferentially utilizes NADPH as the physiologic electron donor, though it can also utilize NADH as a substrate (see below). In contrast to the lack of mitochondrial metabolism during the respiratory burst, it was demonstrated in 1959 (Sbarra and Karnovsky, 1959) that there was a dramatic increase in cytosolic HMP shunt activity, with conversion of $NADP^+$ to NADPH. More recent studies have shown that there is a stoichiometric relationship between oxygen consumed in the burst and CO_2 produced by the HMP shunt (Borregaard *et al.*, 1984a). These studies lend further support to the importance of the HMP shunt as a source of electrons for the respiratory burst and to the role of NADPH as the physiologic electron donor. Thus, all of the cellular activities associated with the burst—nonmitochondrial oxygen consumption, NADPH utilization and subsequent HMP shunt activity, and oxygen radical production—can be attributed to the activation of a single enzyme system, the respiratory burst NADPH oxidase.

2.2. Oxygen Radical Species Produced in the Respiratory Burst

Some stimuli of the respiratory burst appear to cause a dissociation between oxygen uptake and O_2^- formation (Curnutte and Tauber, 1983). However, the vast majority of stimuli of the respiratory burst appear to act solely through the NADPH oxidase to promote the one-electron reduction of oxygen to O_2^-. Thus, with most stimuli of the burst, all additional oxygen that is consumed by the cell is converted to O_2^- before formation of additional oxidants (Root and Metcalf, 1977; Badwey and Karnovsky, 1979; Badwey *et al.*, 1980). The rate of O_2^- generation can be measured by a number of spectrophotometric or fluorometric techniques; one of the more common assays is the superoxide dismutase (SOD)-inhibitable reduction of ferricytochrome *c* (Babior *et al.*, 1973). Once formed, O_2^- may undergo spontaneous dismutation to H_2O_2 according to the reaction

$$O_2^- + O_2^- + 2H^+ \rightarrow H_2O_2 + O_2 \qquad (2)$$

This reaction occurs rapidly in acidic environments (Bielski and Allen, 1977) but is dramatically accelerated in the presence of SOD (Fielden *et al.*, 1974). The existence of a cytochrome copper/zinc-requiring form of this enzyme (McCord and Fridovich, 1969) may serve to protect mammalian cells from the effects of O_2^- by enhancing its dismutation to H_2O_2. The resulting H_2O_2 is efficiently scavenged in mammalian cells by catalase (Olofsson and Olsson, 1977) or by the glutathione peroxidase–reductase system (Reed, 1969; Strauss *et al.*, 1969; Spielberg *et al.*, 1979). While the immediate products of the respiratory burst, O_2^- and H_2O_2, are themselves relatively innocuous, in the extracellular environment they are used by the neutrophil as a substrate to produce more reactive oxidant species, including oxidized halogens, OH·, and ultimately additional oxidants, including lipid peroxides and mono- and dichloroamines.

One major pathway in the production of these additional oxidizing species involves the enzyme myeloperoxidase (MPO). This peroxidase [molecular weight (MW) 150,000] is found exclusively in the azurophilic primary granules of neutrophils, and the spectral characteristics of its iron–chlorin-prosthetic group give these granules their unique green color (Cramer et al., 1985). This enzyme utilizes H_2O_2 produced by the dismutation of O_2^- to generate oxidized halogens:

$$H_2O_2 + Cl^- \rightarrow HOCl + OH^- \tag{3}$$
$$H_2O_2 + Br^- \rightarrow HOBr + Br^- \tag{4}$$

In neutrophils and monocytes, chloride appears to be the physiologic substrate for the peroxidase, although bromide is the preferred substrate for a different peroxidase found in eosinophils (Klebanoff, 1968; Kanofsky et al., 1984; Weiss et al., 1986). The resulting hypohalite ions are highly reactive and can directly oxidize an assortment of biologic substrates as well as combine with ambient amine groups to form a wide variety of chloroamines (Thomas, 1979a; Albrich et al., 1981). These oxidants have a wide range of bioactivities, ranging from highly reactive and short-lived species with a probable role in host defense (Thomas, 1979b) to those which are marginally reactive and long-lived (e.g., taurine chloramine) (Weiss et al., 1982, 1983; Thomas et al., 1985). The more long-lived products of the respiratory burst have been implicated as a persistent source of oxidants in the tissues and may be able to cause damage over much longer distances (e.g., from the cell surface to the nucleus) than more reactive oxidants (Weiss et al., 1983).

Although the role of the neutrophil MPO in host defense is well established, patients known to have complete MPO deficiency are typically asymptomatic or have only minor fungal infections (Kitahara et al., 1981; Parry et al., 1981). MPO-deficient neutrophils are able to kill bacteria but at a reduced rate (Lehrer and Cline, 1969). Thus, the production of MPO-independent oxidant species from O_2^- must also be crucial for neutrophil activity. A variety of oxygen radicals may be derived from O_2^- by such MPO-independent paths and can be sought following the respiratory burst. The species for which there is the most direct evidence is $OH\cdot$, which can be formed by a direct Haber–Weiss reaction between H_2O_2 and O_2^- (Haber and Weiss, 1934; McClune and Fee, 1976):

$$O_2^- + H_2O_2 \rightarrow OH\cdot + OH^- + O_2 \tag{5}$$

In pure metal-free water this reaction is very sluggish, but it is greatly accelerated in the presence of iron (McCord and Day, 1978; Graf et al., 1984):

$$Fe^{3+} + O_2^- \rightarrow Fe^{2+} + O_2 \tag{6}$$
$$Fe^{2+} + H_2O_2 \rightarrow Fe^{3+} + OH\cdot + OH^- + O_2 \tag{7}$$

$$\text{Net: } O_2^- + H_2O_2 \rightarrow OH\cdot + OH^- + O_2 \tag{8}$$

The exact source of the iron that cycles in this reaction, known as the metal-catalyzed Haber–Weiss reaction, remains unknown, although both lactoferrin and ferritin are candidates. Lactoferrin is present in the specific (secondary) granules of neutrophils and can enhance $OH\cdot$ production both in whole cells and in particulate NADPH oxidase systems (Ambruso and Johnston, 1981). However, the ability of HL-60 cells (a myeloid leukemia-derived cell line) to generate $OH\cdot$ in the apparent absence of lactoferrin (Newburger and

Tauber, 1982) suggests that alternate sources of iron may be available. Ferritin has been shown to release Fe^{2+} on exposure to O_2^- and is therefore an attractive candidate for participation in the formation of OH· (Halliwell and Gutteridge, 1986; Rohrer *et al.*, 1987).

The presence of OH· during the respiratory burst can be determined by the use of the spin-trapping agent 5,5-dimethyl-1-pyrroline-1-oxide (DMPO), which combines with OH· to form a long-lived nitroxide adduct DMPO/OH (Rosen and Klebanoff, 1979; Green *et al.*, 1979) by bioassays for OH·-mediated tissue damage (Fantone and Ward, 1982; Braughler *et al.*, 1986; Shanblatt and Revzin, 1987) or by assays for the OH·-dependent production of ethylene from methional (Weiss *et al.*, 1978). However, in each of these assays, the reagents may interact with other oxygen radical species and thus are not truly specific for OH·. Increased specificity for OH· has been claimed by performing these assays in the presence of OH· scavengers such as mannitol, benzoate, thiourea, or the combination of SOD and catalase. The use of these agents when studying stimulated neutrophils has provided evidence suggesting that OH· is produced in the respiratory burst. However, the scavengers themselves are not completely specific for OH·, and therefore it is not possible to definitively state that OH· is produced in the burst. In summary, it appears that potent oxidants that behave much like OH· are being produced following the immediate formation of O_2^- and H_2O_2, but the short half-life of these species makes them difficult to measure.

3. THE ACTIVATED NADPH OXIDASE

The earliest studies of the enzyme responsible for O_2^- production during the respiratory burst were confused by uncertainty about the true products of the burst and its physiologic substrate. In 1964, an H_2O_2-producing, NADPH-requiring enzyme found in the $19,000 \times g$ pellet of neutrophil homogenates was first described (Rossi and Zatti, 1964). Enzyme activity was greater in preparations from phagocytizing cells than in preparations made from unstimulated cells. Enzyme activity was greatly enhanced in the presence of Mn^{2+}, a finding that caused concern because the activity appeared to be nonphysiologic. This concern was allayed with the demonstration that O_2^- and not H_2O_2 was the initial product of the burst (Babior *et al.*, 1973) and that the apparent Mn^{2+} requirement was an artifact resulting from an H_2O_2-producing oxygen radical-mediated chain reaction. This chain reaction was initiated when Mn^{2+} was oxidized to Mn^{3+} by the conjugate acid of O_2^- (HO_2·; pK_a 4.8) with the concomitant formation of H_2O_2, and this Mn^{3+} subsequently reacted with NADPH to regenerate Mn^{2+} and form NADP·, which in turn converted oxygen to O_2^- and was itself oxidized to $NADP^+$ (Patriarca *et al.*, 1975; Curnutte *et al.*, 1976). Removal of Mn^{2+} from the cell preparations demonstrated that O_2^- was the initial product of the respiratory burst and prepared the way for subsequent studies that have accurately described the properties of the respiratory burst NADPH oxidase. These studies have confirmed that the NADPH oxidase described by Rossi and Zatti (1964) is the "true" respiratory burst oxidase; this activity is detectable only in phagocytic cells* and is present only following stimulation of the cell with a

---

*A very low level of respiratory burst oxidase activity has recently been found in B lymphocytes (Volkman *et al.*, 1984).

suitable agent. Only minimal if any activity can be detected in preparations from resting, nonstimulated cells.

3.1. Subcellular Localization of the Oxidase

Once the nature of the oxidants produced during the respiratory burst was recognized, it became possible to localize the activated oxidase through cytochemical techniques. These studies (Briggs *et al.*, 1977; Badwey *et al.*, 1980; Ohno *et al.*, 1982; Wakeyama *et al.*, 1983) relied on the ability of H_2O_2 to react with indicators such as Ce^{3+} to form an insoluble precipitate, visible by microscopy, at the site of oxidant formation. Oxidant production was found to be localized to the neutrophil cell surface or the phagosome following stimulation with soluble or particulate agents, respectively.

The ability to detect oxidase activity in sonicated homogenates from stimulated neutrophils also permitted the use of subcellular fractionation techniques to further localize the NADPH oxidase to the neutrophil plasma membrane. This work (Dewald *et al.*, 1979; Cohen *et al.*, 1980; Bellavite *et al.*, 1982; Gabig *et al.*, 1982; Borregaard *et al.*, 1983) demonstrated that oxidase activity comigrated with the plasma membrane fraction when cell homogenates were sedimented during gradient centrifugation. Oxidase activity was not found in the fractions containing markers for the azurophilic granules (MPO), or the specific granules (lysozyme or cobalamine-binding protein). The addition of increasing concentrations of detergent to particulate membrane fractions obtained from stimulated neutrophils also established that the oxidase in those preparations exhibits latency. In studies using Triton X-100, there was an abrupt increase in NADPH oxidase activity at the critical micellar concentrations (CMC) of the detergent (Babior *et al.*, 1981). The 2.5-fold increase in the rate of O_2^- formation at the detergent CMC suggested that the oxidase is vectorially oriented in the plasma membrane; 40% of the vesicles were oriented with the NADPH-binding site facing the external media, and the remaining vesicles were oriented so that the NADPH-binding site faced the interior of the vesicle and was inaccessible to substrate. As the detergent concentration reached its CMC, the vesicular structure was lost and substrate was able to gain access to all oxidase NADPH-binding sites.

Confirmation of the directional orientation of the oxidase was obtained by additional studies on phagosomal oxidase preparations (Babior *et al.*, 1981). Since all of the phagosomal membranes are oriented in the same direction (as opposed to the neutrophil sonicates, which have a random orientation), oxidase latency should be an all-or-none phenomenon. The complete absence of latency in these preparations established that the NADPH-binding site of the oxidase on phagosomal membranes is directed outward, consistent with an orientation toward the cytoplasm in the intact cell.

3.2. Stoichiometry and Cofactor Requirements of the Active Oxidase

Studies that used detergent to permeabilize phagocytic cells yielded much of the initial information about the catalytic properties of the NADPH oxidase. Upon the addition of sodium deoxycholate to stimulated neutrophils or macrophages, O_2^- formation by the oxidase disappears as a result of loss of cytosolic substrate upon permeabilization of the plasma membrane (Babior and Kipnes, 1977; Bellavite *et al.*, 1981; Nakamura *et al.*, 1981; Berton *et al.*, 1982). Activity can be restored by the addition of exogenous

NADPH, and a stoichiometric correlation exists between the production of O_2^- by the intact cell and by the permeabilized cell preparation following the addition of NADPH. Thus, the permeabilized cell preparation has an absolute requirement for the addition of an exogenous pyridine nucleotide substrate. In such cell preparations, as well as in particulate oxidase preparations from stimulated neutrophils or from activated macrophages, the oxidase demonstrates a K_m for NADPH of 30–80 μM (Babior *et al.*, 1976; Gabig and Babior, 1979; Lew *et al.*, 1981; Seger *et al.*, 1983; Tamura *et al.*, 1988). In contrast, the K_m of these preparations for NADH is substantially greater (250–930 μM) and well above the expected cytosolic concentrations in whole cells. Interestingly, the oxidase activity that can be detected at low levels in unstimulated cells appears to have a low affinity for NADPH ($K_m = 4000$ μM) (Patriarca *et al.*, 1971). Although this value may be due to a second enzyme with different substrate specificities, modulation of the affinity of the oxidase for NADPH has also been observed during the transition of monocytes and macrophages from a dormant to an activated state (Sasada *et al.*, 1983; Tsunawaki and Nathan, 1984; Berton *et al.*, 1985; Cassatella *et al.*, 1985). Upon activation of the macrophage by incubation with gamma interferon (IFN-γ), the level of oxidase activity after stimulation with phorbol esters is much greater (see Section 5.1.2) (Cassatella *et al.*, 1985). Moreover, the K_m for NADPH in cells not treated with IFN-γ was 150–300 μM, as opposed to the value of 30–50 μM seen in IFN-γ-treated macrophages. Similar differences in the affinity of the oxidase for NADPH have been demonstrated between resident murine peritoneal macrophages and those elicited and activated by the intraperitoneal injection of *Corynebacterium parvum* (Berton *et al.*, 1985) or bacterial lipopolysaccharide (Sasada *et al.*, 1983). Modulation of the affinity of the oxidase for NADPH occurs over a physiologic range and may reflect an important mechanism for regulation of oxidase activity.

Oxidase affinity for oxygen has been measured in particulate oxidase preparations and in intact cells (Babior and Kipnes, 1977; Gabig and Babior, 1979). In these studies, the K_m for oxygen was 0.01–0.03 mM, corresponding to a partial pressure of oxygen of 7 mm Hg in an aqueous environment. The oxidase activity in these preparations has a broad pH optimum centered at pH 7.3. This latter value is somewhat surprising in view of the acid pH normally found inside phagosomal vesicles, the presumed physiologic milieu of the oxidase, but this may reflect a requirement for phagocyte O_2^- production over a wide range of tissue environments rather than just in the interior of the phagosome.

4. IDENTIFICATION OF COMPONENTS OF THE OXIDASE

The ability to obtain particulate membrane preparations containing NADPH oxidase activity allowed the initiation of studies that described the catalytic properties of the oxidase and also prepared the way for further efforts to purify the active oxidase. The recovery of soluble oxidase activity from these membrane particles after treatment with a nonionic detergent marked the first step in that purification (Gabig *et al.*, 1978). In these studies, human neutrophils were stimulated with opsonized zymosan, sonicated, and ultracentrifuged to obtain a particulate NADPH oxidase preparation. These particles were then suspended in 0.4% Triton X-100 for 30 min and centrifuged (105,000 × g for 30 min). The supernatant contained NADPH oxidase activity that was not retained by a 300-

kDa-cutoff filter. Additional studies demonstrated that more than 40% of the activity present in the particles could be recovered in a soluble form, and the solubilized oxidase had catalytic properties similar to those of the particulate preparations (Babior and Peters, 1981).

The solubilized oxidase had properties that included (1) marked instability of the preparation, (2) a requirement for phospholipid, and (3) flavin dependence (Gabig and Babior, 1979). The phospholipid requirement could be met by phosphatidylethanolamine but not by phophatidylserine or phosphatidylcholine. A similar requirement for phospholipid was not seen in preparations in which deoxycholate was the solubilizing agent, either because deoxycholate itself met that requirement or because the deoxycholate allowed the formation of a micelle that contained a greater proportion of membrane lipid. The flavin requirement seen in Triton X-100-solubilized preparations was not unexpected; a similar requirement had previously been seen in particulate oxidase preparations treated with permeabilizing concentrations of detergent (Babior and Kipnes, 1977). The instability of the solubilized preparations also posed a number of problems during attempts at purification. The particulate oxidase preparations were known to be stable for many weeks if frozen at $-70°C$, although at $4°C$ the $\tau\frac{1}{2}$ was 1–2 days. In contrast, the solubilized preparations had a markedly shorter $\tau\frac{1}{2}$ at $4°C$ and immediately lost oxidase activity if exposed to even modest concentrations of salt. Therefore, many of the usual techniques of protein purification (e.g., ion exchange chromatography) were fruitless since they resulted in loss of oxidase activity.

4.1. Attempts to Isolate the NADPH Oxidase from Stimulated Cells

Gel filtration of a deoxycholate–Lubrol-solubilized oxidase preparation from guinea pig neutrophils showed that NADPH oxidase activity was eluted in the void volume of the column (exclusion limit, 1.2×10^3 kDa), suggesting that the oxidase was part of a high-molecular-weight complex (Bellavite et al., 1983, 1984; Serra et al., 1984). Cytochrome b_{558} was found to comigrate with oxidase activity in these experiments. The investigators then took the column eluate and, after concentration by ultrafiltration, performed isopycnic sedimentation over a glycerol gradient, followed by precipitation in 0.4 M NaCl of a pellet enriched in oxidase activity. A fivefold enrichment in oxidase activity, to >200 nmol $O_2^-\cdot$ $min^{-1}\cdot mg^{-1}$, was achieved, with a progressive enrichment in a 32-kDa polypeptide that was felt to be a component of the oxidase. There was an identical enrichment of cytochrome b_{558}, suggesting that it also was a component of the enzyme complex comprising the active oxidase.

Human neutrophil NADPH oxidase, solubilized with deoxycholate, was found to be retained by a red dye–agarose (a nucleotide analog) column but could not be eluted by buffers containing nucleotide or salt (Markert et al., 1985; Glass et al., 1986; Babior, 1987). An abrupt change to a buffer devoid of detergent did result in the elution of small amounts of NADPH oxidase; the eluted enzyme was found to have an extremely high specific O_2^--forming activity, 5000–10,000 nmol O_2^- min^{-1} mg^{-1}, and had kinetic properties identical to those described earlier for the particulate oxidase. This purified oxidase preparation contained large amounts of FAD, no ubiquinone, and minimal cytochrome b_{558} (<0.05 moles/mole of active sites). Nondenaturing gel electrophoresis of the final preparation revealed a two closely spaced bands, only the lower of which contained flavin; SDS-PAGE disclosed three major bands of 65–67, 48, and 32 kDa (Markert *et al.*, 1985).

Given the lack of agreement between these two studies about the composition of the oxidase, it is interesting to consider the results of Doussiere and Vignais (1985), who extracted NADPH oxidase activity from PMA-stimulated bovine neutrophils. These investigators used ion-exchange chromatography of this membrane extract, followed by gel filtration and isoelectric focusing (pI of active fraction = 5.0) to obtain a final highly purified preparation. The final preparation was moderately active (≈ 200 nmol O_2^- $\min^{-1}$ mg^{-1}) and, like the particulate oxidase, could directly reduce oxygen to O_2^- in the presence of NADPH (K_m for NADPH = 30 μM) and was inhibitable by Cibachron blue and mersalyl but not cyanide. Surprisingly, they were unable to detect cytochrome b_{558} or flavin in their final preparation. In the initial description of this purification, there was no demonstration of a difference between resting and stimulated membranes in the recovery of this 65-kDa protein and there was a major loss of NADPH activity during solubilization. In subsequent studies (Doussiere *et al.*, 1986), the same investigators showed that an NADPH analog, *N*-4-azido-2-nitrophenyl aminobutyryl NADP$^+$, acted as a competitive inhibitor of NADPH oxidase activity in detergent-solubilized preparations and, when radiolabeled, covalently bound to a number of different polypeptides, including one of 65 kDa. This labeled 65-kDa polypeptide was shown to be the only labeled protein in the final purified oxidase preparation described above. The investigators also reported that the amount of the 65-kDa labeled polypeptide was greatly reduced in the membranes prepared from nonstimulated cells, evidence strongly implicating this 65-kDa protein as the NADPH-binding component of the oxidase.

Isoelectric focusing has also been used to isolated NADPH oxidase activity from an octyl glucoside-solubilized preparation of stimulated neutrophil membranes (Kakinuma *et al.*, 1987). A sharply focused band at pI 5.0 was found to reduce nitroblue tetrazolium (NBT) in the presence of NADPH and, when extracted from the gel, had a K_m for NADPH of 30 μM, contained FAD but not cytochrome *b*, and had an apparent mass of 67 kDa. Although the investigators did report an increase in the amount of this pI 5.0 protein in the membranes of stimulated versus resting cells, the NADPH oxidase activity was insensitive to SOD, suggesting that the isolated enzyme was able to transfer electrons only to cytochrome *c* and not to oxygen. Freeze–thawing of the preparation restored sensitivity to SOD, suggesting that the enzyme was competent to transfer electrons to oxygen under some micellar conditions; however, it is also possible that the investigators were isolating a cytochrome *c* reductase rather than the intact NADPH oxidase. In this context, it is interesting to examine the NADPH-dependent cytochrome *c* reductase of guinea pig neutrophils (Kojima *et al.*, 1987), which was put forth as a potential proximal component of the NADPH oxidase electron transport chain. This enzyme was extracted with Triton X-100 from resting membrane preparations and subjected to ion-exchange chromatography and to affinity chromatography over 2',5'-ADP–agarose and Bio-Gel HTP. The purified enzyme had a mass of 80 kDa and contained equimolar amounts of both FAD and FMN. The K_m for NADPH of the purified enzyme was found to be 3.6 μM, as opposed to a previously reported value of 25 μM (Sakane *et al.*, 1984). Although the sensitivity of the purified reductase to several inhibitors was similar to that described for the NADPH oxidase of particulate human neutrophil preparations, and although the isolated reductase was able to catalyze the reduction of oxygen to O_2^- in the presence of a purified preparation of cytochrome b_{558}, the difference in substrate affinities suggests caution in accepting this enzyme as a component of the respiratory burst oxidase. If it is not a component of the oxidase, its presence in neutrophil membranes certainly suggests caution in the interpretation of labeling studies with NADPH analogs.

The use of other affinity labels to identify a component of the NADPH oxidase has also been reported. First, the 2′,3′-dialdehyde derivative of NADPH (NADPH-dialdehyde) was shown to compete with NADPH for nucleotide-binding sites and function as a substrate for the porcine neutrophil NADPH oxidase with a K_m of ≈ 35 μM (Umei *et al.*, 1986). When incubated over longer periods with oxidase preparations, the NADPH-dialdehyde inhibited the oxidase activity presumably as a result of the formation of a Schiff base between the nucleotide and an amine group at the enzyme-active site. When tritiated sodium cyanoborohydride (NaCNBH$_3$) was added to reduce and stabilize the Schiff base, a 66-kDa polypeptide was found to be labeled. In subsequent experiments, the same group was not able to find a difference between resting and stimulated cells and found no difference in the labeling of this 66-kDa polypeptide in preparations from CGD neutrophils (Umei *et al.*, 1987). Second, diphenylene iodonium (DPI), which had previously been shown to be an inhibitor of NADH : ubiquinone reductase (Ragan and Bloxham, 1977), was also shown to be a rapid inhibitor of O_2^- production by the porcine neutrophil NADPH oxidase in solubilized preparations or in intact cells (Cross and Jones, 1986). Oxidase preparations inhibited by DPI were unable to reduce either cytochrome b_{558} or flavin, suggesting that the point of inhibition was proximal to the binding sites of both of these substances in the postulated electron transport chain of the oxidase and possibly proximal to the putative NADPH-binding flavoprotein. When radioiodinated, DPI was shown to be incorporated into a 45-kDa polypeptide, with less prominent labeling of 70-, 40-, and 36-kDa polypeptides. This incorporation was reduced in the presence of NADPH in soluble oxidase preparations, but no difference was seen in the extent of incorporation in membranes from resting or stimulated cells.

Although it is possible that the disparity between the results of some of these studies is due to extraction or isolation of enzyme systems other than the respiratory burst NADPH oxidase, or may be in part due to interspecies differences, none of these explanations is completely satisfying. A more likely interpretation is that the process of extracting the oxidase from cells with some detergents, and the subsequent steps in the isolation of the "active oxidase" moiety, causes the dissociation of a multicomponent enzyme complex. Under some circumstances, the proximal components of this enzyme complex may be able to effect transfer of electrons from NADPH to various acceptors. In particular, the presence of even modest concentrations of a flavin in an assay may facilitate such a transfer of electrons. This kind of nonphysiologic electron transport is likely to be quite dependent on assay conditions, on the presence of lipids that may serve to stabilize a component, and the concentration of detergent; variations between these factors are common in the "NADPH oxidase" preparations reported in the literature.

4.2. Activation of the Oxidase in Resting Phagocyte Membranes in a Cell-Free System

A dramatic step forward in the understanding of the oxidase occurred with the demonstration that the oxidase could be activated in resting phagocyte membranes in a cell-free system derived from phagocytic cells (Bromberg and Pick, 1984; Heyneman and Vercauteren, 1984; Curnutte, 1985; McPhail *et al.*, 1985). Unexpectedly, this activation process required the simultaneous presence of both plasma membranes (either particulate or solubilized) and cytosol. Simply stated, resting cells could be disrupted and the homogenates fractionated into membrane (sedimenting) and cytosolic (nonsedimenting at $230,000 \times g$) fractions. Membrane oxidase activity could be elicited, in the presence of the cytosolic fraction and Mg^{2+}, with the addition of an activating agent (typically an amphophilic detergent or lipid such as SDS, oleic acid, or arachidonic acid). The rate of O_2^- production increased 2-fold in the presence of guanine nucleotide and 10-fold in the presence of Mg^{2+}. The relevance of these cell-free oxidase activation systems is underscored by the observations that all CGD patients studied thus far have been shown to have a defect in either the membrane or cytosolic component of their neutrophils.

4.2.1. Cytosolic Components of the Oxidase

In the initial descriptions of the cell-free systems, the "cytosolic factor" was found to be unique to phagocytic cells (Ligeti *et al.*, 1988), heat labile, and nondialyzable. Attempts at gel filtration of cytosol suggested that the cytosolic activity was part of a high-molecular-weight complex (mass of >300 kDa) (Curnutte *et al.*, 1987; Fujita *et al.*, 1987); under some circumstances, cytosol factor activity was eluted from columns with lower molecular weights (10–50 kDa) (Fujita *et al.*, 1987). The activation process in the cell-free system was independent of calcium and ATP, although protein kinase C (PKC) and exogenous ATP could substitute for cytosol in some settings (Cox *et al.*, 1985), supporting a role for phosphorylation in the intact cell. However, chromatography of crude cytosol results in the separation of PKC from cytosol activity in the cell-free system (Curnutte *et al.*, 1987).

An examination of the kinetics and cofactor requirements of a fully soluble cell-free system in our laboratory has demonstrated a first-order relationship between the amount of neutrophil membrane added to the activation reaction and the resulting O_2^--forming activity (Babior *et al.*, 1988a). Surprisingly, there was a 2.5-order exponential relationship between the amount of cytosol added to the reaction and the final oxidase activity. This data supported the speculation that there was more than one cytosolic component of the oxidase and suggested that a minimum of three kinetically distinct entities were interacting during the activation process when deoxycholate was present to dissociate any complexes. By testing preparative cytosol fractions in the presence of membranes, Mg^{2+}, and a limiting amount of unfractionated neutrophil cytosol (to provide threshold amounts of all the cytosolic components), the presence of single cytosolic components of the oxidase could be detected by their ability to augment O_2^- formation. This approach has been used with success by three independent groups of investigators to demonstrate the presence of at least three different cytosolic oxidase components when crude cytosol preparations are subjected to ion-exchange chromatography (Nunoi *et al.*, 1988) or four different components when the cytosol is subjected to preparative isoelectric focusing (Curnutte *et al.*, 1989a). Patients with autosomal recessive inheritance cytochrome b_{558}-positive CGD, possessing cytosolic defects in the cell-free system ($\approx 30\%$ of all CGD), can be divided into two categories on the basis of complementation experiments (Nunoi *et al.*, 1988; Volpp *et al.*, 1988). Similarly, in the cell-free assay systems, different preparative fractions obtained by either isoelectric focusing (Curnutte *et al.*, 1989a) or ion-exchange chromatography (Nunoi *et al.*, 1988; Bolscher *et al.*, 1989) are able to restore the defective activity in two different CGD cytosols. The most direct interpretation of these data suggests that the cytosolic components of the oxidase exist in the form of a complex in the cytosol of resting neutrophils, but this complex can be dissociated into its various constituents and the separated components retain the ability to participate in the activation of the oxidase in the cell-free system.

The fact that the cytosolic components of the oxidase normally are part of a larger complex has also been exploited in additional purification attempts. Volpp *et al.* (1988) found that cytosolic components of the oxidase (presumably in the form of a complex) would bind to a GTP–agarose column, and the bound oxidase components could be eluted with buffer containing ATP or GTP. The fractions eluted from the column by either of these nucleotides were still able to participate in the cell-free system, suggesting that all necessary components were binding and eluting together as a complex. The active fractions eluted from this column contained a large number of proteins by SDS-PAGE. Despite the impurities, when the mixture of proteins was used to raise a polyclonal antibody, blotting studies demonstrated that the antibody preferentially bound to two cytosolic polypeptides of 47 and 65 kDa in size; these proteins were found only in the cytosol of phagocytes. Furthermore, the 47-kDa polypeptide was found to be absent in the majority of the patients with autosomal inheritance cytosol-deficient CGD. The 65-kDa polypeptide was absent in the cytosol from one patient with autosomal inheritance CGD who was found to have normal levels of the 47-kDa polypeptide; this patient's cytosol was found to complement that of the patients missing the 47-kDa protein.

In the cell-free system, it was shown that the affinity for NADPH of the oxidase generated in the activation reaction was dependent on the amount of cytosol present in the reaction. At low concentrations of cytosol, the oxidase had a decreased affinity for NADPH ($K_m = 160$ μM); when increased amounts of cytosol are combined with the same amount of resting neutrophil membranes, the K_m for NADPH falls to ≈ 35 μM (Babior *et al.*, 1988a; Curnutte *et al.*, 1989b). The affinity of the oxidase for NADPH in the cell-free system appears to have a bimodal distribution at these two values, rather than varying continuously over a range of values. These observations suggested that a cytosolic component or cofactor is involved in the modulation of the affinity of the oxidase for its NADPH substrate. This observation is particularly interesting in view of the similarity between K_m values observed in the cell-free system and those measured in the monocyte at various states of activation.

Additional evidence linking a cytosolic component with the binding of NADPH to the oxidase was obtained in the cell-free system with the use of the affinity label NADPH-dialdehyde. The active oxidase, as described above, is quite sensitive to inhibition by NADPH-dialdehyde; however, the incubation of resting neutrophil membranes with NADPH-dialdehyde and $NaCNBH_3$ did not alter the ability of those membranes to be activated by normal neutrophil cytosol. In contrast, if cytosol is incubated with NADPH-dialdehyde, it is unable to participate in the cell-free system. The inactivation of a cytosolic component of the oxidase follows first-order kinetics and can be blocked by the addition of unmodified NADPH to the incubation [$K_s = 36$ μM] (Smith *et al.*, 1989a). The k_i for NADPH-dialdehyde is 42 μM, a value which suggests that the catalytic NADPH-binding site of the oxidase may be contained on a cytosolic protein that must then be translocated to the plasma membrane during the process of activation (and assembly) of the oxidase. Experiments using cytosol in which the NADPH-binding component was occupied and inhibited by NADPH-dialdehyde demonstrated that only one cytosolic oxidase component was affected and the other cytosolic oxidase components appeared to remain functional (Smith *et al.*, 1989b). In particular, the NADPH-dialdehyde-inhibited cytosol was still able to restore activity to CGD cytosol deficient in the 47-kDa polypeptide described above. Despite the retention of activity by some components of the oxidase, the cytosol that was inactivated by NADPH-dialdehyde was not able to affect the K_m of the oxidase when added to low concentrations of normal cytosol in an activation reaction. Thus, in addition to being a cytosolic protein, the oxidase NADPH-binding subunit appears to be directly involved in control of the affinity of the oxidase for its physiologic substrate, NADPH.

4.2.2. Identification of Components of the Oxidase Present in Resting Membranes

The use of a cell-free activation system to identify membrane components of the oxidase has not received as much attention as has cytosol. Treatment of a particulate preparation of unstimulated guinea pig macrophage membranes with octyl glucoside allowed solubilization and recovery of membrane-associated oxidase activity assayed in a cell-free system (Pick *et al.*, 1987). Although the soluble membrane fraction contained cytochrome b_{558}, results of attempts to demonstrate anaerobic reduction of the cytochrome with the cell-free system were equivocal. When the components of the soluble membrane preparation were separated over a Cibacron blue–agarose column and an HPLC sizing column, the membrane-associated activity in the cell-free system separated from fractions containing spectrally detectable cytochrome b_{558}. The fractions containing the greatest cell-free oxidase activity were reported to contain a single polypeptide of 29-kDa mass. In contrast to these results, which suggest that the cytochrome b_{558} may not be an important component of the oxidase in the cell-free system, partially purified preparations of cytochrome b_{558} are able to restore missing activity to solubilized membranes from cytochrome b_{558}-deficient CGD patients of both X-linked and autosomal inheritance (Curnutte *et al.*, 1988a). Interestingly, more highly purified preparations of cytochrome b_{558} were not able to reconstitute activity, suggesting the possibility that an additional component of the oxidase had been removed during purification.

5. POSTULATED COMPONENTS OF THE NADPH OXIDASE

The components of the NADPH oxidase and its attendant electron transport chain are unknown, but it is possible to postulate a structure for the oxidase and its cofactors that corresponds to the available data. Although there remains some controversy, it is likely at a minimum that electrons are transported from NADPH through FAD to the heme group of cytochrome b_{558} and then to molecular oxygen to form O_2^- (Figure 2). It is also clear that this electron transport chain is not intact in the resting cell and is only assembled *in toto* following stimulation of the phagocyte. A major question, just now being answered, is what polypeptides serve as structural and/or additional electron-carrying components of

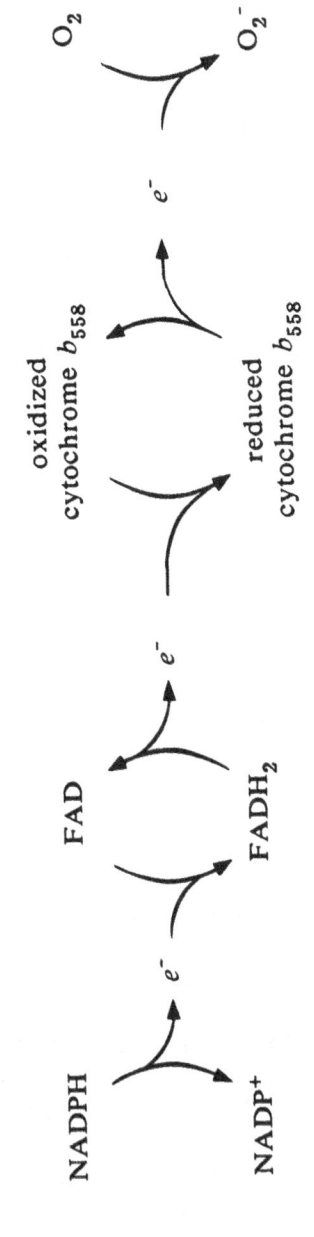

NADPH	FAD	oxidized cytochrome b_{558}		O_2
NADP$^+$	FADH$_2$	reduced cytochrome b_{558}		O_2^-
Midpoint Potential	−208 mV	−245 mV		−330 mV

FIGURE 2. Postulated electron-carrying components of the NADPH oxidase. Electrons are passed from NADPH to FAD, from FADH$_2$ to cytochrome b_{558}, and from the cytochrome to oxygen, to form O_2^-.

the oxidase transport chain. On the basis of the data described above as well as the additional information presented below, we suggest the following components.

5.1. A Membrane-Associated Cytochrome *b*

A unique mammalian *b*-type cytochrome (designated either cytochrome b_{558} to reflect its alpha-band absorption at 558 nm or b_{-245} to describe its midpoint potential of -245 mV) was first described in human neutrophils in 1978 (Segal *et al.*, 1978; Segal and Jones, 1978). A similar cytochrome had previously been detected in horse (Hattori, 1961; Ohta *et al.*, 1966) and rabbit (Shinagawa *et al.*, 1966) neutrophils. Subsequent studies showed that it could be detected only in human cells of phagocytic lineage, including monocytes, macrophages, eosinophils, and neutrophils (Segal *et al.*, 1981). The functional relationship between this cytochrome and the NADPH oxidase was firmly established by studies which demonstrated that cytochrome b_{558} was absent in the membranes of most X-linked inheritance (type I; Table I) CGD patients. The recognition that the oxidase contained a critically important cytochrome had a number of important implications. In particular, it introduced the concept that the NADPH oxidase is a complex multicomponent enzyme system rather than a single catalytic protein.

5.1.1. Cytochrome b_{558} Structure and Function

Cytochrome b_{558} has recently been purified and its polypeptide structure determined (Parkos *et al.*, 1987). On the basis of extensive experimental evidence, cytochrome b_{558} has been shown to be a heterodimer formed by a small nonglycosylated 22-kDa alpha subunit and a larger 91-kDa glycosylated beta subunit (Segal, 1987; Parkos *et al.*, 1987, 1988a); the 22-kDa and 91-kDa subunits are tightly associated in a 1 : 1 stoichiometry (Parkos *et al.*, 1988b). The 91-kDa subunit is highly glycosylated (15–20%) with carbohydrate linkages predominantly of the N-linked high-lactosamine form (Parkos *et al.*, 1987; Segal, 1988, 1989). The deglycosylated polypeptide core of the 91-kDa subunit is estimated to be 50–55 kDa in mass (Parkos *et al.*, 1987). The spectrum of the reduced cytochrome is typical of other *b* cytochromes, with characteristic absorption peaks at 558 nm (alpha), 528 nm (beta), and 428 nm (gamma or Soret).

Purification and separation of the 22-kDa and 91-kDa subunits unfortunately results in displacement of the heme prosthetic group (Parkos *et al.*, 1987). *A priori,* however, the 22-kDa subunit of the cytochrome may be the more likely candidate to contain the heme-binding site based on its size and lack of glycosylation, both of which are more characteristic of the known heme-binding regions of other *b*-type cytochromes. Supporting this idea is a recent report of a purification of the 22-kDa subunit with its heme group attached (Yamaguchi *et al.*, 1989). It is also possible that the heme is bound by both cytochrome b_{558} subunits, similar to hemoglobin. The number of prosthetic heme groups bound by each cytochrome b_{558} heterodimer is uncertain, but a 1 : 1 stoichiometry is unlikely (Parkos *et al.*, 1988a).

Initially, cytochrome b_{558} was found to be distributed between the neutrophil plasma membrane and specific granule (Segal *et al.*, 1978; Segal and Jones, 1978, 1979b, 1980b). Cell fractionation experiments using nitrogen cavitation and Percoll density gradient centrifugation showed that 90% of cytochrome b_{558} in the resting neutrophil can be

Table I
Classification of CGD and Correlation with Defects in Oxidase Components[a]

Type	Mode of inheritance	No. of cases or % of total	Cyto. b_{558} spectrum	% NBT-positive cells	Whole-cell activity (%)	Oxidase kinetics Particulate Km (mM)	Vmax (%)	Defect in cell-free system	Defect (if known)
I									
A	X linked (classic)	65%	Absent	0	0			Membrane	Absence of message for 91-kDa subunit of cyto. b_{558}
B	X linked (variant)	5 cases	Absent	60–90	1–10	0.8–8.4	7–70	Membrane	Unknown
II									
A	Auto.	25–30%	Normal	0	0			Cytosol	Absence of pp47
B	Auto.	1–5%	Normal	0	0			Cytosol	Absence of p65[b]
C	Auto.	1 case	Normal	85	1–4			Membrane	Unknown
III	Auto.	8 cases	0–4%	0.4	0	0.17	85	Membrane	Absence of message for 22-kDa subunit of cyto. b_{558}
IV	X linked	2 cases	Normal	0	0.4			Membrane	Point mutation in 91-kDa subunit of cyto. b_{558}

[a]CGD is subdivided into four types on the basis of the spectroscopic presence or absence of cytochrome b_{558} (cyto. b_{558}) and mode of inheritance, X linked or autosomal recessive (Auto.). CGD patients can be further classified by the site of the oxidase defect in a cell-free system, cytosol or membrane. Although most defects result in a total absence of oxidase activity, in X-linked variant CGD and a rare variant of type II CGD, oxidase activity is present but the enzyme has an abnormal affinity for NADPH.
[b]p65 refers to the 65–kDa cytosolic protein described by Volpp et al. (1988).

localized to the membrane of the specific granule (Borregaard *et al.*, 1983). Following stimulation of the cell, 40–75% of the cytochrome b_{558} was found in association with the plasma membrane, apparently as a consequence of fusion of the specific granule membrane with the plasma membrane during degranulation. This translocation of cytochrome b_{558} is not dependent on intact microfilaments or on microtubules (Garcia and Segal, 1984).

Although there has been debate about whether cytochrome b_{558} plays a structural or an electron transport role in the oxidase, it has several unique biochemical properties that make it an ideal terminal electron carrier for the NADPH oxidase (Table II). First, its redox or midpoint potential (E_m pH 7.0) is -245 mV, a value similar to the redox potential of the oxygen–superoxide couple (Cross *et al.*, 1981); this is the lowest midpoint potential of any known mammalian b-type cytochrome. Second, the cytochrome has been shown to weakly bind carbon monoxide, a property which suggests that it can also bind oxygen directly. This feature is unusual for mammalian b-type cytochromes and has previously been seen only in bacterial o cytochromes (Segal and Jones, 1978; Shinagawa *et al.*, 1966). Third, cytochrome b_{558} has the apparent kinetic capacity to participate in oxidase electron transport. Oxidation of the reduced cytochrome b_{558} by oxygen occurs at a very rapid rate ($t_{1/2}$ of ~5 msec) (Cross *et al.*, 1982a). Although the cytochrome can only be reduced slowly under anaerobic conditions (Segal and Jones, 1979a), in the presence of physiologic oxygen levels, cytochrome b_{558} reduction occurs at a rate comparable to the rate of O_2^- production by the whole cell (Cross *et al.*, 1985; Segal, 1988). Finally, cytochrome b_{558} cannot be reduced directly by either NADH or NADPH (Cross *et al.*, 1981), which suggests another intermediate redox carrier such as FAD. O_2^-· itself cannot reduce cytochrome b_{558} (Segal and Jones, 1980a). All of these characteristics make cytochrome b_{558} the ideal carrier for direct transfer of electrons to molecular oxygen.

Soon after its discovery, cytochrome b_{558} was found to be absent in the neutrophils of some CGD patients (Segal *et al.*, 1978; Segal and Jones, 1980a). The absence of cytochrome b_{558} in CGD neutrophils conclusively established its close relationship to the NADPH oxidase. Studies with some of these cytochrome b_{558}-deficient patients, how-

Table II

Characteristics of the Cytochrome b_{558} of Phagocytes

1. Found overwhelmingly in mammalian phagocytic cells, although small amounts have recently been detected in B lymphocytes. Neutrophils have approximately 100 pmol of cytochrome b_{558}/mg of protein.
2. Integral membrane protein located in specific granules and plasma membrane.
3. Translocation of cytochrome b_{558} to plasma membrane occurs with activation, is dependent on association of cytochrome b_{558} with 47-kDa phosphoprotein family, and is associated with FAD-flavoprotein translocation to plasma membranes.
4. Unique midpoint potential of -245 mV.
5. Ability to bind CO, suggesting capability to bind O_2.
6. Heterodimer composed of a large glycosylated 91-kDa subunit and a smaller 22-kDa subunit. Heme binding site possibly on the 22-kDa subunit.
7. 91-kDa subunit encoded on chromosome Xp21.1; expression is enhanced by IFN-γ.
8. Absent in type I (X-linked) and type III (autosomal recessive) CGD. Intermediate levels are found in X-linked heterozygote carriers.

ever, have yielded a certain amount of evidence calling into question the participation of the cytochrome as an obligatory electron carrier of the respiratory burst oxidase. For example, patients have been reported whose neutrophils, completely lacking in cytochrome b_{558}, can furnish activated particles that manufacture O_2^- at rates up to 40% of control (Seger *et al.*, 1983; Newburger *et al.*, 1986; Ezekowitz *et al.*, 1987). In addition, recent studies have shown that when treated with IFN-γ, neutrophils from certain cytochrome b_{558}-deficient patients will produce normal amounts of O_2^- even though their levels of the cytochrome increase only slightly (Ezekowitz and Newburger, 1988). Finally, a purified oxidase preparation containing almost no heme was able to produce O_2^- at a rate of 5–10 μmol/min^{-1} mg of protein^{-1} (Markert *et al.*, 1985; Glass *et al.*, 1986). Thus, the role of the cytochrome in O_2^- production by the respiratory burst oxidase is not yet quite settled.

The epidemiology of cytochrome b_{558}-deficient CGD has been extensively studied in a large multicenter European study, which connected cytochrome b_{558} deficiency with X-linked CGD (type I; Table I) (Segal *et al.*, 1983); female heterozygotes for the X-linked disease had intermediate levels of cytochrome b_{558}, whereas autosomal recessive CGD patients were found to have normal cytochrome b_{558} levels (type II CGD; Table I). Later, a third group of CGD patients (type III; Table I) was identified who also had an autosomal recessive mode of inheritance but had no detectable cytochrome b_{558} (Segal and Jones, 1980a). Confirmation of three biochemically and genetically distinct forms of CGD was accomplished by fusing monocytes from the three types of CGD and measuring oxidase activity by NBT reduction (Hamers *et al.*, 1984). The finding that fused heterokaryons were able to produce O_2^- provided a substantive background for the classification of CGD patients shown in Table I, which is based on inheritance patterns and on the presence or absence of a cytochrome b_{558} spectrum.

Although a linkage had already been established between CGD and the blood group antigen Xg, located distally on the short arm of the X chromosome, it was not until two patients were identified who had the unfortunate concordance of cytochrome b_{558}-negative CGD, Duchenne's muscular dystrophy, and a cytologically visible interstitial deletion within the Xp21 locus of the X chromosome that accurate mapping of X-linked inheritance (type I) CGD was possible (reviewed in Dinauer and Orkin, 1988). The assignment of the CGD gene to the Xp21.1 locus by formal linkage analysis narrowed the region coding for this form of CGD to about 3000 kb of DNA. Through the elegant application of "reverse genetics," the cDNA encoding the 91-kDa subunit of cytochrome b_{558} was identified even before the protein had been purified (Royer-Pokora *et al.*, 1986). First, a radioactively labeled cDNA pool was prepared from HL-60 cells that had been differentiated through incubation with dimethylformamide, a stimulus that leads to granulocytic differentiation and the appearance of NADPH oxidase activity. This cDNA pool was then subtractively hybridized with RNA obtained from an Epstein-Barr virus-transformed B-cell line derived from one of the patients with an interstitial deletion of Xp21. The subtracted cDNA pool consisted of approximately 500 different clones and was used to screen a library of bacteriophage clones that were known to encode a portion of the Xp21 genomic locus (these were initially prepared in studies of the Duchenne's muscular dystrophy locus). Fortuitously, two overlapping bacteriophage clones were identified, which were then used to screen cDNA libraries prepared from differentiated HL-60 cells. A series of cDNA clones were identified that spanned a single 4.27-kb open reading frame. This transcript was shown to be absent in five of six X-linked cytochrome b_{558}-negative CGD patients. In a sixth patient, a normal transcript appeared to be present, but additional studies established that there was in fact a small genomic deletion that resulted in the loss of the cDNA coding for the C-terminal 41 amino acids of the protein. Although these elegant genetic studies identified the protein encoded by the X-linked CGD gene, the exact function of the encoded protein was not known. Polyclonal antibodies were generated against synthetic peptides corresponding to the C terminus of the protein and against bacterially expressed β-galactosidase fusion proteins; these antibodies were shown to cross-react with both the glycosylated and deglycosylated 91-kDa subunit of cytochrome b_{558} that had been purified by Parkos *et al.* (1987), providing direct evidence that the X-linked CGD gene encodes the 91-kDa subunit of cytochrome b_{558} (Dinauer *et al.*, 1987). Using another approach, an independent group of investigators was able to demonstrate almost complete homology between the N-terminal amino acid sequence (43 residues) of the cytochrome b_{558} 91-kDa subunit and the sequence predicted by the X-linked CGD cDNA (Teahan *et al.*, 1987).

Subsequent studies have provided the full cDNA sequence for the 22-kDa subunit of the cytochrome as well as for the 91-kDa subunit, and thus the deduced amino acid sequence for both subunits has been determined (Royer-Pokora *et al.*, 1986; Parkos *et al.*, 1988a). Neither amino acid sequence subunit demonstrates homology with any other previously characterized mammalian cytochrome. Both subunits have several potential N-glycosylation sites as well as having hydrophobic domains that possibly represent membrane-spanning regions (Dinauer and Orkin, 1988; Parkos *et al.*, 1988a); the latter data are in keeping with earlier experiments using Triton X-114 phase separation, which showed that cytochrome b_{558} was an integral membrane protein (Borregaard and Tauber, 1984). Hydropathy profiles suggest that myoglobin and the 22-kDa cytochrome b_{558} subunit may have similar secondary polypeptide structures (Parkos *et al.*, 1988a).

The most current classification of CGD is summarized in Table I. A rare X-linked cytochrome b_{558}-normal CGD subset (type IV) has been identified, and the defect has been attributed in one case to a dysfunctional cytochrome due to a Pro → His amino acid substitution at position 415, the result of a point mutation in the DNA coding for the 91-kDa cytochrome b_{558} subunit (Dinauer *et al.*, 1989). The most common form of CGD, X-linked cytochrome b_{558} negative (type I), occurs as a result of a mutation in the gene encoding the 91-kDa subunit as described above. Similarly, the rare cytochrome b_{558}-negative autosomal recessive form of CGD (type III) results from an absence of cDNA coding for the 22-kDa subunit of the cytochrome (Parkos *et al.*, 1989). Interestingly, in both forms of cytochrome b_{558}-negative CGD (type I and type III), the stable expression of each cytochrome b_{558} subunit appears to be dependent on the presence of mRNA for both subunits (Parkos *et al.*, 1988a, 1989). Supporting this hypothesis is the observation that in a variety of nonphagocytic cells, the 22-kDa subunit mRNA is present, but in the absence of the 91-kDa subunit mRNA, no 22-kDa subunit protein can be found (Parkos *et al.*, 1988a). The purpose of the constitutive presence of 22-kDa mRNA in nonphagocytic cells remains unknown.

5.1.2. Gamma Interferon and Cytochrome b_{558}

The ability of a cytokine to modify respiratory burst activity was first demonstrated by using IFN-γ. Although IFN-γ cannot stimulate the respiratory burst directly, it has

Table III
Effects of IFN-γ on the Respiratory Burst

Effects on macrophages

1. Priming is dose dependent and occurs at picomolar concentrations.
2. Time dependent, with maximal effect occurring 2–4 days after exposure to IFN-γ.
3. Enhancement persists for up to 6 days with continuous IFN-γ exposure and is completely reversible within 3 days of discontinuing IFN-γ.
4. A 10-fold reduction in K_m of NADPH with no change in V_{max}.
5. Priming unaffected by dexamethasone.

Effects on neutrophils

1. Time-dependent O_2^- augmentation occurring within 1 hr of incubation and maximal by 2–4 hr.
2. Priming after PMA stimulation occurs only with suboptimal concentrations (<10 ng/ml).
3. O_2^- enhancement inhibited by IFN-γ monoclonal antibody, cycloheximide, and actinomycin D but not by polymyxin B.
4. Priming not dependent on adherent cells.
5. No change in oxidase V_{max} or K_m for NADPH.

been shown to enhance or prime the respiratory burst activity of normal human neutrophils or macrophages in response to known stimuli (Nathan *et al.*, 1983; Cassatella *et al.*, 1985, 1988; Berton *et al.*, 1986). As described in Table III, the effects of IFN-γ priming on the respiratory burst are dose and time dependent, occur at picomolar concentrations, and, at least in the neutrophil, appear to be dependent on *de novo* protein synthesis. Unfortunately, a definite mechanism for IFN-γ priming has not been established.

To further understand the effect of IFN-γ on the NADPH oxidase and explore the therapeutic potential in CGD, other investigators have examined the *in vitro* effects of IFN-γ on CGD macrophages and neutrophils (Ezekowitz *et al.*, 1987). PMA-induced O_2^- production was significantly enhanced following *in vitro* IFN-γ treatment of normal human macrophages, as well as X-linked variant* (type IB; Table I) CGD macrophages and neutrophils; IFN-γ had no effect on normal human neutrophils, on classic X-linked cytochrome b_{558}-negative CGD neutrophils, or on autosomal recessive cytochrome b_{558}-positive CGD neutrophils. Also, NBT dye reduction as an indicator of O_2^- formation demonstrated that cells responded homogeneously, either as nonresponsive or as highly responsive cells. In the responsive X-linked variant cytochrome b_{558}-negative (type IB) CGD neutrophils, a twofold increase in the V_{max} of the NADPH oxidase was observed; control neutrophils demonstrated no change in either V_{max} or NADPH K_m. IFN-α and -β had no effect on either neutrophil or macrophage NBT reduction (Newburger *et al.*, 1988).

The first clue regarding a possible mechanism for IFN-γ priming of O_2^- production occurred with the examination of cytochrome b_{558} gene expression. The CGD-responsive neutrophils had a 5- to 10-fold increase in cytochrome b_{558} 91-kDa subunit mRNA; however, cytochrome b_{558} remained undetectable using spectrophotometric techniques. With human macrophages, Newburger and co-workers (1988) demonstrated that IFN-γ increased the level of cytochrome b_{558} 91-kDa subunit mRNA in a dose- and time-

* "Variant" refers to a small subset of CGD patients who have approximately 1–5% of normal superoxide production.

dependent manner that was compatible with priming of the NADPH oxidase. Nuclear run-off experiments suggested an up regulation of 91-kDa subunit gene transcription and not an improvement in the posttranscriptional stability of the mRNA. IFN-γ did not appear to affect the level of 22-kDa subunit mRNA. IFN-α decreased the level of 91-kDa subunit mRNA compatible with its effect on O_2^- production, and IFN-β had no detectable effect on 91-kDa subunit mRNA. Although other investigators had failed to spectrally detect an increase in normal human macrophage cytochrome b_{558} levels with IFN-γ treatment (Cassatella *et al.*, 1985, 1988), it was proposed that IFN-γ induction of 91-kDa subunit gene expression might increase cytochrome b_{558} levels (Newburger *et al.*, 1988).

In vivo administration of IFN-γ as two subcutaneous doses induced a 5- to 10-fold increase in neutrophil and monocyte O_2^- production that persisted for 2–5 weeks (Ezekowitz *et al.*, 1988). Despite the above *in vitro* results, both classic (type IA) and variant (type IB) X-linked cytochrome b_{558}-negative CGD patients appeared to respond to *in vivo* administration of IFN-γ. The increase in cytochrome b_{558} 91-kDa subunit mRNA was confirmed, but surprisingly there was also an increase in spectral levels of cytochrome b_{558}. Spectroscopic assays demonstrated an increase in cytochrome b_{558} levels to 10–50% of normal with IFN-γ therapy (pretreatment levels were 0–14% of normal). In two patients, immunoblotting studies demonstrated an increase in the apparent level of cytochrome 91-kDa subunit. One explanation proposed for the discrepancy between the amount of O_2^- produced and the amount of cytochrome b_{558} detected was a nonlinear relationship between the level of cytochrome b_{558} 91-kDa subunit mRNA, its protein product, and NADPH oxidase activity (Ezekowitz *et al.*, 1987); another was discussed above.

As a result of the above studies, two interesting hyaotheses developed. First, bacterial killing improved significantly in all patients regardless of their change in O_2^- production, suggesting an oxygen-independent effect of IFN-γ on phagocytic cells. Second, in addition to the *in vitro* data demonstrating a IFN-γ priming effect on mature phagocytic cells, which have a very short half-life in the circulation, the sustained effect of *in vivo* IFN-γ suggested an effect on hemapoietic progenitor cells in the marrow or possibly the associated accessory cells that reside in the marrow microenvironment. Recently, IFN-γ and tumor necrosis factor, products of accessory cells, have been shown to have synergistic effects on both cytochrome b_{558} 91-kDa subunit gene expression and NADPH oxidase activity in HL-60 cells (Cassatella *et al.*, 1989).

Sechler *et al.* (1988) subsequently demonstrated improved O_2^- production after IFN-γ in patients with autosomal recessive cytochrome b_{558}-positive CGD that was also associated with increased levels of cytochrome b_{558} 91-kDa subunit on western blots (Sechler, 1988). Using a cell-free system, Curnutte *et al.* (1988b) recently demonstrated augmentation of cytosolic oxidase activity in two patients with autosomal cytochrome b_{558}-positive (type II) CGD following *in vivo* administration of IFN-γ.

If an absence of cytochrome b is the limiting factor in electron transport to molecular oxygen by the oxidase, an increase in the level of cytochrome b_{558} transcript could certainly increase the level of oxidase activity as reflected in the V_{max}. It is less apparent how increased cytochrome b_{558} levels could influence the affinity of the oxidase for NADPH if the cytochrome is functioning only as a redox carrier, particularly since evidence from cell-free oxidase systems has suggested that alterations in the affinity of the oxidase for NADPH are controlled by a cytosolic component (Babior *et al.*, 1988a;

Curnutte *et al.*, 1989b; Smith *et al.*, 1989b). If the carboxy terminus of the 91-kDa subunit of cytochrome b_{558} does serve as an attachment site for the cytosolic components of the oxidase, then increased levels of the cytochrome could lead to increased binding of the cytosolic components of the oxidase that are responsible for controlling the affinity of the oxidase for NADPH. Alternatively, IFN-γ treatment may induce both cytosol and membrane components of the oxidase, with cooperativity between these effects leading to the changes seen in whole-cell experiments or in patients.

5.2. A 47-kDa Phosphoprotein

The involvement of a phosphoprotein in the activation or regulation of the NADPH oxidase had been suggested by a number of lines of reasoning. First, many enzymes are regulated by reversible phosphorylation–dephosphorylation cycles (Krebs and Beavo, 1979; Edelman *et al.*, 1987), and these phosphorylation systems are important in the regulation of cellular metabolism in a number of cells, including platelets, hepatocytes, muscle cells, nerve cells, and adrenocortical cells. Second, the rapid conversion of the NADPH oxidase from its normally dormant state into an active enzyme species suggests the involvement of rapid and reversible processes such as protein phosphorylation. Finally, both agonists (phorbol esters and 1-oleyl-2-acetyl diglyceride) and antagonists (phenothiazines, retinal, H-7, and W-7) of protein kinases, particularly of PKC, have parallel effects on the activation and inhibition of the NADPH oxidase (reviewed in Babior, 1988).

Andrews and Babior were among the first investigators to demonstrate that protein phosphorylation is an important phenomenon during neutrophil stimulation (Andrews and Babior, 1983). After incubation of neutrophils with $^{32}PO_4^{3-}$, the cells were stimulated and lysed, and the lysates were examined by gel electrophoresis. This technique demonstrated the presence of at least 40 distinct radiophosphorylated proteins in the resting neutrophil; dramatic changes in phosphorylation could be consistently induced by stimulation of the respiratory burst in the cells. Stimulation of the neutrophil with either PMA or opsonized zymosan resulted in the incorporation of ^{32}P into proteins* of 80, 69, 55–53, 48, 22, and 13 kDa and a decrease in ^{32}P incorporation into a protein of 20 kDa. Both formylmethionine-leucine-phenylalanine (fMLP) and digitonin caused similar phosphorylation patterns except that pp55-53 and pp13 were not seen consistently. In subsequent experiments, the investigators demonstrated that both O_2^- production and protein phosphorylation in response to PMA or fMLP were inhibited by trifluoperazine, an inhibitor of both calmodulin-dependent and -independent protein kinases (Andrews and Babior, 1984). Washing the neutrophils free of trifluoperazine restored O_2^- production as well as phosphorylation in all phosphoproteins except pp69. This latter phosphoprotein also had a time course of phosphorylation that was inconsistent with participation in activation of the NADPH oxidase, data suggesting that it was not a component of the NADPH oxidase. Similar patterns of protein phosphorylation were seen upon stimulation of neutrophils with opsonized zymosan, sodium fluoride, concanavalin A, A23187, latex beads, or arachidonate (Heyworth and Segal, 1986; Babior, 1988; Heyworth *et al.*, 1989).

Several groups have shown that stimulation of the neutrophil by PMA results in

---

*These proteins will be designated by mass, e.g., pp67 for phosphoprotein of 67 kDa mass.

phosphorylation of many of these proteins with a time course that closely parallels that of O_2^- production (Andrews and Babior, 1983, 1984; Hayakawa *et al.*, 1986; Heyworth and Segal, 1986). Although, the rate of incorporation of ^{32}P into the numerous phosphoproteins was variable, phosphorylation generally began within 12 sec after PMA stimulation and continued for at least several minutes, closely following the PMA time course of O_2^- production in neutrophils (Andrews and Babior, 1983; Segal *et al.*, 1985). Additionally, the proteins in neutrophils from several other species (rabbits, guinea pigs, and cattle) have been shown to contain phosphoproteins of a similar size and to undergo patterns of phosphorylation following stimulation similar to those of human neutrophils (Babior, 1988). Finally, phosphorylation studies performed in normal neutrophils under anaerobic conditions demonstrated that protein phosphorylation occurred in the absence of O_2^- production or production of any other oxygen-derived free radical (Hayakawa *et al.*, 1986).

Although the preceding evidence established an interesting association between protein phosphorylation and the NADPH oxidase, a functional relationship was not established until an abnormality of phosphorylation was demonstrated in certain forms of CGD (Segal *et al.*, 1985). Using one-dimensional PAGE, these studies demonstrated the absence of phosphorylation of a protein of approximately 44 kDa from autosomal cytochrome b_{558}-positive (type II) CGD neutrophils; the neutrophils obtained from X-linked cytochrome b_{558}-negative (type I) CGD patients or from normal subjects both appeared to contain this phosphoprotein when they were stimulated. Phosphorylation of all other phosphoproteins was otherwise identical between CGD and control neutrophils, suggesting that the absence of radiophosphorylated protein was probably not due to a defective phosphorylating system *per se*. Subsequent experiments more accurately identified the molecular weight of this entity as approximately 47–48 kDa* (Heyworth and Segal, 1986; Hayakawa *et al.*, 1986; Okamura *et al.*, 1988a). The importance of this phosphoprotein was highlighted further by additional experiments using various stimuli of the NADPH oxidase (e.g., opsonized zymosan, latex beads, fMLP, sodium fluoride, and the calcium ionophore A23187) which produced different lag times and rates of oxidase activation; in all cases, there was an exact correspondence between oxygen consumption by the neutrophil and both the time course and extent of ^{32}P incorporation into pp47 (Heyworth and Segal, 1986).

Efforts to localize pp47 in resting or stimulated human neutrophils by using subcellular fractionation techniques demonstrated its presence in both plasma membrane and cytosol fractions but not in granule fractions (Hayakawa *et al.*, 1986). Recently, the dual cytosolic and membrane location of the pp47 was confirmed in normal neutrophils, and evidence was provided demonstrating that phosphorylation initially occurs in the cytosol, with subsequent transfer of pp47 to the membrane (Heyworth *et al.*, 1989). Experiments on neutrophils from X-linked cytochrome b_{558}-positive and -negative CGD patients demonstrated an association between the inability to transfer pp47 to the membrane and the absence of cytochrome b_{558} (Okamura *et al.*, 1988b). Intermediate results were obtained with X-linked carriers. The precise mechanism of binding between pp47 and the cytochrome b_{558} subunits in the neutrophil membrane remains to be determined.

Using two-dimensional gel electrophoresis, further experiments demonstrated that

---

*This phosphoprotein will subsequently be referred to as pp47.

Table IV
Features of Neutrophil Respiratory Burst Phosphoprotein Family

1. Mass of 47 kDa.
2. At least six members of phosphoprotein family with pIs of 6.8, 7.3, 7.8, 8.7, 9.5, and 10.0.
3. Identical one-dimensional phosphopeptide maps indicating a family of phosphoproteins encoded by the same gene.
4. Phosphorylation occurs exclusively on serine residues.
5. Abnormalities in phosphorylation identified in 3 of the 4 types of CGD.

the previously described pp47 was in fact a family of phosphoproteins with a mass of 47 kDa and isoelectric points (pI) of 6.8, 7.3, 7.8, 8.3, 9.5, and approximately 10.0 (Hayakawa et al., 1986; Babior et al., 1988b; Okamura et al., 1988a).* The phosphorylation of this family of proteins in response to PMA or fMLP correlated with the production of O_2^-, similar to the findings in one-dimensional electrophoresis. Amino acid mapping experiments revealed that phosphorylation of the pp47 family was exclusively confined to serine residues (Okamura et al., 1988a). After protease digestion of excised spots, one-dimensional peptide mapping of the resulting radiopeptides yielded identical patterns for each of the members of the pp47 family, suggesting that pp47/6.8, pp47/7.3, and pp47/7.8 have similar if not identical polypeptide structures (Okamura et al., 1988a). It has been suggested that these phosphoproteins are derived from a single protein precursor and are therefore encoded by a single gene (Okamura et al., 1988a). Although variations in pI may represent differences in glycosylation or in microheterogeneity of amino acid composition, they are more likely due to differing degrees of phosphorylation of a single polypeptide whose pI decreases as the phosphorylation increases (Babior, 1988). The major characteristics of the neutrophil 47-kDa phosphoprotein family are summarized in Table IV. Interestingly, although the 47-kDa phosphoproteins are easily detected by autoradiography, they are not visible on silver-stained gels, suggesting that they may be present in extremely small amounts (N. Okamura, R. Woodman, and B. Babior, unpublished observations).

In control neutrophils, phosphorylation of pp47/6.8 after activation occurred inconsistently; the significant of this phosphoprotein, therefore, remains uncertain. However, extensive investigations in CGD patients demonstrated abnormalities in phosphorylation of pp47/7.3 and pp47/7.8 in three of the four major types of CGD (Okamura et al., 1988a,b) (Table V). Autosomal and X-linked CGD patients deficient in cytochrome b_{558} (types III and I, respectively) had absent phosphorylation of pp47/7.3 and pp47/6.8 but normal phosphorylation of pp47/7.8 [these findings are thus consistent with the earlier studies using one-dimensional electrophoresis and autoradiography showing normal phosphorylation in whole neutrophil lysates from X-linked CGD patients (Segal et al., 1985]. Neutrophils from different autosomal recessive cytochrome b_{558}-positive (type II) CGD patients did not phosphorylate any of the three 47-kDa proteins. Finally, one patient with the uncommon X-linked cytochrome b_{558}-positive (type IV) CGD had a completely normal 47-kDa phosphorylation pattern. These results suggested that cytochrome b_{558} may be required for phosphorylation of the most acidic 47-kDa phosphoproteins and that

*The abbreviation pp47/6.8 will be used to refer to the phosphoprotein with molecular weight of 47 kDa.

Table V
Phosphorylation Patterns of 47-kDA Phosphoproteins in Neutrophils from Normal Subjects and from Patients with CGD

		Phosphorylation of pp47 of given pI[a]		
Source of cells		pI 6.8	pI 7.3	pI 7.8
Normal subjects		±	+	+
CGD subjects				
Inheritance	Cytochrome b_{558}			
X-linked	Absent	—	—	+
Autosomal	Absent	—	—	+
Autosomal	Present	—	—	—
X-linked	Present	±	+	+

[a]Stimulation of the respiratory burst in neutrophils loaded with ³²P results in the incorporation of radiophosphorus into pp47. The presence or absence of phosphorylation of pp47 at different isoelectric points can be established by 2-dimensional electrophoresis and autoradiography (Okamura *et al.*, 1988a, 1988b). Phosphorylation patterns of the acidic members of the pp47 family differ among the subtypes of CGD and are indicated by + for phosphorylation identified, — for phosphorylation not found, and ± for inconsistent phosphorylation.

phosphorylation of the 47-kDa proteins probably occurs in separate successive steps; the first step is the generation of pp47/7.8 within the neutrophil cytosol, followed by translocation to the plasma membrane possibly through an interaction with cytochrome b_{558}. Finally, further phosphorylation of pp47 occurs at the plasma membrane to form pp47/7.3 and pp47/6.8.

The kinase responsible for phosphorylation of the pp47 family has not been conclusively identified. However, substantial evidence exists favoring PKC. The resting neutrophil has substantial levels of cytosolic PKC activity (Helfman *et al.*, 1983), and the use of inhibitors with specificity for PKC inhibits both O_2^- production and pp47 phosphorylation (reviewed in Tauber, 1987). Furthermore, induction of protein phosphorylation in a cell-free system by treatment either with PMA, to directly stimulate endogenous PKC, or by the addition of PKC and [³²P]ATP produced banding patterns similar to those seen in PMA-stimulated intact neutrophils (Pontremoli *et al.*, 1986). Although many of the stimuli of the respiratory burst are not known to act directly on PKC, well-described alternate pathways exist (e.g., via phosphoinositide-mediated production of diacylglycerol) for PKC activation after ligand binding to receptors.

Though not conclusively demonstrated by the foregoing experiments, there is strong support for the hypothesis that pp47 is absent in the autosomal recessive inheritance, cytochrome b_{558}-positive form of CGD. Since the protein cannot be detected by Coomassie blue or silver staining, it is difficult to be certain that an abnormal protein was not present or that a kinase with narrow specificity for the pp47 family was not defective or absent. The recent demonstration of a missing 47-kDa protein in the cytosol of patients with autosomal recessive cytochrome b_{558}-positive CGD whose neutrophils do not demonstrate pp47 after stimulation raises the obvious question of whether this 47-kDa protein is a substrate for phosphorylation in the neutrophil and is identical to the previously described pp47. Recent experiments confirm that a polyclonal antibody, directed against the cytosolic 47 kDa-peptide that is missing in autosomal CGD, is able to immunoprecipitate a 47-kDa protein that is phosphorylated during the respiratory burst (Nunoi *et al.*, 1989). Blotting studies with the same antibody demonstrate that this 47-kDa protein is translocated from cytosol to membrane during stimulation of the respiratory burst (Clark and Nauseef, 1989). The recent publication of the cDNA sequence coding for this phosphoprotein revealed that the derived amino sequence contains four potential serine phosphorylation sites, consistent with the previous studies of the pp47 family. Furthermore, it is clear that the neutrophil 47-kDa phosphoprotein is unique and is not the same as the 47-kDa phosphoprotein phosphorylated by PKC during platelet activation (Haslam and Lynham, 1977; Lyons and Atherton, 1979; Imaoka *et al.*, 1983). Although there had been earlier speculation that the neutrophil and platelet 47-kDa phosphoproteins might be identical, comparison of their amino acid sequences shows no significant homology (Tyers *et al.*, 1988; Lomax *et al.*, 1989). Thus, it is likely that the cytosolic 47-kDa protein is identical to pp47 and that sequential phosphorylation steps generate the members of the pp47 family. Interestingly, Northern blotting studies of neutrophils from autosomal recessive cytochrome b_{558}-negative CGD patients disclose apparently normal quantities of pp47 message despite the absence of the transcribed protein (Lomax *et al.*, 1989). It is not known yet whether the primary defect in these patients (type IIA CGD) is due to an abnormality in the transcription of pp47 message or whether there is an additional subunit of the oxidase missing whose presence is required for the stable expression of pp47. A complete understanding of the function of the pp47 family in the assembly and catalytic activity of the oxidase awaits further studies; we speculate that pp47 is responsible for controlling the assembly (and disassembly?) of the oxidase. Upon successive phosphorylations, pp47 may gain the ability to be recognized by both the additional cytosolic oxidase components and the membrane oxidase components and thereby serve as a "bridge" to allow the formation of the intact active oxidase. The membrane component to which pp47 binds may be the carboxy terminus of the 91-kDa subunit of cytochrome b_{558}, though this has not yet been conclusively demonstrated. As pp47 itself does not contain the NADPH-binding subunit of the oxidase (Smith *et al.*, 1989b), the additional cytosolic oxidase components that must combine with pp47 include the 65-kDa protein described below and (if different) the cytosolic NADPH-binding subunit.

5.3. A Cytosolic 65-kDa Flavoprotein

It is now well established that the NADPH oxidase is a FAD-dependent enzyme (Babior and Kipnes, 1977; Babior and Peters, 1981; Light *et al.*, 1981). Other flavoproteins (FMN and riboflavin) are ineffective cofactors for restoring oxidase activity and have been detected in only minimal amounts in either subcellular fractions of normal neutrophils (Gabig, 1983; Borregaard and Tauber, 1984) or purified oxidase preparations (Glass *et al.*, 1986). The FAD content of the purified oxidase is 20 pmol/μg of protein (Glass *et al.*, 1986), and for solubilized oxidase preparations the K_m is 60–80 nM (Babior and Peters, 1981). Experiments with the FAD analogs 8-chloro-FAD and 1-deaza-FAD, which can transfer either one or two electrons, are effective oxidase cofactors; however, 5-carba-5-deaza-FAD, which can transfer only two electrons completely, inhibits oxidase activity (Light *et al.*, 1981). These results suggest that FAD acts as a one-electron acceptor within the NADPH oxidase complex. The role of FAD as a redox carrier within the NADPH oxidase complex is also supported by electron spin resonance studies, reduction by NADPH under anaerobic conditions, and its midpoint potential of -208 mV, which is intermediate between the redox potentials of NADPH and cytochrome b_{558} Cross *et al.*, 1984; Kakinuma *et al.*, 1986).

On the basis of following evidence, there appears to be a very interesting relationship between FAD-flavoprotein and cytochrome b_{558}. Subcellular fractionation has localized FAD and cytochrome b_{558} to similar sites (Borregaard *et al.*, 1983; Borregaard and Tauber, 1984b). Approximately 50% of the total resting neutrophil FAD is found within the specific granules; the remainder is distributed between plasma membrane ($\sim$30%) and cytosol ($\sim$20%), with essentially none ($<$3%) in the azurophilic granules. The exact molar ratio of FAD to cytochrome b_{558} in the specific granules and in the activated plasma membrane remains controversial but probably is not $1:1$ at either location (Borregaard and Tauber, 1984). FAD is not bound to cytochrome b_{558}, as both are completely dissociated after Triton X-114 treatment (Borregaard and Tauber, 1984). The FAD appears to be noncovalently (Babior and Kipnes, 1977; Borregaard and Tauber, 1984) but tightly associated ($K_d = < 0.2$ μM) (Light *et al.*, 1981) with its apoprotein. Following PMA stimulation, the FAD-flavoprotein also appears to translocate from the specific granules to the plasma membrane in a fashion paralleling the movement of cytochrome b_{558}.

A deficiency of FAD has been considered in the pathogenesis of CGD. FAD measurements in two patients with X-linked cytochrome b_{558}-negative (type I) CGD demonstrated a 60% reduction in total cellular FAD due entirely to a 80% reduction in specific granule content; no change in plasma membrane or cytosol FAD content was detected (Borregaard and Tauber, 1984). However, in the few CGD patients examined, neutrophil FAD deficiency has usually occurred in conjunction with cytochrome b_{558} deficiency, and thus it is difficult to suggest that a flavin or flavoprotein defect is the primary cause of the diminished $O_2{}^-$ production (Gabig, 1983; Cross *et al.*, 1982b; Gabig and Lefker, 1984b; Borregaard and Tauber, 1984). Furthermore, FAD replacement *in vitro* has not restored oxidase activity (Babior and Kipnes, 1977; Cross *et al.*, 1982b; Gabig, 1983; Gabig and Lefker, 1984b; Borregaard and Tauber, 1984).

One CGD patient has been reported with $<$10% normal FAD content in the particulate oxidase fraction, in the presence of apparently normal cytochrome b_{558} levels (Gabig, 1983; Gabig and Lefker, 1984a). The mechanism for this CGD phenotype could be due to an intrinsic abnormality in cytochrome b_{558} that prevents it from binding the flavin or flavoprotein or could represent a primary defect in the flavoprotein itself.

The exact nature of the flavoprotein that appears to function as a component of the oxidase is unknown. We speculate that the flavoprotein is not an intrinsic membrane protein but rather a cytosolic protein that associates with the membrane through binding with cytochrome b_{558}, a binding that may be mediated through pp47. Thus, cytochrome b_{558} deficiency could lead to a reduction in membrane-associated flavin as a result of an inability to bind cytosolic flavoprotein. Although the exact function of the 65-kDa cytosolic oxidase component is not known, preliminary reports of the cDNA sequence suggest that it potentially contains both NADPH- and flavin-binding domains (Leto *et al.*, 1989). Thus, it is possible that the 65-kDa cytosolic protein first described by Clark and coworkers (Volpp *et al.*, 1988) is a flavoprotein that contains the catalytic NADPH-binding site of the oxidase. Although this hypothesis is partially consistent with the earlier work of Doussiere and associates (Doussiere and Vignais, 1985; Doussiere *et al.*, 1986), confirmation of its validity awaits further studies.

5.4. Cytosolic Nucleotide-Binding Proteins

As described in Section 4.2.1, a number of investigators have used affinity labeling techniques to identify NADPH-binding proteins in preparations of neutrophil or macrophage membranes and have assigned these proteins to the NADPH oxidase. Work in cell-free oxidase systems, now reported by two independent laboratories (Smith *et al.*, 1989a,b; Takasugi *et al.*, 1989), strongly suggests that the subunit containing the catalytic NADPH-binding site of the oxidase exists in the cytosol in the resting cell. Additional support for a cytosolic location for the catalytic NADPH-binding subunit of the oxidase is given by experiments in which it was possible to use an NADP(H) affinity column to attain a 40-fold enrichment in activity of the cytosolic oxidase components (Sha'ag and Pick, 1988). The activity eluted from the affinity column was retained by a 100-kDa-cutoff membrane, possibly suggesting that it consisted of a complex of oxidase components, one of which is the NADPH-binding subunit, rather than a single component of the oxidase. A translocation of a high-affinity NADPH-binding subunit of the oxidase from the cytosol to the membrane presumably marks one of the steps in the assembly and activation of the oxidase. The attachment of the subunit of the oxidase containing such a high-affinity binding site may also account for the drop in K_m of the oxidase for NADPH.

In addition to a cytosolic location for the catalytic NADPH-binding subunit of the oxidase, there is evidence to suggest that there are other cytosolic nucleotide-binding proteins which may function as regulatory elements of the oxidase. In particular, GTP-binding proteins have long been suggested to play a role in regulating oxidase activity and may exist in the cytosol as well as on the membranes of phagocytic cells. The evidence implicating GTP-binding proteins in respiratory burst activation has been gathered primarily through experiments using pertussis toxin. Activation of the respiratory burst oxidase by chemotactic peptides such as fMLP, platelet-activating factor, interleukin 8, opsonized zymosan, and leukotriene B_4 can be inhibited by pertussis toxin, which blocks the activity of GTP-binding proteins by the ADP ribosylation of their alpha subunits (Gilman, 1987; Neer and Clapham, 1988; Thelen *et al.*, 1988). In addition to inhibiting O_2^- production, pertussis toxin also prevents phosphoinositide turnover, intracellular calcium flux, and protein phosphorylation (Ohta *et al.*, 1985; Kikuchi *et al.*, 1986). These observations combined with evidence demonstrating activation of a phosphoinositide specific phospholipase C in the membranes of human neutrophils provide a role for the coupling of chemotactic peptide receptors and GTP-binding proteins in the activation of the NADPH oxidase. Superoxide production by PMA or diacylglycerol (both directly activate PKC), by concanavalin A, and by A23187 is not inhibited by pertussis toxin. Also, concanavalin A-stimulated phosphoinositide turnover and intracellular calcium changes are unaffected by pertussis toxin. These observations would suggest at least two different pathways for activation of the NADPH burst oxidase, one of which involves GTP-binding proteins.

In cell-free oxidase systems, the addition of the GTP analog GTPγS typically results in increased oxidase activity (Seifert and Schultz, 1987; Ligeti *et al.*, 1988). A cytosolic location for this GTP-binding component of the oxidase is suggested, at least in part, by the experiments described earlier in which a GTP–agarose column was used to bind and partly purify a cytosolic complex that could participate in the activation of neutrophil membranes in a cell-free system. At a minimum, this complex is known to contain the 47-kDa phosphoprotein and the 65-kDa (flavo?)protein, which are now recognized as components of the oxidase. It is not known which of these two oxidase subunits might bind GTP, though on the basis of the cDNA sequences available to date, the 65-kDa protein is the more likely.

5.5. Other Oxidase Components or Cofactors

It is possible and even likely that other components of the oxidase exist in addition to those described above. In particular, many of the earlier attempts at purification of the oxidase described a 28–35-kDa polypeptide that appeared to be a component of purified oxidase preparations. The work of Pick and co-workers (1987) might suggest that this polypeptide represents an additional membrane-bound component of the oxidase. Additionally, the 45-kDa polypeptide that is labeled by using DPI (Cross and Jones, 1986) has not yet been assigned a role in the function of the oxidase. Whether these and other polypeptides are a structural or electron-carrying component of the oxidase, or whether they may be involved in early activation steps, remains to be seen. To date, no CGD patients who are missing these putative oxidase components have been described, but this may be because mutations in these components are rare or are lethal in the homozygous genotype.

The association of other electron carriers within the NADPH oxidase complex has been examined by many investigators. Nonheme iron redox carriers have not been detected (Cross *et al.*, 1981). Ubiquinones have occasionally been detected in the neutrophil (Gabig and Lefker, 1985; Glass *et al.*, 1986), but they probably originate from platelet mitochondria contaminating the neutrophil preparation (Lutter *et al.*, 1984) and quinone reduction occurs in both the resting and stimulated neutrophil (Crawford and Schneider, 1982). Overall, there are no experimental data to conclusively implicate other redox carriers in the electron transport chain of the NADPH oxidase.

6. SUMMARY

The precise delineation of components of the oxidase, their sequence, and their role in the function of the oxidase remains an area of intensive investigation. Until many of these studies are completed, the hypotheses presented in this chapter must be considered tentative at best. An increased understanding of the NADPH oxidase, however, offers a threefold promise: the application of this knowledge to the design of drugs for combating inflammation, to improved understanding of human disease processes, and to increased comprehension of basic cellular functions.

ACKNOWLEDGMENTS. This work was supported in part by Public Health Service grant AI-24227 and RR00833. R.M.S. is the recipient of a Parker B. Francis Fellowship in Pulmonary Research and an American Lung Association Research support grant.

7. REFERENCES

Albrich, J. M., McCarthy, C. A., and Hurst, J. K., 1981, Biological reactivity of hypochlorous acid: Implications for microbicidal mechanisms of leukocyte myeloperoxidase, *Proc. Natl. Acad. Sci. USA* **78:** 210–214

Ambruso, D. R., and Johnston, R. B., 1981, Lactoferrin enhances hydroxyl radical production by human neutrophils, neutrophil particulate fractions, and an enzymatic generating system, *J. Clin. Invest.* **67:**352–360

Andrews, P. C., and Babior, B. M., 1983, Endogenous protein phosphorylation by resting and activated human neutrophils, *Blood* **61:**333–340

Andrews, P. C., and Babior, B. M., 1984, Phosphorylation of cytosolic proteins by resting and activated human neutrophils, *Blood* **64:**883–890

Arroyo, C. M., Kramer, J. H., Dickens, B. F., and Weglicki, W. B., 1987, Identification of free radicals in myocardial ischemia/reperfusion by spin trapping with nitrone DMPO, *FEBS Lett.* **221:**101–104

Babior, B. M., 1987, The respiratory burst oxidase, *Trends Biochem. Sci.* **12:**241–243

Babior, B. M., 1988, Protein phosphorylation and the respiratory burst, *Arch. Biochem. Biophys.* **264:**361–367

Babior, B. M., and Kipnes, R. S., 1977, The superoxide-forming enzyme from human neutrophils. Evidence for a flavin requirement, *Blood* **50:**517–524

Babior, B. M., and Peters, W. A., 1981, The O_2^--producing enzyme of human neutrophils. Further properties, *J. Biol. Chem.* **256:**2321–2323

Babior, B. M., Kipnes, R. S., and Curnutte, J. T., 1973, Biological defense mechanisms: The production by leukocytes of superoxide, a potential bactericidal agent, *J. Clin. Invest.* **52:**741–744

Babior, B. M., Curnutte, J. T., and McMurrich, B. J., 1976, The particulate superoxide-forming system from human neutrophils, *J. Clin. Invest.* **58:**989–996

Babior, B. M., Kuver, R., and Curnutte, J. T., 1988a, Kinetics of activation of the respiratory burst oxidase in a fully soluble system from human neutrophils, *J. Biol. Chem.* **263:**1713–1718

Babior, B. M., Curnutte, J. T., and Okamura, N., 1988b, The respiratory burst oxidase and certain members of the 48K phosphoprotein family are associated with the neutrophil cytoskeleton, *Blood* **72:**141A

Babior, G. L., Rosin, R. E., McMurrich, B. J., Peters, W. A., and Babior, B. M., 1981, Arrangement of the respiratory burst oxidase in the plasma membrane of the neutrophil, *J. Clin. Invest.* **67:**1724–1728

Badwey, J. A., and Karnovsky, M. L., 1979, Production of superoxide and hydrogen peroxide by an NADH-oxidase in guinea pig polymorphonuclear leukocytes, *J. Biol. Chem.* **254:**11530–11537

Badwey, J. A., Curnutte, J. T., Robinson, J. M., Lazdins, J. K., Briggs, R. T., Karnovsky, M. J., and Karnovsky, M. L., 1980, Comparative aspects of oxidative metabolism of neutrophils from human blood and guinea pig peritonea: Magnitude of the respiratory burst, dependence upon stimulating agents, and localization of the oxidases, *J. Cell Physiol.* **105:**541–551

Badwey, J. A., Curnutte, J. T., Robinson, J. M., Berde, C. B., Karnovsky, M. J., and Karnovsky, M. L., 1984, Effects of free fatty acids on release of superoxide and on change of shape by human neutrophils: Reversibility by albumin, *J. Biol. Chem.* **259:**7870–7877

Bellavite, P., Berton, G., Dri, P., and Soranzo, M. R., 1981, Enzymatic basis of the respiratory burst of guinea pig resident peritoneal macrophages, *J. Reticuloendothel. Soc.* **29:**47–60

Bellavite, P., Serra, M. C., Davoli, A. and Rossi, F., 1982, Selective enrichment of NADPH oxidase activity in phagosomes from guinea pig polymorphonuclear leukocytes, *Inflammation* **6:**21–29

Bellavite, P., Cross, A. R., Serra, M. C., Davoli, A., Jones, O. T. G., and Rossi, F., 1983, The cytochrome b and flavin content and properties of the O2- forming NADPH oxidase solubilized from activated neutrophils, *Biochim. Biophys. Acta* **746:**40–47

Bellavite, P., Jones, O. T. G., Cross, A. R., Papini, E., and Rossi, F., 1984, Composition of partially purified NADPH oxidase from pig neutrophils, *Biochem. J.* **223:**639–648

Berton, G., Bellavite, P., Dri, P., De Togni, P., and Rossi, F., 1982, The enzyme responsible for the respiratory burst in elicited guinea pig peritoneal macrophages, *J. Pathol.* **136:**273–279

Berton, G., Cassatella, M., Cabrini, G., and Rossi, F., 1985, Activation of mouse macrophages causes no change in expression and function of phorbol diester receptors, but is accompanied by alterations in the activity and kinetic parameters of NADPH oxidase, *Immunology* **54:**371–377

Berton, G., Zeni, L., Cassatella, M. A., and Rossi, F., 1986, Gamma interferon is able to enhance the oxidative metabolism of human neutrophils, *Biochem. Biophys. Res. Commun.* **138:**1276–1282

Bielski, B. H. J., and Allen, A. O., 1977, Mechanism of the disproportionation of superoxide radicals, *J. Phys. Chem.* **81**:1048–1050

Bolscher, B. G. J. M., van Zwieten, R., Weening, R. S., Verhoeven, A. J., and Roos, D., 1989, A phosphoprotein of Mr 47,000, defective in autosomal chronic granulomatous disease, copurifies with one of two soluble components required for NADPH : O$_2$ oxidoreductase activity in human neutrophils, *J. Clin. Invest.* **83**:757–763

Borregaard, N., Schwartz, J. H., and Tauber, A. I., 1984a, Proton secretion by neutrophils: Significance of hexose monophosphate shunt activity as a source of electrons and protons for the respiratory burst, *J. Clin. Invest.* **74**:455–459

Borregaard, N., and Tauber, A. I., 1984b, Subcellular localization of the human neutrophil NADPH-oxidase : b-cytochrome and associated flavoprotein, *J. Biol. Chem.* **259**:47–52

Borregaard, N., Heiple, J. M., Simons, E. R., and Clark, R. A., 1983, Subcellular localization of the b-cytochrome component of the human neutrophil microbicidal oxidase: Translocation during activation, *J. Cell Biol.* **97**:52–61

Braughler, J. M., Duncan, L. A., and Chase, R. L., 1986, The involvement of iron in lipid peroxidation. Importance of ferric to ferrous ratios in initiation, *J. Biol. Chem.* **261**:10282–10289

Briggs, R. T., Karnovsky, M. L., and Karnovsky, M. J., 1977, Hydrogen peroxide production in chronic granulomatous disease. A cytochemical study of reduced pyridine nucleotide, *J. Clin. Invest.* **59**:1088–1098

Bromberg, Y., and Pick, E., 1984, Unsaturated fatty acids stimulate NADPH-dependent superoxide production by cell-free system derived from macrophages, *Cell Immunol.* **88**:213–221

Cassatella, M. A., Della Bianca, V., Berton, G., and Rossi, F., 1985, Activation by gamma interferon of human macrophage capability to produce toxic oxygen molecules is accompanied by decreased Km of the superoxide-generating NADPH oxidase, *Biochem. Biophys. Res. Commun.* **132**:908–913

Cassatella, M. A., Cappelli, R., Della Bianca, V., Grzeskowiak, M., Dusi, S., and Berton, G., 1988, Interferon-gamma activates human neutrophil oxygen metabolism and exocytosis, *Immunology* **63**:499–506

Cassatella, M. A., Hartman, L., Perussia, B., and Trinchieri, G., 1989, Tumor necrosis factor and immune interferon synergistically induce cytochrome b$_{-245}$ heavy-chain gene expression and nicotinamide-adenine dinucleotide phosphate hydrogenase oxidase in human leukemic myeloid cells, *J. Clin. Invest.* **83**:1570–1579

Cerutti, P. A., 1985, Pro-oxidant states and tumor promotion, *Science* **227**:375–381

Clark, R. A., and Nauseef, W. M., 1989, Translocation of cytosolic components of neutrophil NADPH oxidase, *Clin. Res.* **37**:607A

Cohen, H. J., Chovaniec, M. E., and Davies, W. A., 1980, Activation of the guinea pig granulocyte NAD(P)H-dependent superoxide generating enzyme: Localization in a plasma membrane enriched particle and kinetics of activation, *Blood* **55**:355–363

Cohen, H. J., Chovaniec, M. E., Wilson, M. K., and Newburger, P. E., 1982, Con-A-stimulated superoxide production by granulocytes: Reversible activation of NADPH oxidase, *Blood* **60**:1188–1194

Cox, J. A., Jeng, A. Y., Sharkey, N. A., Blumberg, P. M., and Tauber, A. I., 1985, Activation of the human neutrophil nicotinamide adenine dinucleotide phosphate (NADPH)-oxidase by protein kinase C, *J. Clin. Invest.* **76**:1932–1938

Cramer, E., Pryzwansky, K. B., Villeval, J.-L., Testa, U., and Breton-Gorius, J., 1985, Ultrastructural localization of lactoferrin and myeloperoxidase in human neutrophils by immunogold, *Blood* **65**:423–432

Crawford, D. R., and Schneider, D. L., 1982, Identification of ubiquinone-50 in human neutrophils and its role in microbicidal events, *J. Biol. Chem.* **257**:6662–6668

Cross, A. R., and Jones, O. T. G., 1986, The effect of the inhibitor diphenylene iodonium on the superoxide-generating system of neutrophils, *Biochem. J.* **237**:111–116

Cross, A. R., Jones, O. T. G., Harper, A. M., and Segal, A. W., 1981, Oxidation-reduction properties of the cytochrome b found in the plasma-membrane fraction of human neutrophils. A possible oxidase in the respiratory burst, *Biochem. J.* **194**:599–606

Cross, A. R., Higson, F. R., Jones, O. T. G., Harper, A. M., and Segal, A. W., 1982a, The enzymic reduction and kinetics of oxidation of cytochrome b-245 of neutrophils, *Biochem. J.* **204**:479–485

Cross, A. R., Jones, O. T. G., Garcia, R., and Segal, A. W., 1982b, The association of FAD with the cytochrome b-245 of human neutrophils, *Biochem. J.* **208**:759–763

Cross, A. R., Parkinson, J. F., and Jones, O. T. G., 1984, The superoxide-generating oxidase of leucocytes. NADPH- dependent reduction of flavin and cytochrome b in solubilized preparations, *Biochem. J.* **223:**337–344

Cross, A. R., Parkinson, J. F., and Jones, O. T. G., 1985, Mechanism of the superoxide-producing oxidase of neutrophils, *Biochem. J.* **226:**881–884

Curnutte, J. T., 1985, Activation of human neutrophil nicotinamide adenine dinucleotide phosphate, reduced (triphosphopyridine nucleotide, reduced) oxidase by arachidonic acid in a cell-free system, *J. Clin. Invest.* **75:**1740–1743

Curnutte, J. T., and Tauber, A. I., 1983, Failure to detect superoxide in human neutrophils stimulated with latex particles, *Pediatr. Res.* **17:**281–284

Curnutte, J. T., Babior, B. M., and Karnovsky, M. L., 1979, Fluoride-mediated activation of the respiratory burst in human neutrophils, *J. Clin. Invest.* **63:**637–647

Curnutte, J. T., Karnovsky, M. L., and Babior, B. M., 1976, Manganese-dependent NADPH oxidation by granulocyte particles, *J. Clin. Invest.* **57:**1059–1067

Curnutte, J. T., Kuver, R., and Scott, P. J., 1987, Activation of neutrophil NADPH oxidase in a cell-free system. Partial purification of components and characterization of the activation process, *J. Biol. Chem.* **262:**5563–5569

Curnutte, J. T., Scott, P. J., and Mayo, L. A., 1988b, The cytosolic components of the respiratory burst oxidase: Resolution of four components, two of which are missing in complementing types of gamma interferon-responsive chronic granulomatous disease, *Blood* **72:** 144 (Abstract)

Curnutte, J. T., Scott, P. J., and Mayo, L. A., 1989a, The cytosolic components of the respiratory burst oxidase: Resolution of four components, two of which are missing in complementing forms of chronic granulomatous disease, *Proc. Natl. Acad. Sci. USA* **86:**825–836

Curnutte, J. T., Scott, P. J., and Babior, B. M., 1989b, The functional defect in neutrophil cytosols from 2 patients with type II (autosomal recessive cytochrome-positive) chronic granulomatous disease, *J. Clin. Invest.* **83:**1236–1240

Curnutte, J. T., Parkos, C. A., Scott, P. J., and Jesaitis, A. J., 1988a, Reconstitution of NADPH oxidase activity in membranes from patients with Chronic Granulomatous Disease, *Clin. Res.* **36:**403A(Abstract)

Curnutte, J. T., Scott, P. J., and Babior, B. M., 1989c, The functional defect in neutrophil cytosols from 2 patients with type II (autosomal recessive cytochrome-positive) chronic granulomatous disease, *J. Clin. Invest.* **83:**1236–1240

Dewald, B., Baggiolini, M., Curnutte, J. T., and Babior, B. M., 1979, Subcellular localization of the superoxide-forming enzyme in human neutrophils, *J. Clin. Invest.* **63:**21–29

Dinauer, M. C., and Orkin, S. H., 1988, Chronic granulomatous disease: Molecular genetics. *Hematol./Oncol. Clin. North Am.* **2:**225–240

Dinauer, M. C., Orkin, S. H., Brown, R., Jesaitis, A. J., and Parkos, C. A., 1987, The glycoprotein encoded by the X-linked chronic granulomatous disease locus is a component of the neutrophil cytochrome b complex, *Nature* (London) **327:**717–720

Dinauer, M. C., Orkin, S. H., Hurst, J. K., Parkos, C. A., Jesaitis, A. J., Rosen, H., and Curnutte, J. T., 1989, A missence mutation in the neutrophil cytochrome b heavy chain leading to X-linked chronic granulomatous disease, *Clin. Res.* **37:**544A

Doussiere, J., and Vignais, P. V., 1985, Purification and properties of O_2^- generating oxidase from bovine polymorphonuclear neutrophils, *Biochemistry* **24:**7231–7239

Doussiere, J., Laporte, F., and Vignais, P. V., 1986, Photolabeling of a O_2^- generating protein in bovine polymorphonuclear neutrophils by an arylazido NADP+ analog, *Biochem. Biophys. Res. Commun.* **139:**85–93

Edelman, A. M., Blumenthal, D. K., and Krebs, E. G., 1987, Protein serine/threonine kinases, *Annu. Rev. Biochem.* **56:**567–614

Ezekowitz, R. A. B., and Newburger, P. E., 1988, New perspectives in chronic granulomatous disease, *J. Clin. Immunol.* **8:**419–425

Ezekowitz, R. A. B., Orkin, S. H., and Newburger, P. E., 1987, Recombinant interferon gamma augments phagocyte superoxide production and X-chronic granulomatous disease gene expression in X-linked variant chronic granulomatous disease, *J. Clin. Invest.* **80:**1009–1016

Ezekowitz, R. A. B., Dinauer, M. C., Jaffe, H. S., Orkin, S. H., and Newburger, P. E., 1988, Partial correction of the phagocyte defect in patients with X-linked chronic granulomatous disease by subcutaneous interferon gamma, *N. Engl. J. Med.* **319:**146–151

Fantone, J. C., and Ward, P. A., 1982, Role of oxygen-derived free radicals and metabolites in leukocyte-dependent inflammatory reactions, *Am. J. Pathol.* **107:**397–418

Fielden, E. M., Roberts, P. B., Bray, R. C., Lowe, D. T., Mautner, G. N., Rotilio, G., and Calabrese, L., 1974, The mechanism of action of superoxide dismutase from pulse radiolysis and electron paramagnetic resonance, *Biochem. J.* **139:**49–60

Frenkel, K., Chrzan, K., Troll, W., Teebor, G. W., and Steinberg, J. J., 1986, Radiation-like modification of bases in DNA exposed to tumor promoter-activated polymorphonuclear leukocytes, *Cancer Res.* **46:**5533–5540

Fujita, I., Takeshige, K., and Minakami, S., 1987, Characterization of the NADPH-dependent superoxide production activated by sodium dodecyl sulfate in a cell-free system of pig neutrophils, *Biochim. Biophys. Acta* **931:**41–48

Gabig, T. G., 1983, The NADPH-dependent O_2^--generating oxidase from human neutrophils. Identification of a flavoprotein component that is deficient in a patient with chronic granulomatous disease, *J. Biol. Chem.* **258:**6352–3656

Gabig, T. G., and Babior, B. M., 1979, The O_2^--forming oxidase responsible for the respiratory burst in human neutrophils. Properties of the solubilized enzyme, *J. Biol. Chem.* **254:**9070–9079

Gabig, T. G., and Lefker, B. A., 1984a, Deficient flavoprotein component of the NADPH-dependent O_2^--generating oxidase in the neutrophils from three male patients with chronic granulomatous disease, *J. Clin. Invest.* **73:**701–705

Gabig, T. G., and Lefker, B. A., 1984b, Catalytic properties of the resolved flavoprotein and cytochrome b components of the NADPH dependent O_2^- generating oxidase from human neutrophils, *Biochem. Biophys. Res. Commun.* **118:**430–436

Gabig, T. G., and Lefker, B. A., 1985, Activation of the human neutrophil NADPH oxidase results in coupling of electron carrier function between ubiquinone-10 and cytochrome b559, *J. Biol. Chem.* **260:**3991–3995

Gabig, T. G., Kipnes, R. S., and Babior, B. M., 1978, Solubilization of the O2- -forming activity responsible for the respiratory burst in human neutrophils, *J. Biol. Chem.* **253:**6663–6665

Gabig, T. G., Schervish, E. W., and Santinga, J. T., 1982, Functional relationship of the cytochrome b to the superoxide-generating oxidase of human neutrophils, *J. Biol. Chem.* **257:**4114–4119

Garcia, C. C., and Segal, A. W., 1984, Changes in the subcellular distribution of the cytochrome b-245 on stimulation of human neutrophils, *Biochem. J.* **219:**233–242

Gilman, A. G., 1987, G proteins: Transducers of receptor-generated signals, *Annu. Rev. Biochem.* **56:**615–650

Glass, G. A., DeLisle, D. M., DeTogni, P., Gabig, T. G., Magee, B. H., Markert, M., and Babior, B. M., 1986, The respiratory burst oxidase of human neutrophils. Further studies of the purified enzyme, *J. Biol. Chem.* **261:**13247–13251

Graf, E., Mahoney, J. R., Bryant, R. G., and Eaton, J. W., 1984, Iron-catalyzed hydroxyl radical formation, *J. Biol. Chem.* **259:**3620–3624

Green, M. R., Hill, H. A. O., Okolow-Zubkowska, M. J., and Segal, A. W., 1979, The production of hydroxyl and superoxide radicals by stimulated human neutrophils—measurement by EPR spectroscopy, *FEBS Lett.* **100:**23–26

Haber, F., and Weiss, J., 1934, The catalytic decomposition of hydrogen peroxide by iron salts, *Proc. R. Soc. Edinburgh Sect. A* **147:**332–334

Halliwell, B., 1987, Oxidants and human disease: Some new concepts, *FASEB J.* **1:**358–364

Halliwell, B., and Gutteridge, J. M. C., 1986, Iron and free radical reactions: Two aspects of antioxidant protection, *Trends Biochem. Sci.* **11:**372–375

Halliwell, B., Gutteridge, J. M. C., and Blake, D., 1985, Metal ions and oxygen radical reactions in human inflammatory joint disease, *Philos. Trans. R. Soc. London Ser. B* **311:**659–671

Hamers, M. N., de Boer, M., Meerhof, L. J., Weening, R. S., and Roos, D., 1984, Complementation in monocyte hybrids revealing genetic heterogeneity in chronic granulomatous disease, *Nature* (London) **307:**553–555

Haslam, R. J., and Lynham, J. A., 1977, Relationship between phosphorylation of blood platelet proteins and secretion of platelet granule constituents. I. Effects of different aggregating agents, *Biochem. Biophys. Res. Commun.* **77**:714–722

Hattori, H., 1961, Studies on the labile, stable NADI oxidase and peroxidase staining reactions in the isolated particles of horse granulocyte, *Nagoya J. Med. Sci.* **23**:362–378

Hayakawa, T., Suzuki, K., Suzuki, S., Andrews, P. C., and Babior, B. M., 1986, A possible role for protein phosphorylation in the activation of the respiratory burst in human neutrophils, *J. Biol. Chem.* **261**:9109–9115

Helfman, D. M., Appelbaum, B. D., Vogler, W. R., and Kuo, J. F., 1983, Phospholipid-sensitive Ca2+-dependent protein kinase and its substrates in human neutrophils, *Biochem. Biophys. Res Commun.* **111**:847–853

Heyneman, R. A., and Vercauteren, R. E., 1984, Activation of a NADPH oxidase from horse polymorphonuclear leukocytes in a cell-free system, *J. Leukocyte Biol.* **36**:751–759

Heyworth, P. G., and Segal, A. W., 1986, Further evidence for the involvement of the phosphoprotein in the respiratory burst oxidase of human neutrophils, *Biochem. J.* **239**:723–731

Heyworth, P. G., Shrimpton, C. F., and Segal, A. W., 1989, Localization of the 47 kDa phosphoprotein involved in the respiratory-burst NADPH oxidase of phagocytic cells, *Biochem. J.* **260**:243–248

Imaoka, T., Lynham, J. A., and Haslam, R. J., 1983, Purification and characterization of the 47,000-dalton protein phosphorylated during degranulation of human platelets, *J. Biol. Chem.* **258**:11404–11414

Johnson, R. J., Couser, W. G., Chi, E. Y., Adler, S., and Klebanoff, S. J., 1987, New mechanism for glomerular injury. Myeloperoxidase-hydrogen peroxide-halide system, *J. Clin. Invest.* **79**:1379–1387

Jolly, S. R., Kane, W. J., Bailie, M. B., Abrams, G. D., and Lucchesi, B. R., 1984, Canine myocardial reperfusion injury. Its reduction by the combined administration of superoxide dismutase and catalase, *Circ. Res.* **54**:277–285

Kakinuma, K., Kaneda, M., Chiba, T., and Ohnishi, T., 1986, Electron spin resonance studies on a flavoprotein in neutrophil plasma membranes, *J. Biol. Chem.* **261**:9426–9432

Kakinuma, K., Fukuhara, Y., and Kaneda, M., 1987, The respiratory burst oxidase of neutrophils. Separation of an FAD enzyme and its characterization, *J. Biol. Chem.* **262**:12316–12322

Kanofsky, J. R., Wright, J., Miles-Richardson, G. E., and Tauber, A. I., 1984, Biochemical requirements for singlet oxygen production by purified human myeloperoxidase, *J. Clin. Invest.* **74**:1489–1495

Kikuchi, A., Kozawa, O., Kaibuchi, K., Katada, T., Ui, M., and Takai, Y., 1986, Direct evidence for involvement of a guanine nucleotide-binding protein in chemotactic peptide-stimulated formation of inositol bisphosphate and trisphosphate in differentiated human leukemic (HL-60) cells: Reconstitution with G1 or G0 of the plasma membranes ADP-ribosylated by pertussis toxin, *J. Biol. Chem.* **261**:11558–11562

Kitagawa, S., and Johnston, R. B., Jr., 1986, Deactivation of the respiratory burst in activated macrophages: Evidence for alteration of signal transduction, *J. Immunol.* **136**:2605–2612

Kitahara, M., Eyre, H. J., Simonian, Y., Atkin, C. L., and Hasstedt, S. J., 1981, Hereditary myeloperoxidase deficiency, *Blood* **57**:888–893

Klebanoff, S. J., 1968, Myeloperoxidase-halide-hydrogen peroxide antibacterial system, *J. Bacteriol.* **95**:2131–2138

Kojima, H., Takahashi, K., Sakane, F., and Koyama, J., 1987, Purification and characterization of NADPH-cytochrome c reductase from porcine polymorphonuclear leukocytes, *J. Biochem.* **102**:1083–1088

Krebs, E. G., and Beavo, J. A., 1979, Phosphorylation-dephosphorylation of enzymes, *Annu. Rev. Biochem.* **48**:923–959

Lehrer, R. I., and Cline, M. J., 1969, Leukocyte myeloperoxidase deficiency and disseminated candidiasis: The role of myeloperoxidase in resistance to Candida infection, *J. Clin. Invest.* **48**:1478–1488

Leto, T. L., Lomax, K. J., Sechler, J. M. G., Nunoi, H., Gallin, J. I., and Malech, H. L., 1989, Molecular cloning of the 65 kDa cytosolic factor absent in a rare form of autosomal recessive chronic granulomatous disease, *Clin. Res.* **37**:547A

Lew, P. D., Southwick, F. S., Stossel, T. P., Whitin, J. C., Simons, E., and Cohen, H. J., 1981, A variant of chronic granulomatous disease: Deficient oxidative metabolism due to a low-affinity NADPH oxidase, *N. Engl. J. Med.* **305**:1329–1333

Ligeti, E., Doussiere, J., and Vignais, P. V., 1988, Activation of the O_2^--generating oxidase in plasma membrane from bovine polymorphonuclear neutrophils by arachidonic acid, a cytosolic factor of protein nature, and nonhydrolyzable analogues of GTP, *Biochemistry* **27**:193–200

Light, D. R., Walsh, C., O'Callaghan, A. M., Goetzl, E. J., and Tauber, A. I., 1981, Characteristics of the cofactor requirements for the superoxide-generating NADPH oxidase of human polymorphonuclear leukocytes, *Biochemistry* **20**:1468–1476

Lomax, K. J., Leto, T. L., Nunoi, H., Gallin, J. I., and Malech, H. L., 1989, Recombinant 47-kilodalton cytosol factor restores NADPH oxidase in chronic granulomatous disease, *Science* **245**:409–412

Lutter, R., van Zwieten, R., Weening, R. S., Hamers, M. N., and Roos, D., 1984, Cytochrome b, flavins, and ubiquinone-50 in enucleated human neutrophils (PMN cytoplasts), *J. Biol. Chem.* **259**:9603–9606

Lyons, R. M., and Atherton, R. M., 1979, Characterization of a platelet protein phosphorylated during the thrombin-induced release reaction, *Biochemistry* **18**:544–551

Markert, M., Glass, G. A., and Babior, B. M., 1985, Respiratory burst oxidase from human neutrophils: Purification and some properties, *Proc. Natl. Acad. Sci. USA* **82**:3144–3148

Matzner, Y., Brass, L. M., McMurrich, B. J., Peters, W. A., Andre-Schwartz, J., and Babior, B. M., 1982, Expression of a chronic granulomatous disease-like defect by flouride-exhausted neutrophils, *Blood* **60**:822–826

McClune, G. J., and Fee, J. A., 1976, Stopped flow spectrophotometric observation of superoxide generation in aqueous solution, *FEBS Lett.* **67**:294–298

McCord, J. M., and Day, E. D., 1978, Superoxide-dependent production of hydroxyl radical catalyzed by iron-EDTA complex, *FEBS Lett.* **86**:139–142

McCord, J. M., and Fridovich, I., 1969, Superoxide dismutase. An enzymic function for erythrocuprein, *J. Biol. Chem.* **244**:6049–6055

McGuire, W. W., Spragg, R. G., Cohen, A. B., and Cochrane, C. G., 1982, Studies on the pathogenesis of the adult respiratory distress syndrome, *J. Clin. Invest.* **69**:543–553

McPhail, L. C., Shirley, P. S., Clayton, C. C., and Snyderman, R., 1985, Activation of the respiratory burst enzyme from human neutrophils in a cell-free system: Evidence for a soluble cofactor, *J. Clin. Invest.* **75**:1735–1739

Nakamura, M., Baxter, C. R., and Master, B. B. S., 1981, Simultaneous demonstration of phagocytosis-connected oxygen consumption and corresponding NAD(P)H oxidase activity. Direct evidence of NADPH as the predominant electron donor to oxygen in phagocytizing human neutrophils, *Biochem. Biophys. Res. Commun.* **98**:743–751

Nathan, C. F., Murray, H. W., Wiebe, M. E., and Rubin, B. Y., 1983, Identification of interferon-gamma as the lymphokine that activates human macrophage oxidative metabolism and antimicrobial activity, *J. Exp. Med.* **158**:670–689

Neer, E. J., and Clapham, D. E., 1988, Roles of G protein subunits in transmembrane signalling, *Nature* (London) **333**:129–134

Newburger, P. E., and Tauber, A. I., 1982, Heterogenous pathways of oxidizing radical production in human neutrophils and the HL-60 cell line, *Pediatr. Res.* **16**:856–860

Newburger, P. E., Luscinskas, F. W., Ryan, T., Beard, C. J., Wright, J., Platt, O. S., Simons, E. R., and Tauber, A. I., 1986, Variant chronic granulomatous disease: Modulation of the neutrophil by severe infection, *Blood* **68**:914–919

Newburger, P. E., Ezekowitz, R. A. B., Whitney, C., Wright, J., and Orkin, S. H., 1988, Induction of phagocyte cytochrome b heavy chain gene expression by interferon gamma, *Proc. Natl. Acad. Sci. USA* **85**:5215–5219

Nunoi, H., Rotrosen, D., Gallin, J. I., and Malech, H. L., 1988, Two forms of autosomal chronic granulomatous disease lack distinct neutrophil cytosol factors, *Science* **242**:1298–1301

Nunoi, H., Gallin, J. I., and Malech, H. L., 1989, Evidence for interaction of neutrophil 47 kDa phosphoprotein with cytochrome during activation of the respiratory burst, *Clin. Res.* **37**:438A

Ohno, Y., Hirai, K., Kanoh, T., Uchino, H., and Ogawa, K., 1982, Subcellular localization of H_2O_2 production in human neutrophils stimulated with particles and an effect of cytochalasin-B on the cells, *Blood* **60**:253–260

Ohta, H., Takahashi, H., and Hattori, H., 1966, Some oxidative enzymes and cytochrome in the specific granules of neutrophil leukocytes, *Acta Haematol Jpn.* **29**:799–808

Ohta, H., Okajima, F., and Ui, M., 1985, Inhibition by islet-activating protein of a chemotactic peptide-induced early breakdown of inositol phospholipids and Ca^{2+} mobilization in guinea pig neutrophils, *J. Biol. Chem.* **260**:15771–15780

Okamura, N., Curnutte, J. T., Roberts, R. L., and Babior, B. M., 1988a, Relationship of protein phosphorylation to the activation of the respiratory burst in human neutrophils. Defects in the phosphorylation of a group of closely related 48-kDa proteins in two forms of chronic granulomatous disease, *J. Biol. Chem.* **263**:6777–6782

Okamura, N., Malawisti, S. E., Roberts, R. L., Rosen, H., Ochs, H. D., Babior, B. M., and Curnutte, J. T., 1988b, Phosphorylation of the oxidase-related 48K phosphoprotein family in the unusual autosomal cytochrome-negative and x-linked cytochrome-positive types of chronic granulomatous disease, *Blood* **72**:811–816

Olofsson, T., and Olsson, I., 1977, Purification of human granulocyte catalase in chronic myeloid leukemia, *Biochim. Biophys. Acta* **482**:301–308

Parkos, C. A., Allen, R. A., Cochrane, C. G., and Jesaitis, A. J., 1987, Purified cytochrome b from human granulocyte plasma membrane is comprised of two polypeptides with relative molecular weights of 91,000 and 22,000, *J. Clin. Invest.* **80**:732–742

Parkos, C. A., Dinauer, M. C., Walker, L. E., Allen, R. A., Jesaitis, A. J., and Orkin, S. H., 1988a, Primary structure and unique expression of the 22-kilodalton light chain of human neutrophil cytochrome b, *Proc. Natl. Acad. Sci. USA* **85**:3319–3323

Parkos, C. A., Allen, R. A., Cochrane, C. G., and Jesaitis, A. J., 1988b, The quaternary structure of the plasma membrane b-type cytochrome of human granulocytes, *Biochim. Biophys. Acta* **932**:71–83

Parkos, C. A., Dinauer, M. C., Jesaitis, A. J., Orkin, S. H., and Curnutte, J. T., 1989, Absence of both the 91kD and 22kD subunits of human neutrophil cytochrome b in two genetic forms of chronic granulomatous disease, *Blood* **73**:1416–1420

Parry, M. F., Root, R. K., Metcalf, J. A., Delaney, K. K., Kaplow, L. S., and Richar, W. J., 1981, Myeloperoxidase deficiency: Prevalence and clinical significance, *Ann. Intern. Med.* **95**:293–301

Patriarca, P., Cramer, R., Moncalvo, S., Rossi, F., and Romeo, D., 1971, Enzymatic basis of metabolic stimulation in leukocytes during phagocytosis: the role of activated NADPH oxidase, *Arch. Biochem. Biophys.* **145**:255–262

Patriarca, P., Dri, P., Kakinuma, K., Tedesco, F. and Rossi, F., 1975, Studies on the mechanism of metabolic stimulation in polymorphonuclear leukocytes during phagocytosis. I. Evidence for superoxide anion involvement in the oxidation of NADPH, *Biochim. Biophys. Acta* **385**:380–386

Pick, E., Bromberg, Y., Sphungin, S., and Gadba, R., 1987, Activation of the superoxide forming NADPH oxidase in a cell-free system by sodium dodecyl sulfate. Characterization of the membrane-associated component, *J. Biol. Chem.* **262**:16476–16483

Pontremoli, S., Melloni, E., Michetti, M., Sacco, O., Salamino, F., Sparatore, B., and Horecker, B. L., 1986, Biochemical response in activated human neutrophils mediated by protein kinase C and a Ca^{2+}-requiring proteinase, *J. Biol. Chem.* **261**:8309–8313

Ragan, C. I., and Bloxham, D. P., 1977, Specific labeling of a constituent polypeptide of bovine heart mitochondrial reduced nicotinamide-adenine dinucleotide-ubiquinone reductase by the inhibitor diphenyleneiodinium, *Biochem. J.* **163**:605–615

Reed, P. W., 1969, Glutathione and the hexose monophosphate shunt in phagocytizing and hydrogen peroxide-treated rat leukocytes, *J. Biol. Chem.* **244**:2459–2464

Rohrer, J. S., Joo, M., Dartyge, E., Sayers, D. E., Fontaine, A., and Theil, E. C., 1987, Stabilization of iron in a ferrous form by ferritin. A study using dispersive and conventional X-ray absorption spectroscopy, *J. Biol. Chem.* **262**:13385–13387

Root, R. K., and Metcalf, J. A., 1977, H_2O_2 release from human granulocytes during phagocytosis: Relationship to superoxide anion formation and cellular catabolism of H_2O_2: Studies with normal and cytochalasin B treated cells, *J. Clin. Invest.* **60**:1266–1279

Rosen, H., and Klebanoff, S. J., 1979, Hydroxyl radical generation by polymorphonuclear leukocytes measured by electron spin resonance spectroscopy, *J. Clin. Invest.* **64**:1725–1729

Rossi, F., and Zatti, M., 1964, Changes in the metabolic pattern of polymorphonuclear leukocytes during phagocytosis, *Br. J. Exp. Pathol.* **45**:548–559

Royer-Pokora, B., Kunkel, L. M., Monaco, A. P., Goff, S. C., Newburger, P. E., Baehner, R. L., Cole, F. S., Curnutte, J. T., and Orkin, S. H., 1986, Cloning the gene for an inherited human disorder—chronic granulomatous disease—on the basis of its chromosomal location, *Nature* (London) **322**:32–38

Sakane, F., Takahashi, K., and Koyama, J., 1984, Purification and characterization of a membrane-bound NADPH-cytochrome c reductase capable of catalyzing meadione-dependent O_2^- formation in guinea pig polymorphonuclear leukocytes, *J. Biochem.* **96**:671–678

Sasada, M., Pabst, M. J., and Johnston, R. B., 1983, Activation of mouse peritoneal macrophages by lipopolysaccharide alters the kinetic parameters of the superoxide-producing NADPH oxidase, *J. Biol. Chem.* **258**:9631–9635

Sbarra, A. J., and Karnovsky, M. L., 1959, The biochemical basis of phagocytosis: I. Metabolic changes during the ingestion of particles by polymorphonuclear leukocytes, *J. Biol. Chem.* **234**:1355–1362

Sechler, J. M. G., Malech, H. L., White, C. J., and Gallin, J. I., 1988, Recombinant human interferon-gamma reconstitutes defective phagocyte function in patients with chronic granulomatous disease of childhood, *Proc. Natl. Acad. Sci. USA* **85**:4874–4878

Segal, A. W., 1987, Absence of both cytochrome b-245 subunits from neutrophils in X-linked chronic granulomatous disease, *Nature* (London) **326**:88–92

Segal, A. W., 1988, Cytochrome b-245 and its involvement in the molecular pathology of chronic granulomatous disease, *Hematol. Oncol. Clin. North Am.* **23**:213–223

Segal, A. W., 1989, The electron transport chain of the microbicidal oxidase of phagocytic cells and its involvement in the molecular pathology of chronic granulomatous disease, *J. Clin. Invest.* **83**:1785–1793

Segal, A. W., and Jones, O. T. G., 1978, Novel cytochrome b system in phagocytic vacuoles of human granulocytes, *Nature* (London) **276**:515–517

Segal, A. W., and Jones, O. T. G., 1979a, Reduction and subsequent oxidation of a cytochrome b of human neutrophils after stimulation with phorbol myristate acetate, *Biochem. Biophys. Res. Commun.* **88**:130–134

Segal, A. W., and Jones, O. T. G., 1979b, The subcellular distribution and some properties of the cytochrome b component of the microbicidal oxidase system of human neutrophils, *Biochem. J.* **182**:181–188

Segal, A. W., and Jones, O. T. G., 1980a, Absence of cytochrome b reduction in stimulated neutrophils from both female and male patients with chronic granulomatous disease, *FEBS Lett.* **110**:111–114

Segal, A. W., and Jones, O. T. G., 1980b, Rapid incorporation of the human neutrophil plasma membrane cytochrome b into phagocytic vacuoles, *Biochem. Biophys. Res. Commun.* **92**:710–715

Segal, A. W., Jones, O. T. G., Webster, D., and Allison, A. C., 1978, Absence of a newly described cytochrome b from neutrophils of patients with chronic granulomatous disease, *Lancet* **ii**:446–449

Segal, A. W., Garcia, R., Goldstone, H., Cross, A. R., and Jones, O. T. G., 1981, Cytochrome b-245 of neutrophils is also present in human monocytes, macrophages and eosinophils, *Biochem. J.* **196**:363–367

Segal, A. W., Cross, A. R., Garcia, R. C., Borregaard, N., Valerius, N., Soothill, J. F., and Jones, O. T. G., 1983, Absence of cytochrome b-245 in chronic granulomatous disease: A multicenter European evaluation of its incidence and relevance, *N. Engl. J. Med.* **308**:245–251

Segal, A. W., Heyworth, P. G., Cockcroft, S., and Barrowman, M. M., 1985, Stimulated neutrophils from patients with autosomal recessive chronic granulomatous disease fail to phosphorylate a Mr-44,000 protein, *Nature* (London) **316**:547–549

Seger, R. A., Tiefenauer, L., Matsunaga, T., Wildfeuer, A., and Newburger, P. E., 1983, Chronic granulomatous disease due to granulocytes with abnormal NADPH oxidase activity and deficient cytochrome-b, *Blood* **61**:423–428

Seifert, R., and Schultz, G., 1987, Reversible activation of NADPH oxidase in membranes of HL-60 human leukemic cells, *Biochem. Biophys. Res. Commun.* **146**:1296–1302

Serra, M. C., Bellavite, P., Davoli, A., Bannister, J. V., and Rossi, R., 1984, Isolation from neutrophil membranes of a complex containing active NADPH oxidase and cytochrome b-245, *Biochim. Biophys. Acta* **788**:138–146

Sha'ag, D., and Pick, E., 1988, Macrophage derived superoxide-generating NADPH oxidase in an amphiphile-activated, cell-free system; partial purification of the cytosolic component and evidence that it may contain the NADPH binding site, *Biochim. Biophys. Acta* **952**:213–219

Shanblatt, S. H., and Revzin, A., 1987, Interactions of the catabolite activator protein (CAP) at the galactose and lactose promoters of Escherichia coli probed by hydroxyl radical footprinting. The second CAP molecule which binds at gal and the one CAP at lac may act to stimulate transcription in the same way, *J. Biol. Chem.* **262**:11422–11427

Shinagawa, Y., Tanaka, C., and Teraoka, A., 1966, A new cytochrome in neutrophilic granules of rabbit leucocyte, *J. Biochem.* **59**:622–624

Smith, R. M., Curnutte, J. T., and Babior, B. M., 1989a, Affinity labeling of the cytosolic and membrane components of the respiratory burst oxidase by the 2′,3′-dialdehyde derivative of NADPH. Evidence for a cytosolic location of the nucleotide binding site in the resting cell, *J. Biol. Chem.* **264**:1958–1962

Smith, R. M., Curnutte, J. T., Mayo, L. A., and Babior, B. M., 1989b, Use of an affinity label to probe the function of the NADPH binding component of the respiratory burst oxidase of human neutrophils, *J. Biol. Chem.* **264**:12243–12248

Spielberg, S. P., Boxer, L. A., Oliver, J. M., Allen, J. M., and Schulman, J. D., 1979, Oxidative damage to neutrophils in glutathione synthetase deficiency, *Br. J. Haematol.* **42**:215–223

Strauss, R. R., Paul, B. B., Jacobs, A. A., and Sbarra, A. J., 1969, The role of the phagocyte in host-parasite interactions. XIX. Leukocyte glutathione reductase and its involvement in phagocytosis, *Arch. Biochem. Biophys.* **135**:265–271

Takasugi, S., Ishida, K., Takeshige, K., and Minakami, S., 1989, Effect of 2′,3′-dialdehyde NADPH on activation of superoxide-producing NADPH oxidase in a cell-free system of pig neutrophils, *Biochem. J.* **23**:23–23

Tamura, M., Tamura, T., Tyagi, S. R., and Lambeth, J. D., 1988, The superoxide-generating respiratory burst oxidase of human neutrophil plasma membrane. Phosphatidylserine as an effector of the activated enzyme, *J. Biol. Chem.* **263**:17621–17626

Tauber, A. I., 1987, Protein kinase C and the activation of the human neutrophil NADPH-oxidase, *Blood* **69**:711–720

Teahan, C., Rowe, P., Parker, P., Totty, N., and Segal, A. W., 1978, The X-linked chronic granulomatous disease gene codes for the beta-chain of cytochrome b-245, *Nature* (London) **327**:720–721

Thelen, M., Peveri, P., Kernen, P., von Tscharner, V., Walz, A., and Baggiolini, M., 1988, Mechanism of neutrophil activation by NAF, a novel monocyte-derived peptide agonist, *FASEB J.* **2(11)**:2702–2706

Thomas, E. L., 1979a, Myeloperoxidase-hydrogen peroxide-chloride antimicrobial system: Nitrogen-chlorine derivatives of bacterial components in the bactericidal action against *Escherichia coli*, *Infect. Immun.* **23**:522–531

Thomas, E. L., 1979b, Myeloperoxidase-hydrogen peroxide-chloride antimicrobial system: Effect of exogenous amines on antibacterial action against *Escherichia coli*, *Infect. Immun.* **25**:110–116

Thomas, E. L., Grisham, M. B., Melton, D. F., and Jefferson, M. M., 1985, Evidence for a role of taurine in the in vitro oxidative toxicity of neutrophils toward erythrocytes, *J. Biol. Chem.* **260**:3321–3329

Trush, M. A., Seed, J. L., and Kensler, T. W., 1985, Oxidant-dependent metabolic activation of plycyclic aromatic hydrocarbons by phorbol ester-stimulated human polymorphonuclear leukocytes: Possible link between inflammation and cancer, *Proc. Natl. Acad. Sci. USA* **82**:5194–5199

Tsunawaki, S., and Nathan, C. F., 1984, Enzymatic basis of macrophage activation. Kinetic analysis of superoxide production in lysates of resident and activated mouse peritoneal macrophages and granulocytes, *J. Biol. Chem.* **259**:4305–4312

Tyers, M., Rachubinski, R. A., Stewart, M. I., Varrichio, A. M., Shorr, R. G. L., Haslam, R. J., and Harley, C. B., 1988, Molecular cloning and expression of the major protein kinase C substrate of platelets, *Nature* (London) **333**:470–474

Umei, T., Takeshige, K., and Minakami, S., 1986, NADPH binding component of neutrophil superoxide-generating oxidase, *J. Biol. Chem.* **261**:5229–5232

Umei, T., Takeshige, K., and Minakami, S., 1987, NADPH-binding component of the superoxide-generating oxidase in unstimulated neutrophils and the neutrophils from the patients with chronic granulomatous disease, *Biochem. J.* **243**:467–472

Volkman, D. J., Buescher, E. S., Gallin, J. I., and Fauci, A. S., 1984, B-cell lines as models for inherited phagocytic diseases: Abnormal superoxide generation in chronic granulomatous disease and giant granules in Chediak-Higashi syndrome, *J. Immunol.* **133**:3006–3009

Volpp, B. D., Nauseef, W. M., and Clark, R. A., 1988, Two cytosolic neutrophil oxidase components absent in autosomal chronic granulomatous disease, *Science* **242**:1295–1297

Wakeyama, H., Takeshige, K., and Minakami, S., 1983, NADPH-dependent reduction of 2,6,-dichlorophenol-indophenol by the phagocytic vesicles of pig polymorphonuclear leukocytes, *Biochem. J.* **210**:577–581

Weiss, S. J., Rustagi, P. K., and LoBuglio, A. F., 1978, Human granulocyte generation of hydroxyl radical, *J. Exp. Med.* **147**:316–323

Weiss, S. J., Slivka, A., and Wei, M., 1982, Chlorination of taurine by human neutrophils: Evidence for hypochlorous acid generation, *J. Clin. Invest.* **70**:598–607

Weiss, S. J., Lampert, M. B., and Test, S. T., 1983, Long-lived oxidants generated by human neutrophils: Characterization and bioactivity, *Science* **222:**625–628

Weiss, S. J., Test, S. T., Eckmann, C. M., Roos, D., and Regiani, S., 1986, Brominating oxidants generated by human eosinophils, *Science* **234:**200–202

Weitzman, S. A., Weitberg, A. B., Clark, E. P., and Stossel, T. P., 1985, Phagocytes as carcinogens: Malignant transformation produced by human neutrophils, *Science* **227:**1231–1233

Yamaguchi, T., Hayakawa, T., Kaneda, M., Kakinuma, K., and Yoshikawa, A., 1989, Purification and some properties of the small subunit of cytochrome b558 from human neutrophils, *J. Biol. Chem.* **264:**112–118

Chapter 10

Myeloperoxidase: Localization, Structure, and Function

Andreas Tobler and H. Phillip Koeffler

1. INTRODUCTION

Myeloperoxidase (MPO) is a heme-containing enzyme that, in the presence of peroxide and halide ions, is effective in killing various microorganisms; in addition, it exerts a wide variety of extracellular functions. Polymorphonuclear leukocytes (PMN) are the main source of this enzyme. MPO is a myeloid cell-specific enzyme whose synthesis is tightly regulated at the promyelocyte stage of differentiation of myeloid cells. This review will consider recent developments concerning the biochemical and molecular structure, biosynthesis and processing, and various possible functions of MPO.

2. BACKGROUND

MPO belongs to the class of oxyreductases (EC 1.11.1.7). In 1941, K. Agner was the first to detect a green enzyme in tubercular empyema which he called verdoperoxidase (Agner, 1941a). This enzyme, now known as MPO to indicate its source, is highly cationic (isoelectric point > 10) (Agner, 1941b) and was shown to be structurally different from other iron-containing proteins, including hemoglobin, catalase, and myoglobin (Ehrenberg and Agner, 1958; Agner, 1958). The same investigators revealed that MPO is able to inactivate diphtheria and tetanus toxins, suggesting a biological role for MPO (Agner, 1947, 1950). Interest in this enzyme was increased by observations indicating that

Andreas Tobler   Central Hematology Laboratory, University of Berne, Inselspital, Switzerland.
H. Phillip Koeffler   Division of Hematology/Oncology, UCLA School of Medicine, Los Angeles, California 90024, USA.

MPO may play a role in killing of bacteria, fungi, and tumor cells and in modulating a variety of extracellular responses.

PMN are the chief reservoir of MPO, representing 2–5% of cell weight (Agner, 1941; Schultz and Kaminker, 1962). The MPO protein is localized entirely in the primary (azurophilic) granules of PMN (Bainton et al., 1971, Bainton and Farquhar, 1968; Bretz and Baggiolini, 1982; Cramer et al., 1985). MPO is also present in human monocytes but not in tissue macrophages (Lehrer, 1975; Bos and Dirk, 1978). The level of MPO in monocytes is considerably lower than in PMN, being approximately 0.9% of cell weight (Bos and Dirk, 1978).

Investigations of the structure and function of MPO initially were hampered by the available material, requiring large amounts of human buffy coat cells. Studies in the 1960s and early 1970s analyzed the amino acid composition, investigated the nature of the porphyrin of the heme group, the linkage of the peptide chain, its relation to cytochrome oxidase, and the subunit structure with the possible presence of two hemes, and provided evidence for various isoenzymes (Schultz, 1980; Felberg and Schultz, 1972; Schultz et al., 1972; Schultz and Shmukler, 1964). In 1977, the introduction of new isolation procedures, such as cetyltrimethylammonium bromide as a detergent, increased the yield and purity of MPO (Harrison et al., 1977). At the same time, unlimited and well-defined cell material became available by the establishment of various human myeloid leukemic cell lines, such as the promyelocytic HL-60 cell line, to study in more detail the structure, biosynthesis, and function of MPO (Collins et al., 1977; Koeffler, 1983, 1986). The recent isolation, cloning, and characterization of the human MPO gene based on cDNA libraries of HL-60 cells now makes it possible to study this enzyme at the molecular level in health and deficient states (Yamada et al., 1987; Chang et al., 1986; Weil et al., 1987; Johnson et al., 1987a; Morishita et al., 1987a).

3. STRUCTURE

The molecular weight (MW) of the mature human MPO molecule was first determined by Agner in 1958 and calculated to be 149,000 daltons (Da) (Ehrenberg and Agner, 1958). Further studies showed an MW of mature MPO in the range of 120,000–160,000 Da in various species (Schultz, 1980; Felberg and Schultz, 1972; Harrison et al., 1977; Klebanoff, 1980a; Zgliczynski, 1980; Andrews and Krinsky, 1981; Bakkenist et al., 1978; Miyasaki et al., 1986a; Pember et al., 1983; Yamada et al., 1981a, b; Kincade et al., 1983; Desser et al., 1972; Matheson et al., 1981a).

Early studies had already noted the existence multiple components of MPO, reporting 2–10 different forms of the enzyme (Felberg and Schultz, 1972; Schultz et al., 1967; Himmelhoch et al., 1969; Felberg et al., 1969; Strauven et al., 1978). Later investigations agreed on the distinction between three isoenzymes in PMN (I, II, and III) and four in promyelocytic HL-60 cells (IA, IB, II, III) (Miyasaki et al., 1986a; Pember et al., 1982, 1983; Yamada et al., 1981b; Kinkade et al., 1983; Wright et al., 1987; Suzuki et al., 1986). Different methods and various cell sources, such as pooled PMN, PMN from a single individual, murine peritoneal exudate cells, and HL-60 cells, yielded multiple isoenzymes, demonstrating that the isoenzymatic pattern of MPO is not due either to genetic heterogeneity or to different isolation, separation, and digestion procedures used.

There is uncertainty as to the structures of the three isoenzymes. In PMN, MPO I was found to be the largest isoenzyme, followed by MPO II and MPO III, with MWs of 120,000, 115,000, and 110,000 Da, respectively (Miyasaki *et al.*, 1986). The amino acid compositions of the three forms were similar, but major differences were observed in selected residues, such as charged amino acids (Felberg and Schultz, 1972; Miyasaki *et al.*, 1986a; Pember *et al.*, 1983; Yamada *et al.*, 1981b; Strauven *et al.*, 1978; Wright *et al.*, 1987). A relatively high proportion of aspartate, glutamyl, leucyl, and propyl residues and an amino sugar *N*-acetylglucosamine moiety were seen in all isoenzymes (Bakkenist *et al.*, 1978; Miyasaki *et al.*, 1986a; Pember *et al.*, 1983; Matheson *et al.*, 1981a). Tumors of human promyelocytic HL-60 cells grown in athmymic nude mice possessed multiple forms (IA, II, and III) of MPO with MWs of 153,000 Da (Yamada *et al.*, 1981b). One additional form was smaller (calculated MW of 79,000 Da) and made up about 8% of total MPO; the rest consisted of larger forms. The amino acid sequence of the small form was different from that of the large MPO. During differentiation along the granulocytic pathway by dimethyl sulfoxide, both the large and small forms decreased, but the small form was reduced preferentially (Yamada *et al.*, 1981a).

The functional properties of the isoenzymes are a matter of debate. Some investigators found differences in functional characteristics, showing different sensitivities to the peroxidase inhibitor aminotriazole and different heat stabilities (Miyasaki *et al.*, 1986; Straumen *et al.*, 1978; Pember *et al.*, 1981, 1982, 1983). The possibility of distinct functional properties was supported by studies demonstrating differences in the pattern of compartmentalization and in exocytosis (Pember *et al.*, 1983; Kinkade *et al.*, 1983; Pember and Kinkade, 1983). In hypertonic salt solutions, only forms II and III but not form I) were solubilized (Pember *et al.*, 1983). Extraction of peak I required the use of detergent, further suggesting that differences might exist in compartmentalization of the different forms. Exposure of purified human PMN to the synthetic chemotactic peptide fMet-Leu-Phe in the presence of cytochalasin B resulted in specific exocytosis of forms II and III (Pember and Kinkade, 1983). The lower-density granules contained mainly forms II and III, whereas the higher-density granules consisted of all three forms (Kinkade *et al.*, 1983). This heterogeneity of the primary granules might reflect differences in the site of action of the different forms of MPO. Selective exocytosis of MPO II and III from the lower-density granules might be important for extracellular processes. The high-density subpopulation, with its content of MPO and lysozyme, may function primarily in an intracellular environment. Other investigators could not identify additional properties of these isotypes. No difference in K_m and V_{max} values, enzymatic activity, and Reinheitszahl was observed among the three isoenzymes in both normal PMN and HL-60 cells (Wright *et al.*, 1987; Suzuki *et al.*, 1986; Selsted and Novotny, 1988).

The structural basis for the MPO isoenzymes and their possible distinct functional roles are not conclusively evident from the reported data. Recent molecular analyses of MPO composition also did not clarify this issue (Hashinaka *et al.*, 1988). Since the precise primary structure is still unknown, it is impossible to correlate the different isoenzymes with predicted sequences on the basis of different cDNA clones of MPO. A more recent study revealed that peptide maps of the large and small subunits are indistinguishable (Selsted and Novotny, 1988). Endoglycosidase digestion of the large subunits of the various isoenzymes (I, II, and III) resulted in a single electrophoretic species. Enzymatic deglycosylation took place only when the large subunit was denatured. These results suggest that the different isoenzyme forms are the result of differential N-glycosylation and that the different glycosylation patterns may play a role in cytoplasmic trafficking and sorting. Both the precise structures and possible different functions of MPO isoenzymes remain to be elucidated.

The first evidence for a subunit structure of human MPO was provided by treatment with mercaptoethanol in guanidine and analysis by SDS-PAGE (Schultz, 1980; Schultz *et al.*, 1972). Both a heavy chain, which was green in color and had an MW of 60,000 Da, and a light chain with an MW of 10,000 Da were isolated. On the basis of the total MW of 140,000 Da, it was assumed that MPO consisted of two heavy and two light chains that formed a dimer. Numerous studies confirmed the existence of heavy- and light-chain subunits of MPO in both PMN, HL-60 leukemia cells, and human myeloid leukemia cells from patients (Andrews and Krinsky, 1981; Bakkenist *et al.*, 1978; Yamada *et al.*, 1981a, b; Harrison *et al.*, 1977; Atkin *et al.*, 1982; Andersen *et al.*, 1982; Nauseef *et al.*, 1983a; Olsen and Little, 1984; Nauseef and Clark, 1986; Yamada, 1982; Nauseef and Malech, 1986; Koeffler *et al.*, 1985; Nauseef, 1986). The MWs reported for the heavy- and light-chain subunits are in the range of 55,000–65,000 and 10,000–15,000 Da, respectively. In terms of isoenzyme pattern, all three major forms consist of heavy and light peptides (Miyasaki *et al.*, 1986a; Nauseef and Malech, 1986; Olsen and Little, 1983). The light subunits have the same MW, whereas the heavy subunit varies in size. A single disulfide bridge connects the two hemi-MPOs, with the half-enzyme retaining its enzymatic activity (Andrews and Krinsky, 1981). Carbohydrate moieties are attached to the heavy subunit, demonstrating that MPO is a glycosylated protein (Wright *et al.*, 1987; Harrison *et al.*, 1977; Olsen and Little, 1984).

The existence of additional bands of approximately 80,000, 39,000, 22,000 Da was noted by several investigators (Matheson *et al.*, 1981a; Wright *et al.*, 1987; Andersen *et al.*, 1982; Nauseef *et al.*, 1983a; Olsen and Little, 1983, 1984; Nauseef and Malech, 1986; Koeffler *et al.*, 1985; Nauseef, 1986). The electrophoretic enzymatic variants seen are not because of genetic polymorphisms, protease activity, or MPO autoxidation (Klebanoff, 1980a; Zgliczynski, 1980; Pember *et al.*, 1983; Nauseef and Malech, 1986). The major form of these additional peptides is the 39,000-Da unit, whose appearance is dependent on conditions used to study the enzyme. Reduction before heat treatment virtually abolishes this band in favor of the 59,000-Da unit, whereas boiling without reduction results in presence of the 39,000-Da fragment (Matheson *et al.*, 1981a; Nauseef *et al.*, 1983a; Olsen and Little, 1984). One explanation for the 39,000-Da species might be the extreme stability of the MPO enzyme, which shows little unfolding even in the presence of 10 M urea or 5 M guanidium chloride (Olsen and Little, 1983, 1984). Therefore, unless the 59,000-Da unit is reduced before denaturation, heating even in SDS may cause it to fold up tightly rather than open up; once the enzyme is in this state, reduction of the disulfide bridges might be impossible. The two heavy subunits (57,000 and 39,000 Da) are essentially composed of the same polypeptide chain (Olsen and Little, 1984). Peptide mapping of the 59,000- and 39,000-Da forms shows that the two subunits are related but not identical (Nauseef and Malech, 1986). Furthermore, the 39,000-Da subunit can be converted into a 59,000-Da species; deglycosylation of the 59,000-Da unit does not result in the formation of the 39,000-Da fragment (Olsen and Little, 1984; Nauseef and Malech, 1986).

MPO is a heme-containing protein. Two iron atoms bound to the heme prosthetic group are part of the MPO molecule (Ehrenberg and Agner, 1958). The iron content is in the range of 0.069–0.1%, depending on the MPO species studied (Ehrenberg and Agner, 1958; Agner, 1958; Schultz and Shmukler, 1964; Schultz and Rosenthal, 1958). The heme prosthetic group is most likely covalently bound to the heavy subunits and is pivotal for enzymatic activity (Andrews and Krinsky, 1981; Harrison and Schultz, 1978; Arnljots and Olsson, 1987; Andrews et al., 1984). Uncertainty exists as to the relationship between the two iron centers in the dimeric MPO protein. Some studies suggest that the two hemes bind to MPO in a nonequivalent fashion (Agner, 1958; Harrison and Schultz, 1978). Experiments by other investigators using optical spectra and halide-binding properties have not confirmed these findings (Andrews and Krinsky, 1981; Andrews et al., 1984). The two hemi-MPOs still possesses both heme groups and retain full specific activity, indicating that whatever structural differences exist between the two hemes, these differences do not lead to different functions (Andrews and Krinsky, 1981; Andrews et al., 1984).

The exact structure of the heme prosthetic group of MPO is not clear. On the basis of the characteristic absorption spectrum with a relatively intense peak at 700 nm and a red shifted Soret band, it is most likely that the prosthetic group in MPO is a chlorine type of heme (dihydroporphyrin), the chloride binding to the heme (Agner, 1958; Harrison and Schultz, 1978; Babcock et al., 1985; Ikeda-Saito and Prince, 1985). Other studies, by contrast, concluded that one of the two iron atoms exists in a formyl moiety similar to heme a-like chromophore and that the other is a nonheme iron (Harrison and Schultz, 1978; Odajima, 1980).

On the basis of these data, the following model of the quaternary structure for native MPO is proposed. MPO is a tetramer with an MW of 120,00–160,000 Da consisting of two heavy and two light chains (H2 and L2). An iron–chlorine prosthetic group is incorporated into each of the heavy subunits. The two heavy- and light-chain protomers lie side by side along their axis to form the native enzyme. A single disulfide bridge joins the two halves of the native enzyme. Observations with respect to the 39,000-Da unit suggest that a different model is possible, i.e., an asymmetric heavy-chain structure of HH'L2 rather than H2L2 (Nauseef and Malech, 1986).

4. SYNTHESIS

4.1. Biosynthesis, Glycosylation, Further Processing, and Intracellular Transport

Early studies suggested that synthesis of MPO occurs at the promyelocytic stage of differentiation (Bainton et al., 1971). Human myeloid leukemia cell lines blocked at different stages of differentiation provide a suitable model system to study synthesis and processing of MPO (Koeffler, 1983, 1986). The human promyelocytic HL-60 cell line is commonly used. Wild-type cells are at the promyelocytic stage of differentiation; these can be induced to differentiate to either granulocytelike or macrophagelike cells (Koeffler, 1983, 1986; Collins et al., 1977).

The in vivo synthesis and posttranslational processing of MPO was studied by incubating HL-60 cells with [^{35}S]methionine or [^{14}C]leucine. The cells were lysed and immunoprecipitated with anti-MPO antibodies, and the radiolabeled immunoprecipitate was subjected to SDS-PAGE and autoradiography (Yamada, 1982; Koeffler *et al.*, 1985; Nauseef, 1986, 1987; Arnlijots and Olsson, 1987; Yamada and Kurahashi, 1984; Hasilik *et al.*, 1984; Stroemberg *et al.*, 1985; Olsson *et al.*, 1984, 1988; Akin and Kinkade, 1986; Akin *et al.*, 1987). The MPO is precipitated from pulse-labeled cells as a precursor polypeptide of approximately 90,000 Da with a transient intermediate of 82,000 Da (Yamada, 1982; Arnlijots and Olsson, 1987; Yamada and Kurahashi, 1984; Hasilik *et al.*, 1984; Olsson *et al.*, 1984; Stroemberg *et al.*, 1985; Nauseef, 1986, 1987; Akin and Kinkade, 1986; Akin *et al.*, 1987). Another intermediate of approximately 75,000 Da was also observed, which was converted to the heavy and light subunits (Koeffler *et al.*, 1985; Akin and Kinkade, 1986). The stepwise processing and its time course was studied by pulse-chase experiments. Within a few hours of chase, the precursor peptide was processed to the heavy and small subunits. A fraction of the precursor protein remained unprocessed even after a longer chase of up to 100 hr, and a minor fraction of the MPO precursor was found in the medium (Nauseef, 1986; Hasilik *et al.*, 1984; Olsson *et al.*, 1984; Akin and Kinkade 1986). The *in vitro* synthesis of MPO protein was studied by using a rabbit reticulocyte lysate system. The primary translation product of HL-60 RNA was immunoprecipitated with anti-MPO antibody as a single protein of approximately 80,000 Da (Koeffler *et al.*, 1985; Nauseef, 1986; Yamada and Kurahashi, 1984; Hasilik *et al.*, 1984). Taken together, these *in vivo* and *in vitro* results show that MPO is synthesized as a large precursor protein of 90,000–80,000 Da that is processed over time to the mature enzyme consisting of the heavy and small subunits.

Studies using metabolic labeling of HL-60 cells in combination with subcellular fractionation on a Percoll gradient defined more precisely the subcellular localization of MPO processing (Nauseef and Clark, 1986; Nauseef, 1986; Olsson *et al.*, 1984; Akin and Kinkade, 1986; Akin *et al.*, 1987). The 90,000-Da pro-MPO form cosedimented with the light-density organelles, containing endoplasmatic reticulum, Golgi complex, and plasma membranes. The subunits of mature MPO cosedimented with the high-density lysosomal fraction. A mixture of mature and unprocessed MPO proteins was present in fractions in the intermediate area of the gradient corresponding to the Golgi elements. These studies indicate that the proteolytic cleavage of MPO occurs mainly at the prelysosomal level.

MPO is a lysosomal enzyme. Studies using nonmyeloid cells provided the necessary understanding of glycosylation, processing, and intracellular trafficking of lysosomal enzymes (Kornfeld, 1986; Von Figura and Hasilik, 1986; Marshall, 1974; Hasilik and Von Figura, 1984). Lysosomal enzymes are synthesized in a prepro form with amino-terminal extension (Von Figura and Hasilik, 1986; Hasilik and Von Figura, 1984). Initially, the nascent lysosomal protein is glycosylated in the rough endoplasmic reticulum, where the signal prepeptide is excised. The protein is then translocated to the Golgi. Here, lysosomal enzymes undergo an important posttranslational modification by acquisition of phosphomannosyl residues. This structure is responsible for binding to the mannose-6-phosphate (M-6-P) receptor in the Golgi and subsequent transfer to the lysosome. The protein–M-6-P complex is transported to a prelysosomal area, where the complex is dissociated by acidification. The receptor cycles back to the Golgi network to pick up another protein. By this mechanism, lysosomal enzymes remain intracellular, in contrast to secretory proteins. A small portion of lysosomal enzymes fails to bind to the receptor and is secreted together with other secretory proteins into the culture medium. The proteolytic processing and transport of lysosomal enzymes are dependent on an acid intracellular lysosomal environment. Alkalinization of the lysosomes increases secretion of lysosomal enzymes and stops proteolytic maturation (Hasilik *et al.*, 1983). Finally, the proenzyme is proteolytically processed into the mature form before or after being transported to the lysosome and packaged into vesicles forming the primary lysosomes. In contrast to the lysosomal enzymes, secretory proteins are processed to complex-type oligosaccharides and are secreted.

Knowledge of carbohydrate structures and their further trimming is important for understanding whether MPO has structural features in common with other lysosomal enzymes. *In vitro* translation studies showed that without addition of membranes to the translational system, the nascent MPO protein was not glycosylated, but when *in vitro* translation was performed in the presence of cell membranes, a larger form of MPO (87,000 Da) was observed (Hasilik *et al.*, 1984). Experiments using endoglycosidases that modify carbohydrate side chains further elucidated the glycosylation structure of MPO. Endoglycosidase F cleaves both the complex and high-mannose oligosaccharides, whereas endoglycosidase H cleaves only the high-mannose units (Elder and Alexander, 1982). Both endoglycosidases H and F cleaved the 90,000-Da pro-MPO, producing five N-linked high-mannose oligosaccharides and a 79,000-Da protein (Hasilik *et al.*, 1984; Nauseef, 1987). This finding indicates that the pro-MPO contains only high-mannose oligosaccharides and no complex mannose units; the latter are features of lysosomal enzymes. The terminal product of endoglycosidase digestion was 1,000 Da smaller than the primary translation product of 80,000 Da observed in *in vitro* translation studies (Yamada *et al.*, 1981a; Hasilik *et al.*, 1984; Koeffler *et al.*, 1985; Nauseef, 1986). This difference may be due to the cleavage of a signal peptide. Recent amino acid sequence analysis of MPO is consistent with this conclusion, showing that MPO contains N-terminal sequences with features of a signal peptide (Johnson *et al.*, 1987; Morishita *et al.*, 1987b). Further insight into MPO processing was provided by studies using agents that inhibit the action of enzymes responsible for the trimming of oligosaccharides in the endoplasmic reticulum and Golgi apparatus, such as glucosidases and mannosidases (Nauseef, 1987). Glucosidases I and II are located in the endoplasmic reticulum, and mannosidase II is located in the medial cisternae of the Golgi system. In the presence of the glucosidase I inhibitor castanopsermine and the glucosidase II inhibitor bromoconduritol, [^{35}S]methionine-labeled and immunoprecipitated MPO from HL-60 cells showed a pro-MPO of 92,000 Da instead of the previously identified unit of 90,000 Da. The action of glycosidases I and II is probably so rapid that it may occur cotranslationally (Atkinson and Lee, 1983). This might explain why the 92,000 Da protein is detected only in the presence of glucosidase inhibitors. By contrast, when the action of mannosidase II was inhibited by swainsonine, the processing of pro-MPO was not altered. The results of these experiments suggest that pro-MPO leaves the Golgi system proximal to the mannosidases II, which are localized in the medial cisternae of the Golgi system, and are consistent with the results of the endoglycosidase experiments, since formation of complex mannose side chains occurs in the Golgi system.

As mentioned, trafficking of enzymes toward the lysosome is dependent on acidic pH values in lysosomes and the M-6-P receptor complex (Kornfeld, 1986; Von Figura and Hasilik, 1986). Lysosomotropic agents such as ammonium chloride and chloroquine raise the pH, providing a means to study pH dependence of lysosomal transfer. Another agent used to study movement of proteins to the lysosomes is the ionophore monensin, which leads to perturbation of the Golgi complex and increases the pH of lysosomes (Tartakoff, 1983). The increase of intralysosomal pH by ammonium chloride did not alter the intracellular processing and transport of MPO in HL-60 cells (Hasilik *et al.*, 1984; Stroemberg *et al.*, 1985; Nauseef, 1987; Olsson *et al.*, 1988). Contradictory results were obtained when the weak base chloroquine or the proton exchanger monensin was used. No influence of these agents on processing and transport of MPO was reported in one study (Nauseef, 1987), whereas others noted that monensin and chloroquine almost completely inhibited the processing and transport of MPO (Stroemberg *et al.*, 1985; Olsson *et al.*, 1988; Akin *et al.*, 1987). These results suggest that processing and transport of MPO are not necessarily influenced by pH-dependent mechanisms. On the other hand, studies showed that pro-MPO incorporates phosphomannosyl residues similarly to lysosomal enzymes, suggesting an M-6-P-dependent pathway of enzyme transfer (Stroemberg *et al.*, 1985; Olsson *et al.*, 1988; Hasilik *et al.*, 1983). The finding that ammonium chloride does not affect the transport and further processing of MPO indicates an M-6-P-independent pathway involved in MPO trafficking because of pH-sensitive dissociation of the enzyme from the M-6-P receptor. Several lines of evidence suggest that lysosomal enzymes can be transported by mechanisms independent of M-6-P receptors (Kornfeld, 1986; Von Figura and Hasilik, 1986). For example, cells of patients with I-cell disease are characterized by a deficiency of phosphotransferase activity; as a result, these cells are incapable of building phosphomannose residues, which prevents binding to the specific receptor. In fibroblasts of patients with I-cells, lysosomal enzymes are secreted instead of being transported to the prelysosomal area. However, in other cell types of these patients, the activity of many lysosomal enzymes remains normal. These cells must therefore use an M-6-P-independent pathway for transporting lysosomal enzymes. Similarly, lysosomal enzymes such as MPO in PMN of normal individuals may be directed to the lysosomes at least in part by M-6-P-independent mechanisms.

In sum, MPO is a lysosomal enzyme with many of the biosynthesis and processing properties found for other lysosomal enzymes. However, intracellular trafficking seems to be different for MPO than for most other lysosomal enzymes.

4.2. Biosynthesis of the Heme Prosthetic Group

A critical element in the biosynthesis of MPO is the timing of heme incorporation, since the heme prosthetic group is necessary for the catalytic activity of the MPO enzyme. Within the limited sensitivity of the assay used, one study found no spectral evidence for a heme group in the precursor protein, whereas the mature subunits contained the heme group (Nauseef and Clark, 1986). The precursor protein displayed no enzymatic activity. A more recent study showed that the precursor MPO protein incorporates heme, as demonstrated by labeling HL-60 cells with a 5-amino-[^{14}C]levulinic acid (Arnlijots and Olsson, 1987). After a 3-hr labeling with 5-amino-[^{14}C]levulinic acid, the isotope was incorporated into MPO precursors of 90,000 and 82,000 Da. During a subsequent chase for 20 hr, 5-amino-[^{14}C]levulinic acid-labeled MPO was converted to the mature heavy subunit of 62,000 Da. The small subunit of 12,000 Da lacked labeling with 5-amino-[^{14}C]levulinic acid, confirming previous studies that showed no heme moiety in the light subunit (Olsen and Little, 1984). Peptide mapping showed that after proteolytic treatment, cleavage of 5-amino-[^{14}C]levulinic acid was associated with a single radioactive fragment of 23,000 Da, whereas cleavage of the [^{14}C]leucine-labeled MPO generated multiple radioactive fragments of different sizes, indicating that the heme moiety is incorporated at a specific site. The incorporation of heme appears to be independent of the final processing of the peptide. The functional significance of a heme-containing precursor, however, remains to be determined.

5. MOLECULAR BIOLOGY: THE MYELOPEROXIDASE GENE

The human gene for MPO was recently isolated and characterized by several groups (Yamada *et al.*, 1987; Chang *et al.*, 1986; Weil *et al.*, 1987; Johnson *et al.*, 1987a; Morishita *et al.*, 1987a). Full-length cDNA clones from libraries constructed with mRNA from promyelocytic HL-60 leukemia cells were isolated by two groups (Johnson *et al.*, 1987a; Morishita *et al.*, 1987a). The MPO gene has an open reading frame of approximately 2200 bp coding for the primary translation product of 745 amino acids with a calculated MW of approximately 83,000 Da. *In vitro* translation of mRNA selected by cDNA hybridization reveals the synthesis of a 74,000-Da protein, and antibodies specific for either light- or heavy-chain subunits precipitate the 74-kDa protein, indicating that MPO is probably synthesized by a single protein from a single gene (Yamada *et al.*, 1987). This finding confirms those of our previous studies (Koeffler *et al.*, 1985). The human MPO gene is located on the long arm of chromosome 17 in the region of 17q12-23 (Weil *et al.*, 1988; Van Tuinen *et al.*, 1988). The human MPO gene contains 12 exons and 11 introns spanning 10 kb (Morishita *et al.*, 1987b). All splice donor and acceptor sites fit the GT-AT rule for nucleotides flanking exon borders. The heavy chain of MPO is encoded by part of exon 6 and exons 7–12; the light chain is encoded by the 3' half of exon 4, exon 5, and part of exon 6. The remaining exons 1–3 and the 5' part of exon 4 code for the signal peptide and the pro-MPO precursor peptide. The heavy chain consists of 467 amino acids located at the carboxy terminal of the protein (Johnson *et al.*, 1987a; Morishita *et al.*, 1987a; Hashinaka *et al.*, 1988). The light chain consists of 108 amino acid residues (Hashinaka *et al.*, 1988). For the formation of heavy and light chains, 164 amino acids of the prepro sequence are removed; in addition, a small peptide of six amino acids between light and heavy chains and a single amino acid at the carboxy terminus of the heavy chain are removed (Hashinaka *et al.*, 1988). The most likely cleavage site for the signal (pre) sequence is at amino acids 41–42, the alanine–threonine being a likely site for cleavage of pre sequences (Van Heijne, 1986). The protein contains five potential cleavage sites, one in the pro sequence and four in the heavy-chain sequence; none are present in the light chain. These findings, based on sequence information, are in agreement with previous observations showing that only the heavy subunit changes electrophoretic mobility after digestion with endoglycosidase (Nauseef, 1986; Hasilik *et al.*, 1984; Stroemberg *et al.*, 1985). Interestingly, a recent comparison of the sequences of human MPO and human thyroid peroxidase genes shows close to 50% homology, especially in the sequences encoding the subunits of MPO (Kimura and Ikeda-Saito, 1988). This finding suggests that the two human peroxidases are members of the same gene family, derived from the same ancestral gene.

Analyses of the 5' flanking region of the gene by S1 nuclease mapping determined a single transcription initiation site at 180 bp upstream of the ATG initiation codon (Morishita et al., 1987b). Examination of the promoter region did not reveal authentic sequences of CAAT and TATA, which are normally responsible for initiation of transcription by eukaryotic polymerase (Hashinaka et al., 1988). However, Alu sequences and an enhancer core sequence, as well as a phorbol ester responsive-like element, were found, which might represent different classes of initiation sites (Schmid and Jelinek, 1982; Weiher et al., 1983; Angel et al., 1987). Interestingly, sequence homologies occur in the 5' flanking region of MPO and the c-myc proto-oncogene (Morishita et al., 1987b).

The 3' untranslated region consists of approximately 800 bp with a poly(A) tail of 10 bp (Hashinaka et al., 1988). Two polyadenylation signals separated by 613 bp are present (Morishita et al., 1987b; Hashinaka et al., 1988). In total, five classes of polyadenylation sites exist. Two species of MPO mRNA, approximately 3.3 and 2.6 kb in length, are seen (Chang et al., 1986; Weil et al., 1987; Johnson et al., 1987a; Morishita et al., 1987a). The different sizes of the MPO mRNAs can be explained by the two different polyadenylation signals, which are separated by approximately 600 bp.

A survey of various human myeloid leukemia cell lines blocked at different stages of differentiation confirmed that gene expression is confined to the myeloid lineage; promyelocytic HL-60 cells expressed high levels of MPO mRNA, whereas more immature myeloblastic leukemia cells (KG and KG-1A) did not express detectable MPO mRNA (Tobler et al., 1988). Upon differentiation of HL-60 cells to macrophagelike or granulocytelike cells, expression of MPO mRNA decreased to undetectable levels (Johnson et al., 1987a; Tobler et al., 1988; Sagoh and Yamada, 1988). The decrease in MPO mRNA levels was inversely proportional to an increase in the number of differentiated cells and increased expression of a differentiation-specific mRNA (CD11b/CD18) (Tobler et al., 1988; Rosmarin et al., 1989). The temporal expression of MPO mRNA during myeloid differentiation was studied by exposing murine hematopoietic cells to hematopoietic growth factors. In the presence of granulocyte colony-stimulating factor, MPO mRNA rapidly increased within 24–48 hr of exposure, reached a maximum at day 6, and was no longer detectable at day 9. This increase was paralleled by an increase in MPO-positive cells in culture (Valtieri et al., 1987). A similar maximal expression of MPO mRNA was observed during the peak proliferation of murine immature myeloid cells in the presence of granulocyte–macrophage colony-simulating factor (Jaffe et al., 1988). MPO protein synthesis closely followed accumulation of MPO mRNA.

The decrease in steady-state MPO mRNA with myeloid differentiation could not be explained by either a selective decrease in MPO gene transcription or a change in MPO mRNA stability, as shown by in vitro nuclear run-on and RNA stability experiments (Tobler et al., 1988). Possibly an alteration in posttranscriptional processing of MPO RNA occurs with induction of differentiation, causing a large percentage of MPO RNA to be degraded before reaching a stable pool of RNA. For example, decreased gene expression due to inhibition of posttranscriptional processing was demonstrated in mRNA coding for class I major histocompatibility complex protein in adenovirus-transformed cells (Vaessen et al., 1987). MPO is a fairly stable mRNA with a half-life of approximately 5 hr, compared with the short-lived (< 30 min) mRNA of c-myc (Tobler et al., 1988; Dani et al., 1984).

We studied the molecular changes associated with regulation of expression of MPO in different myeloid and nonmyeloid cells (Lubbert et al., submitted for publication).

Analysis of DNA methylation of the MPO gene by digestion with methylation-sensitive restriction enzymes revealed a pattern of demethylation in MPO-expressing myeloid cells that was most distinct in the 5' region of the gene. This pattern was unchanged with terminal differentiation. Stepwise demethylation preceded the onset of transcription of MPO in less mature myeloid cells. Further studies suggested that hypomethylation of MPO is necessary but not sufficient for expression of this gene. The chromatin structure of the MPO gene was evaluated by DNase I treatment of isolated nuclei and Southern blot analysis. Three sites of DNase I hypersensitivity sites were identified in the MPO gene of cells in which this gene is expressed. These sites mapped to the regions of the putative promoter and 5' upstream region. They were strongly reduced with induction of terminal differentiation of myeloid cells. Only one of these sites was present in early myeloid cells that did not express MPO, and none were present in lymphoid cells. These studies suggest that several structural features in the region of the MPO gene are modulated not only with regulation of expression of MPO but also during earlier steps of myeloid differentiation.

The precise role of MPO in normal and leukemic hematopoiesis has yet to be defined. MPO cytochemical activity is an important diagnostic tool in classifying acute leukemias according to the French-American-British (FAB) system (Bennett et al., 1976, 1985). Acute myeloblastic leukemia cells possess cytochemical MPO activity, whereas acute lymphoblastic leukemias have no activity. However, examples occur in which a tissue- and maturation-related gene is or is not expressed in neoplastic cells at the same stage of differentiation as the normal cells (Greaves et al., 1986; McCulloch, 1987). In this context, a recent observation of MPO expression in lymphoblastic leukemia cells is of interest (Ferrari et al., 1988). Leukemic cells from patients classified as having acute lymphoblastic leukemia by morphological, cytochemical, immunohistochemical, and molecular criteria expressed high levels of normal-size MPO mRNA. No MPO protein was detected in these cells by Western blotting, and neither a gene rearrangement nor a gene amplification of MPO was seen.

Similar to promyelocytic HL-60 leukemia cells, cells from patients with M3 (FAB classification) leukemia express high levels of MPO mRNA (Chang et al., 1986; Tobler et al., 1988). Cytogenetic analyses show a consistent specific translocation involving chromosomes 15 and 17 in acute promyelocytic leukemia (APL) [t(15;17)(q22;q11.2)] (Rowley, 1978). Indications for a possible role of MPO in the pathogenesis of promyelocytic leukemia (M3) were recently provided (Weil et al., 1988). In all five patients studied, translocation of the MPO gene to chromosome 15 was seen. Genomic DNA analyses by Southern blotting demonstrated a rearrangement of MPO in two of four cases. However, further studies by us were unable to document a rearrangement of MPO in 13 leukemic samples from patients with APL having the t(15;17) (Miller et al., 1989). In addition, a number of genes intervene between the APL breakpoint and MPO locus, making rearrangement of the MPO gene unlikely in APL.

The isolation and characterization of the MPO gene provide the tools to study at the molecular level the mechanisms underlying hereditary MPO deficiency. Previous studies demonstrated the presence of a 89,000-Da protein in MPO-deficient PMN (Nauseef et al., 1983b). This finding suggested that a defect in posttranslational processing might lead to defective gene expression in these individuals. More recent studies by the same investigator supported this hypothesis (Nauseef, 1989). Neutrophils from three completely MPO-deficient individuals contained normal-size MPO mRNA. Restriction endonuclease digests of genomic DNA with $BglII$ showed an additional fragment not present in control persons; also, two different Bgl restriction patterns of MPO were observed in the MPO-deficient individuals, who were biochemically and phenotypically identical. Other digests resulted in the same restriction pattern in MPO-deficient and control persons. By contrast, we recently showed that a completely MPO-deficient individual displayed no MPO protein on Western blots and no mature MPO mRNA on Northern blots (Tobler *et al.*, 1989). Using an intron-specific probe for MPO, we could detect heterogeneous RNA in mononuclear bone marrow cells. No aberrant restriction pattern of the MPO gene was seen. Our results indicate that a failure to process MPO RNA might underlie MPO deficiency in this subject. The data obtained to date suggest that no major deletion of the MPO gene occurs in hereditary deficiency. One can further assume that, similar to thalassemia, MPO deficiency is a heterogeneous genetic disorder with various mechanisms leading to defective gene expression. Most genotypic abnormalities will probably be found to be point mutations in the MPO region.

6. FUNCTION

6.1. Mechanism of Action of the Myeloperoxidase System

Phagocytosis is accompanied by a burst of oxidative reactions. During this respiratory burst, both O_2 consumption and O_2^- and H_2O_2 production increase, and chemiluminescence is observed. MPO can be released into both the phagocytic vacuole and the extracellular environment during the degranulation process (Klebanoff, 1970, 1980a, Klebanoff, 1988; Clark, 1983). MPO itself has no direct effect on microorganisms. In a peroxidatic reaction, ferric MPO ($MP3^+$) together with peroxide (H_2O_2) forms compound I ($MP^{3+} + H_2O_2$) (Table I). In an alternative reaction, MPO together with superoxide anion (O_2^-) forms oxymyeloperoxidase (compound III; $MP^{2+} + O_2$). As a third possibility, the MPO–H_2O_2–halide system may generate singlet oxygen ($^1O_2$).

The components of the complete MPO system are the enzyme itself, peroxide, and an oxidizable cofactor such as a halide. The H_2O_2 is formed by the leukocytes during the respiratory burst from O_2 via a superoxide anion intermediate. At least 28% of the oxygen consumed by stimulated normal human PMN is converted into active chlorination agents (Foote *et al.*, 1983). MPO reacts with H_2O_2 to form the short-lived compound I, which is highly reactive with halides (Schultz, 1980; Zgliczynski, 1980; Klebanoff, 1968; Sbarra *et*

Table I
Reactions of MPO[a]

$MP^{3+} + H_2O_2$	→	$MP^{3+} + H_2O_2$	(1)
$MP^{3+} + H_2O_2 + Cl$	→	$MP^{3+} + HOCl + OH^-$	(2)
$MP^{3+} + O_2$	→	$MP^{2+} + O_2$	(3)
$MP^{2+} + O_2 + MP^{2+} + H_2O_2$	→	$2MP^{3+} + O_2 + 2OH^-$	(4)
$MP^{2+} + O^2 + O_2^- + 2H$	→	$MP^{3+} + H_2O_2 + O_2$	(5)
$MP^{2+} + H_2O_2 + H_2O_2$	→	$MP^{3+} + O_2^- + _{H2}O$	(6)

[a]MP^{2+}, Ferrous MPO; MP^{3+}, ferric MPO; $MP^{3+} + H_2O_2$, compound I; $MP^{2+} + H_2O_2$, compound II; $MP^{3+} + O_2^- = MP^{2+} + O_2$, compound III.

al., 1976). By reacting with other electron donors, compound I may be converted into compound II, which is inactive in reacting with halides (Table I) (Schultz, 1980). Among the halides used in the peroxidatic MPO reaction are chloride, bromide, iodide, and the pseudohalide thiocyanate. *In vivo,* the halide requirement is probably met by chloride. The intracellular chloride concentration is approximately 75 meq/liter and the extracellular concentration is about 103 meq/liter (Zgliczynski, 1980). Complete killing is seen at an oxidized chloride concentration as low as 0.5 meq/liter (5×10^{-4} M) (Zgliczynski, 1980; Lehrer, 1975; Klebanoff, 1980a). Oxidants generated from chloride are chlorine (Cl_2), chloridum ion (Cl^+), and hypochlorous acid. Hypochlorous acid is probably the primary and most important germicidal agent (Klebanoff, 1980a; Zgliczynski *et al.,* 1977). MPO-mediated chlorination can be carried out over a wide range of pH provided adjustment is made between the concentrations of H_2O_2 and chloride (Klebanoff, 1980a; Zgliczynski, 1980; Klebanoff *et al.,* 1984). The optimum for MPO activity occurs at an acidic pH of approximately 5 (Klebanoff, 1968; McRipley and Sbarra, 1967). Some chlorination and bactericidal activity can be observed at neutral pH (Sbarra *et al.,* 1976).

The production of hypochlorites by the peroxidatic reaction is considered the classical mechanism in the cell-free system. However, there is doubts as to whether this system functions as efficiently *in vivo* in the phagolysosome as it does *in vitro* (Winterbourn *et al.,* 1985; Cech and Lehrer, 1984; Hamers *et al.,* 1984). Quantitative spectrophotometric methods for assessing the intravacuolar pH show that, in contrast to earlier findings (Mandell, 1976; Jensen and Bainton, 1973), the vacuoles are alkaline, with a pH of 7.8 during the initial phase of the respiratory burst and phagocytosis, and remain neutral to slightly alkaline during the first 15–30 min (Cech and Lehrer, 1984; Segal *et al.,* 1981). Subsequently, the pH falls over 60 min to an acidic value of approximately 5.5. Several enzymes have their reaction optima at neutral pH. For instance, the NAD(P)H oxidase that generates the superoxide is maximally active at neutral pH (Gabig and Babior, 1979; Tauber and Goetzl, 1979; Suzuki and Lehrer, 1980). Data also suggest that H_2O_2 production is maximal at pH values of 7.0–7.8, as measured indirectly by enhanced O_2 consumption (Cech and Lehrer, 1984; Segal *et al.,* 1981). After 1 hr of phagocytosis, when the pH becomes acidic, the H_2O_2 production is reduced by 90% compared with the value at the beginning. Since high concentrations of H_2O_2 inhibit the chlorinating activity of the MPO system (Winterbourn *et al.,* 1985; Svensson *et al.,* 1987), the high concentrations observed during the initial phase of phagocytosis could therefore be restrictive for optimal chlorination activity.

The MPO system may circumvent inactivation during the initial phases of phagocytosis by an alternative pathway of reacting with superoxide anion. Superoxide (O_2^-) reacts with ferric MPO (MP^{3+}) to generate the oxy derivative, compound III (Table I) (Winterbourn *et al.,* 1985; Odajima and Yamazaki, 1972; Kettle and Winterbourn, 1988). Whether compound III has the ability for halogenation remains controversial. Some studies suggested that compound III is not able to halogenate (Morrison and Schonbaum, 1976; Cuperus *et al.,* 1987; Zgliczynski *et al.,* 1977); others showed that compound III is directly and indirectly involved in the halogenation mechanism (Winterbourn *et al.,* 1985; Kettle and Winterbourn, 1988; Cuperus *et al.,* 1986). Compound III, in the presence of Cl^-, H_2O_2, and ferric MPO, is able to chlorinate at the same rate as does compound I (Winterbourn *et al.,* 1985). In addition, compound III can be directly converted to compound I, which is active in chlorination (Cuperus *et al.,* 1986). On the basis of these observations, a cycle can be proposed between the native (ferric) MPO, compound III, and compound II, with MPO acting in a combined superoxide dismutase–catalase system (Kettle and Winterbourn, 1988). The ferric MPO by reacting with O_2^- forms compound III, which by reacting with H_2O_2 forms compound II; compound II in turn reacts with O_2^-, leading to the native ferric MPO, which is active in chlorination. By this proposed mechanism, O_2^- may help to maintain MPO in its active form and prevent inhibition by high concentrations of H_2O_2 in the early phase of phagocytosis, with MPO functioning predominantly as catalase rather than as peroxidase. At later phases of phagocytosis, when both pH and H_2O_2 concentrations decrease, the peroxidatic activity of MPO might predominate, helping to eliminate microorganisms.

The MPO system, in a third reaction, may take part in the generation of singlet oxygen ($^1O_2$), as suggested by observations that during phagocytosis the MPO–halide system emits light (chemiluminescence) (Zgliczynski, 1980; Badway and Karnovsky, 1980; Suematsu et al., 1988; Wymann et al., 1987; Allen, 1975). Singlet oxygen is powerful in causing cell damage. The in vivo significance of this system remains to be determined.

6.2. The Myeloperoxidase-Mediated Antimicrobial System

The MPO system is involved in the oxygen-dependent killing of a number of microorganisms, including bacteria, viruses, parasites, and fungi (Clark, 1983; Klebanoff, 1980b, 1988). The antimicrobial activities of other granule proteins (e.g., defensin), the role of other oxygen products, and nonoxidative microbial mechanisms have been reviewed recently (Klebanoff, 1988; Ganz et al., 1986; Elsbach and Weiss, 1983). The exact mechanism by which the MPO-mediated system exerts its antimicrobial properties is unknown. Binding of MPO to the microbial surface is believed to be the first step. Electron microscopic examination showed that intraphagolysosomal MPO is adsorbed to the surface of ingested microorganisms (Klebanoff, 1970). Preincubation of bacteria with MPO enhances the bactericidal capacity of the MPO system (Selvaraj et al., 1978). Binding of MPO was studied in depth with Actinobacillus actinomycetemconcomitans, a gram-negative coccobacillus involved in juvenile periodontitis whose killing requires MPO (Miyasaki et al., 1986b). Binding of MPO occurs very rapidly, reaching maximal effect within 1 min, with a binding capacity of 4500 sites per cell (Miyasaki et al., 1987). The MPO retains chlorination activity after binding to the target. The MPO retains chlorination activity after binding to the target. The MPO also binds to mannan polysaccharides isolated from the cell wall of Candida albicans, again with retention of peroxidatic activity (Wright et al., 1984a). Addition of soluble mannan reduced MPO-mediated killing of yeast cells (Wright et al., 1983). The interaction between the cationic MPO enzyme and the anionic polysaccharide is probably ionic rather than lectinlike (Agner, 1941; Wright et al., 1984b).

These studies suggest that one step in MPO-mediated microbicidal killing involves binding of the enzyme to the cell wall. This binding provides close proximity of the chlorination process with the cells and may avoid dilution of the oxidative agents. The phenomenon of MPO binding to microorganisms may be of utmost importance in the killing of microorganisms that cannot be readily phagocytosed because of their shape, such as fungi and parasites. Several studies showed that stimulated PMN and the cell-free

MPO system can attach to fungi and induce extracellular killing (Wright *et al.*, 1983; Wright and Nelson, 1988; Wagner *et al.*, 1986; Diamond *et al.*, 1980). The importance of this mechanism is supported by observations that individuals with MPO deficiency are markedly impaired in their ability to kill fungi such as *C. albicans* (Cech *et al.*, 1979; Nauseef, 1988; Lehrer and Cline, 1969; Kitahara *et al.*, 1981; Parry *et al.*, 1981).

Several mechanisms are thought to be associated with the MPO-mediated killing of bacteria. Bacteria commonly lose their ability to divide within minutes of encountering phagolysosomes in leukocytes (Elsbach, 1973). Hypochloric acid itself is able to attack microorganisms at a variety of chemical sites. Potential targets are unsaturated carbon bonds, sulfhydryl groups, amino groups, nucleic acids, and heme enzymes [reviewed in Root and Cohen, 1981, and Klebanoff, 1988; see also Section 6.3]. Hypochlorous acid can further react with nitrogenous substances to form chloramines that may contribute to the MPO-mediated toxicity (Klebanoff, 1980a; Zgliczynski *et al.*, 1977; Grisham *et al.*, 1984a, b). Chloramines take part in decarboxylation reactions and breakage of peptide bonds. The degradation of chloramines further results in the formation of aldehydes, which are directly toxic to the cells (Zgliczynski *et al.*, 1968).

The potential biological activity of hypochlorous acid was tested with compounds that might serve as models for bacterial cellular components (Albrich *et al.*, 1981). Hypochlorous acid was able to bleach iron sulfur proteins extremely rapidly, followed in decreasing order by beta-carotene, nucleotides, porphyrins, and heme proteins. These results suggest that iron–sulfur proteins, cytochromes, and nucleotides are possible targets for the MPO-mediated cell damage. Exposure of *Escherichia coli* to the MPO system actually leads to a loss of iron content (Rosen and Klebanoff, 1982). The release of iron was mainly due to MPO-mediated oxidation of iron–sulfur centers in intact *E. coli*, which was followed by the loss of viability of the cells (Rosen and Klebanoff, 1985).

The MPO-mediated oxidative damage of iron–sulfur centers was studied in greater detail by assessing the activity of the iron–sulfur center containing succinate dehydrogenase (Rosen *et al.*, 1987). Succinate dehydrogenase is a component of the succinate oxidase system that is involved in the electron transport system. Within 30 min of exposure to MPO, the dehydrogenase activity decreased to undetectable levels. Further studies demonstrated oxidative damage of other proteins of *E. coli*, such as glutamine synthetase (Fucci *et al.*, 1983).

Another oxidative mechanism of cell damage is the alteration of cell membrane permeability. Earlier morphological studies suggest that oxidative damage of the cell envelope is followed by changes in membrane permeability that might allow leakage of essential small metabolites (Bringmann, 1953). Both the cell-free MPO system and isolated PMN are able to increase membrane permeability, as indicated by the observation of increased hydrolysis of galactopyranoside in mutant *E. coli* strains that lack lactose permease (Sips and Hamers, 1981; Hamers *et al.*, 1984). In these mutant cells, galactopyranoside enters the cell only when the cell is perforated. Membrane proton conductivity and permeability remain unchanged in the presence of hypochloric acid, indicating that the cytoplasmic membrane of *E. coli* retains its generalized impermeability toward small molecules and ions (Albrich *et al.*, 1986). These studies show that the MPO system might affect the viability of bacterial cells by inducing alterations in membrane permeability and by oxidation of iron–sulfur centers of enzymes.

The *in vivo* role of the MPO system in bacterial killing cannot be pivotal, since

Table II
Targets and Effects of the MPO System

Target	Effects
Mammalian cells	
Neutrophils, erythrocytes	
Platelets	
Lymphocytes	Impairment of function or cytotoxicity
Natural killer cells	
B and T lymphocytes	
Tumor cells	
Platelets	Activation
Microorganisms	
Fungi, viruses, protozoa, helminths, bacteria	Toxicity resulting in killing
Soluble mediators	
Bacterial toxins	
Clostridia, diphtheria, pneumolysin	
Chemoattractants	Inactivation resulting in decreased inflammation
C5a, fMet-Leu-Phe	
Arachidonic metabolites	
Granule enzymes	
Lysosomal enzymes, vitamin B_{12}-binding protein, MPO	
Latent collagenases and gelatinases	Activation resulting in increased inflammation
Alpha-1-proteinase inhibitor	Inactivation resulting in increased inflammation
Macroglobulins	Formation of immune complexes resulting in increased inflammation
Connective tissue	Crosslinking of protein complexes resulting in increased inflammation
Lung	Emphysema
Kidney	Glomerulonephritis
DNA	Binding to DNA resulting in protection of DNA from breakage

individuals with complete MPO deficiency do not have an apparent increase in suscepti-bility to bacterial infections (Nauseef, 1988; Lehrer and Cline, 1969; Kitahara *et al.*, 1981; Parry *et al.*, 1981). MPO-deficient PMN kill staphylococci and *Serratia mar-cescens* three to four times more slowly than do PMN from normal individuals, but ultimately these MPO-deficient cells are able to eliminate the bacteria. Therefore, MPO-deficient PMN are able to kill bacteria by an MPO-independent mechanism that is slower but sufficient. This is in contrast to the case with chronic granulomatous disease, which is associated with repeated and severe bacterial infections beginning in childhood (Klebanoff, 1980b). These patients are unable to activate the respiratory burst and there-fore are unable to generate superoxide and peroxide.

6.3. Extracellular Effects of the Myeloperoxidase System

The MPO–hydrogen peroxide–halide system has not only intracellular but probably also extracellular effects. Selective extracellular release of granule enzymes occurs during phagocytosis (Henson, 1980). Soluble agents such as bacterial chemoattractants (Becker *et al.*, 1974), the complement fragment C5a (Goldstein *et al.*, 1973), formylmethionine peptides (Showell *et al.*, 1976), and pyrogens from leukocytes (Klempner *et al.*, 1978) are capable of inducing selective release of granule enzymes. Several studies provide evi-dence for an extracellular role of the MPO system. When bacteria are ingested by neu-trophils *in vitro*, 52% of MPO activity remains intracellular, 8% is detected in the extracellular fluid, and 40% is lost (Bradley *et al.*, 1982). MPO activity can be measured in the extracellular fluid of infected lesions containing neutrophils (Bradley and Erickson, 1981). Following stimulation with formylmethionine peptides or opsonized zymosan par-ticles, neutrophils also excrete H_2O_2 (Test and Weiss, 1984). Extracellular peroxide is detected 40 sec after stimulation, and the peak concentration is reached within 3 and 8 min for fMet-Leu-Phe- and zymosan-triggered cells, respectively. These studies suggest that extracellular chloride, H_2O_2, and MPO may constitute an extracellular microbicidal sys-tem. In addition, extracellular superoxide anion might participate by oxidation of extra-cellular reductants such as ascorbate and sulfhydryl components (Thomas *et al.*, 1988). By eliminating these oxidants, superoxide helps the MPO–halide system exert its toxicity.

6.3.1. Cytotoxicity

Cellular injury to normal and malignant cells can be mediated by neutrophils by two mechanisms: nonoxidatively through release of neutrophilic enzymes such as elastase, defensins, and cathepsin G (Lichtenstein *et al.*, 1986; Smedly *et al.*, 1986; Clark *et al.*, 1976) and by reactive oxygen intermediates, including peroxide (Nathan *et al.*, 1979; Simon *et al.*, 1981; Harlan *et al.*, 1981; Weiss *et al.*, 1981).

Klebanoff and Smith (1970) were the first to recognize that the MPO system might be cytotoxic to mammalian cells, showing that spermatozoa were sensitive to the MPO–peroxidase system. The MPO system has an effect on normal hematopoietic cells. Lysis of red cells is seen when the cells are exposed either to purified MPO, H_2O_2, and chloride or to isolated neutrophils (Klebanoff and Clark, 1975; Baehner *et al.*, 1971; Dallegri *et al.*, 1986). Lysis can be prevented by azide or catalase. When hypochlorous acid reacts with endogenous ammonia (NH_4^+), monochloramine is generated (Grisham *et al.*,

1984a). Monochloramine is very potent in oxidizing erythrocyte hemoglobin, up to 10 times more active than hypochlorous acid and H_2O_2. However, hypochlorous acid is more potent in cell lysis. Oxidation is lowered by eliminating H_2O_2 by catalase. On the other hand, studies show that red cells are well equipped to escape oxidative damage (Stern, 1985; Agar et al., 1986; Cantin et al., 1987; Dallegri et al., 1986). This protection includes the hexose monophosphate shunt, superoxide dismutase, catalase, and glutathione peroxidase. Erythrocytes are able efficiently to remove H_2O_2, which prevents OH^- production and generation of MPO-dependent hypochlorous acid. This suggests that erythrocytes are able to protect tissue from oxidative damage.

Exposure of platelets to the MPO system results in platelet activation. Isolated neutrophils trigger the release of serotonin and adenine (Clark and Klebanoff, 1979a). Studies in the cell-free MPO system show that MPO, in combination with halide and H_2O_2, stimulates release of serotonin but not of adenine, suggesting that MPO is able to initiate secretion of specific granules by a nonlytic process (Clark and Klebanoff, 1980). In addition to its role in the MPO system, superoxide is able to enhance platelet aggregation and release (Handin et al., 1977).

Autoinactivation of neutrophils seems to be mediated either by the MPO system or by H_2O_2 alone (see also Section 6.3.2) (Clark and Klebanoff, 1977; Tsan and Denison, 1980). Autotoxicity mediated by phorbol ester-stimulated neutrophils is inhibited by catalase but not by azide, indicating a predominant role of peroxide in autoinactivation of neutrophils. Taken together, these in vitro studies of effects of the MPO system on normal blood cells indicate that interaction between blood cells and the MPO system might play a role in various inflammatory processes and perhaps also in blood coagulation via platelets.

The MPO system is cytotoxic to some tumor cells (reviewed in Klebanoff, 1970; Klebanoff et al., 1976). Recent studies on malignant cells performed by using activated peripheral blood neutrophils suggest that different targets might respond differently to oxidative injury and that the susceptibility might also depend on environmental conditions. Inhibitors of MPO did not prevent lysis of teratocarcinoma cells by freshly obtained murine inflammatory neutrophils (Lichtenstein, 1986). By contrast, catalase and superoxide dismutase can inhibit cell lysis, indicating that in these tumor cells, peroxide and hydroxyl radicals rather than the MPO system are lytically active (Lichtenstein, 1986; English and Lukens, 1983). Synergism in cytolytic effects on erythroleukemia cells was observed when peroxide was combined with defensin, another enzyme located in the primary granules (Lichtenstein et al., 1988).

Oxidative cellular damage might also involve RNA and DNA (Badway and Karnovsky, 1980). For instance, stimulated neutrophils are able to break their own DNA strands (Bradley and Erickson, 1981), and oxidants produced by activated neutrophils trigger malignant transformation of C3H 10T $\frac{1}{2}$ mouse fibroblasts (Weitzman et al., 1985). While cell lysis might support a defense mechanism, inducing mutations in DNA strands might cause malignant transformation. The role of extracellular H_2O_2 in inducing DNA strand break was shown in various targets cells (Bradley and Erickson, 1981; Birnboim, 1982; Schraufstaetter et al., 1988). Strand breakages of DNAs are induced at low concentrations (< 100 μM) of H_2O_2. In contrast to the extracellular DNA-damaging effect of peroxide, ingested E. coli minicells containing plasmid DNA of 10 kb show only a very limited effect of peroxide. This is seemingly in contrast to the finding that endogenous DNA of neutrophils becomes damaged (Cheh et al., 1980). This discrepancy could be explained by the huge differences in target size of genomes of plasmids and humans, the human genome being approximately 100,000 times greater than the plasmid genome.

The MPO system, by contrast, seems to have a protective role against DNA damage. Concentrations of hypochlorous acid that induce cytolysis do not induce DNA breaks in whole cells or in isolated DNA (Schraufstaetter *et al.*, 1988). When MPO is added to the cells together with peroxide and chloride, a decrease in DNA damage is observed. This may be explained by conversion of H_2O_2 to hypochlorous acid by MPO. Consistent with a protective effect of MPO against DNA breaks is a recent study showing that MPO may be a nuclear antigen with DNA-binding properties (Murao *et al.*, 1988). Addition of MPO to an oxygen radical-generating reaction results in 50–80% decreased damage of plasmid DNA. The interaction of intranuclear MPO with DNA suggests that MPO might be able to exert a protective function depending on its localization.

6.3.2. Modulation of Immunoinflammatory Responses

6.3.2.a. Interaction between Lymphocytes and Neutrophils. The coexistence of lymphocytes and activated neutrophils in tumor masses and inflammatory tissues suggests the possibility of interactions between lymphocytes and neutrophil products such as MPO. Several reports provided evidence for a role of the MPO system in modulation of the immune system by mediating interactions between these two cell types. The MPO system can compromise the function of natural killer (NK) cells (El-Hag and Clark, 1984, 1987; El-Hag *et al.*, 1986; Dallegri *et al.*, 1985; Seaman *et al.*, 1981, 1982). Suppression of NK function is dependent on each compound of the MPO–halide–peroxidase system (El-Hag and Clark, 1984). The MPO alone has no effect. With increasing concentrations of peroxide, the NK suppression becomes independent of MPO, and omission of chloride diminishes the immunosuppressive effect of neutrophils. NK suppression is reversible after 24 hr of exposure and lymphocyte viability is not affected, suggesting that the MPO system exerts a functional rather than a cytotoxic effect (El-Hag and Clark, 1984, 1987; El-Hag *et al.*, 1986).

Not only NK cells but all three classes of lymphocytes are susceptible to the inhibitory effects of the MPO system (El-Hag *et al.*, 1986). Immunoglobulin (Ig)-secreting B cells are the most sensitive, NK cells are intermediate, and T lymphocytes are the least sensitive cells. Lymphocyte function is protected by catalase but not by superoxide dismutase (El-Hag and Clark, 1987). This is consistent with the finding that removal of monocytes from the mononuclear cell population increased the susceptibility of NK cells to oxidative functional impairment, probably because of the high catalase content of monocytes (El-Hag *et al.*, 1986; Meerhof and Roos, 1980). Neutrophils from patients with chronic granulomatous disease, which fail to produce peroxide, do not suppress lymphocyte functions, suggesting the importance of peroxide in the MPO-mediated modulation of lymphocytes (El-Hag and Clark, 1987). Neutrophils from patients with MPO deficiency do not suppress the function of lymphocytes as strongly as do neutrophils from normal individuals. Nevertheless, the suppression by MPO-deficient neutrophils is less than that by neutrophils from chronic granulomatous disease. This might be explained by the enhanced H_2O_2 production observed in MPO-deficient neutrophils as a result of their increased respiratory burst (Klebanoff and Pincus, 1971; Berkow and Dodson, 1987; Stendahl *et al.*, 1984). The partial impairment in cells of MPO-deficient patients is corrected by addition of purified MPO. These studies further support the observation that increased influx of H_2O_2 renders the oxidatively induced impairment of lymphocyte function less dependent on MPO.

6.3.2.b. Inflammatory Responses. MPO is implicated in both the inactivation and activation of humoral mediators and enzymes. One of the major circulating plasma proteins involved in the inflammatory tissue reaction is human alpha-1-proteinase inhibitor (alpha-1-antitrypsin), which is the most important serine protease inhibitor. This enzyme is capable of passing through vascular membranes into tissues and is pivotal in controlling tissue proteolysis. Imbalance between proteolytic activity and protease inhibitors leads to elastolytic breakdown of tissue, e.g., in the lungs, leading to pulmonary emphysema (Janoff, 1985). The importance of this enzyme is demonstrated by individuals who genetically lack alpha-1-antitrypsin. These individuals are prone to develop severe emphysema in adult life, and some may manifest hepatic cirrhosis (Morse, 1978). Several studies reveal that the MPO system may diminish the action of the alpha-1-proteinase inhibitor (Matheson and Travis, 1985; Matheson *et al.*, 1979, 1981a; Shock and Baum, 1988; Clark *et al.*, 1981). The inactivation is due to oxidation of methionine residues of alpha-1-proteinase inhibitor (Matheson *et al.*, 1979; Matheson and Travis, 1985). Hypochlorous acid alone inactivates alpha-1-proteinase inhibitor at concentrations in the micromolar range, causing inactivation of up to six methionine residues and one tyrosine residue, whereas the MPO system is able to oxidize up to six methionine residues, but no tyrosine moieties are modified. Studies using alpha-1-proteinase inhibitor purified from the synovial fluid of rheumatic arthritis patients show oxidation of four methionine residues, suggesting that *in vivo* four residues might be modified (Wong and Travis, 1980). The pH optimum of the alpha-1-proteinase inhibitor inactivation by MPO is 6.2 in the presence of 0.16 M NaCl. The inhibitory activity of the MPO system is influenced by initial concentrations of alpha-1-proteinase inhibitor and H_2O_2 (Matheson *et al.*, 1979, 1981a, b). At H_2O_2 concentrations of greater than 95 mM, the rate of inactivation progressively decreases, in accordance with previous studies showing that high concentrations of H_2O_2 inhibit MPO activity (Zgliczynski, 1980). Inactivation of alpha-1-proteinase inhibitor by both purified neutrophils and human alveolar macrophages is seen. H_2O_2 alone shows neither an effect on alpha-1-proteinase inhibitor activity nor oxidation of amino acids (Matheson and Travis, 1985). In the context of the MPO-mediated inactivation of antiproteases, an observation of note is that neutrophils from cigarette smokers show significantly higher MPO activity than do those from nonsmoking controls (Matheson *et al.*, 1981a; Shock and Baum, 1988).

The MPO system might induce inflammatory responses not only by inactivating enzymes such as proteinase inhibitors but also by activating latent enzymes of PMN such as the metalloproteins collagenase and gelatinase. Collagenase is a latent metalloenzyme that attacks specifically interstitial collagen; gelatinase has been shown to attack denatured collagen and to potentiate the activity of collagenase. Activated human neutrophils are able to degrade gelatin by releasing and activating the metalloenzyme gelatinase (Peppin and Weiss, 1986). The inhibitory effects of catalase, azide, and methionine on gelatinolysis suggest that the MPO system plays an important role in the activation of gelatinase. The same group of investigators showed that zymosan-activated neutrophils released large amounts of latent collagenase, which was activated by hypochlorous acid (Weiss *et al.*, 1985).

Both isolated neutrophils and the cell-free MPO system achieve oxidation of lysine side chains of elastin, a reaction which is blocked by azide, cyanide, and catalase (Clark *et al.*, 1986). Oxidation of lysine side chains leads to lysine-derived crosslinking. Crosslinking between the polypeptide chains of elastin and collagen results in the insoluble fibrous connective tissue form of the proteins. These studies suggest a role of MPO in connective tissue damage.

Another crosslinking reaction mediated by the MPO system was observed by Jasin (1988). Incubation with both human neutrophils and the cell-free MPO system resulted in covalent crosslinking of protein complexes between IgG–anti-IgG and type II collagen–anticollagen antibodies.

The precise nature of chemical bonds responsible for the neutrophil mediated crosslinking is unknown. The main effect of H_2O_2-mediated oxidation on amino acids is decarboxylation and deamination, yielding aldehyde derivatives of the original amino acids (Sbarra *et al.*, 1976). The same reactive groups generated by the action of lysyl oxidase on lysine and hydroxylysine are responsible for covalent crosslinking of collagen fibers. Crosslinking of lysine-containing proteins by a similar oxidative mechanism might be achieved by the MPO–H_2O_2–halide system, leading to crosslinking of immune complexes to themselves and to structural macromolecules, such as collagen and basal membrane (Jasin, 1983; Stahmann *et al.*, 1977). The MPO system is further capable of generating large amounts of IgG aggregates (Stahmann and Spencer, 1977). These complexes behave like typical immune complexes because they consume complement and are detected by the Raji cell and solid-phase Clq assays. These observations would explain the detection of circulating immune complex-like material in disorders without apparent immunopathogenesis.

Further evidence for a role of the MPO system in tissue inflammation is provided by studies showing MPO-mediated glomerular damage (Johnson *et al.*, 1987a, b, 1988; Vissers and Winterbourn, 1986; Shah *et al.*, 1987). In several experiments, small quantities of MPO were infused into rat renal arteries in combination with H_2O_2 and chloride. This resulted in proteinuria, endothelial cell swelling, and effacement of the epithelial cell foot process as well as halogenation of glomerular structures (Johnson *et al.*, 1987b, 1988), similar to the glomerular injury found in immune complex glomerulonephritis in rats (Johnson *et al.*, 1987c). The highly cationic MPO enzyme released from primary granules binds to the anionic glomerular structures (Johnson *et al.*, 1987b). The H_2O_2 alone does not induce glomerular damage (Johnson *et al.*, 1988). Furthermore, activated neutrophils may adhere to basement membranes coated with IgG, which results in production of oxidants, degranulation, and release of proteinases (Vissers and Winterbourn, 1986). The oxidants, such as hypochlorous acid, may render the basement membrane proteins more susceptible to proteolytic actions of enzymes, as demonstrated by increased hydroxyproline solubilization. This suggests that an interplay between neutrophil enzymes may lead to glomerular injury. The MPO system may produce glomerular damage either directly by halogenation and oxidation or indirectly by activation of latent metalloenzymes and inhibition of antiproteinases. On the other hand, released proteolytic neutrophilic enzymes may be inactivated by the neutrophils themselves (see below) (Voetman *et al.*, 1981; Clark and Borregaard, 1985; Vissers and Winterbourn, 1987). Taken together, these studies provide evidence for an activating role of the MPO system in inflammatory processes.

However, studies also show several mechanisms by which MPO might control and down regulate inflammatory responses, such as by inactivation of arachidonic acid metabolites, soluble chemoattractants, lysosomal enzymes, and bacterial toxins and by autoinactivation. The MPO system interacts with arachidonic acid metabolites. Among them are the leukotrienes generated by the lipooxygenase pathway and the prostaglandins generated by the cyclooxygenase pathway. Stimulated neutrophils generate both leukotriene A_4, which is metabolized to leukotriene C_4, a potent slow-reacting substance of anaphylaxis, and leukotriene B_4, a potent chemoattractant comparable to C5a. Both a cell-free MPO system and isolated neutrophils are able to down regulate rapidly the leukotriene-mediated inflammatory response by oxidative modifications of leukotrienes B_4 and C_4 (Lee et al., 1983; Henderson et al., 1982). The oxidized product has less than 5% the spasmogenic activity of the parent compound (Lee et al., 1983). Leukotriene C_4 is inactivated by sulfur oxidation as a result of interaction with the MPO system (Lee et al., 1983; Henderson et al., 1982). A second mechanism of leukotriene C_4 inactivation might involve hydroxyl radicals, as shown with activated neutrophils from MPO-deficient patients (Henderson and Klebanoff, 1983). The influence of the MPO system on the inflammatory activity of prostaglandins is suggested by the ability of activated human PMN and the cell free MPO system to transform prostaglandin fα (PGFα), $PGF_2α$, and PGE_2 to unidentified products (Paredes and Weiss, 1982).

The MPO system might further contribute to the down regulation of phagocytic activities of neutrophils by inactivation of soluble chemotactic factors, including complement-derived and synthetic peptide agents (Clark and Klebanoff, 1979b; Clark et al., 1980; Clark, 1982; Clark and Szot, 1982). Both the cell-free MPO system and isolated, stimulated neutrophils inactivate both C5a and formylmethionine peptide. Patients with MPO deficiency and chronic granulomatous disease were unable to inactivate these factors unless MPO and H_2O_2, respectively, were added. The molecular mechanism of MPO-catalyzed inactivation was shown to be oxidation of the thioether group of methionine (Clark et al., 1980, 1982). The biochemical inactivation of the chemoattractant resulted in a decreased affinity of the peptide for its membrane receptors. Furthermore, the chemoattractant fMet-Leu-Phe could induce sufficient secretory stimuli to trigger autoinactivation of the neutrophil (Clark, 1982), suggesting that not only exogenous agents such as latex particles (Tsan and Denison, 1981) or zymosan or phorbol esters (Clark and Szot, 1982) but also chemotactic factors per se were able to induce inactivation. Taken together, these data indicate that proteases such as cathepsin and elastase as well as products of the oxidative metabolism of neutrophils are able to inactivate neutrophil attractants (Gallin et al., 1978).

The MPO-mediated oxidative system might attack the neutrophil itself. Several studies showed that neutrophil-generated oxidants regulate the activity of granule enzymes and thus provide a self-defense system by autoinactivating secretory products. After phagocytosis of opsonized zymosan particles, neutrophils from patients with chronic granulomatous disease, which are not able to generate oxidative agents, released two to three times more lysozymes and β-glucuronidase than did normal neutrophils (Voetman et al., 1981). In normal PMN, the β-glucuronidase is probably inactivated mainly in the cells and not in the medium. The MPO requirement is equivocal. Vitamin B_{12}-binding protein, which is a marker of granule enzyme secretion, is inactivated extracellularly; this inactivation can be prevented by azide, suggesting a role of the MPO system (Clark and Borregaard, 1985). Neutral neutrophil proteinases are also susceptible to MPO-dependent oxidation (Vissers and Winterbourn, 1987). Almost complete oxidative inactivation of β-glucuronidase, collagenase, and gelatinase can occur, along with an approximately 50% inhibition of lysozyme.

Observations that the MPO system inactivates diphtheria and tetanus toxins were the first evidence for a possible role of MPO in biological responses (Agner, 1950, 1958). Later studies demonstrated an inhibitory effect of the cell-free MPO system and isolated PMN on the toxic microbial metabolite aflatoxin and the cytosolic toxin of *Clostridium difficile* (Odajima, 1981; Ooi *et al.*, 1984). The latter is implicated in antibiotic-associated colitis. Inactivation of *Clostridium* toxin suggests a protective role of MPO in inflammatory bowel diseases.

Of interest is a study showing that another cytolytic bacterial toxin, pneumolysin (a hemolytic toxin from *Streptococcus pneumoniae*), is inactivated by stimulated PMN and the MPO system (Clark, 1986). Neutrophils from patients with chronic granulomatous disease and MPO deficiency were unable to inactivate pneumolysin. These studies indicate that the MPO system might help in detoxifying oxygen-sensitive cytolysins.

MPO itself could be inactivated by both hypochlorous acid and H_2O_2 alone. Inactivation of MPO was the result of oxidation of methionine and tyrosine residues (Matheson and Travis, 1985). The down-regulatory effect of MPO on neutrophil activity was also indirectly suggested by studies using MPO-deficient neutrophils and monocytes (Stendahl *et al.*, 1984; Locksley *et al.*, 1983; Nauseef *et al.*, 1983a). MPO-deficient cells had increased phagocytosis of IgG- and C3b-opsonized yeast particles and a prolonged fMet-Leu-Phe-mediated stimulation (Paredes and Weiss, 1982). In these cells, both oxygen consumption and the formation and release of superoxide and peroxide were greater than in normal PMN (Locksley *et al.*, 1983). Using anti-MPO antibodies with normal neutrophils, this supermetabolic state was shown to be due to increased duration of superoxide production rather than to accelerated production or decreased metabolism (Edwards and Swan, 1986). The underlying mechanism of MPO-induced modulation of the respiratory burst was postulated to be MPO-mediated inactivation of the superoxide-producing NADPH oxidase system.

Taken together, the numerous studies suggest a role of MPO in modulating acute and chronic inflammatory responses and demonstrate that a complex interplay exists between neutrophils and their products at the site of inflammation. The balance between damaging and protective mechanisms determines the net effect of the MPO system in immunoinflammatory responses.

ACKNOWLEDGEMENTS. We would like to thank Heidi Haag, Elaine Epstein, Margery Goldberg, and Elisa Weiss for their excellent secretarial help. This research was supported by grants of the Swiss National Science Foundation (3100-009141.87), the Bernische Krebsliga, and the Roche Research Foundation (AT); and in part by U.S. Public Health Service grants CA26038, CA33936, and CA00975 from the National Institutes of Health and by the Marilyn Levine and Erwin Memorial Fund (HPK).

REFERENCES

Agar, N. S., Sadradeh, S. M. H., Hallaway, P. E., Eaton, J. W., 1986, Erythrocyte catalase. A somatic oxidant defense, *J. Clin. Invest.* **79**:319–321

Agner, K., 1941a, Verdoperoxidase. A ferment isolated from leucocytes, *Acta Physiol. Scand.* **2**(Suppl. 8):1–62

Agner, K., 1941b, Detoxicating effect of verdoperoxidase on toxins, *Nature (London)* **159**:271–272

Agner, K., 1950, Studies on peroxidative detoxification of purified diphtheria toxin, *J. Exp. Med.* **92**:337–342

Agner, K., 1958, Crystalline myeloperoxidase, *Acta Chem. Scand.* **12**:89–94

Akin, D. T., and Kinkade, J. M., 1986, Processing of a newly identified intermediate of human myeloperoxidase in isolated granules occurs at neutral pH, *J. Biol. Chem.* **261**:8370–8375

Akin, D., Kinkade, J. M., and Parmley, R. T., 1987, Biochemical and ultrastructural effects of monensin on the processing, intracellular transport, and packaging of myeloperoxidase into low and high density compartments of human leukemia (HL-60) cells, *Arch. Biochem. Biophys.* **257**:451–463

Albrich, J. M., McCarthy, C. A., and Hurst, J. K., 1981, Biological reactivity of hypochlorus acid: Implication for microbicidal mechanisms of myeloperoxidase from leukocytes, *Proc. Natl. Acad. Sci. USA* **78**:210–214

Albrich, J. M., Gilbaugh, J. H., III, Callahan, K. B., and Hurst, J. K., 1986, Effects of the putative neutrophil-generated toxin, hypochlorus acid, on membrane permeability and transport systems of Escherichia coli, *J. Clin. Invest.* **78**:177–184

Allen, R. C., 1975, Halide dependence of the myeloperoxidase-mediated antimicrobial system of the polymorphonuclear leucocyte in the phenomenon of electronic excitation, *Biochem. Biophys. Res. Commun.* **63**:675–683

Andersen, M. R., Atkin, C. L., and Eyre, H. J., 1982, Intact form of myeloperoxidase from normal human neutrophils, *Arch. Biochem. Biophys.* **214**:273–283

Andrews, P. C., and Krinsky, N. I., 1981, The reductive cleavage of myeloperoxidase in half, producing enzymatically active hemimyeloperoxidase, *J. Biol. Chem.* **256**:4211–4218

Andrews, P. C., Parnes, C., and Krisnky, N. I., 1984, Comparison of myeloperoxidase and hemi-myeloperoxidase with respect to catalysis, regulation, and bactericidal activity, *Arch. Biochem. Biophys.* **228**:439–442

Angel, P., Imagawa, M., Chiu, R., Stein, B., Imbra, R. J., Rahmsdorf, H., Carsten, J., Herrlich, P., and Karin, M., 1987, Phorbol ester-inducible genes contain a common *cis* element recognized by a TPA-modulated *trans*-acting factor, *Cell* **49**:729–739

Arnlijots, K., and Olsson, I., 1987, Myeloperoxidase precursors incorporate heme, *J. Biol. Chem.* **262**:10430–10433

Atkin, C. L., Andersen, M. R., and Eyre, H. J., 1982, Normal neutrophil myeloperoxidase from a patient with chronic myelocytic leukemia, *Arch. Biochem. Biophys.* **214**:284–292

Atkinson, P. H., and Lee, J. T., 1983, Cotranslational excision of alpha-glucose and alpha-mannose in nascent vesicular stomatitis G proteins, *EMBO J.* **2**:823–832

Babcock, G. T., Ingle, R. T., Oertling, W. A., Davis, J. C., Averill, B. A., Hulse, C. L., Stufkens, D. J., Bolscher, B. G. J. M., Wevers, R., and Raman, P., 1985, Characterization of human myeloperoxidase and bovine spleen hemoprotein. Insight into chromophore structure and evidence that the chromophores of myeloperoxidases are equivalent, *Biochim. Biophys. Acta* **828**:58–66

Badway, J. A., and Karnovsky, M. L., 1980, Active oxygen species and the functions of phagocytic leukocytes, *Annu. Rev. Biochem.* **49**:695–726

Baehner, R. L., Nathan, D. G., and Castle, W. B., 1971, Oxidant injury of Caucasian glucose-6-phosphate dehydrogenase-deficient red blood cells by phagocytosing leukocytes during infection, *J. Clin. Invest.* **50**:2466–2473

Bainton, D. F., and Farquhar, M. G., 1968, Differences in enzyme content of azurophil and specific granules of polymorphonuclear leukocytes, *J. Cell Biol.* **39**:299–317

Bainton, D. F., Ullyot, J. L., and Farquar, M. G., 1971, The development of neutrophilic polymorphonuclear leukocytes in human bone marrow, *J. Exp. Med.* **134**:907–933

Bakkenist, A. R. J., Wever, R., Vulsma, T., Plat, H., and Van Gelder, F., 1978, Isolation procedure and some properties of myeloperoxidase from human leukocytes, *Biochim. Biophys. Acta* **524**:45–54

Bakkenist, A. R. J., De Boer, J. E. G., Plat, H., and Wever, R., 1980, The halide complexes of myeloperoxidase and the mechanism of the halogenation reactions, *Biochim. Biophys. Acta* **613**:337–348

Becker, B. N., Henson, P., Showell, H. J., and Hsu, L. S., 1974, The ability of chemotactic factors to induce lysosomal enzyme release. I. The characteristics of the release, the importance of surfaces and the relation of enzyme release to chemotactic responsiveness, *J. Immunol.* **112**:2047–2054

Bennett, J. M., Catovsky, D., Daniel, M. T., Flandrin, G., Galton, D. A. G., Gralnick, H. R., and Sultan, C., 1976, Proposal for the classification of acute leukemia, *Br. J. Haematol.* **33**:451–458

Bennett, J. M., Catovsky, D., Daniel, M. T., Flandrin, G., Galton, D., Gralnick, H. R., and Sultan, C., 1985, Proposed revised criteria for the classification of acute myeloid leukemia. A report of the French-American-British cooperative group, *Ann. Intern. Med.* **103**:626–629

Berkow, R. L., and Dodson, R. W., 1978, Functional analysis of the marginating pool of human polymorphonuclear leukocytes, *Am. J. Hematol.* **24**:47–54

Birnboim, H. C., 1982, DNA strand breakage in human leukocytes exposed to a tumor promotor, phorbol myristate acetate, *Science* **215**:1247–1249

Bos, A., Wever, R., and Dirk, R., 1978, Characterization and quantification of the peroxidase in human monocytes, *Biochim. Biophys. Acta* **525**:37–44

Bradley, M., and Erickson, L. C., 1981, Comparison of the effects of hydrogen peroxide and x-rays irradiation on toxicity, mutation, and DNA damage/repair in mammalian cells (V-79), *Biochim. Biophys. Acta* **654**:135–141

Bradley, P. P., Christensen, R. D., and Rothstein, G., 1982, Cellular and extracellular myeloperoxidase in pyogenic inflammation, *Blood* **60**:618–622

Bretz, U., and Baggiolini, M., 1974, Biochemical and morphological characterization of azurophil and specific granules of human neutrophilic polymorphonuclear leukocytes, *J. Cell Biol.* **63**:251–269

Bringmann, G., 1953, Elektronenmikroskopische Befunde zur Wirkung von Chlor, Brom, Jod, Kupfer, Silber und Wasserstoffsuperoxide auf E. coli, *Z. Hyg. Infektionskr.* **138**:155–166

Cantin, A. M., North, S. L., Fells, G. A., Hubbard, R. C., and Crystal, R. G., 1987, Oxidant-mediated epithelial cell injury in idiopathic pulmonary fibrosis, *J. Clin. Invest.* **79**:1665–1673

Cech, P., and Lehrer, R. I., 1984, Phagolysosomal pH of human neutrophils, *Blood* **63**:88–95

Cech, P., Stalder, H., Widman, J. J., Rohrer, A., and Miescher, P. A., 1979, Leukocyte myeloperoxidase deficiency and diabetes mellitus associated with Candida albicans liver abscess, *Am. J. Med.* **66**:149–153

Chang, S. K., Trujillo, J. M., Cook, R. G., and Stass, S. A., 1986, Human myeloperoxidase gene: Molecular cloning and expression in leukemic cells, *Blood* **68**:1411–1414

Cheh, A. M., Skochdopole, J., Koski, P., and Cole, L., 1980, Nonvolatile mutagens in drinking water. Production by chlorination and destruction of sulfide, *Science* **207**:90–92

Clark, R. A., 1982, Chemotactic factors trigger their own oxidative inactivation by human neutrophils, *J. Immunol.* **129**:2725–2728

Clark, R. A., 1983, Extracellular effects of the myeloperoxidase-hydrogen peroxide-halide system, in *Advances in Inflammation Research* (G. Weissmann, ed.), pp. 107–146, Raven Press, New York.

Clark, R. A., 1986, Oxidative inactivation of pneumolysin by the myeloperoxidase system and stimulated human neutrophils, *J. Immunol.* **136**:4617–4622

Clark, R. A., and Borregaard, N., 1985, Neutrophils autoinactivate secretory products by myeloperoxidase-catalyzed oxidation, *Blood* **65**:375–381

Clark, R. A., and Klebanoff, S. J., 1977, Myeloperoxidase-H$_2$O$_2$-halide system: Cytotoxic effect on human blood leukocytes, Blood **50**:65–70

Clark, R. A., and Klebanoff, S. J., 1979a, Myeloperoxidase-mediated platelet release reaction, *J. Clin. Invest.* **63**:177–183

Clark, R. A., and Klebanoff, S. J., 1979b, Chemotactic factor inactivation by the myeloperoxidase-hydrogen peroxide-halide system, *J. Clin. Invest.* **64**:913–920

Clark, R. A., and Klebanoff, S. J., 1980, Neutrophil-platelet interaction mediated by myeloperoxidase and hydrogen peroxide, *J. Immunol.* **124**:399–407

Clark, R. A., and Szot, S., 1982, Chemotactic factor inactivation by stimulated human neutrophils mediated by myeloperoxidase-catalyzed methionine oxidation, *J. Immunol.* **128**:1507–1513

Clark, R. A., Olsson, I., and Klebanoff, S. J., 1976, Cytotoxicity for tumor cells of cationic proteins from human neutrophil granules, *J. Cell Biol.* **70**:719–723

Clark, R. A., Szot, S., Venkatasubramanian, K., and Schiffmann, E., 1980, Chemotactic factor inactivation by myeloperoxidase mediated oxidation of methionine, *J. Immunol.* **124**:2020–2029

Clark, R. A., Stone, P. J., El Hag, A., Calore, J. D., and Franzblau, C., 1981, Myeloperoxidase-catalyzed inactivation of alpha-1 protease inhibitor by human neutrophils, *J. Biol. Chem.* **256**:3348–3353

Clark, R. A., Szot, S., Williams, M. A., and Kagan, H. M., 1986, Oxidation of lysine side-chains of elastin by the myeloperoxidase system and by stimulated neutrophils, *Biochem. Biophys. Res. Commun.* **135**:451–457

Collins, S. J., Gallo, R. C., and Gallagher, R. E., 1977, Continuous growth and differentiation of human myeloid leukemia cells in suspension culture, *Nature (London)* **270**:347–349

Cramer, E., Pryzwansky, K. B., Villeval, J. L., Testa, U., and Breton-Gorius, J., 1985, Ultrastructural localization of lactoferrin and human myeloperoxidase in human neutrophils by immunogold, *Blood* **65**:423–432

Cuperus, R. A., Muijers, A. O., and Wever, R., 1986, The superoxide dismutase activity of myeloperoxidase; formation of compound III, *Biochim. Biophys. Acta* **871**:78–84

Cuperus, R. A., Hoogland, H., Wever, R., and Muijers, A. O., 1987, The effect of D-penicillamine on myeloperoxidase: Formation of compound III and inhibition of the chlorinating activity, *Biochim. Biophys. Acta* **912**:124–131

Dallegri, F., Patrone, F., Frumento, G., Ballestro, A., and Sacchetti, C., 1985, Downregulation of natural Killer cell activity by neutrophils, *Blood* **65**:571–577

Dallegri, F., Patrone, F., Ballestrero, A., Frumento, G., and Sacchetti, C., 1986, Inhibition of neutrophil cytolysin production by target cells, *Blood* **67**:1265–1272

Dani, C., Blanchard, M., Piechaczyk, M., El Sabouty, S., Marty, L., and Janteur, P., 1984, Extreme instability of *myc* mRNA in normal and transformed human cells, *Proc. Natl. Acad. Sci. USA* **81**:7046–7050

Desser, R. K., Himmelhoch, S. R., Evans, W. H., Januska, M., Mage, M., and Shelton, E., 1972, Guinea pig heterophil and eosinophil peroxidase, *Arch. Biochem. Biophys.* **148**:452–465

Diamond, R. D., Clark, R. A., and Haudenschild, C. C., 1980, Damage to Candida albicans hyphae and pseudohyphae by the myeloperoxidase system and oxidative producers of neutrophil metabolism in vitro, *J. Clin. Invest.* **66**:908–917

Edwards, E. W., and Swan, T. F., 1986, Regulation of superoxide generation by myeloperoxidase during the respiratory burst of human neutrophils, *Biochem. J.* **237**:601–604

Ehrenberg, A., and Agner, K., 1958, The molecular weight of myeloperoxidase, *Acta Chem. Scand.* **12**:95–100

Elder, J. H., and Alexander, S., 1982, Endo-beta-N-acetylglucosaminidase F: Endoglycosidase from Flavobacterium meningosepticum that cleaves both high-mannose and complex glycoproteins, *Proc. Natl. Acad. Sci. USA* **79**:4540–4544

El-Hag, A., and Clark, R. A., 1984, Down-regulation of human natural killer activity against tumors by the neutrophil myeloperoxidase system and hydrogen peroxide, *J. Immunol.* **133**:3291–3297

El-Hag, A., and Clark, R. A., 1987, Immunosuppression by activated human neutrophils. Dependence on the myeloperoxidase system, *J. Immunol.* **139**:2406–2413

El-Hag, A., Lipsky, P. E., Bennett, M., and Clark, R. A., 1986, Immunomodulation by neutrophil myeloperoxidase and hydrogen peroxide: Differential susceptibility of human lymphocyte functions, *J. Immunol.* **136**:3420–3426

Elsbach, P., 1973, On the interaction between phagocytes and microorganisms, *N. Engl. J. Med.* **16**:846–852

Suematsu, M., Oshio, C., Mura, S., Suzuki, M., Houzawa, S., and Tsuchiya, M., 1988, Luminol-dependent photoemission from single neutrophil stimulated by phorbol ester and calcium ionophore—role of degranulation and myeloperoxidase, *Biochem. Biophys. Res. Commun.* **155**:106–111

Elsbach, P., and Weiss, J. A., 1983, A reevaluation of the roles of the O_2-dependent and O_2-independent microbicidal systems of the phagocytes, *Rev. Infect. Dis.* **5**:843–853

English, D., and Lukens, J. N., 1983, Regulation of neutrophil inflammatory mediator release: Chemotactic peptide activation of stimulus-dependent cytotoxicity, *J. Immunol.* **130**:850–860

Felberg, N. T., and Schultz, J., 1972, Evidence that myeloperoxidase is composed of isoenzymes, *Arch. Biochem. Biophys.* **148**:407–413

Felberg, N. T., Putterman, G. J., and Schultz, J., 1969, Myeloperoxidase X: Comparison of normal human leukocyte myeloperoxidase prepared with and without the use of trypsin, *Biochem. Biophys. Res. Commun.* **37**:213–218

Ferrari, S., Mariano, M. T., Tagliafico, E., Sarti, M., Ceccherelli, G., Selleri, L., Merli, F., Narni, F., Donelli, A., Torelli, G., and Torelli, U., 1988, Myeloperoxidase gene expression in blast cells with lymphoid phenotype in cases of acute lymphoblastic leukemia, *Blood* **72**:873–876

Foote, C. S., Goyne, T. E., and Lehrer, R. I., 1983, Assessment of chlorination by human neutrophils, *Nature* **301**:715–716

Fucci, L., Oliver, C. N., Coon, M. J., and Stadtman, E. R., 1983, Inactivation of key metabolic enzymes by mixed-function oxidation reaction: Possible implication in protein turnover and aging, *Proc. Natl. Acad. Sci. USA* **80**:1521–1525

Gabig, T. G., and Babior, B. M., 1979, The O_2^--forming oxidase responsible for the respiratory burst in human neutrophils, *J. Biol. Chem.* **254**:9070–9074

Gallin, J. I., Wright, D. G., and Schiffmann, E., 1978, Role of secretory events in modulating human neutrophil chemotaxis, *J. Clin. Invest.* **62**:1364–1374

Ganz, T., Selsted, M. E., and Lehrer, R. I., 1986, Antimicrobial activity of phagocyte granule proteins, *Semin. Respir. Infect.* **1**:107–117

Goldstein, I., Hoffstein, S., Gallin, J., and Weissmann, G., 1973, Mechanisms of lysosomal enzyme release from human leukocytes: Microtubule assembly and membrane fusion induced by a component of complement, *Proc. Natl. Acad. Sci. USA* **70**:2916–2920

Greaves, M. F., Chan, L. C., Furley, A. J. W., Watt, S. M., and Molgaard, H. V., 1986, Lineage promiscuity in hematopoietic differentiation and leukemia, *Blood* **67**:1–11

Grisham, M. B., Jefferson, M. M., and Thomas, E. L., 1984a, Role of monochloramine in the oxidation of erythrocyte hemoglobin by stimulated neutrophils, *J. Biol. Chem.* **259**:6757–6765

Grisham, M. B., Jefferson, M. M., Melton, D. F., and Thomas, E. L., 1984b, Chlorination of endogenous amines by isolated neutrophils: Ammonia-dependent bactericidal, cytotoxic, and catalytic activities of the chloramines, *J. Biol. Chem.* **259**:10404–10413

Hamers, M. N., Bot, A. A. M., Weening, R. S., Sips, H. J., and Ross, D., 1984, Kinetics and mechanism of the bactericidal action of human neutrophils against Escherichia coli, *Blood* **64**:635–641

Handin, R. I., Karabin, R., and Boxer, G. J., 1977, Enhancement of platelet function by superoxide anion, *J. Clin. Invest.* **59**:959–965

Harlan, J. M., Killer, P. D., Harker, L. A., Striker, G. E., and Wright, D. G., 1981, Neutrophil mediated injury in vitro. Mechanism of cell detachment, *Clin. Invest.* **68**: 1394–1403

Harrison, J. E., and Schultz, J., 1978, Myeloperoxidase: Confirmation and nature of heme-binding inequivalence. Resolution of a carbonyl-substituted heme, *Biochim. Biophys. Acta* **536**:341–349

Harrison, J. E., Pabalan, S., and Schultz, J., 1977, The subunit structure of crystalline canine myeloperoxidase, *Biochim. Biophys. Acta* **493**:247–259

Hasilik, A., and Von Figura, K., 1984, Processing of lysosomal enzymes in fibroblasts, in *Lysosomes in Biology and Pathology* (J. T. Doyle, R. T. Dean, and W. Sly, eds.), pp. 3–26, Elsevier Science Publications, New York

Hasilik, A., Pohlmann, R., Steckel, F., Gieselmann, V., Von Figura, K., Olsen, R., and Waheed, A., 1983, Biosynthesis and transport of lysosomal enzymes, in *13th Lindstrom-Lang Conference on Translational and Post-Translational Events*, pp. 349–369, The Humana Press, Inc., Clifton, N.J.

Hasilik, A., Pohlmann, R., Olsen, R. L., and Von Figura, K., 1984, Myeloperoxidase is synthesized as a larger phosphorylated percursor, *EMBO J.* **3**:2671–2676

Hashinaka, K., Nishio, C., Hur, S.-J., Sakiyama, F., Tsunasawa, S., and Yamada, M., 1988, Multiple species of myeloperoxidase messenger RNAs produced by altered splicing and differential polyadenylation, *Biochemistry* **27**:5906–5914

Henderson, W. R., and Klebanoff, S. J., 1983, Leukotriene B4, C4, D4 and E4 inactivation by hydroxyl radicals, *Biochem. Biophys. Res. Commun.* **110**:266–272

Henderson, W. R., Joerg, A., and Klebanoff, S. J., 1982, Eosinophil peroxidase-mediated inactivation of leukotrienes B4, C4, and D4, *J. Immunol.* **128**:2609–2613

Henson, P. M., 1980, Mechanisms of exocytosis in phagocytic inflammatory cells, *Am. J. Pathol.* **101**: 494–511

Himmelhoch, R. S., Evans, W. H., Mage, M. G., and Peterson, E. A., 1969, Purification of myeloperoxidase from the bone marrow of the guinea pig, *Biochemistry* **8**:914–921

Ikeda-Saito, M., and Prince, R. C., 1985, The effect of chloride on the REDO and EPR properties of myeloperoxidase, *J. Biol. Chem.* **260**:8301–9305

Jaffe, B. D., Sabath, D. E., Johnson, G. D., Moscinski, L. C., Johnson, K. R., Rovera, G., Nauseef, W. M., and Prystowsky, M. B., 1988, Myeloperoxidase and oncogene expression in GM-CSF induced bone marrow differentiation, *Oncogene* **2**:167–174

Janoff, A., 1985, Elastases and emphysema. Current assessment of the protease-antiprotease hypothesis, *Am. Rev. Respir. Dis.* **132**:417–433

Jasin, H. E., 1983, Generation of IgG aggregates by the myeloperoxidase-hydrogen peroxide system, *J. Immunol.* **130**:1918–1923

Jasin, H. E., 1988, Oxidative cross-linking of immune complexes by human polymorphonuclear leukocytes, *J. Clin. Invest.* **81**:6–15

Jensen, M. S., and Bainton, D. F., 1973, Temporal changes in pH within the phagocytic vacuole of the polymorphonuclear neutrophilic leukocyte, *J. Cell Biol.* **56**:379–338

Johnson, K. R., Nauseef, M., Care, A., Wheelock, M. J., Shane, S., Hudson, Koeffler, H. P., Selsted, M., Miller, C., and Rovera, G., 1987a, Characterization of cDNA clones for human myeloeproxidase: Predicted amino acid sequence and evidence for multiple mRNA species, *Nucleic Acids Res.* **15**:2013–2028

Johnson, R., Couser, W. G., Chi, E. Y., Adler, S., and Klebanoff, S. J., 1987b, New mechanism for glomerular injury. Myeloperoxidase-hydrogen peroxide-halide system, *J. Clin. Invest.* **79**:1379–1387

Johnson, R. J., Klebanoff, S. J., Ochi, R. F., Adler, S., Baker, P., Sparks, L., and Couser, W. G., 1987c, Participation of the myeloperoxidase-H_2O_2-halide system in immune complex nephritis, *Kidney Int.* **32**:342–349

Johnson, R. J., Guggenheim, S. J., Klebanoff, S. J., Ochi, R. F., Wass, A., Baker, P., Schulze, M., and Couser, W. G., 1988, Morphologic correlates of glomerular oxidant injury induced by the myeloperoxidase-hydrogen peroxide-halide system of the neutrophil, *Lab. Invest.* **5**:294–301

Kettle, A. J., and Winterbourn, C. C., 1988, Superoxide modulates the activity of myeloperoxidase and optimizes the production of hypochlorus acid, *Biochem. J.* **252**:529–536

Kimura, S., and Ikeda-Saito, M., 1988, Human Myeloperoxidase and thyroid peroxidase, two enzymes with separate and distinct physiological functions, are evolutionary related members of the same gene, *Proteins* **3**:113–120

Kinkade, J. M. J., Pember, S. O., Barnes, K. C., Shapira, R., Spitznagel, J. R., and Martin, L. E., 1983, Differential distribution of distinct forms of myeloperoxidase in different azurophilic granule subpopulations from human neutrophils, *Biochem. Biophys. Res. Commun.* **114**:296–303

Kitahara, M., Eyre, H. J., Simonian, Y., Atkin, C. L., and Hasstedt, S. J., 1981, Hereditary myeloperoxidase deficiency, *Blood* **57**:88–893

Klebanoff, S. J., 1968, Myeloperoxidase-halide-hydrogen peroxide antimicrobial system, *J. Bacteriol.* **95**:2131–2138

Klebanoff, S. J., 1970, Myeloperoxidase-mediated antimicrobial systems and their role in leukocyte function, in *Biochemistry of the Phagocytic Process: Localization and the Role of Myeloperoxidase and the Mechanism of the Halogenation Reaction* (J. Schultz, ed.), pp. 89–110, Elsevier/North-Holland, Amsterdam

Klebanoff, S. J., 1980a, Myeloperoxidase-mediated cytotoxic systems, in *The Reticuloendothelial System,* Vol. 2. *Biochemistry and Metabolism* (A. J. Sbarra and R. R. Strauss, eds.), pp. 279–308, Plenum Press, New York

Klebanoff, S. J., 1980b, Oxygen metabolism and the toxic properties of phagocytes, *Ann. Intern. Med.* **93**: 480–489

Klebanoff, S. J., 1988, Phagocytic cells: Products of oxygen metabolism, in *Inflammation: Basic Principles and Clinical Correlates* (J. I. Gallin and R. Snyderman, eds.), pp. 391–444, Raven Press, New York

Klebanoff, S. J., and Clark, R. A., 1975, Hemolysis and iodination of erythrocyte components by a myeloperoxidase-mediated system, *Blood* **45**:699–707

Klebanoff, S. J., and Pincus, S. H., 1971, Hydrogen peroxide utilization in myeloperoxidase-deficient leukocytes: A possible microbicidal control mechanism, *J. Clin. Invest.* **50**:2226–2229

Klebanoff, S. J., and Smith, D. C., 1970, The source of H_2O_2 for the uterine fluid-mediated sperm-inhibitory system, *Biol. Reprod.* **3**:236–242

Klebanoff, S. J., Clark, R. A., and Rosen, H., 1976, Myeloperoxidase-mediated cytotoxicity, in *Cancer Enyzymology* (J. Schultz and F. Ahmad, eds.), pp. 267–288, Academic Press, New York

Klebanoff, S. J., Waktersdirogm, A. M., and Rosen, H., 1984, Antimicrobial activity of myeloperoxidase, *Methods Enzymol.* **105**:399–423

Klempner, M. S., Dinarello, C. A., and Gallin, J. I., 1978, Human leukocyte pyrogen induces release of specific granule contents from human neutrophils, *J. Clin. Invest.* **61**:1330–1336

Koeffler, H. P., 1983, Induction of differentiation of human acute myelogenous leukemia cells: Therapeutic implications, *Blood* **62**:709–721

Koeffler, H. P., 1986, Human acute myeloid leukemia lines: Models of leukemogenesis, *Semin. Hematol.* **23**:223–236

Koeffler, H. P., Ranyard, J., and Pertchek, M., 1985, Myeloperoxidase: Its structure and expression during myeloid differentiation, *Blood* **65**:484–491

Kornfeld, S., 1986, Trafficking of lysosomal enzymes in normal and disease states, *J. Clin. Invest.* **77**:1–6

Lee, C. W., Lewis, R. A., Tauber, A. I., Mehrotra, M., Corey, E. J., and Austen, K. F., 1983, The myeloperoxidase-dependent metabolism of leukotrienes C4, D4, and E4 to 6-trans-leukotriene B4 diastereoisomers and the subclass-specific S-diastereoisomeric sulfoxides, *J. Biol. Chem.* **258**:15004–15010

Lehrer, R. I., 1975, The fungicidal mechanisms of human monocytes. I. Evidence for myeloperoxidase-linked and myeloperoxidase-independent candidicidal mechanisms, *J. Clin. Invest.* **55**:338–346

Lehrer, R. I., and Cline, M. J., 1969, Leukocyte myeloperoxidase deficiency and disseminated candidiasis: The role of myeloperoxidase in resistance to candida infection, *J. Clin. Invest.* **48**:1478–1488

Lichtenstein, A., 1986, Spontaneous tumor cytolysis mediated by inflammatory neutrophils: Dependence upon divalent cations and reduced oxygen intermediates, *Blood* 67:657–665

Lichtenstein, A., Ganz, T., Selsted, M. E., and Lehrer, R. I., 1986, In vitro tumor cell cytolysis mediated by peptide defensins of human and rabbit granulocytes, *Blood* **68**:1407–1410

Lichtenstein, A. K., Ganz, T., Selsted, M. E., and Lehrer, R. I., 1988, Synergistic cytolysis mediated by hydrogen peroxide combined with peptide defensins, *Cell. Immunol.* **114**:104–116

Locksley, R. M., Wilson, C. B., and Klebanoff, S. J., 1983, Increased respiratory burst in myeloperoxidase-deficient monocytes, *Blood* **62**:902–909

Lubbert, M., Miller, C. W., and Koeffler, H. P., submitted for publication, Methylation and chromatin changes in the human myeloperoxidase gene during myeloid differentiation (submitted for publication)

Mandell, G. L., 1976, Intraphagosomal pH of human polymorphonuclear neutrophils, *Proc. Soc. Exp. Med.* **134**:447–449

Marshall, R. D., 1974, Glycoproteins, *Annu. Rev. Biochem.* **41**:673–702

Matheson, N. R., and Travis, J., 1985, Differential effects of oxidizing agents on human plasma alpha-1-proteinase inhibitor and human neutrophil myeloperoxidase, *Biochemistry* **24**:1941–1945

Matheson, N. R., Wong, D. S., and Travis, J., 1979, Enzymatic inactivation of human alpha-1-proteinase inhibitor by neutrophil myeloperoxidase, *Biochem. Biophys. Res. Commun.* **88**:402–409

Matheson, N. R., Wong, P. S., and Travis, J., 1981a, Isolation and properties of human neutrophil myeloperoxidase, *Biochemistry* **20**:325–330

Matheson, N. R., Wong, P. S., Schuler, M., and Travis, J., 1981b, Interaction of human alpha-1-proteinase inhibitor with neutrophil myeloperoxidase, *Biochemistry* **20**:331–336

McCulloch, E. A., 1987, Lineage infidelity or lineage promiscuity, *Leukemia* **1**:135–140

McRipley, R. J., and Sbarra, A. J., 1967, Role of the phagocyte in host-parasite interactions, *J. Bacteriol.* **94**:1425–1430

Meerhof, L. J., and Roos, D., 1980, An easy, specific and sensitive assay for the determination of catalase activity of human blood cells, *J. Reticuloendothel. Soc.* **28**:419–425

Miller, C. W., Rovera, G., VanTuinen, P., Kitchingman, G., Bernstein, I., and Koeffler, H. P., 1989, Myeloperoxidase gene in acute promyelocytic leukemia, *Science* **244**:823–826

Miyasaki, K. T., Wilson, M. E., Brunetti, A. J., and Genco, R. J., 1986a, Oxidative and nonoxidative killing of *Actinobacillus actinomycetemconcomitans* by human neutrophils, *Infect. Immun.* **53**:154–160

Miyasaki, K. T., Wilson, M. E., Cohen, E., Jones, P. C., and Genco, R. J., 1986b, Evidence for and partial characterization of three major and three minor chromographic forms of human neutrophil myeloperoxidase, *Arch. Biochem. Biophys.* **246**:751–764

Miyasaki, K. T., Zambon, J. J., Jones, C. A., and Wilson, M. E., 1987, Role of high-avidity binding of human neutrophil myeloperoxidase in the killing of *Actinobacillus actinomycetemconcomitans*, *Infect. Immun.* **55**:1029–1036

Morishita, K., Kubota, N., Asano, S., Kaziro, Y., and Nagata, S., 1987a, Molecular cloning and characterization of cDNA for human myeloperoxidase, *J. Biol. Chem.* **262**:3844–3851

Morishita, K., Tschiya, M., Asano, S., Kaziro, Y., and Nagata, S., 1987b, Chromosomal gene structure of human myeloperoxidase and regulation of its expression by granulocyte colony-stimulating factor, *J. Biol. Chem.* **262**:15208–15313

Morrison, M., and Schonbaum, G. R., 1976, Peroxidase-catalyzed halogenation, *Annu. Rev. Biochem.* **45**:861–888

Morse, J. O., 1978, Alpha-1-antitrypsin deficiency, *N. Engl. J. Med.* **299**:1045–1048, 1099–1105

Murao, S. I., Stevens, F. J., Ito, A., and Huberman, E., 1988, Myeloperoxidase: A myeloid nuclear antigen with DNA-binding properties, *Proc. Natl. Acad. Sci. USA* **85**:1232–1236

Nathan, C. F., Brukner, L., Silverstein, S., and Cohn, Z. A., 1979, Extracellular cytolysis by activated macrophages and granulocytes. II. Hydrogen peroxide as a mediator of cytotoxicity, *J. Exp. Med.* **149**:84–99

Nauseef, W. M., 1986, Myeloperoxidase biosynthesis by a human promyelocytic leukemia cell line: Insight into myeloperoxidase deficiency, *Blood* **67**:865–872

Nauseef, W., 1987, Posttranslational processing of a human lysosomal protein, myeloperoxidase, *Blood* **70**:1143–1150

Nauseef, W. M., 1988, Myeloperoxidase deficiency. *Hematol. Oncol. Clin. North Am.* **2**:135–158

Nauseef, W. M., 1989, Aberrant restriction endonuclease digests of DNA from subjects with hereditary myeloperoxidase deficiency, *Blood* **73**:290–295

Nauseef, W. M., and Clark, R. A., 1986, Separation and analysis of sub cellular organelles in human promyelocytic leukemia cell line, HL-60: Application to the study of myeloid lysosomal enzyme synthesis and processing, *Blood* **68**:442–449

Nauseef, W. M., and Malech, H. L., 1986, Analysis of the peptide subunits of human neutrophil myeloperoxidase, *Blood* **67**:1504–1507

Nauseef, W. M., Root, R. K., and Malech, H. L., 1983a, Biochemical and immunologic analysis of hereditary myeloperoxidase deficiency, *J. Clin. Invest.* **71**:1297–1307

Nauseef, W. M., Metcalf, J. A., and Root, R. K., 1983b, Role of myeloperoxidase in the respiratory burst of human neutrophils, *Blood* **61**:483–492

Odajiama, T., 1980, Myeloperoxidase of the leukocyte of normal blood. Nature of the prosthetic group of myeloperoxidase, *J. Biochem.* **87**:379–391

Odajima, T., 1981, Oxidative destruction of microbial metabolite aflatoxin by the myeloperoxidase-hydrogen peroxide-chloride system, *Arch. Oral Biol.* **26**:339–340

Odajima, T., and Yamazaki, I., 1972, Myeloperoxidase of the leukocyte of normal blood. III. The reaction of ferric myeloperoxidase with superoxide anion, *Biochim. Biophys. Acta* **284**:355–359

Olsen, R. L., and Little, C., 1983, Purification and some properties of myeloperoxidase and eosinophil peroxidase from human blood, *Biochem. J.* **209**:781–787

Olsen, R. L., and Little, C., 1984, Studies on the subunits of human myeloperoxidase, *Biochem. J.* **222**:701–709

Olsson, I., Persson, A. M., and Stroemberg, K., 1984, Biosynthesis, transport and processing of myeloperoxidase in the human leukemic promyelocytic cell line HL-60 and normal bone marrow cells, *Biochem. J.* **223**:911–920

Olsson, I., Lantz, M., Persson, A. M., and Arnljots, K., 1988, Biosynthesis and processing of lactoferrin in bone marrow cells, a comparison with processing of myeloperoxidase, *Blood* **71**:441–447

Ooi, W., Levine, H. G., LaMont, J. T., and Clark, R. A., 1984, Inactivation of Clostridium difficile cytotoxin by the neutrophil myeloperoxidase system, *J. Infect. Dis.* **149**:215–219

Paredes, J. M., and Weiss, S. J., 1982, Human neutrophils transform prostaglandins by a myeloperoxidase-dependent mechanism, *J. Biol. Chem.* **257**:2738–2740

Parry, M. F., Root, R. K., Metcalf, J. A., Delaney, K. K., Kaplow, L. S., and Richard, W. J., 1981, Myeloperoxidase deficiency. Prevalence and clinical significance, *Ann. Intern. Med.* **95**:293–301

Pember, S. O., and Kinkade, J. M., Jr., 1983, Differences in myeloperoxidase activity from neutrophilic polymorphonuclear leukocytes of differing density: Relationship to selective exocytosis of distinct forms of the enzyme, *Blood* **61**:1116–1124

Pember, S. O., Kellar, K. L., Winton, E. F., and Kinkade, F. M., 1981, Chromatographic isolation of two murine leukocyte peroxidases distinct from eosinophil peroxidase, isoenzymes or cell line-specific proteins, *J. Reticuloendothel. Soc.* **29**:451–458

Pember, S. O., Fuhrer-Kruesi, S. M., Barnes, K. C., and Kinkade, J. M., 1982, Isolation of three native forms of myeloperoxidase from human polymorphonuclear leukocytes, *FEBS Lett.* **140**:103–108

Pember, S. O., Shapira, R., and Kinkade, J. M. J., 1983, Multiple forms of myeloperoxidase from human neutrophilic granulocytes: Evidence for differences in compartmentalization, enzymatic activity, and subunit structure, *Arch. Biochem. Biophys.* **221**:391–403

Peppin, G. J., and Weiss, S. J., 1986, Activation of endogenous metalloproteinase, gelatinase, by triggered human neutrophils, *Proc. Natl. Acad. Sci. USA* **83**:4322–4326

Root, R. K., and Cohen, M. S., 1981, The microbicidal mechanism of human neutrophils and eosinophils, *Rev. Infect. Dis.* **3**:565–598

Rosen, H., and Klebanoff, S. J., 1982, Oxidation of Escherichia coli iron centers by the myeloperoxidase-mediated microbicidal system, *J. Biol. Chem.* **257**:13731–13725

Rosen, H., and Klebanoff, S. J., 1985, Oxidation of microbial iron-sulfur centers by the myeloperoxidase-H_2O_2-halide antimicrobial system, *Infect. Immun.* **47**:613–618

Rosen, H., Rakita, R. M., Waltersdorph, A. M., and Klebanoff, S. J., 1987, Myeloperoxidase-mediated damage to the succinate oxidase system of Escherichia coli, *J. Biol. Chem.* **242**:15004–15010

Rosmarin, A. G., Weil, S. C., Rosner, G. L., Griffin, J. D., Arnout, M. A., and Tenen, D. G., 1989, Differential expression of CD11b/CD18 (Mol) and myeloperoxidase genes during myeloid differentiation, *Blood* **73**:131–136

Rowley, J. D., 1978, General report on the first International Workshop on chromosomes in leukemia, *Int. J. Cancer* **21**:307–308

Sagoh, T., and Yamada, M., 1988, Transcriptional regulation of myeloperoxidase gene expression in myeloid leukemia HL-60 cells during differentiation into granulocytes and macrophages, *Arch. Biochem. Biophys.* **262**:599–604

Sbarra, A. J., Selvaraj, R. J., Paul, B. B., Zgliczynski, J. M., Poskitt, K. F., Mitchell, G. W., and Loui, F., 1976, Chlorination, decarboxylation, and bactericidal activity mediated by the MPO-H_2O_2-Cl-system, *Adv. Exp. Med. Biol.* **73**:191–203

Schmid, C. W., and Jelinek, W. R., 1982, The Alu family of dispersed repetitive sequences, *Science* **216**:1065–1070

Schraufstaetter, I., Hyslop, P. A., Jackson, J. H., and Cochrane, C. G., 1988, Oxidant-induced DNA damage of target cells, *J. Clin. Invest.* **82**:1040–1050

Schultz, J., 1980, Myeloperoxidase, in *The Reticuloendothelial System*, Vol. 2. *Biochemistry and Metabolism* (A. J. Sbarra and R. R. Strauss, eds.), pp. 231–254, Plenum Press, New York

Schultz, J., and Kaminker, K., 1962, Myeloperoxidase of the leukocyte of normal human blood. I. Content and localization, *Arch. Biochem. Biophys.* **98**:465–471

Schultz, J., and Rosenthal, S., 1958, Iron (II) inactivation of myeloperoxidase, *J. Biol. Chem.* **234**:2486–2490

Schultz, J., and Shmukler, H. W., 1964, Myeloperoxidase of the leukocyte of normal human blood. II. Isolation, spectrophotometry, and amino acid analysis, *Biochemistry* **3**:1234–1238

Schultz, J., Felberg, N., and John, S., 1967, Myeloperoxidase. VIII. Separation into the components by free-flow electrophoresis, *Biochem. Biophys. Res. Commun.* **28**:543–549

Schultz, J., Snyder, H., Wu, N. C., Berger, N., and Bonner, M. J., 1972, Chemical nature and biological activity of myeloperoxidase, in *The Molecular basis of Electron Transfer* (J. Schultz and B. F. Cameron, eds.), pp. 301–321, Academic Press, New York

Seaman, W. E., Gindhart, T. D., Blackman, M. A., Dalal, B., Talal, N., and Werb, P., 1981, Suppression of natural killing in vitro by human peripheral blood cells. Suppression of killing in vitro by tumor promoter diesters, *J. Clin. Invest.* **67**:1324–1333

Seaman, W. E., Gindhart, T. D., Blackman, M. A., Dalal, B., Talal, N., and Werb, P., 1982, Suppression of natural killing in vitro by monocytes and polymorphonuclear leukocytes. Requirement for reactive metabolites of oxygen, *J. Clin. Invest.* **69**:876–888

Segal, A. W., Geisow, M., Garcia, R., Harper, A., and Miller, R., 1981, The respiratory burst of phagocytic cells is associated with a rise in vacuolar pH, *Nature (London)* **290**:406–409

Selsted, M. E., and Novotny, M. I., 1988, The isoenzyme forms of human myeloperoxidose result from differential N-glycosylation, *Blood* **72**(s):152a

Selvaraj, R. J., Zgliczynski, J. M., Paul, B. B., and Sbarra, A. J., 1978, Enhanced killing of myeloperoxidase-coated bacteria in the myeloperoxidase-H_2O_2-Cl system, *J. Infect. Dis.* **137**:481–485

Shah, S. V., Baricos, W. H., and Basci, A., 1987, Degradation of human glomerular basement membrane by stimulated neutrophils, *J. Clin. Invest.* **79**:25–31

Shock, A., and Baum, H., 1988, Inactivation of alpha-1-proteinase inhibitor in serum by stimulated human polymorphonuclear leucocytes. Evidence for a myeloperoxidase-dependent mechanism, *Cell Biochem. Funct.* **6**:13–23

Showell, H. J., Freer, R. J., Zigmond, S. H., Schiffmann, E., Aswanikumar, S., Corcoran, B., and Becker. E. L., 1976, The structure activity relations of synthetic peptides as chemotactic factors and inducers of lysosomal enzyme secretion for neutrophils, *J. Exp. Med.* **143**:1154–1169

Simon, R. H., Scoggin, C. H., and Patterson, D., 1981, Hydrogen peroxide causes the fatal injury to human fibroblasts exposed to oxygen radicals, *J. Biol. Chem.* **256**:7180–7186

Sips, H. J., and Hamers, M. N., 1981, Mechanism of bactericidal action of myeloperoxidase: Increased permeability of the *Escherichia coli* cell envelope, *Infect. Immun.* **31**:11–16

Smedly, L. A., Tonnesen, M. G., Sandhans, R. A., Haslett, C., Guthrie, A., Johnston, R. B., Henson, P. M., and Worthen, G. S., 1986, Neutrophil-mediated injury to endothelial cells. Enhancement by endotoxin and essential role of neutrophil elastase, *J. Clin. Invest.* **77**:1233–1243

Stahmann, M. A., and Spencer, A. K., 1977, Deamination of protein lysyl epsilon-amino groups by peroxidase in vitro, *Biopolymers* **16**:1299–1306

Stahmann, M. A., Spencer, A. K., Honold, G. R., 1977, Crosslinking of proteins in vitro by peroxidase, *Biopolymers* **16**:1307–1318

Stendahl, O., Coble, B. I., Dahlgren, C., Hed, J., and Molin, L., 1984, Myeloperoxidase modulates the phagocytic activity of polymorphonuclear neutrophil leukocytes. Studies with cells from a myeloperoxidase-deficient patient, *J. Clin. Invest.* **73**:366–373

Stern, A., 1985, Red cell oxidative damage, in *Oxidative Stress* (H. Sies, ed.), pp. 321–330, Academic Press, Orlando, Fla.

Strauven, T. A., Armstrong, D., James, G. T., and Austin, J. H., 1978, Separation of leukocyte peroxidase isoenzymes by agarose acrylamide disc electrophoresis, *Age* **1**:111–117

Stroemberg, K., Persson, A. M., and Olsson, I., 1985, The processing and intracellular transport of myeloperoxidase. Modulation by lysosomotropic agents and monensin, *Eur. J. Cell Biol.* **39**: 424–431

Suematsu, M., Oshio, C., Mura, S., Suzuki, M., Houzawa, S., and Tsuchiya, M., 1988, Luminol-dependent photoemission from single neutrophil stimulated by phorbol ester and calcium ionophore—role of degranulation and myeloperoxidase, *Biochem. Biophys. Res. Commun.* **155**: 106–111

Suzuki, Y., and Lehrer, R., 1980, NAD(P)H oxidase activity in human neutrophils stimulated by phorbol myristate acetate, *J. Clin. Invest.* **66**:1409–1418

Suzuki, K., Yamada, M., Akashi, K., and Fujikura, T., 1986, Similarity of kinetics of three types of myeloperoxidase from human leukocytes and four types from HL-60 cells, *Arch. Biochem. Biophys.* **245**:167–163

Svensson, B. E., Domeij, K., Lindvall, S., and Rydell, G., 1987, Peroxidase and peroxidase-oxidase activities of isolated human myeloperoxidases, *Biochem. J.* **242**:673–680

Tartakoff, A. M., 1983, Perturbation of vesicular traffic with the carboxylic ionophore monensin. *Cell* **32**:1026–1028

Tauber, A. I., and Goetzl, E. J., 1979, Structural and catalytic properties of the solubilized superoxide-generating activity of human polymorphonuclear leukocytes. Solubilization, stabilization, and partial characterization, *Biochemistry* **18**:5576–5584

Test, S. T., and Weiss, S. J., 1984, Quantitative and temporal characterization of the extracellular H_2O_2 pool generated by human neutrophils, *J. Biol. Chem.* **259**:399–405

Thomas, E. L., Learn, D. B., Jefferson, M. M., and Weatherred, W., 1988, Superoxide-dependent oxidation of extracellular reducing agents by isolated neutrophils, *J. Biol. Chem.* **263**:2178–2186

Tobler, A., Miller, C. W., Johnson, K. R., Selsted, M. E., Rovera, G., and Koeffler, H. P., 1988, Regulation of gene expression of myeloperoxidase during myeloid differentiation, *J. Cell. Physiol.* **136**: 215–225

Tobler, A., Selsted, M. E., Miller, C. W., Johnson, K. R., Novotny, M. J., Rovera, G., and Koeffler, H. P., 1989, Evidence for a pretranslational defect in hereditary and acquired myeloperoxidase deficiency, *Blood* **73**:1980–1986

Tsan, M. F., and Denison, R. C., 1980, Phorbol myristate acetate-induced neutrophil autotoxicity. A comparison with H_2O_2 toxicity, *Inflammation* **4**:371–380

Tsan, M. F., and Denison, R. C., 1981, Oxidation of n-formyl methionyl chemotactic peptide by human neutrophils, *J. Immunol.* **126**:1387–1389

Vaessen, R., Houweling, A., and Vonder, E. A., 1987, Post transcriptional control of class 1 MHC mRNA expression in adenovirus 120 transformed cells, *Science* **235**:1486–1488

Valtieri, M., Tweardy, D. J., Caracciolo, D., Johnson, K., Mavilio, F., Altmann, S., Santoli, D., and Rovera, G., 1987, Cytokine-dependent granulocytic differentiation. Regulation of proliferative and differentiative responses in a murine progenitor cell line, *J. Immunol.* **138**:3829–3835

Van Heijne, G., 1986, A new method for predicting signal sequence cleavage site, *Nucleic Acids Res.* **14**:4683–4690

Van Tuinen, P., Johnson, K. R., Ledbetter, S. A., Nussbaum, R. L., Rovera, G., and Ledbetter, D. H., 1988, Localization of myeloperoxidase to the long arm of human chromosome 17: Relationship to the 15;17 translocation of acute promyelocytic leukemia, *Oncogene* **1**:319–322

Vissers, M. C. M., and Winterbourn, C. C., 1986, The effect of oxidants on neutrophil-mediated degradation of glomerular basement membrane-collagen, *Biochim. Biophys. Acta* **889**:277–286

Vissers, M. C. M., and Winterbourn, C. C., 1987, Myeloperoxidase-dependent oxidative inactivation of neutrophil neutral proteinases and microbicidal enzymes, *Biochem. J.* **245**:277–280

Voetman, A. A., Weening, R. S., Hamers, N. N., Meerhof, L. J., Bot, A. A. A. M., and Roos, D., 1981, Phagocytosing human neutrophils inactivate their own granular enzymes, *J. Clin. Invest.* **67**:1541–1549

Von Figura, K., and Hasilik, A., 1986, Lysosomal enzymes and their receptors, *Annu. Rev. Biochem.* **55**:167–193

Wagner, D. K., Collins-Lech, C., and Sohnle, P. G., 1986, Inhibition of neutrophil killing of *Candida albicans* pseudohyphae by substances which quench hypochlorous acid and chloramines, *Infect. Immun.* **51**:731–735

Weiher, H., Koenig, M., and Gruss, P., 1983, Multiple point mutations affecting the simian virus 40 enhancer, *Science* **219**:626–631

Weil, C., Rosner, G. L., Reid, M. S., Chisholm, R. L., Farber, N. M., Spitznagel, J. K., and Swanson, M. S., 1987, cDNA cloning of human myeloperoxidase: Decrease of myeloperoxidase mRNA upon induction of HL-60 cells, *Proc. Natl. Acad. Sci. USA* **84**:2057–2061

Weil, S. C., Rosner, G. L., Reid, M. S., Chisholm, R. L., Lemons, R. S., Swanson, M. S., Carrino, J. J., Diaz, M. O., and Le Beau, M. M., 1988, Translocation and rearrangement of myeloperoxidase gene in acute promyelocytic leukemia, *Science* **249**:790–792

Weiss, S. J., Young, J., LoBuglio, A., Slivka, G. E., and Nimeh, N. F., 1981, Role of hydrogen peroxide in neutrophil mediated destruction of cultured endothelial cells, *J. Clin. Invest.* **68**:714–721

Weiss, S. J., Peppin, G., Ortiz, X., Ragsdale, C., and Test, S. T., 1985, Oxidative autoactivation of latent collagenase by human neutrophils, *Science* **227**:747–749

Weitzman, S., Weitberg, A. B., Clark, E. P., and Stossel, T. P., 1985, Phagocytes as carcinogens: Malignant transformation produced by human neutrophils, *Science* **227**:1231–1233

Winterbourn, C. C., Garcia, R. C., and Segal, A. W., 1985, Production of the superoxide adduct of myeloperoxidase (compound III) by stimulated human neutrophils and its reactivity with hydrogen peroxide and chloride, *Biochem. J.* **228**:583–592

Wong, P. S., and Travis, J., 1980, Isolation and properties of oxidized alpha-1-proteinase inhibitor from human rheumatoid synovial fluid, *Biochem. Biophys. Res. Commun.* **96**:1449–1454

Wright, C. D., and Nelson, R. D., 1988, Candidacidal activity of myeloperoxidase: Characterization of myeloperoxidase-yeast complex formation, *Biochem. Biophys. Res. Commun.* **154**:809–817

Wright, C. D., Bowie, J. U., Gray, G. R., and Nelson, R. D., 1983, Candidacidal activity of myeloperoxidase: Mechanisms of inhibitory influence of soluble cell wall mannan, *Infect. Immun.* **42**:76–80

Wright, C. D., Bowie, J. U., and Nelson, R. D., 1984a, Influence of yeast mannan on release of myeloperoxidase by human neutrophils: Determination of structural features required for formation of myeloperoxidase-mannan-neutrophil complex, *Infect. Immun.* **43**:467–471

Wright, C. D., Herron, M. J., Gray, G. R., Holmes, B., and Nelson, R. D., 1984b, Influence of yeast mannan on human neutrophil function: inhibition of release of myeloperoxidase related to carbohydrate-binding property of the enzyme, *Infect. Immun.* **43**:467–471

Wright, J., Yoshimoto, S., Offner, G. D., Blanchard, R. A., Troxler, R., and Tauber, A. I., 1987, Structural characterization of the isoenzymatic forms of human myeloperoxidase, *Biochim. Biophys. Acta* **915**:68–76

Wymann, M. P., von Tscharner, V., Deranleau, D. A., and Baggiolini, M., 1987, Chemiluminescence detection of H_2O_2 produced by human neutrophils during the respiratory burst, *Anal. Biochem.* **165**:371–378

Yamada, M., 1982, Myeloperoxidase precursors in human myeloid leukemia HL-60 cells, *J. Biol. Chem.* **257**:5980–5982

Yamada, M., and Kurahashi, K., 1984, Regulation of myeloperoxidase gene expression during differentiation of human myeloid leukemia HL-60 cells, *J. Biol. Chem.* **259**:3021–3025

Yamada, M., Mori, M., and Sugimura, T., 1981a, Myeloperoxidases in cultured human promyelocytic leukemia cell line HL-60, *Biochem. Biophys. Res. Commun.* **98**:219–226

Yamada, M., Mori, M., and Sugimura, T., 1981b, Purification and characterization of small molecular weight myeloperoxidase from human promyelocytic leukemia HL-60 cells, *Biochemistry* **20**:766–771

Yamada, M., Hur, S. J., Hasinaka, K., Tsuneoka, K., Saeki, T., Nishio, C., Sakiyama, F., and Tsunasawa, S., 1987, Isolation and characterization of cDNA coding for human myeloperoxidase, *Arch. Biochem. Biophys.* **255**:147–155

Zgliczynski, J. M., 1980, Characteristics of myeloperoxidase from neutrophils and other peroxidases from different cell types, in *The Reticuloendothelial System*, Vol. 2. *Biochemistry and Metabolism* (A. J. Sbarra and R. R. Strauss, eds.), pp. 255–278, Plenum Press, New York

Zgliczynski, J. M., Stelmaszynska, T., Ostrowski, W., Naskalsi, J., and Sznaid, J., 1968, Myeloperoxidase of human leukemic leukocytes. Oxidation of amino acids in the presence of hydrogen peroxide, *Eur. J. Biochem.* **4**:540–547

Zgliczynski, J. M., Selvaraj, R., Paul, B. B. Stelmaszynska, T., Poskitt, K., and Sbarra, A. J., 1977, Chlorination by myeloperoxidase-H_2O_2-Cl-antimicrobial system at acid and neutral pH, *Proc. Soc. Exp. Biol. Med.* **154**:418–422

Chapter 11

Mechanisms of Oxidase Activation in Neutrophils
Importance of Intracellular Calcium and Cytoskeletal Interactions

Futwan A. Al-Mohanna and Maurice B. Hallett

1. IMPORTANCE OF THE NEUTROPHIL OXIDASE IN PHYSIOLOGY AND PATHOLOGY

1.1. Overview

Polymorphonuclear neutrophilic leukocytes, or neutrophils, play a key role in combating infection by killing the infecting microorganisms. These cells also play a role in wound healing by removing cell debris and extracellular material from the wound site. Neutrophils leave the circulation and move through the tissues (by a process of chemotaxis) to the inflammatory site, where they accumulate in high density (up to $10^8/ml$). The neutrophils at this site phagocytose the infecting microbes and kill them within phagosomes. This is achieved after the fusion of granules containing degradative and hydrolytic enzymes with the phagosome. Accompanying these events is activation of a non-mitochondrial oxidase system, which generates superoxide ions (see Section 2), which dismutate to peroxide. The most abundant enzyme within the phagolysosome is the enzyme myeloperoxidase, which in the presence of the formed peroxide and chloride ions catalyzes the production of hypochlorite (Table I). Hypochlorites as well as superoxide and peroxide are extremely toxic to the microbes within the vacuoles, which are consequently killed. Other reactions that also lead to reactive and hence toxic oxygen metabolites have been postulated (Table I). The crucial role of the activation of the oxidase

Futwan A. Al-Mohanna and Maurice B. Hallett   University Department of Surgery, University of Wales College of Medicine, Cardiff CF4 4XN, United Kingdom.

Table I
Reactive Oxygen Metabolites Generated
by the Neutrophil Oxidase[a]

Reaction	Enzyme
$O_2 + e \rightarrow O_2^-$	Oxidase
$O_2^- + O_2^- + 2H^+ \rightarrow H_2O_2$	Superoxide dismutase
$H_2O_2 + Cl^- \rightarrow OCl^- + H_2O$	Myeloperoxidase
$O_2^- + H_2O_2 \rightarrow O_2 + OH^-$	
$OCl^- + H_2O_2 \rightarrow {}^1O_2 + Cl^- + H_2O$	
$H_2O_2 + Fe^{2+} \rightarrow Fe^{3+} + OH^- + OH^{\bullet}$	

[a]Some reactions that generate reactive metabolites of oxygen, including O_2^- (superoxide), OCl^- (hypochlorite), $OH^{\bullet}$ (hydroxyl radical), and ${}^1O_2$ (singlet oxygen).

system is dramatically demonstrated in the genetic disease chronic granulomatous disease, in which the oxidase is absent. Children with this condition are prone to various infections and rarely survive to adulthood. Activation of the oxidase therefore represents a key event in the physiological response to injury and infection (Figure 1).

The toxic potential of reactive oxygen metabolites leads to the possibility that if somehow misdirected, neutrophils will cause damage to the extracellular matrix and neighboring cells. Thus, in inflammatory diseases such as rheumatoid arthritis and Crohn's disease, it has been speculated that the presence of neutrophils at the inflammatory site mediates tissue damage by producing extracellular oxygen metabolites. As with bacterial killing, hypochlorite may be of particular importance. Apart from a direct cytotoxic effect, this ion has also been demonstrated to inhibit α-1-antitrypsin and to activate the latent neutrophil collagenase. Thus, inappropriate activation of the neutrophil

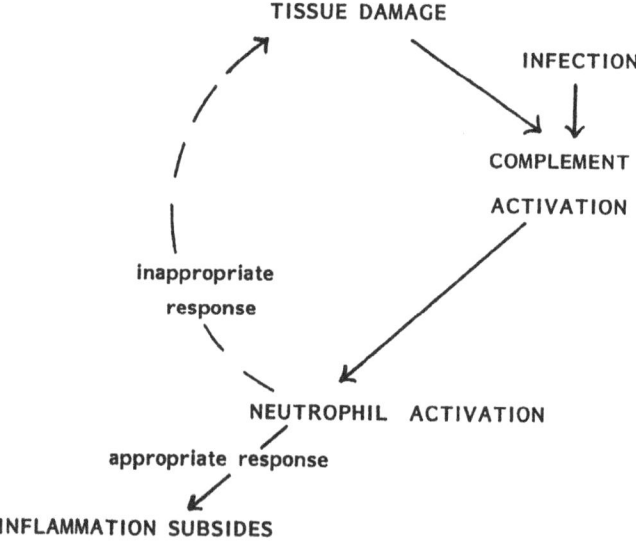

FIGURE 1. Pivotal role of the neutrophil in inflammation. This flow diagram illustrates how the inappropriate response of neutrophils can produce chronic rather than acute inflammation.

leads to a switch from the physiological course of inflammation to a cycle of chronic tissue-damaging inflammation (Figure 1). Once the cycle of neutrophil-mediated tissue is in place, the initial trigger of the inflammation is no longer needed for perpetuation of the tissue-damaging process. It may therefore be argued that the key event in determining the course of inflammation is the response of the neutrophils, and in particular the mechanisms that control appropriate and inappropriate oxidase activation in these cells.

1.2. Appropriate and Inappropriate Oxidase Activation

The appropriate activation of the neutrophil oxidase, which occurs during phagocytosis (Figure 2), leads to the generation of reactive oxygen metabolites within the phagosomal vacuole. It would be appropriate for the oxidase only in the membrane of the phagolysosome to be activated, as activation at other cellular sites would serve no beneficial purpose. Activation is thus restricted to particular sites within the cell where phagolysosome formation has occurred after fusion of the granular membrane with the phagosomal membrane (Figure 2). In contrast, inappropriate activation of the oxidase may result in the extracellular production of reactive metabolites. This can occur after fusion of the granular membrane with the plasma membrane (Figure 2). Under these conditions, the events normally restricted to within the phagolysosome occur extracellularly and are thus potentially damaging to the extracellular matrix and neighboring cells.

Furthermore, unlike the appropriate activation following phagocytosis, which is restricted to subcellular sites, the conditions that induce inappropriate activation *in vitro* lead to whole-cell events which show no such restrictions. Although the final event is common to both the appropriate and inappropriate routes of activation, that is, fusion of the granular membrane with the plasma membrane (either as part of the cell periphery or in the form of an isolated phagolysosomal vacuole), the intracellular mechanisms that

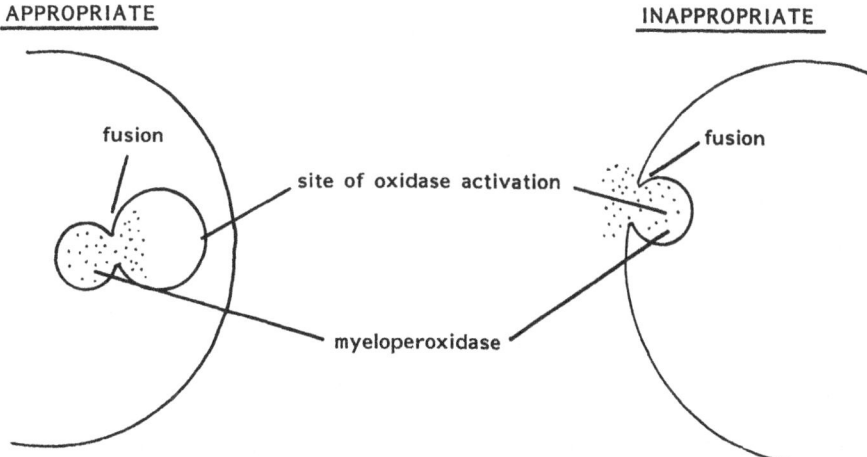

FIGURE 2. Appropriate and inappropriate locations of oxidase activity. Appropriate oxidase activation occurs during phagocytosis, when products of the oxidase are generated in the phagosome. Inappropriate activity occurs after exocytosis of myeloperoxidase-containing granules when products of the oxidase are generated extracellularly. In both responses, fusion occurs between the granular membrane and the plasma membrane (or phagosome formed from the plasma membrane).

determine which type of activation occurs are different at a number of levels. In particular, the roles played by calcium ions and the actin microfilament network are critical for determining the end response to a particular neutrophil stimulus.

2. THE NEUTROPHIL OXIDASE

2.1. Molecular Properties

The burst of oxygen consumption accompanying phagocytosis was first demonstrated in 1933 (Ado et al., 1933; Balridge and Gerald, 1933) and used as evidence for the energy requirement of phagocytosis. However, in 1959, when Sharra and Karnovsky discovered that mitochondrial inhibitors such as cyanide failed to prevent this oxygen consumption, it became obvious that the neutrophil oxidase was of a different class than the mitochondria. Evidence has accumulated that the increase in the hexose monophosphate shunt associated with the respiratory burst provides NADPH as the electron source for oxygen reduction (Rossi, 1986), and the oxidase has therefore been called the NADPH oxidase. At least two components of the oxidase system have been identified, cytochrome b_{-245} and flavoprotein.

2.1.1. Cytochrome b_{-245}

The crucial component of the neutrophil oxidase is a b-type cytochrome, the midpoint potential of which is -245 mV (Cross et al., 1981). This potential is lower than that of any other mammalian b-type cytochrome and is sufficiently low to allow direct reduction of oxygen to superoxide. The cytochrome was first identified in human neutrophil phagocytic vacuoles (Segal and Jones, 1978) but was probably the same cytochrome reported earlier by Japanese workers (Hattori, 1961; Ohta et al., 1966; Shinagawa et al., 1966; see also Segal, 1988) in rabbit and horse. It has since been identified in the neutrophils of a number of other mammalian species as well as fish (Higson and Jones, 1984).

Cytochrome b_{-245} (also known as cytochrome b_{558}, from a characteristic absorbance peak at 558 nm) is composed of two subunits, a small subunit of 23 kDa and a larger subunit that migrates on SDS-PAGE with a molecular weight of 60–90 kDa (Harper et al., 1985). The subunit is a glycoprotein and is probably a structural protein. The subunit is similar in size to other b-type cytochromes and is probably a heme-binding structure. The genes for both subunits have been cloned, and the amino acid sequence for the subunit has been determined (Parker et al., 1988; Royer-Pokora et al., 1986).

2.1.2. Flavoproteins

It has long been speculated that the NADPH oxidase contains a flavorprotein dehydrogenase that acts as an electron carrier between NADPH and the cytochrome. The evidence for this proposal arises for the ability of a flavin analog, 5-carbo-deaza-FAD, to inhibit the activity of solubilized preparations of the oxidase (Light et al., 1981) and a diphenylene iodonium, a potential flavoprotein inhibitor, to inhibit the oxidase in intact cell (Cross and Jones, 1986). The latter component also specifically labels a polypeptide component of the oxidase with a molecular weight of 45 kDa. FAD has also been identified in neutrophils associated with cytochrome b_{-245} (Cross *et al*, 1982), and in solubilized preparations of the oxidase, FAD is reduced by NADPH (Cross *et al.*, 1984; Gabig and Lefker, 1984).

2.1.3. Quinones

Although the possibility that ubiquinone is a further component in the electron transport chain has been raised (Crawford and Schneider, 1982; Gabig and Lefker, 1985), it would seem unnecessary from theoretical considerations. This possibility is further undermined by the failure of some workers to identify nonmitochondrial ubiquinone in neutrophil preparations (Cross *et al.*, 1983; Luther *et al.*, 1984; Bellavite *et al.*, 1984).

2.2. Subcellular Location

The subcellular location of cytochrome b_{-245} has not been entirely resolved. The original reports from Japanese workers showed a location within the granular fraction (Hattori, 1961; Ohta *et al.*, 1966; Shinagawa *et al.*, 1966). More detailed analysis suggested that most of the cytochrome is located in the membrane of specific granules, which was translocated to the plasma membrane during stimulation (Borregaard *et al.*, 1983). However, in human neutrophils there seems to be an equal distribution between the specific granules and the plasma membrane (Segal and Jones, 1979; Garcia and Segal, 1984).

FAD has been identified in the plasma membrane of human neutrophils in approximately $1:1$ ratio with cytochrome b_{-245} (Cross *et al.*, 1982). Interestingly, translocation of the cytochrome from the plasma membrane to phagocytic vesicles resulted in incorporation of FAD to give an FAD/cytochrome b ratio in these vesicles of greater than $2:1$. It is not clear whether the addition of FAD resulted in greater oxidase activity.

2.3. Superoxide Production

The oxidase is thus a short electron transport chain in the membrane of neutrophil granules and plasma membrane (Figure 3). NADPH is the electron donor to a flavoprotein

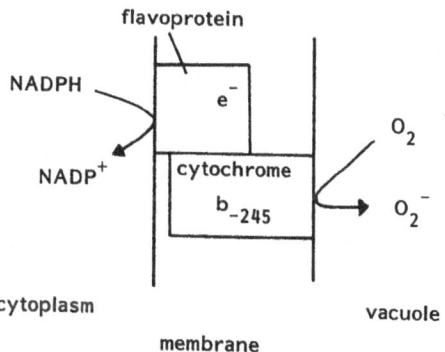

FIGURE 3. The neutrophil oxidase. Shown is the relationship of the cytochrome, flavoprotein, and the electron donor and acceptor, NADPH and O_2.

dehydrogenase on the cytoplasmic side of the membrane. The electrons are subsequently transported through cytochrome b_{-245} to oxygen to generate superoxide on the extracellular side of the membrane, normally within the phagosome, but also possibly extracellularly (see Section 1.1). The result of this vectorial transfer of electrons is the generation of transmembrane potential, which in turn is associated with proton channel opening (Henderson *et al.*, 1987).

3. STIMULUS–OXIDASE ACTIVATION COUPLING

From a teleological viewpoint, it can be argued that the two routes leading to oxidase activation (outlined in Section 1.1) would differ in their requirement for cytoplasmic messengers. Whereas a membrane-associated oxidase may be activated in the phagosomal membrane without the need for cytoplasmic messengers, the activation of oxidase molecules throughout the whole cell by surface receptor occupancy requires that intracellular messengers be released into the cytoplasm. The evidence presented in this section suggests that this is indeed the case. Cytoplasmic Ca^{2+} acts as the cytoplasmic messenger, and lipid products, perhaps diacylglycerol (DAG) or fatty acids, mediate the restricted activation that accompanies phagocytosis. However, the end result of either route of stimulation, oxidase activation, may be the same molecular event and involve phosphorylation.

3.1. Role of Calcium

The first experimental indication that a rise in the concentration of intracellular calcium could trigger a response in neutrophils was provided by the demonstration that the toxin leucocidin produced both in accumulation of calcium within the cell and secretion of proteins (Woodin and Wienke, 1963). Later, with the development of more specific methods for increasing cell membrane permeability for calcium with ionophores, it was confirmed that raised cytoplasmic calcium concentrations triggered not only secretion but also oxidase activation (Schell-Frederick, 1974; Romeo *et al.*, 1975). Measurement of Ca^{2+} fluxes across the plasma membrane of the neutrophil showed that an increase in plasma membrane permeability for Ca^{2+} was also accompanied by chemotactic and secretory activity (Boucek and Snyderman, 1976; Naccache *et al.*, 1977). It was therefore widely accepted that, as in other cells, intracellular calcium acted as the sole intracellular trigger for neutrophil responses. However, it was not until the development of techniques for direct measurement of intracellular Ca^{2+} within living neutrophils that the complexity of the role of this ion was fully realized.

There is agreement from the direct measurement of intracellular Ca^{2+} in neutrophils from a number of species that the resting intracellular Ca^{2+} concentration is approximately 0.1 μM (Table I). The effects on this resting level of a number of oxidase-activating stimuli have been determined. For convenience, the effects can be classified as shown in Table II. Some stimuli, such as the experimental ionophores A23187 and ionomycin, elevate cytoplasmic Ca^{2+} and stimulate the oxidase, the intracellular Ca^{2+} concentration required for activation being as high as 15 μM. Other agents, however, such as the receptor agonists, activate the oxidase but elevate intracellular Ca^{2+} only to

Table II
Requirement of the Action of Some Stimuli
for Intracellular Ca^{2+}

Group	Agent	Ca^{2+} conc. required for activation
0	Unstimulated	75–150 nM
I	Latex beads	
	C3bi-opsonized particles	No rise required
	PMA	
II	fMLP	
	ConA	
	C5a	0.1–1 μM
	Leukotriene B$_4$	
	IgG	
III	A23187	
	Ionomycin	
	Terminal attack complex of complement	1–10 μM
	Sendai virus	
	Saponin	
	Digitonin	

about 1 μM. Yet other agents are capable of activating the oxidase system without any elevation of intracellular Ca^{2+}. In light of these data, it may be asked whether Ca^{2+} plays any role at all. To answer this question, it is necessary to look in more detail at the mechanisms by which these agents stimulate.

3.1.1. Ionophores

The experimental Ca^{2+} ionophores A23187 and ionomycin provide a means of artificially increasing the permeability of the membrane to Ca^{2+} ions. In the presence of extracellular Ca^{2+} concentrations in the millimolar range, intracellular Ca^{2+} rises rapidly to high levels. The intracellular Ca^{2+} concentration is sufficient to saturate the fluorescent probe quin-2 (Pozzan et al., 1983; Lew et al., 1984; White et al., 1983a) and is estimated by use of the Ca^{2+}-activated photoprotein obelin to be approximately 10 μM (Hallett and Campbell, 1982a, 1984; Campbell and Hallett, 1983). When the conditions are manipulated, either by reducing the extracellular concentration of Ca^{2+} or by reducing the concentration of ionophore, lower levels of intracellular Ca^{2+} can be achieved. In these conditions, the intracellular requirement for Ca^{2+} can be titrated. It was found that oxidase activation in both rat and human neutrophils required intracellular Ca^{2+} in excess of 1 μM for maximum stimulation (Pozzan et al., 1983; Hallett and Campbell, 1984a; Al-Mohanna and Hallett, 1988a). The terminal attack complex of complement elevated intracellular Ca^{2+} to micromolar levels and stimulated the oxidase to an extent predicted from ionophore titration (Hallett and Campbell, 1984a). A similar dependence on intracellular Ca^{2+} exists for the secretion of β-glucuronidase when stimulated by ionomycin, no secretion occurring at below 2 μM intracellular Ca^{2+} (Lew et al., 1986). There is thus a clearly demonstrable causal relationship between cytoplasmic Ca^{2+} and oxidase activation by these agents.

3.1.2. Phorbol Esters

Phorbol esters, especially phorbol myristate acetate (PMA) (tetradeconoyl phorbol acetate) are potent activators of the neutrophil oxidase. However, in extreme contrast to the ionophores, stimulation by these agents occurs without a requirement for an elevation of intracellular Ca^{2+} (Sha'afi et al., 1983; DiVirgilio et al., 1984; Al-Mohanna and Hallett, 1988a). The possibility therefore exists that these agents act at a site closer to oxidase activation than the elevation in Ca^{2+} seen with the ionophores. Furthermore, they allow for the possibility that some physiological stimuli may also act without elevating cytoplasmic Ca^{2+}.

3.1.3. Receptor Agonists

The receptor agonist formylmethionyl-leucyl-phenylalanine (fMLP) stimulates chemotaxis, oxidase activation, and (in the presence of cytochalasin B) secretion. However, this agent and other receptor agonists (White et al., 1983a; Lew et al., 1987) cause only a small rise in intracellular Ca^{2+} concentration to a maximum of approximately 0.8 μM (Table II). Furthermore, the rise to this level is transient (although a sustained but lower elevation can persist for several minutes) (Hallett and Campbell, 1982a; Al-Mohanna and Hallett, 1988a). This peak of elevated intracellular Ca^{2+} is far less than that found to be necessary for stimulation by ionophores. However, this small intracellular Ca^{2+} is essential for triggering the response. When intracellular Ca^{2+} is "clamped" by intracellular EGTA (Campbell and Hallett, 1983) or by high intracellular concentrations of quin-2 (Lew et al., 1984), fura-2 (Al-Mohanna and Hallett, 1988a), or BAPTA (Cooke et al., 1989), responsiveness to the peptide is lost. By increasing the Ca^{2+}-buffering capacity of the neutrophil cytoplasm, the relationship between oxidase activation and the peptide-induced cytoplasmic Ca^{2+} rise can be titrated. Under these conditions, the Ca^{2+} concentration required for half-maximal oxidase activation was 600 nM (Al-Mohanna and Hallett, 1988a), whereas ionomycin-induced Ca^{2+} of similar magnitude did not significantly activate.

There are several possibilities for the discrepancy in the magnitudes of intracellular Ca^{2+} rises necessary for stimulation by peptide and ionophores. One possibility is that the intracellular Ca^{2+} induced by peptide was in reality much higher than that measured by the available techniques. This would occur if the magnitude of an intracellular Ca^{2+} change were restricted to a small part of the cytoplasm or to a small number of cells. Intracellular Ca^{2+} measurements performed on cell populations would significantly underestimate this local rise, since this approach reports only a "population average" Ca^{2+}-dependent signal. This possibility, however, would seem to be unlikely because both Ca^{2+}-activated photoprotein and the Ca^{2+} chelators quin-2 and fura-2 produce similar intracellular Ca^{2+} rises. The error associated with underestimation due to restriction of elevated Ca^{2+} would be different for each method. In the submicromolar range, the former technique produces a signal that is proportional to $(Ca^{2+})^n$ where $n = 2$–3 (Campbell et al., 1985), whereas for the latter technique of signal would be proportional to $\log (Ca^{2+})$ (Hallett et al., 1989). If an intracellular Ca^{2+} rise of x-fold were restricted to a fraction of the cytoplasm (or a fraction of the cells), f, the methods would underestimate the local change by f^{-n} for photoprotein (Hallett and Campbell, 1982b) and $x^{(f-1)}$

for the fluorescent probes (Hallett *et al.*, 1989). Since both methods report approximately equivalent intracellular Ca^{2+} changes, f must be close to unity; i.e., the intracellular Ca^{2+} change is not significantly restricted.

The second possibility is that the peptide generates a second intracellular messenger that was not generated by ionophores, the effect of which was to increase the affinity of Ca^{2+}-sensitive machinery within the cell for Ca^{2+}. The existence of this non-Ca^{2+} intracellular messenger was clearly demonstrated by Pozzan *et al.*, (1983). In the absence of extracellular Ca^{2+}, they showed that ionomycin elevated intracellular Ca^{2+} by a small extent that was insufficient to activate the cells. Subsequent stimulation by the peptide, however, resulted in activation of the cells without any further intracellular Ca^{2+} change. It was therefore concluded that the peptide had generated a non-Ca^{2+} intracellular messenger or messengers that triggered the cells. Although the identities of these noncalcium messengers have not been unequivocally established, further clues to their nature have accumulated (see Section 3.2).

3.1.4. Phagocytic Stimuli

After phagocytosis and fusion of the phagosome and granule membranes, the oxidase is activated. Although there had been much speculation about the role of calcium in these processes, it was difficult to interpret some early experiments. For example, the significance of the dependency of phagocytosis of some particular particles on extracellular calcium is not clear, since these experiments do not distinguish between the requirements of Ca^{2+} for binding of the particle to the membrane and as an intracellular messenger. Direct measurement of intracellular Ca^{2+} in neutrophils during phagocytosis, however, has revealed that the relationship between intracellular Ca^{2+} and phagocytosis is complex and that there is no absolute requirement for a change in intracellular Ca^{2+}. In neutrophil–erythrocyte hybrids loaded with the Ca^{2+}-activated photoprotein obelin, no intracellular Ca^{2+} rise accompanying phagocytosis-stimulated oxidase activation was detectable (Hallett and Campbell, 1982a; Campbell and Hallett, 1984a). The possibility that a rise in intracellular Ca^{2+} had actually occurred but was too localized or too small to be detectable was excluded by the demonstration that intracellular EGTA in concentration sufficient to inhibit the detectable intracellular Ca^{2+} rise induced by the peptide fMet-Leu-Phe did not prevent the latex bead-induced activation of the cells (Campbell and Hallett, 1983). No rise in intracellular Ca^{2+} in rat neutrophils was detected by the fluorescent Ca^{2+} indicator fura-2. (Al-Mohanna and Hallett, 1990). Similarly, phagocytosis of *Candida albicans* hyphae by human neurtrophils can also occur without a rise in cytoplasmic Ca^{2+} (Levitz *et al.*, 1987; Murata *et al.*, 1987) although rises in cytoplasmic Ca^{2+} can occur (Sawyer *et al.*, 1985). Furthermore, phagocytosis of unopsonized latex particles by the rat neutrophils occurs even when the intracellular Ca^{2+} concentration is clamped at a low level by using BAPTA (Cooke *et al.*, 1989). A similar lack of intracellular Ca^{2+} rise detectable by Ca^{2+}-activated photoprotein has been demonstrated during phagocytosis of unopsonized latex beads by rat macrophages (Hallett and Campbell, 1984b), by immunoglobulin G (IgG)-coated red blood cells by human macrophages (McNeil *et al.*, 1986), and by HL-60 cells (DiVirgilio *et al.*, 1988).

However, it is interesting that the intracellular free Ca^{2+} estimated by quin-2 flourescence rose in the macrophagelike cell line J774 after phagocytosis of opsonized red blood cells (Young *et al.*, 1984), and in neutrophils quin-2 indicated an intracellular Ca^{2+} rise after phagocytosis of either IgG- or C3bi-coated yeast particles (Lew *et al.*, 1985). However, phagocytosis of the C3bi particles was not inhibited by suppression of this Ca^{2+} rise, suggesting that two mechanisms for phagocytosis coexist in neutrophils, one dependent on and the other independent of a rise in intracellular Ca^{2+}. Oxidase activation after phagocytosis, unlike that accompanying receptor activation, can thus occur without a rise in intracellular Ca^{2+}.

3.2. Role of Protein Kinases

The involvement of protein kinases in neutrophil activation is well documented. However, although several studies agree that phosphorylation of a number of proteins accompanies stimulation, there is little agreement as to the molecular weights of these substrates (Table III). A feature of protein kinases is that although phosphorylation occurs at specific sites on the protein, e.g., serine and threonine, there is little specificity for particular proteins. Presumably, in the cell specificity is provided by the location or accessibility of the enzyme or its substrate. However, in cell homogenates, where these constraints are not present, kinases have been shown to catalyze the phosphorylation of a number of proteins. Although the physiological significance of a number of these observations remains to be established, it is of obvious importance that one substrate is the neutrophil oxidase. It has been demonstrated that phosphorylation of the partially purified inactive oxidase can occur *in vitro*, with the result that the oxidase becomes activated (Cox *et al.*, 1986; Papini *et al.*, 1985). Furthermore, a 44–47-kDa protein that becomes phosphorylated during stimulation by PMA, fMLP, and latex beads in normal cells (Heyworth and Segal, 1986; Hayakawa *et al.*, 1986; Gennaro *et al.*, 1985) but not autosomal recessive granulomatous cells (Segal *et al.*, 1985) may be identified as a component responsible for activation of the oxidase (Badwey *et al.*, 1984; Bolscher *et al.*, 1989). These demonstrations may provide the underlying molecular basis for the activation of the oxidase.

Table III
Activators and Inhibitors of c-Kinase

Activators	Inhibitors
Phorbol myristate acetate	Trifluoperazine
Mezerein	Retinal
Bryostatin	C1
1-Oleoyl-2-acetate glycerol	H9
Retinal	Polymyxin B
Phorbol dibutyrate	Gossypol
1,2-dioctanoyl glycerol	Tamoxifen
Dansyl phorbol acetate	Adriamycin
	K-252b
	Staurosporin
	Sphingosine
	Sphinganine

3.2.1. c-Kinase

Perhaps the first indication that c-kinase played a role in activation of the oxidase was provided by the discovery that the receptor for the phorbol ester PMA was the enzyme c-kinase (Niedel *et al.*, 1983). Since it has been known for many years that PMA is a potent activator of the oxidase system (DeChatelet *et al.*, 1976), the possibility existed that c-kinase played a key role in stimulus–oxidase activation coupling. In fact, neutrophils have particularly high c-kinase activity, human and rat neutrophils containing c-kinase activity equivalent to, respectively, 38,000 and 21,000 pmol of phosphate transferred to histone per min per g of cells (Helfman *et al.*, 1983a,b; Cooke *et al.*, 1989). As will be seen from considering some properties of c-kinase, a role for this enzyme is also consistent with the existence of non-Ca^{2+} intracellular messengers that increase the affinity of the intracellular machinery for Ca^{2+}.

3.2.1a. Some Properties of c-Kinase. c-Kinase was originally described by Nishizuka (1984) as "the Ca^{2+}-activated phospholipid-dependent kinase." This name implied that the enzyme was activated by Ca^{2+} but that its activity was also dependent on binding to phospholipid. In fact, only phosphatidylserine was capable of producing full Ca^{2+}-dependent activation. Under these conditions, activation occurred with Ca^{2+} concentrations in the range of 1–10 μM. However, the affinity of the enzyme for Ca^{2+} was greatly increased by certain diacylglycerols such that the enzyme could be fully activated by submicromolar concentrations of Ca^{2+}. This particular feature would provide a molecular explanation for the intracellular Ca^{2+} requirement of the three types of stimuli described in Table II (if DAG were also generated to provide the Ca^{2+} affinity shift seen with group II receptor-operated agonists). It must also be assumed that the ionophores do not produce the DAG and hence activate by Ca^{2+} alone, requiring high intracellular Ca^{2+} for stimulation. Furthermore, if c-kinase is the sole mediator of oxidase activation, then it also follows that phagocytic stimuli produce sufficient DAG or other c-kinase activator (possibly fatty acids; see Section 3.3) to enable activation of the cell in the absence of a rise in intracellular Ca^{2+}.

Another route by which c-kinase can be activated is limited proteolysis. Although much recent attention has focused on the lipid activators of c-kinase as potential physiological activators of this enzyme, its activation was first described following proteolysis. The proteolytically cleaved kinase (PKC-M) has no requirement for phospholipid and no sensitivity to Ca^{2+} and is consequently released from the membrane to become free in the cytoplasm. Proteolysis can be catalyzed by the enzyme calpain, which is activated by micromolar Ca^{2+} concentrations. This route of c-kinase activation has been demonstrated to occur in neutrophils (Melloni *et al.*, 1986) and may play a role in oxidase activation, since PKC-M has been shown to phosphorylate and activate the oxidase *in vitro* (Tauber et al., 1989). At high intracellular Ca^{2+} concentrations, this mechanism of activation may result in c-kinase becoming irreversibly activated and gaining access to both membrane-associated and cytoplasmic proteins. This may thus represent a route for inappropriate and hence potentially pathogenic activation of the oxidase (see Section 1.1). However, the affinity of the proteinase may possibly be regulated and allow activated at lower Ca^{2+} (Pontremoli *et al.*, 1988). Thus, this route may also play a role in physiological activation.

3.2.1b. Experimental Activators of C-Kinase. A number of experimental agents have been shown to activate c-kinase (Table III). Many are phorbol esters or have phorbol-like structures. Others are diacylglycerols, and yet others are structurally distinct. All, however, activate the neutrophil oxidase with a potency that is markedly correlated with their ability to activate c-kinase (Cooke et al., 1989). The most potent activator by far is PMA. It can be calculated that since 1 ng can maximally activate the oxidase of 2.10^7 cells (assuming a maximum of one c-kinase molecule activated per molecule of PMA), then a mere 1.46×10^5 c-kinase molecules per cell are required for activation. This must be compared with the estimated c-kinase content of 4×10^5 PKC molecules per neutrophil (Gordon and Weinberg, 1982). It can also be estimated that since the maximum rate of oxygen consumption was 2.5 nmol of O_2 consumed per 10^6 cells per min and the maximum rate for the oxidase was 13 moles of O_2^- produced per sec per mole of cytochrome b (Cross et al., 1985), then 1 molecule of c-kinase must activate at least 14 molecules of cytochrome b. This may suggest an efficient coupling of oxidase activation of c-kinase.

The experimental activators of c-kinase have been shown to have the expected synergistic effect with Ca^{2+} ionophores (Dale and Penfield, 1987). However, with use of these agents to investigate the physiological situation, a more complex picture for the involvement of c-kinase has emerged. Cells pretreated with PMA (in order to fully activate c-kinase) could not be further stimulated by unopsonized latex beads, whereas fMLP was capable of provoking further activation (Cooke and Hallett, 1985). This suggested that whereas both latex beads and PMA produced activation of the oxidase by acting at a common site (presumably c-kinase), the peptide also acted at a site that was independent of PMA and thus unlikely to be c-kinase. The possibility exists that since the peptide under these conditions generated the Ca^{2+} transient (Cooke et al., 1989), another Ca^{2+}-sensitive site was involved. One possibility was calmodulin-dependent kinase (see Section 3.2.2).

3.2.1c. Experimental Inhibitors of c-Kinase.

Although inhibitors of particular intracellular events are often useful for establishing the role played by the event, no agent has been shown to have totally specific inhibitory activity against c-kinase. Many agents, however, have the ability to inhibit, among other kinases, c-kinase and calmodulin-dependent kinase (Kimura et al., 1985; Mazzei et al., 1982; O'Brian et al., 1985; Su et al., 1985, Taffet et al., 1983; Wilson et al., 1986; Wright and Hoffman, 1986). Although these agents, such as trifluoperazine, are strong inhibitors of oxidase activation (Naccache et al., 1980), they are of little use in distinguishing the roles played by these different kinases. Other agents, such as sphinganine and sphingosine, which were claimed to be selective for phospholipid-dependent enzymes (and thus selective inhibitors of c-kinase), may also have other, nonspecific effects (Bazzi and Nelsesteun, 1987; Pittet et al., 1987).

With these limitations in mind, it is still relevant that a number of agents which inhibit c-kinase have the ability to selectively inhibit activation by PMA but not by fMLP (e.g., Naccache et al., 1985) (Table III). This differential inhibition suggests that fMLP utilizes a pathway that is independent of that activated by PMA and thus probably independent of the activity of c-kinase. For example, retinal, a partial agonist of c-kinase, produces a small but detectable stimulation and profound inhibition to subsequent stimulation by c-kinase activators but fails to inhibit subsequent activation by fMLP (Cooke and Hallett, 1985). In contrast, subsequent activation by unopsonized latex beads is inhibited. This suggests that whereas c-kinase is involved in the mechanism of activation used by unopsonized particles, it is not involved in the route by which the receptor agonist, fMLP, activates. It must be emphasized again that no demonstration that these inhibitors act solely on c-kinase has been provided. The differential effect of inhibition does, however, again point to fundamental differences in the mechanisms of oxidase activation by different agents.

3.2.2. Calmodulin-Dependent Kinase

The possibility that Ca^{2+}-activated, calmodulin-dependent kinase plays a role in physiological stimulation of the oxidase system must also exist. Calmodulin has been identified in human and rabbit neutrophils (Jones et al., 1982; Chafouleas et al., 1979), the concentration in human neutrophils being approximately 15 μM. Although inhibitors of calmodulin-dependent kinase prevent oxidase activation, as mentioned in Section 3.2.1b, the specificity is often poor with accompanying inhibition of c-kinase. However, the differential effect of some inhibitors is suggestive of a role for calmodulin-dependent kinase in oxidase activation. For example, activation of the oxidase by the peptide fMLP is not inhibited by retinal (and some other c-kinase inhibitors; see Section 3.2.1c) but is inhibited by trifluoperazine (Cooke et al., 1985). This suggests that a key step is trifluoperazine sensitive but not c-kinase dependent. Since trifluoperazine is an inhibitor of calmodulin-dependent kinase (as well as c-kinase), an interpretation of this result is that the key step was calmodulin dependent. Interestingly, only the stimuli that depend on a rise in cytoplasmic Ca^{2+} show the calmodulin dependency (Cooke et al., 1985, 1989). This suggests that whereas c-kinase plays a role in Ca^{2+}-independent stimulation, the rise in cytoplasmic Ca^{2+} also activates calmodulin.

Calmodulin-dependent phosphorylation has been demonstrated in neutrophils (Hirayama and Kato, 1982) but has not been shown to have a direct stimulatory effect on th isolated oxidase system in the same way as c-kinase. However, in vitro, calmodulin enhances the activity of stimulated oxidase (Jones et al., 1982). Whether this phenomenon plays a role in the cellular response is unclear. The role of calmodulin-dependent NAD-kinase, found in neutrophils (Williams and Jones, 1985), is also unclear. However, a specific calmodulin-dependent kinase, myosin light-chain kinase, may have a role. It has been shown to play a role in contractile processes in nonmuscle cells. Phosphorylation of the myosin light chain produces an active molecule capable of assembly and interaction with actin filament to generate contraction (Scholey et al., 1980; Southwick and Stossel, 1983). While contraction may be important for chemotaxis, its role in phagocytosis is not established (Rickard and Sheterline, 1989). Furthermore, while the polymerization of actin may play a role in oxidase activation (see Section 5.2), the relationship between myosin-mediated contraction and oxidase activity is unclear.

3.3. Role of Phospholipid-Derived Messengers

A key event associated with stimulation of neutrophils is an enhancement in the turnover of phospholipids (Cockcroft, 1989). Interest has focused most closely on the activity of phospholipase C, which cleaves phosphatidylinositol 4,5-bisphosphate to generate two potential intracellular messengers, DAG and inositol 1,4,5-triphosphate (IP_3). The former is an activator of c-kinase (see Section 3.2.1), and the latter has been shown to release Ca^{2+} from a store of calcium within neutrophils (Prentki et al., 1984), which may play an important role in neutrophil triggering (see Section 3.1). Phospholipids are also a source for the generation of fatty acids, which may play a role in c-kinase activation (Section 3.3.3). One fatty acid, arachidonic acid, also can generate other important bioactive molecules such as prostaglandins, leukotrienes, and eicosanoids (Naccache *et al.*, 1989).

3.3.1. Diacylglycerol

The possibility that DAG produced during phospholipase C activation allows activation of c-kinase at resting cytoplasmic Ca^{2+} concentrations or acts synergistically with Ca^{2+} has been suggested. However, there was no detectable increase in DAG concentration immediately after stimulation by fMLP, although an increase was observed later (Cockcroft and Allan, 1984). Furthermore, this late production of DAG did not have the fatty acid composition expected for that derived from phosphatidylinositol and therefore may not have arisen from the action of phospholipase C on phosphoinositide 4,5-bisphosphate (PIP_2). The production of phosphatidic acid, however, rises sharply after stimulation by fMLP, reaching five times its resting concentration within 10 sec (Cockcroft and Allan, 1984). Since this rise in phosphitadic acid probably resulted from the phosphorylation of DAG, DAG may have been generated with a time course appropriate for it to act as a intracellular messenger but rapidly phosphorylated by a DAG-kinase. This enzyme has been detected in the neutrophil plasma membrane (Cockcroft *et al.*, 1984) and can be inhibited by the agent R59022. In platelets, R59022 has been shown to elevate DAG concentrations after stimulation and to increase phosphorylation of a 40-kDa protein (de Chaffoy de Courcelles *et al.*, 1985). An important test of the physiological importance of DAG in neutrophil oxidase activation was therefore the determination of the effect of elevation of DAG by inhibition of DAG-kinase.

The DAG-kinase inhibitor R59022 (0.1–20 μM) had no significant effect on the resting level of oxidase activity, but the response to an optimal concentration of fMLP, 1 μM, was enhanced by up to fourfold (Muid *et al.*, 1987; Cooke *et al.*, 1987). This finding suggested that DAG generated by this agonist was the limiting factor in determining the extent of the response. Furthermore, the enhancement was probably mediated by c-kinase, since it could be prevented by c-kinase inhibitors (Cooke *et al.*, 1987). Interestingly, the response to the calcium inophore A23187 was also enhanced, 20 μM R59022 producing an approximately 20-fold enhancement (Cooke *et al.*, 1987), whereas responses induced by PMA and unopsonized latex beads were not. This may reflect the intracellular Ca^{2+} concentrations existing in the cells under these conditions. Whereas stimulation by fMLP and ionophores is totally dependent on a rise in intracellular Ca^{2+} (Lew *et al.*, 1984; Hallett and Campbell, 1984a), activation by unopsonized latex beads (Campbell and Hallett, 1983; Hallett and Campbell, 1984a) does not depend on such a rise. However, lack of enhancement of activation by latex beads suggests one of two possibilities: (1) DAG played no part in signal transduction by this route or (2) the DAG concentration generated was already sufficiently high to maximally activate c-kinase at the existing Ca^{2+} concentration (approximately 0.1 μM). Since a membrane-bound intracellular messenger such as DAG would be restricted to the phagosomal membrane, the latter possibility cannot be excluded. This was also suggested by the fact that a characteristic of the responses to both fMLP and latex beads after pretreatment with R59022 was a reduction in the rate at which the oxidase activation returned to baseline (Cooke *et al.*, 1987). This may have resulted from a prolongation of a stimulatory level of the intracellular signal in these conditions. Interestingly, the breakdown of DAG to fatty acids by a DAG-lipase seems to play an insignificant part in reducing the concentration of DAG after activation, DAG-lipase inhibitor having little effect on oxidase activation (Dale and Penfield, 1987). DAG has also been suggested to have a role in modulating the association of actin to the plasma membrane, although the evidence for this role in neutrophils has not yet been provided.

3.3.2. Inositol Phosphates

Associated with the generation of DAG in neutrophils is the production of inositol phosphates. Perhaps the most important is IP_3, which has been shown to be capable of releasing Ca^{2+} from an intracellular store of calcium in permeabilized neutrophils (Prentki *et al.*, 1984). Subcellular fractionation has suggested that the IP_3-sensitive store called calcisomes, is probably located near the plasma membrane (Volpe *et al.*, 1988). Other inositol phosphates may have physiological roles. For example, inositol 1,3,4,5-tetraphosphate generated by phosphorylation of IP_3 may open Ca^{2+} channels in the plasma membrane (Irvine and Moor, 1986). The possibility that cyclic IMP can act as an intracellular messenger has also been suggested (Majerus *et al.*, 1986). Furthermore, since the activities of both profilin (Lassing and Linberg, 1985) and gelsolin (Janmey and Stossel, 1987) are both by influenced phosphatidylinositol 4,5-bisphosphate, it has also been suggested that this phospholipid has importance in regulating actin polymerization.

3.3.3. Fatty Acids

Since fatty acids activate c-kinase *in vitro,* they also play a physiological role in neutrophil activation (McPhail *et al.*, 1984b). Arachidonic acid and other fatty acids generated during stimulation of neutrophils (Naccache *et al.*, 1989) may therefore serve as intracellular messengers activating c-kinase. It is interesting that phagocytosis of latex particles, which does elevate cytoplasmic Ca^{2+}, results in a transient production of fatty acid (Elsbach and Farrow, 1969; Elsbach *et al.*, 1972). Although many studies have also shown that extracellularly added fatty acids can activate neutrophils (Kakinuma, 1974; Badwey *et al.*, 1981; Bromberg and Pick, 1983), it is likely that this activation resulted from the increased Ca^{2+} influx that accompanies insertion of the fatty acid into the plasma membrane.

Arachidonic acid also serves as a source for other lipid-derived bioactive molecules, such as prostaglandins and leukotrienes (Naccache *et al.*, 1989). The role of these agents as intracellular messengers in oxidase activation remains unclear.

3.4. Role of Other Intracellular Messengers

3.4.1. GTP

Both of the classical G proteins, Ns and Ni, have been identified in neutrophils (Lad *et al.*, 1984; Okajima and Ui, 1984; Becker *et al.*, 1985). Cholrea toxin, which ADP ribosylates and activates Ns, elevates cyclic AMP (cAMP) and subsequently inhibits responses to chemotactic stimuli (Rivkin *et al.*, 1975; Hill *et al.*, 1975). Pertussis toxin, which ADP ribosylates and activates Ni, has no effect on cAMP (Becker *et al.*, 1985) but does inhibit a number of neutrophil responses, including oxidase activation (Goldman *et al.*, 1985; Verghese *et al.*, 1985). This inhibition may result from the inhibition of Ca^{2+} mobilization (Goldman *et al.*, 1985) and phosphatidylinositol breakdown (Smith *et al.*, 1986). It is therefore interesting that phagocytosis which occurs without a requirement for a rise in intracellular Ca^{2+} is unaffected by pertussis toxin (Lad *et al.*, 1986). It may therefore be suggested that pertussis toxin acts on a crucial "transducinlike" molecule that provides the link to the rise in intracellular Ca^{2+}. Thus, it may not be unexpected that microinjection of nonhydrolyzable analogs of GTP can activate neutrophils by reducing the dependence of phospholipase C for Ca^{2+} (Cockcroft and Gomperts, 1985; Smith *et al.*, 1986; Cockcroft, 1986; Bradford and Rubin, 1986); DAG and IP_3 may thus be generated and so activate c-kinase. However, a second route for activation has also been suggested via a novel G protein that causes activation independently of Ca^{2+} (Barrowman *et al.*, 1986). The introduction of nonhydrolyzable analogs of GTP into permeabilized neutrophils may also evoke secretion by increasing the affinity of a key component of the secretory machinery for Ca^{2+} by approximately 10-fold (Smolen and Stoehr, 1986). In this latter experiment, GTP acts in a manner that would explain the lowered Ca^{2+} sensitivity of physiological stimuli for oxidase activation (see Section 3.1).

3.4.2. Membrane Potential Change

The membrane potential of neutrophils changes dramatically during stimulation, transiently depolarizing from a resting potential of -60 mV (Gallin and McKinney, 1989). This represents a transient collapse of a voltage field across the membrane of 1.2×10^5 V/cm and may have a nonconformational effect on proteins, such as the oxidase or phospholipases, within the membrane. The possibility that such effects play a role in signal transduction in neutrophils has remained unexplored.

3.5. Priming

A phenomenon that may have great importance to our understanding of the mechanisms of oxidase activation is that seen when neutrophils are exposed to more than one stimulus. Cells prestimulated with a receptor agonist, e.g., fMLP, respond poorly if at all to second exposure to the same stimulus. This is probably due to a decrease in the number of receptors or their function (Nelson *et al.*, 1980; English *et al.*, 1981; Donabedian and Gallin, 1981; Seligmann *et al.*, 1982). However, if the second stimulus is different from the first, the second response may be greatly enhanced. Furthermore, when the first stimulus can be given at a concentration insufficient to trigger the oxidase, enhancement of response to the second stimulus may still occur. This effect of the first stimulus on the cells has been described as priming. The oxidase or coupling mechanism may thus exist in one of three possible states in the cell, inactive, active, and primed. A number of agonists have been shown to be able to prime (Table IV) and include agonists of both Ca^{2+}-dependent and -independent classes (groups I and II Table II).

Two possibilities exist for the mechanism. The first is that there is a molecular state

Table IV
"Primers" of Oxidase Activation

Agent	Intracellular Ca^{2+} dependency
f-Met-Leu-Phe	$\checkmark$
Con A	$\checkmark$
Phorbol myristate acetate	$\times$
Phorbol dibutyrate	$\times$
Tumor necrosis factor	$\times$
Lipopolysaccharide	?
Gamma interferon	?
Diacylglycerols	$\times$
Alkylacylglycerols	?

A dependency upon an intracellular Ca^{2+} rise is signified by a "$\checkmark$," independence by an "$\times$" and an unestablished relationship to cytoplasmic Ca^{2+} by a "?".

corresponding to the primed oxidase and that normal signal transduction processes lead to the activation of a more efficient oxidase system. Although there is little direct evidence for this in neutrophils, it may be noteworthy, however, that pretreatment of macrophages with agents such as γ interferon and lipopolysaccharide results in an alteration in the affinity of the oxidase for NADPH. The oxidases in macrophages pretreated in this way have a decrease K_m for NADPH and show an enhanced response to stimulation (Sasada *et al.*, 1983; Berton *et al.*, 1985, 1986; Cassatella *et al.*, 1985). The relevance of these observations to neutrophil priming is unresolved but may provide a potential molecular basis for the first possibility of a primed oxidase state.

The second possibility is that priming results from an alteration in an event earlier in the signal transduction pathway. The intracellular messengers that trigger such a change would require different characteristics than the messengers involved in triggering oxidase activity. First, the primed state is long-lived, lasting many minutes (McPhail *et al.*, 1984a), whereas the activated state is transient. For example, the maximum oxidase activity following stimulation with the peptide fMLP occurs for approximately 60 sec, the accompanying Ca^{2+} transient lasts for only a comparable time (Figure 4), and the cytochalasin B-sensitive state (see Section 5.2) lasts for only 6 sec (Figure 4). Second, the concentration of agonist, and hence the percentage receptor occupancy required for priming, is less than 1/10 that required for triggering. It is interesting that with these lower receptor occupancies, intracellular messengers are generated that may induce actin polymerization (F. A. Al-Mohanna and M. B. Hallett, unpublished data; see Section 5.4.1). Third, some agents, such as tumor necrosis factor, are primers but are not capable of triggering oxidase directly (Berkow *et al.*, 1985). It is therefore possible that the intracellular messenger which induces priming is not involved in oxidase triggering. Rossi and colleagues have eliminated an involvement of Ca^{2+} or phosphotidylinositol products in PMA priming (Grzeskowiak *et al.*, 1986; Della Bianca *et al.*, 1986; Rossi *et al.*, 1986) by demonstrating that priming by PMA can occur in neutrophils depleted in Ca^{2+} and without activation of phospholipase C. Actin polymerization also does not seem to be involved in priming, whereas the microfilament inhibitor cytochalasin B has profound

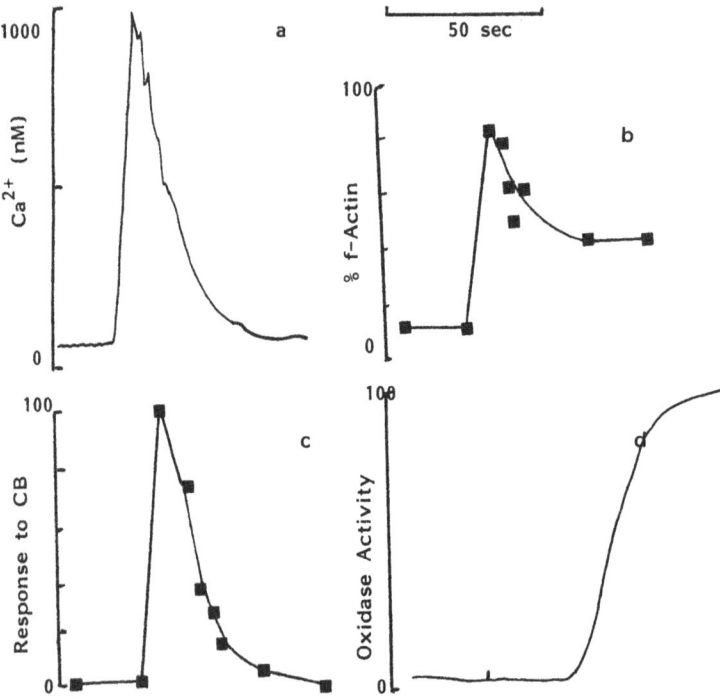

FIGURE 4. Changes in cytoplasmic Ca^{2+} (a) and actin polymerization actin (b) after stimulation of neutrophils with fMLP. (c) Time dependency of the enhancement caused by cytochalasin B (CB; 5μg/ml). (d) Oxidase activation as determined by luminol-dependent chemiluminescence. The time scale is common for all four parameters, and the peptide fMLP (1μM) was added at the first deflection on each graph.

effects on oxidase activation (see Section 4), it has no effect on the ability for the cell to be primed (McPhail *et al.*, 1984a).

4. CYTOSKELETAL COMPONENTS AND NEUTROPHIL FUNCTIONS

In the late 1950s, electron microscopy revealed that the viscous protoplasm of Heilbrunn (1956) contains numerous organelles, including at least three morphologically distinct intracellular fibers, that comprise the cytoskeleton. It is now recognized that the neutrophil cytoskeleton is made up of four components: microfilaments, microtubules, intermediate filaments, and microtrabeculae.

4.1. Microfilaments

4.1.1. Overview

Microfilaments are dynamic structures made up of actin polymers. Microfilaments exists in a reversible equilibrium with a pool of monomeric actin (Sheterline, 1983; Pollard and Cooper, 1986). In resting rat neutrophils, about 44% of total cellular actin

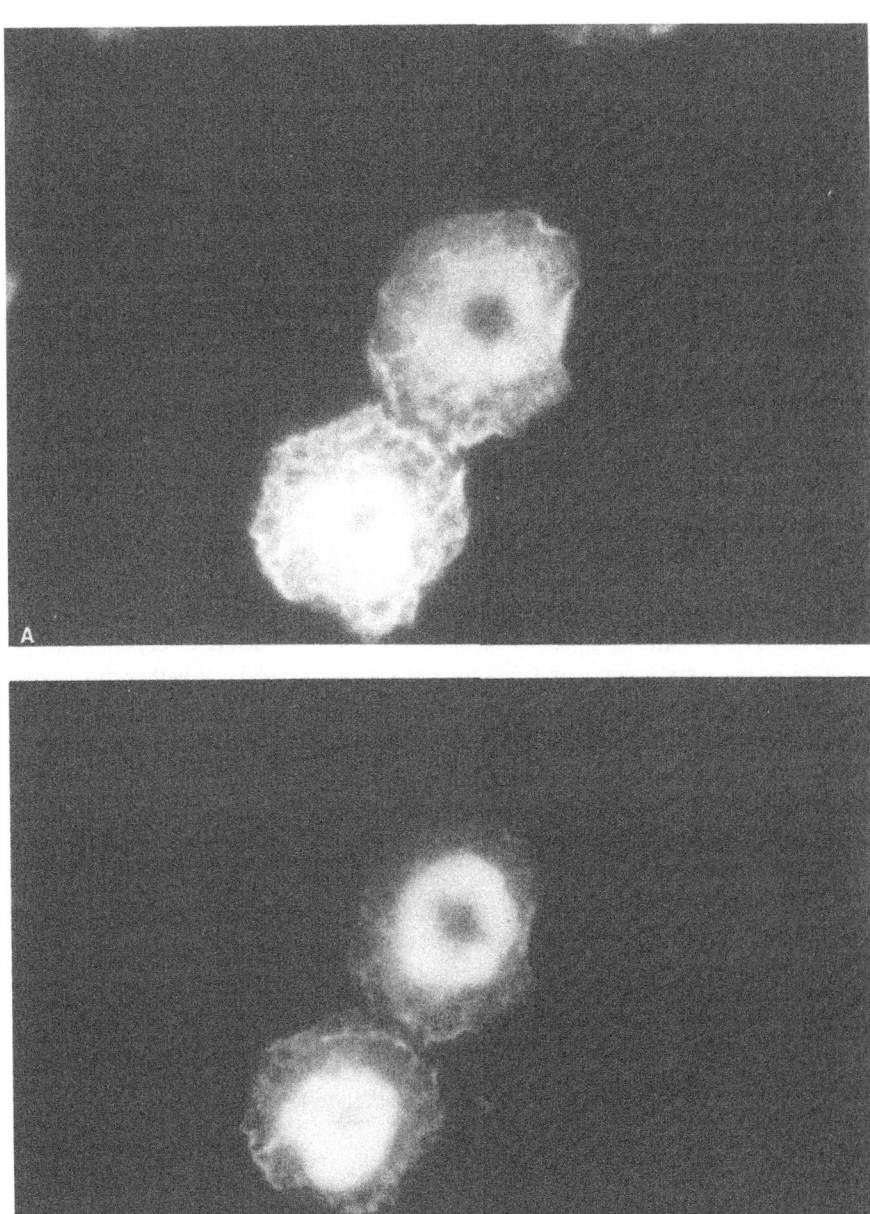

FIGURE 5. Fluorescence micrographs of actin filaments distribution in rat neutrophils. Spreading cells were fixed and doubly stained with rhodamine-labeled goat anti-mouse immunoglobulin antibodies (A) and iodoacetamidofluorescein (B).

exists in the polymeric state, i.e., F-actin (Al-Mohanna and Hallett, 1987). Upon stimulation of neutrophils with formylated peptides, the amount of polymerized actin increases and reaches a maximum of about 90% within 5 sec before returning to near prestimulation levels. Polymerized actin is found mainly in regions underlying the plasma membrane and in regions of phagocytosis and pseudopod formation. As neutrophils adhere to solid surfaces, actin-containing filaments form throughout the cell (Figure 5).

Microfilaments are found primarily in regions underlying the plasma membrane and in regions of active phagocytosis (Korn et al., 1974) and pseudopod formation. They also extend down all microprojections that form as neutrophils adhere to surfaces (Figure 5). Actin is one of the most highly conserved proteins throughout evolution (Korn, 1982); only 6% of residues are different in actins from sources as evolutionarily distant as *Acanthamoeba castillanii* and rabbit skeletal muscle (Korn, 1978). It is a single polypeptide with a molecular weight of about 42 kDa (Sheterline, 1983). In neutrophils it represents about 12% of total cellular proteins, i.e., 6 pg per cell (Al-Mohanna and Hallett, 1987). The polymerization of actin and its reorganization into highly structured microfilaments is governed by a number of actin-binding proteins (Table V).

Actin-binding proteins have varying degrees of dependency on intracellular Ca^{2+} levels (Table VI) and phospholipids. Of particular importance are profilin, gelsolin, bundling proteins such as α-actinin, and anchorage proteins such as vinculin. Profilin binds to monomeric globular actin (G-actin) in a 1 : 1 stoichiometry (Dinubile and Southwick, 1985). The complex (profilactin) has an apparent dissociation constant of 3 μM. *In vitro*, dissociation of the complex is mediated by PIP_2 and leads to concomitant polymerization of the actin monomers (Lassing and Lindberg, 1985). Whether this occurs *in vivo* is yet to be demonstrated. It is, however, possible, especially since phospholipid turnover (Prentki et al., 1984; DiVirgilio et al., 1985) and actin polymerization (Varani et al., 1983; White et al., 1983b; Sheterline et al., 1984a, b; Yassin et al., 1985; Al-Mohanna and Hallett, 1987) are among the earliest events in cell activation.

The other important actin-binding protein is gelsolin, which constitutes about 1% of total cellular protein (Southwick, 1985). This 90kDa protein has three effects on actin (Yin, 1987): (1) severs preformed filaments, (2) inhibits elongation of newly formed filaments (Yin and Stossel, 1979), and (3) increases the equilibrium monomer concentration (concentration of G-actin that is required before polymerization ensues) and thus

Table V
Organization of Neutrophil Cytoskeleton

Cytoskeletal component	Diameter (nm)	Subunits (monomers)	State	Reference
Microfilaments	6	Actin	Dynamic equilibrium between filaments and monomer pools	Sheterline, 1983
Intermediate filaments	10–12	Vimentin	Static	Parysek and Eckert, 1984
Microtrabecular lattice	7–17	Many, but actin predominates	Dynamic?	Bershadsky and Vasiliev, 1988
Microtubules	25	Tubulin (dimer)	Dynamic	Lackie, 1986

Table VI

Properties of Actin-Binding Proteins

Protein	Mol. wt. (kDa)	Ca²⁺ sensitivity	PIs sensitivity	Function	Reference(s)
Profilin	16	?	+	Maintains a pool of G-actin	Korn, 1982; Korn et al., 1987; Lassing and Lindberg, 1985; Pollard and Cooper, 1986
Gelsolin	90,95	+	+	Severs filaments, nucleates growth of new filaments, inhibits elongation by capping the barbed end of filaments	Yin and Stossel, 1979, 1980; Janmey and Stossel, 1987; Yin, 1987; Kwiatkowski and Yin, 1987
α-Actinin	210	+	+	Cross-linker and spacer between actin filaments	Schook et al., 1978; Burridge and Feramisco, 1981; Burn et al., 1985
Vinculin	130	+		Bundles filaments and decreases polymerization of preformed nuclei	Elzinga et al., 1976; Geiger, 1979; Burridge and Feramisco, 1982
Filamin	520	−	−	Causes angular branching of filaments	Hartwig and Stossel, 1975; Heizmann and Hauptle, 1977; Hartwig and Stossel, 1981
Acumentin	65	−	−	Caps pointed ends of filaments	Southwick and Hartwig, 1982; Mariyana, 1971; Southwick and Stossel, 1981

decreases the length distribution of actin filaments. Gelsolin must bind Ca^{2+} ($K_d = 0.9$ µM) (Yin and Stossel, 1980) for the F-actin-severing effect of this protein. Gelsolin binds two actin monomers on F-actin cooperatively (Kwiatkowski *et al.*, 1985; Chaponnier *et al.*, 1986). It is suggested that Ca^{2+} is required for occupation of the first actin-binding site at the COOH terminus of gelsolin. Occupation of this site exposes the second actin-binding site on the NH_2 terminus of gelsolin. Occupation of the site becomes inherently independent of Ca^{2+} (Kwiatkowski and Yin, 1987). Treatment of gelsolin-bound actin with EGTA causes dissociation of the one actin monomer from the first binding site (Bryan and Kurth, 1984; Kurth and Bryan, 1984) and leaves an "EGTA-resistant" complex that has higher affinity for actin filament blocking. This complex can be further dissociated by either PIP_2 or PIP (phosphatidylinositide monophosphate) to gelsolin and actin monomers (Janmey and Stossel, 1987).

α-Actinin binds DAG *in vitro* and forms in microfilamentlike structures that resemble those occurring in living cells (Burn *et al.*, 1985). In contrast to muscle α-actinin, nonmuscle α-actinin is Ca^{2+} dependent (Burridge and Feramisco, 1981). Anchorage of the actin network to plasma membrane is thought to be mediated by another Ca^{2+}-sensitive actin-binding protein, vinculin (Geiger, 1979). It is interesting that upon transformation of fibroblasts, vinculin becomes phosphorylated, leading to the collapse of the actin network (Elzinga *et al.*, 1976).

4.1.2. Microfilaments and Neutrophil Functions

An intact actin network is necessary for the normal functions of neutrophils. Impairment of filament formation by pharmacological agents such as cytochalasins (Cooper, 1987) or actin dysfunction syndrome (Boxer *et al.*, 1974) leads to defective chemotaxis and phagocytosis and changes the neutrophil into a secretory cell. In actin dysfunction syndrome, patients have increased susceptibility to bacterial infection (Boxer *et al.*, 1974). Cytochalasin B inhibits actin polymerization into filamentous F-actin by blocking the barbed end of F-actin (Maclean-Fletcher and Pollard, 1980). The effect of cytochalasin B on neutrophils is attributed to its effects on actin polymerization (Al-Mohanna and Hallett, 1987) and not on glucose and nucleotide uptake (Maclean-Fletcher and Pollard, 1980). Furthermore, botulinum C2 toxin produces effects on neutrophils similar to those evoked by cytochalasin B (Al-Mohanna *et al.*, 1987). C2 toxin acts on monomeric actin and inhibits its polymerization by ADP ribosylation (Aktories *et al.*, 1986).

4.1.2a. Chemotaxis. Formylated peptides are characteristics of prokaryotic proteins and are extremely potent inducers of neutrophil chemotaxis (Schiffman *et al.*, 1975). Since peptide sequences which begin with formyl are always nonmammalian, this may be important in recognition of microbial proteins. The formyl peptide receptor on neutrophil membrane has been identified as an integral membrane protein with a molecular weight of 60 kDa (Painter *et al.*, 1984). The number of receptors per cell varies with species. Part of this variation may be attributed to down regulation and recycling of receptors by the cells (Zigmond and Sullivan, 1981) and by the possibility that some receptors are switched from low to high affinity during chemotaxis (Snyderman, 1983). Nevertheless, the accepted figure is in the range of 50,000 per cell in rabbit and human neutrophils (Cochrane, 1984).

The ligand–receptor interactions lead to generation of intracellular messengers such as cAMP, DAG, phosphatidylinositides, and Ca^{2+}. Neutrophils move rapidly into tissues in response to inflammatory stimuli. They are probably the fastest-moving mammalian cells (10–15 μm/min) (Lackie, 1986). They change shape during locomotion from regions of low to regions of high chemoattractant concentration. The movement starts with the formation of pseudopods, followed by contraction of the tail part of the body. The pseudopod is rich in actin and actin-binding proteins (Stendahl *et al.*, 1980; Heath, 1981; Sheterline *et al.*, 1984b). It is suggested that the formation of F-actin in the pseudopod pushes the membrane to the exterior, therefore causing its extension (Pollard and Weshing, 1974; Bershadsky and Vasiliev, 1988). Pseudopod extension can also be explained by the contraction of the cortical actin, which may exert hydrostatic pressure on the internal fluid cytoplasm and lead to its extrusion. It is also possible to explain this extension in terms of swelling of cortical actin.

It was postulated that actin polymerization in pseudopods is due to intracellular generation of Ca^{2+} gradients (Southwick and Stossel, 1983; Lackie, 1986) in which actin moves from regions of high to regions of low Ca^{2+} concentration (Southwick and Stossel, 1983). After the formation of the pseudopod, adhesion sites are formed (Lackie, 1986). These are patches of integral membrane proteins into which the actin filaments are inserted. The filaments are stabilized by lateral interactions with vinculin at the insertion sites and along the filaments through filamin, α-actinin, and other actin-binding proteins.

It has been suggested that propulsion of the rest of the cytoplasm toward the leading site is a result of a rise in intracellular Ca^{2+} (Lackie, 1986; Bershadsky and Vasiliev, 1988). When intracellular Ca^{2+} is sufficiently high to activate actin-binding proteins such as gelsolin, the gel matrix would be weaker, and this would permit myosin to interact effectively with F-actin bundles to cause a concentration of the gel. In the presence of ATP, myosin is released from the actin filaments; after sequestration of Ca^{2+}, the rigidity of the system is restored. The neutrophil is capable of detecting a 1% gradient of chemoattractant across its dimensions (Zigmond, 1977). For continual movement, the cell may need to clear the occupied receptors or insert new receptors into the plasma membrane. The down regulation and appearance of new receptors have been demonstrated (Niedel *et al.*, 1979; Sullivan and Zigmond, 1980). Movement continues by formation of new pseudopods and contraction of the rest of the cytoplasm in the direction of pseudopods until the cell can no longer detect a chemotactic gradient. This whole process (chemotaxis) is inhibited by cytochalasins (Wessels *et al.*, 1971; Becker *et al.*, 1972) and C2 toxin (Norgauer *et al.*, 1988).

4.1.2b. Phagocytosis. Phagocytosis is the engulfment of particles by the flow of hyaline pseudopodia. The neutrophil surrounds the particle with an organelle-free peripheral cytoplasm (hyaline ectoplasm or hyaloplasm). Phagocytosis commences with recognition, followed by internalization of the particle. The process is also concerned with removal of cellular and extracellular debris. Recognition is enhanced by opsonization with opsonins, e.g., IgG and the C3b fragment of the complement system. The neutrophil has cell surface receptors to regions of immunoglobulins and the C3b fragment of complement (Griffin, 1977). The hyaloplasm forms the pseudopod, which attaches to the free surface of the particle, and the process is repeated until the particle is completely surrounded by the membrane of the pseudopod. Regions of active phagocytosis have higher concentrations of actin and actin-binding proteins than do other areas of the cell (Stendahl *et al.*, 1980; Sheterline *et al.*, 1984b). Cytochalasin B and C2 toxin are potent inhibitors of phagocytosis (Allison, 1973; Al-Mohanna *et al.*, 1987). Furthermore, actin dysfunction syndrome is manifested as defective phagocytosis by neutrophils (Boxer *et al.*, 1974). It is not clear whether internalization is achieved by contraction of the actin network in the surrounding pseudopod, which may pull the particle into the cortex, or whether the particle is held while cytoplasm flows around it (Sheterline and Rickard, 1989).

4.1.2c. Secretion.   Secretion involves the translocation of secretory granules to the plasma membrane and fusion of granule and plasma membrane, with the subsequent release of granule content to the outside by exocytosis. In resting neutrophils, the granules are separated from the plasma membrane by a hyaline cortex within which the granules are embedded. It is therefore necessary for the secretory granule to pass through this hyaline cortex before fusion can take place. It has been suggested that the secretory granules are actively transported through the concerted action of the cellular actin–myosin contractile machinery (Burgoyne and Cheek, 1987). Secretion requires an external signal, which is then transduced across the plasma membrane; through second messenger signals, the process commences. The actin network is suggested to play a crucial part in preventing secretion until the appropriate signal is received by the cell (Cheek & Burgoyne, 1986) Evidence for this view is drawn from the fact that disruption of the actin network by cytochalasin B enhances secretion by neutrophils in response to formylated peptides and Ca^{2+} ionophores (Lew *et al.*, 1986). Furthermore, actin filament stabilizers such as phalloidin inhibit secretion in permeabilized cells (Lelkes *et al.*, 1986). Involvement of the actin-binding protein fodrin in secretion by adrenal chromaffin cells has been reported (Perrin *et al.*, 1987). Although actin depolymerization has been demonstrated in some secretory cells after stimulation (Cheek and Burgoyne, 1986), this does not occur in neutrophils.

ADP-ribosylated actin was suggested to act as a capping protein for preformed actin filaments (Wegner and Aktories, 1988), which may ultimately lead to actin filament disassembly. Cytochalasin B inhibits neutrophil oxidase activation by particulate stimuli (e.g., latex beads) by inhibiting phagocytosis, which is essential for oxidase activation by this route (Hallett and Campbell, 1983, 1984a). Both cytochalasm B and C2 toxin inhibit the transient increase in actin polymerization after fMLP stimulation (Al-Mohanna and Hallett, 1987; Al-Mohanna *et al.*, 1987). This inhibition is accompanied by enhanced oxidase activation by fMLP. Both temporal and quantitative data suggest a strong correlation between inhibition of actin polymerization and enhancement of oxidase activation. These effects are limited to the Ca^{2+}-dependent route of oxidase activation.

4.2. Microtubules

Microtubules are universal structures of all mammalian cells. In resting human neutrophils there are 11.9 ± 0.8–22.3 ± 2.0 microtubules per cell (Hoffstein, 1980). They are approximately 25 nm in diameter (Bershadsky and Vasiliev, 1988) with a wall width of about 5 nm and are composed of tubulin subunits. Tubulin is a highly conserved dimeric protein that has a basic polypeptide chain and a less basic second chain, each with a molecular weight of about 50 kDa. Microtubules are polar structures, filament growth being biased toward one end (designated the plus end). Microtubules are associated with proteins that control their dynamic state, e.g., microtubule-associated proteins (MAP) and microtubule-organizing centers. In the neutrophil, about 35–40% of total cellular tubulin exists in the assembled state at rest, but the remainder can be transiently assembled after stimulation (Sherline and Mundy, 1977).

Microtubules and Neutrophil Functions

Chemotaxis.   An involvement of microtubules in chemotaxis has been suggested. Although some reports have shown an inhibition of chemotaxis by some microtubule-disrupting agents such as colchicine, colcemid, vinblastine and vincristine (Caner, 1965; Edelson and Fudenberg, 1973; Bandmann et al., 1974; Gallin and Rosenthal, 1974; Gallin et al., 1975), others report no effect (Chang, 1975) or inhibition only at very high concentrations of these agents (Ramsey and Harris, 1973). This discrepancy may be attributed to effects, other than on microtubules, by these agents on cellular functions. Some investigators have suggested that inhibition resulted from the prevention of release of chemotactic factors from migrating cells (Rydgren et al., 1976). Although chemotactic stimulation of neutrophils causes transient assembly of microtubules, an event that persists even in the presence of cytochalasin B (Goldstein et al., 1973), it seems unlikely that microtubule assembly is essential for chemotaxis. Cytokineplasts and cytoplasts (fragments of neutrophils that are devoid of nuclei and cytoplasmic organelles) lack a microtubule system yet are still capable of directional movement (Malawista and Chevance, 1982).

Phagocytosis.   The demonstration that phagocytosis can occur in the presence of microtubule-disrupting agents such as colchicine and vinblastine at concentrations of up 250 and 50 μM, respectively, rules out a requirement for microtubule assembly (Malawista and Bensch, 1967; Malawista and Bodel, 1967; Malawista, 1971). Some reports, however, suggested that microtubules are involved with particles (Klebanoff and Clark, 1978).

Secretion.   There is some evidence that microtubules are involved in neutrophil secretion. For example, (1) colchicine and vinblastine inhibit secretion even in cytochalasin B-treated neutrophils (Weissmann et al., 1973); (2) D_2O, which promotes microtubule assembly, enhances secretion in cells treated with cytochalasin B (Weissmann et al., 1973; Zurier et al., 1974); (3) some secretory stimuli also increase microtubule assembly (Weissmann et al., 1975); and (4) there is a morphological association between microtubules and the sites of exocytosis (Klebanoff and Clark, 1978).

The mechanism of microtubule involvement in secretion remains unclear. There is however, a general contention that microtubules are only partly involved in secretion. Total disassembly of microtubules by colchicine causes a mere 40% inhibition of enzyme release (Weissmann et al., 1971), while the Ca^{2+} ionophore A23187 causes a massive increase in secretion with virtually no change in microtubule assembly (Nacchache et al., 1977; Hoffstein and Weissmann, 1978). It has also been suggested that microtubule assembly is required for the physical translocation of phagosomes from the neutrophil periphery to the central regions of the cytoplasm (Hoffstein, 1980).

Oxidase Activation and Intracellular Killing.   The anti-inflammatory effects of colchicine have been known for sometime and have led to its use in the treatment of gouty arthritis (Malawista and Bodel, 1967). In vitro, colchicine inhibits particle-induced oxidase activation as measured by oxygen uptake (Malawista and Bodel, 1967; Malawista, 1971).

Microtubular dysfunction may underlie Chédiak-Higashi syndrome (CHS). This disease is an autosomal recessive genetic condition characterized by oculocutaneous albinism, enlarged granules in many cells, and an increased susceptibility to bacterial infection (Baehner and Boxer, 1979). Neutrophils from patients with CHS share many characteristics with colchicine-treated neutrophils (Baehner and Boxer, 1979), including concanavalin A (ConA) capping (Oliver *et al.*, 1973). Although untrastructural studies revealed fewer microtubules in CHS neutrophils (Oliver 1976), it has been suggested that the microtubule dysfunction does not result from alterations in tubulin structure or quantity but may be due to an aberrant control mechanism (Klebanoff and Clark, 1978). The increased susceptibility to bacterial infection in CHS was found to be a result of delayed rather than defective intracellular killing (Stossel *et al.*, 1972) and may be attributed to impairment of fusion between the phagocytic vacuole and lysosome (Boxer *et al.*, 1976). There is some disagreement about the effect of microtubule disruption on activation by soluble stimuli of the neutrophil oxidase (Klebanoff and Clark, 1978). However, colchicine has no effect on oxygen uptake stimulated by chemotactic peptide in rat neutrophils (Al-Mohanna and Hallett, 1987).

4.3. Intermediate Filaments

Unlike microfilaments and microtubules, intermediate filaments are not universal components of the eukaryotic cytoskeleton. They are, however, found in neutrophils (Pryswansky *et al.*, 1983). Their contribution to total cellular protein and many of their characteristics remain largely unknown. In fibroblasts and epidermal cells, they constitute 2–4% and about 30% of total cellular protein, respectively (Bershadsky and Vasiliev, 1988). Human neutrophils contain intermediate filaments of the vimentin type (Parysek and Eckert, 1984); the desmin and keratin forms have not yet been detected.

Vimentin is a rod-shaped molecule with a molecular weight of about 57 kDa (Bershadsky and Vasiliev, 1988) that polymerizes readily under physiological conditions to give a very stable filament of about 10–12 nm in diameter. Harsh treatment of the cytoskeletal components, including those that remove microfilaments and microtubules, leaves intermediate filaments intact. Intermediate filaments appear to exist in the filamentous form only, and in neutrophils they form an open network of single filaments in the perinuclear space (Pryzwansky *et al.*, 1983).

The role of intermediate filaments in neutrophil function is not yet established. Intermediate filaments are located by immunofluorescence and transmission electron microscopy in areas between the nucleus and the trailing end of the cell and in uropodal areas of chemoattractant-treated neutrophils (Parysek and Eckert, 1984). The question of whether these locations are crucial for cell locomotion remains unanswered. However, since intermediate filaments appear to be stable and unable to move independently of other cellular structures, it is possible that their main function in neutrophils is to confer a mechanical network through which peripheral and central regions of the cell are connected.

4.4. Microtrabeculae

The microtrabecular lattice (MTL) is a highly organized component of the neutrophil cytoskeleton (Pryzwansky, 1987). It is composed of an intricate network of fine strands within which all cytoplasmic components are embedded. The strands are variable in diameter (7–17 nm) (Pryzwansky *et al.*, 1983) and are made up to at least 400 different polypeptides, of which actin is the predominant species (Bershadsky and Vasiliev, 1988). These structures are highly dynamic and appear to be reversibly sensitive to many experimental manipulations such as altered ionic strength, altered Ca^{2+} and Mg^{2+} levels, and low temperature (Bershadsky and Vasiliev, 1988).

It has been suggested that the MTL may play a role in phagocytosis. Pryzwansky (1987) reported that the neutrophil granules are distributed at different depths within the MTL and suggested that the MTL may provide an organizational compartmentation within which granule and phagosome fusion proceed more efficiently. The possibility also exists that the MTL is involved in directional movement (Pryzwansky, 1987).

4.5. Interaction between Cytoskeletal Components

The neutrophil cytoskeleton is composed of at least four distinct structures: microfilaments, microtubules, intermediate filaments, and the MTL. The MTL seems to be the interconnecting ground substance through which the other components of the cytoskeleton may interact. Interactions can also occur through specific proteins such as MAP that bind actin filaments.

5. ACTIN POLYMERIZATION AND OXIDASE ACTIVATION

5.1. Characteristics of Actin Polymerization Accompanying Oxidase Activation

In the resting spheroid neutrophils, at least half of the actin within the cell exists as unpolymerized globular monomers (G-actin) (Sheterline *et al.*, 1984a; White *et al.*, 1983b; Al-Mohanna and Hallett, 1987). The pool of unpolymerized actin can be rapidly and reversibly polymerized to form oligomeric filaments (F-actin). Since total actin represents 5–10% of total cell protein (Sheterline and Rickard, 1989), this transition represents a major change in the cytoplasmic structure and has been described in earlier literature as sol-to-gel transformation. However, the mechanism of the transition remains unestablished and its effect on other intracellular events poorly understood.

In the resting neutrophil, the polymerized actin as detected by phalloidin binding is mainly in the cortical layer of cytoplasm, with very little general cytoplasmic F-actin being detectable. Transmission electron microscopy of Triton-insoluble cytoskeletal "ghosts" suggests that this cortical layer extends only 100–200 nm into the cytoplasm (Bray *et al.*, 1986). Sheterline and Rickard (1989) have suggested that if the layer is composed of actin filaments angled between 10° and 90° to the plasma membrane, then each filament would be composed of only 30–100 subunits. This cortical network may be static, with little turnover of actin, and is insensitive to cytochalasin D (Sheterline *et al.*, 1986) at concentrations known to prevent actin assembly.

Phagocytosis of particles is accompanied by a transient assembly of actin into the cortical network at the site of pseudopod formation (Sheterline *et al.*, 1984a,b). This assembly is inhibited by cytochalasins with apparent $K(0.5)$ values of approximately 3 μM (Al-Mohanna and Hallett, 1987; Sheterline and Rickard, 1989). In contrast, phorbol esters stimulate a massive increase in actin polymerization throughout the cell (Sheterline *et al.*, 1986) and produce a characteristic shape change (Sheterline and Rickard, 1989). Both events are inhibited by 1 μM cytochalasin D. The peptide fMLP also provokes a large transient polymerization of actin, 1 μm peptide provoking almost total actin polymerization in rat neutrophils (Al-Mohanna and Hallett, 1987). This event is very rapid, reaching a peak within 5 sec and subsiding over the following 60 sec to reach a slight plateau. Cytochalasins and botulinium C2 toxin inhibit the rise (Al-Mohanna and Hallett, 1987; Al-Mohanna *et al.*, 1987). The ionophore A23187 also gives rise to a slow actin polymerization (Al-Mohanna and Hallett, 1986). Thus, stimuli from each of the groups shown in Table II provoke both polymerization of actin and oxidase activation. In the next section, the relationship between the two events is examined.

5.2. Interaction of Actin Polymerization with Oxidase Activation

Evidence that actin polymerization plays a role in oxidase activation is restricted to that provided by inhibitors. Cytochalasins, which have a inhibitory effect on actin polymerization, also influence the extent of oxidase activation by some stimuli. Stimulation by the phagocytic route is abolished, probably as a result of inhibition of phagocytosis. This suggests that oxidase activation follows internalization or phagosome–lysosome fusion, rather than particle binding. Stimulation of the oxidase by PMA is, however, unaffected by pretreatment with cytochalasins and hence leads to two possible conclusions: either (1) activation of the oxidase by direct activation of c-kinase is not influenced by actin polymerization or (2) no change in actin polymerization occurs before oxidase activation by the stimulus.

With the peptide fMLP, however, the effect of cytochalasin B is dramatic. There is a massive enhancement of oxidase activity after pretreatment with cytochalasins; up to five times the oxygen consumption rate occurs. The evidence that this effect of cytochalasin B is mediated by inhibition of actin polymerization is that (1) the concentration required for the enhancement is the same as that required for inhibition of actin polymerization (Al-Mohanna and Hallett, 1987), (2) the concentration required for enhancement is the same as that required for inhibition of other "microfilament-dependent" cellular processes (Al-Mohanna and Hallett, 1986), and (3) other cytochalasins and the dihydro derivative of cytochalasin B also produce the same enhancement. Furthermore, botulinum C2 toxin, which has been shown to ADP ribosylate nonmuscle G actin and prevent polymerization of actin (Aktories *et al.*, 1986; Wegner and Aktories, 1988) has a similar enhancing effect on oxidase activation (Al-Mohanna *et al.*, 1987; Norgauer *et al.*, 1988). As with the cytochalasins, botulinum C2 toxin enhanced activation by the Ca^{2+}-dependent stimuli fMLP-lu-pe and ConA but not by PMA (Al-Mohanna *et al.*, 1987; Norgauer *et al.*, 1988). This leads to the conclusion that with Ca^{2+}-dependent stimuli, the polymerization of actin which accompanies oxidase activation exerts an inhibitory influence on the magnitude of oxidase activity. In neutrophils in which actin polymerization is prevented, the Ca^{2+} signal is diverted (from producing chemotaxis) to oxidase activation. The mechanisms of the inhibitory influence of polymerized actin and of the intracellular signal diversion remain unestablished.

Two possibilities can be suggested. The first is that the cortical actin network presents a physical barrier to the movement of granules, micelles, or other structures necessary for oxidase activation. The second possibility is that actin polymerization controls or limits some biochemical changes in the cell, for example, Ca^{2+} homeostasis (see Section 5.3) or phospholipase C activity. There is evidence for both examples, although the effect on phospholipid metabolism may be small (Tau and Stjernholm, 1975; Bennet et al., 1980). It is striking that the cytochalasin-sensitive state (and hence polymerized actin inhibitory state) is transient after stimulation with fMLP, as would be predicted from the transience of actin polymerization (Al-Mohanna and Hallett, 1987) (see Figure 4). However, the effect of cytochalasin B has another unexpected time characteristic which may give a clue to the mechanism of enhancement. In the absence of cytochalasin, there is a marked lag period between stimulation and onset of oxidase activation of about 40 sec. This lag cannot be accounted for by time delays in the measuring systems, by mixing and receptor binding, or by intrinsic properties of the oxidase (Hallett et al., 1987); this lag therefore originates at the level of coupling receptor occupancy to oxidase activation, and has been shown to be related to the cytoplasmic Ca^{2+} concentration (Campbell and Hallett, 1983; Hallett and Campbell, 1984a). In neutrophils pretreated with cytochalasins, the lag is reduced to 6 sec (Al-Mohanna and Hallett, 1986), but there are no differences in the peptide-induced Ca^{2+} signal (Al-Mohanna and Hallett, unpublished data). This finding suggests that as well as limiting the extent of oxidase activation, actin polymerization limits the rate at which post-Ca^{2+} signal transduction to the oxidase occurs. Interestingly, when cytochalasin is added to neutrophils after fMLP, a lag of 6 sec is again observed. However, although the magnitude of the cytochalasin-induced response declines as the interval between peptide and cytochalasin is increased (Figure 4), the lag of 6 sec is constant. This time may thus represent a basic limitation of the coupling process and is reminiscent of the retardation period observed in chemotaxing neutrophils, during which time the cell is insensitive to external stimuli. The retardation time is approximately 20 sec (in non-cytochalasin-treated cells) and may originate from an "internal clock" with a characteristic time of 40 sec (Gruler, 1984, 1989). The biochemical basis for such a clock, however, remains obscure.

5.3. Actin Polymerization and Ca^{2+} Homeostasis

Since a transient rise in intracellular free Ca^{2+} and actin polymerization are among the earliest events following neutrophil stimulation with fMLP, the possibility of a correlation between the two events was investigated. In quin-2- (Treves et al., 1987) and fura-2-loaded neutrophils, cytochalasin B causes a dose-dependent transient rise in intracellular Ca^{2+} (Figure 6). Removal of extracellular Ca^{2+} and the presence of 1 mM EGTA suppress the cytochalasin B (10 μM)-induced transient Ca^{2+} rise in rat neutrophils by approximately 33%, indicating that the Ca^{2+} transient is largely due to release from intracellular stores. The cytochalasin effect is totally abolished by pretreatment of the cells with EGTA (1 mM) and ionomycin (2 μM) (Treves et al., 1987; Al-Mohanna and Hallett, unpublished data). This suggests that the cytochalasin B-sensitive Ca^{2+} store is membrane bound (Lew et al., 1984; Treves et al., 1987). It can be irreversibly emptied by 10 μM cytochalasin B (Al-Mohanna and Hallett, unpublished data). Cytochalasin B does not affect the extent of Ca^{2+} rise by fMLP, but the return to prestimulatory levels is retarded (Treves et al., 1987; Al-Mohanna and Hallett, unpublished data), perhaps as a consequence of the effect on the Ca^{2+} store.

Cytochalasins bind to three sites in neutrophils and lymphocytes (Parker et al., 1976;

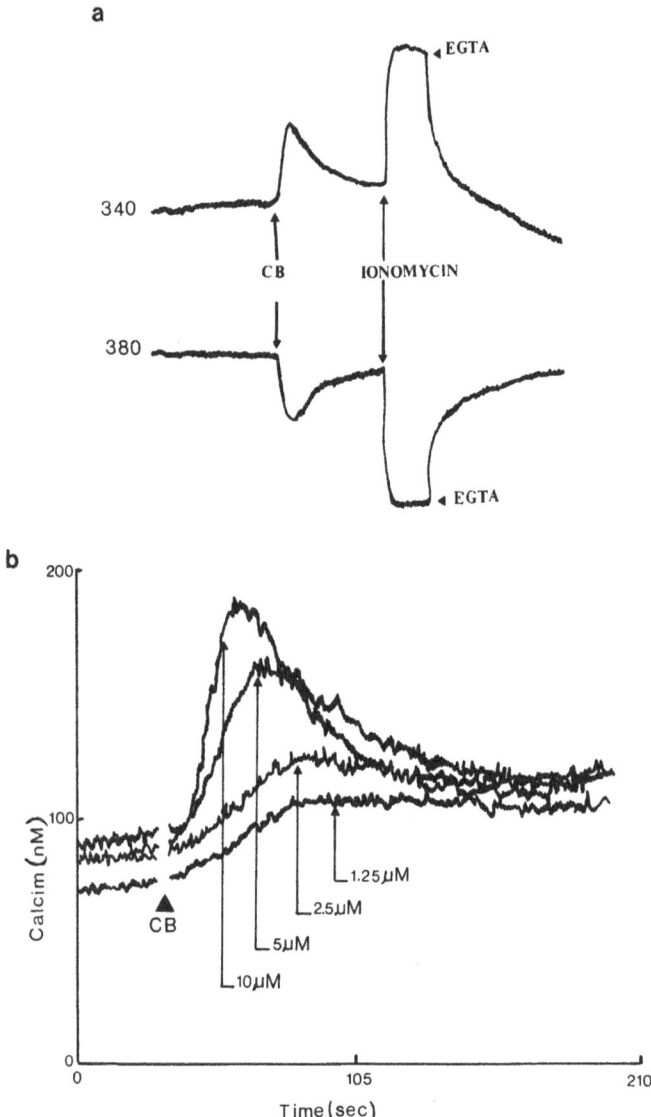

FIGURE 6. Effect of cytochalasin B (CB) on Ca^{2+}. (a) Fluorescence signals at 340 and 380 nm from fura-2-loaded rat neutrophils treated with cytochalasin B. (b) From the fluorescence ratio, the cytoplasmic free calcium was calculated and is shown after treatment of the cells with the concentration of cytochalasin B indicated.

Mookerjee *et al.,* 1981): a high-affinity binding site that is closely associated with the plasma membrane, a medium-affinity binding site that is the glucose transporter, and a low-affinity binding site that is the actin network, the proportions of the cytochalasin B binding to these sites being 7 : 8 : 85%, respectively (Mookerjee *et al.,* 1981). The possibility that the Ca^{2+} rise was mediated by a nonactin effect of cytochalasin B was eliminated by the use of botulinum C2 toxin. This toxin inhibits actin polymerization by ADP ribosylation of G-actin and also causes an increase in intracellular Ca^{2+} (Al-Mohanna and Hallett, unpublished data). Interestingly, DNase I, which depolymerizes actin filaments, causes a marked influx of Ca^{2+} into bovine adrenal medullary cells when introduced into the cytoplasm by liposomal fusion (Harish *et al.*, 1984). It was speculated that the Ca^{2+} influx was mediated by actin connected to the surface membrane that is associated with membrane calcium channels. The possibility also exists that the mechanism of intracellular Ca^{2+} release involves phosphoinositide breakdown (Bradford and Rubin 1985).

5.4. Intracellular Triggers for Actin Polymerization

5.4.1. Calcium

Actin polymerization accompanies activation of neutrophils by stimuli from each of the three groups shown in Table II (see Section 5.1). Thus, there is a clear dissociation from an elevation in cytoplasmic Ca^{2+} with some stimuli. For example, phagocytosis, which depends on actin polymerization, is not prevented by increasing the buffering capacity of the cytoplasm with quin-2 (Lew *et al.*, 1985), fura-2 (DiVirgilio *et al.*, 1988), or BAPTA (Cooke *et al.*, 1989). Actin polymerization accompanying treatment with PMA occurs without a rise in cytoplasmic Ca^{2+} and is also not prevented by buffering cytoplasmic Ca^{2+} (Sheterline *et al.*, 1986; Sheterline and Rickard, 1989; Sha'afi *et al.*, 1986b). With stimuli that elevate cytoplasmic Ca^{2+}, such as the peptide fMLP, it has been speculated that the Ca^{2+} may play a role in triggering actin polymerization. However, with a low concentration of the peptide, actin polymerization is triggered without an accompanying Ca^{2+} rise (Al-Mohanna and Hallett, 1989). Also, buffering the Ca^{2+} rise at higher peptide concentrations does not prevent actin polymerization. Buffering intracellular calcium to less that 25 nM by the calcium chelator fura-2 (2 mM) fails to inhibit actin polymerization (Al-Mohanna and Hallett, 1988b); this treatment unexpectedly increases the peak by about 25% and produces a rise that is sustained for at least 2 min. The evidence thus suggests that calcium plays no role in triggering actin polymerization.

There may, however, be a role for calcium in regulating the extent and cellular location of polymerization as well as contractility. The major calcium-regulated protein that interacts with actin is gelsolin (Table VI). This protein, which has an apparent K_d for Ca^{2+} of 1.1 μM (Yin and Stossel, 1980), caps actin filaments and prevents their further growth. The transient nature of the polymerization of actin seen with the peptide fMLP may thus in part be due to the activation of gelsolin. In the absence of the calcium signal, as with PMA, actin may be permanently locked into the polymerized form.

For the actin to generate contraction, it is apparent that a crosslinked actin gel is not appropriate. Only if the actin is in shorter filaments with the possibility for "sliding" is conventional myosin-mediated contraction possible. Stossel has suggested that gelsolin may thus play a role in cleaving actin filaments and provide the potentiality for contraction (Yin and Stossel, 1979, 1982; Southwick and Stossel, 1983). Hence, cytoplasmic Ca^{2+} may be the key signal for contraction acting through gelsolin (to cleave crosslinked immobile actin filaments) and myosin light-chain kinase (allowing myosin assembly). If this is so, then we must conclude that chemotaxis, which is stimulated by agents that elevate cytoplasmic Ca^{2+} (group II), involves contraction of the actin cytoskeleton and that phagocytosis, which can occur without an elevation in cytoplasmic Ca^{2+}, does not. The latter conclusion may seem unlikely, especially in view of the presence of both actin and myosin at the site of phagocytosis (Valerius *et al.*, 1981). However, there are two distinct possibilities for the mechanism of phagocytosis (Sheterline and Rickard, 1989). The first is that the particle is pulled into the cell and hence requires contraction; the second is that the particle is surrounded by cytoplasmic protusions and requires no contraction. It is difficult at present to resolve which of these two possible events actually occurs, especially since phagocytosis is often accompanied by locomotion toward the particle.

5.4.2. Protein Phosphorylation

It has been reported (Ohta *et al.*, 1987) that protein kinase C (PKC) phosphorylates muscle and nonmuscle G-actin *in vitro* and that this phosphorylation increases G-actin polymerizability. cAMP-dependent protein kinase also phosphorylates G-actin but at a different site and, in contrast to PKC phosphorylation, leads to weaker polymerizability. It is not clear whether these phenomena play any role within the living neutrophil. For example, protein phosphorylation can be dissociated from actin polymerization by the use of phorbol didecanoate (White *et al.*, 1984), which induces actin polymerization in the absence of any detectable phosphorylation. In addition, in the U937 human monocytelike cell line, PMA does not evoke actin polymerization (Banks *et al.*, 1988) even though protein phosphorylation occurs. Also, inhibitors of protein kinases have no effect on actin polymerization induced by these two agents in neutrophils (White *et al.*, 1984).

5.4.3. GTP

The neutrophil membrane contains a family of GTP-binding proteins, including G, and G_s, which are, respectively, the inhibitory and stimulatory components of adenylate cyclase and transducin (Dickey *et al.*, 1987), and G, which has a stimulatory effect on phospholipase C. Huang and Devanney (1986) demonstrated that GTPγS caused the dissociation of actin and myosin from rabbit neutrophil membranes, an effect that was independent of pertussis toxin, cytochalasin B, and calcium (100 μM). GTPγS (1 mM) caused approximately 37–58% solubilization of membrane-associated actin and myosin. It may be postulated that there are two distinct mechanisms for actin depolymerization in neutrophils, both dependent on GTP, either acting directly with actin-binding proteins to cause actin depolymerization or binding to a G protein to cause actin depolymerization through activation of phospholipase C.

5.4.4. Cyclic Nucleotides

The involvement of cyclic nucleotides in actin polymerization and depolymerization may be suggested on the basis of their effects on actin-dependent functions. cAMP and its analogs have been shown to inhibit chemotaxis and phagocytosis (Rivkin and Becker, 1972; Patriarca *et al.*, 1973). In addition, agents that are know to increase intracellular cAMP, such as β-adrenergic agents, cause similar inhibition (Anderson *et al.*, 1977; Ignarro *et al.*, 1974). However, although increased cAMP concentrations are observed at sites of active phagocytosis (Pryzwansky *et al.*, 1981), neither cAMP nor cGMP influences the actin states of the neutrophil membrane (Huang and Devanny, 1986). Cyclic nucleotides thus play no role in triggering actin polymerization, although a modulatory role cannot be ruled out (McRobbie and Newell, 1984).

5.4.5. Intracellular pH

The increase in actin polymerization in stimulated cells is usually accompanied by changes in intracellular pH (pH_i). For instance, the addition of the chemotactic factor fMLP (1 μM) to neutrophils causes a rapid drop in pH_i, followed by a substantial increase (Molski *et al.*, 1980). It is, however, unlikely that this plays a triggering role for actin polymerization. In fMLP-treated cells, actin polymerization precedes the rise in pH_i, suggesting that alkalinization does not trigger actin polymerization in these cells (Yassin *et al.*, 1985); the Na^+/H^+ antiport inhibitor amiloride (1 mM), which abolishes pH_i increases, has no effect on the amount of actin associated with the cytoskeleton (Yassin *et al.*, 1985). Cytoplasmic acidification, however, causes a rapid and transient increase in actin polymerization (Yuli and Oplatka, 1987), and Sha'afi and Molski (1987) have proposed that receptor-induced actin polymerization may be mediated by low pH_i. However, pH_i measurements by fluorescein demonstrated that acidification occurs approximately 40 sec after stimulation by fMLP in the presence of cytochalasin B (5 μg/ml), perhaps as a result of activation of the hexose monophosphate shunt, and is too late to play a part as an intracellular signal for actin polymerization (Cooke *et al.*, 1989). In neutrophils in which the cytoplasmic pH_i is manipulated by nigericin/K^+, there is no correlation with actin polymerization (Figure 7). A role of pH_i in triggering actin polymerization thus appears unlikely.

5.4.6. Polyphosphoinositides

One of the earliest events of neutrophil activation by fMLP is a rise in PIP_2 hydrolysis by phospholipase C, resulting in the concomitant formation of the two intracellular

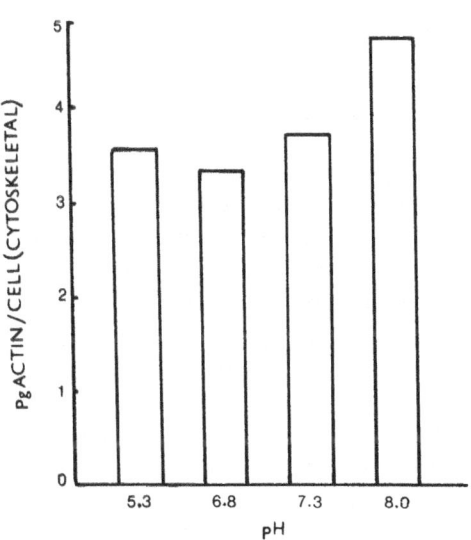

FIGURE 7. Effect of pH_i on the association of actin with cytoskeleton. Rat neutrophils were treated with nigericin (13.6 μM), and the intracellular pH was adjusted and measured. The incubation medium contained 130 mM KCl, 5 mM NaCl, 1.3 mM $CaCl_2$, 1.2 mM KH_2PO_4, 25mM HEPES, and 0.1% bovine serum albumin. The amount of actin was determined as 40-kDa protein associated with the Triton X-100-insoluble fraction. This was achieved by SDS-page of the insoluble cytoskeletal matrix after extensive boiling.

second messengers inositol IP_3, which is released into the aqueous cytoplasm and acts on Ca^{2+} stores, and DAG, the membrane-bound activator of PKC at resting levels of intracellular calcium (0.1 μM). DAG has also been shown to form microfilamentlike structures in the presence of α-actinin and F-actin (Burn et al., 1985). In addition, two key actin-binding proteins have been shown to be regulated by PIP_2. Gelsolin, which severs actin filaments in the presence of micromolar calcium and increases critical monomer concentrations, binds to two actin monomers on the actin filament (Yin, 1987). Removal of calcium causes the dissociation of one of the actin monomers (see Section 4.1). The remaining gelsolin–actin complex is dissociated by PIP_2. Profilactin (the profilin–actin complex) provides a pool of readily polymerizable G-actin by its PIP_2-dependent dissociation into actin and profilin. The modification of the activity of these actin-binding proteins by inositol phosphates provides a mechanism for controlling the actin states (Korn et al., 1987). Hartwig et al., (1989) provided ultrastructural and immunohistochemical data demonstrating the transient association of gelsolin with actin filaments and the plasma membrane in activated platelets and macrophages. The association with the plasma membrane was thought to be mediated by the binding site for phosphoinositides on gelsolin (Yin et al., 1988). In thrombin-activated platelets, gelsolin translocation occurs within 15 sec. Within this time many other cellular responses occur, including a transient intracellular Ca^{2+} rise, gelsolin–actin complex formation, cell shape changes, and a fall and rise in PIP_2 concentration (Hartwig et al., 1989). It has been proposed that gelsolin binding to membrane PIP_2 causes release of actin oligomers from gelsolin (Janmey et al., 1987) and therefore provides active nuclei for rapid polymerization. However, this proposal does not account for the calcium independence of actin polymerization in neutrophils. Since the effects of gelsolin are calcium-dependent, such a mechanism would impose a calcium dependency on actin polymerization.

5.4.7. Osmolarity

Although hyperosmolarity has been shown to inhibit the fMLP-induced actin polymerization (Yassin et al., 1985), it has been suggested that this inhibition resulted from receptor uncoupling rather than a direct effect on actin polymerization.

6. CONCLUSION

In this review, we have attempted to highlight the ways in which stimulation of neutrophils leads to oxidase activation and cytoskeletal changes. The interaction between these two events has not yet been fully explored. However, the evidence presented here suggests that there may be a fundamental interaction between these two which lies at the heart of determining whether the outcome of a particular stimulus to the neutrophil is beneficial or pathogenic. An understanding of the intracellular events associated with these phenomena may thus provide a key to solving the puzzle of inflammatory disease.

ACKNOWLEDGMENTS. We are grateful to the Medical Research Council, the Arthritis and Rheumatism Council, and the Saudi Educational Office, which have supported our work presented here.

REFERENCES

Aktories, K., Barmann, M., Ohishi, I., Tsuyama, S., Jacobs, K. H., and Habermann, E. 1986. Botulinum C2 toxin ADP-ribosylates actin, *Nature (London)* **322**:390–392

Allison, A. C., 1973, The role of microfilaments and microtubules in cell movement, endocytosis and exocytosis, *Ciba Found. Symp.* **14**:109–148

Al-Mohanna, F. A., and Hallett, M. B., 1987, Actin polymerization modifies stimulus-oxidase coupling in rat neutrophils, *Biochem. Biophys. Acta.* **927**:366–371

Al-Mohanna, F. A., and Hallett, M. B., 1988a, The use of fura 2 to determine the relationship between cytoplasmic free Ca^{2+} and oxidase activation in rat neutrophils, *Cell Calcium* **9**:17–26

Al-Mohanna, F. A. A., and Hallett, M. B., 1988b, Is actin polymerization in neutrophils regulated by intracellular Ca^{2+}?, *Eur. J. Clin. Invest.* **18**:A41.

Al-Mohanna, F. A. and Hallett, M. B., 1989, Actin polymerization in neutrophils is triggered without a requirement for a rise in cytoplasmic Ca^{2+}, *Biochem J.* **266**:669–674

Al-Mohanna, F. A., Ohishi, I., and Hallett, M. B., 1987, Botulinum C toxin potentiates activation of the neutrophil oxidase; further evidence for a role for actin polymerization, *FEBS Lett.* **219**:40–44

Anderson, R., Glover, A., and Robson, A. R., 1977, The in vitro effects of histamine and metiamide on neutrophil motility and their relationship to intracellular cyclic nucleotide levels, *J. Immunol.* **118**:1690–1696

Andrews, P. C., and Babior, B. M., 1983, Endogenous protein phosphorylation by resting and activated human neutrophils, *Blood* **61**:333–342

Badwey, J. A., Curnutte, J. T., and Karnovsky, M. L., 1981, *cis*-Polyunsaturated fatty acids induce high levels of superoxide production by human neutrophils, *J. Biol. Chem.* **256**:12640–12643

Badwey, J. A., Heyworth, P. G., and Karnovsky, M. L., 1984, Phosphorylation of both 47 and 49-KDa proteins accompanies superoxide release from neutrophils, *Biochem. Biophys. Res. Commun.* **158**:1029–1035

Baehner, R. L., and Boxer, L. A., 1979, Disorders of polymorphonuclear leukocyte functions related to alterations in integrated reactions of cytoplasmic constituents with the plasma membrane, *Semin. Hematol.* **16**:148–161

Balridge, C. W., and Gerald, R. W., 1933, The extra respiration of phagocytosis, *Am. J. Physiol.* **103**:235–239

Bandmann, U., Norberg, B., and Rydgren, L., 1974, Polymorphonuclear leucocyte chemotaxis in Boyden chambers, Effect of 305–312

Banks, P., Barker, M. D., and Burton, D. R., 1988, Recruitment of actin to the cytoskeleton of human monocyte-like cells activated by complement fragment C5a. Is protein kinase C involved?, *Biochem. J.* **252**:765–768

Barrowman, M. M., Cockroft, S., and Gomperts, B. D., 1986, Two roles for guanine nucleotides in the stimulus-secretion coupling of neutrophils, *Nature* (London) **319**:504–507

Bazzi, M. D., and Nelsesteun, G. L., 1987, Mechanism of protein kinase inhibition by sphingosine, *Biochem. Biophys. Res. Commun.,* **146**:203–207

Becker, E. L., Davis, A. T., Estensen, R. D., and Quie, P. G., 1972, Cytochalasins B. IV. Inhibition and stimulation of chemotaxis of rabbit and human polymorphonuclear leukocytes, *J. Immunol.* **108**:396–402

Becker, E. L., Kermode, J. C., Naccache, P. H., Yassin, R., Marsh, M. L., Munoz, J. J., and Sha'afi, R. I., 1985, The inhibition of neutrophil granule enzyme secretion and chemotaxis by pertussis toxin, *J. Cell Biol.* **100**:1641–1646

Bellavite, P., Jones, O. T. G., Cross, A. P., Papini, E., and Rossi, F., 1984, Composition of partially purified NADPH oxidase from pig neutrophils, *Biochem. J.* **223**:639–648

Bennet, J. P., Cockcroft, S., and Gomperts, B. D., 1980, Use of cytochalasin B to distinguish between early and late events in neutrophil activation, *Biochim. Biophys. Acta* **601**:584–591

Berkow, R. L., and Kraft, A. S., 1985, Bryostatin a non-phorbol macrocyclic lactone activates intact human polymorphonuclear leukocytes and binds to the phorbol ester receptor, *Biochem. Biophys. Res. Commun.* **131**:1109–1116

Bershadsky, A. D., and Vasiliev, J. M., 1988, *Cytoskeleton*, Plenum Press, New York

Berton, G., Castella, M., Cabrini, G. and Rossi, F., 1985, Activation of mouse macrophages causes no change in expression and function of phorbol diester receptors but is accompanied by alterations in activity and kinetics of NADPH oxidase, *Immunology* **54**:371–379

Berton, G., Zeni, L., Castella, M. A., and Rossi, F., 1986, Interferon-γ is able to enhance the oxidative metabolism human neutrophils, *Biochem. Biophys. Res. Commun.* **138**:1276–1282

Bolscher, B. G. J. M., van Zwieten, R., Kramer, I. M., Weening, R. S., Verhoeven, A. J., and Ross, D., 1989, A phosphoprotein of Mr 47,000 defective in autosomal chronic granulomatous disease, copurifies with one of two soluble components required for NADPH O oxidoreductase activity in human neutrophils, *J. Clin. Invest.* **83**:757–763

Borregaard, N., Heiple, J. M., Simons, E. R., and Clark, R. A., 1983, Subcellular localisation of the b-cytochrome component of the human microbicidal oxidase: Translocation during activation, *J. Cell Biol.* **97**:52–61

Boucek, M. M., and Snyderman, R., 1976, Calcium influx requirement for human neutrophil chemotaxis: Inhibition by lathanum ions, *Science* **193**:905–907

Boxer, L. A., Hedley-Whyte, E. T., and Stossel, T. P. 1974, Neutrophil actin dysfunction and abnormal neutrophil behavior, *New Eng. J. Med.* **291**:1093–1099

Boxer, L. A., Wantanake, A. M., and Rister, M., 1976, Correction of leukocyte function in Chediak-Higashi syndrome by ascorbate, *N. Engl. J. Med.* **295**:1041–1045

Bradford, P. G., and Rubin, R. P., 1985, Characterisation of formylmethionyl-leucyl-phenylalanine stimulation of inositol triphosphate accumulation in rabbit neutrophils, *Mol. Pharmacol.* **27**:74–78

Bradford, P. G., and Rubin, R. P., 1986, Guanine nucleotide regulated phospholipase C activity in permeabilized rabbit neutrophils, *Biochem. J.* **239**:97–102

Bray, D., Heath, J., and Moss, D., 1986, The membrane-associated 'cortex' of animal cells: Its structure and mechanical properties. *J. Cell. Sci. Suppl.* **4**:71–88

Bromberg, Y., and Pick, E., 1983, Unsaturated fatty acids as second messengers of superoxide generation in macrophages. *Cell Immunol.* **79**:240–252

Bryan, J., and Kurth, M. C., 1984, actin-gelsolin interactions: Evidence for two actin-binding sites, *J. Biol. Chem.* **259**:7480–7487

Burgoyne, R. D., and Cheek, T. R., 1987, Cytoskeleton: Role of fodrin in secretion, *Nature* (London) **326**:448

Burn, P., Rotman, A., Meyer, R. K., and Burger, M. M., 1985, Diacylglycerol in large-actinin/actin complexes and in the cytoskeleton of activated platelets, *Nature* (London) **31**:469–472

Burridge, K., and Feramisco, J. R., 1981, Non-muscle α-actinins are calcium-sensitive actin-binding proteins, *Nature* (London) **294**:565–567.

Burridge, K., and Feramisco, J. R., 1982, Alpha-actinin and vinculin for non-muscle cells: calcium sensitive interactions with actin, *Cold Spring Harbor Symp.* **46**:587–597

Campbell, A. K., and Hallett, M. B., 1983, Direct measurement of intracellular free calcium and oxygen radicals in polymorphonuclear leucocyte-erythrocyte "ghost" hybrids, *J. Physiol.* **338**:537–550

Campbell, A. K., Daw, R. A., Hallett, M. B., and Luzio, J. P., 1981, Direct measurement of the increase in intracellular free calcium in concentration in response to the actin of complement. *Biochem. J.* **194**:551–560

Campbell, A. K., Dormer, R. L., and Hallett, M. B. 1985, Coelenterate photoproteins as indicators of cytoplasmic free Ca^{2+} in small cells, *Cell Calcium* **6**:69–82

Caner, J. E., 1965, Colchicine inhibits chemotaxis, *Arthritis Rheum.* **8**:757–760

Castanga, M., Takai, Y., Kaibuchi, K., Sano, K., Kikkawa, U., and Nishizuka, Y., 1982, Direct activation of calcium-activated phospholipid-dependent protein kinase by tumor-promoting phorbol esters, *J. Biol. Chem.* **257**:7847–7857

Cassatella, M., Della Bianca, V., Berton, G., and Rossi, F., 1985, Activation by interferon-γ of human macrophage capability to produce toxic oxygen molecules is accompanied by decreased Km of NADPH oxidase, *Biochem. Biophys. Res. Commun.* **132**:908–914

Chafouleas, J. G., Dedman, J. R., Munjaal, R. P., and Means, A. R., 1979, Calmodulin; development and application of a sensitive radio immunoassay. *J. Biol. Chem.* **254**:10262–10271

Chang, Y. H., 1975, Mechanism of action of colchicine. II. Effects of colchicine and its analogs on phagocytosis and chemotaxis in vitro. *J. Pharmacol. Exp. Ther.* **194**:159–164

Chaponnier, C., Janmey, P. A., and Yin, H. L., 1986, The actin filament-severing domain of plasma gelsolin. *J. Cell Biol.* **103**:1473–1481

Cheek, T. R., and Burgoyne, R. D., 1986, Nicotine-evoked disassembly of cortical actin filaments in adrenal chromaffin cells. *FEBS Lett.* **207**:110–114

Clarke, M., and Spudich, J. A., 1977, Non-muscle contractile proteins: The role of actin and myosin in cell motility and shape determination. *Annu. Rev. Biochem.* **46**:797–822

Cochrane, C. G., 1984, Mechanisms of coupling stimulation and function in leukocytes, *Fed. Proc.* **43**:2729–2731

Cockroft, S., 1986, The dependence on Ca^{2+} of the guanine nucleotide-activated polyphosphoinositide phosphodiesterase in neutrophil plasma membranes, *Biochem. J.* **240**:503–507

Cockroft, S., and Allen, D., 1984, The fatty acid composition of phosphatidylinositol, phosphatidate and 1,2,diacylglycerol in stimulated human neutrophils, *Biochem. J.* **222**:557–561

Cockroft, S., and Gomperts, B. D., 1985, Role of guanine nucleotide binding protein in the activation of polyphosphoinositide phosphodiesterase, *Nature* (London) **314**:534–536

Cockroft, S., Baldwin, J. M., and Allan, D., 1984, The Ca^{2+}-activated polyphosphoinositide phosphodiesterase of human and rabbit neutrophil membranes, *Biochem. J.* **221**:447–481

Cooke, E., and Hallett, M. B., 1985, The role of c-kinase in the physiological activation of the neutrophil oxidase; evidence from pharmacological manipulation of c-kinase activity in intact cells, *Biochem. J.* **232**:323–327

Cooke, E., Al-Mohanna, F. A., and Hallett, M. B., 1987, Diacylglycerol inhibitor, R59022, potentiates neutrophil oxidase activation by Ca^{2+}-dependent stimuli, *Biochem. Pharmacol.* **36**:3459–3463

Cooke, E., Al-Mohanna, F. A., and Hallett, M. B., 1989, Ca^{2+}-dependent and independent mechanisms in neutrophil activation; roles of Kinase C, diacylglycerol and unidentified intracellular messengers, in *The Neutrophil: Cellular Biochemistry and Physiology* (M. B. Hallett ed.), pp. 219–242, CRC Press, Boca Raton, Fla.

Cooper, J. A., 1987, Effects of cytochalasin and phalloidin on actin, *J. Cell Biol.* **105**:1473–1478

Cox, J. A., Jeng, A. J., Sharkey, N. A., Blumberg, J. M., and Tauber, A. I., 1986, Activation of human neutrophil NAPH-oxidase by protein kinase C, *J. Clin. Invest.* **76**:1932–1939

Crawford, D. R., and Schneider, D. L., 1982, Identification of ubiquinone-50 in human neutrophils and its role in microbicidal events, *J. Biol. Chem.* **257**:6662–6668

Cross, A. R., and Jones, O. T. G., 1986, The effect of the inhibitor diphenylene iodonium on the superoxide generating system of neutrophils, *Biochem J.* **237**:111–116

Cross, A. R., Jones, O. T. G., Harper, A. M., and Segal, A. W., 1981, Oxidation-reduction properties of the cytochrome b found in the plasma-membrane fraction of human neutrophils. A possible oxidase in the respiratory burst, *Biochem. J.* **194**:599–606

Cross, A. R., Jones, O. T. G., Garcia, R., and Segal, A. W., 1982, The association of FAD with the cytochrome b-245 of human neutrophils, *Biochem. J.* **208**:759–763

Cross, A. R., Jones, O. T. G., Garcia, R. C., and Segal, A. W., 1983, The subcellular localization of ubiquinone in human neutrophils, *Biochem. J.* **216**:765–768

Cross, A. R., Parkinson, J. F., and Jones, O. T. G., 1984, The superoxide generating oxidase of leukocytes. NADPH-dependent reduction of flavin and cytochrome b in solubilised preparations, *Biochem. J.* **223**:337–344

Cross, A. R., Parkinson, J. F., and Jones, O. T. G., 1985, Mechanism of the superoxide-producing oxidase of neutrophils; oxygen is necessary for fast reduction of cytochrome b-245, *Biochem. J.* **226**:881–884

Dale, M. M., and Penfield, A., 1987, Comparison of the effect of indomethacin, RHC80267 and R59022 on superoxide production by 1,oleoyl-2, acetyl glycerol and A23187 in human neutrophils, *Br. J. Pharmacol.* **92**:63–68

de Chaffoy De Courcelles, D., Roevens, P., and Van Belle, H., 1985, R59022, a diacylglycerol kinase inhibitor; its effect on diacylglycerol and thrombin-induced c-kinase activation in intact platelets, *J. Biol. Chem.* **260**:15762–15770

DeChatelet, L. R., Shirley, P. S., and Johnson, R. B., Jr., 1976, Effect of phorbol myristate acetate on the oxidative metabolism of human polymorphonuclear leukocytes, *Blood* **47**:545–554

Della Bianca, V., Grzeskowiak, M., Castella, M., Zeni, L., and Rossi, F., 1986, PMA potentiates the respiratory burst while it inhibits phosphoinositide hydrolysis and calcium metabolism by Fmlp in human neutrophils, *Biochem. Biophys. Res. Commun.* **135**:556–565

Dickey, B. F., Pyun, H. Y., Williamson, K. C., and Navarro, J., 1987, Identification and purification of a novel G protein from neutrophils, *FEBS. Lett.* **219**:289–292

Dinubile, M. T., and Southwick, F. S., 1985, Effect of macrophage profilin on actin in the presence and absence of acumentin and gelsolin, *J. Biol. Chem.* **260**:7402–7409

DiVirgilio, F., Lew, D. P., and Pozzan, T., 1984, Protein kinase activation of physiological processes in human neutrophil at vanishingly small cytosolic Ca^{2+} levels, *Nature*, (London) **310**:691–693

DiVirgilio, F., Vicentini, L.M., Treves, S., Riz, G., and Pozzan, T., 1985, Inositol phosphate formation in f-met-leu-phe stimulated human neutrophils does not require an increase in the cytosolic free Ca^{2+} concentration, *Biochem. J.* **229**:361

DiVirgilio, F., Meyer, B. C., Greenberg, S., and Silverstein, S. C., 1988, Fc-receptor mediated phagocytosis occur in macrophages at exceedingly low cytosolic Ca^{2+} levels, *J. Cell Biol.* **106**:657–666

Donabedian, H., and Gallin, J. F., 1981, Deactivation of human neutrophil chemotaxis by chemoattractants: Effect on receptors for the chemotactic factor f-met-leu-phe, *J. Immunol.* **127**:839–844

Edelson, P. J., and Fudenburg, H. F., 1973, Effect of vinblastin on chemotactic responsiveness of normal human neutrophils, *Infect. Immun.* **8**:127–129

Elsbach, P. and Farrow, S., 1969, Cellular triglyceride as a source of fatty acid for lecithin synthesis during phagocytosis, *Biochim. Biophys. Acta* **176**:438–442

Elsbach, P., Patriarca, P., Pettis, P., Stossel, T. P., Mason, R. J. and Vaughan, M., 1972, The appearance of lecithin ^{32}P, synthesis from bysolecithin ^{32}P, in phagosomes of polymorphonuclear leukocytes, *J. Clin. Invest,* **51**:1910–1919

Elzinga, M., Maron, B. J., and Adelstein, R. S. 1976, Human heart and platelet actins are products of different genes, *Science* **191**:94–95

English, D., Roloff, J. S., and DuKens, J. N., 1981, Regulation of human polymorphonuclear leukocyte superoxide release by cellular responses to chemotactic peptides, *J. Immunol.* **126**:165–171

Gabig, T. G., and Lefker, B. A., 1984, Catalytic properties of the resolved flavoprotein and cytochrome b components of the NADPH dependent O$_2$-generating oxidase of human neutrophils. *Biochem. Biophys. Res. Commun.* **118**:430–436

Gallin, E. K., and McKinney, L. C., 1989, Ion transport in phagocytes, in *The Neutrophil: Cellular Biochemistry and Physiology* (M. B. Hallett, ed.), pp. 243–260, CRC Press, Boca Raton, Fla.

Gallin, J. I., and Rosenthal, A. S., 1974, The regulatory role of divalent cations in human granulocyte chemotaxis. Evidence for an association between calcium exchanges and microtubule assembly, *J. Cell Biol.* **62**:594–609

Gallin, J. I., Durocher, J. R., and Kaplan, A. P., 1975, Interaction of leukocyte chemotactic factors with the cell surface. I. Chemotactic factor-induced changes in human granulocyte surface change, *Clin. Invest.* **55**:967–974

Garcia, R. C., and Segal, A. W., 1984, Changes in the subcellular distribution of the cytochrome b-245 on stimulated human neutrophils, *Biochem. J.* **219**:233–242

Geiger, B., 1979, A 130 K protein from chicken gizzard: its localization at the termini of microfilament bundles in cultured chicken cells, *Cell* **18**:193–205

Gennaro, R., Florio, C., and Romeo, D., 1985, Activation of protein kinase C in neutrophil cytoplasts. Localization of protein substrates and possible relationship with stimulus-response coupling, *F.E.B.S. Lett* **180**:185–189

Goldman, D. W., Chang, F. H., Gifford, L. A., Goetzl, E. J., and Bourne, H. R., 1985, Pertussis toxin inhibition of chemotactic factor-induced calcium mobilization and function in human polymorphonuclear leukocytes, *J. Exp. Med.* **162**:145–156

Goldstein, I., Hoffstein, S., Gallin, J. I., and Wiessmann, G., 1973, Mechanisms of lysosomal enzyme release from human leukocytes: Microtubule assembly and membrane fusion induced by a component of complement, *Proc. Natl. Acad. Sci. USA* **70**:2916–2920

Gordon, B. J., and Weinberg, J. B., 1982, Receptor-mediated modulation of human monocyte, neutrophil, lymphocyte and platelet function by phorbol diesters, *J. Clin. Invest.* **70**:699–707

Griffin, F. M., 1977, Opsonization, in *Biological Amplification Systems in Immunobiology,* (N. K. Day, and R. A. Good, eds.), pp. 85–114, Plenum Press, New York

Gruler, H., 1984, Physical basis for defining terms used to describe cell movement phenomena, *Blood Cells* **10**:123–131

Gruler, H., 1989, Biophysics of leukocytes; neutrophil chemotaxis, characteristics and mechanisms, in *The Neutrophil: Cellular Biochemistry and Physiology* (M. B. Hallett, ed.), pp. 63–96, CRC Press, Boca Raton, Fla.

Grzeskowiak, M., Della Bianca, V., Cassatella, M. A., and Rossi, F., 1986, Complete dissociation between the activation of phosphoinositide turnover and NADPH oxidase by Fmlp in human neutrophils depleted of Ca^{2+} and primed with sub-threshold doses of PMA, *Biochem. Biophys. Res. Commun.* **135**:785–794

Hallett, M. B., and Campbell, A. K., 1982a, Measurement of changes in cytoplasmic free calcium in fused cell hybrids, *Nature* (London) **295**:155–158

Hallett, M. B., and Campbell, A. K., 1982b, Applications of coelenterater luminescent proteins, In *Clinical and Biochemical Luminescience. Clinical and Biochemical analysis*, Vol. 12, (L. J. Kricka and T. J. N. Carter, eds.), pp. 89–133, Marcel Dekker, Inc., New York

Hallett, M. B., and Campbell, A. K., 1983, Two distinct mechanisms for stimulation of oxygen-radical production by polymorphonuclear leucocytes, *Biochem. J.* **216**:459–465

Hallett, M. B., and Campbell, A. K., 1984a, Is intracellular Ca^{2+} the trigger for oxygen radical production by polymorphonuclear leucocytes?, *Cell Calcium* **5**:1–19

Hallett, M. B., and Campbell, A. K., 1984b, Direct measurement of intracellular free Ca^{2+} in rat peritoneal macrophages; correlation with oxygen radical production. *Immunology* **50**:487–495

Hallett, M. B., Dormer, R. L., and Campbell, A. K., 1990, Cytoplasmic free calcium; measurement and manipulation in small living cells, in *Peptide Hormones: A Practical Approach* (K. Siddle and J. C. Hutton, ed.), (in press), IRL Press, Cambridge.

Hallett, M. B., Edwards, S. W., and Campbell, A. K., 1987, Control of oxygen radical production by polymorphonuclear leukocytes monitored by luminol-dependent chemiluminescence: the roles of intracellular Ca^{2+}, oxygen concentration and redox components. *Cellular Chemiluminescence* Vol I (K. Van-Dyke, ed.). pp. 173–192, CRC Press, Boca Raton, FL.

Harper, A. M., Chaplin, M. F., and Segal, A. W., 1985, Cytochrome b-245 from human neutrophils is a glycoprotein, *Biochem. J.* **227**:783–788

Harish, O. E., Levy, R., Rosenheck, K., and Oplatka, A., 1984, Possible involvement of actin and myosin in Ca^{2+} transport through the plasma membrane of chromaffin cells, *Biochem. Biophys. Res. Commun.* **119**:652–655

Hartwig, J. H., and Stossel, T. P., 1975, Isolation and properties of actin, myosin and a new actin-binding protein in rabbit alveolar macrophages, *J. Biol. Chem.* **250**:5696–5705

Hartwig, J. H., and Stossel, T. P., 1987, Structure of macrophage actin-binding protein in solution and with actin-filaments, *J. Mol. Biol.* **145**:563–581

Hartwig, J. H., Chambers, K. A., and Stossel, T. P., 1989, Association of gelsolin with actin filaments and cell membranes of macrophages and platelets, *J. Cell Biol.* **106**:805–812

Hattori, H., 1961, Studies on the labile, stable NADH oxidase and peroxidase staining reaction in the isolated particles of horse granulocyte, *Nagova J. Med. Sci.* **23**:362–378

Hayakawa, T., Suzuki, K., Suzuki, S., Andrews, P. C., and Babior, B. M., 1986, A possible role for protein phosphorylation in the activation of the respiratory burst in human neutrophils. Evidence from studies with cells from patients with chronic granulomatous disease, *J. Biol. Chem.* **261**:9101–9114

Heath, J. P., 1981, Arcs: Curved microfilament bundles beneath the sorsal surface of the leading lamellae of moving chick embryo fibroblasts, *Cell Biol. Int. Rep.* **5**:975–980

Heilbrunn, L. V., 1956, *The Dynamics of Living Protoplasm*, Academic Press, New York

Heizmann, C. W., and Hauptle, M. T., 1977, The purification, characterization and localization of a parvalbumin-like protein from chicken leg muscle, *Cur. J. Biochem.* **80**:443–457

Helfman, D. M., Appelbaum, B. D., Vogler, N. R., and Kuo, J. F., 1983a, Phospholipid-sensitive calcium-dependent protein kinase and its substrates in human neutrophils, *Biochem. Biophys. Res. Commun.* **111**:847–852

Helfman, D. M., Barnes, K. C., Kinade, J. M., Vogler, W. R., Shoji, M., and Kuo, F. F., 1983b, Phospholipid-sensitive and Ca^{2+}-dependent protein phosphorylating system in various types of leukemic cells from patients and from leukemic cell-line HL60 and K562, *Cancer Res.* **43**:2955–2962

Henderson, L., Chappell, J. B., and Jones, O. T. P., 1987, The superoxide-generating NADPH oxidase is electrogenic and associated with an H^+ channel, *Biochem. J.* **246**:325–329

Heyworth, P., and Segal, A. W., 1986, Further evidence for the involvement of the phosphoprotein in the respiratory burst oxidase of human neutrophils. *Biochem. J.* **239**:723–727

Higson, F. K., and Jones, O. T. G., 1984, The generation of active oxygen species by stimulated rainbow trout leukocytes in whale blood, *Comp. Biochem. Physiol.* **77**:583–587

Hill, H. R., Estensen, R. D., Quie, P. G., Hogan, N. A., and Goldberg, N. D., 1975, Modulation of human neutrophil chemotosis responses by cAMP, *Metabolism* **24**:447–452

Hirayama, T., and Kato, I., 1982, Two types of Ca^{2+}-dependent phosphorylation in rabbit leukocytes, *FEBS Lett.* **146**:209–213

Hoffstein, S., 1980, Intra and extracellular secretion from polymorphonuclear leukocytes, in *Handbook of*

Inflammation 2. The Cell Biology of Inflammation (G. Weissmann ed.), pp. 387–430, Elsevier/North-Holland Biomedical Press, Amsterdam

Hoffstein, S., and Weissmann, G., 1978, Microfilaments and microtubules in calcium ionophore-induced secretion of lysosomal enzymes from human polymorphonuclear leukocytes, *J. Cell Biol.* **78**:769–781

Hoffstein, S., Zurier, R. B., and Weissmann, G., 1974, Mechanisms of lysomal enzyme release from human leucocytes. III. Quantitative morphologic evidence from an effect of cyclic nucleotides and colchicine on degranulation, *Clin. Immunol. Immunopathol.* **3**:201–217

Hoffstein, S., Goldstein, I. M., and Weissmann, G., 1977, Role of microtubule assembly in lysosomal enzyme secretion from human polymorphonuclear leukocytes. A re-evaluation, *J. Cell Biol.* **73**:242–256

Howard, T. H., and Oresajo, C. O., 1985a, A method for quantifying F-actin in chemotactic peptide activated neutrophils: Study of the effect of tBoc peptides, *Cell Motil.* **5**:545–557

Howard, T. H., and Oresajo, C. O., 1985b, The kinetics of chemotactic peptide-induced changes in f-actin content, f-actin distribution, and the shape of neutrophils, *J. Cell Biol.* **101**:1078–1085

Huang, C. K., and DeVanney, J. F, 1986, GTP γ-5 induced solubilization of actin and myosin from rabbit peritoneal neutrophil membrane, *F.E.B.S. Lett.* **202**:41–44

Ignarro, L. J., Lunt, T. F., and George, W. J., 1974, Hormonal control of lysosomal enzyme release from human neutrophils. Effect of autonomic agents on enzyme release, phagocytosis, and cyclic nucleotide levels, *J. Exp. Med.* **139**:1395–1414

Irvine, R. F., and Moor, R. M., 1980, Microinjection of inositol 1,3,4,5 tetrakisphosphate activates sea urchin eggs by a mechanism that depends on external Ca^{2+}, *Biochem. J.* **240**:917–923

Janmey, P. A., and Stossel, T. P., 1987, Modulation of gelsolin function by phosphortidylinositol 4,5-biphosphate, *Nature* (London) **325**:362–364

Janmey, P. A., Iida, K., Yin, H. L., and Stossel, T. P., 1987, Polyphosphoinositide micelles and poly-phosphoinositide-containing vesicles dissociate endogenous gelsolin-actin complexes and promote actin assembly from the fast-growing end of actin filaments blocked by gelsolin, *J. Biol. Chem.* **262**:12228–12236

Jones, H. P., Ghai, G., Petrone, W. F., and McCord, J. M., 1982, Calmodulin-dependent stimulation of the NADPH-oxidase in human neutrophils, *Biochim. Biophys. Acta* **714**:152–159

Kakinuma, K., 1974, Effects of fatty acids on oxidative metabolism of leukocytes, *Biochim. Biophys. Acta* **348**:76–87

Kimura, K., Saberada, K., and Katoh, N., 1985, Inhibition by gossypol of phospholipid-sensitive calcium-dependent protein kinase from pig testis, *Biochim. Biophys. Acta* **839**:276–280

Klebanoff, S. J., and Clark, R. A., 1978, *The Neutrophil: Function and Clinical Disorders*, North-Holland Publishing Co., Amsterdam

Korn, E. D., 1982, Actin polymerization and its regulation by proteins from nonmuscle cells, *Physiol. Rev.* **62**:672–737

Korn, E. D., 1978, Biochemistry of actomyosin-dependent cell motility, *Proc. Natl. Acad. Sci. USA* **75**:588–599

Korn, E. D., Bowers, B., Batzri, S., Simmons, S. R. R., and Victoria, E. J., 1974, Endocytosis and exocytosis: Role of microfilaments and involvement of phospholipids in membrane fusion, *J. Supramol. Struct.* **2**:517–528

Korn, E. D., Carlier, M.-F., and Pantaloni, D., 1987, Actin polymerization and ATP hydrolysis, *Science* **238**:638–644

Kurth, M. C., and Bryan, J., 1984, Platelet activation induces the formation of a stable gelsolin-actin complex from monomeric gelsolin, *J. Biol. Chem.* **259**:7473–7379

Kwiatkowski, D. J., and Yin, H. L., 1987, Molecular biology of gelsolin: A calcium-regulated actin filament severing protein, *Biorheology* **24**:643–647

Kwiatkowski, D. J., Janmey, P. A., Mole, J. E., and Yin, H. L., 1985, Isolation and properties of two actin-binding domains in gelsolin, *J. Biol. Chem.* **260**:15232–15238

Lackie, J. M., 1986, *Cell Movement and Cell Behavior*, Allen & Unwin, London

Lad, P. M., Glovsky, M. M., Smiley, P. A., Klempner, M., Reissinger, D. M., and Richards, J. M., 1984, The β-adrenergic receptor in the human neutrophil plasma membrane; receptor cyclase coupling is associated with amplified GTP activation, *J. Immunol.* **132**:1466–1471

Lad, P. M., Olson, C. V., and Grewal, I. S., 1986, A step sensitive to pertussis toxin and phorbol esters in neutrophils regulates chemotaxis and capping but not phagocytosis, *FEBS Lett.* **200**:91–96

Lassing, I., and Lindberg, U., 1985, Specific interaction between phosphotidylinositol 4,5 biphosphate and profilactin, *Nature* (London) **314:**472–474

Lelkes, P. I., Friedman, J. E., Rosenheck, K., and Oplatka, A., 1986, Destabilization of the actin filaments as a requirement for the secretion of catecholamines from permeabilised chromaffin cells, *FEBS Lett.* **208:**357–363

Levitz, S. M., Lyman, C. A., Murata, T., Sullivan, J. A., Mandell, G. L., and Diamond, R. D., 1987, Cytosolic calcium changes in individual neutrophils stimulated by opsonized and unopsonized *Candida albicans* hyphae, *Infect. Immun.* **55:**2783–2788

Lew, P. D., Wollheim, C. B., Waldvogel, F. A., and Pozzan, T., 1984, Modulation of cytosolic free calcium transcience by changes in intracellular calcium buffering capacity: Correlation with cytosis and O production in human neutrophils, *J. Cell Biol.* **99:**1212–1221

Lew, D. P., Andersson, T., Divirgilio, F., Pozzan, T., and Stendahl, O., 1985, Ca^{2+}-dependent and Ca^{2+}-independent phagocytosis in human neutrophils, *Nature* (London) **315:**509–511

Lew, D. P., Monod, A., Waldvogel, F. A., Dewald, B., Bagglioni, M., and Pozzan, T., 1986, Quantitative analysis of the cytosolic free calcium dependency of exocytosis from three subcellular components in intact human neutrophils, *J. Cell Biol.* **102:**2197–2204

Lew, D. P., Monod, A., Waldvogel, F. A., and Pozzan, T., 1987, Role of cytosolic free calcium and phospholipase C in leukotriene B4 stimulated secretion in human neutrophils; comparison with the chemotactic peptide f-mlp, *Eur. J. Biochem.* **162:**161–168

Light, D. R., Walsh, C., O'Callaghan, A. M. Goetzl, E. J., and Tauber, A. L., 1981, Characteristics of the cofactor requirements for the superoxide-generating NADPH oxidase of human polymorphonuclear leukocytes, *Biochemistry* **20:**1468–1476

Luther, R., Van Zwieten, R., Weening, R. S., Hamers, M. N., and Roos, D., 1984, Cytochrome b, flavins and ubiquinone in enucleated human neutrophils, *J. Biol. Chem.* **259:**9603–9606

Maclean-Fletcher, S., and Pollard, T. D., 1980, Mechanism of action of cytochalasin B on actin, *Cell* **20:**329–341

Majerus, P. W., Connolly, T. M., Deckmyn, H., Ross, T. S., Bross, T. E., Malawista, S. E., 1971, Vinblastin; colchicine-like effects on human blood leukocytes during phagocytosis, *Blood* **37:**519–529

Malawista, S. E., and Bensch, K. G., 1967, Human polymorphonuclear leukocyes; demonstration of microtubules and the effect of colchicine, *Science* **15:**521–522

Malawista, S. E., and Bodel, P. T., 1967, The dissociation by colchicine of phagocytosis from increased oxygen consumption in human leukocytes, *J. Clin. Invest.* **46:**786–796

Malawista, S. E., and Chevance, A. D. B., 1982, The cytokineplast: Purified, stable and functional motile machinery from human blood polymorphonuclear leukocytes. Possible formative role of heat-induced centrasomal dysfunction, *Cell Biol.* **95:**960–973

Maruyama, K. B., 1971, A study of β-actinin, microfibrillar protein from rabbit skeletal muscle, *J. Biochem.* **69:**369–386

Mazzei, G. J., Katoh, N., and Kuo, J. F., 1982, Polymixin B is a more selective inhibitor for phospholipid-sensitive calcium-dependent protein kinase than for calmodulin-dependent kinase, *Biochem. Biophys. Res. Commun.* **109:**1129–1133

McNeil, P. L., Swanson, J. A., Wright, J. D., Silverstein, S. C., and Taylor, D. L., 1986, Fc-mediated phagocytosis occurs in macrophages without an increase in average Ca^{2+}, *J. Cell Biol.* **102:**1586–1592

McPhail, L. C., Clayton, C. C., and Snyderman, R., 1984a, The NADPH oxidase of human polymorphonuclear leukocytes: Evidence for multiple signals, *J. Biol. Chem.* **259:**5768–5775

McPhail, L. C., Clayton, C., and Snyderman, R., 1984b, A potential second messenger role for unsaturated fatty acids: Activation of Ca^{2+}-dependent protein kinase, *Science* **224:**622–625

McRobbie, S. J., and Newell, P. C., 1984, A new model for chemotactic signal transduction in *Dictyotelium discoideum*, *Biochem. Biophys. Res. Comm.* **123:**1076–1083

Melloni, E., Pontremoli, S., Michetti, M., Sacco, O., Sparatore B., Miyake, R., Tanaka, Y., Tsuda, T., Kaibuchi, K., Kikkkawa, U., and Nishisuka, Y., 1984, Activation of protein kinase C by a non-phorbol tumor promotor mezerein, *Biochem. Biophys. Res. Commun.* **121:**648–656

Melloni, E., Pontremoli, S., Michetti, M., Sacco, O., Sparatore, B., and Horecker, B. L., 1986, The involvement of calpain in the activation of protein kinase C in neutrophils stimulated with phorbol myristate acetate, *J. Biol. Chem.* **261:**4101–4105

Molski, T. F. P., Naccache, P. H., Volpi, M., Wolpert, L., and Sha'afi, R. I., 1980, Specific modulation of the intracellular pH of rabbit neutrophils by chemotactic factor, *Biochem. Biophys. Res. Commun.* **94:**508–514

Mookerjee, B. K., Cuppoletti, J., Rampal, A. L., and Jung, C. Y., 1981, The effects of cytochalasins on lymphocytes. Identification of distinct cytochalasin-binding sites in relation to mitogenic response and hexose transport, *J. Biol. Chem.* **256:**1290–1300

Muid, R. E., Penfield, A., and Dale, M. M., 1987, The diacylglycerol kinase inhibitor R59022, enhances the superoxide generation from human neutrophils induced by stimulation of the fmlp, IgG and C3b receptors, *Biochem. Biophys. Res. Commun.* **143:**630–637

Murata, T., Sullivan, J. A., Sawyer, D. W., and Mandell, G. L., 1987, Influence of type and opsonization of ingested particles on intracellular free calcium distribution of superoxide production by human neutrophils, *Infect. Immun.* **55:**1784–1791

Naccache, P. H., Showell, H. J., Becker, E. L., and Sha'afi, R. I., 1977, Transport of sodium, potassium and calcium across rabbit polymorphonuclear leukocyte membranes. Effect of chemotactic factor, *J. Cell Biol.* **73:**428–444

Naccache, P. H., Molski, T. F. P., Becker, E. L., Showell, H. J., and Sha'afi, R. I., 1980, Calmodulin inhibitors block neutrophil degranulation at a step distal from the mobilization of calcium, *Biochem. Biophys. Res. Commun.* **97:**62–68

Naccache, P. H., Molski, T. F. P., and Sha'afi, R. I., 1985, Polymixin B inhibits PMA but not chemotactic peptide induced effects in rabbit neutrophils, *FEBS Lett.* **193:**227–232

Naccache, P. H., Sha'afi, R. I., and Borgeat, P., 1989, Mobilization, metabolism, and biological effects of eicosanoids in polymorphonuclear leukocytes, in *The Neutrophil: Cellular* Biochemistry and Physiology (M. B. Hallett, ed.) pp. 113–140, CRC Press, Boca Raton, Fla.

Nelson, R. D., Fiegel, V. D., Herron, M. J., and Simmons, R. C., 1980, Chemotactic deactivation of human neutrophils: Relationship of loss of cytotoxin receptor function and temporal nature of phenomenon, *J. Reticuloendothel. Soc.* **28:**285–294

Niedel, J. E., Kahane, I., and Cuatrecassas, P., 1979, Receptor-mediated internalisation of fluorescent chemotactic peptide by human neutrophils, *Science* **205:**1412

Niedel, J. E., Kuhn, L. J. T., and Vanderbark, G. R., 1983, Phorbol diester receptor copurifies with protein kinase C, *Proc. Natl. Acad. Sci. USA* **80:**34–40

Nishizuka, Y., 1984, The role of protein kinase C in cell surface signal transduction and tumor promotion, *Nature* (London) **308:**693-698

Norgauer, J., Kownatzki, E., Seifert, R., and Aktories, K., 1988, Botulinum C2 toxin ADP-ribosylates actin and enhances O_2^- production and secretion but inhibits migration of activated human neutrophils, *J. Clin. Invest.* **82:**1376–1382

O'Brian, C. A., Liskamp, R. M., Solomon, D. H., and Weinstein, I. B., 1985, Inhibition of protein kinase C by tamoxifen, *Cancer Res.* **45:**2469–2475

Ohta, H., Takahashi, H., Hattori, H., Yamada, H., and Takikawa, K., 1966, Some oxidative enzymes and cytochrome in the specific granules of neutrophil leukocytes, *Acta Haematol. Jpn.* **29:**799–808

Ohta, Y. Akiyama, T., Nishida, E., and Sakai, H., 1987, Protein kinase C and cyclic AMP-dependent kinase induces opposite effects on cation polymerizability, *FEBS Lett.* **222:**305–310

Okajima, F., and Ui, M., 1984, ADP-ribosylation of the specific membrane by islet-activating factor, pertussis toxin, associated with inhibition of chemotactic peptide-induced arachidonate release in neutrophils, *J. Biol. Chem.* **259:**13863–13871

Oliver, J. M., 1976, Impaired microtubule function correctable by cyclic GMP and cholinergic agonists in Chediak-Higashi syndrome, *Am. J. Pathol.* **85:**395–418

Oliver, J. M., Zurier, R. B., and Berlin, R. D., 1973, Concanavalin A cap formation on polymorphonuclear leukocytes of normal and Beige (Chediak-Higashi) mice, *Nature,* (London) **253:**471–473

Painter, R. G., Sklar, L. A., Tesaitis, A. J., Schmitt, M., and Cochrane, C. G., 1984, Activation of neutrophils by N-formyl chemotactic peptides, *Fed. Proc.* **43:**2737–2742

Papini, E., Grzeskowiak, M., Bellavite, P., and Rossi, F., 1985, Protein kinase C phosphorylates a component of NADPH oxidase of neutrophils, *FEBS Lett.* **190:**204–208

Parker, C. W., Creene, W. C., and MacDonald, H. H., 1976, Cytochalasin binding in lymphocytes and polymorphonuclear leukocytes, *Exp. Cell Res.* **103:**99–108

Parkos, C. A., Dinauer, M. C., Walker, L. E., Allen, R. A., Jesaitis, A. J., and Orkin, S. H., 1988, Primary structure and unique expression of the 22-kilodalton, light chain of human neutrophil cytochrome b, *Proc. Natl. Acad. Sci. USA* **85:**3319–3323

Parysek, L. M., and Eckert, B. S., 1984, Vimentin filaments in spreading, randomly locomoting, and f-met-leu-phe-treated neutrophils, *Cell Tissue Res.* **235:**575–581

Patriarca, P., Cramer, R., Dri, P., Sorcenzo, M. R., and Rossi, F., 1973, Biochemical studies on the effect of papaverine on polymorphonuclear leukocytes, *Biochem. Pharmacol.* **22:**3257–3266

Pevin, D., Langley, O.K., and Amis, D. 1987, Anti-α-fodrin inhibits secretion from permeabilized chromaffin cells, *Nature* **376:**498–507

Pittet, D., Krause, K. H., Wollheim, C. B., Bruzzone, R. and Lew, D. P., 1987, Non-selective inhibition of neutrophil functioning by sphinganine, *J. Biol. Chem.* **262:**10072–10076

Pollard, T. D., and Cooper, J. A., 1986, Actin and actin binding proteins. A critical evaluation of mechanisms and functions, *Annu. Rev. Biochem.* **55:**987–1035

Pollard, T. D., and Weihing, R. R., 1974, Actin and myosin and cell movement, *Crit. Rev. Biochem.* **2:**1–65

Pontremoli, S., Melloni, E., Salamino, F., Sparatore, B., Michetti, M., Sacco, O., and Horecker, B. L., 1986, Phosphorylation of proteins in human neutrophils activated with phorbol myristate acetate or with chemotactic factor, *Arch. Biochem. Biophys.* **250:**23–29

Pontremoli, S., Melloni, E., Michetti, M., Salamino, F., Sparatore, B., and Horecker, B. L., 1988, An endogenous activator of Ca^{2+}-dependent proteinase of human neutrophils that increases its affinity for Ca^{2+}, *Proc. Natl. Acad. Sci. USA* **85:**1749–1743

Porter, K. R., 1984, The cytomatrix: A short history of its study, *J. Cell Biol.* **99:**35–125

Pozzan, T., Lew, D. P., Wollheim, C. B., and Tsien, R. Y., 1983, Is cytosolic calcium the trigger for neutrophil activation?, *Science* **221:**1413–1415

Prentki, M., Wollheim, C. B., and Lew, D. P., 1984, Ca^{2+} homeostasis in permeabilized human neutrophils: Characterization of Ca^{2+} sequestering pools and the action of inositol 1,4,5 triphosphate, *J. Biol. Chem.* **259:**13777–13782

Pryzwansky, K. B., 1987, Human leukocytes as viewed by stereo high-voltage electronmicroscopy, *Blood Cells* **12:**505–530

Pryzwansky, K. B., Steiner, A. L., Spitznagel, J. K., and Kapoor, C. L., 1981, Compartmentation of cyclic AMP during phagocytosis by human neutrophilic granulocytes, *Science* **211:**407

Pryzwanksky, K. B., Schliwa, M., and Porter, K. R., 1983, Comparison of the three-dimensional organisation of unextracted and triton-extracted human neutrophilic polymorphonuclear leukocytes, *Eur. J. Cell Biol.* **30:**112–125

Ramsey, W. S., and Harris, A., 1973, Leucocyte locomotion and its inhibition by antimitic drugs, *Exp. Cell Res.* **82:**262–270

Rickard, J. E., and Sheterline, P. E., 1985, Evidence that phorbol ester interfers with stimulated Ca^{2+} redistribution by activating Ca^{2+} efflux in neutrophil leucocytes, *Biochem J.* **231:**632–628

Rivkin, I., Rosenblatt, J., and Becker, E. L., 1975, The role of cAMP in the chemotactic responsiveness and spontaneous motility of rabbit peritoneal neutrophils. The inhibition of neutrophil movement and the elevation of cAMP levels by catecholamines, prostaglandins, theophylline and cholera toxin, *J. Immunol.* **115:**1114–1134

Romeo, D., Zabucchi, G., Miani, N., and Rossi, F., 1975, Ion movement across leukocyte plasma membrane and excitation of their metabolism, *Nature* (London) **253:**542–544

Rossi, F., 1986, The O_2^--forming NADPH oxidase of the phagocytes: Mechanisms of activation and function, *Biochim. Biophys. Acta* **53:**65–89

Rossi, F., Grzeskowiak, M., and Della Bianca, V., 1986, Double stimulation with Fmlp and con A restores the activation of the respiratory burst but not of the phosphoinositide turnover in Ca^{2+}-depleted human neutrophils: A further example of dissociation between stimulation of the NADPH oxidase and phosphoinositide turnover, *Biochem. Biophys. Res. Commun.* **140:**1–11

Royer-Pokora, B., Kunkel, L. M., and Monaco, A. P., 1986, Cloning the gene for an inherited human disorder—chronic granulomatous disease—on the basis of chromosomal location, *Nature* (London) **322:**32–38

Rydgren, L., Simmingskold, G., Bandmann, U., and Narberg, B., 1976, The role of cytoplasmic microtubules in polymorphonuclear leucocyte chemotaxis. Evidence for the release hypothesis by means of time-lapse analysis of PMN movement relative to dot-like attractants, *Exp. Cell Res.* **99:**207–220

Salamino, F., and Horecker, B. L., 1985, Binding of protein kinase C to neutrophil membranes in the presence of Ca and its activation by a Ca^{2+}-requiring proteinase, *Proc. Natl. Acad. Sci. USA* **82:**6435–6439

Sasada, M., Dabst, M. J., and Johnston, R. B., 1983, Activation of mouse peritoneal macrophages by lipopolysaccharide alters the Kinches parameters of the superoxide producing NADPH-oxidase, *J. Biol. Chem.* **258:**9631–9635

Sawyer, D. W., Sullivan, J. A., and Mandell, G. L., 1985, Intracellular free calcium localised in neutrophils during phagocytosis, *Science* **230:**663–666

Schell-Frederick, E., 1974, Stimulation of the oxidative metabolism of polymorphonuclear leukocytes by the calcium ionophore A23187, *FEBS Lett.* **48:**37–40

Schell-Frederick, E., 1984, A comparison of the effects of soluble stimuli on free cytoplasmic and membrane bound calcium in human neutrophils, *Cell Calcium* **5:**237–251

Schiffmann, E., Corcoran, B. A., and Wahl, S. M., 1975, N-formyl-methionyl peptides as chemoattractants for leukocytes. *Proc. Natl. Acad. Sci. USA* **782:**1059–1064

Scholey, J. M., Taylor, K. A., and Kendrick-Jones, J., 1980, Regulation of non-muscle myosin assembly by calmodulin-dependent light chain kinase, *Nature* (London) **287:**233–235

Schook, W., Ones, C., and Puzzkin, S., 1978, Isolation and properties of α-Actinin, *Biochem. J.* **175:**63–72

Segal, A. W., 1988, The molecular and cellular pathology of chronic granulomatous disease, *Eur. J. Clin. Invest.* **18:**433–443

Segal, A. W., and Jones, O. T. G., 1978, Novel cytochrome b system in phagocytic vacuoles from human granulocytes, *Nature* (London) **276:**515–517

Segal, A. W., and Jones, O. T. G., 1979, The subcellular distribution and some properties of the cytochrome b component of the microbicidal oxidase system of human neutrophils, *Biochem J.* **182:**181–188

Segal, A. W., Heyworth, P. G., Cockroft, S., and Barrowman, M. M., 1985, Stimulated neutrophils from patients with autosomal recessive chronic granulomatous disease fail to phosphorylate a Mr 44,000 protein, *Nature* (London) **316:**547–549

Seligmann, B. E., Fletcher, M. P., and Gallin, J. P., 1982, Adaption of human neutrophil responsiveness to the chemoattractant N-formyl-methionyl-leucyl-phenylalanine, *J. Biol. Chem.* **257:**6280–6286

Sha'afi, R. I., and Molski, T. F. P., 1987, Signalling for increased cytoskeletal actin in neutrophils, *Biochem. Biophys. Res. Commun.* **145:**934–941

Sha'afi, R. I., White, J. R., Molski, T. F. P., Schefcyk, J., Volpi, M., Naccache, P. H., and Feinstein, M. B., 1983, Phorbol-12-myristate 13-acetate activates rabbit neutrophils without an apparent rise in the level of intracellular free calcium, *Biophys. Biochem. Res. Commun.* **114:**638–645

Sha'afi, R. I., Molski, T. F. P., Huang, C.-K., and Naccache, P. H., 1986a, The inhibition of neutrophil responsiveness caused by phorbol esters is blocked by the protein kinase inhibitor H7, *Biochem. Biophys. Res. Commun.* **137:**50–60

Sha'afi, R. I., Shetcyk, J., Yassin, R., Molski, T. F. P., Volpi, M., Naccache, P. H., White, J. R., Feinstein, M. B., and Becker, E. C., 1986b, Is a rise in intracellular concentration of free calcium necessary or sufficient for stimulated cytoskeletal-associated actin?, *J. Cell Biol.* **102:**1459–1463

Sherline, P., and Mundy, G. R., 1977, Role of the tubulin microtubule system in lymphocye activation, *J. Cell Biol.* **74:**371–376

Sheterline, P., 1983, *Mechanisms of Cell Motility, Molecular Aspects of contractility*, Academic Press, New York

Sheterline, P., and Rickard, J. E., 1989, The cortical actin network of neutrophilic leukocytes during phagocytosis and chemotaxis, in *The Neutrophil: Cellular Biochemistry and Physiology* (M.B. Hallett, ed.), pp. 141–166, CRC Press, Boca Raton, Fla.

Sheterline, P., Rickard, J. E., Boothroyd, B., and Richards, R. C., 1986, Phorbol ester induces rapid actin assembly in neutrophil leucocytes independently of change in Ca²⁺ and pH, *J. Muscle Res. Cell Motil.* **7:**405–416

Sheterline, P., Rickard, J. E., and Richards, R. C., 1984a, Fc receptor-directed phagocytic stimuli induce transient actin assembly at an early stage of phagocytosis in neutrophil leukocytes, *Eur. J. Cell Biol.* **34:**80–87

Sheterline, P., Rickard, J. E., and Richards, R. C., 1984b, Involvement of cortical actin network of neutrophilic leukocytes during phagocytosis, *Biochem. Soc. Trans.* **12:**983–987

Shinagawa, Y., Shinagawa, Y., Tanaka, C., and Teraoka, A., 1966, Electron microscopic and biochemical study of the neutrophilic granules from leukocytes, *J. Electron Microsc.* **15:**81–85

Smith, C. D., Lane, B. C., Kusaka, I., Verghese, M. W., and Snyderman, R., 1985, Chemoattractant receptor-induced hydrolysis of phosphoinositol 4,5-biphosphate in human polymorphonuclear leukocyte membranes, *J. Biol. Chem.* **260:**5875–5878

Smith, C. D., Cox, C. C., and Snyderman, R., 1986, Receptor-coupled activation of phosphoinositide-specific phospholipase C by N protein, *Science*, **232**:97–100

Smolen, J. E., and Stoehr, S. J., 1986, Guanine nucleotides reduce the free calcium requirement for granule of granule constituents from permeabilized human neutrophils, *Biochem. Biophys. Acta* **889**:171–178

Snyderman, R., 1983, Chemoattractant receptor affinity reflects its ability to transduce different biological responses, in *Leucocyte Locomotion and Chemotaxis* (H. U. Keller, and G. O. Till, eds.), pp. 323–336, Birkhauser, Basel.

Southwick, F. S., and Hartwig, J. H., 1982, Acumentin, a protein in macrophages which caps the 'pointed' end of actin filaments, *Nature* **297**:303–307

Southwick, F. S., and Stossel, T. P., 1981, Isolation of an inhibitor of actin polymerization from human polymorphonuclear leukocytes, *J. Biol. Chem.* **256**:3030–3036

Southwick, F. S., and Stossel, T. P., 1983, Contractile proteins in leukocye function, *Semin. Hematol.* **20**:305–321

Southwick, F. S., 1985, Regulation of phagocyte movement, in *Inflammation, Basic Mechanisms, Tissue Injuring Principles and Clinical Models* (P. Venge and A. Lindbom, eds.), pp. 139–160, Almqvist and Wiksell International, Stockholm.

Stendahl, O. I., Hartwig, J. H., Brotschi, E. A., and Stossel, T. P., 1980, Distribution of actin-binding protein and myosin in macrophages during spreading and phagocytosis, *J. Cell Biol.* **84**:215–224

Stossel, T. P., Root, R. R., and Vaughn, M., 1972, Phagocytosis in chronic granulomatous disease and Chediak-Higashi syndrome, *N. Engl. J. Med.* **286**:120–123

Su, H.-D., Mazzei, G. J., Vogler, W. R., and Kuo, J. F., 1985, Effect on tamoxifen, a nonsteroidal antiestrogen, on phospholipid/calcium dependent protein kinase and phosphorylation of its endogenous substrate protein from the rat brain and ovary, *Biochem. Pharmacol.* **34**:3649–3653

Sullivan, S. J., and Zigmond, S. H., 1980, Chemotactic peptide receptor modulation in polymorphonuclear leukocytes, *J. Cell Biol.* **85**:703–711

Taffet, S., Greenfield, A. R. L., and Haddox, M. K., 1983, Retinal inhibits TPA activated calcium-activated phospholipid-dependent protein kinase ("C kinase"), *Biochem. Biophys. Res. Commun.* **114**:1194–1199

Taui, J.-S., and Stjernholm, R., 1975, Cytochalasin B: Effect on phospholipid metabolism and lysosomal enzyme release by leukocytes, *Biochim. Biophys. Acta* **392**:1-12

Tauber, A. I., Cox, J. A., Curnutte, J. T., Carrol, P. M., Nakakuma, H., Warren, B., Gilbert, H., and Blumberg, P. M., 1989, Activation of human neutrophil NADPH-oxidase in vitro by the catalytic fragment of protein kinase-C, *Biochem. Biophys. Res. Commun.* **158**:884–890

Treves, S. D., DiVirgilio, F., Vasselli, G. M., and Pozzan, T., 1987, Effects of cytochalasins on cytosolic free Ca^{2+} concentrations and phosphoinositide metabolism in leukocytes, *Exp. Cell Res.* **168**:285–298

Valerius, N. H., Stendahl, O., Hartwig, J. H., and Stossel, T. P., 1987, Distribution of actin binding protein and myosin in polymorphonuclear leukocytes during locomotion and phagocytosis, *J. Cell, Biol.* **68**:195–202

Varani, J., Wass, J. A., and Rao, M. K., 1983, Actin changes in normal human and rat leukocytes and in transformed leukocytes, *J. Natl. Cancer Inst.* **70**:805–809

Verghese, M. W., Smith, C. D., and Snyderman, R., 1985, Potential role for a guanine nucleotide regulatory protein in chemoattractant receptor mediated polyphosphoinositide metabolism, Ca^{2+} mobilization and cellular responses by leukocytes, *Biochem. Biophys. Res. Commun.* **127**:450–457

Volpe, P., Krause, K. H., Hashimoto, S., Pozzan, T., Meldolesi, J., and Lew, D. P., 1988, "Calcisome," a cytoplasmic organelle; the inositol 1,4,5 trisphosphate-sensitive Ca^{2+} store of non-muscle cells?, *Proc. Natl. Acad. Sci. USA* **85**:1091–1095

Wegner, A., and Aktories, K., 1988, ADP-ribosylated actin caps the barbed ends of actin filaments, *J. Biol. Chem.* **263**:13739–13742

Weissmann, G., Zurier, R. B., and Hoffstein, S., 1973, Leukocytes as secretory organs of inflammation, *Agents Actions* **3**:370–379

Weissmann, G., Goldstein, I., Hoffstein, S., Chasavet, G., and Robineaux, R., 1975, Yin/Yang modulation of lysosomal enzyme release from polymorphonuclear leukocytes by cyclic nucleotides, *Ann. N.Y. Acad. Sci.* **256**:222–232

Weissmann, G., Zurier, R. B., Spieler, P. J., and Goldstein, I. M., 1971, Mechanisms of lysosomal enzyme release from leukocytes exposed to immune complexes and other particles, *J. Exp. Med.* **134**:149s–165s

Wessells, N. K., Spooner, B. S., Ash, J. F., Bradley, M. O., Luduena, M. A., Taylor, E. L., Wrenn, J. T., and Yamada, K. M., 1971, Microfilaments in cellular and developmental processes, *Science* **171**:135–143

White, J. R., Naccache, P. H., Molski, T. F. P., Borgeat, P., and Sha'afi, R. I., 1983a, Direct demonstration of increased intracellular concentration of free calcium in rabbit and human neutrophils following stimulation by chemotactic factors, *Biophys. Res. Commun.* **113**:44–50

White, J. R., Naccache, P. H., and Sha'afi, R. I., 1983b, Stimulation by chemotactic factor of actin association with the cytoskeleton in rabbit neutrophils, *J. Biol. Chem.* **258**:14041–14047

White, J. R., Huang, C. K., Hill, J. M., Naccache, P. H., Becker, E. L., and Sha'afi, R. I., 1984, Effect of phorbol 12-myristate 13-acetate and its analogue 4-phorbol 12,13-didecanoate on protein phosphorylation and lysosomal enzyme release in rabbit neutrophils, *J. Biol. Chem.* **259**:8605–8611

Williams, M. B., and Jones, H. P., 1985, Calmodulin-dependent NAD-kinase of human neutrophils, *Arch. Biochem. Biophys.* **237**:80–87

Wilson, E., Olcott, M. C., Bell, R. M., Merrill, A. H., and Lambeth, J. D., 1986, Inhibition of oxidative burst in human neutrophils by shingoid long-chain bases; role of protein kinase C in activation of the burst, *J. Biol. Chem.* **261**:12616–12623

Woodin, A. M., and Wienke, A. A., 1963, The accumulation of calcium by polymorphonuclear leucocytes treated with staphylococcal leucocidin, *Biochem. J.* **87**:487–495

Wright, C. D., and Hoffman, M. D., 1986, The protein kinase inhibitors H7 and H9 fail to inhibit human neutrophil activation, *Biochem. Biophys. Res. Commun.* **135**:749–755

Yassin, R., Shefeyk, J., White, J. R., Tao, W., Volpi, M., Molski, T. F. P., Naccache, P. H., Feinstein, M. B., and Sha'afi, R. I., 1985, Effects of chemotactic factors and other agents on the amount of actin and a 65,000 MW protein associated with the cytoskeleton of rabbit and human neutrophils, *J. Cell Biol.* **101**:182–188

Yin, H. L., 1987, Gelsolin: Calcium and polyphosphoinositide-regulated actin-modulating protein, *BioEssays* **7**:176–179

Yin, H. L., and Stossel, T. P., 1979, Control of cytoplasmic actin gel-sol transformation by gelsolin, a calcium-dependent regulatory protein, *Nature* (London) **281**:583–385

Yin, L. Y., and Stossel, T. P., 1980, Purification and structural properties of gelsolin, a Ca^{2+}-activated regulatory protein of macrophages, *J. Biol. Chem.* **255**:9490–9493

Yin, H. L., and Stossel, T. P., 1982, The mechanism of phagocytosis, in *Phagcytosis—Past and Future* (M. L. Karnovsky and L. Bolis, eds.), pp 13–28. Academic Press, New York

Yin, H. L., Iida, K., and Janmey, P. A., 1988, Identification of polyphosphoinositide-modulated domain in gelsolin which binds to the sides of actin filaments, *J. Cell Biol.* **106**:805–812

Young, T. D. E., Ko, S. S., and Cohn, Z. A., 1984, The increase in intracellular free calcium associated with IgG/Fc receptor-ligand interactions: Role in phagocytosis, *Proc. Natl. Acad. Sci. USA* **81**:5430–5434

Yuli, I., and Oplatka, A., 1987, Cytosolic acidification as an early transductory signal of human neutrophil chemotaxis, *Science* **235**:240–342

Zigmond, S. H., 1977, Ability of polymorphonuclear leukocytes to orient in gradients of chemotactic factors, *J. Cell Biol.* **75**:606–616

Zigmond, S. H., and Sullivan, S. J., 1981, Receptor modulation and its consequences for the response to chemotactic peptides, in *Biology of the Chemotactic Response* (J. M. Lackie and P. C. Wilkinson, eds.), pp. 73–87, Cambridge University Press, Cambridge

Zurier, R. B., Hoffstein, S., and Weissmann, G., 1973a, Mechanisms of lysomal enzyme release from human leukocytes: Effect of cyclic nucleotides and colchicine, *J. Cell Biol.* **58**:27–41

Zurier, R. B., Hoffstein, S., and Weissman, G., 1973b, Cytochalasin B: effect on lysomal enzyme release from human leukocytes, *Proc. Natl. Acad. Sci. USA* **70**:844–848

Zurier, R. B., Weissman, G., Hoffstein, S., Kammerman, S., and Tai, H. H., 1974, Mechanisms of lysosomal enzyme release from human leukocytes. II. Effects of cAMP and CGMP autonomic agonists and agents which affect microtubule function, *J. Clin. Invest.* **53**:297–309

Chapter 12

Neutrophil and Eosinophil Granules as Stores of "Defense" Proteins

Renato Gennaro, Domenico Romeo, Barbara Skerlavaj, and Margherita Zanetti

1. INTRODUCTION

The accumulation of phagocytic cells at the site of infection represents one of the most important defense mechanisms for protecting the host from noxious microorganisms, viruses, and parasites. The defense cells, also known as professional phagocytes, include neutrophils, eosinophils, monocytes, and a variety of macrophage phenotypes.

The efficiency of the defense mechanisms is ensured by the engulfment of noxious agents within the microenvironment of the phagocytic vacuoles, coupled with the concomitant discharge of a variety of toxic factors into the vacuoles. These factors can be divided into two systems. One microbicidal system utilizes toxic O_2 derivatives, such as superoxide anion, hydrogen peroxide, and hydroxyl radical, whose production is triggered upon stimulation of the so-called respiratory burst, as well as myeloperoxidase (for a review, see Klebanoff, 1988; see the chapter by Smith *et al.* and the chapter by Tobler and Koeffler, this volume). The other system relies on an arsenal of preformed cytotoxic polypeptides, stored inside the cytoplasmic granules of the phagocytic cells. These poly-

The abbreviations used are: PMA, phorbol myristate acetate; MBP, major basic protein; ECP, eosinophil cationic protein; EDN, eosinophil derived neurotoxin; EPO, eosinophil peroxidase; BPI, bactericidal/permeability increasing protein; IM, inner membrane; OM, outer membrane; PMSF, phenylmethylsulphonyl fluoride; LPS, lipopolysaccharide; LBP, lipopolysaccharide binding protein; KDO, 2-keto-3-deoxyoctanoate; CAP, cationic antimicrobial protein; CHS, Chédiak-Higashi syndrome; SGD, specific granule deficiency; CGD, chronic granulomatous disease.

Renato Gennaro   Institute of Biology, University of Udine, 33100 Udine, Italy.   **Domenico Romeo, Barbara Skerlavaj, and Margherita Zanetti**   Department of Biochemistry, Biophysics and Macromolecular Chemistry, University of Trieste, 34127 Trieste, Italy.

peptides, which have been shown to exert, at least *in vitro,* a cytotoxic activity toward various invading organisms, are the subject of this chapter.

Although a historical overview of the subject is beyond the purpose of our review, it is worth noting that the first studies on the oxygen-independent bactericidal activity of phagocytes were carried out by Hirsch (1956) with a crude protein fraction, named phagocytin, prepared from acid-extracted rabbit polymorphonuclear leukocytes. The active components of this fraction, toxic against both gram-positive and gram-negative microorganisms, were subsequently shown to be associated with the cytoplasmic granules of the neutrophils (Cohn and Hirsch, 1960a,b).

All of the antibacterial polypeptides thus far isolated from phagocytes are localized inside cytoplasmic granules. The first part of this chapter will thus be concerned with the biogenesis, content, and species specificity of granule types present in these cells and with the conditions leading to the discharge of their content. A description of the features of the various polypeptides exerting antibacterial activity will follow. Where known, their intracellular localization, biosynthesis and processing, physicochemical properties, activity spectrum, mechanism of action, and potential role *in vivo* will also be discussed.

2. THE VACUOLAR APPARATUS OF GRANULOCYTES

The vacuolar apparatus of professional phagocytes has been studied in great detail in neutrophils and eosinophils. The granules present in the cytoplasm assemble at various stages of cell maturation in the bone marrow during a relatively narrow lapse of time, and the storage proteins, mainly hydrolytic enzymes and antibacterial factors (Table I), are thus synthesized and packaged according to a precise temporal scheme. Once released from the bone marrow, the mature granulocytes exhibit a much lower biosynthetic capacity and have a short life span. If engaged in phagocytic events, they use up all of the weapons stored in the granules, without replenishing them. They thus differ from mononuclear phagocytes, which during their long life span are able to continuously synthesize proteins, including those stored into the granules, and can thus perform repeated pathogen-neutralizing phagocytic events (Schnyder and Baggiolini, 1978; McCarthy *et al.,* 1982).

2.1. Neutrophils

Several investigations on the biogenesis, content, and function of cytoplasmic granules have been conducted, initially in rabbit neutrophils (Bainton and Farquhar, 1966; Baggiolini *et al.,* 1969) and then in humans and other species (Bainton *et al.,* 1971; Bretz and Baggiolini, 1974; Spitznagel *et al.,* 1974; Gennaro *et al.,* 1983b).

The pioneering studies of Cohn and Hirsch (1960a) allowed the isolation by differential centrifugation of a granule-rich fraction sharing many properties with lysosomes. A few years later, ultrastructural studies showed that rabbit neutrophils contain two main populations of organelles, named azurophil or primary granules and specific or secondary granules, whose biogenesis is sequential and can be traced to the promyelocyte and the myelocyte stages of maturation, respectively (Bainton and Farquhar, 1966). This discovery prompted further investigations aimed at resolution of the two populations. This was

Table I
Constituents of Neutrophil Granules

Protein	Constituent		
	Azurophil granules	Specific granules	Other granules
Antimicrobial proteins	BPI Defensins Cathepsin G Myeloperoxidase Lysozyme	Lysozyme	Bactenecins[a]
Neutral proteinases	Elastase Cathepsin G Proteinase 3	Collagenase (type 1)	Gelatinase[b] Heparanase[b]
Acid hydrolases	β-Glycerophosphatase β-Glucuronidase N-Acetyl-β-glucosaminidase α-Mannosidase Arylsulfatase β-Galactosidase α-Fucosidase Cathepsins B and D		

Other		Lactoferrin
		Cytochrome b
		Vitamin B_{12}-binding protein
		Receptors for:
		  fMet-Leu-Phe
		  C3bi
		  Laminin
			Lactoferrin[a]

[a]Present in the large granules of the bovine neutrophils.
[b]Present in a type of granule identified in human neutrophils distinct from azurophils and specific granules (see text).

achieved by subjecting a postnuclear supernatant of homogenized cells to zonal differential sedimentation and to isopycnic equilibration (Baggiolini *et al.*, 1969; Bretz and Baggiolini, 1974; Spitznagel *et al.*, 1974). These techniques led to a fairly complete resolution of azurophil and specific granules, thereby permitting the study of their contents. Table I summarizes the results obtained in a number of investigations on constituents of neutrophil granules.

Since the first fractionation experiments, evidence suggesting the presence of granular populations other than the two major ones has accumulated. Furthermore, the use of high-resolution gradient media, such as colloidal silica (Percoll), has shown that the azurophil and specific granules can be resolved into at least five and eight subpopulations, respectively. These are heterogeneous with respect to both size and density, and possibly content, suggesting functional differences (Rice *et al.*, 1986).

A novel subcellular secretory compartment, resolved from azurophil and specific granules, has been identified in human neutrophils by differential sedimentation (Dewald *et al.*, 1982). Although still morphologically unidentified, this organelle is the store of gelatinase, a metalloproteinase specifically acting on denatured collagen, and very likely of heparanase, which catalyzes the degradation of glycosaminoglycans of the extracellular matrix (Matzner *et al.*, 1985). The high responsiveness of this secretory compartment to stimuli suggests that its content may play a role in the early events of neutrophil mobilization, such as diapedesis.

A striking difference in the vacuolar apparatus has been found to occur in the neutrophils of a number of ruminants and has been studied particularly in the cow (Gennaro *et al.*, 1983b). The neutrophils of this family contain three major distinct populations of cytoplasmic granules. Two of them correspond to the well-known azurophil and specific granules described above; the third one, which is prominent relative to the others, is characteristic of ruminants and has not been found in the neutrophils of the other animal species thus far studied. These granules are denser, larger, and more numerous than azurophil and specific granules. Furthermore, they lack the typical enzymatic activities stored into the azurophils (myeloperoxidase, neutral proteases, and various types of acid hydrolases), share lactoferrin, but not the vitamin B_{12}-binding protein, with the specific granules, and are the store of a group of highly cationic proteins capable of exerting a potent oxygen-independent antibacterial activity (bactenecins; see Section 3.5). Finally, they are truly secretory granules, since their contents can be released upon neutrophil stimulation with either opsonized phagocytosable particles or soluble agonists (Gennaro *et al.*, 1983b).

The study of bovine neutrophil maturation in bone marrow specimens has led to the conclusion that this novel granular population is formed at an intermediate stage with respect to the early-appearing azurophils and the late-appearing specific granules (Baggiolini *et al.*, 1985).

As pointed out above, the contents of the granules of the vacuolar apparatus of granulocytes are secretory, and their main destination is the phagocytic vacuole and/or the extracellular space (for a review, see Henson *et al.*, 1988).

Considering that these organelles are heterogeneous in content, it seems obvious that they may be functionally distinct both with respect to their release from and their utilization by the cell. The secondary granules and the gelatinase-containing granules appear to fuse with the phagosome earlier and also to release extracellularly their contents more easily than the azurophil granules (Henson, 1971; Bainton, 1973; Wright *et al.*, 1977; Dewald *et al.*, 1982). The ease of mobilization of these two compartments thus suggests that some of their constituents may be required for diapedesis and maximal chemotactic response in order for the neutrophils to migrate from the blood vessels to the infection site, and possibly for other early events in the defense reactions. This view is supported by the finding that these compartments contain hydrolytic enzymes, such as gelatinase (Dewald *et al.*, 1982), heparanase (Matzner *et al.*, 1985), and collagenase (Murphy *et al.*, 1977), which may help digest the extracellular matrix during cell migration toward the site of inflammation. In addition, they include receptors for laminin (Yoon *et al.*, 1987), for the chemoattractant fMet-Leu-Phe (Jesaitis *et al.*, 1982; Fletcher and Gallin, 1983), and for the opsonin C3bi (O'Shea *et al.*, 1985), which may play a role in cell adherence, chemotaxis, and phagocytosis. Moreover, the specific granules contain cytochrome *b* (Segal and Jones, 1979; Borregaard *et al.*, 1983), a component of the NADPH oxidase, responsible for the respiratory burst.

Conversely, the function of the primary granules appears to be mainly related to microbial killing and some late events in the defense reactions, such as digestive processes. This is suggested by the presence of many potentially toxic agents, such as myeloperoxidase and bactericidal proteins, and by the great number of acid hydrolases stored inside this compartment (Table I; for a review, see Henson *et al.*, 1988).

The release of the contents of neutrophil granules can be induced *in vitro* by a variety of stimuli, either physiological or not. Some of these stimuli, such as phagocytosable opsonized particles and immune complexes, induce exocytosis of primary, secondary, and gelatinase-containing granules, whereas others, such as chemoattractants (C5a, fMet-Leu-Phe), calcium ionophores (A23187, ionomycin), lectins (concanavalin A), and phorbol esters [phorbol myristate acetate (PMA)], promote, primarily or exclusively, secretion of the latter two granule populations (for references, see Henson *et al.*, 1988). Furthermore, as outlined above, under almost all experimental conditions thus far tested, mobilization of the contents of these two compartments is more pronounced and occurs more rapidly than that of the azurophils. Observations on the selective secretion of specific granules have also been made *in vivo* during inflammation (Wright and Gallin, 1979).

A selective release of the contents of the different granular populations is also supported by the quantitative analysis of exocytosis as a function of the cytosolic free calcium concentration. In fact, investigations have shown that the gelatinase-storing vesicles require the lowest, and the azurophil granules the highest, increase in cytosolic Ca^{2+} to liberate their contents in the extracellular medium (Lew *et al.*, 1986).

2.2. Eosinophils

In a normal subject, eosinophils usually account for less than 2% of total blood leukocytes. Investigations on this cell type have thus been more difficult than those on neutrophils and have usually been performed with cells obtained from human donors with moderate to extreme eosinophilia. The recent introduction of Percoll gradients in cell separation technology has made possible the purification of over 90% pure eosinophils from the blood of various animal species (Gartner, 1980; Mottola *et al.*, 1980; Roberts and Gallin, 1985), thus leading to a great advance in our knowledge of biochemical composition and function of this cell type.

These studies have shown that the eosinophil is a potent proinflammatory cell that plays a crucial role in the body's defense mechanisms, particularly against helminths in parasitic infections, and is associated with many hypersensitivity diseases, very likely being involved in host tissue damage (for a review, see Gleich and Adolphson, 1986) (see also Chapter 9 in Volume 2 of this series).

In analogy with neutrophils, eosinophils rely on the capacity of reducing oxygen to radicals and of releasing cytotoxic proteins stored inside their cytoplasmic granules for exerting toxic effects on parasites.

The eosinophil granules, like those of neutrophils, are assembled during cell differentiation in the bone marrow (Bainton and Farquhar, 1970; Zucker-Franklin, 1980). At the promyelocyte stage, a population of small and uniformly electron-dense granules appears. As in neutrophils, these are called primary granules and contain peroxidase and a number of hydrolytic enzymes. At the myelocyte stage, other granules, containing in most species an electron-dense crystalline lattice core surrounded by a matrix, begin to appear. These, known as eosinophil specific granules, (not to be confused with the specific granules of the neutrophils), are very likely a mature form of the primary granules, generated when the core protein achieves a critical concentration and crystallizes.

Furthermore, the cytosol of the peripheral eosinophil contains another population of smaller granules, shown to contain arylsulfatase and acid phosphatase (Parmley and Spicer, 1974).

A number of investigations have been carried out on the specific granules. These have been purified and shown to contain several basic proteins, as might be predicted from their avidity for the acidic dye eosin. Among these proteins, four have been characterized: the major basic protein (MBP), which constitutes the large crystalline core (Lewis *et al.*, 1978; Peters *et al.*, 1986), the eosinophil cationic protein (ECP), the eosinophil-derived neurotoxin (EDN), and the eosinophil peroxidase (EPO), all localized by immunological techniques in the matrix (Bainton and Farquhar, 1970; Peters *et al.*, 1986). All of these proteins are cytotoxic to parasites or mammalian cells, either in the presence of H_2O_2, as for the EPO, or independently of oxygen derivatives, as for MBP and ECP, whose characteristics will be described in detail below (see Sections 3.6 and 3.7).

Eosinophils, much like neutrophils, can be induced to discharge the content of their granules into phagocytic vacuole or, under some conditions, into the extracellular milieu by a number of stimuli. These include immunoglobulin G (IgG)-coated parasites (Butterworth *et al.*, 1979a,b), immune complexes (Archer *et al.*, 1969; Takenaka *et al.*, 1977), endotoxin-activated serum, concanavalin A, the two tetrapeptides released by mast cells and basophils known as eosinophil chemotactic factors of anaphylaxis (Sher and Wadee, 1981), and the calcium ionophore A23187 (Henderson *et al.*, 1983; Fukuda *et al.*, 1985).

3. DEFENSE POLYPEPTIDES OF GRANULOCYTES

Several granule-associated polypeptides, active *in vitro* against various agents noxious to the host, have been purified to homogeneity from neutrophils and eosinophils of various animal species.

These polypeptides are synthesized and processed during cellular maturation in the bone marrow and are stored within the granules, where they constitute a ready-to-use dormant weapon that can be mobilized as soon as the cell interacts with microorganisms. The functional integration between the phagocytic capacity of the defense cells and the delivery system of these factors is expressed in the generation of a microenvironment, the phagosome, where the antimicrobial substances are discharged at high concentration on the surface of, and can act synergically on, the noxious agent. Furthermore, in contrast to the oxygen-dependent killing system, these agents can exert their action also in the hypoxic environment sometimes found at the inflammatory site.

3.1. Bactericidal/Permeability-Increasing Protein

A protein called bactericidal/permeability-increasing (BPI) protein was purified first from rabbit (Weiss *et al.*, 1975; Elsbach *et al.*, 1979) and then from human neutrophils (Weiss *et al.*, 1978) more than a decade ago. The name accounts for its capacity to both increase permeability of the microbial envelope and inactivate susceptible microorganisms.

The two BPIs constitute approximately 0.5–1% of the total protein content of neutrophils (Elsbach and Weiss, 1985), are characterized by a high net positive charge (pI >9.6), and have molecular masses of about 50 and 58 kDa for rabbits and humans, respectively (Weiss *et al.*, 1978; Elsbach *et al.*, 1979). In addition, they exhibit immunological cross-reactivity, very similar antimicrobial activity (see below), and 80% homology in the first 17 amino acid residues so far sequenced (Ooi *et al.*, 1987). It is therefore very likely that the two BPIs are homologous proteins.

Immunotechniques, using antisera to human BPI, have permitted Weiss and Olsson (1987) to establish that this protein (1) is a specific product of neutrophils, absent from all other peripheral blood leukocytes; (2) is stored within the primary granules; (3) is synthesized during the promyelocyte stage of cell maturation; and (4) is also present in the promyelocytic leukemic cell line HL-60, although in a lower amount than in mature normal neutrophils. When human neutrophils are challenged with a secretory dose of the chemotactic peptide fMet-Leu-Phe, unlike the case for myeloperoxidase and elastase, also associated with the azurophil granules, only tiny amounts of BPI are released into the extracellular medium. This suggests that BPI is bound to the granule membrane, in keeping with the observation that granule treatment with salt or weak acid solutions solubilizes practically all myeloperoxidase, elastase, and lactoferrin but little BPI. Further support for this hypothesis has come from investigations on the structural organization of BPI as deduced from both limited proteolysis experiments and cDNA sequencing.

Recently, it has been shown that all of the cytotoxic activity of human BPI is associated to a 25-kDa amino-terminal fragment, which can be obtained from the purified holoprotein upon prolonged incubation at neutral pH and at 37°C (Ooi *et al.*, 1987). This fragmentation, whose occurrence *in vivo* and physiological relevance remain to be established, depends on the action of a neutral serine proteinase, as indicated by phenylmethylsulfonyl fluoride (PMSF) inhibition. This activity could be due either to a minor copurifying contaminant or to an as yet unknown (auto)proteolytic activity possessed by the BPI itself. Interestingly, the amino acid composition of this N-terminal domain has shown that it is enriched in charged and polar residues, with respect to the holoprotein. Moreover, from these data it has been possible to deduce that the remaining C-terminal portion is enriched in hydrophobic residues, suggesting that it might serve to anchor BPI to the granule membrane.

The organization of BPI into at least two functional domains has been confirmed by the amino acid sequence derived from a cDNA clone, encoding human BPI, isolated from a library prepared from HL-60 cells (Gray *et al.*, 1989). Analysis of the sequence has shown that a 31-amino-acid signal sequence is followed by the 456-residue mature protein, containing two potential glycosylation sites. The glycosylation of these sites may account for the difference between the mass predicted from the sequence (50.6 kDa) and that estimated for purified BPI (58 kDa). As expected, the amino-terminal portion, corresponding to the 25-kDa active fragment, carries a net positive charge of $+16$, whereas the C-terminal portion is slightly negatively charged and contains many hydrophobic residues.

A hydrophobicity plot of the BPI sequence shows that the amino-terminal antibacterial fragment contains alternating hydrophobic and hydrophilic segments, a pattern already observed in membrane-penetrating proteins such as the pore-forming C9 complement component (DiScipio *et al.*, 1984). On the other hand, the carboxy-terminal portion contains three highly hydrophobic regions that could mediate BPI binding to the granule membrane. Furthermore, the region connecting these two parts is particularly rich in hydrophobic and proline residues (Gray *et al.*, 1989). Similar regions have already been described both as linking hinges between domains and as preferred sites of proteolytic cleavage (Perham *et al.*, 1987). Interestingly, in this linking region there is a sequence corresponding to a potential cleavage site for elastase, a protease present in the azurophil granules, which might play a role in the generation of the 25-kDa antibacterial fragment.

A computer search to identify proteins homologous to BPI has shown that the cholesteryl ester transfer protein (Drayna *et al.*, 1987) shares 22% identity and 30% conservative substitutions with BPI. Interestingly, both proteins bind hydrophobic ligands associated with macromolecular complexes such as lipopolysaccharide (LPS) in the bacterial envelope for BPI (see below) and cholesterol esters, associated with lipoproteins, for the cholesteryl ester transfer protein.

An even more significant homology has been recognized by Tobias and co-workers (1988) between BPI and an LPS-binding protein (LBP) present in the acute-phase sera of humans, rats, mice, and rabbits, from which it has been purified (Tobias *et al.*, 1986). The N-terminal amino acid sequence of this protein is homologous to the corresponding amino-terminal portion of BPI, and an anti-LBP antiserum cross-reacts with BPI. These observations provide grounds for the hypothesis that BPI also is an LBP (Weiss *et al.*, 1980, 1984). Despite their sequence and immunological similarities and capacity to bind LPS, the two proteins very likely have different functions, since LBP, in contrast to BPI, is devoid of antibacterial activity (Tobias *et al.*, 1988).

A very potent bactericidal activity is displayed by BPIs from rabbits and humans at concentrations $\geq 10^{-8}$ M. This activity is restricted to several enteric gram-negative bacteria, with the rabbit protein being up to fivefold more active than the human one, at least with some strains (Weiss *et al.*, 1975, 1978; Elsbach *et al.*, 1979). The bactericidal activity is maximal at neutral pH and quickly drops both at acidic and alkaline values (Weiss *et al.*, 1978), suggesting that BPI acts very early in the defense process, as the phagosome pH turns gradually to acidic values upon internalization of the microorganisms (Mandell, 1970; Segal *et al.*, 1981).

Many gram-negative strains of a single susceptible species, whether rough or smooth, are inactivated. However, the smooth strains generally require up to 20-fold-higher BPI concentrations than the rough ones. In contrast, other gram-negative bacteria, such as *Serratia marcescens* and some *Proteus* species, and all of the gram-positive organisms and fungi (*Candida albicans* and *C. parapsilosis*) thus far tested are resistant even to very high BPI concentrations (Weiss *et al.*, 1975, 1978; Elsbach *et al.*, 1979).

A series of elegant studies has been devoted to examining the effects of BPI on susceptible bacteria. Early observations have shown that the protein causes a rapid increase in bacterial permeability to normally impermeant hydrophobic molecules, such as the antibiotic actinomycin D, whereas only negligible effects have been reported, for at least 1 hr after treatment, on bacterial macromolecular synthesis and potassium ion transport (Weiss *et al.*, 1978, 1984; Elsbach *et al.*, 1979). This finding indicates that during this time interval, the functions associated with the bacterial cytoplasmic membrane remain practically unaffected. However, very recently, inhibition of the oxygen consumption by *Escherichia coli* treated with a lethal dose of BPI has been reported (see below) (In't Veld *et al.*, 1988). In contrast, alterations are rapidly produced at the outer membrane level, which loses its ability to act as a barrier to the entrance of small hydrophobic molecules such as actinomycin D. Furthermore, it has also been shown that BPI can rapidly promote bacterial phospholipid hydrolysis by activation of bacterial phospholipases A, located predominantly in the outer membrane of gram-negative micro-organisms, and by certain exogenous phospholipases A_2 (Weiss *et al.*, 1976, 1978, 1979; Weiss and Elsbach, 1977).

The immediate interaction of the protein with surface structures of the bacterial envelope is also suggested by the effect of the divalent cations Mg^{2+} or Ca^{2+}. These ions, at supraphysiological concentrations, release almost totally and selectively BPI molecules bound to the bacteria (Weiss *et al.*, 1983). This finding suggests that the positively charged N-terminal moiety of BPI interacts with the negatively charged phosphate and 2-keto-3-deoxyoctanoate (KDO) residues, clustered at the base of the polysaccharide chains (O antigen) and in close proximity to the lipid A in the LPS molecules (Figure 1). This would explain the competitive action on BPI activity by Mg^{2+} and Ca^{2+}, which likely bind to these negative groups and stabilize the LPS molecules of the bacterial

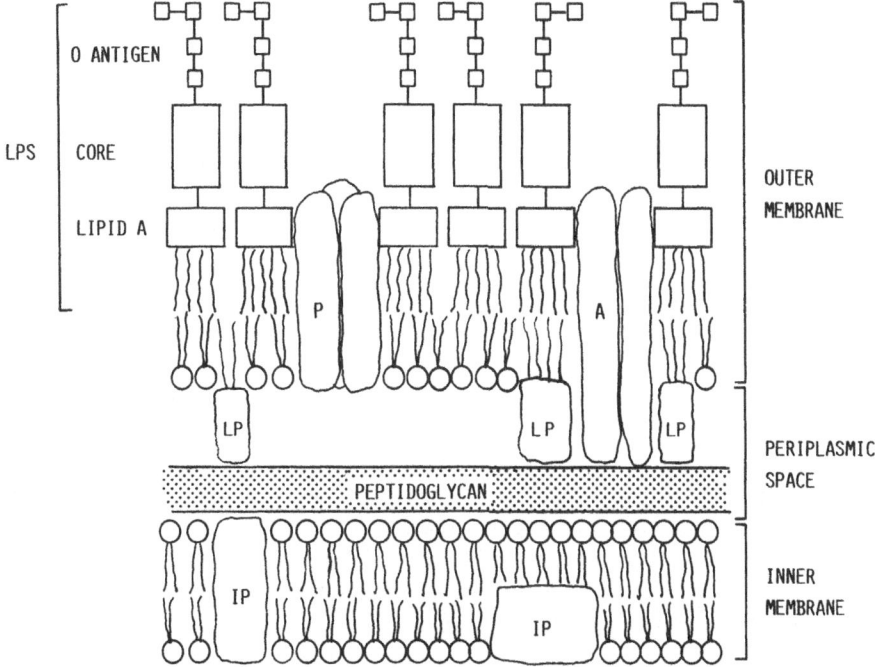

FIGURE 1. Schematic model of the molecular organization of the cell envelope (outer membrane, peptidoglycan, and inner membrane) of *Enterobacteriaceae*. The asymmetric bilayer of the outer membrane has the outer leaflet made up of the lipid portion (lipid A) of the LPS and the inner leaflet occupied by phospholipids. The outwardly directed polysaccharide chains of LPS are constituted by the core region, rich in negative charges potentially binding divalent cations, and by the O antigen. This is a polymer of repeated carbohydrate units, each three to six residues long, whose number can vary from none to more than 40. For the sake of simplicity, only one of these units has been drawn. The outer layer is covalently attached to the peptidoglycan via lipoproteins (LP). The drawing also includes other proteins of the outer membrane, such as the pore proteins LamB (P) and OmpA (A) and some integral proteins (IP) of the inner membrane.

outer membrane into an ordered and cohesive layer (for a review on the gram-negative bacteria outer membrane, see Lugtenberg and van Alphen, 1983). As described above, further evidence in favor of the BPI–LPS interaction has been recently provided by the homology between BPI and an acute-phase LBP (Tobias *et al.*, 1988; Gray *et al.*, 1989).

The role played by the organization of the gram-negative bacterial outer membrane (OM) in the BPI–microorganism interaction is further supported by experiments in which the susceptibility to BPI was tested with the same smooth strain of *E. coli* grown under conditions leading to different polysaccharide chain lengths in the LPS (Weiss *et al.*, 1986). The results showed that the longer the chain, the higher the amount of BPI required to obtain a comparable killing. As expected, the sensitivity of rough strains was not affected by growth conditions. The most likely explanation of these observations is that the length of the LPS polysaccharide moiety, by masking the negative charges of phosphate and KDO residues, hinders the BPI interaction with the bacterial surface. This is confirmed by measurements of BPI binding to susceptible microorganisms. Although both rough and smooth strains bind at saturation the same amount of BPI, the protein concentration necessary to reach this value increases in parallel with the polysaccharide chain length (Weiss *et al.*, 1980). Recently, by using biologically active ^{125}I-labeled human BPI, it has been possible to measure precisely its binding to the *E. coli* strain J5 (Mannion *et al.*, 1989). The results obtained indicate that binding is saturable and that there are approximately 2.2×10^6 binding sites per bacterium, with an apparent dissociation constant of 23 nM. Moreover, the binding of BPI to bacteria is unaffected by lysozyme or by BPI-depleted crude rabbit neutrophil extracts and only slightly inhibited by myeloperoxidase and by cathepsin G, even when added in large excess with respect to BPI. This implies that other neutrophil proteins, which may be present inside the phagocytic vacuole, do not interfere with BPI binding to target bacteria.

The charge interaction and the subsequent envelope alterations produced by BPI are necessary but not sufficient to explain its effects. In fact, displacement of BPI from the bacterial surface, even as early as 15–30 sec after contact, is not sufficient to reverse the effect on bacterial viability, although rapid repair of the envelope alterations is promoted. Experimental evidence suggests that a postbinding step(s), dependent on the hydrophobic properties of the outer membrane and of unknown nature, is necessary for the irreversible killing of the microorganism (Weiss *et al.*, 1983). In this respect, the hypothesis has been advanced that a small subpopulation of BPI penetrates the OM and reaches the cytoplasmic membrane (very likely at the adhesion zones between the two membranes), where it exerts its bactericidal action by an as yet unknown mechanism.

To determine whether BPI actually interacts with the bacterial inner membrane (IM) and subsequently impairs membrane-dependent functions, investigations have been carried out with cytoplasmic membrane vesicles prepared from *E. coli* and also from the gram-positive *Bacillus subtilis*. These studies have shown that BPI, once added to energized vesicles from both microorganisms, rapidly inhibits oxygen consumption and amino acid active transport, with similar dose dependence (In't Veld *et al.*, 1988). These data thus provide evidence that once the external peptidoglycan layer is removed, BPI may also affect membrane-associated functions of gram-positive organisms, which are normally resistant to its toxic effects. That BPI can perturb the cytoplasmic membrane is further suggested by the marked reduction in oxygen consumption by intact *E. coli* cells exposed to a lethal dose of the bactericidal protein (In't Veld *et al.*, 1988). However, although respiration is reduced by about 90% after 20 min, there is apparently no effect on the cell energy charge. In fact, only a minor decrease in the incorporation of labeled amino acids into acid-precipitable protein is observed even 60 min after treatment. It remains to be established whether BPI reaches the metabolic machinery of the inner membrane of the susceptible organisms and to what extent this interaction contributes to the bactericidal action of the protein.

3.2. Cationic Antimicrobial Proteins 57 and 37

By using a procedure very similar to that adopted by Weiss *et al.* (1978) to obtain BPI, Spitznagel and co-workers have purified from granule extracts of human granulocytes two antibacterial proteins, named cationic antimicrobial proteins 37 (CAP37) and 57 (CAP57) from their respective masses of 37 and 57 kDa (Shafer *et al.*, 1984). These two polypeptides differ substantially in amino acid composition and spectrum of antibacterial activity and are considered to be two different entities.

Of the two proteins, CAP57 was studied more extensively. In particular, two monoclonal antibodies raised against this protein have been used to investigate its cellular localization. Immunocytochemical studies have shown that CAP57 is associated with cytoplasmic granules of neutrophils and of some of their bone marrow precursors and is not present in other blood cells (Spitznagel *et al.*, 1987). In addition, these antibodies do not react with CAP37.

Several lines of evidence, including amino acid composition and antibacterial properties, indicate that CAP57 is very likely identical to BPI (Shafer *et al.*, 1984). In fact, like BPI, CAP57 is active against various gram-negative species and is virtually inactive against gram-positive microorganisms. In addition, its bactericidal activity is a function of the LPS structure, with deep rough mutants being more susceptible than the smooth parental strains (Shafer *et al.*, 1984). In this respect, the use of biologically active radiolabeled CAP57 has led to the conclusions that its binding to bacteria is relatively specific and saturable and that the amount of CAP57 bound by the various strains correlates directly with their susceptibility to the antibacterial protein (Farley *et al.*, 1987, 1988).

Farley and co-workers (1988) have demonstrated that CAP57, like BPI, interacts both electrostatically and hydrophobically with the microbial surface. In fact, the polymyxin B-resistant mutant strain SH7426 of *Salmonella typhimurium* (mutant *pmrA*), characterized by decreased electronegativity of LPS as a result of increased substitution of phosphate residues of the core oligosaccharide and of lipid A with 4-amino-4-deoxy-L-arabinose (Vaara *et al.*, 1981), is more resistant to the CAP57 bactericidal activity than its *prmA*[+] parental strain SH9178. In competition experiments, excess polymyxin B competes with CAP57 for binding to SH9178, whereas a polymyxin B-derived polycationic peptide lacking the hydrophobic portion of the antibiotic is much less capable of competing with CAP57 (Farley *et al.*, 1988). This result points to the importance of hydrophobic interactions, in addition to the electrostatic ones, for CAP57 action.

As seen with other neutrophil antimicrobial factors, the susceptibility of microorganisms to CAP57 appears to require an active metabolic state of the bacteria. In fact, anaerobic conditions inhibiting the oxidative metabolism of *Neisseria gonorrhoeae* strongly decrease its susceptibility to the killing action of CAP57. In contrast, a facultative anaerobe such as *S. typhimurium* is readily killed both in the presence and in the absence of oxygen (Casey *et al.*, 1985).

As mentioned above, the procedure developed to obtain CAP57 also led to the purification of CAP37, an antibacterial protein that, though less potent than CAP57, is active against many gram-negative microorganisms in neutral as well as in acidic (pH 5.5) media (Shafer *et al.*, 1986a). Further investigation is needed to better characterize this protein.

3.3. Cathepsin G

In 1974, Olsson and Venge resolved a granule extract prepared from human granulocytes into two groups of cationic proteins more basic than lysozyme (Olsson and Venge, 1974). The most cationic components 1–4 exhibited apparent molecular masses of 25.5–28.5 kDa, virtually identical amino acid composition, and immunochemical identity. Similarly, the less cationic components 5–7 could be pooled in a common subset on the base of their immunological identity. Subsequently, the components of the latter group were localized into the eosinophil granules and shown to constitute a family of isoforms named eosinophil cationic protein (ECP) (described in Section 3.7). In contrast, the former group was localized in the azurophil granules of neutrophils (Ohlsson *et al.*, 1977) and was shown to consist of several isozymes, very likely differing in carbohydrate content, which displayed at neutral pH a chymotrypsinlike esterolytic activity and were thus classified as cathepsin G (Odeberg *et al.*, 1975).

A clone, apparently coding for the complete amino acid sequence of this enzyme, has been isolated by using a cDNA library constructed from the U937 human leukemic cell line (Salvesen *et al.*, 1987). The protein appears to be synthesized as pre-procathepsin G with a 18 residues N-terminal signal peptide. This is followed by the dipeptide Gly-Glu, not found in the mature protein isolated from neutrophils, which contains a Ile-Ile sequence at its amino terminus. This dipeptide, as seen for other short pro-portions in various serine proteinases (Isackson *et al.*, 1984), likely enables the sorting and targeting of this polypeptide in an inactive form, thereby avoiding undesired proteolytic events. The mature protein is 235 residues long, has three disulfide bonds, a potential N-linked glycosylation site, and reveals homology with the rat mast cell proteinase II and with the proteinase CCP I, a putative product of activated mouse cytotoxic T lymphocytes.

Cathepsin G, in the micromolar range of concentration, exerts *in vitro* a significant antimicrobial activity against both gram-positive and gram-negative bacteria (Odeberg and Olsson, 1975; Thorne *et al.*, 1976; Shafer *et al.*, 1986b) and fungi (Drazin and Lehrer, 1977) and is cytotoxic to mammalian cells (Clark *et al.*, 1976). All of these activities are heat stable and appear to be unrelated to the heat-sensitive enzymatic function. Further, the bactericidal effect is inhibited at pH values below neutrality and by divalent cations and decreases with increasing ionic strength, suggesting a charge interaction with the target microorganisms (Odeberg and Olsson, 1975, 1976a, Shafer and Onunka, 1989).

By investigating the effects of cathepsin G, Odeberg and Olsson (1976a) have shown that it rapidly impairs oxygen consumption and active transport as well as incorporation of labeled precursors into RNA, DNA, and protein of susceptible bacteria. Furthermore, $^{86}Rb^+$-preloaded *E. coli* and *Candida* spp., but not *Staphylococcus aureus*, rapidly release the cation once exposed to cathepsin G, which thus appears to alter the permeability properties of the plasma membrane. More recently its has been found that *Neisseria gonorrhoeae* susceptibility to cathepsin G depends on the carbohydrate content of the LPS core region. In fact, a mutant strain which has lost sugars in this region, binds cathepsin G and is killed much better than its isogenic parental strain (Shafer *et al.*, 1986c). This observation suggests that gonococcal susceptibility may be controlled by LPS-masking of cathepsin G-binding sites, which have been tentatively identified with at least three of the OM proteins (Shafer, 1988).

Finally, it has been found that a mutant strain of *S. aureus* deficient in polymeric teichoic acid is more resistant to the cidal activity than its parental strain, suggesting that binding of cathepsin G to gram positive bacteria is mediated by electrostatic interactions with teichoic acid (Shafer and Onunka, 1989).

3.4. Defensins

Defensins are a well-characterized family of relatively small antimicrobial peptides that are active against various types of bacteria, fungi, and viruses and are also cytotoxic to mammalian cells. The history of research on these peptides can be traced to the early observation of the presence of an antimicrobial activity (see Section 1) in extracts of rabbit leukocytes (Hirsch, 1956). Some years later, the pioneering investigations by Zeya and Spitznagel (1963, 1966a–c) on the purification of neutrophil antibacterial factors showed that the main antimicrobial components of rabbit and of guinea pig leukocyte extracts were several low-molecular-weight cationic peptides. These early observations have been greatly extended in the past decade by Lehrer and co-workers, who have named these peptides defensins on the basis of their putative role in host defense (Ganz *et al.*, 1985).

Thus far fifteen distinct components of the defensin family have been purified to homogeneity. The first peptides were obtained from rabbit alveolar macrophages and were designated MCP-1 and MCP-2 (Patterson-Delafield *et al.*, 1980). Subsequently, six components of the family were obtained from rabbit peritoneal granulocytes NP-1, -2, -3a, -3b, -4 and -5 (Selsted *et al.*, 1984), and it was shown that there was a total identity between NP-1 and MCP-1 and between NP-2 and MCP-2 (Selsted *et al.*, 1985a). Later, four defensins have been purified from human neutrophils: HNP-1, -2 and -3 (Ganz *et al.*, 1985), and the minor component HNP-4 (Singh *et al.*, 1988), two defensin from guinea pig neutrophils: GPNP (Selsted and Harwig, 1987), and GNCP-2 (Yamashita and Saito, 1989), and three defensins from rat neutrophils: RatNP-1, -3 and -4 (Eisenhauer *et al.*, 1989).

Defensins are localized in cytoplasmic granules of the neutrophils, as demonstrated by both immunogold electron microscopy (Ganz *et al.*, 1985) and subcellular fractionation (Rice *et al.*, 1987). By using high-resolution Percoll gradients, it was possible to demonstrate that these peptides are mainly stored within the five granule subpopulations collectively constituting the azurophil granules (see Section 2.1). Moreover, among these five subpopulations, it was demonstrated that the largest and the densest (fraction H5) is particularly rich in defensins and relatively deficient in other azurophil markers, such as myeloperoxidase, elastase, and β-glucuronidase (Rice *et al.*, 1987). These defensin-rich granules show a peculiar ultrastructural morphology, with an uneven distribution of myeloperoxidase, which concentrates at the electron-dense rim, in contrast to the other azurophil granule subpopulations. The uniqueness of composition and morphology of the granules collected in the H5 fraction suggests that they may play a specialized role in the antimicrobial function of the human neutrophil.

Determinations of amino acid composition and sequence have shown that the defensins are rich in cysteine (6 residues) and arginine (4–10 residues) and are 29–34 amino acids long. The sequence homology among the defensins of different species is relatively low (27–45%), suggesting a rapid evolution of these peptides. However, when appropriately aligned, they display eight invariant residues: Arg-15, Gly-18, and the six cysteines (Selsted et al., 1983, 1985a,c; Selsted and Harwig, 1987; Singh et al., 1988; Eisenhauer et al., 1989; Yamashita and Saito, 1989). The latter residues are linked in three disulfide bridges, which appear to be essential for preservation of structure and function. In fact, their cleavage virtually abolishes the biological activity of these peptides (Selsted and Harwig, 1989). The three disulfide bridges, as determined for HNP-2, are arranged according to a motif not yet found in other peptides (Selsted and Harwig, 1989). On the basis of the invariance of the cysteine residues, it can be argued that this motif is very likely conserved in all members of the defensin family. The link between the first and the last cysteine residues, present at the N and C termini in the case of HNP-2 and near these termini in all other peptides, constrains HNP-2 and very likely also the other defensins to a cyclic structure (see also discussion of the structure of the dodecapeptide bactenecin in Section 3.5).

The high density of disulfide links, shared with many toxic proteins ranging from snake and spider neurotoxins (Chicheportiche et al., 1975; Sheumack et al., 1985; Brown et al., 1988) to bacterial enterotoxin (Gariépy et al., 1986), suggests that defensins have a compact structure. This might help explain their high resistance to low pH and to proteolytic enzymes, thereby allowing them to exert their biological activity in the acidic and protease-rich microenvironment of the phagolysosome.

Recently, by applying two-dimensional nuclear magnetic resonance techniques, a solution conformation of the rabbit NP-5 defensin has been obtained. The overall folding of the peptide backbone is well defined, is stabilized by the three disulfide bonds (in keeping with the above observations), and contains a β hairpin encompassing residues 19–28 (Bach et al., 1987; Pardi et al., 1988). Further analysis aimed at the achievement of high-resolution conformations in a set of defensins should help find structure–function relationships and explain the differences in biological activity among the various components of this family of antimicrobial peptides (see below). In this respect, some information might derive from high-resolution determination of the structures of the available crystals of the rabbit NP-2 and human HNP-1 defensins (Westbrook et al., 1984; Stanfield et al., 1988).

Screening of an HL-60 human promyelocytic leukemia cDNA library has allowed the isolation of four clones encoding HNP-1 and HNP-3 defensins (Daher et al., 1988). The amino acid sequences derived from these clones have confirmed that these two 30-residue-long peptides differ only at their N-terminal residue (Ala in HNP-1 and Asp in HNP-3). The failure of identifying a clone encoding HNP-2 suggests that this peptide could derive, by degradation or processing, from either HNP-1 or HNP-3. In fact, HNP-2 lacks only the N-terminal amino acid; the rest of its sequence is identical to that of the other two congeners (Selsted et al., 1985c).

Analysis of these clones has indicated that the peptides are synthesized as pre-prodefensins, consisting of a 19-residue signal peptide and a 45-residue pro segment, which must be removed to generate the mature form of the 30-residue-long HNP-1 and -3. Since no evidence of prodefensins has been found in mature neutrophils, it is very likely that the pro portion is cleaved before the mature peptides are stored within the granules. Southern blot analysis of genomic DNA has revealed a single 3.7-kb *Eco*RI fragment, suggesting that either the two defensin genes are closely linked or the flanking *Eco*RI sites have been conserved as a result of a recent duplication. Finally, defensin mRNA has been found in bone marrow cells (in addition to HL-60 cells) but not in mature circulating normal neutrophils.

Defensins are able to inactivate various types of bacteria, fungi, and viruses as well as to exert cytotoxic activity against both normal and malignant mammalian cells. The potency of these biological effects varies considerably among individual peptides despite their common structural features. This suggests, as mentioned above, that minor structural differences might be responsible for the variable degrees of potency of the various defensins.

Killing activity *in vitro* is displayed against a broad range of both gram-positive and gram-negative organisms, the former generally being more susceptible than the latter (Patterson-Delafield *et al.*, 1980; Lehrer *et al.*, 1983; Selsted *et al.*, 1984; Ganz *et al.*, 1985; Shafer *et al.*, 1988). With respect to bactericidal activity, there is only a partial correlation between cationicity and bactericidal potency of individual defensins, suggesting that factors other than net charge contribute to their activity. For instance, although the rabbit NP-4 and NP-5 (net charges of +6 and +4) are much less effective than the most positively charged NP-1 and NP-2 (net charges of +9 and +8), they are also less effective than HNP-1 or HNP-2 (net charge of +3) (Selsted *et al.*, 1984; Ganz *et al.*, 1985).

The inactivation of bacteria is time and concentration dependent and requires neutral or slightly alkaline pH as well as low ionic strength. Addition of salt to the medium decreases the *in vitro* antibacterial activity, which is totally abolished at 50–100 mM NaCl (Lehrer *et al.*, 1983; Selsted *et al.*, 1984; Ganz *et al.*, 1985). Furthermore, the assay medium has to be supplemented with substrates, different for different microorganisms, that can support bacterial metabolic activity (Ganz *et al.*, 1985). This suggests that the susceptible microorganisms have an active rather than a passive role *vis-à-vis* the bactericidal action of defensins (see also below).

The mode of action of defensins has not yet been clearly established. However, it is known that exposure of *E. coli* to a lethal dose of HNP-1, in the presence of nutrients, rapidly inhibits bacterial respiration, with a concomitant permeabilization of both the OM and IM. This has been shown by the loss of crypticity of the periplasmic β-lactamase and of the cytoplasmic β-galactosidase, for the OM and IM respectively (Lehrer *et al.*, 1989b). OM permeabilization, which can be at least partially dissociated from that of IM, shows a minimal lag time of 8–10 min after defensin addition and is rapidly followed by disruption of IM integrity with a minimal lag time of 12–15 min. This immediately produces an inhibition of the synthesis of the macromolecules (protein, DNA, and RNA) and loss of bacterial viability. All these effects appear to be a consequence of the IM permeabilization, with its attendant leak of cellular metabolites, altered ionic cellular environment, and collapse of transmembrane potential or proton motive force. Finally, all these effects can be abolished either by omission of nutrients from the incubation medium, or by addition of metabolic inhibitors, thereby supporting the notion that defensins initially require an actively metabolizing bacterium to exert their bactericidal activity (Lehrer *et al.*, 1988a and 1989b).

Defensins also exert a potent bactericidal activity against various fungi (Patterson-Delafield *et al.*, 1980, 1981). Investigations, carried out mainly with *C. albicans*, have shown that this activity is optimal at conditions very similar to those reported above for bacteria, such as low ionic strength and active cellular metabolism. *C. albicans* can be protected from the toxic effects of defensin by inhibitors of mitochondrial activity (Lehrer *et al.*, 1988b) or by Ca^{2+} and, to a lesser extent, by Mg^{2+} (Selsted *et al.*, 1985b).

With fungi, as for bacteria, there is variability in the activity of the individual defensins. The fungicidal activity parallels the capacity of defensins to bind to the *Candida* surface (Lehrer *et al.*, 1985b). The binding shows a biphasic kinetic, with a rapid primary phase inhibitable by addition of salt and relatively unaffected by calcium or low temperature (0°C). Conversely, the secondary phase occurs only under candidacidal conditions and is inhibited by millimolar concentrations of calcium ions or low temperature. This phase is not observed when sublethal doses of defensins are tested. According to these data, binding of the peptides to external accessible sites of the cell takes place during the primary rapid phase, whereas the secondary phase, essential for the subsequent fungicidal effect, represents binding to additional sites that become accessible after membrane permeabilization. The fungicidal mechanisms of defensins thus appear to be similar to those proposed for other neutrophil microbicidal peptides (see Sections 3.1 and 3.2).

Some members of the defensin family are able to directly neutralize several types of enveloped viruses, such as herpes simplex virus types 1 and 2, vesicular stomatitis virus, and influenza virus, but are inactive against the nonenveloped echovirus and reovirus (Lehrer *et al.*, 1985a; Daher *et al.*, 1986; Selsted and Harwig, 1987). In particular, the three human, the guinea pig, and the rabbit NP-1 and NP-2 defensins show considerable *in vitro* neutralizing activity, whereas the other four rabbit congeners and the rat defensins are relatively inactive (Eisenhauer *et al.*, 1989). The virucidal activity is optimal at slightly alkaline pH, unaffected by ionic strength, and completely abolished by reduction and alkylation of the disulfide bridges of defensins, indicating that it depends on peptide conformation and cannot be simply attributed to their cationicity (Lehrer *et al.*, 1985a; Daher *et al.*, 1986). The finding that only enveloped viruses are susceptible to defensins and that the neutralizing activity requires temperatures higher than 15°C suggests that perturbation of the lipid envelope could determine the virion neutralization.

Defensins also exert cytotoxic activity *in vitro* against both normal and tumor cells in culture (Lichtenstein *et al.*, 1986). Cell lysis is detected after at least 3–4 hr of incubation, indicating that rapid generation of plasma membrane lesions is not the primary effect of defensins. As seen with *C. albicans*, treatment of mammalian cells with metabolic inhibitors blocks the lytic process, suggesting that the target cells play an active role in their own lysis (Lichtenstein *et al.*, 1988). This role might consist of an energy-dependent translocation or internalization of defensin molecules, as found for some bacterial and plant toxins (Sandvig and Olsnes, 1982).

Recently, to the polyfunctionality of defensins was added the capacity to inhibit corticotropin-stimulated rat adrenal cell corticosterone production (Zhu *et al.*, 1988; Singh *et al.*, 1988) the activity of protein kinase C (Charp *et al.*, 1988), and to stimulate histamine release from rat mast cells (Yamashita and Saito, 1989).

As pointed out above, the broad spectrum toxic activity of defensins against viruses, microorganisms, and eukaryotic cells is very likely associated with their capacity to interact with and perturb lipid membranes. All of these activities are exerted at the relatively high concentration of 10^{-6}–10^{-5} M. Furthermore, both the antibacterial and fungicidal activities are practically abolished at physiological ionic strength, which might arise some doubts as to the effective role of defensins *in vivo*. However, these reservations may be easily overcome by considering the abundance of the defensins within the neutrophils, where they account for approximately 5–10% of the total cell protein. This represents an average cellular concentration that exceeds 10 mg/ml. The exclusive localization of defensins in the cytoplasmic granules allows for an even higher local concentration, which could be sufficient for toxic activity upon discharge of the granule content into the phagocytic vacuole. Conversely, a potential toxicity of defensins in the extracellular milieu is less convincing, considering that upon cell stimulation by PMA or opsonized zymosan, only very tiny amounts of these peptides are found extracellularly (Ganz, 1987).

3.5. Bactenecins

Investigations on the antibacterial factors associated with the large granules of bovine neutrophils (see Section 2.1) conventionally began with assays of bactericidal activity of granule acid extracts and of protein fractions derived therefrom (Gennaro *et al.*, 1983a, 1983b; Savoini *et al.*, 1984). These extracts were then used as starting material for the purification of potent antimicrobial peptides, which were named bactenecins from the Latin words *bacterium* and *necare* (to kill).

The first bactenecin described is a dodecapeptide with the amino acid sequence Arg-Leu-Cys-Arg-Ile-Val-Val-Ile-Arg-Val-Cys-Arg (Romeo *et al.*, 1988), which is maintained in a cyclic structure by a disulfide bond between the two cysteine residues (see also the discussion of the structure of defensins in Section 3.4). A model conformation for this peptide was computer generated by exploiting the restraint offered by the disulfide link. In the conformation proposed, the peptide chain adopts an antiparallel extended structure, which reverses at residue 7, forming a bend containing three α-carbon atoms (γ turn) (Romeo *et al.*, 1988).

Two more bactenecins were subsequently purified and named Bac5 and Bac7 from their apparent molecular masses of 5 and 7 kDa, respectively. Their unique feature is a high content of proline (>45%) and arginine (>20%) (Gennaro *et al.*, 1989). Despite the very similar amino acid compositions, Bac5 and Bac7 have totally different sequences. In particular, the former is characterized by Arg-Pro-Pro and the latter by Pro-Arg-Pro repeated sequences, spaced by single hydrophobic amino acid residues (R. W. Frank *et al.*, 1990).

Immunofluorescence and Western blot analyses have provided evidence that the two latter bactenecins are not present in bovine blood cells other than neutrophils. In addition, they are absent from human neutrophils, which lack the large granules, and from a variety of cell lines, including a bovine leukemia cell line and the human HL-60 line. These experiments, conducted with rabbit antibodies against Bac5 and Bac7, have also indicated that in bovine neutrophils, the antibacterial polypeptides exist as probactenecins of 15 and 20 kDa, respectively. These precursor molecules are detected by Western blot analysis in neutrophils precipitated (fixed) with trichloroacetic acid and are converted to the respective bactenecin when the cells are solubilized with the nonionic detergent Triton X-100 at pH 7.4. If cell solubilization is carried out in the presence of the proteinase inhibitors DFP and PMSF, or at pH 4.0, the processing of the precursor molecules into Bac5 and Bac7 is markedly prevented. This suggests that in solubilized bovine neutrophils, the cleavage of the probactenecins is catalyzed by a neutral proteinase(s) (M. Zanetti *et al.*, 1990). The 15- and 20-kDa precursors may thus be the granule storage forms, which would be converted into the active Bac5 and Bac7 during secretion into the phagocytic vacuoles. In fact, the bovine neutrophil granules contain a large variety of neutral proteinases (Gennaro *et al.*, 1983a), which might catalyze the generation of Bac5 and Bac7 from their precursors.

Purified proBac5 and proBac7 have been found to be inactive on Bac5- and Bac7-susceptible bacteria (see below). In the stimulated neutrophil the activation of probactenecins may thus occur only after their discharge into the phagocytic vacuoles, whose pH remains slightly alkaline for a few minutes after sealing of the boundary membrane (Segal *et al.*, 1981). Here they would meet neutral protease(s), such as cathepsin G and elastase, which are known to be concomitantly released from the azurophils upon phagocytosis.

The probactenecins can be immunoprecipitated from [^{3}H]leucine-labeled bovine bone marrow cells, where they are very likely synthesized concomitantly with the assembly of the large granules, but not from mature peripheral neutrophils. As shown by cell-free translation experiments with poly(A)$^+$ RNA prepared from bone marrow cells enriched in neutrophil precursors, the primary translation products have approximate molecular masses of 21 kDa for Bac5 and 24 kDa for Bac7. These preprobactenecins are cotranslationally cleaved to polypeptides of about 16 and 20 kDa, respectively, thus indicating that the signal peptides removed could be about 35–40 residues long. The 16-kDa polypeptide is then processed to the 15-kDa probactenecin, as shown by pulse-chase experiments (Zanetti *et al.*, 1990).

The cyclic Cys-containing bactenecin exhibits potent bactericidal activity against both *E. coli* and *S. aureus* at 1–30 μg/ml (10^{-5}–10^{-6} M) (Romeo *et al.*, 1988). On the contrary, Bac5 and Bac7 appear to be active only on gram-negative organisms. In particular, at 2–50 μg/ml they efficiently kill *E. coli*, *S. typhimurium*, and *Klebsiella pneumoniae*. They also arrest the growth of *Enterobacter cloacae*. Finally, Bac7 suppresses the growth of *Pseudomonas aeruginosa* and neutralizes human herpes simplex virus but not rhinovirus (Zerial *et al.*, 1987; Gennaro *et al.*, 1989).

Recent experiments have indicated that the mechanism of action of Bac5 and Bac7 could be linked to a perturbation of the inner membrane of the susceptible microorganisms. In fact, bacterial respiration, which depends on the integrity of the inner membrane, is markedly decreased by both polypeptides. In this respect, *E. coli* respiration is inhibited by 60–75% as soon as 5 min after addition of bactericidal amounts of Bac5 or Bac7. The inhibitory effect on respiration is coupled to a concomitant drop in ATP levels as well as in uridine and leucine incorporation into RNA and proteins (B. Skerlavaj *et al.*, unpublished data). On the contrary, the oxygen consumption of Bac5- and Bac7-resistant *S. aureus* is completely unaffected by both bactenecins. In addition, as reported for defensins (see 3.4), Bac 5 and Bac7 cause accessibility of substrates to periplasmic β-lactamase and cytosolic β-galactosidase, indicating a permeabilization of both OM and IM (Skerlavaj *et al.*, 1990).

The bactericidal activity of Bac5 and Bac7 is optimally manifested at or above neutral pH, decreases with increasing ionic strength of the assay medium, and is significantly impaired by millimolar concentrations of Ca^{2+} or Mg^{2+} (Skerlavaj *et al.*, 1990). This implies that Bac5 and Bac7, in common with other antibacterial polypeptides (see Sections 3.1, 3.3, and 3.4), have an initial interaction with negatively charged groups in the outer membrane before reaching their target on the cytoplasmic membrane.

As with other antibacterial polypeptides of neutrophils, it remains to be established whether the three bactenecins potentially exert their effects *in vivo* as whole molecules or as split products. The likelihood that they play a physiological role in the defense process is supported by the yield of pure peptides from the granule extracts (5 μg for the dodecapeptide and 125 μg each for Bac5 and Bac7 per 10^9 neutrophils). In fact, it has been calculated that if only 10% of the granule contents were discharged into the phagocytic vacuoles, the intraphagolysosomal concentration of bactenecins would largely exceed the *in vitro* effective concentrations and may also provide a margin of safety of antibacterial activity against possible antagonists such as negatively charged macromolecules or low pH.

3.6. Eosinophil Major Basic Protein

MBP has been purified from the eosinophils of various mammalian species (guinea pig, human, rat, and horse) (Gleich *et al.*, 1973, 1976; Lewis *et al.*, 1976). It accounts for roughly half of the total protein of the eosinophil specific granules, where it constitutes the typical crystalloid core structure (see Section 2.2) (Gleich *et al.*, 1973; Lewis *et al.*, 1978; Peters *et al.*, 1986). MBP has been found, although in much lower amount than eosinophils, in the histamine-containing granules of basophils (Ackerman *et al.*, 1983) and in the HL-60 promyelocytic leukemia cell line when induced to differentiate to eosinophil-like cells (Fischkoff *et al.*, 1984). A protein reacting with MBP antibodies has also been localized by immunofluorescence in human placenta, where it is present in the placental X cells and in the giant cells (Maddox *et al.*, 1984).

MBP has a relatively low molecular mass (9–14 kDa), is highly cationic (pI > 10), and is characterized by a relatively high content of arginine (>13%) and hydrophobic residues and by several cysteines, some in free sulfydryl form (Gleich *et al.*, 1974, 1976). The sulfydryl groups undergo rapid oxidation when the protein is present in highly concentrated solution, thus leading to formation of reversible disulfide-linked insoluble aggregates with reduced biological activity (Gleich *et al.*, 1974).

Low solubility, coupled to unusual amino acid composition, has complicated the sequence determination of this protein. Only recently has the complete amino acid sequence of human MBP been obtained by conventional methods (Wasmoen *et al.*, 1988) as well as by sequencing of cDNA clones obtained from an HL-60 library (Barker *et al.*, 1988; McGrogan *et al.*, 1988).

Analyses of cDNA clones have shown that the protein is synthesized as a preproprotein of 222 residues, with an initial 15-amino-acid putative signal followed by 90 residues not found in the mature protein and by a C-terminal portion of 117 residues representing MBP. These data suggest that the preproprotein is likely cotranslationally modified to proMBP by signal sequence cleavage and then posttranslationally processed to the mature form by removal of the 90 residues of the pro fragment.

The 90-residue sequence contains 26 strongly acidic and 7 strongly basic amino acids, with a predicted pI of 3.7–3.9. In contrast, mature MBP has only one strongly acidic and 17 strongly basic amino acids, with a predicted pI of 10.9–11.1. Similar calculations for the proMBP form give a slightly acidic pI value of 6.0–6.2. Together, these data suggest that the acidic residues of the pro portion, by interacting with the basic amino acids of the MBP portion, could function to mask the toxic effects of mature MBP, thereby protecting eosinophils from damage during protein processing.

It remains to be established when after translation the acidic portion is split off from proMBP to obtain the mature product. The proteolytic cleavage might be carried out (1) during proprotein transport from the endoplasmic reticulum to the granules, (2) immediately before crystallization in the granules, or (3) concomitantly with secretion once the cell has been activated (see discussion bactenecin processing in Section 3.5 . Since no evidence of proMBP has been found in mature eosinophils, the latter hypothesis is very unlikely. The answer to this question could come from investigations on MBP processing by use of pulse-chase experiments in actively MBP-synthesizing cells, followed by immunoprecipitation and autoradiography.

By using restriction analysis and Southern blot hybridization, McGrogan and coworkers (1988) have provided preliminary results on the structure of the MBP gene, which appears to be less than 3 kb long and to contain at least one intron.

Prediction methods applied to the MBP sequence strongly suggest the possibility that segments 1–19, 31–49, and 107–115 assume a β-sheet conformation, whereas there is no evidence of helical structure.

MBP is known to be cytotoxic to several parasites, such as *Schistosoma mansoni*, *Trichinella spiralis*, *Fasciola hepatica*, and the trypomastigote form of *Trypanosoma cruzi* (Butterworth *et al.*, 1979a; Wassom and Gleich, 1979; Duffus *et al.*, 1980; Kierszenbaum *et al.*, 1981). The toxic effect is observed at concentrations of 10^{-4}–10^{-6} M, which could easily be achieved *in vivo*. In fact, MBP is present within the eosinophil granules in high amount and is delivered during degranulation directly on the surface either of an extracellular parasite, to which the cell tightly adheres (Glauert *et al.*, 1978), or of a parasite sequestered into a phagocytic vacuole (Villalta and Kierszenbaum, 1984).

MBP is cytotoxic *in vitro* to various mammalian cells at concentrations similar to those exerting antihelminthic activity (Gleich *et al.*, 1979). This effect has been implicated in the *in vivo* tissue damage associated with eosinophil infiltrates, where MBP accumulates, as in the lung epithelium of asthma-affected patients (Filley *et al.*, 1982). Furthermore, MBP can induce histamine release from basophils and mast cells, thereby generating a possible mechanism of eosinophil involvement in immune hypersensitivity reactions (O'Donnell *et al.*, 1983).

A common notion is that MBP exerts only weak antibacterial and antifungal activity (Gleich *et al.*, 1974). However, very recently human MBP has been shown to display potent bacteri- and fungicidal activity against *E. coli*, *Streptococcus faecalis*, and *C. albicans*, at least when tested at relatively low ionic strength and slightly acidic pH (5.5–6.0) (McGrogan *et al.*, 1988). These observations have been extended to *S. aureus*, which is killed optimally at neutral or slightly alkaline pH, and, at variance with *E. coli*, is equally susceptible to MBP in stationary and logarithmic phase of growth (Lehrer et al., 1989a).

The mechanism by which MBP damages susceptible parasites and mammalian cells remains elusive, although some membrane alteration certainly occurs, as shown by ^{51}Cr or [^{3}H]uridine release from preloaded cells (Butterworth *et al.*, 1979a; Villalta and Kierszenbaum, 1984). In this respect, as for other antimicrobial cationic proteins (see Sections 3.1, 3.2, 3.4, and 3.5), it has been hypothesized that the first step is the interaction of the positively charged groups of MBP with negative charges on the cell surface. This would be immediately followed by insertion of MBP apolar residues, which are likely exposed to external interactions, into the membrane lipid environment (Wasmoen *et al.*, 1988). Evidence supporting this hypothesis has been recently gained by demonstrating that MBP causes OM and IM permeabilization of the E. coli ML-35 strain (Lehrer *et al.*, 1989a).

3.7. Eosinophil Cationic Protein

ECP was originally purified by Olsson and Venge (1974) from a granule extract of leukocytes obtained from a patient with chronic myeloid leukemia. As described in Section 3.3, this extract was resolved into seven cationic proteins. Within each of the protein subsets 1–4 and 5–7 there was great similarity in amino acid composition and complete immunological identity. Later it was shown that the latter subset, which was named ECP, is a major constituent of matrix of the eosinophil granules (Olsson *et al.*, 1977) Peters *et al.*, 1986), and represents a microheterogeneous family of at least four zinc-containing extremely basic proteins (pI of about 11), with an almost identical amino acid composition, a high content of arginine (15%) and apparent molecular masses ranging from about 17 to 21 kDa (Olsson *et al.*, 1986; Gleich *et al.*, 1986).

Digestion of the purified components with endoglycosidase F has shown that the heterogeneity displayed by ECP can be almost totally ascribed to N-linked oligosaccharides, added post-translationally to the protein (see below). In fact, the enzymatic treatment reduces the mass of the largest species to about 18 kDa. Since the apparent molecular mass of the smallest form purified is about 17 kDa, there is a small mass excess in the digested forms that remains to be explained. This excess could be accounted for by other posttranslational modifications such as addition of an O-linked oligosaccharide. Moreover, these investigations have suggested that the N-linked oligosaccharides are of the so-called complex type, since digestion with endoglycosidase H has no effect on purified ECP (Gleich *et al.*, 1986). The biological significance, if any, of this microheterogeneity remains to be established.

The conclusion that the differences in charge and size observed among the various ECP subspecies depend on posttranslational glycosylation is further supported by the complete identity showed by the N-terminal sequences (through residue 59) of two ECP forms. These observations have been confirmed and extended by examination of a full-length cDNA clone for ECP, recently isolated in two independent laboratories (Rosenberg *et al.*, 1989; Barker *et al.*, 1989). The cDNA codes for a preprotein of 160 amino acids consisting of a 27 amino-terminal putative signal sequence followed by the mature 133 residue-long polypeptide which contains three potential N-linked glycosylation sites. This analysis has also revealed an about 70% identity with the eosinophil-derived neurotoxin (EDN), which shares with ECP the localization (Peters et al., 1986), the presence of a common antigenic epitope (Tai *et al.*, 1984), and the capacity to promote, once injected intrathecally, a neurological syndrome known as Gordon phenomenon (Durack *et al.*,

1981; Fredens *et al.*, 1982). Despite the close homology between the two proteins, ECP has a much higher isoelectric point than EDN, with calculated values of 10.9 and 8.9, respectively. This reflects the net increase in the number of Arg residues from 8 in EDN to 19 in ECP.

Furthermore, both proteins possess similarities to RNases from various animal species. In fact, when compared to the sequence of human pancreatic ribonuclease, EDN and ECP show conservation of the eight Cys residues and a high conservation of residues known to participate in substrate-binding and catalysis (Gleich *et al.*, 1986). This is in keeping with the observation that both eosinophil proteins possess RNase activity, with ECP displaying 100-fold less activity than EDN with a yeast RNA as substrate (Slifman *et al.*, 1986). The role, if any, played by this ribonucleolytic activity in the cytotoxic effect of ECP remains to be established (see below).

A posttranslational modification or conformational rearrangement of ECP might also occur during secretion of the protein from stimulated eosinophils. In fact, Tai *et al.* (1984) have shown that whereas two anti-ECP monoclonal antibodies, EG1 and EG2, react with a secreted form of the cytotoxic protein, only EG1 is able to recognize the stored form.

In addition to the studies on the structure of purified ECP, investigations have also been carried on the biosynthesis of this protein. To this end, [^{14}C]leucine-labeled bone marrow cells from patients with eosinophilia were immunoprecipitated with anti-ECP antibodies. This led to the yield of a 22-kDa polypeptide that does not correspond to any of the ECP forms thus far purified and very likely represents an ECP precursor. In fact, the 22-kDa form is processed, although with an unusually slow kinetic, to lower-molecular-mass forms of 18–19 kDa, whose formation is prevented by treatment of the cells with monensin, a known inhibitor of transport of secretory or lysosomal proteins (Olsson *et al.*, 1986).

ECP lacks any antibacterial activity (Olsson *et al.*, 1977), unless this is measured in low ionic strength media. Under these conditions it is particularly potent against stationary phase *S. aureus,* whereas *E. coli* is killed only if mid-logartithmic phase bacteria are used or if the assay medium is enriched with nutrients (Lehrer *et al.*, 1989a). At concentrations as low as 10^{-7} M, ECP exerts a potent toxic effect against parasites (McLaren *et al.*, 1981; Ackerman *et al.*, 1985). The antihelminthic activity has been tested mainly against the schistosomula of *Schistosoma mansoni,* and the results obtained indicate that ECP is approximately 10-fold more active than MBP (Ackerman *et al.*, 1985). However, when considering the possible relative contribution of these two proteins to the cytotoxic effect of the eosinophil *in vivo,* it should be kept in mind that the amount of MBP inside the granules is much higher than that of ECP.

These studies also indicate that ECP differs from MBP not only in potency but also in the mechanim of action on the schistosomula. In fact, ECP inhibits the motility of, and promotes dye uptake by, the parasite within 1 hr, far more rapidly than MBP, without any evident disorganization of the membrane structure (Ackerman *et al.*, 1985). At least initially, ECP is likely to produce discrete damage to the membrane, that subsequently leads to fragmentation of the tegumental membrane and extrusion of internal contents of the parasite (McLaren *et al.*, 1981; Ackerman *et al.*, 1985). In this respect, it has been demonstrated that purified human ECP at cytotoxic concentrations, but not MBP or EDN, can form functional channels into lipid vesicles, making them permeable to ions, sucrose, and fluorescent markers (Young *et al.*, 1986b). Formation of the pores requires no specific lipids and occurs more easily at 37°C than at lower temperatures, suggesting that it depends on membrane fluidity or temperature-induced changes in protein conformation.

The capacity of ECP to form channels may represent the mechanism by which this protein exerts its cytotoxic effect on target cells once it has been discharged directly onto their surface by activated eosinophils. Grounds for this hypothesis are provided by the observation that ECP displays hemolytic activity and produces a rapid and sustained membrane depolarization in all of the nucleated cells thus far tested (Young *et al.*, 1986b). The ability to form channels relates ECP to a number of different cytolytic proteins such as the terminal membrane attack complex of complement (Muller-Eberhard, 1986; Young *et al.*, 1986c), the perforins of cytotoxic T and natural killer cells (Podak *et al.*, 1985; Young *et al.*, 1986a), and a number of hole-producing cytotoxins from protozoa (Young *et al.*, 1982), yeasts (Kagan, 1983), and bacteria (Fussle *et al.*, 1981).

This capacity to perforate the membrane might explain the synergism that has been observed *in vitro* between ECP and cytotoxic oxygen derivatives in killing of schistosomula (Yazdanbakhsh *et al.*, 1987). In fact, when eosinophils closely adhere to the parasite, they concomitantly discharge their granule contents and activate the production of toxic oxygen derivatives. In the cell–parasite contact zone, the holes made by the released ECP could thus facilitate diffusion of superoxide anion and hydrogen peroxide, which might thereby exert their toxic effect directly inside the parasite.

4. CONCLUSIONS

A wealth of information on the structures and properties of the granulocyte antimicrobial polypeptides has recently accumulated. However, additional studies on their mechanism of action appear to be required to reach the level of knowledge gained for some bacterial (Konisky, 1982; Neville and Hudson, 1986), arthropod (Boman and Hultmark, 1987), amphibian (Zasloff, 1987; Marion *et al.*, 1988; Terry *et al.*, 1988), and pore-forming mammalian (Muller-Eberhard, 1986; Young *et al.*, 1986c) cytotoxins.

Common features shared by the granulocyte antimicrobial factors are the net positive charge, conferred by a relatively high content in Arg residues, and the ability to engage in hydrophobic interactions. These features appear to be essential for interaction with their first target, i.e., the plasma membrane surrounding prokaryotic and eukaryotic cells, as well as enveloped viruses. In bacteria, this interaction is complicated by the presence of other layers external to the cytoplasmic membrane, such as the OM of gram-negative microorganisms. These outer layers can act as a barrier for the antimicrobial factor–microorganism interaction (see discussion of the peptidoglycan of gram-positive microorganisms with respect to the action of BPI, Bac5, and Bac7 in Sections 3.1 and 3.5), or act as first binding site before interaction of the factor with the cytoplasmic membrane takes place (see discussion of BPI–LPS interaction in Section 3.1). It is likely that all of the granulocyte bactericidal polypeptides partially or totally penetrate into the surface membrane, thereby producing a perturbation in structure and function that may lead to a more or less rapid impairment of energy production (bacteria) and transport of metabolites. In this connection, further investigations on the effects of the antimicrobial polypeptides other than the simple loss of colony-forming ability should be pursued.

Elucidation of the molecular basis of cytotoxicity would greatly benefit from knowledge of the three-dimensional structure of these polypeptides. Despite the lack of homology among primary structures of the various antimicrobial factors described above, these studies might reveal common structural motifs not predictable on the basis of amino acid sequence. They could also lead to recognition of specific domains involved in different functions, such as binding to and penetration into the membrane, as already observed with some well-known bacterial toxins such as bacteriocins and diphtheria toxin (Konisky, 1982; Neville and Hudson, 1986). Such a function-related structural dissection might take advantage of limited proteolysis experiments, as already reported for BPI (see Section 3.1). All of these structure–function relationships would have to take into account the fact that interaction of these polypeptides with the hydrophobic environment of the membrane might promote profound conformational transitions, as observed for pore-forming proteins (Muller-Eberhard, 1986; Young *et al.*, 1986c) and colicins (Escuyer *et al.*, 1986).

Another point that needs further clarification is the potential of these polypeptides to act as antimicrobial agents in intact neutrophils or eosinophils. In fact, most of the studies reported in this chapter were carried out with purified components *in vitro*. Convincing evidence in support of their potential role *in vivo* might derive from studies of defined inborn errors. Recently, deficiencies in neutrophil granule components of the oxygen-independent microbicidal system have been reported in patients known to easily incur in infections, such as those affected by either Chédiak-Higashi syndrome (CHS) or specific granule deficiency (SGD).

Patients affected by the rare inherited CHS suffer from recurrent and severe pyogenic infections induced mainly by *Staphylococcus aureus* but also by *Streptococcus pneumoniae, Streptococcus pyogenes, Haemophilus influenzae,* and fungi (for a review, see Rotrosen and Gallin, 1987). The CHS neutrophils contain giant granules that derive predominantly from the fusion of azurophil granules and, less frequently, of specific granules (White and Clawson, 1980). Although these cells have normal phagocytic capacity and display normal or even enhanced respiratory burst, they show reduced microbicidal activity, which has been ascribed to delayed discharge of the granule contents into the phagocytic vacuole (Root *et al.*, 1972).

Recently, it has been recognized that in both humans CHS and the equivalent murine syndrome, the neutrophils lack elastase and cathepsin G but show a normal content of defensins (Takeuchi *et al.*, 1986; Ganz *et al.*, 1988b). As reported in Section 3.3, cathepsin G displays potent *in vitro* antibacterial activity, and thus its absence in CHS neutrophils may contribute substantially to their impaired antimicrobial capacity. The concomitant absence of elastase might exacerbate the situation, since it has been observed *in vitro* that this proteinase, although not antibacterial *per se*, potentiates the killing action of cathepsin G (Odeberg and Olsson, 1976b). The hypothesis that the absence of cathepsin G and elastase may contribute to the defect in postphagocytic bacterial killing by CHS neutrophils, although appealing, requires further investigation, since these cells manifest several defects other than those in granule morphology and composition.

Investigations recently performed on two patients affected by the very rare congenital SGD have added new insight into the role played by the oxygen-independent system in the killing of pathogens. These patients are characterized by the absence or aborted development of specific granules in their neutrophils, which, on the contrary, show a normal content of azurophil granules. The SGD subjects suffer from recurrent bacterial infections, with no preference for a particular class of microorganisms (for a review, see Rotrosen and Gallin, 1987).

The SGD neutrophils exert *in vitro* a markedly reduced bactericidal activity against both *E. coli* and *S. aureus,* although they can mount a normal respiratory burst and contain a normal amount of myeloperoxidase (Ambruso *et al.,* 1984). These observations indicate that the oxygen-related killing activity is not impaired. Although the presence of the azurophil granules would suggest that the main microbicidal components of the oxygen-independent system (BPI, cathepsin G, and defensins) are present, the SGD neutrophils actually contain about $\frac{1}{10}$ the mean defensin content of normal cells, whereas other azurophil granule components, such as elastase, cathepsin G, and myeloperoxidase, are present in normal amounts (Ganz *et al.,* 1988b).

The greatly reduced content of defensins in SGD neutrophils may help explain the microbicidal defect observed both *in vitro* and *in vivo.* As for the cathepsin G deficiency in CHS neutrophils, attribution of the impairment in the postphagocytic killing capacity of the SGD neutrophils to the defensin deficit also requires further investigation, since many other granular constituents are either absent or markedly reduced.

Another approach, aimed at clarifying the role *in vivo* of the granulocyte antimicrobial polypeptides, has been followed by Weiss and co-workers (1982), with particular reference to the contribution of these factors to the overall bactericidal activity of intact neutrophils *vis-à-vis* the oxygen-dependent antimicrobial system. In the attempt to evaluate the role of BPI in intact cells, they have measured the bactericidal activity of normal rabbit peritoneal or human blood neutrophils under aerobic and anaerobic conditions, as well as of human neutrophils isolated from the blood of chronic granulomatous disease (CGD) patients. CGD is a group of inherited conditions impairing the capacity to reduce oxygen to reactive cytotoxic derivatives. In this manner, it has been possible to at least partially point to the relative contributions of the oxygen-dependent and oxygen-independent systems to the bactericidal capacity of leukocytes.

The results obtained can be summarized as follows: (1) the inactivation by rabbit neutrophils of the BPI-resistant gram-positive *Staphylococcus epidermidis* is markedly impaired under anaerobic conditions, whereas ingested *Salmonella typhimurium,* which is BPI susceptible, is killed equally well in the presence and in the absence of oxygen by both human and rabbit neutrophils; (2) ingested *S. typhimurium* and *E. coli* are killed by neutrophils from CGD patients almost as effectively as those obtained from normal donors; and (3) whole homogenates and crude acid extracts from rabbit and from normal or CGD human neutrophils exert antibacterial activity toward *S. typhimurium* or *E. coli,* which can be ascribed to their BPI content since this activity is nearly abolished by anti-BPI but not by preimmune goat IgG-rich fractions (Weiss *et al.,* 1982, 1985). Together, these investigations, although not proving directly the role of BPI in the killing activity of intact neutrophils, provide circumstantial evidence that it may play a remarkable role in O_2-independent killing reactions against several enteric gram-negative species *in vivo.*

At any rate, the two antimicrobial systems present in neutrophils and eosinophils should not be considered alternate pathways for microbicidal activity; all the microbicidal components, once released into the phagocytic vacuole, may play a complementary role, on the basis of their different specificities toward the various noxious agents, and/or a synergic role, also involving other granule components (see discussion above of the potentiating effect of elastase on cathepsin G antibacterial activity). This would render the antimicrobial polypeptides active at concentrations significantly lower than those used for the *in vitro* assays, performed with single purified components, as well as significantly amplify their spectrum of activity. Finally, the oxygen-independent antimicrobial system might enable the leukocytes to effectively inactivate microbes, whose ingestion fails to trigger reactive oxygen derivatives production, or exert microbicidal activity when environmental conditions, such as hypoxia or acidosis, preclude the onset of the respiratory burst (Gabig *et al.*, 1979).

5. REFERENCES

Ackerman, S. J., Kephart, G. M., Habermann, T. M., Greipp, P. R., and Gleich, G. J., 1983, Localization of eosinophil granule major basic protein in human basophils, *J. Exp. Med.* **158**:946–961

Ackerman, S. J., Gleich, G. J., Loegering, D. A., Richardson, B. A., and Butterworth, A. E., 1985, Comparative toxicity of purified human eosinophil granule cationic proteins for schistosomula of *Schistosoma mansoni, Am. J. Trop. Med. Hyg.* **34**:735–745

Ambruso, D. R., Sasada, M., Nishiyama, H., Kubo, A., Koiyama, A., and Allen, R. H., 1984, Defective bacterial activity and absence of specific granules in neutrophils from a patient with recurrent bacterial infections, *J. Clin. Immunol.* **4**:23–30

Archer, G. T., Nelson, M., and Johnston, J., 1969, Eosinophil granule lysis in vitro induced by soluble antigen-antibody complexes, *Immunology* **17**:777–787

Bach, A. C., Selsted, M. E., and Pardi, A., 1987, Two-dimensional NMR studies of the antimicrobial peptide NP-5, *Biochemistry* **26**:4389–4397

Baggiolini, M., Hirsch, J. G., and de Duve, C., 1969, Resolution of granules from rabbit heterophil leukocytes into distinct populations by zonal sedimentation, *J. Cell Biol.* **40**:529–541

Baggiolini, M., Horisberger, U., Gennaro, R., and Dewald, B., 1985, Identification of three types of granules in neutrophils of ruminants. Ultrastructure of circulating and maturing cells, *Lab. Invest.* **52**:151–158

Bainton, D. F., 1973, Sequential degranulation of the two types of polymorphonuclear leukocyte granules during phagocytosis of microorganisms, *J. Cell Biol.* **58**:249–264

Bainton, D. F., and Farquhar, M. G., 1966, Origin of granules in polymorphonuclear leukocytes: Two types derived from opposite faces of the Golgi complex in developing granulocytes, *J. Cell Biol.* **28**:277–301

Bainton, D. F., and Farquhar, M. G., 1970, Segregation and packaging of granule enzymes in eosinophilic leukocytes, *J. Cell Biol.* **45**:54–73

Bainton, D. F., Ullyot, L. J., and Farquhar, M. G., 1971, The development of neutrophilic polymorphonuclear leukocytes in human bone marrow: Origin and content of azurophil and specific granules, *J. Exp. Med.* **134**:907–934

Barker, R. L., Gleich, G. J., and Pease, L. R., 1988, Acidic precursor revealed in human eosinophil granule major basic protein cDNA, *J. Exp. Med.* **168**:1493–1498

Barker, R. L., Loegering, D. A., Ten, R. M., Hamann, K. J., Pease, L. R., and Gleich, G. J., 1989, Eosinophil cationic protein cDNA. Comparison with other toxic cationic proteins and ribonucleases, *J. Immunol.* **143**:952–955

Bashford, C. L., Alder, G. M., Menestrina, G., Micklem, K. J., Murphy, J. J., and Pasternak, C. A., 1986, Membrane damage by hemolytic viruses, toxins, complement, and other cytotoxic agents. A common mechanism blocked by divalent cations, *J. Biol. Chem.* **261**:9300–9308

Boman, H. G., and Hultmark, D., 1987, Cell-free immunity in insects, *Annu. Rev. Microbiol.* **41**:103–126

Borregaard, N., Heiple, J. M., Simons, E. R., and Clark, R. A., 1983, Subcellular localization of the b-cytochrome component of the human neutrophil microbicidal oxidase: Translocation during activation, *J. Cell Biol.* **97**:52–61

Bretz, U., and Baggiolini, M., 1974, Biochemical and morphological characterization of azurophil and specific granules of human neutrophilic polymorphonuclear leukocytes, *J. Cell Biol.* **63**:251–269

Brown, M. R., Sheumack, D. D., Tyler, M. I., and Howden, M. E. H., 1988, Amino acid sequence of versutoxin, a lethal neurotoxin from the venom of the funnel-web spider *Atrax versutus, Biochem. J.* **250**:401–405

Butterworth, A. E., Wassom, D. L., Gleich, G. J., Loegering, D. A., and David, J. R., 1979a, Damage to schistosomula of *Schistosoma mansoni* induced directly by eosinophil major basic protein, *J. Immunol.* **122:**221–229

Butterworth, A. E., Vadas, M. A., Wassom, D. L., Dessein, A., Hogan, M., Sherry, B., Gleich, G. J., and David, J. R., 1979b, Interactions between human eosinophils and schistosomula of *Schistosoma mansoni*. II. The mechanism of irreversible eosinophil adherence, *J. Exp. Med.* **150:**1456–1471

Casey, S. G., Shafer, W. M., and Spitznagel, J. K., 1985, Anaerobiosis increases resistance of *Neisseria gonorrhoeae* to O_2-independent antimicrobial proteins from human polymorphonuclear granulocytes, *Infect. Immun.* **47:**401–407

Charp, P. A., Rice, W. G., Raynor, R. L., Reimund, E., Kinkade, J. M., Jr., Ganz, T., Selsted, M. E., Lehrer, R. I., and Kuo, J. F., 1988, Inhibition of protein kinase C by defensins, antibiotic peptides from human neutrophils, *Biochem. Pharmacol.* **37:**951–956

Chicheportiche, R., Vincent, J. P., Kopeyan, C., Schweitz, H., and Lazdunski, M., 1975, Structure-function relationship in the binding of snake neurotoxins to the torpedo membrane receptor, *Biochemistry* **14:**2081–2091

Clark, R., Olsson, I., and Klebanoff, S. J., 1976, Cytotoxicity for tumor cells of cationic proteins from human neutrophil granules, *J. Cell Biol.* **70:**719–723

Cohn, Z. A., and Hirsch, J. G., 1960a, The isolation and properties of the specific cytoplasmic granules of rabbit polymorphonuclear leukocytes, *J. Exp. Med.* **112:**983–994

Cohn, Z. A., and Hirsch, J. G., 1960b, The influence of phagocytosis on the intracellular distribution of granule-associated components of polymorphonuclear leukocytes, *J. Exp. Med.* **112:**1015–1022

Daher, K., Selsted, M. E., and Lehrer, R. I., 1986, Direct inactivation of viruses by human granulocyte defensins, *J. Virol.* **60:**1068–1074

Daher, K., Lehrer, R. I., Ganz, T., and Kronenberg, M., 1988, Isolation and characterization of human defensin cDNA clones, *Proc. Natl. Acad. Sci. USA* **85:**7327–7331

Dewald, B., Bretz, U., and Baggiolini, M., 1982, Release of gelatinase from a novel secretory compartment of human neutrophils, *J. Clin. Invest.* **70:**518–525

DiScipio, R. G., Gehring, M. R., Podack, E. R., Kan, C. C., Hugli, T. E., and Fey, G. H., 1984, Nucleotide sequence of cDNA and derived amino acid sequence of human complement component C9, *Proc. Natl. Acad. Sci. USA* **81:**7298–7302

Drayna, D., Jarnagin, A. S., McLean, J., Henzel, W., Kohr, W., Fielding, C., and Lawn, R., 1987, Cloning and sequencing of human cholesteryl ester transfer protein cDNA, *Nature* (London) **327:**632–634

Drazin, R. E., and Lehrer, R. I., 1977, Fungicidal properties of a chymotrypsin-like cationic protein from human neutrophils: Adsorption to *Candida parapsilosis, Infect. Immun.* **17:**382–388

Duffus, W. P. H., Thorne, K., and Oliver, R., 1980, Killing of juvenile *Fasciola hepatica* by purified bovine eosinophil proteins, *Clin. Exp. Immunol.* **40:**336–344

Durack, D. T., Ackerman, S. J., Loegering, D. A., and Gleich, G. J., 1981, Purification of human eosinophil-derived neurotoxin, *Proc. Natl. Acad. Sci. USA* **78:**5165–5169

Eisenhauer, P. B., Harwig, S. S. L., Szklarek, D., Ganz, T., Selsted, M., and Lehrer, R. I., 1989, Purification and antimicrobial properties of three defensins from rat neutrophils, *Infect. Immun.* **57:**2021–2027

Elsbach, P., and Weiss, J., 1985, Oxygen-dependent and oxygen-independent mechanisms of microbicidal activity of neutrophils, *Immunol. Lett.* **11:**159–163

Elsbach, P., Weiss, J., Franson, R. C., Beckerdite-Quagliata, S., Schneider, A., and Harris, L., 1979, Separation and purification of a potent bactericidal/permeability-increasing protein and a closely associated phospholipase A_2 from rabbit polymorphonuclear leukocytes. Observations on their relationship, *J. Biol. Chem.* **254:**11000–11009

Escuyer, V., Boquet, P., Perrin, D., Montecucco, C., and Mock, M., 1986, A pH-induced increase in hydrophobicity as a possible step in the penetration of colicin E3 through bacterial membranes, *J. Biol. Chem.* **261:**10891–10898

Farley, M. M., Shafer, W. M., and Spitznagel, J. K., 1987, Antimicrobial binding of a radiolabeled cationic neutrophil granule protein, *Infect. Immun.* **55:**1536–1539

Farley, M. M., Shafer, W. M., and Spitznagel, J. K., 1988, Lipopolysaccharide structure determines ionic and hydrophobic binding of a cationic antimicrobial neutrophil granule protein, *Infect. Immun.* **56:**1589–1592

Filley, W. V., Holley, K. E., Kephart, G. J., and Gleich, G. J., 1982, Identification by immunofluorescence of eosinophil granule major basic protein in lung tissues of patients with bronchial asthma, *Lancet* **ii:**11–15

Fischkoff, S. A., Pollak, A., Gleich, G. J., Testa, J. R., Misawa, S., and Reber, T. J., 1984, Eosinophilic differentiation of the human promyelocytic leukemia cell line, HL-60, *J. Exp. Med.* **160:**179–196

Fletcher, M. P., and Gallin, J. I., 1983, Human neutrophils contain an intracellular pool of putative receptors for the chemoattractant N-formylmethionylleucylphenylalanine, *Blood* **62:**792–799

Frank, R. W., Gennaro, R., Schneider, K., Przybylski, M., and Romeo, D., 1990, Amino acid sequences of two proline-rich bactenecins, antimicrobial peptides of bovine neutrophils, *J. Biol. Chem.* **265** (in press)

Fredens, K., Dahl, R., and Venge, P., 1982, The Gordon phenomenon induced by eosinophil cationic protein and eosinophil protein X, *J. Allergy Clin. Immunol.* **70:**361–366

Fukuda, T., Ackerman, S. J., Reed, C. E., Peters, M. S., Dunnette, S. L., and Gleich, G. J., 1985, Calcium ionophore A23187 calcium-dependent cytolytic degranulation in human eosinophils, *J. Immunol.* **135:**1349–1356

Fussle, R., Bhakdi, S., Sziegoleit, A., Tranumjensen, J., Kranz, T., and Wellensiek, H. J., 1981, On the mechanism of membrane damage by *Staphylococcus aureus* α-toxin, *J. Cell Biol.* **91:**83–94

Gabig, T. G., Bearman, S. I., and Babior, B. M., 1979, Effects of oxygen tension and pH on the respiratory burst of human neutrophils, *Blood* **53:**1133–1139

Ganz, T., 1987, Extracellular release of antimicrobial defensins by human polymorphonuclear leukocytes, *Infect. Immun.* **55:**568–571

Ganz, T., Selsted, M. E., Szklarek, D., Harwig, S. S. L., Daher, K., Bainton, D. F., and Lehrer, R. I., 1985, Defensins. Natural peptide antibiotics of human neutrophils, *J. Clin. Invest.* **76:**1427–1435

Ganz, T., Selsted, M. E., and Lehrer, R. I., 1988a, Defensins: Antimicrobial/cytotoxic peptides of phagocytes, in *Bacteria-Host Cell Interaction* (M. Horwitz and M. Lovett, eds.), pp. 3–14, Alan R. Liss, Inc., New York

Ganz, T., Metcalf, J. A., Gallin, J. I., Boxer, L. A., and Lehrer, R. I., 1988b, Microbicidal/cytotoxic proteins of neutrophils are deficient in two disorders: Chédiak-Higashi syndrome and "specific" granule deficiency, *J. Clin. Invest.* **82:**552–556

Gariépy, J., Lane, A., Frayman, F., Wilbur, D., Robien, W., Schoolnik, G. K., and Jardetzky, O., 1986, Structure of the toxic domain of the *Escherichia coli* heat-stable enterotoxin ST I, *Biochemistry* **25:**7854–7866

Gartner, I., 1980, Separation of human eosinophils in density gradients of polyvinylpyrrolidone-coated silica gel (Percoll), *Immunology* **40:**133–136

Gennaro, R., Dewald, B., Horisberger, U., Gubler, H. U., and Baggiolini, M., 1983a, A novel type of cytoplasmic granule in bovine neutrophils, *J. Cell Biol.* **96:**1651–1661

Gennaro, R., Dolzani, L., and Romeo, D., 1983b, Potency of bactericidal proteins purified from the large granules of bovine neutrophils, *Infect. Immun.* **40:**684–690

Gennaro, R., Skerlavaj, B., and Romeo, D., 1989, Purification, composition, and activity of two bactenecins, antibacterial peptides of bovine neutrophils, *Infect. Immun.* **57:**3142–3146

Glauert, A. M., Butterworth, A. E., Sturrock, R. F., and Houba, V., 1978, The mechanism of the antibody-dependent, eosinophil-mediated damage to schistosomula of *Schistosoma mansoni in vitro:* A study by phase contrast and electron microscopy, *J. Cell Sci.* **34:**173–192

Gleich, G. J., and Adolphson, C. R., 1986, The eosinophilic leukocyte: Structure and function, *Adv. Immunol.* **39:**177–253

Gleich, G. J., Loegering, D. A., and Maldonado, J. E., 1973, Identification of a major basic protein in guinea pig eosinophil granules, *J. Exp. Med.* **137:**1459–1471

Gleich, G. J., Loegering, D. A., Kueppers, F., Bajaj, S. P., and Mann, K. G., 1974, Physiochemical and biological properties of the major basic protein from guinea pig eosinophil granules, *J. Exp. Med.* **140:**313–332

Gleich, G. J., Loegering, D. A., Mann, K. G., and Maldonado, J. E., 1976, Comparative properties of the Charcot-Leyden crystal protein and the major basic protein from human eosinophils, *J. Clin. Invest.* **57:**633–640

Gleich, G. J., Frigas, E., Loegering, D. A., Wassom, D. L., and Steinmuller, D., 1979, Cytotoxic properties of the eosinophil major basic protein, *J. Immunol.* **123:**2925–2927

Gleich, G. J., Loegering, D. A., Bell, M. P., Checkel, J. L., Ackerman, S. J., and McKean, D. J., 1986, Biochemical and functional similarities between human eosinophil-derived neurotoxin and eosinophil cationic protein: Homology with ribonuclease, *Proc. Natl. Acad. Sci. USA* **83:**3146–3150

Gray, P. W., Flaggs, G., Leong, S. R., Gumina, R. J., Weiss, J., Ooi, C. E., and Elsbach, P., 1989, Cloning of the cDNA of a human neutrophil bactericidal protein. Structural and functional correlations, *J. Biol. Chem.*, **264:**9505–9509

Henderson, W. R., Chi, E. Y., Jorg, A., and Klebanoff, S. J., 1973, Horse eosinophil degranulation induced by the ionophore A23187. Ultrastructure and role of phospholipase A$_2$, *Am. J. Pathol.* **111:**341–349

Henson, P. M., 1971, The immunologic release of constituents from neutrophil leukocytes. II. Mechanisms of release during phagocytosis and adherence to non phagocytosable surfaces, *J. Immunol.* **107:**1547–1557

Henson, P. M., Henson, J. E., Fittschen, C., Kimani, G., Bratton, D. L., and Riches, D. W. H., 1988, Phagocytic cells: degranulation and secretion, in *Inflammation: Basic Principles and Clinical Correlates* (J. I. Gallin, I. M. Goldstein, and R. Snyderman, eds.), pp. 363–390, Raven Press, New York

Hirsch, J. G., 1956, Phagocytin: A bactericidal substance from polymorphonuclear leukocytes, *J. Exp. Med.* **103:**589–592

In't Veld, G., Mannion, B., Weiss, J., and Elsbach, P., 1988, Effects of the bactericidal/permeability-increasing protein of polymorphonuclear leukocytes on isolated bacterial cytoplasmic membrane vesicles, *Infect. Immun.* **56:**1203–1208

Isackson, P. J., Ullrich, A., and Bradshaw, R. A., 1984, Mouse 7S nerve growth factor: complete sequence of a cDNA coding for the α-subunit precursor and its relationship to serine proteases, *Biochemistry* **23:**5997–6002

Jesaitis, A. J., Naemura, J. R., Painter, R. G., Sklar, L. A., and Cochrane, C. G., 1982, Intracellular localization of N-formyl chemotactic receptor and Mg^{2+} dependent ATPase in human granulocytes, *Biochim. Biophys. Acta* **719:**556–568

Kagan, B. L., 1983, Mode of action of yeast killer toxins: Channel formation in lipid bilayer membranes, *Nature* (London) **302:**709–711

Kierszenbaum, F., Ackerman, S. J., and Gleich, G. J., 1981, Destruction of bloodstream forms of *Trypanosoma cruzi* by eosinophil granule major basic protein, *Am. J. Trop. Med. Hyg.* **30:**775–779

Klebanoff, S. J., 1988, Phagocytic cells: Products of oxygen metabolism, in *Inflammation: Basic Principles and Clinical Correlates* (J. I. Gallin, I. M. Goldstein, and R. Snyderman, eds.), pp. 391–444, Raven Press, New York

Konisky, J., 1982, Colicins and other bacteriocins with established modes of action, *Annu. Rev. Microbiol.* **36:**125–144

Lehrer, R. I., Selsted, M., Szklarek, D., and Fleischmann, J., 1983, Antibacterial activity of microbicidal cationic proteins 1 and 2, natural peptide antibiotics of rabbit lung macrophages, *Infect. Immun.* **42:**10–14

Lehrer, R. I., Daher, K., Ganz, T., and Selsted, M. E., 1985a, Direct inactivation of viruses by MCP-1 and MCP-2, natural peptide antibiotics from rabbit leukocytes, *J. Virol.* **54:**467–472

Lehrer, R. I., Szklarek, D., Ganz, T., and Selsted, M., 1985b, Correlation of binding of rabbit granulocyte peptides to *Candida albicans* with candidacidal activity, *Infect. Immun.* **49:**207–211

Lehrer, R. I., Ganz, T., and Selsted, M. E., 1988a, Oxygen-independent bactericidal systems: Mechanisms and disorders, *Hematol. Oncol. Clin. N. Am.* **2:**159–169

Lehrer, R. I., Ganz, T., Szklarek, D., and Selsted, M., 1988b, Modulation of the *in vitro* candidacidal activity of human neutrophil defensins by target cell metabolism and divalent cations, *J. Clin. Invest.* **81:**1829–1835

Lehrer, R. I., Szklarek, D., Barton, A., Ganz, T., Hamann, K. J., and Gleich, G. J., 1989a, Antibacterial properties of eosinophil major basic protein and eosinophil cationic protein, *J. Immunol.* **142:**4428–4434

Lehrer, R. I., Barton, A., Daher, K. A., Harwig, S. S. L., Ganz, T., and Selsted, M. E., 1989b, Interaction of human defensins with *Escherichia coli*. Mechanism of bactericidal activity, *J. Clin. Invest.* **84:**553–561

Lew, P. D., Monod, A., Waldvogel, F. A., Dewald, B., Baggiolini, M., and Pozzan, T., 1986, Quantitative analysis of the cytosolic free calcium dependency of exocytosis from three subcellular compartments in intact human neutrophils, *J. Cell Biol.* **102:**2197–2204

Lewis, D. M., Loegering, D. A., and Gleich, G. J., 1976, Isolation and partial characterization of a major basic protein from rat eosinophil granules, *Proc. Soc. Exp. Biol. Med.* **152:**512–515

Lewis, D. M., Lewis, J. C., Loegering, D. A., and Gleich, G. J., 1978, Localization of the guinea pig eosinophil major basic protein to the core of the granule, *J. Cell Biol.* **77:**702–713

Lichtenstein, A. K., Ganz, T., Selsted, M. E., and Lehrer, R. I., 1986, *In vitro* tumor cell cytolysis mediated by peptide defensins of human and rabbit granulocytes, *Blood* **68:**1407–1410

Lichtenstein, A. K., Ganz, T., Nguyen, T. M., Selsted, M. E., and Lehrer, R. I., 1988, Mechanism of target cytolysis by peptide defensins. Target cell metabolic activities, possibly involving endocytosis, are crucial for expression of cytotoxicity, *J. Immunol.* **140:**2686–2694

Lugtenberg, B., and van Alphen, L., 1983, Molecular architecture and functioning of the outer membrane of *Escherichia coli* and other gram-negative bacteria, *Biochim. Biophys. Acta* **737**:51–115

Maddox, D. E., Kephart, G. M., Coulam, C. B., Butterfield, J. H., Benirschke, K., and Gleich, G. J., 1984, Localization of a molecule immunochemically similar to eosinophil major basic protein in human placenta, *J. Exp. Med.* **160**:29–41

Mandell, G. L., 1970, Intraphagosomal pH of human polymorphonuclear neutrophils, *Proc. Soc. Exp. Biol. Med.* **134**:447–449

Mandell, G. L., 1974, Bactericidal activity of aerobic and anaerobic polymorphonuclear neutrophils, *Infect. Immun.* **9**:337–341

Mannion, B. A., Kalatzis, E. S., Weiss, J., and Elsbach, P., 1989, Preferential binding of the neutrophil cytoplasmic granule-derived bactericidal/permeability increasing protein to target bacteria. Implications and use as a means of purification, *J. Immunol.* **142**:2807–2812

Marion, D., Zasloff, M., and Bax, A., 1988, A two-dimensional NMR study of the antimicrobial peptide magainin 2, *FEBS Lett.* **227**:21–26

Matzner, Y., Bar-Ner, M., Yahalom, J., Ishai-Michaeli, R., Fuks, Z., and Vlodavsky, I., 1985, Degradation of haparan sulfate in the subendothelial extracellular matrix by a readily released heparanase from human neutrophils: Possible role in invasion through basement membranes, *J. Clin. Invest.* **76**:1306–1313

McCarthy, K. M., Musson, R. A., and Henson, P. M., 1982, Protein synthesis dependent and protein synthesis independent secretion of lysosomal hydrolases from rabbit and human macrophages, *J. Reticuloendothelial. Soc.* **31**:131–144

McGrogan, M., Simonsen, C., Scott, R., Griffith, J., Ellis, N., Kennedy, J., Campanelli, D., Nathan, C., and Gabay, J., 1988, Isolation of a complementary DNA clone encoding a precursor to human eosinophil major basic protein, *J. Exp. Med.* **168**:2295–2308

McLaren, D. J., McKean, J. R., Olsson, I., Venge, P., and Kay, A. B., 1981, Morphological studies on the killing of schistosomula of *Schistosoma mansoni* by human eosinophil and neutrophil cationic proteins *in vitro*, *Parasite Immunol.* **3**:359–373

Mottola, C., Gennaro, R., Marzullo, A., and Romeo, D., 1980, Isolation and partial characterization of the plasma membrane of purified bovine neutrophils, *Eur. J. Biochem.* **111**:341–346

Muller-Eberhard, H. J., 1986, The membrane attack complex of complement, *Annu. Rev. Immunol.* **4**:503–528

Murphy, G., Reynolds, J. J., Bretz, U., and Baggiolini, M., 1977, Collagenase is a component of the specific granules of human neutrophil leukocytes, *Biochem. J.* **162**:195–197

Neville, D. M., Jr., and Hudson, T. H., 1986, Transmembrane transport of diphtheria toxin, related toxins, and colicins, *Annu. Rev. Biochem.* **55**:195–224

Odeberg, H., and Olsson, I., 1975, Antibacterial activity of cationic proteins from human granulocytes, *J. Clin. Invest.* **56**:1118–1124

Odeberg, H., and Olsson, I., 1976a, Mechanisms for the microbicidal activity of cationic proteins of human granulocytes, *Infect. Immun.* **14**:1269–1275

Odeberg, H., and Olsson, I., 1976b, Microbicidal mechanisms of human granulocytes: Synergistic effects of granulocyte elastase and myeloperoxidase or chymotrypsin-like cationic protein, *Infect. Immun.* **14**:1276–1283

Odeberg, H., Olsson, I., and Venge, P., 1975, Cationic proteins of human granulocytes. IV. Esterase activity, *Lab. Invest.* **32**:86–90

O'Donnell, M. C., Ackerman, S. J., Gleich, G. J., and Thomas, L. L., 1983, Activation of basophil and mast cell histamine release by eosinophil granule major basic protein, *J. Exp. Med.* **157**:1981–1991

Ohlsson, K., Olsson, I., and Spitznagel, J. K., 1977, Localization of chymotrypsin-like cationic protein, collagenase and elastase in azurophil granules of human neutrophilic polymorphonuclear leukocytes, *Hoppe-Seyler's Z. Physiol. Chem.* **358**:361–366

Olsson, I., and Venge, P., 1974, Cationic proteins of human granulocytes. II. Separation of the cationic proteins of the granules of leukemic myeloid cells, *Blood* **44**:235–246

Olsson, I., Venge, P., Spitznagel, J. K., and Lehrer, R. I., 1977, Arginine-rich cationic proteins of human eosinophil granules. Comparison of the constituents of eosinophilic and neutrophilic leukocytes, *Lab. Invest.* **36**:493–500

Olsson, I., Persson, A. M., and Winqvist, I., 1986, Biochemical properties of the eosinophil cationic protein and demonstration of its biosynthesis in vitro in marrow cells from patients with an eosinophilia, *Blood* **67**:498–503

Ooi, C. E., Weiss, J., Elsbach, P., Frangione, B., and Mannion, B., 1987, A 25-kDa NH$_2$-terminal fragment carries all the antibacterial activities of the human neutrophil 60-kDa bactericidal/permeability-increasing protein, *J. Biol. Chem.* **262**:14891–14894

O'Shea, J. J., Brown, E. J., Seligman, B. E., Metcalf, J. A., Frank, M. M., and Gallin, J. I., 1985, Evidence for distinct intracellular pools of receptors for C3b and C3bi in human neutrophils, *J. Immunol.* **134**:2580–2587

Pardi, A., Hare, D. R., Selsted, M. E., Morrison, R. D., Bassolino, D. A., and Bach, A. C., II, 1988, Solution structures of the rabbit neutrophil defensin NP-5, *J. Mol. Biol.* **201**:625–636

Parmley, R. T., and Spicer, S. S., 1974, Cytochemical and ultrastructural identification of a small type granule in human late eosinophils, *Lab. Invest.* **30**:557–567

Patterson-Delafield, J., Martinez, R. J., and Lehrer, R. I., 1980, Microbicidal cationic proteins in rabbit alveolar macrophages: A potential host defense mechanism, *Infect. Immun.* **30**:180–192

Patterson-Delafield, J., Szklarek, D., Martinez, R. J., and Lehrer, R. I., 1981, Microbicidal cationic proteins of rabbit alveolar macrophages: Amino acid composition and functional attributes, *Infect. Immun.* **31**:723–731

Perham, R. N., Packman, L. C., and Radford, S. E., 1987, 2-Oxo acid dehydrogenase multi-enzyme complexes: In the beginning and halfway there, *Biochem. Soc. Symp.* **54**:67–81

Peters, M. S., Rodriguez, M., and Gleich, G. J., 1986, Localization of human eosinophil granule major basic protein, eosinophil cationic protein, and eosinophil-derived neurotoxin by immunoelectron microscopy, *Lab. Invest.* **54**:656–662

Podack, E. R., Young, J. D. E., and Cohn, Z. A., 1985, Isolation and biochemical and functional characterization of perforin 1 from cytolytic T-cell granules, *Proc. Natl. Acad. Sci. USA* **82**:8629–8633

Rice, W. G., Kinkade, J. M., Jr., and Parmley, R. T., 1986, High resolution of heterogeneity among human neutrophil granules: Physical, biochemical, and ultrastructural properties of isolated fractions, *Blood* **68**:541–555

Rice, W. G., Ganz, T., Kinkade, J. M., Jr., Selsted, M. E., Lehrer, R. I., and Parmley, R. T., 1987, Defensin-rich dense granules of human neutrophils, *Blood* **70**:757–765

Roberts, R. L., and Gallin, J. I., 1985, Rapid method for isolation of normal human peripheral blood eosinophils on discontinuous Percoll gradients and comparison with neutrophils, *Blood* **65**:433–440

Romeo, D., Skerlavaj, B., Bolognesi, M., and Gennaro, R., 1988, Structure and bactericidal activity of an antibiotic dodecapeptide purified from bovine neutrophils, *J. Biol. Chem.* **263**:9573–9575

Root, R. K., Rosenthal, A. S., and Balestra, D. J., 1972, Abnormal bactericidal, metabolic, and lysosomal functions of Chédiak-Higashi syndrome leukocytes, *J. Clin. Invest.* **51**:649–665

Rosenberg, H. F., Ackerman, S. J., and Tenen, D. G., 1989, Human eosinophil cationic protein. Molecular cloning of a cytotoxin and helminthotoxin with ribonuclease activity, *J. Exp. Med.* **170**:163–176

Rotrosen, D., and Gallin, J., 1987, Disorders of phagocyte function, *Annu. Rev. Immunol.* **5**:127–150

Salvesen, G., Farley, D., Shuman, J., Przybyla, A., Reilly, C., and Travis, J., 1987, Molecular cloning of human cathepsin G: structural similarity to mast cell and cytotoxic T lymphocyte proteinases, *Biochemistry* **26**:2289–2293

Sandvig, K., and Olsnes, S., 1982, Entry of the toxic proteins abrin, modeccin, ricin, and diphtheria toxin into cells. II. Effect of pH, metabolic inhibitors, and ionophores and evidence for toxin penetration from endocytotic vesicles, *J. Biol. Chem.* **257**:7504–7513

Savoini, A., Marzari, R., Dolzani, L., Serranò, D., Graziosi, G., Gennaro, R., and Romeo, D., 1984, Wide-spectrum antibiotic activity of bovine granulocyte polypeptides, *Antimicrob. Agents Chemother.* **26**:405–407

Schnyder, J., and Baggiolini, M., 1978, Secretion of lysosomal hydrolases by stimulated and non-stimulated macrophages, *J. Exp. Med.* **148**:435–450

Segal, A. W., and Jones, O. T. G., 1979, The subcellular distribution and some properties of the cytochrome b component of the microbicidal oxidase system of human neutrophils, *Biochem. J.* **182**:181–188

Segal, A. W., Geisow, M., Garcia, R., Harper, A., and Miller, R., 1981, The respiratory burst of phagocytic cells is associated with a rise in vacuolar pH, *Nature* (London) **290**:406–409

Selsted, M. E., and Harwig, S. S. L., 1987, Purification, primary structure, and antimicrobial activities of a guinea pig neutrophil defensin, *Infect. Immun.* **55**:2281–2286

Selsted, M. E., and Harwig, S. S. L., 1989, Determination of the disulfide array in the human defensin HNP-2. A covalently cyclized peptide, *J. Biol. Chem.* **264**:4003–4007

Selsted, M. E., Brown, D. M., DeLange, R. J., and Lehrer, R. I., 1983, Primary structures of MCP-1 and MCP-2, natural peptide antibiotics of rabbit lung macrophages, *J. Biol. Chem.* **258**:14485–14489

Selsted, M. E., Szklarek, D., and Lehrer, R. I., 1984, Purification and antibacterial activity of antimicrobial peptides of rabbit granulocytes, *Infect. Immun.* **45**:150–154

Selsted, M. E., Brown, D. M., DeLange, R. J., Harwig, S. S. L., and Lehrer, R. I., 1985a, Primary structures of six antimicrobial peptides of rabbit peritoneal neutrophils, *J. Biol. Chem.* **260**:4579–4584

Selsted, M. E., Szklarek, D., Ganz, T., and Lehrer, R. I., 1985b, Activity of rabbit leukocyte peptides against *Candida albicans*, *Infect. Immun.* **49**:202–206

Selsted, M. E., Harwig, S. S. L., Ganz, T., Schilling, J. W., and Lehrer, R. I., 1985c, Primary structures of three human neutrophil defensins, *J. Clin. Invest.* **76**:1436–1439

Shafer, W. M., 1988, Lipopolysaccharide masking of gonococcal outer-membrane proteins modulates binding of bactericidal cathepsin G to gonococci, *J. Gen. Microbiol.* **134**:539–545

Shafer, W. M., and Onunka, V. C., 1989, Mechanism of staphylococcal resistance to non-oxidative antimicrobial action of neutrophils: importance of pH and ionic strength in determining the bactericidal action of cathepsin G, *J. Gen. Microbiol.* **135**:825–830

Shafer, W. M., Martin, L. E., and Spitznagel, J. K., 1984, Cationic antimicrobial proteins isolated from human neutrophil granulocytes in the presence of diisopropyl fluorophosphate, *Infect. Immun.* **45**:29–35

Shafer, W. M., Martin, L. E., and Spitznagel, J. K., 1986a, Late intraphagosomal hydrogen ion concentration favors the in vitro antimicrobial capacity of a 37-kilodalton cationic granule protein of human neutrophil granulocytes, *Infect. Immun.* **53**:651–655

Shafer, W. M., Onunka, V. C., and Martin, L. E., 1986b, Antigonococcal activity of human neutrophil cathepsin G, *Infect. Immun.* **54**:184–188

Shafer, W. M., Onunka, V. C., and Hitchcock, P. J., 1986c, A spontaneous mutant of *Neisseria gonorrhoeae* with decreased resistance to neutrophil granule proteins, *J. Infect. Dis.* **153**:410–417

Shafer, W. M., Engle, S. A., Martin, L. E., and Spitznagel, J. K., 1988, Killing of *Proteus mirabilis* by polymorphonuclear leukocyte granule proteins: Evidence for species specificity by antimicrobial proteins, *Infect. Immun.* **56**:51–53

Sher, R., and Wadee, A. A., 1981, Eosinophil degranulation. Monitoring by interference contrast microscopy, *Inflammation* **5**:37–53

Sheumack, D. D., Claassens, R., Whiteley, N. M., and Howden, M. E. H., 1985, Complete amino acid sequence of a new type of lethal neurotoxin from the venom of the funnel-web spider *Atrax robustus*, *FEBS Lett.* **181**:154–156

Singh, A., Bateman, A., Zhu, Q., Shimasaki, S., Esch, F., and Solomon, S., 1988, Structure of a novel human granulocyte peptide with anti-ACTH activity, *Biochem. Biophys. Res. Commun.* **155**:524–529

Skerlavaj, B., Romeo, D., and Gennaro, R., 1990, Rapid membrane permeabilization and inhibition of vital functions of gram negative bacteria by bactenecins, *Infect. Immun.* **58** (in press).

Slifman, N. R., Loegering, D. A., McKean, D. J., and Gleich, G. J., 1986, Ribonuclease activity associated with human eosinophil-derived neurotoxin and eosinophil cation protein, *J. Immunol.* **137**:2913–2917

Spitznagel, J. K., Dalldorf, F. G., Leffell, M. S., Folds, J. D., Welsh, I. R. H., Cooney, M. H., and Martin, L. E., 1974, Character of azurophil and specific granules purified from human polymorphonuclear leukocytes, *Lab. Invest.* **30**:774–785

Spitznagel, J. K., Pereira, H. A., Martin, L. E., Guzman, G. S., and Shafer, W. M., 1987, A monoclonal antibody that inhibits the antimicrobial action of a 57 kD cationic protein of human polymorphonuclear leukocytes, *J. Immunol.* **139**:1291–1296

Stanfield, R. L., Westbrook, E. M., and Selsted, M. E., 1988, Characterization of two crystal forms of human defensin neutrophil cationic peptide 1, a naturally occurring antimicrobial peptide of leukocytes, *J. Biol. Chem.* **263**:5933–5935

Tai, P. C., Spry, C. J. F., Peterson, C., Venge, P., and Olsson, I., 1984, Monoclonal antibodies distinguish between storage and secreted forms of eosinophil cationic protein, *Nature* (London) **309**:182–184

Takenaka, T., Okuda, M., Kawabori, S., and Kubo, K., 1977, Extracellular release of peroxidase from eosinophils by interaction with immune complexes, *Clin. Exp. Immunol.* **28**:56–63

Takeuchi, K., Wood, H., and Swank, R. T., 1986, Lysosomal elastase and cathepsin G in beige mice, *J. Exp. Med.* **163**:665–677

Terry, A. S., Poulter, L., Williams, D. H., Nutkins, J. C., Giovannini, M. G., Moore, C. H., and Gibson, B. W., 1988, The cDNA sequence coding for prepro-PGS (prepro-magainins) and aspects of the processing of this prepro-polypeptide, *J. Biol. Chem.* **263**:5745–5751

Thorne, K. J. I., Oliver, R., and Barrett, A. J., 1976, Lysis and killing of bacteria by lysosomal proteinases, *Infect. Immun.* **14**:555–563

Tobias, P. S., Soldau, K., and Ulevitch, R. J., 1986, Isolation of a lipopolysaccharide-binding acute phase reactant from rabbit serum, *J. Exp. Med.* **164**:777–793

Tobias, P. S., Mathison, J. C., and Ulevitch, R. J., 1988, A family of lipopolysaccharide binding proteins involved in responses to gram-negative sepsis, *J. Biol. Chem.* **263**:13479–13481

Vaara, M., Vaara, T., Jensen, M., Helander, I., Nurminen, M., Rietschel, E. T., and Makela, P. H., 1981, Characterization of the lipopolysaccharide from the polymyxin-resistant *pmrA* mutants of *Salmonella typhimurium*, *FEBS Lett.* **129**:145–149

Villalta, F., and Kierszenbaum, F., 1984, Role of inflammatory cells in Chagas' disease. I. Uptake and mechanism of destruction of intracellular (amastigote) forms of *Trypanosoma cruzi* by human eosinophils, *J. Immunol.* **132**:2053–2058

Wasmoen, T. L., Bell, M. P., Loegering, D. A., Gleich, G. J., Prendergast, F. G., and McKean, D. J., 1988, Biochemical and amino acid sequence analysis of human eosinophil granule major basic protein, *J. Biol. Chem.* **263**:12559–12563

Wassom, D. L., and Gleich, G. J., 1979, Damage to *Trichinella spiralis* newborn larvae by eosinophil major basic protein, *Am. J. Trop. Med. Hyg.* **28**:860–863

Weiss, J., and Elsbach, P., 1977, The use of a phospholipase A-less *Escherichia coli* mutant to establish the action of granulocyte phospholipase A on bacterial phospholipids during killing by a highly purified granulocyte fraction, *Biochim. Biophys. Acta* **466**:23–33

Weiss, J., and Olsson, I., 1987, Cellular and subcellular localization of the bactericidal/permeability-increasing protein of neutrophils, *Blood* **69**:652–659

Weiss, J., Franson, R. C., Beckerdite, S., Schmeidler, K., and Elsbach, P., 1975, Partial characterization and purification of a rabbit granulocyte factor that increases permeability of *Escherichia coli*, *J. Clin. Invest.* **55**:33–42

Weiss, J., Franson, R. C., Schmeidler, K., and Elsbach, P., 1976, Reversible envelope effects during and after killing of *Escherichia coli* W by a highly-purified rabbit polymorphonuclear leukocyte fraction, *Biochim. Biophys. Acta* **436**:154–169

Weiss, J., Elsbach, P., Olsson, I., and Odeberg, H., 1978, Purification and characterization of a potent bactericidal and membrane active protein from the granules of human polymorphonuclear leukocytes, *J. Biol. Chem.* **253**:2664–2672

Weiss, J., Beckerdite-Quagliata, S., and Elsbach, P., 1979, Determinants of the action of phospholipases A on the envelope phospholipids of *Escherichia coli*, *J. Biol. Chem.* **254**:11010–11014

Weiss, J., Beckerdite-Quagliata, S., and Elsbach, P., 1980, Resistance of gram-negative bacteria to purified bactericidal leukocyte proteins. Relation to binding and bacterial lipopolysaccharide structure, *J. Clin. Invest.* **65**:619–628

Weiss, J., Victor, M., Stendhal, O., and Elsbach, P., 1982, Killing of gram-negative bacteria by poly-morphonuclear leukocytes. Role of an O_2-independent bactericidal system, *J. Clin. Invest.* **69**:959–970

Weiss, J., Victor, M., and Elsbach, P., 1983, Role of charge and hydrophobic interactions in the action of the bactericidal/permeability-increasing protein of neutrophils on gram-negative bacteria, *J. Clin. Invest.* **71**:540–549

Weiss, J., Muello, K., Victor, M., and Elsbach, P., 1984, The role of lipopolysaccharides in the action of the bactericidal/permeability-increasing neutrophil protein on the bacterial envelope, *J. Immunol.* **132**:3109–3114

Weiss, J., Kao, L., Victor, M., and Elsbach, P., 1985, Oxygen-independent intracellular and oxygen-dependent extracellular killing of *Escherichia coli* S15 by human polymorphonuclear leukocytes, *J. Clin. Invest.* **76**:206–212

Weiss, J., Hutzler, M., and Kao, L., 1986, Environmental modulation of lipopolysaccharide chain length alters the sensitivity of *Escherichia coli* to the neutrophil bactericidal/permeability-increasing protein, *Infect. Immun.* **51**:594–599

Westbrook, E. M., Lehrer, R. I., and Selsted, M. E., 1984, Characterization of two crystal forms of neutrophil cationic protein NP2, a naturally occurring broad-spectrum antimicrobial agent from leukocytes, *J. Mol. Biol.* **178**:783–785

White, J. G., and Clawson, C. C., 1980, The Chédiak-Higashi syndrome: The nature of the giant neutrophil granules and their interactions with cytoplasm and foreign particulates, *Am. J. Pathol.* **98**:151–167

Wright, D. G., and Gallin, J. I., 1979, Secretory responses of human neutrophils: Exocytosis of specific (secretory) granules by human neutrophils during adherence in vitro and during exudation in vivo, *J. Immunol.* **123**:285–294

Wright, D. G., Bralove, D. A., and Gallin, J. I., 1977, The differential mobilization of human neutrophil granules: Effects of phorbol myristate acetate and ionophore A23187, *Am. J. Pathol.* **87**:273–284

Yamashita, T., and Saito, K., 1989, Purification, primary structure, and biological activity of guinea pig neutrophil cationic peptides, *Infect. Immun.* **57**:2405–2409

Yazdanbakhsh, M., Tai, P. C., Spry, C. J. F., Gleich, G. J., and Roos, D., 1987, Synergism between eosinophil cationic protein and oxygen metabolites in killing of *Schistosoma mansoni*, *J. Immunol.* **138**:3443–3447

Yoon, P. S., Boxer, L. A., Mayo, L. A., Yang, A. Y., and Wicha, M. S., 1987, Human neutrophil laminin receptors: Activation-dependent receptor expression, *J. Immunol.* **138**:259–265

Young, J. D. E., Young, T. M., Lu, L. P., Unkeless, J. C., and Cohn, Z. A., 1982, Characterization of a membrane pore-forming protein from *Entamoeba histolytica*, *J. Exp. Med.* **156**:1677–1690

Young, J. D. E., Hengartner, H., Podack, E. R., and Cohn, Z. A., 1986a, Purification and characterization of a cytolytic pore-forming protein from granules of cloned lymphocytes with natural killer activity, *Cell* **44**:849–859

Young, J. D. E., Peterson, C. G. B., Venge, P., and Cohn, Z. A., 1986b, Mechanism of membrane damage mediated by human eosinophil cationic protein, *Nature* (London) **321**:613–616

Young, J. D. E., Cohn, Z. A., and Podack, E. R., 1986c, The ninth component of complement and the pore-forming protein (perforin 1) from cytotoxic T cells: Structural, immunological, and functional similarities, *Science* **233**:184–190

Zanetti, M., Litteri, L., Gennaro, R., Horstmann, H., and Romeo, D., 1990, Bactenecins, defense polypeptides of bovine neutrophils, are generated from precursor molecules stored in the large granules, *J. Cell Biol.* **111** (in press)

Zasloff, M., 1987, Magainins, a class of antimicrobial peptides from *Xenopus* skin: Isolation, characterization of two active forms, and partial cDNA sequence of a precursor, *Proc. Natl. Acad. Sci. USA* **84**:5449–5453

Zerial, A., Skerlavaj, B., Gennaro, R., and Romeo, D., 1987, Inactivation of herpes simplex virus by protein components of bovine neutrophil granules, *Antiviral Res.* **7**:341–352

Zeya, H. I., and Spitznagel, J. K., 1963, Antibacterial and enzymic basic proteins from leukocyte lysosomes: Separation and identification, *Science* **142**:1085–1087

Zeya, H. I., and Spitznagel, J. K., 1966a, Cationic proteins of polymorphonuclear leukocytes. I. Resolution of antibacterial and enzymatic activities, *J. Bacteriol.* **91**:750–754

Zeya, H. I., and Spitznagel, J. K., 1966b, Cationic proteins of polymorphonuclear leukocytes. II. Composition, properties and mechanisms of antibacterial action, *J. Bacteriol.* **91**:755–762

Zeya, H. I., and Spitznagel, J. K., 1966c, Antimicrobial specificity of leukocyte lysosomal cationic proteins, *Science* **154**:1049–1051

Zhu, Q., Hu, J., Mulay, S., Esch, F., Shimasaki, S., and Solomon, S., 1988, Isolation and structure of corticostatin peptides from rabbit fetal and adult lung, *Proc. Natl. Acad. Sci. USA* **85**:592–596

Zucker-Franklin, D., 1980, Eosinophil structure and maturation, in: *The Eosinophil in Health and Disease* (A. A. F. Mahmoud and F. K. Austen, eds.), pp. 43–59, Grune & Stratton, New York

Chapter 13

Membrane Glycoproteins of Mast Cells and Basophils

Eric F. Rimmer and Michael A. Horton

1. INTRODUCTION

Mast cells (MC) and basophils are morphologically similar, heavily granulated cells of bone marrow origin, basophils being a circulating cell type and MC a numerically minor component of many tissues. Both cells are characterized by their granular content of bioactive amines, enzymes, and other mediators (reviewed in Schwartz and Austen, 1984).

The most conspicuous biological role of MC/basophils is in anaphylaxis, the immediate induction of smooth muscle spasm and vasopermeability due to activation of MC/basophils. Stimulated release of granule-stored mediators results in immediate inflammatory effects, and MC/basophil activation may also lead to *de novo* synthesis of bioactive arachidonic acid metabolites. The most important physiological routes to degranulation of MC/basophils are (1) via the interaction of specific antigen with the membrane complex of immunoglobulin E (IgE) and its MC/basophil-specific receptor and (2) by the action of soluble cleavage products of the complement cascade, although other physiological and nonphysiological stimuli are known (reviewed in Galli *et al.*, 1984, Ishizaka and Ishizaka, 1984).

MC/basophils may be of considerable importance in the initiation of inflammatory lesions. Their products include vasoactive agents, leukocyte chemoattractants and activating factors, and enzymes able to generate active molecules from soluble precursors (reviewed in Schwartz and Austen, 1984). These products facilitate movement of effector cells and molecules from the bloodstream into the tissue spaces at inflammatory sites.

Eric F. Rimmer and Michael A. Horton   Imperial Cancer Research Fund Haemopoiesis Research Group, Department of Haematology, St. Bartholomew's Hospital, London EC1A 7BE, United Kingdom.

MC and basophils, although not closely related cell types, are functionally similar and may be regarded as the two arms of the immediate inflammatory response. Circulating basophils are able to initiate or participate in an inflammatory response wherever they encounter appropriate stimuli. Although their mediator content is small, they are comparatively numerous. MC, in contrast, are "sentinel" cells, responding only to stimuli arising near their home site. It is evident that MC themselves divide into two classes, characterized by the tendency to locate either in connective tissue and fluid exudates or near mucosal sites; hence they are called connective tissue and mucosal mast cells (CTMC and MMC), respectively. CTMC and MMC differ in mediator and enzyme content and in granule proteoglycan type. MMC proliferate *in situ* in response to T-lymphocyte-derived growth factors, leading to pronounced mastocytoses in many mucosal immune reactions, such as those provoked by helminth parasites (reviewed in Jarrett and Haig, 1984). MMC and CTMC are most readily distinguishable in rodents, but studies of lung MC mediator and proteoglycan content suggests that the distinction is also valid in humans (Stevens *et al.*, 1988).

In most cases, the molecules discussed in this review, although present on MC/basophils, have been studied only on other cell types. Others, such as receptors for the complement components C3a and C5a, have not been formally demonstrated on MC/basophils but may be inferred from the presence of the corresponding bioactivity. The data are also drawn from studies in several species, notably rat, mouse, guinea pig, and human. This is unavoidable, since MC/basophils have largely been studied for their functional role in inflammation and for the convenience of the IgE–receptor system as a model of receptor–ligand interaction and receptor-mediated control of secretion. The rat basophilic leukemia (RBL) cell line has been invaluable in such studies. Data on MC/basophil surface biochemistry must, therefore, be assembled from disparate sources. The aim of this review is to describe membrane glycoproteins of MC/basophils and to comment on their contribution to the effector role of these two cell types.

2. RECEPTORS FOR COMPLEMENT ANAPHYLATOXINS ON MAST CELLS AND BASOPHILS

2.1. The Anaphylatoxins

2.1.1. Overview

Proteolytic cleavage of the C3 or C5 complement component liberates diffusible, bioactive minor polypeptide fragments, C3a and C5a, respectively. Both polypeptides induce degranulation of tissue MC/basophils, immediate inflammatory effects of which include smooth muscle contraction and increased vascular permeability (Dias da Silva *et al.*, 1967; Cochrane and Muller-Eberhard, 1968; Hook *et al.*, 1975). This bioactivity, historically known as anaphylatoxicity, is central to the participation of MC/basophils in tissue inflammatory responses. Direct and secondary (via MC/basophil-produced chemotactic factors) effects of anaphylatoxins on other leukocytes lead to accumulation of inflammatory cells at sites of complement activation or anaphylatoxin administration (Snyderman *et al.*, 1971; Kay and Austen, 1972; Kay *et al.*, 1973; Fernandez *et al.*, 1978; Shaw *et al.*, 1978; Stimler *et al.*, 1981; Yancey *et al.*, 1985). Thus anaphylatoxin responses of MC/basophils contribute both to initiation and to maintenance of the inflammatory lesion.

Cell surface receptor molecules specific for C5a have been identified on human neutrophils and myeloid cell lines, and their interaction with C5a has been extensively characterized. A distinct C3a receptor from guinea pig platelets has also been identified and partly characterized. Neither receptor type has yet been identified on either MC or basophils.

There is, however, considerable functional evidence for specific interaction of anaphylatoxins with MC/basophils (see below). Several features of the MC/basophil anaphylatoxin response, notably stimulus-specific tachyphylaxis (Chenoweth and Huey et al., 1986), unresponsiveness to either molecular bereft of terminal arginine (desArg derivatives; Bokisch and Muller-Eberhard, 1970; Gerard and Hugli, 1981; Schulman et al., 1981; Hugli, 1981), response to C3a oligopeptides (Hugli, 1981), and dose response to purified anaphylatoxins (Huey et al., 1986), are consistent with the presence on MC/basophils of specific receptor molecules. This evidence is described in detail below.

2.1.2. Biological Activities of C3a and C5a

Both anaphylatoxins are immediately effective in *in vitro* or *in vivo* bioassays of smooth muscle contraction or increased vascular permeability (reviewed in Hugli, 1981). Tissue fragments, rodent peritoneal MC, or human blood leukocytes release histamine and other MC mediators when challenged *in vitro* with either anaphylatoxin (Johnson et al., 1975; Siraganian and Hook, 1976; Glovsky et al., 1979; Regal, 1982; Stimler et al., 1983; Yancey et al., 1985). Antihistamines and inhibitors of cyclooxygenase or lipooxygenase metabolism oppose smooth muscle effects of C5a and C3a, evidence that C5a and C3a induce both release of preformed mediators and *de novo* synthesis of prostaglandins and leukotrienes by MC/basophils (Regal, 1982; Stimler et al., 1983).

The distinctive property of C5a is leukocyte chemoattractant activity, observed in both *in vivo* and *in vitro* bioassays of leukocyte migration or accumulation at challenge sites (Snyderman et al., 1971; Kay et al., 1973; Fernandez et al., 1978; McCarthy and Henson, 1979; Yancey et al., 1985). C3a, in contrast, lacks chemoattractant activity (Fernandez et al., 1978). C5a desArg, produced following deletion of the C-terminal arginine by serum carboxypeptidase B, retains residual chemoattractant activity and stimulates neutrophil granule exocytosis (Webster et al., 1980), although C5a of most species lacks MC/basophil-degranulating activity (see below). Release of leukocyte-derived effector molecules probably contributes to tissue damage at inflammatory sites (Henson et al., 1979; Shaw et al., 1980).

Within seconds of C5a challenge, membrane-bound vesicles containing intracellular stores of leukocyte function-associated antigens (LFAs) (see Section 5.1.1) fuse with the plasma membrane of myeloid cells, including basophils, greatly enhancing adhesiveness (Springer et al., 1984; Springer and Anderson, 1985). We suggest that this phenomenon reflects a coordinating role for anaphylatoxins in local inflammation. Thus, anaphylatoxin-stimulated release of mediators from MC/basophils causes vasoconstriction and vascular permeability. Vasoconstriction slows blood flow, allowing leukocytes to contact the endothelium; C5a-enhanced LFA expression and chemotactic activity then facilitate leukocyte adhesion and extravasation.

Anaphylatoxin responsiveness may not be a general property of mucosal MC. The rat MMC-like cell line RBL is unresponsive to both anaphylatoxins, nor are all interleukin 3 (IL3)-dependent, mouse bone marrow-derived MMC-like cell lines and clones responsive (Huey *et al.*, 1986). Isolated human lung MC, the granule proteoglycan profile of which suggests equivalence to rodent MMC (Stevens *et al.*, 1988), are unresponsive to C5a (Schulman *et al.*, 1981). Unresponsiveness of MMC may limit inflammation at vulnerable mucosal sites.

Table I summarizes the biological activities of C3a and C5a on MC/basophils.

2.2. Characterization of Anaphylatoxin Receptors

2.2.1. Evidence for Distinct C3a and C5a Receptors

The distinct biological activities of C3a and C5a originally suggested that they bound to different cellular receptors. In bioassays of their effects on MC/basophils, both molecules demonstrate specific tachyphylaxis (induction of stimulus-specific unresponsiveness to further exposure) but not cross-tachyphylaxis (Johnson *et al.*, 1975; *Hugli*, 1981; Huey *et al.*, 1986). Thus, anaphylatoxin activities are independent and additive. Similarly, specific tachyphylaxis characterizes neutrophil responses to C5a.

Analyses of C5a and C3a receptors in whole leukocytes or membrane preparations demonstrates distinct patterns of cellular expression. ^{125}I-C5a binds with high affinity to polymorphonuclear cells, mononuclear phagocytes, and some myeloid cell lines (Chenoweth and Hugli, 1978; Chenoweth *et al.*, 1982). ^{125}I-C3a does not inhibit C5a binding to its receptor on neutrophils (Chenoweth and Hugli, 1978). Fluorescein isothiocyanate (FITC)-conjugated C5a binds to almost all polymorphs and most monocytes, as assessed by fluorescence-activated cell sorting. Within the limits of sensitivity of this technique, C3a does not bind to blood leukocytes (Van Epps and Chenoweth, 1984).

Prior to identification of candidate receptor molecules, the most compelling evidence for distinct C5a and C3a receptors arose from studies using synthetic peptide fragments of the two molecules (Glovsky *et al.*, 1979; Hugli, 1981; Huey *et al.*, 1986). Peptides representing the COOH-terminal end of C3a retain spasmogenic activity, the smallest effective peptide being Leu-Gly-Leu-Ala-Arg-COOH (residues 73–77 of C3a). Modification of C3a peptides identifies three crucial residues; substitution of either Leu-73 or Leu-75, deletion of Arg-77, and addition of an extra COOH-terminal residue all abrogate bioactivity (Hugli, 1981). Longer human C3a peptides, containing residues 56–77, are equipotent to native C3a. Circular dichroism revealed that residues 56–71 formed an organized α helix (Lu *et al.*, 1984), suggesting that secondary structure of the C3a molecule enhances interaction of the COOH-terminal receptor-binding portion with the receptor.

Peptides corresponding to the C5a COOH terminus have little or no anaphylatoxic activity, indicating that binding sites elsewhere on C5a are important (Hugli, 1981). The COOH-terminal arginine of C5a is not crucial to activity, although C5a desArg is much less biologically active than is native C5a (Cochrane and Muller-Eberhard, 1968; Gerard and Hugli, 1981; Webster *et al.*, 1979). Pig C5a bereft of both COOH-terminal Arg-74 and Gly-73 retains residual activity; removal of Leu-72 abolishes bioactivity (Gerard and Hugli, 1981).

Table I

Biological Activities of C3a and C5a on MC/Basophils

Cell type	Bioactivity		Effective dose range (M)	
	C3a	C5a	C3a	C5a
Rat mast cells (*in vitro*)	Degranulation, *de novo* mediator synthesis	As for C3a	$10^{-6}-10^{-5}$	$4 \times 10^{-7}-4 \times 10^{-6}$
Human skin MC (*in vivo*)	Inflammation	As for C3a	$10^{-13}-10^{-10}$	$10^{-14}-10^{-13}$
Human basophils (*in vitro*)	Degranulation	Degranulation, chemotaxis	$2 \times 10^{-7}-2 \times 10^{-6}$	$10^{-8}-2 \times 10^{-7}$
Human neutrophils (*in vitro*)		Chemotaxis, enhanced phagocytosis, enzyme release		$10^{-10}-0.5 \times 10^{-7}$

It is noteworthy that C5a of pig, human, and rat cells all terminate in the tripeptide Leu-Gly-Arg-COOH and that C3a of all three species terminate in Leu-Ala-Arg-COOH (Hugli, 1981). The fact that for both C3a and C5a, COOH-terminal arginine and a leucine two residues NH_2 terminal are both important for bioactivity implies structural similarity between C5a and C3a receptors.

2.2.2. C5a Receptor

2.2.2a. Interaction of C5a with Its Cellular Receptor.   Nonspecific interaction of C5a with surface-exposed heparin proteoglycan greatly exceeds specific binding to isolated MC, precluding characterization of their C5a receptors (Huey *et al.*, 1986). Functional and biochemical characterization of a C5a receptor has, however, been achieved by using human neutrophils as a source (Chenoweth and Hugli, 1978; Huey and Hugli, 1985; Rollins and Springer, 1985). C5a biological activities against neutrophils and MC/basophils are consistent with identity of their C5a receptors. All three cell types respond almost immediately to very low concentrations of C5a but are much less responsive to C5a desArg and become refractory within minutes to further contact (Johnson *et al.*, 1975; Fernandez *et al.*, 1978; Glovsky *et al.*, 1979; Webster *et al.*, 1979; Gerard and Hugli, 1981; ·Yancey *et al.*, 1985). The rapid development of tachyphylaxis is receptor mediated rather than due to cell fatigue following maximal stimulation. Rat peritoneal MC exposed to C5a *in vitro* at 5°C, a temperature that does not permit degranulation, show diminished histamine release by comparison with control cells on subsequent C5a challenge at 37°C (Johnson *et al.*, 1975). C5a-treated neutrophils remain normally responsive to chemoattractants other than C5a.

The rapid onset of desensitization to C5a observed in bioassays of MC or neutrophil function (Johnson *et al.*, 1975; Henson *et al.*, 1978; Yancey *et al.*, 1985) reflects the *in vivo* importance of tachyphylaxis. C5a is a potent inflammatory stimulus, immediately active on effector cell populations: desensitization prevents prolonged activation of inflammatory cells.

C5a receptors so far identified on myeloid cells other than neutrophils are of comparable affinity to neutrophil receptors and again are a single affinity class (Chenoweth *et al.*, 1982; Johnson and Chenoweth, 1987). It is likely that most leukocyte C5a receptors are of this type. The closely related basophils, monocytes and neutrophils, might be expected to express the same C5a receptor molecule, and the actions of C5a on neutrophils, basophils, and MC are consistent with identity of their C5a receptors. On this assumption, we will consider the properties of the myeloid cell C5a receptors.

The neutrophil C5a receptor binds C5a with high affinity, K_d for the interaction being about $10^{-9}M$ (estimates range from 2×10^{-12} to 2×10^{-9} M; Chenoweth and Hugli 1978; Johnson and Chenoweth, 1985; Huey and Hugli, 1985; Rollins *et al.*, 1988). Excess unlabeled C5a specifically inhibits binding of ^{125}I-C5a; excess unlabeled C3a does not (Chenoweth and Hugli, 1978). Scatchard analysis of competitive binding assays on neutrophils and the U937 myeloid cell line reveal a single class of receptors, the determined affinity of which coincides with the range of concentrations over which C5a affects neutrophil function (Chenoweth and Hugli, 1978; Henson *et al.*, 1978). *In vitro* degranulation of rat peritoneal MC is apparently optimal at slightly higher C5a concentrations (Johnson *et al.*, 1975), but nonspecific interactions with heparin proteoglycan and associated proteolysis (Schwartz *et al.*, 1983; Gervasoni *et al.*, 1986) probably combine to reduce the effective C5a concentration at the mast cell surface. High-affinity receptors are clearly present on human CTMC; C5a induces the classical wheal-and-flare response of type I hypersensitivity on injection of human and rodent skin at concentrations as low as 10^{-13} M (Hugli, 1981; Yancey *et al.*, 1985).

Estimates of receptor number per human neutrophil range from 5×10^4 to 2×10^5, with considerable individual variation (Huey *et al.*, 1986). Interaction with the ligand regulates receptor expression. At 37°C, binding to viable neutrophils peaks within 20–30 min. From 60 to 100 min, the amount of ^{125}I-C5a bound diminishes to an equilibrium as a result of endocytosis and degradation of bound C5a. This suggests that desensitization to C5a results from receptor down regulation following binding (Huey *et al.*, 1986).

C5a effects on neutrophil secretion indicate that onset of desensitization is very rapid. Dose-dependent inhibition of the neutrophil secretory response to further C5a challenge develops within 2 to 5 min of C5a pretreatment at 37°C (Henson *et al.*, 1978). Interestingly, cytochalasin B both enhances the human neutrophil response to C5a and opposes the inhibitory effect of C5a pretreatment if administered within 2 min of pretreatment (Henson *et al.*, 1978). It may be that by preventing actin microfilament assembly, cytochalasin B inhibits receptor modulation following C5a contact (Goldstein *et al.*, 1973). Alternatively, the C5a receptor may normally associate with actin microfilaments, perturbation of this interaction increasing receptor mobility, and hence the effective receptor concentration within the membrane.

2.2.2b. Biochemistry of the C5a Receptor. Nonspecific interactions of C5a with surface proteoglycans on MC (Gervasoni *et al.*, 1986) prevent biochemical characterization of their C5a receptor (Huey *et al.*, 1986). A 40–45-kDa putative receptor has been identified by covalent crosslinking of ^{125}I-C5a to membrane structures of neutrophil or myeloid cell lines (Huey and Hugli, 1985; Johnson and Chenoweth, 1985; Rollins and Springer, 1985). Excess "cold" C5a inhibits ^{125}I-C5a crosslinking to this putative receptor, demonstrating its C5a specificity. Reassuringly, the same 40–45-kDa species is consistently isolated by using a variety of crosslinking methods and reagents (Huey and Hugli, 1985; Johnson and Chenoweth, 1985; Rollins and Springer, 1985; Johnson and Chenoweth, 1987; Rollins *et al.*, 1988). Larger-molecular-size bands were also occasionally observed (Rollins and Springer, 1985).

Solubilization of intact neutrophil C5a receptor complexes in β-dodecylmaltoside permits further elucidation of receptor composition (Rollins *et al.*, 1988). Binding activity for ^{125}I-C5a occurs in fractions eluting in molecular size ranges of 150–200 and 30–70 kDa by gel filtration. With use of membrane preparations from cells pretreated with ^{125}I-C5a, however, significant amounts of radioactivity are recovered only in association with the 150–200-kDa fraction. This implies that the low-molecular-size fraction represents the 40–45-kDa C5a-binding protein and that a proportion of this protein occurs free in the membrane, complexing with other proteins upon ligand binding. Supporting this contention, both the 40–45-kDa C5a-binding protein and a 95-kDa band are readily identifiable by SDS-PAGE following C5a crosslinking to solubilized intact receptor preparations (Rollins *et al.*, 1988).

Several features of the neutrophil response to C5a suggest involvement of a GTP-binding protein in signal transduction from the C5a receptor. C5a-induced chemotaxis and degranulation of human and rabbit neutrophils are inhibited by pertussis or cholera toxin

(Goldman *et al.*, 1985; Lad *et al.*, 1985; Feltner *et al.*, 1986). A membrane GTPase activity, of K_m comparable to that of some GTP-binding proteins, is enhanced on C5a treatment of human or rabbit neutrophils (Feltner *et al.*, 1986). A protein reportedly associated with the 40–45-kDa C5a-binding unit and the putative third component of the 95-kDa intact receptor complex are both estimated at about 40 kDa (Johnson and Chenoweth, 1985), a molecular weight similar to that of the GTP-binding protein alpha subunits (Okajima and Ui, 1984; Bokotch *et al.*, 1984). Thus, the data support a model of the C5a receptor consisting of one 40–45-kDa binding subunit associated with an integral membrane GTP-binding protein, the complex binding C5a in a 1 : 1 molar ratio. The presence of both free and complexed binding units in the absence of C5a suggests that this molecular association is either transient or dynamic. Binding of C5a apparently stabilizes the receptor complex. Given that the intact receptor complex elutes at 150–200 kDa on gel filtration (Rollins *et al.*, 1988), the complex may include further components.

Solubilization of intact receptor complexes offers a possible route to overcoming the nonspecific interactions with heparin-linked proteoglycans which have defeated attempts to investigate MC C5a receptors. Availability of solubilized receptor would also permit specific antisera or monoclonal antibodies to be raised against the C5a receptor. Such reagents would be valuable in establishing the identity (or otherwise) of C5a receptors on MC/basophils, neutrophils, and other leukocytes and in functional, structural, and molecular analysis of the C5a receptor. Since the 40–45-kDa C5a-binding protein is expressed and functional in free monomeric form, it would seem to be a promising candidate for cDNA cloning by the ligand affinity technique (Seed and Aruffo, 1987).

2.2.3. C3a Receptor

The immediate effects of C3a demonstrate its action on MC/basophils (Dias da Silva, 1968; Johnson *et al.*, 1975; Glovsky *et al.*, 1979). Few other cell types are directly affected by C3a. Guinea pig platelets are reportedly stimulated to aggregate (Grossklaus *et al.*, 1976) and to release 5-hydroxytryptamine (5-HT); it is uncertain whether human platelets respond to C3a (Fukuoka and Hugli, 1988).

As with C5a, nonspecific interactions with cell surface heparin-linked proteoglycans and subsequent proteolysis preclude characterization of the C3a receptor of MC/basophils. A guinea pig platelet C3a receptor was recently identified, however (Fukuoka and Hugli, 1988), and its properties are consistent with C3a bioactivities against MC. Thus, ^{125}I-C3a desArg shows no specific binding to guinea pig platelets, nor does C3a desArg induce 5-HT secretion from these cells. Scatchard analysis of the binding of ^{125}I-C3a to guinea pig platelets, although complicated by nonspecific interactions, reveals a high-affinity receptor of K_d 8×10^{-10}M, present at an average 1200 sites per cell (Fukuoka and Hugli, 1988).

Crosslinking of human C3a to guinea pig platelets reveals a specific interaction of C3a with distinct molecular species of 105 and 115 kDa, readily inhibitable by excess cold C3a or by the C3a C-terminal peptide 57–77 but not by C3a desArg. On SDS-PAGE of ^{125}I-C3a crosslinked to guinea pig platelet membrane preparations, the two bands are present in apparently equimolar amounts. It is possible that distinct receptor molecules are present, although Scatchard analysis suggests a single affinity class of receptor. Given the molecular weight of C3a (about 10 kDa), it is possible that the receptor is bivalent, the two species representing receptor crosslinked to either one or two molecules of ^{125}I-C3a. Alternatively, the receptor may be a heterodimer. The molecular sizes of the two bands are consistent with membership of the integrin family (Hynes, 1987), which includes leukocyte receptors for C3b and C3bi, although the Arg-Gly-Asp motif characteristic of many integrin ligands is absent from C3a (Hugli, 1981).

Attempts to crosslink ^{125}I-C3a to the same molecular species on rat MC were unsuccessful. It must be stressed, though, that nonspecific binding of C3a to MC surface proteoglycans results in rapid proteolysis (Gervasoni et al., 1986), and extracellular proteolysis by secreted MC proteases is also reported (Schwartz et al., 1983). On the guinea pig platelet membrane, by contrast, unaltered C3a persists for at least 60 min (Fukuoka and Hugli, 1988), potentially permitting a significant degree of specific binding to the receptor. It may be for this reason that the receptor is demonstrable on guinea pig platelets but not on MC (Fukuoka and Hugli, 1988).

It remains possible that C3a effects on MC are mediated via nonspecific interactions. The lack of spasmogenic activity of either C3a desArg (Bokisch and Muller-Eberhard, 1970) or C3a C-terminal peptides lacking Arg-77 (Hugli, 1981), however, is strong evidence for a specific C3a receptor on MC. A promising approach to MC C3a receptor characterization is the use of C-terminal oligopeptide analogs substituted with a helix-promoting residue, such as 2-aminobutyric acid or 2-aminoisobutyric acid (Huey et al., 1986). Enhancement of the α-helical structure of C3a, with consequent improvement in receptor binding relative to nonspecific binding, may permit characterization of the MC C3a receptor.

Alternatively, solubilization of the intact receptor from MC membranes may aid both functional and biochemical characterization of the C3a receptor and could provide sufficient receptor protein for immunochemical studies. It may also be possible to clone C3a receptor cDNA by the ligand affinity technique (Seed & Aruffo, 1987).

3. IMMUNOGLOBULIN RECEPTORS ON MAST CELLS AND BASOPHILS

3.1. FcεR1, the High-Affinity Receptor for IgE Fc

3.1.1. Interaction of Mast Cells and Basophils with IgE Fc

MC/basophils may be sensitized for degranulation by incubation with nonspecific IgE, indicating the presence of specific IgE receptors (Ishizaka et al., 1970a,b; Tigelaar et al., 1971). Binding of IgE to human basophils was directly demonstrated by immunoelectron microscopy (Sullivan et al., 1971), and iodinated IgE was shown to bind to MC/basophils in vitro. This interaction was dose dependent, saturable, and inhibitable by unlabeled IgE (Kulzycki et al., 1974; Kulzycki and Metzger, 1974). The determined affinities of IgE Fc receptors on RBL, rat and mouse peritoneal MC, and human basophils were all similar, K_d approximately 10^{-9}M (Ishizaka et al., 1973; Kulzycki et al., 1974; Kulzycki and Metzger, 1974; Conrad et al., 1975; Sterk and Ishizaka, 1982; Lee et al., 1985).

MC/basophil high-affinity IgE receptors are univalent. Thus, fluorescein- and rhodamine-labeled IgE preparations on MC membranes were not coaggregated by antisera to either fluorochrome (Schlessinger et al., 1976). ^{131}I-rat IgE and ^{131}I-mouse IgE were independently isolated from RBL cell membranes as receptor–IgE complexes by affinity purification on antigen or specific anti-IgE (Kanellopoulos et al., 1980). The molecular size of the IgE–receptor complex, estimated by gel filtration, is too small to include more than one IgE molecule (Newman et al., 1977).

Binding of IgE monomers to MC/basophils has no obvious physiological effect. Receptor–IgE complexes are stable on the membrane for several days, reflecting the low dissociation constant of the system (Cass et al., 1968). Any treatment causing crosslinking of IgE Fc receptors immediately activates MC/basophils. Aggregated or dimerized/oligomerized IgE, macromolecular antigen, antisera to IgE or IgE receptors, lectins, and divalent haptens all effectively stimulate immediate degranulation of MC/basophils, though potency increases with increasing molecular size of stimulus (Segal et al., 1977; Ishizaka and Ishizaka, 1978; Kagey-Sobotka, 1981). The efficacy of dimeric and divalent activators suggests that comparatively few receptor crosslinks suffice for cellular activation.

3.1.2. Biochemistry of FcεR1

IgE-binding proteins, seen on SDS-PAGE as a broad band spanning 45–60 kDa, were originally identified both by affinity purification on IgE–Sepharose and by immunoprecipitation of IgE preincubated with RBL membrane preparations (Kulzycki et al., 1976; Conrad and Froese, 1976, 1978). The IgE-binding species, now known to be the α chain of FcεR1, biosynthetically incorporates both [³H]leucine and [³H]glucosamine (Kulzycki et al., 1976; Pecoud et al., 1981) and binds to lectin affinity columns (Helm et al., 1979). FcεR1α isolated from RBL cells cultured with tunicamycin has a molecular weight 30% smaller than that of the native molecule (Hempstead et al., 1981). The molecular weight of FcεR1α estimated by denaturing gel filtration in 6 M guanidine hydrochloride is 36 kDa (Kumar and Metzger, 1982), which is reduced to 27.3 kDa following endoglycosidase treatment. This peptide core molecular weight is similar to the size of the target IgE-binding species demonstrated by radiation inactivation of RBL membrane preparations (Fewtrell et al., 1981) and to that predicted from cDNA (see below).

FcεR1α may be extrinsically iodinated on intact RBL cells prior to affinity purification on IgE, demonstrating that it is surface exposed. A minor IgE-associated species of about 30–35 kDa by SDS-PAGE was identified in ³H-amino acid biosynthetically labeled RBL membrane preparations (Holowka et al., 1985). This molecule, FcεR1β, was not extrinsically iodinated on intact RBL cells but was labeled by a lipophilic probe, iodoacetylnaphthyl, covalently crosslinked to FcεR1α at a constant 1 : 1 molecular ratio (Holowka and Metzger, 1982) and failed to incorporate [³H]glucosamine. cDNA analysis of β (see below) is consistent with the presence of cytoplasmic regions. These properties identify the 30–35-kDa species as an integral membrane subunit of FcεR1.

Having identified a β chain, it was apparent that α and β dissociated during detergent lysis. Addition of phospholipids to preparative buffers, and maintenance of the detergent–lipid ratio within an empirically defined range, allowed reproducible purification of intact receptors (Rivnay et al., 1984). This strategy also revealed a third subunit, a 20-kDa disulfide-bonded homodimer, FcεR1γ, dissociating from α as a unit with β (βγ₂; Perez-

Montfort *et al.*, 1983). V8 protease mapping showed γ to be dissimilar to both α and β. Both β and γ are efficiently iodinated on permeabilized RBL cell membrane vesicles, indicating that both are exposed at the cytoplasmic aspect of the membrane (Holowka and Baird, 1984).

The nature of interactions with membrane lipid, suggested by the detergent sensitivity of FcεR1 and its enhanced stability in the presence of exogenous lipid, have been investigated (Kinet *et al.*, 1985a,b). Anomalously, IgE-prebound FcεR1 was resistant to oxidative iodination (Conrad and Froese, 1976). This was not due to effects of β or γ, since physicochemical conditions originally used for IgE receptor isolation dissociate β and γ from α–IgE complexes. Brief washing of affinity column-bound intact FcεR1 with lipid-free micellar detergent and transfer to lipid-free submicellar detergent, however, maintains FcεR1 integrity and permits iodination of IgE-bound FcεR1. The presence of either micellar detergent or exogenous lipid in the final preparation inhibited FcεR1 iodination (Kinet *et al.*, 1985b). Thus, failure of IgE-bound FcεR1 to iodinate was due to loosely associated lipid masking tyrosine residues not already blocked by IgE. The FcεR1 complex is soluble at low ratios of micellar detergent to micellar lipid but stable for prolonged periods at considerably higher detergent–lipid ratios. Together with the stability of FcεR1 in lipid-free submicellar detergent, this observation suggests the presence of tightly bound lipids on FcεR1 (Kinet *et al.*, 1985b).

RBL cells cultured with [^{3}H]palmitic acid rapidly incorporate radioactivity into β and γ. Treatment of β and γ bands excised from SDS-PAGE gels to cleave peptide ester bonds liberates [^{3}H]methylpalmitate, and extensive detergent washing releases only a minority of β- and γ-associated counts, indicating covalent linkage of fatty acid to these subunits. Incorporation of fatty acid into β and γ was quantitatively similar in cells cultured in the presence or absence of IgE. α, β, and γ insert into the membrane and turn over coordinately (see below). Since IgE-bound remains on the membrane, esterification of β and γ must follow their insertion to the membrane (Kinet *et al.*, 1985b).

FcεR1 alone cannot be successfully incorporated into liposomes, whereas the intact αβγ$_2$ complex incorporates efficiently (Rivnay *et al.*, 1984). Upon chromatography on a hydrophobic support (C$_{18}$ Sep-pak), both β and γ remain bound to the matrix at propanol concentrations which elute bound α, showing that both have a greater hydrophobic surface than does α (Alcaraz *et al.*, 1984). The marked effective hydrophobicity of β and γ and their association as a unit is consistent with a role in membrane anchorage of the FcεR1 complex. It would seem feasible that the ability of β and γ in particular to associate with lipid enhances their ability to stabilize FcεR1 within the lipid bilayer. Receptor-associated phospholipids might also be involved in signal transduction from the receptor.

3.1.3. Biosynthesis and Turnover of FcεR1 Subunits

Analysis of [^{3}H]leucine incorporation at intervals during metabolic labeling of RBL cells showed that α, β, and γ labeled at similar, constant rates (Quarto *et al.*, 1985). At each time point considered, the ratio of radioactivity incorporated into α, β, and γ remained constant (Quarto *et al.*, 1985). Thus, α, β, and γ are coordinately synthesized and are each present at a constant integral ratio with respect to the FcεR1 complex, as would be predicted for subunits of an oligomeric functional complex.

The average receptor number per RBL cell is inversely related to culture growth rate,

Table II
Subunits of FcεR1, the MC/Basophil High-Affinity IgE Receptor

| Subunit | IgE binding | Mol. wt. (kDa) | | Molecules/ receptor complex | Orientation |
		Mature protein[a]	Peptide core[b]		
α	+	46–60	26.1	1	Transmembrane
β	–	30–35	27	1	Integral membrane, 4 transmembrane sequences
γ	–	10	7.1	2	Integral membrane homodimer

[a]Assessed by SDS-PAGE.
[b]Predicted from cDNA.

and receptor accumulation continues in stationary cultures (Isersky *et al.*, 1975). Surface expression and turnover of FcεR1 on RBL depends on the presence or otherwise of receptor-bound IgE (Coutts *et al.*, 1980). Binding of IgE profoundly inhibits receptor turnover but not receptor synthesis or membrane insertion of newly synthesized FcεR1 (Furuichi *et al.*, 1985). IgE-treated RBL express higher levels of FcεR1 than do cells cultured without IgE. Similarly, basophils of atopic human patients have a higher receptor density than do those of controls, suggesting a correlation between receptor number and serum IgE level (Malveaux *et al.*, 1978).

These phenomena are likely to be relevant to *in vivo* functions of MC/basophils. FcεR1 concentrates and retains IgE on the cell surface, conferring antigen-specific effector function upon the cell. The stability of receptor–IgE complexes within the membrane is consistent with this role, as is the observation that IgE binding does not inhibit further incorporation of FcεR1 into the membrane. Any such negative-feedback process would limit the rate of IgE acquisition and hence the number of antigen-specific reactivities acquired by the cell.

The subunits of FcεR1 are characterized in Table II.

3.2. Molecular Cloning of FcεR1 Subunits

3.2.1. cDNA Cloning of FcεR1α

Three independent reports describe cDNA clones encoding rat FcεR1 (Kinet *et al.*, 1987; Shimizu *et al.*, 1988; Liu *et al.*, 1988). In each case, tryptic peptides of α were sequenced, and corresponding oligodeoxyribonucleotides were used to screen RBL plasmid or λgt11 cDNA libraries. cDNA for the human homolog was identified in a KU812 library by hybridization using rat FcεR1α cDNA as probe (Shimizu *et al.*, 1988).

Full-length rat FcεR1α cDNA clones predict a mature protein of 220 amino acids. One reported sequence (Kinet *et al.*, 1987) encodes five novel C-terminal residues and differs at three residues from other predicted sequences. Overall, however, reported cDNA sequences predict the same features, notably (1) 180 extracellular residues, including seven potential N-linked glycosylation sites (Asn-Xaa-Ser/Thr), (2) a 19-residue hydrophobic transmembrane sequence, and (3) a 21-residue (or 26-residue) hydrophilic

COOH-terminal cytoplasmic tail. All sequences have an aspartic acid at position 218, in the transmembrane portion. The presence of a positively charged residue in the transmembrane hydrophobic region may explain the inability of FcεR1α to incorporate into liposomes when dissociated from β and γ.

The predicted protein sequence includes sequences determined from tryptic peptides, has a molecular weight (26.1 kDa) near to that of endoglycosidase-treated FcεR1α, and has an amino acid composition similar to that determined by direct analysis, confirming that the cDNA-encoded protein is FcεR1α.

FcεR1α is homologous to the mouse macrophage IgG1/IgG2b receptor, FcγRα (Ravetch et al., 1986); there is 32% amino acid identity (49% nucleotide identity) between the coding regions of FcεR1α and FcγRα. Homology is particularly marked at the extracellular N-terminal region (10 of the first 14 residues in common) and in a run of 8 identical residues around Asp-218 in the transmembrane region. There is no homology between FcεR1α and previously described RBL (Albrandt et al., 1987) or T-lymphocyte (Martens et al., 1985) IgE-binding proteins, nor with the human leukocyte low-affinity IgE receptor (FcεR2α; Ikuta et al., 1987), recognized by CD23 monoclonal antibodies.

FcεR1α has two extracellular regions of 26% amino acid internal homology, spanning residues 65–102 and 148–182. COOH proximally in each internal homology region are cysteine residues preceded by the motif Asp-Xaa-Gly-Xaa-Tyr-Xaa. This feature identifies the second cysteine partner in intradomain disulfide bonds in several proteins of the immunoglobulin family, including FcγRα. The two homologous regions would thus be disulfide bonded between, respectively, Cys-49–Cys-91 and Cys-130–Cys-174.

The presence of disulfide-bonded homology regions and the high homology to FcγRα at both protein and nucleotide levels establish FcεR1α as a member of the immunoglobulin gene superfamily. Other immunoglobulin receptors which are themselves members of the immunoglobulin superfamily are the mouse macrophage IgG receptors, FcγRα and FcγRβ (Ravetch et al., 1986), the epithelial cell receptor for polymeric IgA/IgM (Mostov et al., 1984), and the neonatal rat intestinal IgG Fc receptor, FcRn (Simister and Mostov, 1989).

Analysis of cDNA encoding human FcεR1α predicts a protein of 257 residues, homologous to both rat FcεR1α and FcγRα but not to other IgE-binding proteins (Shimizu et al., 1988). Again, the predicted human FcεR1α sequence would be consistent with the presence of two extracellular disulfide-bonded Ig-like domains. Extracellular amino acids 140–146 were common to both rat and human FcεR1α but not conserved on either FcγRα or FcγRβ, suggesting that this region may be important in conferring IgE specificity on FcεR1α. Transmembrane sequences of rat and human FcεR1α are similar, including 10 conserved amino acids around Asp-218 (Val-212–Leu-221), suggesting that this region is important either in interaction with other subunits or in signal transduction. Rat and human FcεR1α cytoplasmic sequences are dissimilar.

Rat FcεR1α cDNA fails to restore either IgE-binding activity or reactivity with BC4, a monoclonal antibody to FcεR1α, upon transfection into a mouse MC clone deficient in FcεR1 expression (Shimizu et al., 1988). These cells may lack other FcεR1 subunits. Curiously, FcεR1α cDNA transfected to COS-7 is transcribed, but not translated to protein (Kinet et al., 1987). Rat FcεR1α cDNA probes used to screen a range of hemopoietic and other cell lines for mRNA expression detect the corresponding mRNA only in lines of MC/basophil origin (Kinet et al., 1987; Shimizu et al., 1988).

3.2.2. FcεR1α Structural Heterogeneity: Evidence from cDNA Variants

Extensive screening of an RBL cDNA library, using as probe a 30-mer FcεR1α oligodeoxyribonucleotide, identifies three classes of FcεR1α cDNA variants, summarized as follows:

1. cDNAs incorporating a novel 5′ untranslated region lacking the previously reported signal peptide-encoding sequence, the first in-frame initiation codon being that for Met-3 of the coding sequence. The coding region is otherwise identical to previously published cDNA sequences, although novel 3′ untranslated sequence, including a third polyadenylation site, is present (type clone R8-2b).
2. cDNAs possessing a major internal deletion of 163 bp (nucleotides 537–699) but otherwise identical to previously reported sequences. The deletion would alter the reading frame, encoding a truncated 137-residue protein product with 17 novel C-terminal amino acids but lacking transmembrane and cytoplasmic sequences.
3. cDNAs encoding both the major internal deletion and novel 5′ untranslated sequence, thus lacking signal, transmembrane, and cytoplasmic sequences.

A 305-base RNA probe derived from the 5′ end of an R8-2b restriction fragment, spanning 140 bases of the novel 5′ untranslated sequence and 165 bases of FcεR1α-coding region, protected 305- and 140-base RBL mRNAs in an RNase protection assay. An 83-base RNA probe, encompassing 50 bases of FcεR1α-coding sequence and the first 33 bases of the major internal deletion, protected RBL mRNA species of 83 and about 50 bases. These data establish that the three cDNA variants represent detectable cellular mRNA species, presumably translated to protein in RBL and other MC-like cells. Variant mRNAs, or their protein products, may function in regulation of FcεR1 expression. Variants encoding the major internal deletion probably result from differential mRNA splicing. Their protein product would lack the hydrophobic transmembrane sequence and so would be soluble. If IgE-binding activity was retained, these molecules might participate in intracellular processing of IgE, a role previously suggested for an unrelated novel IgE-binding protein of RBL (Albrandt et al., 1987).

3.2.3. cDNA Cloning of FcεR1β

cDNA clones encoding FcεR1β were isolated from RBL pUC9 and λgt11 cDNA libraries, using as probes oligodeoxyribonucleotides based on FcεR1β tryptic peptide sequences (Kinet et al., 1986). The full-length FcεR1β cDNA has two in-frame potential initiation codons and predicts a protein product of either 243 or 246 amino acids without protein or nucleotide homology to any reported sequence. The outstanding feature of the predicted protein was the presence of four 19-residue hydrophobic sequences, implying that FcεR1β crosses the cell membrane four times.

The predicted molecular weight of the cDNA product is only about 27 kDa, rather than the 33 kDa of native FcεR1β (Holowka and Metzger, 1982). In vitro transcription of the cDNA and cell-free translation, however, yields a protein identical in electrophoretic mobility to native FcεR1β and reactive with antisera and monoclonal antibodies [MAb(JRK) and MAb(NB)] to FcεR1β (Kinet et al., 1988). Electrophoretic mobility of both recombinant and native β is unaffected by reduction.

Three potential N-linked glycosylation sites are predicted, although β does not incorporate tritiated sugars.

Subcloning and expression of cDNA restriction fragments encoding residues 18–148 and 149–243 in *Escherichia coli* provide further information on FcεR1β topology. The latter peptide is recognized by MAb(NB), locating the epitope for this MAb toward the C terminus. MAb(JRK) recognizes neither recombinant peptide, suggesting specificity for an epitope incorporating, or consisting of, N-terminal residues. Both MAbs react with RBL sonicates; neither reacts with intact cells. Together with the epitope localization data, this suggests that both NH_2 and COOH termini are on the cytoplasmic face of the membrane.

FcεR1β cDNA hybridizes to mRNA species of 2.7, 1.75, and 1.2 kb from RBL but not to mRNA of cell lines from other lineages. This is significant, since both FcγRα and mouse FcεR2 reportedly associate with proteins of molecular weight 33 kDa or immunoreactivity with MAb(JRK) (Finbloom and Metzger, 1983), supporting speculation that Ig Fc receptors other than FcεR1 associate with β-like molecules.

Cotransfection of COS-7 cells with both FcεR1α and -β cDNAs fails to reconstitute FcεR1 expression.

3.2.4. Molecular Cloning of FcεR1γ

cDNAs encoding the rat FcεR1γ subunit have recently been identified by probing a λgt11 library of RBL with oligonucleotides based on tryptic peptide sequences (Blank *et al.*, 1989). A mature peptide of 68 residues is predicted (molecular weight 7.1 kDa), including five extracellular NH_2-terminal residues, a 21-residue hydrophobic transmembrane sequence, and 42 cytoplasmic residues towards the COOH terminus. This topology is consistent with known properties of γ, notably its ability to be oxidatively iodinated on inside-out RBL cell membrane vesicles and to phosphorylate on threonine *in vivo*. The 42-residue segment has two tyrosines and four threonines; the 5-residue segment has neither amino acid.

β and the $γ_2$ dimer dissociate from FcεR1α as a unit and may disulfide bond to each other. It is suggested that these bonds form between γ Cys-7 and β Cys-80, both located at the cytoplasmic end of transmembrane segments and so likely to be physically close. γ–γ disulfide bonds would thus form between the Cys-26 residues. The predicted sequence agrees with previous compositional analyses of γ, two cysteines but no methionine or tryptophan being present in the mature peptide.

A 300-bp restriction fragment of γ cDNA in the pSVL expression vector was cotransfected into COS-7 cells along with similar, separate constructs containing FcεR1α or -β cDNA fragments, in each case containing the entire in-frame coding sequence. DNA uptake by cells was assessed by probing Northern blots for corresponding mRNA production. High-affinity IgE binding was assessed by rosetting of cells with rat IgE-conjugated red cells. IgE-binding cell clones expressed mRNAs for all three subunits. Together with the demonstration that α, β, and γ are coordinately synthesized, this finding unequivocally establishes β and γ as true subunits of FcεR1 and demonstrates that all three subunits are required for receptor expression.

3.3. Mast Cell and Basophil Receptors for IgG Fc

In several experimental systems, antibodies of IgG isotype sensitize MC/basophils for antigen-specific degranulation. IgG Fc receptors were demonstrated by binding assays on RBL cells and rat peritoneal MC and by immunofluorescence on human basophils (Halper and Metzger, 1976; Stain *et al.*, 1987).

For each species examined, a particular IgG subclass is associated with sensitization; IgG1 of mouse, IgG2a of rat, and IgG4 of human (Tigelaar *et al.*, 1971; Halper and Metzger, 1976; Vijay and Perelmutter, 1977). IgG sensitization of MC/basophils *in vivo*, unlike IgE sensitization, persists for only a few hours.

In systems in which IgG inhibits *in vivo* IgE-mediated anaphylaxis, or binds to MC/basophils *in vitro*, IgG aggregation or immune complex is invariably necessary (Halper and Metzger, 1976; Moller and Konig 1980). A single report described binding of monomeric mouse hybridoma-derived IgG to RBL and suggested that clonal characteristics of IgG determine ability to bind to the cell (Hall and Rittenberg, 1986).

Intact mouse IgG1 antibodies from sera of appropriate anti-H-2 allospecificity induce histamine release from mouse MC, whereas their $F(ab')_2$ fragments are ineffective (Daeron and Voisin, 1979). The mechanism of IgG-induced degranulation would appear to be crosslinking of IgG Fc receptors to each other or to other surface structures.

IgG aggregates do not inhibit IgE binding or IgE-mediated degranulation in the presence of IgE, preadministration of IgG being essential. RBL cells bear distinct populations of IgG and high-affinity IgE receptors, as shown by the differential turnover and processing of cell-bound IgE and IgG. Thus, it seems unlikely that IgG binds to FcεR1 but possible that steric effects of cell-bound IgG aggregates may significantly hinder binding of IgE to FcεR1. This phenomenon is unlikely to be relevant to MC/basophil function *in vivo*.

The nature of the IgG receptors on most MC/basophil types has not been established. Human basophils react with CDw32 MAbs (Stain *et al.*, 1987), recognizing FcγR2 (the human homolog of murine FcγRβ), but human MC do not. Neither cell type bears either the low-affinity receptor, $FcγR_{lo}$ (CD16), or the monocyte-associated high-affinity FcγR1(CD64; Rimmer and Horton, 1987; Stain *et al.*, 1987).

The human low-affinity IgE Fc receptor (FcεR2/CD23) is serologically undetectable on MC or basophils.

4. IMMUNOGLOBULIN E ON MAST CELLS AND BASOPHILS

4.1. Interaction of IgE Fc with Mast Cells and Basophils

The interaction of IgE with specific high-affinity receptors on MC/basophils is characterized by the lack of cellular response to binding of monomeric IgE and the persistence on the cell of receptor–IgE complexes (Cass and Anderson, 1968; Stanworth *et al.*, 1967). Receptor-bound IgE retains antigen-binding capacity; bridging of adjacent IgE receptors by multivalent antigen, anti-IgE, antiserum to the receptor, or IgE dimers or oligomers results in cellular activation and exocytosis (Segal *et al.*, 1977). IgE, therefore, functions as an acquired cellular receptor for antigen, arming the cell for participation in antigen-specific inflammatory responses.

IgE is the least abundant serum immunoglobulin, and original characterization was possible only through analyses of IgE myeloma proteins. Extensive biochemical characterizations and amino acid sequence analysis have been reviewed elsewhere (Kochwa *et al.*, 1971; Dorrington and Bennich, 1978). For this reason, we will describe only briefly the structure and biochemistry of IgE and concentrate on more recent functional and molecular studies.

4.2. Structure and Biochemistry of IgE

Early studies of human IgE myeloma protein identified an immunoglobulin of apparent molecular weight 190 kDa, the heavy chain, ϵ, of which was antigenically distinct from other Ig isotypes (Johansson and Bennich, 1967). The carbohydrate content, including both simple and complex carbohydrate residues, was approximately 12% (Johansson and Bennich, 1967; Nezlin *et al.*, 1973; Mendez *et al.*, 1973; Baenziger *et al.*, 1974). The molecular weight of the peptide core is similar to that of μ (IgM) but exceeds by about 8–11 kDa those of α (IgA), δ (IgD), and γ (IgG). This discrepancy suggested that ϵ, like μ, had five domains, later confirmed by amino acid sequencing and cDNA cloning (Dorrington and Bennich, 1978; Kenten *et al.*, 1982; Hellmann *et al.*, 1982; Liu *et al.*, 1982).

Complete amino acid sequence analysis of a human IgE myeloma revealed that ϵ consisted of 547 amino acid residues, comprising an NH_2-terminal VH region and four constant-region domains (Cϵ1–Cϵ4) (Dorrington and Bennich, 1978). Of the 15 cysteine residues per ϵ chain, 10 participate in typical intradomain disulfide bonds, one in the inter-heavy–light-chain disulfide bond (Cys-138 in Cϵ1) and two in the bonds between the paired chains (Cys-241 and Cys-328 (both in Cϵ2). Cysteines at residues 139 and 225 form an additional disulfide bridge within Cϵ1, a characteristic feature of human IgE. Each domain of ϵ is homologous to the corresponding domain (where present) of other Ig isotypes. The cDNA-predicted amino acid sequence of mouse ϵ suggests structural differences from human ϵ (Liu *et al.*, 1982). The Cϵ1 intrachain bond of human is absent, and an extra cysteine, with no human homolog, may form a third inter-ϵ-chain bond in Cϵ2.

Early studies established that IgE interacts with its high-affinity receptor via the Fc portion (Stanworth *et al.*, 1967; Ishizaka *et al.*, 1970a). The proteolytic fragment Fc$_\epsilon$, roughly equivalent to the paired S═S-bonded Cϵ2–Cϵ3–Cϵ4 regions, is equipotent to native IgE in *in vitro* and *in vivo* bioassays (Perez-Montfort and Metzger, 1982). Aggregated Fc$_\epsilon$ is an MC/basophil-degranulating signal (Ishizaka *et al.*, 1970b); IgE F(ab')$_2$ fragment is not. The irreversible loss of IgE bioactivity on prolonged heating at 56°C correlates to exposure of aromatic residues in Cϵ3 and Cϵ4, detected by CD spectral analysis (Dorrington and Bennich, 1978). Similarly, progressive loss of human IgE bioactivity on treatment with increasing concentrations of mild reducing agents (e.g., dithiothreitol in the range of 1–10 mM) correlates with the degree of reduction of disulfide bonds in Fcϵ (Takatsu *et al.*, 1975). A significant increment of loss of activity follows reduction of the inter-ϵ-chain bond at Cys-241, suggestive of marked conformational change. The extent of observed loss of activity on reduction varies between studies and between bioassays. Residual bioactivity may remain after reduction so extensive as to cause loss of Fc$_\epsilon$ antigenicity, and rodent IgEs retain unimpaired bioactivity even after cleavage of all inter-ϵ-chain S═S bonds to yield ϵ–light-chain heterodimers (Perez-Montfort and Metzger, 1982).

A full understanding of IgE–receptor interaction depended on accurate localization of the receptor binding site on IgE. Early data suggested that the $C\epsilon2$–$C\epsilon3$ junction was important, since proteolytic cleavage near this point, to yield the fragment called Fc'_ϵ, inactivates IgE (Dorrington and Bennich, 1978). Rodent IgE, trypsin-cleaved in mid-$C\epsilon3$, retains bioactivity, however, whereas the $C\epsilon4$-containing cleavage product is inactive. A site near the $C\epsilon2$–$C\epsilon3$ junction is significantly less trypsin sensitive on cell-bound than on free IgE (Perez-Montfort and Metzger, 1983). Given that a fragment containing $C\epsilon2$ and part of $C\epsilon3$ is bioactive whereas Fc'_ϵ (a $C\epsilon2$ dimer) is not, it was apparent that the receptor binding site lay at or near the $C\epsilon2$–$C\epsilon3$ junction.

4.3. Conformation of Receptor-Bound IgE

The binding of monomeric IgE to its high-affinity receptor on MC/basophils is perhaps the best characterized of all cellular receptor–ligand systems. The unique feature is the role of receptor-bound IgE as a cellular receptor for antigen. It was of interest, therefore, to determine the configuration of receptor-bound IgE.

Plainly, receptor-bound IgE combining sites are accessible to antigen and so presumably project clear of the cell membrane. Bound IgE must adopt one of three approximate conformations: either perpendicular or angled relative to the membrane or tangential to the membrane.

A series of elegant studies, using measurement of resonance energy transfer between fluorescent probe donors attached to IgE and amphipathic acceptor probes inserting into the plasma membrane (Holowka and Baird, 1983a), elucidated the spatial relationship of receptor-bound IgE to the cell membrane. IgE, labeled at specific sites with donor probes, was bound to RBL-derived membrane-bound vesicles. In early studies, FITC was shown to conjugate to the Fab region of rat IgE, whereas the reduced, alkylated inter-heavy-chain disulfide bonds in $C\epsilon2$ labeled with $N[4-\{7(\text{diethylamino})-4-\text{methylcoumarin-3-yl}\}\text{phenyl}]$maleimide (CPM) (Holowka and Baird, 1983b). The average distance from fluorescent donor on modified, receptor-bound IgE to the plane of membrane-embedded acceptors may be estimated as a function of fluorescence quenching (due to donor–acceptor energy transfer) if the density of acceptor in the membrane is known (Holowka and Baird, 1983a). Because different regions of IgE are preferentially or specifically labeled, the distance of particular regions from the cell surface may be determined. Thus, the FITC sites of Fab'_ϵ are further from the membrane (86–91 Å) than are CPM-labeled sites on the $C\epsilon2$ S=S interchain bonds (35–52 Å; Holowka and Baird, 1983b). Furthermore, the determined distance of $C\epsilon2$ from the membrane is much less than the combined length of $C\epsilon3$ and $C\epsilon4$. Assuming a perpendicular orientation, therefore, a considerable portion of the COOH-terminal end would be buried in the membrane. This seems unlikely, especially since IgE does not interact with β and γ, the integral membrane components of the receptor.

Technical refinements permitted more extensive mapping of receptor-bound IgE. Donor-labeled antigen [dinitrophenol-L-lysine) was used to estimate the combining site-to-membrane distance for specific IgE, about 100–120 Å (Baird and Holowka, 1985). A $C\epsilon1$ site that specifically labeled with CPM under nonreducing conditions (CPM site) was shown to lie 75–87 Å from the membrane (Holowka *et al.*, 1985). Techniques for specific labeling of COOH- or NH_2-terminal residues of antibodies were developed. MAbs B5 and

A2, specific for epitopes on Fab and Fc, respectively, were labeled in this way and used to analyze distances between their epitopes on free and bound IgE and from epitope to membrane on bound IgE. Earlier results were confirmed, the A2 epitope on Fc being considerably nearer the membrane (54 Å) than the B5 epitope on Fab'_ε (78–87 Å). Energy transfer between donor (FITC)-labeled B5 and acceptor iodoacetamido-tetramethylrhodamine-labeled A2 was much less efficient when the MAbs were bound to receptor-complexed IgE, as opposed to free IgE in solution. Both MAbs bind equally well to IgE in either phase, suggesting a change in the spatial relationship of Fc_ε and $F(ab')_{2\varepsilon}$ following binding to the receptor. B5 may bind to either face of Fab'_ε, implying that in binding simultaneously to receptor-bound IgE, B5 and A2 must bind to opposite sides of the molecule. This would effectively increase the distance between probes on MAbs attached to bound, as opposed to unbound, IgE.

In summary, the antigen-combining sites of receptor-bound IgE are the part furthest from the cell, other sites on Fab'_ε (the FITC sites, the B5 epitopes, and the CPM site on $C\varepsilon1$) being significantly closer to the cell. The inter-ε-chain disulfide bonds in $C\varepsilon2$ are yet closer, being about as far from the membrane as the A2 epitope. These data are most conveniently interpreted as representing IgE bound at the $C\varepsilon2$–$C\varepsilon3$ junction to its receptor, the Fc portion lying roughly parallel to the cell and the two Fab'_ε, bent from the twofold axis of symmetry of the molecule, projecting from the cell at a steep angle. The inefficiently of energy transfer between probe-labeled A2 and B5 would be consistent with this "bent IgE" model.

4.4. Molecular Cloning of ε

ε cDNA clones, encoding peptides with full and appropriate bioactivity and immu-noreactivity, are available for mouse (Liu *et al.*, 1982), rat (Hellmann *et al.*, 1982), and human (Kenten *et al.*, 1982) species. Analysis of these clones has confirmed or corrected details of the structure of IgE, such as the primary sequence of residues 112–136 of human ε, not determined by N-terminal sequencing.

More interesting is the use of recombinant peptides encoding selected regions of human ε to localize the receptor interaction site. The expressed products of a cDNA restriction fragment encoding human $C\varepsilon2$–$C\varepsilon3$–$C\varepsilon4$ (Kenten *et al.*, 1984; Liu *et al.*, 1984; Ishizaka *et al.*, 1986) are almost as potent as myeloma IgE in *in vivo* and *in vitro* assays of MC/basophil sensitization (Geha *et al.*, 1985; Coleman *et al.*, 1985). Peptides expressed from vectors containing still smaller restriction fragments, each encoding some part of the $C\varepsilon2$–$C\varepsilon3$–$C\varepsilon4$ region (see Table II), were assessed for their ability to sensitize human basophils to anti-ε *in vitro* and to compete with myeloma IgE for binding sites on skin MC *in vivo* (Helm *et al.*, 1988). Only peptides spanning the $C\varepsilon2$–$C\varepsilon3$ junction are bioactive. The shortest peptide tested, containing residues 301–376, is equipotent to native IgE in both bioassays. IgE activity requires the presence of amino acids of the $C\varepsilon2$–$C\varepsilon3$ junction in unbroken sequence. Peptides equivalent to $C\varepsilon2$ and $C\varepsilon3$–$C\varepsilon4$ (rE2 and rE3-4), comprising sequences Gln-301–Val-336 and Leu-340–Arg-376 respectively, are inactive. The possibility that residues 337–339, absent from both peptides, are essential for binding can be discounted, since favored models of IgE three-dimensional structure predict that these residues are not exposed to the milieu (Pumphrey, 1986; Helm *et al.*, 1988). Structural considerations disqualify other parts of the sequence Gln-301–Arg-376 from consideration as receptor interaction sites. On the native molecule, the sequence of interest is bracketed by carbohydrate residues on Asn-265 and Asn-371. These would effectively obscure the structure at both its NH_2-proximal end and COOH-terminal solvent-exposed residues. The probable location of the receptor interaction site was suggested to lie within a cleft in the molecule, opening to the surface between Cε2 and Cε3. One side of the cleft is delineated by two runs of amino acids of Cε2 and the other by a section of Cε3. FcεR1α probably inserts into, and binds within, the cleft. Peptides rE2 and rE3-4 contain, between them, all of the residues lining the cleft yet are inactive. This argues against the existence of individual crucial residues, overall structure of the cleft being the essential feature, and further implies that contacts of FcεR1α with both sides of the cleft are necessary for high-affinity binding.

The authors note that the sequences Gln-301–Arg-376 on the paired chains of native IgE are 1.1 nm apart at their closest approach (Helm et al., 1988). Given that the monomeric peptides are bioactive, as are unpaired rodent chains, and that FcεR1α is monovalent, the IgE molecule probably binds to receptor via a single site, which may be on either ε chain.

The potency of rE2'-3' and rE2-3 also demonstrates clearly that Cε4 is not involved in receptor binding. It was noted, though, that peptides possessing the inter-ε-bonding cysteine residues, but lacking Cε4, did not readily dimerize. Cε4 may initiate ε–ε dimerization; the homologous Cγ3 domain reportedly has a similar function in γ-chain pairing during IgG assembly.

5. HEMOPOIETIC LINEAGE-ASSOCIATED MEMBRANE GLYCOPROTEINS OF MAST CELLS AND BASOPHILS

5.1. Serologically Defined Membrane Glycoproteins

Abs are available to a wide range of membrane glycoproteins associated with distinct human hemopoietic cell lineages (McMichael et al., 1987; Knapp et al., 1989). Immunochemical characterization of these glycoproteins has facilitated investigation of their roles in leukocyte function and of patterns of expression during cellular differentiation and between leukocyte populations. Molecular cloning of some of these glycoproteins and of their nonhuman homologs has also been possible. Membrane immunophenotypes of hemopoietic cell populations thus indicate both differentiated functions and lineage affinities.

Data on reactivity of MAbs with leukocyte antigens have been collated by the International Workshops on Human Leukocyte Differentiation Antigens (Reinherz et al., 1986; McMichael et al., 1987; Knapp et al., 1989). Each well-defined antigenic specificity receives the designation "CD" and a unique code number; MAbs of defined reactivity are then assigned to the appropriate CD.

Human MC/basophil immunophenotypes are dissimilar, particularly in expression of myeloid lineage-associated glycoproteins (Horton and O'Brien, 1983; Rimmer and Horton, 1987; Stain et al., 1978; Valent et al., 1988a). The immunophenotype of basophils, although characteristic (Stain et al., 1987), is comparable to those of granulocytes and monocytes, establishing the origin of basophils from a common myeloid precursor.

The range of glycoproteins shared with other myeloid cells is consistent with known functions of basophils and with their role in inflammatory responses. All three LFA α chains and their common β chain (Sanchez-Madrid *et al.*, 1983) are present on basophils (Horton and O'Brien, 1983; Stain *et al.*, 1987; McIntyre *et al.*, 1987). These adhesive glycoproteins are perhaps the most important functional molecules of myeloid cells (see below). Another adhesive glycoprotein of basophils is the 220-kDa C3b receptor [complement receptor type 1 (CR1)], detected by CD35 MAbs (Stain *et al.*, 1987). Since basophils are not phagocytic, this receptor may be involved in immune adhesion to deposited C3b at sites of inflammation, and perhaps also on membranes of target cells or parasites. Significantly, basophils lack both the low-affinity IgG Fc receptor CD16/$Fc\gamma R_{lo}$ (Stain *et al.*, 1987), which mediates Fc-dependent phagocytosis in neutrophils (Fleit *et al.*, 1982), and the 75-kDa high-affinity IgG receptor, $Fc\gamma R1$ (CD64). The 40-kDa IgG Fc receptor ($Fc\gamma R2$/CDw32) is present on basophils; as with CR1, a role in immune adhesion seems feasible.

Basophils express gp150/CD13, a "myeloid-associated" but widely expressed marker, apparently identical to aminopeptidase N (Look *et al.*, 1989). This membrane metalloprotease preferentially degrades small peptides and may modulate cellular responses to low-molecular-weight peptide hormones. Basophils retain the 67-kDa/CD33 structure, which is lost from other myeloid lineages during terminal differentiation (Griffin *et al.*, 1984).

The outstanding phenotypic feature of basophils is expression of antigens associated with cellular activation, absent from other myeloid cells (Stain *et al.*, 1987). Basophils

Table III
Hemopoietic Lineage-Associated Glycoproteins
on Human MC/Basophils[a]

		Reactivity[b] with:	
CD Number	Specificity	MC	Basophils
1–8	T-cell antigens	−	−
9	Activation marker/24 kDa	+	+
11a	LFA-1α/180 kDa	−	+
11b	Mac-1α/170 kDa	−	+
11c	gp150/150 kDa	−	+
13	Aminopeptidase N/150 kDa	−	+
16	Low-affinity FcγR/50–70 kDa	−	−
18	LFA-β/95 kDa	−	+
23	FcεR2/45 kDa	−	−
25	IL2Rα/55 kDa	−	+
32	FcγR2/40 kDa	−	+
33	gp67 of myeloid precursors	+	+
35	CR1 (C3b receptor)/220 kDa	−	+
38	OKT10 activation marker/43 kDa	−	+
45	Leukocyte common antigen	+	+
54	ICAM-1/85 kDa	−	+
64	FcγR1/75 kDa	−	−

[a]Data taken from Rimmer and Horton (1987), Stain *et al.* (1987), and Valent *et al.* (1988b).
[b]Assessed by indirect immunofluorescence.

constitutively express CD25, the IL2 receptor α chain (see below). The 24-kDa CD9 antigen, associated with lymphoid precursors and activated B and T lymphocytes, is expressed on both basophils and MC (Stain *et al.*, 1987; Valent *et al.*, 1988b). This molecule reportedly has intrinsic serine kinase activity and mitogenic activity for some cell lines (Zeleznik *et al.*, 1987). Its biological role, however, is undefined. Another activation antigen of basophils is the 45-kDa T10 antigen (CD38), absent from other myeloid cells but expressed by plasma cells and activated B and T lymphocytes and by some bone marrow cells (Janossy *et al.*, 1981).

MC, in contrast, generally lack antigens typical of other lineages, though pan-hemopoietic and some less well-defined antigens are presented (Horton and O'Brien, 1983; Rimmer and Horton, 1987; Valent *et al.*, 1989b; Horton *et al.*, 1989). This immunophenotype (Table III), with other data on surface biochemistry (Rimmer *et al.*, 1986) and MC development *in vitro* and *in vivo* (Sonoda *et al.*, 1983), suggests that MC diverge from other hemopoietic lineages early in their development.

Table III outlines a phenotypic definition of human MC and basophils but is not intended to be an exhaustive list of antigens present on the two cell types.

5.1.1. Leukocyte Function-Associated Antigens on Basophils

5.1.1a. Biology.   Human basophils express LFAs, a family of heterodimeric glycoproteins defined by a common 95-kDa β subunit and restricted to the myeloid and lymphoid lineages (Kurzinger *et al.*, 1983). Three distinct LFA α chains are recognized: the 180-kDa LFA-1, the 170-kDa Mac-1, and p150 (Sanchez-Madrid *et al.*, 1983).

A range of leukocyte functions are inhibitable by MAbs to LFAs or to their common β chain. Anti-LFA-1 MAbs preferentially inhibit *in vitro* adhesion-dependent functions of T cells, notably adhesion to endothelia and antigen-presenting cells and adhesion of cytolytic T cells to target cells (Krensky *et al.*, 1983; Hildreth *et al.*, 1983). Inhibition of myeloid cell functions by subunit-specific MAbs indicates that Mac-1 and p150 are the major LFAs of myeloid cells. Mac-1 mediates myeloid cell adhesion to endothelia (Springer and Anderson, 1985), chemotaxis (Beatty *et al.*, 1983), and phagocytosis of opsonized particles (Beller *et al.*, 1982; Arnaout *et al.*, 1983), whereas the p150,95 complex may be specifically associated with adhesion, chemotaxis, and extravasation of monocytes (Keizer *et al.*, 1987). LFAs on basophils probably mediate the same functions, with the exception of phagocytosis, as on other myeloid cells.

LFAs are probably the primary tissue infiltration system of circulating cells. Both Mac-1 and p150,95 are rapidly mobilized from intracellular stores upon exposure to inflammatory stimuli, such as C5a and leukotriene B_4 (Springer *et al.*, 1984; Springer and Anderson, 1985), and all three LFAs are highly expressed by myeloid cells. The rare human LFA deficiency syndrome is characterized by severe chronic bacterial infection (Springer *et al.*, 1984; Dana *et al.*, 1984). The underlying defect is partial or complete failure to produce LFA β chain, decreasing or abolishing LFA expression (Springer *et al.*, 1984). Leukocytes from such patients are defective in functional assays, including the *in vivo* "skin window" test (basophils are prominent in the normal cellular infiltrate), phagocytosis of opsonized particles, chemotaxis assays, and substrate adhesion and leukoaggregation assays (reviewed in Springer and Anderson, 1985). Similarly, in *in vivo* systems, anti-LFA MAbs prevent tissue infiltration by leukocytes. Anti-LFA-β MAbs reduce inflammatory tissue injury in dog myocardial infarcts (Simpson *et al.*, 1988) and prevent accumulation of peritoneal macrophages in response to inflammatory stimuli in mice and rabbits.

N-terminal partial sequences of the three α chains show considerable homology (Miller *et al.*, 1986). MAbs to the individual α subunits are generally specific, however, permitting their classification into three groups: CD11a, CD11b and CD11c, reactive with LFA-1, Mac-1, and p150, respectively (Hogg and Horton, 1987). CD18 MAbs, recognizing the common β chain, identify all three LFA complexes.

The definitive ligand of LFA-1 is intracellular adhesion molecule 1 (ICAM-1) (Marlin and Springer, 1987), a widely expressed cell surface molecule homologous to immunoglobulins. Endothelial ICAM-1 expression is rapidly enhanced by IL1 at inflammatory sites (Dustin *et al.*, 1986), which may promote adhesion and localization of leukocytes to such sites. ICAM-1 is probably also involved in cell–cell interactions between leukocytes; basophils express ICAM-1. Mac-1 and p150,95 are receptors for C3bi, the proteolytically inactivated form of C3b (Arnaout *et al.*, 1983). Mac-1 also reportedly interacts with fibrinogen, with bacterial surface lipopolysaccharides, with a 63-kDa membrane glycoprotein of *Leishmania donovani*, and possibly with other ligands (Russell and Wright, 1988). Ligand specificities of LFAs are consistent with their demonstrated biological roles, that is, LFA-1 in cell–cell interaction in immune and inflammatory responses and Mac-1 and p150 in adhesion to surfaces prior to phagocytosis or extravasation. In each case, ligand binding requires the presence of divalent metal cations (Martz, 1980).

5.1.1b. Molecular Cloning of Subunits. cDNA clones for both Mac-1 and p150 have been characterized (Corbi *et al.*, 1987, 1988; Arnaout *et al.*, 1988). As suggested by N-terminal partial sequence data, the two molecules show both a high degree of sequence homology and overall similarity in their predicted properties. Comparison with reported sequences reveals strong homology of both Mac-1 and p150 to members of the integrin family of adhesive proteins (Hynes, 1987), including human cellular receptors for fibronectin and vitronectin, platelet glycoprotein IIb-IIIa, and *Drosophila* "position-specific antigen" PS2 (Poncz *et al.*, 1987; Argraves *et al.*, 1987; Suzuki *et al.*, 1987). This establishes LFAs as a subgroup of the integrin family, defined by CD18, also designated integrin β_2.

Both α- chains are 1136 amino acids long, although p150 has 10 fewer extracellular, and thus 10 more cytoplasmic, residues than does Mac-1 (Corbi *et al.*, 1987, Arnaout *et al.*, 1988). Mac-1 has 18 potential N-linked glycosylation sites and p150 has 10, so the higher molecular weight of expressed Mac-1 is probably due to more extensive glycosylation.

The predicted structures show distinctive common features. Tandem repeats of about 60 residues are present in the extracellular portion of both proteins, three such repeats in p150, four in Mac-1. Within each repeat is a sequence homologous to divalent metal ion-binding-site sequences of proteins such as calmodulin and parvalbumin (EF-hand sequences; Szebenyi *et al.*, 1982), having the consensus sequence DXDXDGXXDXXE (single-letter code). Three putative metal-binding sites of Mac-1, and all three of p150, show strong homology to the consensus sequence. As in other divalent cation-dependent integrins, however, the otherwise invariant glutamine at position 12 of the putative cation-binding sequence is not conserved.

Both Mac-1 and p150 possess unique and highly homologous (57% amino acid identity and much conservative substitution) 187-residue extracellular regions NH$_2$ terminal to the metal-binding domains (Corbi *et al.*, 1987, 1988; Arnaout *et al.*, 1988). These regions may include the binding site for the definitive ligand, C3bi. Since other integrins lack a comparable region, however, it is tempting to suggest that the 187-residue sequences confer further ligand specificities. Several biological functions in which Mac-1 and p150 are important could involve binding to a second ligand or homo/heterophilic interactions. Significantly, not all putative Mac-1 ligands contain the Arg-Gly-Asp motif associated with integrin ligands.

cDNA cloning of the LFA common β chain (Law *et al.*, 1987; Kishimoto *et al.*, 1987) predicts a 747-residue protein, including 56 extracellular cysteines. Reduction of intrachain disulfide bonds may explain the decreased electrophoretic mobility of integrin β$_2$ under reducing conditions. A particularly cysteine-rich sequence, spanning residues 445–631, includes four tandem repeats containing similar eight-cysteine motifs. Integrin β$_2$ is strongly homologous to "band 3", the β chain of the avian extracellular matrix receptor, integrin. Overall amino acid identity to band 3 is 45%; particularly striking is the conservation of all but one cysteine residue between integrin β$_2$ and band 3.

Band 3 is known to link to talin (Horwitz *et al.*, 1986), a cytoskeletal protein, suggesting that LFAs might also link to the cytoskeleton via the cytoplasmic tail of β$_2$. Cytochalasin B inhibits several LFA-mediated leukocyte functions (Martz, 1986), providing evidence for linkage to cytoskeletal elements.

No member of the LFA family is detectable on human MC by immunological techniques (Rimmer and Horton, 1987; Valent *et al.*, 1989b). MAbs with interspecies cross-reactivity fail either to bind to dog MC or to prevent their adhesion to bovine tracheal epithelium *in vitro* (Varsano *et al.*, 1988).

The high level of LFA expression on circulating leukocytes and the rapid enhancement of LFA expression by inflammatory stimuli reflect the crucial role of LFAs in leukocyte adhesion and infiltration at inflammatory sites. The specificity of Mac-1 and p150 for C3bi newly generated at inflammatory sites is consistent with this view. MC do not circulate and so have no need of LFAs. Other molecular systems presumably mediate adhesion and "homing" of MC precursors into the tissues (see below).

5.1.2. Unique Glycoproteins of Mast Cells and Basophils

Surface glycoproteins associated with MC/basophils, or distinguishing them from other hemopoietic lineages, have been identified by MAbs.

The MAb Bsp-1 defines a 45-kDa species on membranes of human basophils and their precursors but not on MC or other myeloid cells (Bodger *et al.*, 1987, Bodger *et al.*, 1990). MAb YB5.B8 detects mature human MC, a small percentage of bone marrow cells, and blasts from some acute myeloid leukemias (Gadd and Ashman, 1985; Mayrhofer *et al.*, 1987). YB5.B8 inhibits growth factor-stimulated myeloid colony formation from bone marrow precursors *in vitro* (Cambareri *et al.*, 1988) (Fig. 1). This MAb defines a 150-kDa species, a molecular size comparable to those of murine GM-CSF and human IL3 receptor molecules (Park *et al.*, 1986a,b; Isfort *et al.*, 1988a,b). Specificity for a growth factor receptor would be consistent with the restriction of YB5.B8 reactivity to early myeloid cells and MC.

Katz *et al.* (1983) used a panel of rat MAbs to define mouse MMC and CTMC. One of these, B54.2, reacts with all mouse mature MC, detecting a glycoprotein of 130 kDa

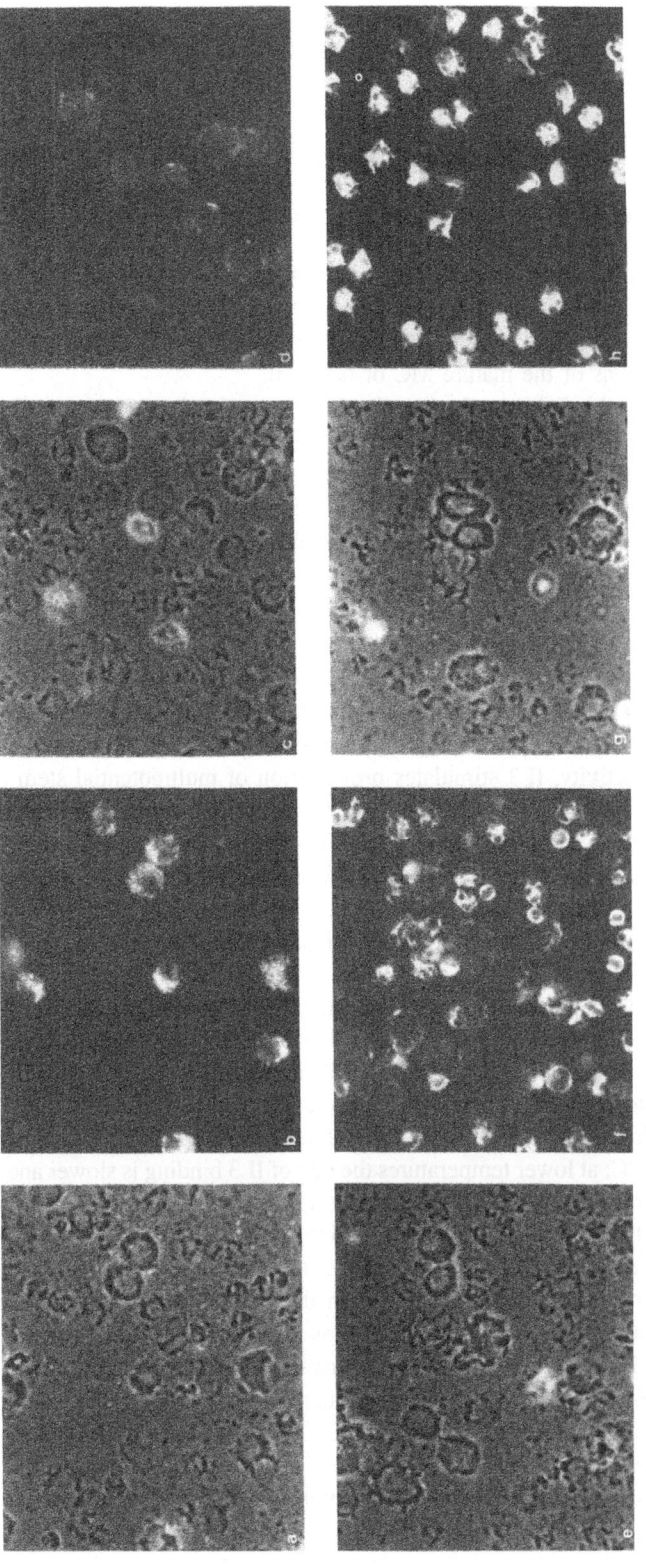

FIGURE 1. (a) Human mastocytosis MC cytocentrifuge preparation from human bone marrow. Phase contrast; ×1200. (b) Same field as 1a, indirect immunofluorescence with MCG35 (anti-MC granule; Rimmer *et al.*, 1984). (c) Human mastocytosis MC cytocentrifuge preparation. Phase contrast; ×1200. (d) Same field as panel c, indirect immunofluorescence with YB5.B8. (e) Human mastocytosis MC cytocentrifuge preparation. Phase contrast; ×1200. (f) Same field as panel e, indirect immunofluorescence with F10-89-4 (CD45). (g) Human mastocytosis MC cytocentrifuge preparation. Phase contrast; ×1200. (h) Same field as panel g, indirect immunofluorescence with MHM23 (CD18).

with a 93-kDa peptide core (Katz and Austen, 1987). A 110-kDa incompletely glycosylated precursor is also detected.

5.2. Receptors for Hemopoietic Growth Factors

5.2.1. Growth Factors in Mast Cell and Basophil Biology

MC/basophil differentiation from bone marrow precursors is influenced by protein growth factors active on hemopoietic cells (Nagao *et al.*, 1981; Schrader, 1981; Tertian *et al.*, 1981; Haig *et al.*, 1982; Valent *et al.*, 1989a). Such factors may also influence differentiated functions of the mature MC or basophil.

Many reports in this field simply describe the effects of defined (or even unidentified) hemopoietic growth factors on MC, basophils, or cell lines of these lineages. Rarely have specific growth factor receptors been characterized on MC/basophils. In general, therefore, the presence or otherwise of receptors is implied by the bioactivity of the factor on MC/basophils, but the possibilities of nonspecific effects or "helper" cell activity should be considered.

5.2.2. Receptors for Interleukin 3

IL3 was originally identified as a T-cell-derived murine MC growth factor (Nagao *et al.*, 1981; Tertian *et al.*, 1981; Schrader, 1981). Later studies showed that, in addition to its MC-promoting activity, IL3 stimulates proliferation of multipotential stem cells and precursors of most hemopoietic lineages, including granulocytes, mononuclear phagocytes, and lymphoid cells (Hapel *et al.*, 1981). Murine IL3 is ineffective in human systems, but the gene encoding a simian IL3 was isolated by DNA transfection into COS-7 cells from a gibbon T-cell line (Yang *et al.*, 1986). IL3 efficiently stimulates human hemopoiesis but is a potent stimulus for basophil rather than MC development (Valent *et al.*, 1989a).

Continuously growing, IL3-dependent mouse myeloid or MC lines have been invaluable tools in characterization of the murine IL3 receptor. Equilibrium binding studies using ^{125}I-IL3 on a range of IL3-dependent cell lines suggest the presence of a single high-affinity receptor class, K_d about 10^{-11}M, present at about 500–5000 receptor sites per cell (Palazynski and Ihle, 1984). ^{125}I-IL3 binding to these cells *in vitro* is maximal within 60 min at 37°C; at lower temperatures the rate of IL3 binding is slower and receptor saturation is not achieved. There is no significant nonspecific binding of ^{125}I-IL3 to cells; binding is inhibited by excess unlabeled IL3, and the radioligand does not bind detectably to IL3-unresponsive cell lines.

Putative IL3 receptor molecules have been detected on several murine cell lines, including some MC lines. Chemical crosslinking of ^{125}I-IL3 bound to intact cells or membrane preparations generates a 65–70-kDa major band and a 140-kDa minor band, as assessed by SDS-PAGE (Park *et al.*, 1986a; Sorensen *et al.*, 1986; Nicola and Peterson, 1986). Both species bind wheat germ agglutinin; neither shows altered electrophoretic mobilty upon reduction.

The presence of two distinct IL3 receptor molecules is inconsistent with demonstrations of a single affinity class of receptor. Extensive studies (Isfort *et al.*, 1988a,b) suggest that the 140-kDa species (gp140) is the native receptor and that the 65–70-kDa molecule (gp65) is a proteolytic fragment with ligand-binding activity. A gp140 phosphorylates on cytoplasmic tyrosine (and also on threonine) residues following binding of IL3 to the cell (Koyasu et al., 1987; Isfort et al., 1988a,b). ^{125}I-IL3 bound to cells may subsequently be affinity isolated from lysates upon reaction with solid-phase antiphosphotyrosine MAb, demonstrating that IL3 binds firmly to a cellular phosphoprotein (Isfort et al., 1988a,b). Covalent crosslinking and SDS-PAGE of the phosphoprotein fraction of IL3-treated cells reveals a major band, equivalent to IL3 (28 kDa), crosslinked to a 140-kDa species (Isfort et al., 1988a,b). Similarly, anti-IL3 antibodies immunoprecipitate a 140-kDa species from affinity-purified, IL3-induced phosphoproteins (Isfort et al., 1988a,b). Finally, on molecular size analysis of IL3-induced phosphoproteins by glycerol gradient centrifugation, ^{125}I-IL3 migrates to a position equivalent to 170 kDa (Isfort et al., 1988a,b), consistent with association of a single IL3 molecule with a 140-kDa protein.

gp65 is unlikely to be a distinct component of an IL3–receptor complex, since neither chemical crosslinking nor glycerol gradient sizing provides evidence for IL3-containing complexes larger than 170 kDa. It is probable that gp65 is an active fragment of gp140. Other cellular receptors, including those for human growth hormone (Leung et al., 1987), GM-CSF (Park et al., 1986b), and polymeric IgA/IgM (Mostov et al., 1986), yield ligand-binding fragments.

gp140 is the only transmembrane glycoprotein shown to phosphorylate upon induction of cells by IL3 (Isfort et al., 1988a). It is not constitutively phosphorylated, nor is it detectable on cell lines that fail to bind ^{125}I-IL3 (Isfort et al., 1988a). The rapidity of gp140 phosphorylation, within 30 sec of IL3 treatment, raises the possibility that gp140 has IL3-inducible intrinsic tyrosine kinase activity. This accords with the observation that introduction of an oncogene tyrosine kinase to IL3-dependent cells abolishes IL3 dependence.

IL3 is probably involved in the "T-cell-dependent" mastocytoses observed at sites of inflammation or parasite infestation (see Jarrett and Haig, 1984, for references). The demonstration of its basophil-promoting activity in humans is too recent to support speculation on the role of IL3 in human basophil biology.

IL3 has not been shown to affect the differentiated functions of the mature mast cell. Recombinant IL3 does not directly induce degranulation of human leukemic basophils in vitro but may prime normal basophils for mediator release (Valent et al., 1989c).

5.2.3. Interleukin 2 Receptors on Basophils

CD25 MABs, detecting the α chain of the IL2 receptor (IL2Rα; gp55; Tac protein), stain a proportion of both normal and leukemic human basophils by indirect immunofluorescence (Stain et al., 1987) and immunoprecipitate gp55 from a cell preparation containing >95% basophils. Hybridization of RNA blots of a similar basophil-enriched preparation revealed 1.5- and 3.5-kb bands, as seen in a T-cell RNA control, demonstrating active synthesis of IL2Rα by basophils (Stockinger et al., 1988).

Equilibrium binding of ^{125}I-IL2 to basophil-enriched preparations indicated 8–16,000 low-affinity sites per cell, K_d 6.3 $\times$ 10^{-8}M, typical for IL2 binding to IL2Rα, and <100 high-affinity sites (K_d 4.8 $\times$ 10^{-11}M) per cell. Assuming that these data are not explicable by the presence of undetected activated T cells, this is a significant finding.

IL2 is known to stimulate T-cell proliferation by binding to a high-affinity receptor complex, comprising IL2Rα and a 75-kDa protein, IL2Rβ (Tsudo *et al.*, 1986; Sharon *et al.*, 1986). IL2 binding to IL2Rα alone is not activating; IL2Rα is not expressed by the resting T cell but is induced following contact with IL2 (Siegel *et al.*, 1987). The role of IL2Rα may simply be to permit formation of high-affinity IL2 receptors. The apparent presence of high-affinity sites on basophils suggests that their expression is physiologically relevant.

The IL2Rα molecule was originally identified by MAbs raised against hyperexpressing human T-cell leukemia virus type I-infected T cells (Leonard *eta l.*, 1984). The heavily glycosylated mature protein has an apparent molecular weight of 55–60 Da, which is only slightly diminished when IL2Rα is derived from tunicamycin-treated cells (see Leonard *et al.*, 1984, and Nikaido *et al.*, 1984, for references).

cDNA cloning predicts a protein of 251 residues with a molecular weight of 28.6 kDa (Leonard *et al.*, 1984; Nikaido *et al.*, 1984; Cosman *et al.*, 1984). Thus, about 50% of the molecular weight of the mature protein is accounted for by carbohydrate. There are only two potential sites for N-linked glycosylation, and since distinct precursor species of 33 and 35 kDa have been identified, the bulk of the carbohydrate is presumably O linked to the numerous extracellular serine/threonine residues. The predicted IL2Rα protein has two 19-residue hydrophobic sequences. One, at residues 220–238, is a typical transmembrane hydrophobic sequence. The other, residues 118–135, is less markedly hydrophobic, dissimilar to reported transmembrane sequences, and may be involved in binding to IL2, itself a partly hydrophobic molecule.

The other partner in the high-affinity IL2 receptor, the 75-kDa IL2Rβ, was identified by crosslinking of ^{125}I-IL2 to T-cell membrane preparations but remains largely uncharacterized (Tsudo *et al.*, 1986; Sharon *et al.*, 1986). Native IL2Rβ binds IL2 with an affinity intermediate between those of IL2Rα and the high-affinity receptor complex. IL2Rβ, but not IL2Rα, is constitutively expressed by subpopulations of peripheral blood resting T cells and large granular lymphocytes LGLs). These cell populations respond to IL2 *in vitro* by proliferation, enhanced natural killer cell activity, development of lymphokine-activated killing activity, enhanced adhesiveness, and synthesis of both IL2Rα and IL2Rβ (Siegel *et al.*, 1987). IL2Rβ-negative cell populations do not respond to IL2, suggesting that IL2Rβ mediates the initial effects of IL2 on T cells and LGLs.

Unfortunately, the lack of appropriate MAbs or cDNA probes precludes direct examination of basophils for IL2Rβ, complicating interpretation of the role of IL2 receptors on basophils. Previous studies of IL2Rα expression by myeloid cells are not helpful (Herrman *et al.*, 1985; Holter *et al.*, 1986; Visani *et al.*, 1987). No other myeloid cell type constitutively expresses IL2Rα, which is, however, induced on normal human monocytes, chronic myelogenous leukemia (CML) blasts, and some monocytic cell lines by 18 hr *in vitro* suspension culture. Culture-induced expression of IL2Rα is much enhanced by gamma interferon on monocytes, though not on basophils, but unaffected by exposure to IL2. IL2 does not affect colony formation by IL2Rα-positive CML bone marrow cells, nor does it generate basophils from human hemopoietic cells *in vitro*. This suggests that IL2 receptors on basophils and other myeloid cells are not important in stimulation of proliferation. Even so, a functional role cannot be excluded. IL2 stimulation, perhaps in concert with other stimuli, is a possible route by which basophils could be activated during T-cell-mediated inflammatory or cytotoxic reactions. Certainly, basophil infiltration is a prominent feature of some cell-mediated hypersensitivity reactions in humans and other species (Galli *et al.*, 1984). Unlike basophils, human mast cells lack immunoreactive IL2Rα (Rimmer and Horton, 1987; Valent *et al.*, 1989b).

5.2.4. Interactions of Mast Cells and Basophils with Other Hemopoietic Growth Factors

Several reports describe actions of growth factors and immunomodulatory hormones on MC/basophils. IL4 maintains limited growth of IL3-dependent mouse MC lines and synergizes with IL3 to enhance their proliferation (Smith and Rennick, 1986). Recombinant human IL4, however, has no MC or basophil-promoting activity on human bone marrow cells, either alone or in concert with recombinant IL3 (Saito *et al.*, 1988). Similarly, erythropoietin (Epo) maintains viability and growth of a mouse MC line in the absence of IL3 (Tsao *et al.*, 1988). The number of Epo-binding sites per cell is inversely related to IL3 concentration. Membrane receptor molecules for both IL4 and Epo await biochemical characterization.

IL1 directly induces mediator release from both human MC and blood basophils (Subramanian and Bray, 1987), the only lymphokine shown to do so. The IL1 receptor is a 557-residue transmembrane structure of 80 kDa. The cytoplasmic tail is 217 residues long, sufficient to possess intrinsic enzymic or signal-transducing activity, suggesting that the 80-kDa species may constitute the entire functional IL1 receptor (Sims *et al.*, 1988).

Finally, there are reports of undefined growth factors in lymphocyte or cell line-conditioned medium stimulating MC/basophil growth from hemopoietic tissues. With hindsight, most of these reports are consistent with IL3 activity. One interesting study, however, indicated the existence of a lymphocyte-produced murine CTMC growth factor distinct from IL3 (Nakahata *et al.*, 1986).

5.3. Cytoadhesive Glycoproteins of Mast Cells

Although MC lack LFAs, several observations suggest the presence of adhesive molecules on MC. Thus, MMC accumulate in inflamed mucosae, and MC infiltrate tissue in pathological mastocytoses. Mouse IL3-dependent MC interaction with fibroblast monolayers *in vitro* induces differentiation to CTMC phenotype, and coculture with fibroblasts maintains viability of *ex vivo* CTMC (Levi-Schaffer *et al.*, 1986; Dayton *et al.*, 1988). Glycoproteins mediate adhesion-dependent phenomena on other cell types, prompting investigation of MC for similar molecules.

A study of adhesion of dog mastocytoma MC to tracheal epithelium *in vitro* is informative in this context (Varsano *et al.*, 1988). These MC adhere specifically to tracheal epithelial cells in frozen sections or in culture, but not significantly to basement membrane, connective tissue, or extracellular matrix. Few of these MC bind to pure fibronectin, nor is there significant binding to collagen type I or IV. Adhesion occurs rapidly, does not require metabolic energy or divalent cations, and is abolished by treatment of MC with proteinase K but not trypsin. These data are inconsistent with adhesion due to integrins, which require energy and divalent cations for function (Charo *et al.*, 1985). Binding of cell surface proteoglycans to membrane proteins on either or both cell populations might explain this result. Proteoglycans are abundant in MC and epithelial cell membranes, and both heparin- and chondroitin sulfate-linked proteoglycans have been shown to bind to protein substrates (Laterra *et al.*, 1980; Ruohslahti and Engvall, 1980; Yamada *et al.*, 1980; Sakashita *et al.*, 1980). Such interactions participate in fibroblast adhesion to epithelia and may be generally important in maintaining the "architecture" of epithelial layers (Saunders and Bernfield, 1988). Integrins, in contrast, function largely in the initial stages of cellular adhesion. We have assessed human mastocytoma MC for expression of integrins by immunofluorescence (unpublished observations). Both α- and β-subunit-specific MAbs fail to detect significant expression of vitronectin receptor (integrin α_V/β_3); glycoproteins IIb-IIIa (integrin α_{IIb}/β_3); or very late activation (VLA) antigens, a group including the cellular receptors for laminin, collagens, and fibronectin (integrins α_1-α_6/β_1). A recent report (Thompson *et al.*, 1989) demonstrates that mouse mast cells adhere to laminin via the 67 (32) kDa laminin–elastin receptor; this suggests that non-integrin receptors may be involved in mast cell interactions with extracellular matrix and should be investigated further.

ACKNOWLEDGMENTS. We wish to thank Drs. L. K. Ashman, K. F. Austen, P. Bettelheim, M. P. Bodger, and J. N. Ihle for their generosity in sharing unpublished data and supplying preprints.

6. REFERENCES

Albrandt, K., Orida, N. K., and Liu, F. T., 1987, An IgE binding protein with a distinct repetitive sequence and homology to an IgG receptor, *Proc. Natl. Acad. Sci. USA* **84**:6859–6863

Alcaraz, G., Kinet, J.-P., Kumar, N., Wank, S. A., and Metzger, H., 1984, Phase separation of receptor for IgE and its subunits in Triton X-114, *J. Biol. Chem.* **259**:14922–14927

Argraves, W. S., Suzuki, S., Arai, H., Thompson, K., Piersbacher, M. D., and Ruoslahti, E., 1987, Amino acid sequence of human fibronectin receptor, *J. Cell Biol.* **105**:1183–1190

Arnaout, A., Todd, R. F., Dana, N., Melamed, J., Schlossman, S. F., and Colten, H. R., 1983, Inhibition of phagocytosis of complement C3 or immunoglobulin G-coated particles and of C3bi binding by mabs to a monocyte/granulocyte membrane glycoprotein (Mo1), *J. Clin. Invest.* **72**:171–179

Arnaout, M. A., Gupta, S. K., Pierce, M. W., and Tenen, D. G., 1988, Amino acid sequence of the alpha subunit of human leucocyte adhesion receptor Mo1 (complement receptor type 3), *J. Cell Biol.* **106**:2153–2158

Baenziger, J., Kornfeld, S., and Kochwa, S., 1974, Structure of carbohydrate units on IgE, *J. Biol. Chem.* **249**:1889

Baird, B., and Holowka, D., 1985, Structural mapping of Fc-receptor bound IgE, *Biochemistry* **24**:6252–6259

Beatty, P. G., Ledbetter, J. A., Martin, P. J., Price, T. H., and Hansen, J. A., 1983, Definition of a common leucocyte surface antigen (Lp 95-150) associated with diverse cell mediated funtions, *J. Immunol.* **131**:2913–2918

Beller, D. L., Springer, T. A., and Schreiber, R. D., 1982, Anti-Mac-1 selectively inhibits mouse and human type 3 complement receptors, *J. Exp. Med.* **156**:1000–1009

Blank, U., Ra, C., Miller, L., White, K., Metzger, H., and Kinet, J.-P., 1989, Complete structure and expression in transfected cells of high affinity IgE receptor, *Nature* (London) **337**:187–189

Bodger, M. P., Rimmer, E. F., Horton, M. A., Yuen, E., and Brown, R. D., 1990, Human mast cells do not express the basophil antigen detected by monoclonal antibody Bsp-1 (submitted for publication)

Bokisch, V. A., and Muller-Eberhard, H. J., 1970, Anaphylatoxin inactivator of human plasma; its isolation and characterisation as a carboxypeptidase, *J. Clin. Invest.* **49**:2427–2436

Bokotch, G. M., Katada, T., Northup, J. K., Ui, M., and Gilman, A., 1984, Purification and properties of the inhibitory guanine-binding regulatory component of adenylate cyclase, *J. Biol. Chem.* **259**:3560–3567

Cambareri, A. C., Ashman, L. K., Cole, S. R., and Lyons, A. B., 1988, A monoclonal antibody to a human myeloid leukaemia-specific antigen binds to normal haemopoietic cells and inhibits colony formation *in vitro*, *Leuk. Res.* **12**:929–939

Cass, R. M., and Anderson, R. R., 1968, The disappearance rate of skin-sensitising antibody, *J. Allergy Clin. Immunol.* **42**:29–35

Charo, I., Yuen, C., and Goldstein, I. M., 1985, Adherence of human polymorphonuclear cells to endothelial monolayers, *Blood* **65**:473

Chenoweth, D. E., and Hugli, T. E., 1978, Demonstration of specific C5a receptors on intact human polymorphonuclear leucocytes, *Proc. Natl. Acad. Sci. USA* **75**:3943–3947

Chenoweth, D. E., Goodman, M. G., and Weigle, W. O., 1982, Demonstration of a specific receptor for C5a on murine macrophages. *J. Exp. Med.* **156**:68

Cochrane, C. G., and Muller-Eberhard, H. J., 1968, The derivation of two distinct anaphylatoxin activities from the third and fifth components of human complement, *J. Exp. Med.* **127**:371–386

Coleman, J. W., Helm, B. A., Stanworth, D. R., and Gould, H., 1985, Inhibition of mast cell sensitisation in vitro by human E-chain fragment synthesised in *E. coli*, *Eur. J. Immunol.* **15**:966–969

Conrad, D. H., and Froese, A., 1976, Characterisation of the target cell receptor for IgE II, *J. Immunol.* **116**:319–326

Conrad, D. H., and Froese, A., 1978, Characterisation of the target cell receptor for IgE IV, *Immunochemistry* **15**:283–288

Conrad, D. H., Bazin, H., Sehon, A. H., and Froese, A., 1975, Binding parameters of interaction between rat IgE and rat mast cell receptors, *J. Immunol.* **114**:1688–1691

Corbi, A. L., Miller, L. J., O'Connor, K., Larson, R. S., and Springer, T. A., 1987, cDNA cloning and complete primary structure of the alpha subunit of a leucocyte adhesion protein, p150,95, *EMBO J.* **6**:4023–4028

Corbi, A. L., Kishimoto, K. T., Miller, L. J., and Springer, T. A., 1988, The human leucocyte adhesion glycoprotein Mac-1 (complement receptor type 3, CD11b) alpha-subunit, *J. Biol. Chem.* **263**:12403–12411

Cosman, D., Ceretti, D. P., Larsen, A., Park, L., March, D., Dower, S., Gillis, S., and Urdal, D., 1984, Cloning, sequence and expression of human interleukin-2 receptor, *Nature* (London) **312**:769–771

Coutts, S. M., Nehring, R. E., and Jariwala, N. U., 1980, Purification of rat peritoneal mast cells; occupation of IgE receptors by IgE prevents loss of receptors, *J. Immunol.* **124**:2309–2315

Daeron, M., and Voisin, G. A., 1979, Mast cell membrane antigens and Fc receptors in anaphylaxis, *Immunology* **38**:447–458

Dana, N., Todd, R. F., Pitt, J., Springer, T. A., and Arnaout, M. A., 1984, Deficiency of a surface membrane glycoprotein (Mo1) in man, *J. Clin. Invest.* **73**:153–159

Dayton, E. T., Pharr, P., Ogawa, M., Serafin, W. E., Austen, K. F., Levi-Schaffer, F., and Stevens, R. L., 1988, 3T3 fibroblasts induce IL-3-dependent mast cells to resemble connective tissue mast cells, *Proc. Natl. Acad. Sci. USA* **85**:569–572

Dias da Silva, W., Eisele, J. W., and Lepow, H., 1967, Complement as a mediator of inflammation, *J. Exp. Med.* **126**:1027–1048

Dorrington, K. J., and Bennich, H., 1973, Thermally-induced structural changes in immunoglobulin E, *J. Biol. Chem.* **248**:8378–8384

Dorrington, K. J., and Bennich, H. H., 1978, Structure-function relationships in human immunoglobulin E, *Immunol. Rev.* **41**:3–25

Dustin, M. L., Rothlein, R., Bhan, A. K., Dinarello, C. A., and Springer, T. A., 1986, Induction by IL-1 and interferon, tissue distribution, biochemistry and function of a natural adherence molecule (iCAM-1), *J. Immunol.* **137**:245–254

Feltner, D. E., Smith, R. H., and Marasco, W. A., 1986, Characterisation of the plasma membrane GTPAse from rabbit neutrophils, *J. Immunol.* **137**:1961–1970

Fernandez, H. N., Henson, P. M., Otani, A., and Hugli, T. E., 1978, Chemotactic response to human C3a and C5a anaphylatoxins, *J. Immunol.* **120**:109–115

Fewtrell, C., Kempner, E., Poy, G., and Metzger, H., 1981, Unexpected findings from target analysis of IgE and its receptor, *Biochemistry* **20**:6589–6594

Finbloom, D. S., and Metzger, H., 1983, Isolation of crosslinked IgE receptor complexes from rat macrophages, *J. Immunol.* **130**:1489–1491

Fleit, H., Wright, S. D., and Unkeless, J. C., 1982, Human neutrophil Fc-gamma receptor distribution and structure, *Proc. Natl. Acad. Sci. USA* **79**:3275–3279

Fukuoka, Y., and Hugli, T. E., 1988, Demonstration of a specific C3a receptor on guineapig platelets, *J. Immunol.* **140**:3496–3501

Furuichi, K., Rivera, J., and Isersky, C., 1985, The receptor for immunoglobulin E on rat basophilic leukaemic cells; effect of ligand binding on receptor expression, *Proc. Natl. Acad. Sci. USA* **82**:1522–1525

Gadd, S. J., and Ashman, L. K., 1985, A murine monoclonal antibody specific for a cell surface antigen expressed by a subgroup of human myeloid leukaemias, *Leuk. Res.* **9**:1329–1336

Galli, S. J., Dvorak, A. M., and Dvorak, H. F., 1984, Basophils and mast cells; morphological insights into their biology, secretory patterns and function, *Prog. Allergy* **34**:1–141

Geha, R., Helm, B., and Gould, H., 1985, Inhibition of Prausnitz-Kustner reaction by an immunoglobulin E-chain fragment synthesized in *E. coli, Nature* (London) **315**:577–578

Gerard, C., and Hugli, T. E., 1981, Identification of classical anaphylatoxin as the desArg form of C5a, *Proc. Natl. Acad. Sci. USA* **78**:1833–1837

Gervasoni, J. E., Conrad, D. H., Hugli, T. E., Schwartz, L. B., and Ruddy, S., 1986, Degradation of human anaphylatoxin C3a by rat peritoneal mast cells, *J. Immunol.* **136**:285–292

Glovsky, M. M., Hugli, T. E., Ishizaka, T., Lichtenstein, L. M., and Erickson, B. W., 1979, Anaphylatoxin-induced histamine release from human leucocytes, *J. Clin. Invest.* **64**:804–811

Goldman, D. W., Chang, F. H., Gifford, L. A., Goetzl, E. J., and Bourne, H. R., 1985, Pertussis toxin inhibition of chemotactic factor-induced calcium mobilisation and function in human polymorphonuclear leucocytes, *J. Exp. Med.* **162**:145–156

Goldstein, I., Hoffstein, S., Gallin, J., and Weissmann, G., 1973, Mechanisms of lysosomal enzyme release from human leucocytes, *Proc. Natl. Acad. Sci. USA* **70**:2916–2920

Griffin, J. D., Linch, D., Sabbath, K., Larcom, D., and Schlossman, S. F., 1984, A monoclonal antibody reactive with normal and leukaemic human myeloid progenitor cells, *Leuk. Res.* **9**:521–527

Grossklaus, C., Damerau, B., Lemgo, E., and Vogt, W., 1976, Induction of platelet aggregation by C3a and C5a, *Arch. Pharmacol.* **295**:71–76

Haig, D. M., McKee, T., Jarrett, E. E. E., Woodbury, R., and Miller, H. R. F., 1982, Generation of mucosal mast cells stimulated in vitro by factors derived from T cells of helminth-infected rats, *Nature* (London) **300**:188–190

Hall, T. J., and Rittenberg, M. B., 1986, Interaction of monomeric and complexed mouse monoclonal IgG with rat basophilic leukaemia cells, *J. Immunol.* **137**:2331–2338

Halper, J., and Metzger, H., 1976, Interaction of IgE with rat basophilic leukaemia cells. VI, *Immunochemistry* **13**:907–913

Hapel, A., Lee, J. C., Farrar, W. L., and Ihle, J. N., 1981, Establishment of continuous cultures of T-cells using purified IL-3, *Cell* **25**:179

Harlan, J. M., 1988, Neutrophil-mediated vascular injury, *Acta Med. Scand. Suppl.* **715**:123–129

Hellmann, L., Petterson, U., and Bennich, H., 1982, Characterisation and molecular cloning of mRNA for the heavy chain of rat immunoglobulin E. *Proc. Natl. Acad. Sci. USA* **79**:1264–1268

Helm, R. M., Conrad, D. H., and Froese, A., 1979, Lentil lectin affinity chromatography of surface glycoproteins and the receptor for IgE from RBL, *Int. Arch. Allergy Appl. Immunol.* **58**:90–98

Helm, B., Marsh, P., Verselli, D., Padlan, E., Gould, H., and Geha, R., 1988, The mast cell binding site on human immunoglobulin E, *Nature* (London) **331**:180–183

Hempstead, B. L., Parker, C. W., and Kulzycki, A., 1981, The cell surface receptor for IgE. *J. Biol. Chem.* **256**:10717–10723

Henson, P. M., McCarthy, K., Larsen, G. L., Webster, R. O., Giclas, P. C., Dreisin, R. B., King, T. E., and Shaw, J. O., 1979, Complement fragments, alveolar macrophages and alveolitis, *Am. J. Pathol.* **97**:93–105

Herrman, F., Cennistra, S., Levine, H., and Griffin, J. D., 1985, Expression of interleukin-2 receptors by gamma-interferon induced leukaemic and normal monocytic cells, *J. Exp. Med.* **162**:1111–1116

Hildreth, J., Gotch, F. M., Hildreth, P. D. K., and McMichael, A. J., 1983, A human lymphocyte associated antigen involved in cell-mediated cytolysis, *Eur. J. Immunol.* **13**:202–208

Hogg, N., and Horton, M. A., 1987, Myeloid antigens; new and previously-defined clusters, in *Leucocyte Typing III: White Cell Differentiation Antigens* (A. McMichael et al., eds.), pp. 576–602, Oxford University Press, Oxford

Holowka, D., and Baird, B., 1983a, Structural studies on membrane bound IgE-receptor complex I; characterisation of plasma membrane vesicles from RBL and insertion of amphipathic fluorescent probes, *Biochemistry* **22**:3466–3474

Holowka, D., and Baird, B., 1983b, Structural studies on membrane bound IgE-receptor complex II; mapping of distances between sites on IgE and membrane surface, *Biochemistry* **22**:3475–3484

Holowka, D., and Baird, B., 1984, Lactoperoxide-catalysed iodination of the receptor for IgE at the cytoplasmic side of the plasma membrane. *J. Biol. Chem.* **259**:3720–3728

Holowka, D., and Metzger, H., 1982, Further characterisation of the B-component of the receptor for IgE. *Mol. Immunol.* **19**:219–227

Holowka, D., Conrad, D. H., and Baird, B., 1985, Structural mapping of membrane bound immunoglobulin E-receptor complexes, *Biochemistry* **24**:6260–6267

Holter, W., Grunow, R., Stockinger, H., and Knapp, W., 1986, Recombinant interferon-gamma induces interleukin-2 receptors on human peripheral blood monocytes, *J. Immunol.* **136**:2171–2175

Horton, M. A., and O'Brien, H. A. W., 1983, Characterisation of human mast cells in long-term culture, *Blood* **62**:1251–1260

Horton, M. A., Rimmer, E. F., Valent, P., and Bettelheim, P., 1989, The use of the myeloid workshop panel for the identification of surface-membrane and cytoplasmic antigens expressed in human mast cells, In Knapp, W., et al., eds. *Leucocyte Typing IV*, pp. 914–915, Oxford University Press, Oxford

Horwitz, A., Duggan, K., Buck, C., Beckerle, M. C., and Burridge, K., 1986, Interaction of plasma membrane fibronectin receptor with talin—a transmembrane linkage, *Nature* (London) **320**:531–533

Hugli, T. E., 1981, Structural basis for anaphylatoxin and chemotactic functions of C3a, C4a and C5a, *Crit. Rev. Immunol.* **4**:321–366

Huey, R., and Hugli, T. E., 1985, Characterisation of a C5a receptor on human polymorphonuclear leucocytes, *J. Immunol.* **135**:2063–2068

Huey, R., Fukuoka, Y., Hoeprich, P. D., and Hugli, T. E., 1986, Cellular receptors to anaphylatoxins C3a and C5a, *Biochem. Soc. Symp.* **51**:69–81

Hynes, R. O., 1987, Integrins; a family of cell surface receptors, *Cell* **48**:549–554

Ihle, J. N., Keller, J., Orozslan, S., Henderson, L. E., Copeland, T., Fitch, F., Prystowsky, M. B., Goldwasser, E., Schrader, J., Palazynski, E., Dy, M., and Lebel, B., 1983, Biological properties of homogeneous interleukin-3, *J. Immunol.* **131**:282–287

Ikuta, K., Takami, M., Kim, C. W., Honjo, T., Miyoshi, T., Tagaya, Y., Kawabe, T., and Yodoi, J., 1987, Human lymphocyte Fc receptor for IgE; sequence homology of cDNA with animal lectins, *Proc. Natl. Acad. Sci. USA* **84**:819–823

Isersky, C., Metzger, H., and Buell, D. N., 1975, Cell cycle associated changes in receptors for IgE during growth and differentiation of rat basophilic leukaemia, *J. Exp. Med.* **141**:1147–1162

Isfort, R. J., Abraham, R., May, A. S., Stevens, D. A., Frackleton, A. R., and Ihle, J. N., 1988a, Mechanisms in IL-3 dependent growth of factor-dependent myeloid leukaemia cell lines, in: *Growth Factors and Their Receptors; Genetic Control and Rational Applications* (R. Ross, J. T. Burgess, and T. Hunter, eds.), pp. 229–238, Alan R. Liss, Inc., New York

Isfort, R. J., Stevens, D., May, W. S., and Ihle, J. N., 1988b, Interleukin-3 binds to a 140kDa phosphotyrosine-containing cell surface protein, *Proc. Natl. Acad. Sci. USA* **85**:7982–7986

Ishizaka, K., Ishizaka, T., and Lee, E. H., 1970a, Biological function of Fc fragments of E myeloma protein, *Immunochemistry* **7**:687

Ishizaka, T., and Ishizaka, K., 1978, Triggering of histamine release from rat mast cells by divalent antibodies against IgE receptor, *J. Immunol.* **136**:623–627

Ishizaka, T., and Ishizaka, K., 1984, Activation of mast cells for mediator release through IgE receptors, *Prog. Allergy* **34**:188–235

Ishizaka, T., Ishizaka, K., Orange, R. P., and Austen, K. F., 1970b, The capacity of IgE to mediate the release of histamine and SRS-A from monkey lung, *J. Immunol.* **104**:335–343

Ishizaka, T., Soto, C., and Ishizaka, K., 1973, Mechanisms of passive sensitisation, *J. Immunol.* **111**:500–511

Ishizaka, T., Helm, B., Hakimi, J., Niebyl, J., Ishizaka, K., and Gould, H., 1986, Biological properties of a human immunoglobulin E-chain fragment, *Proc. Natl. Acad. Sci. USA* **83**:8323–8327

Janossy, G., Tidman, N., Papageorgiou, E. S., Kung, P. C., and Goldstein, G., 1981, Distribution of T-lymphocyte subsets in human bone marrow and thymus, *J. Immunol.* **126**:1608–1613

Jarrett, E. E. E., and Haig, D. M., 1984, Mucosal mast cells in vivo and in vitro, *Immunol. Today* **5**:115–119

Johansson, S. G. O., and Bennich, H., 1967, Immunological studies of an atypical myeloma immunoglobulin, *Immunology* **13**:381–394

Johnson, A. R., Hugli, T. E., and Muller-Eberhard, H. J., 1975, Release of histamine from rat mast cells by C3a and C5a. *Immunology* **28**:1067–1080

Johnson, R. J., and Chenoweth, D. E., 1985, Labelling the granulocyte C5a receptor with a unique photoreactive probe, *J. Biol. Chem.* **260**:7161–7164

Johnson, R. J., and Chenoweth, D. E., 1987, Synthesis of a photoreactive C5a analog that permits identification of the ligand binding component of the granulocyte C5a receptor, *Biochem. Biophys. Res. Commun.* **148**:1330–1337

Kagey-Sobotka, A., Dembo, M., Goldstein, B., Metzger, H., and Lichtenstein, L. M., 1981, Qualitative characteristics of histamine release from human basophils by cross-linked IgE, *J. Immunol.* **127**:2285–2291

Kanellopoulos, J. M., Liu, T. Y., Poy, G., and Metzger, H., 1980, Composition and subunit structure of the cell receptor for IgE, *J. Biol. Chem.* **255**:9060–9066

Katz, H. R., and Austen, K. F., 1987, Biochemical characterisation of a mast cell plasma membrane antigen, *J. Immunol.* **113**:1196–1200

Katz, H. R., LeBlanc, P. A., and Russell, S. W., 1983, Two classes of mouse mast cells delineated by monoclonal antibodies, *Proc. Natl. Acad. Sci. USA* **80**:5916–5918

Kay, A. B., and Austen, K. F., 1972, Chemotaxis of human basophils, *Clin. Exp. Immunol.* **11**:557–563

Kay, A. B., Shin, H. S., and Austen, K. F., 1973, Selective attraction of eosinophils and synergism between eosinophil chemotactic factor and a fragment cleaved from the fifth component of complement (C5a), *Immunology* **24**:969–976

Keizer, G. D., TeVelde, A. A., Schwarting, R., Figdor, C. G., and DeVries, J. E., 1987, Role of p150,95 in adhesion, migration, chemotaxis and phagocytosis of human monocytes, *Eur. J. Immunol.* **17**:1317–1322

Kenten, J. H., Molgaard, H. V., Houghton, M., Derbyshire, R. B., Viney, J., Bell, L. O., and Gould, H., 1982, Cloning and sequence determination for the human immunoglobulin E gene expressed in a myeloma cell line, *Proc. Natl. Acad. Sci. USA* **79**:6661–6662

Kenten, J., Helm, B., Ishizaka, T., Cattini, P., and Gould, H., 1984, Properties of a human immunoglobulin E-chain fragment synthesised in *E. coli*. *Proc. Natl. Acad. Sci. USA* **81**:2955–2959

Kinet, J.-P., Alcaraz, G., Leonard, G., Wank, S. A., and Metzger, H., 1985a, Dissociation of the receptor for IgE in mild detergents, *Biochemistry* **24**:4117–4124

Kinet, J.-P., Quarto, R., Perez-Montfort, R., and Metzger, H., 1985b, Noncovalently and covalently bound lipid on the receptor for IgE, *Biochemistry* **24**:7342–7348

Kinet, J.-P., Metzger, H., Hakimi, J., and Kochan, J., 1987, A cDNA presumptively coding for the alpha-subunit of the receptor with high affinity for IgE, *Biochemistry* **26**:4605–4610

Kishi, K., 1985, A new leukaemic cell line with Philadelphia chromosome characterised as basophil precursors, *Leuk. Res.* **9**:381–390

Kishimoto, K., O'Connor, K., Lee, A., Roberts, T. M., and Springer, T. A., 1987, Cloning of beta subunit of leucocyte adhesion proteins, *Cell* **48**:681–690

Knapp, W., Dörken, B., Gilks, W. R., Reiber, E. P., Schmidt, R. E., Stein, H. L., Von der Borne, A. E. G., eds., 1989, in: *Leucocyte Typing IV: White cell differentiation antigens*, pp. 1–1182, Oxford University Press, Oxford

Kochwa, S., Terry, W. D., Capra, J. D., and Young, N. L., 1971, Structural studies of immunoglobulin E, *Ann. N.Y. Acad. Sci.* **140**:49–70

Koyasu, S., Tojo, A., Miyajima, A., Akiyama, T., Kasuga, M., Urabe, A., Schreurs, J., Arai, K. I., Takaku, F., and Yahara, I., 1987, Interleukin-3 dependent phosphorylation of a membrane glycoprotein of Mr 150,000 in multi-factor dependent myeloid cells lines, *EMBO J.* **6**:3979–3984

Krensky, A. M., Sanchez-Madrid, F., Robbins, E., Nagy, J., Springer, T. A., and Burakoff, S. J., 1983, Functional significance, distribution and structure of LFA-1, LFA-2 and LFA-3; cell surface antigens associated with CTL-target interactions, *J. Immunol.* **131**:611–616

Kulzycki, A., and Metzger, H., 1974, Interaction of rat IgE with rat basophilic leukaemia II, *J. Exp. Med.* **140**:1676–1695

Kulzycki, A., Isersky, C., and Metzger, H., 1974, Interaction of rat IgE with rat basophilic leukaemia cells I, *J. Exp. Med.* **139**:600

Kulzycki, A., McKearney, T. A., and Parker, C. W., 1976, The rat basophilic leukaemia cell receptor for IgE, *J. Immunol.* **117**:661–665

Kumar, N., and Metzger, H., 1982, Gel filtration in 6M guanidine hydrochloride of the alpha-subunit of the receptor for IgE, *Mol. Immunol.* **19**:1561–1567

Kurzinger, K., Reynolds, T., Germain, R. N., Davignon, D., Martz, E., and Springer, T. A., 1983, A novel lymphocyte function-associated antigen (LFA-1); cellular distribution, quantitative expression and structure, *J. Immunol.* **127**:596–602

Lad, P. M., Olson, C. V., and Smiley, P. A., 1985, Association of the N-formyl-Met-Leu-Phe receptor in human neutrophils with a GTP-binding protein, *Proc. Natl. Acad. Sci. USA* **82**:869–873

Laterra, J., Ansbacher, R., and Culp, L. A., 1980, Glycosaminoglycans that bind cold insoluble globulin in cell-substratum adhesion sites of murine fibroblasts, *Proc. Natl. Acad. Sci. USA* **77**:6662–6666

Law, S. K. A., Gagnon, J., Hildreth, J. E. K., Wells, C. E., Willis, A. C., and Wong, A. J., 1987, Primary structure of beta subunit of cell surface adhesion glycoproteins LFA-1, CR3 and p150,95 and its relationship to FnR, *EMBO J.* **6**:915–919

Lee, D. G., Sterk, A., Ishizaka, T., Bienenstock, J., and Befus, A. D., 1985, Number and affinity of receptors for IgE on isolated rat intestinal mast cells, *Immunology* **55**:363–368

Leonard, W. J., Depper, J. M., Crabtree, G. R., Rudikoff, S., Pumphrey, J., Robb, R. J., Kronke, M., Svetlik, P. B., Peffer, M. J., Waldmann, T. A., and Greene, W. C., 1984, Molecular cloning and expression of cDNAs for human interleukin-2 receptor, *Nature* (London) **311**:626–631

Levi-Schaffer, F., Austen, K. F., Gravallese, P. M., and Stevens, R. L., 1986, Coculture of IL-3-dependent mouse mast cells with fibroblasts results in phenotypic change, *Proc. Natl. Acad. Sci. USA* **83**:6485–6488

Liu, F. T., Albrandt, K., Sutcliffe, J. G., and Katz, D. H., 1982, Cloning and nucleotide sequence of mouse immunoglobulin E-chain cDNA, *Proc. Natl. Acad. Sci. USA* **79**:7852–7856

Liu, F. T., Albrandt, K., Bry, C. G., and Ishizaka, T., 1984, Expression of a biologically-active fragment of human IgE in *E. coli*, *Proc. Natl. Acad. Sci. USA* **81**:5369–5373

Look, A. T., Ashmun, R. A., Shapiro, L. H., and Peiper, S. C., 1989, The human myeloid plasma membrane glycoprotein CD13 (gp150) is identical to aminopeptidase N, *Tissue Histocompatibility Immunogenet.* **33**:228

Lu, Z.-X., Fok, K.-F., Erickson, B. W., and Hugli, T. E., 1984, Conformational analysis of the COOH-terminal segment of human C5a, *J. Biol. Chem.* **259**:7367–7370

MacIntyre, E. A., Jones, H. M., Gray, A. G., Morgan, J., Tidman, N., Leung, K. B. P., Newland, A. C., and Linch, D. C., 1987, Analysis of peripheral blood basophils from patients with chronic granulocytic leukaemia, in: *Leucocyte Typing III* (A. McMichael, P. C. L. Beverly, S. Cobbold, M. J. Grumpton, W. Gilks, F. M. Gorch, N. Hogg, M. Horton, N. Ling, I. C. M. McLennan, D. M. Mason, C. Milstein, D. O. Spiegelhalter, H. Waldmann, eds.), pp. 723–725, Oxford University Press, Oxford

Malveaux, F. J., Conroy, M. C., Adkinson, N. F., and Lichtenstein, L. M., 1978, IgE receptors on human basophils; relationship to serum IgE concentration, *J. Clin. Invest.* **62**:176–181

Marlin, J. D., and Springer, T. A., 1987, Purified intercellular adhesion molecule-1 is a ligand for lymphocyte function-associated antigen-1, *Cell* **51**:813–819

Martens, C. L., Huff, T. F., Jardieu, P., Trounstine, M. L., Coffman, R. L., Ishizaka, K., and Moore, K. W., 1985, cDNA clones encoding IgE-binding factors from a rat-mouse T-cell hybridoma, *Proc. Natl. Acad. Sci. USA* **82**:2460–2464

Martz, E., 1980, Immune lymphocyte to tumour cell adhesion; magnesium sufficient, calcium insufficient. *J. Cell Biol.* **84**:584–598

Martz, E., 1986, LFA-1 and other accessory molecules functioning in adhesion of T and B lymphocytes, *Hum. Immunol.* **18**:3–37

Mayrhofer, G., Gadd, S. J., Spargo, L. D. J., and Ashman, L. K., 1987, Specificity of a mouse monoclonal antibody for mast cells in several human mucosal and connective tissues, *Immunol. Cell. Biol.* **65**:241–250

McCarthy, K., and Henson, P. M., 1979, Lysosomal enzyme secretion by alveolar macrophages in response to C5a and C5a desArg, *J. Immunol.* **123**:2511–2517

McMichael, A., *et al.* (eds.), 1987, *Leucocyte Typing III: White Cell Differentiation Antigens*, Oxford University Press, Oxford

Mendez, E., Frangione, B., and Kochwa, S., 1973, Chemical typing of human immunoglobulins E and D, *FEBS Lett.* **33**:4–6

Miller, L. J., Wiebe, M., and Springer, T. A., 1986, Alpha subunit N-terminal sequences of human Mac-1 and p150,95 leucocyte adhesion proteins, *J. Immunol.* **138**:2381–2383

Moller, G., and Konig, W., 1980, Binding characteristics of aggregated IgGa to rat basophilic leukaemia cells and rat mast cells, *Immunology* **141**:605–615

Mostov, K. E., Friedlander, M., and Blobel, G., 1984, The receptor for transepithelial transport of IgA and IgM contains multiple Ig-like domains, *Nature* (London) **308**:37–43

Nagao, K., Yokoro, K., and Aaronson, S., 1981, Continuous lines of basophil/mast cells derived from normal bone marrow, *Science* **212**:333–335

Nakahata, T., Kobayashi, T., Ishiguro, A., Tsuji, K., Nagamuna, K., Ando, O., Yagi, Y., Tadokoro, K., and Akabane, T., 1986, Extensive proliferation of mature connective tissue mast cells in vitro, *Nature* (London) **324**:65–67

Newman, S. A., Rossi, G., and Metzger, H., 1977, Molecular weight and valence of cell surface receptor for IgE, *Proc. Natl. Acad. Sci USA* **74**:869–872

Nezlin, R. S., Zagyansky, Y. A., Kaivaraenen, A. I., and Stefani, D. V., 1973, Properties of myeloma immunoglobulin E (Yu), *Immunochemistry* **10**:681–688

Nicola, N. A., and Peterson, L., 1986, Identification of distinct receptors for two haemopoietic growth factors by chemical crosslinking, *J. Biol. Chem.* **261**:12384–12389

Nikaido, T., Shimizu, A., Ishida, N., Sabe, H., Teshigawa, K., Maeda, M., Uchiyama, T., Yodoi, J., and Honjo, T., 1984, Molecular cloning of cDNA encoding human interleukin-2 receptor, *Nature* (London) **311**:631–635

Okajima, F., and Ui, M., 1984, ADP-ribosylation of specific membrane protein by pertussis toxin associated with inhibition of chemotactic peptide-induced arachidonate release in neutrophils, *J. Biol. Chem.* **259**:13863–13871

Palazynski, E. W., and Ihle, J. N., 1984, Evidence for specific receptor for interleukin-3 on lymphokine-dependent cell lines, *J. Immunol.* **132**:1872–1878

Park, L. S., Friend, D., Gillis, S., and Urdal, D. L., 1986a, Characterisation of cell surface receptor for a multilineage colony-stimulating factor, *J. Biol. Chem.* **261**:205–210

Park, L. S., Friend, D., Gillis, S., and Urdal, D. L., 1986b, Characterisation of cell surface receptor for granulocyte/monocyte colony-stimulating factor, *J. Biol. Chem.* **261**:4177–4183

Pecoud, A. R., Ruddy, S., and Conrad, D. H., 1981, Functional and partial chemical characterisation of the carbohydrate moieties of the IgE receptor, *J. Immunol.* **126**:1624–1629

Perez-Montfort, R., and Metzger, H., 1983, Proteolysis of soluble IgE-receptor complexes; sites on IgE which interact with the receptor, *Mol. Immunol.* **19**:1113–1125

Perez-Montfort, R., Kinet, J.-P., and Metzger, H., 1983, A previously-unrecognised subunit of the receptor for IgE. *Biochemistry* **22**:5722–5728

Poncz, M., Eisman, R., Heidenreich, R., Silver, S. M., Vilaire, G., Surrey, S., Schwartz, E., and Bennett, J. S., 1987, Structure of the platelet membrane glycoprotein IIb, *J. Biol. Chem.* **262**:8476–8482

Pumphrey, R., 1986, Computer models of human immunoglobulins, *Immunol. Today* **7**:174–178

Quarto, R., Kinet, J.-P., and Metzger, H., 1985, Coordinate synthesis and degradation of alpha, beta and gamma subunits of the receptor for IgE, *Mol. Immunol.* **22**:1045–1051

Ravetch, J. V., Luster, A. D., Weinshank, R., Kochan, J., Pavlovec, A., Portnoy, D. A., Hulmes, J., Pan, Y.-C., and Unkeless, J. C., 1986, Structural heterogeneity and functional domains of murine IgG-Fc receptors, *Science* **234**:718–725

Regal, J. F., 1982, C5a-induced aortic contraction; effects of antihistamines and inhibitors of arachidonate metabolism, *J. Pharm. Exp. Ther.* **220**:102–107

Reinherz, E. L., Haynes, B. F., Nadler, L. M., and Bernstein, I. D. (eds.), 1986, *Leucocyte Typing II*, Vol. 3. *Human Myeloid and Haemopoietic Cells*, Springer-Verlag, Berlin

Rimmer, E. F., Turberville, C., and Horton, M. A., 1984, Human mast cells detected by monoclonal antibodies, *J. Clin. Pathol.* **37**:1249–1255

Rimmer, E. F. and Horton, M. A. 1987, Origin of human mast cells studied by dual immunofluorescence, *Clin. Exp. Immunol.* **68**: 712–718.

Rimmer, E. F., Turberville, C., and Horton, M. A., 1986, Cell membrane glycoproteins of human mast cells; a biochemical comparison with basophils, *Exp. Hematol.* **14**:809–811

Rivnay, B., Rossi, G., Henkart, M., and Metzger, H., 1984, Reconstitution of the receptor for IgE into liposomes, *J. Biol. Chem.* **259**:1212

Rollins, T. E., and Springer, M. S., 1985, Identification of the polymorphonuclear leucocyte C5a receptor, *J. Biol. Chem.* **260**:7157–7160

Rollins, T. E., Siciliano, S., and Springer, M. S., 1988, Solubilisation of functional C5a receptor from human polymorphonuclear leucocytes, *J. Biol. Chem.* **263**:520–526

Ruohslahti, E., and Engvall, E., 1980, Complexing of fibronectin, glycosaminoglycans and collagen. *Biochim. Biophys. Acta* **631**:350–358

Russell, D. G., and Wright, S. D., 1988, Complement receptor type 3 (CR3) binds to an Arg-Gly-Asp-containing region of the major surface glycoprotein of *Leishmania* promastigotes, *J. Exp. Med.* **168**:279–292

Saito, H., Hataka, K., Dvorak, A. M., Leiferman, K. M., Donnenberg, A. D., Arai, N., Ishizaka, K., and Ishizaka, T., 1988, Selective differentiation and proliferation of haemopoietic cells induced by recombinant human interleukins, *Proc. Natl. Acad. Sci. USA* **85**:2288–2292

Sakashita, S., Engvall, E., and Ruohslahti, E., 1980, Basement membrane laminin binds to heparin, *FEBS Lett.* **116**:243–246

Sanchez-Madrid, F., Nagy, J., Robbins, E., Simons, P., and Springer, T. A., 1983, A human leucocyte differentiation antigen family with distinct alpha- subunits and a common beta-subunit; the lymphocyte function-associated antigen (LFA-1), C3bi complement receptor (OKM1/Mac-1) and the p150,95 molecule, *J. Exp. Med.* **158**:1785–1803

Saunders, S., and Bernfield, M., 1988, Cell surface proteoglycan binds mouse mammary epithelial cells to fibronectin and behaves as a receptor for interstitial matrix, *J. Cell Biol.* **106**:423–430

Schlessinger, J., Webb, W. W., Elson, E. L., and Metzger, H., 1976, Distribution and valency of receptor for IgE on rodent mast cells, *Nature* (London) **264**:548–550

Schrader, J. W., 1981, The in vitro production of the P-cell, a bone marrow-derived cell that expresses H-2 and Ia, has mast cell-like granules and is regulated by T cells, *J. Immunol.* **126**:425–458

Schwartz, L. B., and Austen, K. F., 1984, Structure and function of chemical mediators of mast cells, *Prog. Allergy* **34**:271–321

Schwartz, L. B., Kawahara, M. S., Hugli, T. E., Vik, D., Fearon, D. T., and Austen, K. F., 1983, Generation of C3a anaphylatoxin from human C3 by mast cell tryptase, *J. Immunol.* **130**:1891–1901

Seed, B., and Aruffo, A., 1987, Molecular cloning of the CD2 antigen by a rapid immunoselection procedure, *Proc. Natl. Acad. Sci. USA* **84**:3365–3369

Segal, D. M., Taurog, J. D., and Metzger, H., 1977, Dimeric IgE serves as a unit signal for mast cell degranulation, *Proc. Natl. Acad. Sci. USA* **74**:2993–2997

Sharon, M., Klausneer, R. D., Cullen, B. R., Chizzonite, R., and Leonard, W. J., 1986, Novel interleukin-2 receptor subunit detected by crosslinking under high-affinity conditions, *Science* **224**:859–863

Shaw, J., Henson, J., Philips, D., and Henson, P. M., 1978, Lung injury induced by intratracheal administration of a chemotactic fragment from the fifth component of complement, *Am. Rev. Respir. Dis.* **117**:Supplement 1 (Abstract) 80

Shaw, J. O., Henson, P. M., and Webster, R. O., 1980, Lung inflammation induced by complement-derived chemotactic fragments, *Lab. Invest.* **42**:547–558

Shimizu, A., Tepler, I., Benfey, P. N., Berenstein, P. H., Siraganian, R. P., and Leder, P., 1988, Human and rat mast cell high affinity IgE receptors; characterisation of putative a-chain gene products, *Proc. Natl. Acad. Sci. USA* **85**:1907–1911

Siegel, J. P., Sharon, M., Smith, P. L., and Leonard, W. J., 1987, The IL-2 receptor receptor beta-chain; role in mediating LAK, NK and proliferative abilities, *Science* **238**:75–78

Simister, M. E., and Mostov, K. E., 1989, An Fc-receptor structurally related to MHC class I antigens, *Nature* (London) **337**:184–187

Simpson, P. J., Todd, R. F., Fantone, J. C., Mickelson, J. K., Griffin, J. D., and Lucchesi, B. R., 1988, Reduction of experimental canine reperfusion injury by a monoclonal antibody (anti-Mo1/ antiCD11b) that inhibits leucocyte adhesion, *J. Clin. Invest.* **81**:624

Sims, J. E., March, C., Cosman, D., Widman, M. B., McDonald, H. R., McMahon, C. J., Grubin, C. E., and Wignall, J. M., 1988, cDNA expression cloning of IL-1 receptor, a member of the immunoglobulin superfamily, *Science* **241**:585–589

Siraganian, R. P., and Hook, W. A., 1976, Complement-induced histamine release from human basophils, *J. Immunol.* **116**:639–646

Smith, C., and Rennick, D. M., 1986, Characterisation of a murine lymphokine distinct from IL-2 and IL-3, possessing a T cell growth factor activity, and a mast cell growth factor activity that synergises with IL-3, *Proc. Natl. Acad. Sci. USA* **83**:1857–1867

Snyderman, R., Philips, J., and Mergenhagen, S. E., 1971, Biological activity of complement *in vivo*, *J. Exp. Med.* **134**:1131–1143

Sonoda, T., Kitamura, Y., Haku, Y., Hara, H., and Mori, K. L., 1983, Mast cell precursors in haemopoietic colonies of mice invitro and in vivo, *Br. J. Haematol.* **53**:611–620

Sorensen, P., Farber, N. M., and Krystal, G., 1986, Identification of the interleukin-3 receptor using an iodinatable cleavable photoreactive crosslinking agent. *J. Biol. Chem.* **261**:9094–9097

Springer, T. A., and Anderson, D. C., 1985, Leucocyte and adhesion proteins in the inflammatory response; insights from an experiment of nature, *Biochem. Soc. Symp.* **51**:47–57

Springer, T. A., Thompson, W. S., Miller, L. J., Schmalstieg, F. C., and Anderson, D. C., 1984, Inherited deficiency of Mac-1, LFA-1 and p150,95 glycoprotein family and its molecular basis. *J. Exp. Med.* **160**:1901–1918

Stain, C., Stockinger, H., Scharf, M., Jager, U., Gossinger, H., Lechner, K., and Bettelheim, P., 1987, Human blood basophils display a unique phenotype including activation-linked membrane structures, *Blood* **70**:1872–1879

Stanworth, D. R., Humphrey, J. H., Bennich, H., and Johansson, S. H. O., 1967, Inhibition of Praustnitz-Kustner reaction by cleavage fragments of a human myeloma protein, *Lancet* **ii**:17–18

Sterk, A., and Ishizaka, T., 1982, Binding properties of IgE receptors on normal mouse mast cells, *J. Immunol.* **128**:838–843

Stevens, R. L., Fox, C. C., Lichtenstein, L. M., and Austen, K. F., 1988, Identification of chondroitin sulfate E proteoglycans and heparin proteoglycans in secretory granules of human lung mast cells, *Proc. Natl. Acad. Sci. USA* **85**:2284–2287

Stimler, N. P., Brocklehurst, W. E., Hugli, T. E., and Bloor, C. M., 1981, Anaphylatoxin-mediated contraction of guineapig lung strips, *J. Immunol.* **126**:2258–2261

Stimler, N. P., Bloor, C. M., and Hugli, T. E., 1983, C3a-induced contraction of guineapig lung parenchyma; role of cyclooxygenase metabolites. *Immunopharmacology* **5**:251–257

Subramanian, N., and Bray, M. A., 1987, Interleukin-1 releases histamine from human basophils and mast cells in vitro, *J. Immunol.* **138**:271–275

Sullivan, A. L., Grimley, P. M., and Metzger, H., 1971, Electron microscope localisation of IgE on the membrane of human basophils, *J. Exp. Med.* **134**:1403–1416

Suzuki, S., Argraves, W. G., Arai, H., Languino, L. R., Piersbacher, M. D., and Ruohslahti, E., 1987, Amino acid sequence of vitronectin receptor alpha subunit and expression of adhesion receptor mRNAs, *J. Biol. Chem.* **262**:14080–14085

Szebenyi, D. M., Obendorf, S. K., and Moffat, K., 1982, Structure of vitamin-D-dependent calcium binding protein from bovine intestine, *Nature* (London) **294**:327–332

Takatsu, K., Ishizaka, T., and Ishizaka, K., 1975, Biological significance of disulphide bonds in human IgE molecules, *J. Immunol.* **114**:1838–1845

Tertian, G., Yung, Y.-P., Guy-Grand, D., and Moore, M. A. S., 1981, Long term in vitro culture of murine mast cells, *J. Immunol.* **127**:788–794

Thompson, H. L., Burbelo, P. D., Segue-real, B., Yamada, Y., and Metcalfe, D. D., 1989, Laminin promotes mast cell attachment, *J. Immunol.* **143**:2323–2327

Tigelaar, R. E., Vaz, N. M., and Ovary, Z., 1971, Immunoglobulin receptors on mouse mast cells, *J. Immunol.* **106**:661–672

Tsao, C. J., Tojo, A., Fukimachi, H., Kitamura, T., Saito, T., Urabe, A., and Takaku, F., 1988, Expression of functional erythropoietin receptors on interleukin-3 dependent murine cell lines, *J. Immunol.* **140**:89–93

Tsudo, M., Kozak, R. W., Goldman, C. K., and Waldmann, T. A., 1986, Demonstration of a non-Tac peptide that binds interleukin-2, *Proc. Natl. Acad. Sci. USA* **83**:9694–9698

Valent, P., Schmidt, G., Besemer, J., Mayer, P., Liehl, E., Hinterberger, W., Lechner, K., Maurer, D., and Bettelheim, P., 1989a, Interleukin-3 is a differentiation factor for human basophils *Blood* **73**:1763–1769

Valent, P., Ashman, L. K., Hinterburger, W., Eckersburger, F., Majdic, O., Lechner, K., and Bettelheim, P., 1989b, Mast cell typing; demonstration of a distinct haemopoietic lineage and evidence for immunophenotypic relationship to mononuclear phagocytes *Blood* **73**:1778–1785 (submitted for publication)

Valent, P., Besemer, J., Liehl, E., Mayer, P., Schmidt, G., Maurer, D., Majdic, O., Knapp, W., and Bettelheim, P., 1989c, Induction of basophil differentiation antigens by recombinant human IL-3, *Tissue Histocompatibility Immunogenet.* **33**:220

Van Epps, D. E., and Chenoweth, D. E., 1984, Analysis of binding of fluorescent C5a and C3a to human blood leucocytes, *J. Immunol.* **132**:2862–2867

Varsano, S., Lazarus, S. C., Gold, W. M., and Nadel, J. A., 1988, Selective adhesion of mast cells to tracheal epithelial cells, *J. Immunol.* **140**:2184–2192

Vijay, H. M., and Perelmutter, L., 1977, Inhibition of reagin-mediated PCA reactions by human IgG₄ subclass, *Int. Arch. Allergy Appl. Immunol.* **53**:78–87

Visani, G., Delwel, R., Touw, I., Bott, F., and Lowenberg, B., 1987, Membrane receptor for interleukin-2 on haemopoietic precursors in chronic myeloid leukaemia, *Blood* **69**:1182–1187

Webster, R. O., Hong, S. R., Johnson, R. B., and Henson, P. M., 1980, Biological effects of human complement fragments C5a and C5a desArg on neutrophil functions, *Immunopharmacology* **2**:201–219

Yamada, K., Kennedy, D. W., Kimita, K., and Pratt, R. M., 1980, Characterisation of fibronectin interaction with glycosaminoglycan, *J. Biol. Chem.* **255**:6055–6063

Yancey, K. B., Hammer, C. H., Narvath, L., Renfer, L., Frank, M. M., and Lawley, T. J., 1985, Studies of human C5a as a mediator of inflammation in normal human skin, *J. Clin. Invest.* **75**:486–495

Yang, Y. C., Ciarletta, A. B., Temple, P. A., Chung, M. P., Korazu, S., Witek-Gianotti, J. A., Leary, A. C., Kriz, R., Donahue, R. E., Wong, G. G., and Clark, S. C., 1986, Human interleukin-3; identification by expression cloning of a novel growth factor related to murine IL-3, *Cell* **47**:3–10

Zeleznik, N. J., Hollingsworth, M. A., and Metzgar, R. S., 1987, Studies on the CD9 antigen defined by monoclonal antibody Du-All-1, in: *Leucocyte Typing III* (A. McMichael *et al.*, eds.), pp. 389–392, Oxford University Press, Oxford

Index

The manufacturer's authorised representative in the EU is Springer
Nature Customer Service Centre GmbH, Europaplatz 3, 69115 Heidelberg,
Germany. If you have any concerns regarding our products, please
contact ProductSafety@springernature.com

Printed and bound by CPI Group (UK) Ltd, Croydon, CR0 4YY

23/04/2026

02095624-0017